D1761134

DRILL HALL LIBRARY
MEDWAY

MW5050S
11006670

# NUTRITIONAL GENOMICS

## The Impact of Dietary Regulation of Gene Function on Human Disease

4646617

# NUTRITIONAL GENOMICS

## The Impact of Dietary Regulation of Gene Function on Human Disease

Edited by

### Wayne R. Bidlack, Ph.D.

Professor, Department Human Nutrition and Food Science
California State Polytechnic University, Pomona, California

### Raymond L. Rodriguez, Ph.D.

Director, Center for Excellence in Nutritional Genomics
University of California, Davis, California

**CRC Press**
Taylor & Francis Group
Boca Raton   London   New York

CRC Press is an imprint of the
Taylor & Francis Group, an **informa** business

CRC Press
Taylor & Francis Group
6000 Broken Sound Parkway NW, Suite 300
Boca Raton, FL 33487-2742

© 2012 by Taylor & Francis Group, LLC
CRC Press is an imprint of Taylor & Francis Group, an Informa business

No claim to original U.S. Government works

Printed in the United States of America on acid-free paper
Version Date: 20110811

International Standard Book Number: 978-1-4398-4452-6 (Hardback)

This book contains information obtained from authentic and highly regarded sources. Reasonable efforts have been made to publish reliable data and information, but the author and publisher cannot assume responsibility for the validity of all materials or the consequences of their use. The authors and publishers have attempted to trace the copyright holders of all material reproduced in this publication and apologize to copyright holders if permission to publish in this form has not been obtained. If any copyright material has not been acknowledged please write and let us know so we may rectify in any future reprint.

Except as permitted under U.S. Copyright Law, no part of this book may be reprinted, reproduced, transmitted, or utilized in any form by any electronic, mechanical, or other means, now known or hereafter invented, including photocopying, microfilming, and recording, or in any information storage or retrieval system, without written permission from the publishers.

For permission to photocopy or use material electronically from this work, please access www.copyright.com (http://www.copyright.com/) or contact the Copyright Clearance Center, Inc. (CCC), 222 Rosewood Drive, Danvers, MA 01923, 978-750-8400. CCC is a not-for-profit organization that provides licenses and registration for a variety of users. For organizations that have been granted a photocopy license by the CCC, a separate system of payment has been arranged.

**Trademark Notice:** Product or corporate names may be trademarks or registered trademarks, and are used only for identification and explanation without intent to infringe.

**Library of Congress Cataloging-in-Publication Data**

Nutritional genomics : the impact of dietary regulation of gene function on human disease / editors, Wayne R. Bidlack and Raymond L. Rodriguez.
    p. ; cm.
  Includes bibliographical references and index.
  ISBN 978-1-4398-4452-6 (hardcover : alk. paper)
  1. Nutrition--Genetic aspects. 2. Diet in disease. 3. Genomics. 4. Diet therapy. I. Bidlack, Wayne R. II. Rodriguez, Raymond L. III. Title.
  [DNLM: 1. Nutrigenomics. 2. Gene Expression Regulation--genetics. 3. Nutrition Therapy. 4. Nutritional Physiological Phenomena--genetics. QU 145]

  QP144.G45N8872 2011
  572.8'6--dc23                                                                                      2011027310

**Visit the Taylor & Francis Web site at**
**http://www.taylorandfrancis.com**

**and the CRC Press Web site at**
**http://www.crcpress.com**

# Contents

## SECTION I  Transforming Dietary Signals into Gene Expression: Epigenetics—The Epigenetic Transfer of Phenotype

## SECTION I  Transforming Dietary Signals into Gene Expression: Systems Biology—Cell Signaling Regulation of Gene Expression

## SECTION II  Gene-Linked Networks: Gene-Linked Chronic Disease

## SECTION II  Gene-Linked Networks: Gene-Linked Cancer Prevention

## SECTION II  Gene-Linked Networks: Intestinal Microbiome- Linked Gene Expression

# Preface

Until recently, the etiology of chronic diseases had been described in terms of biochemical and physiological changes in tissue function. Unfortunately, by the time changes are diagnosed, metabolic dysfunctions already exist, and therapy is designed primarily to reduce the progression of the disease state. Selective dietary regimens and pharmacologic agents have been used to block or slow the progression and alleviate some of the side effects of these diseases. Population groups having similar chronic disease characteristics have consistently been noted to contain responders, nonresponders, and negative responders to each therapeutic regimen.

The future of health care may increasingly depend on identification of personalized nutrition targets based on nutritional genomic markers, enhancing intervention outcomes chosen to benefit the population as a whole. Nutritional products would be safe for all to consume but by characterization of individual genomic health needs, appropriate guidelines could enable consumers to better select nutrient sources to meet their individual genomic health needs. As such, it is imperative that scientists in agriculture, food science, nutrition, genomics, and clinical therapy all work together to assure safe and efficacious application of nutritional genomics to enhance health.

The goal of this book is to provide an opportunity for researchers to establish an integrated perspective for successful development of nutritional genomics. The content of the chapter material is not all encompassing, but in fact has targeted specific areas of interest within the field of nutritional genomics that will rely on further integration to complete the vision for utilization of nutritional genomics as an additional tool for improved health care.

Since the sequencing of the human genome, it has become more apparent that many gene variants exist, which may present a predisposition to a specific disease. Whether or not that genotypic potential will be expressed depends upon a complex interplay among external factors, diet, and environment, and the molecular components that regulate gene expression affecting the physiologic processes they control. The potential integration of dietary components on the phenotypic outcome has led to the development of the field of nutritional genomics.

Nutritional genomics is the study of the interactions between our genetic makeup and the foods we consume, and the health outcomes that may occur. The goal of nutritional genomics is to provide a platform from which a dietary regimen can be tailored to affect individual genetic profiles, potentially increasing the risk of developing a chronic disease or cancer, with the expectation that the onset of these conditions may be delayed or prevented if intervention occurs early enough. The role of bioactive food components, nutrient and non-nutrient, may be to affect multiple mechanisms related to cell signaling, transduction, and transcription factors that regulate gene expression. The regulatory efforts include the use of dietary components in altering chronic diseases and cancer development.

Advances in genomics have increased our understanding of the inherited basis of disease, including both genetic and epigenetic, as well as our understanding of how people differ in their response to their diet and its composition (Chapter 1). The role of epigenetic mechanisms in the etiology of cancer has been recognized in recent years, and more recently it has been associated with diabetes and other metabolic disorders (Chapters 1–3). Diet and other environmental factors may prove to be significant regulators of epigenetic mechanisms.

Different bioactive food components have been shown to exert protective effects through alterations in DNA methylation of CpG islands in promoter and other genomic regions, chromatin silencing, and post-translational modification of histone tail domains. Alteration of these epigenetic marks has been associated with modulation of numerous cellular processes associated with carcinogenesis, including differentiation, inflammation, apoptosis, cell cycle control/proliferation, carcinogen metabolism, and angiogenesis, among others. Epigenetic silencing of tumor suppressor and proapoptosis genes in cancer cells, unlike genetic mutations, can be reversed by the use of DNA demethylating agents (to remove methylation marks) and HDAC inhibitors (to retain histone acetylation marks). Characterization of dietary effects depends on quantity, timing of exposure, chemical form, digestion, and bioavailability (Chapter 1).

Plant-based foods are a natural source of modifiers of the epigenetic machinery. For example, lunasin is a dietary peptide in soy foods, which survives digestion, undergoes absorption into the circulation, and becomes bioavailable to the cells. This peptide causes upregulation of chemopreventive gene expression by specific epigenetic modifications and provides another bioactive food component with the potential to reduce cancer risk (Chapter 2). Both clinical and *in vitro* studies have clearly demonstrated the importance of maintaining a good diet and glycemic control to prevent or slow the progression of diabetic complications. Elevated plasma glucose can induce epigenetic modifications in target cells, resulting in activation of various signaling pathways and genes associated with these complications (Chapter 3).

Type 2 diabetes is a complex disease involving several risk factors including insulin resistance (IR), obesity, dyslipidemia, and genetics. With elevated glucose, each of these risk factors has the potential to induce epigenetic changes in the chromatin structure affecting gene expression patterns in various target tissues. These risk factors participate in the epigenetic mechanisms responsible for metabolic memory as they relate to diabetic complications, when they might be expressed, and how they might be prevented.

Diet has profound effects on short-term and long-term changes in biologic systems. Recent advances in high throughput technologies, such as genomic, transcriptomic, proteomic, and metabolomic analysis, have made measurement of these effects possible. However, the volume of data generated requires mathematical models and use of large computers to pool the data into systems to clarify the molecular basis of human disease (Chapters 4–9). The multiple layers of information need to be integrated into networks to identify the common intersections (hubs) controlling the signaling pathways (Chapters 4 and 12).

N-3 polyunsaturated fatty acids (n-3 PUFA) decreased the risk factors for atherosclerosis in most human studies (Chapter 6). The n-3 PUFA diets containing both eicosapentaenoic acid (EPA) and docosahexaenoic acid (DHA), resulted in decreased fasting and postprandial plasma triglycerides, large VLDL particles, total LDL and small dense LDL particles, and RLP-C, and an increase in large LDL particles. Many studies also indicated anti-inflammatory and lowering of blood pressure (BP) effects. The reduction in BP and the number of small dense LDL particles appeared to respond best to DHA. These results suggest that conflicting data from published research may occur due to a need for optimizing individual doses for EPA and DHA and detailed evaluation of their use together.

Cellular communication is responsible for transduction of signals from external signals and transduction pathways, nuclear receptors (NRs) and coactivators to regulate transcription (Chapters 5, 7, 8, and 13). Fatty acids are believed to regulate gene expression through several transcription factors (TFs), including PPAR, HNF4α, and RXR, which act as fatty acid sensors, selectively binding specific fatty acids to targeted nuclear receptors. The lipid bound TF binds to a responsive element in the promoter region of the target genes and attracts protein coactivators to form the requisite complex for activation of transcription. Thus, NR-mediated transcription and the coactivators have been found to play important roles in many different human disease states related to inflammation, atherosclerosis and those involved in energy metabolism (Chapters 7 and 8).

Other nutritive and non-nutritive bioactive components found in foods have been shown to modify gene expression, affect the inherited genotype, and alter the expressed phenotype. Plant polyphenols, especially the flavonoids, have been most actively studied (Chapters 5 and 9). The green tea polyphenol epigallocatechin gallate has been effective against cancer in animal models. Curcumin polyphenols and cinnamon polyphenols provide anticancer, anti-inflammatory properties, blocking of angiogenesis, and induction of apoptosis. Resveratrol alters global patterns of gene regulation in animal models, perhaps acting through SIRT1, which slows aging and associated physiologic decline, decreases oxidative stress and cancer, and alters the effects of obesity on insulin sensitivity. Quercetin provides a strong antioxidant effect and acts synergistically with resveratrol.

Enormous challenges remain to clarify the beneficial effects of polyphenols. More than 8000 distinct polyphenolic molecules have been identified, yet little understanding exists of their functional similarities and differences, or their cumulative effects over time. For example, the original study that identified resveratrol as a SIRT1 activator also identified 16 other structurally related polyphenols possessing similar activity (Chapter 9). Among the few compounds that have been studied in relative detail, similar effects have been identified *in vitro* and *in vivo*. All have antioxidant capacity. Resveratrol and curcumin bear little structural resemblance to one another, yet both show potent anticancer and anti-inflammatory effects, block angiogenesis, activate glutathione-S-transferases, inhibit cyclooxygenase, induce apoptosis in tumor cells, and are rapidly metabolized. The nature by which the polyphenolic compounds participate in the complex cellular signaling system and in gene expression needs to be characterized in greater detail. Many of the polyphenolic compounds pass through the plasma membrane to bind with nuclear receptors and act as TFs to signal gene expression (Chapter 5). Indeed, if diet can be used to alter phenotypic expression, then nutrition and food scientists can work together to design food products and personalized nutrition guidelines to affect the onset of chronic diseases, such as obesity, hypertension, type 2 diabetes, and cardiovascular disease (Chapters 10, 11, and 16).

Elucidation of the mechanisms by which obesity leads to increased morbidity and mortality has been the focus of intense research. There appears to be a strong link between inflammation and these chronic diseases. In the adipose tissue, expansion of the fat mass with increasing obesity initiates the inflammation cascade, caused by poor delivery of nutrients and limited delivery of oxygen to the core of the expanding adipose mass, increased in fatty acid flux, increased cytokine and chemokine secretion, and adipocyte death. All changes lead to recruitment of macrophages to the site of inflammation and contribute to insulin resistance (Chapter 10). Proinflammatory pathways are initiated in the neighboring insulin target cells, hepatocytes, adipocytes, and myocytes, which release additional cytokines and adipokines, exacerbating general inflammation and enhancing insulin resistance. When occurring together, these events contribute to IR, T2DM, and the metabolic syndrome.

Vitamin D metabolism is well established for regulation of calcium absorption and utilization. Recently, hypovitaminosis vitamin D (25-dihydroxy vitamin D) has been associated with increased risk of cardiovascular disease, elevated BP, and increased cardiohypertrophy, while the vitamin D receptor-agonist (VDR-A) participated in the inhibition of key steps in the inflammation process, angiogenesis and small muscle cell proliferation (Chapter 11). The recognition of genomic-mediated effects of 1,25-dihydroxy vitamin $D_3$ and VDR-A within the cardiovascular system has initiated new insight into strategic nutritional interventions to treat cardiovascular disease.

Vitamin D deficiency produced an increase in renin levels, which was reversed by an increase in vitamin D intake. Research, using vitamin D receptor knock-out (VDR KO) or 25-hydroxy vitamin D 1$\alpha$-hydroxylase KO, indicated that the lack of 1,25-dihydroxy vitamin $D_3$ or the VDR, suppressed the renin gene expression independent of the calcium metabolic pathway. The resultant conclusion is that 1,25-dihydroxy vitamin $D_3$ acts as an independent endocrine regulator of the renin–angiotensin–aldosterone system allowing a decrease in blood pressure (Chapter 11).

Due to the multiple levels of cell signaling involved in the regulation of gene expression, a new methodology, network analysis, has evolved (Chapters 12–15). The analysis is based on comparison of normal cell metabolism to dysfunctional cell metabolism using the same probes (external perturbation) and measuring changes over time in the transcription profiles used to compute and identify hub genes that control the intersection of those pathways.

For example, use of normal breast cells (MCF10A) and inflammatory breast cancer cells (SUM 149) in a network assessment identified the cancer cell hub genes IL-1A and IL-1B, cytokine synthesis, and NFκB were activated. Using EGFR as the probe, these markers were upregulated in the SUM 149 cells but not affected in the normal MCF10A cells. The blockade of the EGFR effect leads to growth arrest and loss of viability in the cancer cells, indicating a control site that could be targeted in functional nutritional genomics (Chapter 12).

Targeting of specific cancer genes, signaling proteins and transcription factors is now considered to have the most effective potential to prevent cancer. Nutrients and other dietary factors, such as, epigallocatechin gallate (EGCG), [6]-gingerol, resveratrol, and various flavonoids, including kaempherol, quercetin, and myricetin, are of interest because of their potential to affect protein kinases and/or transcription factors. Transcription factor proteins bind to specific DNA gene sequences to regulate transcription. Activator protein-1 (AP-1), nuclear factor-kappaB (NF-κB), p53, nuclear factor of activated T cells (NFAT), and cAMP response element binding (CREB) protein have been shown to play a critical role in carcinogenesis and are all regulated by the mitogen activated protein kinase (MAPK) cascades. Activation results in transcription of genes that encode proteins that regulate a multitude of cellular responses including apoptosis, differentiation, development, inflammation, and proliferation (Chapters 9 and 13). A large number of proteins have been screened to identify specific binding sites for the natural polyphenols. Importantly, those identified had binding constants at potential physiologic levels, and demonstrated numerous specific regulatory effects on protein targets. For example, EGCG binds to vimentin, which inhibits AP1 activation and inhibits the p65-CREB-NFκB pathway.

The components of green tea extracts have been shown to inhibit a variety of carcinogenic events *in vitro* cell culture and *in vivo* animal models. The polyphenols, including EGCG, have been examined. The green tea polyphenols were found to play an active preventive role in photoinduced carcinogenesis in animal models (Chapter 14). UVB causes skin damage, initiating DNA damage and stimulating an inflammation response. From green tea polyphenol intake, EGCG stimulates DNA repair and nucleotide excision repair of UVB skin damage. They also decrease production and release of inflammatory cytokines.

Certain bioactive nutrients produce a "U"-shaped health curve indicating an increased incidence of cancer at low and high concentrations (Chapter 15). A low intake of vitamin D is associated with an increased colorectal cancer risk while a similar population with recommended levels of vitamin D indicated a decreased risk. Using meta-analysis, low vitamin D status was also found to be related to a greater risk in breast, skin, and prostate cancer.

In addition to the kidney, multiple cells contain the mitochondrial 1α-hydroxylase enzyme. Active 1, 25-dihydroxy vitamin D stimulates cell differentiation and proliferation. The function of vitamin D is closely linked to the vitamin D receptor (VDR), which has a large number of polymorphisms. One of them, the Fok1 polymorphism, in conjunction with a low calcium level is related to an increased incidence of colorectal cancer (Chapter 15). Inclusion of more dietary vitamin D and calcium reversed the cancer risk of this group. On the other hand, elevated vitamin D has also been correlated with an increased risk of prostate cancer, pancreatic cancer, breast cancer, and esophageal cancer. This effect may be cell specific and/or dependent on precancerous development, such as initiated by smoking. The need then is to specifically quantify the dose, the cell types, and the regulation mechanisms by which vitamin D protects and may increase cancer development.

Similarly, low dietary intake of folate might increase the risk for colorectal cancer, as well as breast, ovary, pancreas, brain, lung, and cervical cancer. Animal studies indicate the regulatory pathways for cancer are expressed through the Wnt cascade and are p53 dependent. The result is a decrease in thymidylate synthesis and a decrease in methylation of cytosine in DNA. A protective effect, 40% reduction in cancer risk, was related to an elevated dietary folate intake and increased plasma levels. While evidence has indicated inadequate folate with cancer risk, increased folate levels in subjects with a history of neoplasms or existing polyps indicate an increase in cell growth and proliferation. Thus, elevated folate could also be a concern for individuals already predisposed to certain cancers (Chapter 15).

The gastrointestinal immune system reflects an array of intricate biological systems, including a highly regulated and carefully controlled inflammatory process and a diverse community of microbes entitled the human microbiome (Chapter 16). Individual inflammatory responses are genetically diverse. The gastrointestinal tract and the microbiome reflect a significant barrier to protect the body from challenges entering the gastrointestinal system. Selected bacterial colonization can alter intestinal physiology by gene modulation that impacts nutrient absorption, mucosal defenses, and xenobiotic metabolism. Thus, dietary choices can affect the inflammatory defense system.

The ultimate goal of nutritional genomics is to use whole foods in our diet to delay or prevent many of the catastrophic health outcomes that are currently overtaking our children, and impact their future health and longevity throughout their life time. Enhancement of nutrient and/or bioactive content in agricultural crops by genomic alteration is one of the important means to achieve this goal. The breeding challenge is to prevent loss of existing positive parameters while making nutritional gains. Specific agricultural breeding efforts have been used to increase calcium uptake into crops (Chapter 17) and genomic efforts to improve the nutritional content of lettuce (Chapter 18). Targeted crops are those consumed in large quantities, but are lacking essential nutrient value. Thus, small gains in content can provide significant impact on dietary intake status.

Another means to enhance nutritional value is to utilize food technology to enhance bioactive food ingredients in our food supply. Naturally occurring anthocyanins, in blue carrots and blueberries, and the carotenoids in other vegetables can contribute varying levels of bioactive components (Chapter 19). If used as an additive in processed food, the amount would contribute to the overall dietary intake of the bioactive profile.

Current manufacturing practices have continued to improve food safety and enhance preservation of nutritional quality and assure bioavailability of functional foods (Chapter 20). Novel processing techniques and product structuring at the nano-, micro-, and macrolevels include utilization of microwave, high hydrostatic pressure, high-intensity pulsed light, high-intensity pulsed electric fields, ultrasonics, and others. Discussion of their use indicates their successful retention of nutritional and bioactive ingredients.

New restructured whole foods have been designed to more effectively deliver bioactive components (Chapter 21). A processing method has been created to liquefy whole grains in a manner to assure delivery of all nutrients, fiber, and bioactive components into a very palatable "grain milk" which can also be added to sauces to enhance delivery of needed dietary fiber to meals or ready-to-eat (RTE) food products. The successful use of these methods may lead to enhancing the delivery of bioactive components from foods designed by agricultural intervention or by the creative processing of our foods contributing to enhanced bioavailability.

The closing chapter assesses the future of food labeling in the United States that will impact nutritional genomic foods and consumer selection (Chapter 22). The discussion examines the existing regulations of labeling and suggests areas needing broader review. Efforts to make the label easier to use are examined, and evaluated in light of FDA responses and decisions. The use of the product label must be considered an essential communication tool to aid consumers in selection of foods to meet individualized nutritional and bioactive component needs.

With this understanding, opportunities will occur to develop personalized nutritional interventions, using evidence-based genomic diets that enable individuals to implement dietary recommendations very early in life and maintain them as a lifelong regimen designed to achieve optimal health. Nutritional genomics provides an opportunity to link future health status for each of us and our families by clear identification of the diet–gene paradigm, enhancing health by personal decisions and commitment. The focus on personalized nutrition will have important implications for the agricultural and food industries to deliver food products to meet individual genomic needs. The future application of nutritional genomics may well bring revolutionary changes to health care and nutrition, changing how we prevent and treat disease, and how food is selected, grown, processed, and provided to the discerning health conscious consumer.

**Wayne R. Bidlack, PhD**
*Editor*

# Acknowledgments

We express our sincere gratitude to each of the authors who provided their time and expertise to enhance the scholarship of this book. A special thank you is extended to Deanna Stewart, College of Agriculture Office of Development at Cal Poly Pomona, for designing the logo used on the book cover.

We also extend our deep appreciation to Randy Brehm, Joette Lynch, Kari Budyk, and the rest of the editorial staff at Taylor & Francis Group, CRC Press, for their encouragement, eternal patience, and the quality of their editorial contributions, as well as to Dennis Troutman and the production team at diacriTech for their valuable and timely assistance in the preparation of this book and bringing this work to publication.

In conclusion we acknowledge the following organizations and their representatives for their sponsorship of the 2009 Nutritional Genomics Conference held at the California State Polytechnic University, Pomona, California, November 16–18, 2009, that led to this publication:

PEPSICO: Robert C. Brown, PhD, MPH, PepsiCo Director of Nutrition; Mark Pirner, MD, PhD, PepsiCo Director of Clinical and Scientific Development; David W. Still, PhD, Coordinator, Cal Poly Pomona ARI; Joe A. Bezerra, Executive Director, California State University ARI, Fresno. Southern California Institute Food Technologist Section: C. M. Barnes, SCIFT; Dan Rosson, Mane Company. NIMHD Center of Excellence in Nutritional Genomics: Raymond L. Rodriguez, PhD, Director, University of California, Davis. ET Horn: Lisa Alley-Zarkades, Vice President of Sales and Food Technology.

# Editors

**Dr. Wayne R. Bidlack**, PhD, is a professor in the Department of Human Nutrition and Food Science at California State Polytechnic University, Pomona, and is actively involved in teaching upper division advanced nutrition and nutritional genomics courses for undergraduate and graduate students. Previously, he served as dean of the College of Agriculture for 12 years (1995–2007).

After receiving his PhD in biochemistry from the University of California, Davis (1972), he completed a postdoctoral fellowship with Dr. Paul Hochstein in pharmacology at the University of Southern California School of Medicine, and joined the faculty in 1974. Dr. Bidlack served as professor of pharmacology and nutrition at the University of Southern California (USC) School of Medicine from 1974–1992. In addition, he served as chair of the department (1990–1992), and as the assistant dean of medical student affairs (1988–1991). During his tenure at USC he developed the nutrition curriculum for the department's MS/PhD program; required nutrition lectures (37 hours) were incorporated into year 1 and year 3 of the medical school curriculum. In 1992, he joined Iowa State University as professor and chair of the Department of Food Science and Human Nutrition, and as director of the Center for Designing Foods to Improve Nutrition at Iowa State University (1992–1995).

Dr. Bidlack's research interests include nutrition, biochemistry, pharmacology, and toxicology. His specific interests include biochemical mechanisms of hepatic drug metabolism and conjugation, lipid peroxidation reactions, metabolism of vitamin A, E, and C, and iron, and copper and zinc interactions. Currently, he has developed course materials and research areas in nutritional genomics. He has published numerous articles and book chapters, and edited the phytochemicals series published with CRC Press, Boca Raton, Florida.

Dr. Bidlack is a member of the American Society of Nutrition and the American Society of Pharmacology and Experimental Therapeutics. He is a professional member of the Institute of Food Technologists and was elected a fellow in 1998. He served as a member of "Functional Foods: Opportunities and Challenges" (Expert Report, 2005) and chaired the Expert Report Committee "Making Decisions about the Risks of Chemicals in Foods with Limited Scientific Information" (2009). Dr. Bidlack served as a member of the California Department of Food and Agriculture Board (2006–2008). He continues to serve the food industry as a consultant in a number of areas (protein modification, lipids, nutritional genomics, food and product safety, and others), and on scientific advisory boards for food companies assessing product quality and safety.

**Dr. Raymond L. Rodriguez**, PhD, is a professor in the section of molecular and cellular biology and director of the Center of Excellence in Nutritional Genomics at the University of California, Davis.

After receiving his PhD at the University of California, Santa Cruz in 1974, he was an A. P. Giannini Foundation Postdoctoral Fellow in the laboratory of Herbert W. Boyer at the University of California, San Francisco (UCSF) Medical Center. While at UCSF, Dr. Rodriguez developed molecular cloning technologies that now serve as the foundation of the modern biotechnology industry.

Dr. Rodriguez joined the faculty at the University of California, Davis in 1977 and is actively involved in research and teaching at the undergraduate and graduate level. In 1988, Dr. Rodriguez was a distinguished visiting professor in the International Center for Biotechnology at Osaka University, Japan, and in 1991, he was a visiting scientist with the Human Genome Project at the Lawrence Livermore National Laboratory.

From 1989 to 1992, Dr. Rodriguez founded and chaired the International Rice Genome Organization, a group that helped establish the framework for the sequencing of the rice genome. In 2000, Dr. Rodriguez cofounded the Laboratory for High-Performance Computing and Informatics in the Section of Molecular and Cellular Biology at UC Davis, and in 2003 he became director of the NIH Center of Excellence for Nutritional Genomics.

Dr. Rodriguez is a member of numerous scientific organizations and committees and he has served as an advisor to the NIH since 1988. He has published numerous articles and books on molecular biology and biotechnology and currently holds 17 U.S. patents. He previously coedited the book, *Nutritional Genomics: Discovering the Path to Personalized Nutrition* (New York, Wiley & Sons, 2006). His current research focus is nutritional epigenomics or the study of how plant-based dietary factors alter human gene activity by chromatin modification.

# Contributors

**Yuriko Adkins, PhD**
USDA/ARS Western Human Nutrition
 Research Center
University of California Davis
Davis, California

**Jorge N. Artaza, PhD**
Department of Medicine
David Geffen School of
 Medicine UCLA

and

Division of Endocrinology,
 Internal Medicine
Charles Drew University of Medicine &
 Science
Los Angeles, California

**Joseph A. Baur, PhD**
Institute for Diabetes, Obesity
 and Metabolism
Department of Physiology
University of Pennsylvania
 School of Medicine
Philadelphia, Pennsylvania

**Wayne R. Bidlack, PhD**
Department of Human Nutrition and
 Food Science
California State Polytechnic
 University Pomona
Pomona, California

**Ann M. Bode, PhD**
The Hormel Institute
University of Minnesota
Austin, Minnesota

**Evelyn D. Cadman**
Bioscience Translation and
 Application
Westminster, Colorado

**Roger A. Clemens, PhD**
ET Horn
La Mirada, California

**Ann Marie D. Craig, PhD**
Independent Food Consultant
Weatherby Lake, Missouri

**Kevin Dawson, MD**
Department of Information Sciences
City of Hope
Duarte, California

**T. Gregory Dewey, PhD**
University of La Verne
La Verne, California

**Zigang Dong, MD, PhD**
The Hormel Institute
University of Minnesota
Austin, Minnesota

**Ryan J. Elias, PhD**
Department of Food Science
The Pennsylvania State University
University Park, Pennsylvania

**Stephen P. Ethier, PhD**
Department of Oncology
Wayne State University School of Medicine
Detroit, Michigan

**Dawn M. Fedor**
USDA/ARS Western Human Nutrition
 Research Center
University of California Davis
Davis, California

**John D. Floros, PhD**
Department of Food Science
The Pennsylvania State University
University Park, Pennsylvania

**Alfredo F. Galvez, PhD**
Soy Labs, LLC
Missouri Plant Science Center
Mexico, Missouri

**Kendal D. Hirschi, PhD**
USDA, ARS, Children's Nutrition
    Research Center
Baylor College of Medicine
Houston, Texas

**Liping Huang, PhD**
USDA/ARS Western Nutritional
    Research Center
University of California Davis
Davis, California

**Santosh K. Katiyar, PhD**
Department of Dermatology
Comprehensive Cancer Center,
    Nutrition Obesity Research Center
University of Alabama Birmingham
Birmingham, Alabama

**Darshan S. Kelley, PhD**
USDA/ARS Western Human Nutrition
    Research Center
University of California Davis
Davis, California

**Sander Kersten, PhD**
Nutrition, Metabolism and
    Genomics Group
Division of Human Nutrition
Wageningen University
Wageningen, the Netherlands

**David M. Lonard, PhD**
Department of Molecular and
    Cellular Biology
Baylor College of Medicine
Houston, Texas

**Mark J. M. Magbanua, PhD**
Comprehensive Cancer Center
University of California San
    Francisco
San Francisco, California

**John A. Milner, PhD**
Nutritional Science Research Group
Division of Cancer Prevention
National Cancer Institute, National
    Institutes of Health
Rockville, Maryland

**Cheryl R. Mitchell, PhD**
Creative Research Management
Stockton, California

**Somen Nandi, PhD**
Center of Excellence in Nutritional
    Genomics
Department of Molecular and
    Cellular Biology
University of California at Davis
Davis, California

**Rama Natarajan, PhD**
Department of Diabetes
Beckman Research Institute
    of City of Hope
Duarte, California

**Keith C. Norris, MD**
Charles Drew University
Lynwood, California

**Jerrold M. Olefsky, MD**
Department of Medicine
University of California San Diego
La Jolla, California

**Bert W. O'Malley, MD**
Department of Molecular and
    Cellular Biology
Baylor College of Medicine
Houston, Texas

**Olivia Osborn, PhD**
Department of Medicine
University of California San Diego
La Jolla, California

**Raymond L. Rodriguez, PhD**
Center of Excellence in Nutritional
    Genomics
Department of Molecular and
    Cellular Biology
University of California Davis
Davis, California

**Sharon A. Ross, PhD, MPH**
Nutritional Science Research Group
Division of Cancer Prevention
National Cancer Institute
Bethesda, Maryland

**Simon Schenk, PhD**
Department of Medicine
University of California San Diego
La Jolla, California

**Sumeet Sharma**
USDA/ARS Western Human
    Nutrition Research Center
University of California Davis
Davis, California

**David W. Still, PhD**
Department of Plant Sciences
California State Polytechnic
    University Pomona
Pomona, California

**Katie L. Streicher, PhD**
Department of Oncology
Wayne State University School of Medicine
Detroit, Michigan

**Sean M. Thompson**
Department of Soil and Crop Sciences
Texas A&M University
College Station, Texas

**Zhidong Tu, PhD**
Merck Research Laboratories
Quincy, Massachusetts

**Behzad Varamini, PhD**
The University of Pennsylvania
    School of Medicine
Philadelphia, Pennsylvania

**Louisa M. Villeneuve, PhD**
Division of Diabetes
Beckman Research Institute
    of City of Hope
Duarte, California

**Sandra F. Williams, MD**
Division of Endocrinology
Cleveland Clinic Florida
Weston, Florida

**Xia Yang, PhD**
Sage Bionetworks
Seattle, Washington

**Jun Zhu, PhD**
Sage Bionetworks
Seattle, Washington

# Section I

## Transforming Dietary Signals into Gene Expression

### Epigenetics—The Epigenetic Transfer of Phenotype

# 1 Epigenetics
## *Molecular Targets for Diet and Cancer Prevention*

*Sharon A. Ross*

## CONTENTS

## INTRODUCTION

Cancer is a leading cause of death in the United States and the number of cancers is projected to increase in the coming decades due to a growing and aging population. Although cancer is several different diseases, it can be generally stated that cancer is influenced by both genetic and environmental factors. Importantly, dietary habits are recognized to be modifiable factors influencing cancer risk and prevention. The World Cancer Research Fund/American Institute of Cancer Research (WCRF/AICR 2007) summary of the available epidemiological evidence on food, nutrition, physical activity, and the prevention of cancer clearly support the suggestion that cancer incidence and death are potentially avoidable by modification of the diet as well as by physical activity. In addition to such observational evidence, both cell culture and animal experimental studies have suggested that several bioactive food components, including phytochemicals found in plants (Manson et al. 2007), zoochemicals such as conjugated linoleic acid found in dairy foods (Kelley et al. 2007), omega-3 fatty acids (Wendel and Heller 2009) present in certain types of fish, fungochemicals found in mushrooms (Adams et al. 2008), and bacteriochemicals (Geier et al. 2006) formed from food fermentation (pre) and those resulting from intestinal flora (pro) are likely to alter susceptibility to cancer. In fact, both essential nutrients and nonessential bioactive food components, have been implicated in altering many of the cellular processes of cancer, including apoptosis, cell cycle control, differentiation, inflammation, angiogenesis, DNA repair, and carcinogen metabolism (Davis and Milner 2007). In cancer, these processes are altered via deregulation of key genes, resulting in an altered cellular phenotype (Hanahan and Weinberg 2000). Such abnormal gene expression may result from genetic disruption, that is, mutation, or from epigenetic modulation by silencing genes that should be active or activating genes that should be silent. Diet and bioactive food components may impact both processes.

The emerging field of nutrigenomics utilizes high-throughput technologies to provide detailed information about the composition and functions of the genome, mechanisms for regulation of gene expression, and the influence of nutrients on gene and protein expression in these cellular processes. Nutrigenomic approaches in diet and cancer research recognize the potential of nutrients to act as signals to influence cell behavior and also should provide insights about variation in biological response. For example, variation in genetic influences on the diet due to gene polymorphisms (nutrigenetics) and/or dietary influences on gene expression (nutritional transcriptomics), DNA methylation, and other epigenetic events (nutritional epigenetics) may account for inconsistencies from study to study in the diet and cancer prevention literature (Trujillo et al. 2006). Understanding variation in response with the help of these "omics" approaches will also assist in the discovery of molecular targets. Identification and characterization of the critical molecular targets, including epigenetic targets, for bioactive components in cancer prevention has the potential to lead to evidence based dietary intervention strategies for preventing cancer.

Many studies provide significant evidence that part of the cancer inhibiting properties associated with several dietary components may relate to changes in epigenetic processes. These processes include DNA methylation of the cytosine phosphate guanine dinucleotide (CpG) islands in promoters as well as other regions of the genome, chromatin remodeling and higher order chromatin structural alterations, and post-translational ATP-dependent modifications of histone tail domains including methylation, acetylation, ubiquitination, and phosphorylation. Epigenetic marks and mechanisms as molecular targets for diet and cancer prevention as well as linkages between diet, epigenetics, and cancer prevention will be explored in this chapter by providing examples of as well as highlighting areas for further research.

## EPIGENETICS AND CANCER

Epigenetics refers to the study of meiotically and/or mitotically heritable changes in gene function that are not attributable to a change in the DNA sequence (Holliday 1987; Urnov and Wolffe 2001). The impact of epigenetic mechanisms in the etiology of cancer and other chronic diseases has been increasingly recognized in recent years. The cellular epigenetic machinery consists of chromatin, which contains a histone protein-based foundation around which DNA is wrapped, histone post-translational modifications, and covalent modifications of a methyl group to cytosines existing in the dinucleotide sequence cytosine guanine in DNA (McGowan et al. 2008). Specific epigenetic marks, typically working coordinately, will determine the accessibility of the transcriptional apparatus to the genome and therefore influence gene expression: genes are switched off when the chromatin is closed (heterochromatin), and they are switched on when the chromatin is open (euchromatin) (Rodenhiser and Mann 2006). Additionally, several proteins or protein complexes have recently been recognized for their ability to regulate chromatin structure and dynamics, including ATP-dependent chromatin remodeling factors (Varga-Weisz and Becker 2006). Noncoding RNAs (ncRNAs) have also been shown to provide an additional level of epigenetic regulation in the nucleus (Kim et al. 2008).

### DNA METHYLATION

The most widely studied epigenetic modification in humans involves the covalent addition of a methyl group ($CH_3$) to the 5'position of a cytosine that precedes a guanosine in the CpG dinucleotide (Esteller 2005). This reaction is catalyzed by DNA methyltransferases (DNMTs) and

S-adenosylmethionine (SAM) serves as the universal methyl donor (Ross 2003). Three DNMTs—DNMT1, DNMT3A, and DNMT3B—have been characterized for their involvement in developmental processes and in cancer (Hermann et al. 2004). Regions rich in CpG dinucleotides, termed CpG islands, often occur in the promoter regions of genes (Esteller 2007). These regions are usually unmethylated, which is associated with the ability of CpG-island-containing genes to be transcribed in the presence of the required transcriptional activators. Methylation at these critical sites, however, inhibits the binding of transcription factors to their recognition elements, recruits DNA binding proteins such as methyl CpG binding protein 2 (MeCP2) and methyl-CpG-binding domain protein 2 (MBD2) to the gene, and activates chromatin modification enzymes such as histone deacetylases (HDACs), which in turn introduce histone modifications, resulting in chromatin silencing (Li 2002). Interestingly, the transcriptional silencing of tumor-suppressor genes by CpG-island-promoter hypermethylation is at least as common as DNA mutations as a mechanism for inactivation of classical tumor suppressor genes in human cancer (Jones and Baylin 2002; Tsou et al. 2002). This aberrant methylation of genes, including tumor suppressor genes, appears to occur early in tumor development and increase progressively, eventually leading to the malignant phenotype (Fearon and Vogelstein 1990; Kim and Mason 1995). Widespread global DNA hypomethylation (Ehrlich 2002) and increased DNA methyltransferase activity (Kautiainen and Jones 1986) are also common characteristics of tumor cells. Thus, aberrant DNA methylation in cancer cells may result in inappropriate under- and over-expression of specific genes, which has been suggested as a mechanism to promote malignant transformation and progression. Importantly, DNA methylation changes are thought to be inherited mitotically in somatic cells, providing a potential mechanism whereby environmental factors, including dietary exposures, can have long-term effects on gene expression (Wolffe 1994).

## HISTONE POST-TRANSLATIONAL MODIFICATIONS

Histones, as the chief protein components of chromatin, undergo a wide range of post-translational modifications, most of which are reversible, that trigger structural changes in the chromatin to confer functional properties. These modifications, which usually occur at certain amino acids that constitute the N-terminal tails of histones, can either facilitate or hinder the association of DNA repair proteins and transcription factors with chromatin and include methylation (Jenuwein 2001), phosphorylation (Oki et al. 2007), acetylation (Wade et al. 1997), sumoylation (Shiio and Eisenman 2003) ubiquitination (Shilatifard 2006), and biotinylation (Kothapalli et al. 2005). For example, histone acetylation neutralizes the positive charge on histones and disturbs the electrostatic interactions between DNA and histone proteins, which promotes chromatin unfolding and has been associated with gene expression (Zhang and Dent 2005). In contrast, deacetylation and condensation typically suppress transcription (Shukla et al. 2008). The specific pattern of histone modifications have been proposed to form a "histone code," which is thought to explain how regions of the genome are expressed at a given point in time in a given cell type (Jenuwein and Allis 2001). It has been hypothesized that similar to a genetic mutation; a change in the post-translational modification(s) of histone tails around a regulatory region of a gene can silence an active gene, or activate a silent gene. Such modifications may also enhance or impair the extent of gene expression in the absence of complete gene silencing or activation. It is therefore not surprising that aberrant histone post-translational modifications have been associated with cancer. For example, a common characteristic of human tumor cells is the loss of monoacetylation and trimethylation of histone H4 (Fraga et al. 2005).

Enzymes that add or remove histone modifications affect a wide range of chromatin-based events including transcription, replication, recombination, and repair, as well as chromosome condensation and nuclear organization (Hake et al. 2004). For example, the acetylation state of histone tails is controlled by the antagonistic action of two enzyme families: histone acetyltransferases (HATs), which transfer an acetyl group from acetyl CoA to an epsilon-amino group of lysine residues of histones, and HDACs, which catalyze the removal of the acetyl group. Abnormal activities of both

HATs and HDACs have been linked to the pathogenesis of cancer (Atsumi et al. 2006; Marks and Dokmonovic 2005; Zhu et al. 2004; Gibbons 2005; Gayther et al. 2000). Although inactivating mutations of HATs and over-expression of HDACs have been described (Mahlknecht and Hoelver 2000), another functional link between HDACs and cancer progression comes from models of acute promyelocytic leukemias, where chromosomal translocations result in chimeric proteins that alter transcriptional events and thereby interfere with normal cell growth, differentiation, and apoptosis (Lin et al. 1998). For example, chromosomal translocations between retinoic acid receptor $\alpha$ (RAR$\alpha$) and the promoyleocytic leukemia protein leads to inappropriate recruitment of co-repressors and HDACs, which abolishes the ability of the RAR$\alpha$ to promote myelocytic differentiation (Atsumi et al. 2006). Aberrant HDAC activity has also been associated with transcriptional silencing of tumor-suppressor genes, including *p21*, which encodes a cyclin-dependent kinase inhibitor that blocks cell cycle progression from $G_1$ into S phase (Gibbons 2005). Moreover, the expression of *p21* has been found to be reduced in many different tumors, allowing uncontrolled cell growth. Interestingly, HDAC inhibitors have been shown to reactivate *p21* expression, thereby inhibiting tumor cell proliferation (Gibbons 2005). This HDAC inhibitor-induced expression of *p21* has also been correlated with an increase in the acetylation of histones associated with the *p21* promoter region. These discoveries have led to the development of HDAC inhibitors as chemotherapeutic agents in clinical trials (Rosato and Grant 2003).

## Chromatin Proteins and Remodeling Complexes

Several proteins or protein complexes have recently been recognized for their ability to regulate chromatin structure and dynamics. For example, ATP-dependent chromatin remodeling factors define accessibility to the transcription machinery via altering the position of nucleosomes around the transcription start site (Varga-Weisz and Becker 2006). One example of an ATP-dependent chromatin remodeling complex is the DNA helicase/ATPase-containing complex termed nucleosome remodeling and deacetylase corepressor complex (NuRD), which represses transcription through chromatin remodeling (Xue et al. 1998). NuRD has recently been found to direct aberrant gene repression and transmission of epigenetic repressive marks in acute promyelocytic leukemia (Morey et al. 2008).

Another type of chromatin protein complex includes the polycomb group (PcG) proteins, containing at least two distinct complexes, PcG complex 1 and 2, which function as transcriptional repressors to silence specific sets of genes through coordinated chromatin modification (Sparmann and van Lohuizen 2006; Takihara 2008; Bracken and Helin 2009; Gieni and Hendzel 2009). PcG complex 2, composed of several factors/proteins, including histone methylase activity, is first recruited to silence chromatin with concomitant methylation of histone H3 at lysine 27 (K27me3). This is followed by PcG complex 1 recruitment through recognition of this histone mark, which then triggers ubiquitination of histone H2A and/or inhibits chromatin remodeling to maintain the silenced state of the locus (Takihara 2008; Bracken and Helin 2009). Knockout mice lacking components of the PcG complex have offered evidence that these chromatin repressive complexes are essential for sustaining stem cell activity. These proteins also contribute to the pathogenesis of cancer (Sparmann and van Lohuizen 2006; Bracken and Helin 2009; Gieni and Hendzel 2009). Enrichment of polycomb repressive complexes has been correlated with cancer progression and prognoses as well as cancer stem cell activity (Simon and Lange 2008). Recent findings have shown that the PcG targeted genes in normal cells may be more likely to acquire aberrant promoter hypermethylation in cancers (Widshwendter et al. 2007; Mohammad et al. 2009). This observation along with evidence of interactions between DNMT1/DNMT3B and several PcG complex subunits (e.g., EZH2 and BMI1) (Vire et al. 2006), suggest clues about a possible mechanism for gene silencing. Additional research is needed to better understand how and in what context these and other epigenetic mechanisms and regulatory factors may interact to regulate chromatin structure, dynamics, and gene expression as well as how they become deregulated during cancer development, so that this knowledge can be exploited to understand how to prevent cancer.

## NONCODING RNA

The role of ncRNA molecules in the regulation of gene expression is an emerging area of research. ncRNAs, including microRNAs, small interfering RNAs, PIWI-interacting RNAs, and various classes of long ncRNAs, perform critical roles as transcriptional and post-transcriptional regulators and in directing chromatin-modifying complexes (Taft et al. 2010). For example, microRNAs have been shown to modulate post-transcriptional silencing through targeted degradation of mRNAs. There is considerable scientific interest in studying deregulation of microRNAs in various diseases, including cancer (Fabbri et al. 2008). In addition, epigenetic modifications have recently been found to be induced and directed by ncRNA molecules in human cells (Hawkins and Morris 2008; Malecová and Morris 2010). In this capacity, ncRNAs are thought to act like the exogenous small inhibitory RNAs (siRNAs) in gene inactivation. In fact, the endogenous small RNA mediated transcriptional gene silencing described by Hawkins and Morris was suggested to be correlated with changes in chromatin structure (including modulation of histone marks and DNA methylation) at specific sites in promoter regions (Hawkins and Morris 2008). Another type of ncRNA, long ncRNAs, have been shown to have a variety of functions, but one of their primary roles appears to be as epigenetic regulators of protein-coding gene expression (Taft et al. 2010).

Bioactive food components have been shown to impact epigenetic mechanisms, including DNA methylation, histone posttranslational modifications, chromatin remodeling, and other chromatin factors. The impact of dietary components on these processes and their role in cancer development and prevention are highlighted in this review. Little is known about the influence of dietary constituents on expression and function of ncRNA that mediate transcriptional gene silencing or that act to alter chromatin structure and function, and therefore it cannot be discussed in this chapter. Ultimately, it will be important to understand how each of these epigenetic processes and other newly identified epigenetic components interact to regulate gene expression and influence cancer risk and prevention.

## DIET AND DNA METHYLATION: IMPORTANCE OF EARLY EXPOSURE AND RISK OF CHRONIC DISEASE

Recent evidence from preclinical models suggests that prenatal and early postnatal diet may alter DNA methylation of important genes or genomic regions which could influence the risk of developing disease, such as cancer, later in life (Wolff et al. 1998; Waterland and Jirtle 2003; Cropley et al. 2006; Dolinoy et al. 2006; Waterland et al. 2006a,b). For example, maternal nutrient restriction impacted global methylation during fetal baboon development which resulted in long-term programming effects (Unterberger et al. 2009). These effects were organ-specific and gestational-age specific. Moreover, a recent study found that individuals who were prenatally exposed to famine during the Dutch Hunger Winter in 1944–1945 had, six decades later, less DNA methylation of the imprinted insulin-like growth factor II (*IGF2*) and increased methylation of *IL10*, *LEP*, *ABCA1*, *GNASAS*, and *MEGF* compared with their unexposed, same-sex siblings (Tobi et al 2009; Heijmans et al. 2008). Additionally, a significant interaction with gender was observed for a number of these genes (*INSIGF*, *LEP*, and *GNSAS*). These data indicate that persistent alterations in DNA methylation may be a common consequence of prenatal famine exposure and that these changes depend on the gender of the exposed individual as well as the gestational timing of the exposure. Although these investigators did not link these findings with health and disease outcomes later in life, the results support the fetal basis or developmental origins of the adult-onset disease hypothesis. This hypothesis implies that an organism can adapt to environmental signals in early life, but that these adaptations may also increase the risk of developing chronic diseases, including cancer, later in life when there is incongruity between the perceived environment and that which is encountered in adulthood.

There is also evidence to suggest that environmental exposures at other vulnerable periods can influence phenotype later in life through epigenetic changes. Investigators using the Netherlands Cohort Study on diet and cancer reported that severe caloric restriction during adolescence was associated with a lower risk of developing colorectal cancer later in life (Dirx et al. 2003; Hughes et al. 2010). Recently, these investigators also explored the association between early life energy restriction (individuals exposed to severe famine during the Hunger Winter) and risk of subsequent colorectal cancer characterized by the (promoter) CpG island methylation phenotype (CIMP) (Hughes et al. 2009). This phenotype has a distinct characteristic of epigenetic instability in colorectal cancer and is characterized by numerous promoter CpG island hypermethylated tumor suppressor genes and DNA repair genes (Boland et al. 2009). Interestingly, it was found that individuals exposed to severe famine in adolescence and early adulthood during the Hunger Winter had a decreased risk of developing a tumor characterized by CIMP compared to those not exposed (HR 0.65, 95%CI: 0.45–0.92). Further categorizing individuals using a methylation index suggested that the degree of hypermethylation was inversely associated with exposure to energy restriction. These results suggest that exposure to a transient environmental condition early in life may result in persistent epigenetic changes that later influence colorectal cancer development.

## Yellow Agouti (A$^{vy}$) Mouse Model

Some of the best evidence for the influence of in utero dietary exposure on DNA methylation and subsequent phenotype comes from studies utilizing the yellow agouti (A$^{vy}$) mouse model. An endogenous retrovirus-like transposon sequence is inserted close to the gene coding for the agouti protein (Duhl et al. 1994) in this mouse model. When this retrotransposon is active, a cryptic promoter in its 5′ long terminal repeat (LTR) takes over transcriptional control of the agouti signaling protein (Duhl et al. 1994). This ectopic agouti expression results in a yellow coat color and obesity (mice over-expressing agouti are hyperphagic), as well as increased susceptibility to chronic diseases, including cancer. When the retrotransposon is silent the agouti protein is expressed in its normal pattern giving the wild-type agouti coat color, which is termed pseudoagouti. Dietary supplementation with a combination of folic acid, vitamin B12, choline, betaine, and zinc to yellow agouti dams in utero has been shown to alter DNA methylation and phenotype of their offspring (Wolff et al. 1998; Waterland and Jirtle 2003; Cropley et al. 2006; Cooney et al. 2002). In initial experiments, supplementation of maternal diets was associated with a change in coat color from a yellow to a pseudoagouti coat in the offspring (Wolff et al. 1998). This phenotype change is typically associated with a lower risk of cancer, diabetes, obesity and prolonged life in this model (Wolff et al. 1998; Cooney et al. 2002). Furthermore, representative yellow mice displayed more hypomethylated long terminal repeats 5′ of the agouti gene compared to the mice with the agouti coat color (Cooney et al. 2002). Another group of investigators verified and expanded these findings when pregnant dams were fed a diet supplemented with folic acid, vitamin B12, choline, and betaine, but not zinc, and found that coat color changes were directly associated with alterations in DNA methylation and there was a distribution shift toward increased CpG methylation at the A$^{vy}$ locus with methyl supplementation (Waterland and Jirtle 2003). Moreover, the coat color phenotype and A$^{vy}$ methylation relationship persisted into adulthood as indicated by a comparison of tail DNA at 21 days and liver DNA at 100 days. These studies demonstrated that maternal methyl donor supplementation during gestation could alter the phenotype of the offspring by methylation changes in the epigenome. It is not yet evident, however, which of the dietary factors are necessary or sufficient for the DNA methylation and phenotypic change.

To make the agouti story more complex, a recent detailed examination of methylation at the A$^{vy}$ locus found that the promoter is incompletely methylated when it is transcriptionally silent and that in utero exposure to supplemental methyl donors did not increase the density of CpG methylation in the silent LTR compared with methylation densities in mice that had never received supplementation (Cropley et al. 2010). The results suggest that methyl donor supplementation acts through a mechanism other than DNA promoter methylation to silence the A$^{vy}$, possibly though histone modifications, the binding of chromatin proteins or involving epigenetic or other mechanisms operating at other loci. Additional studies are needed to clarify the role of methyl supplementation on epigenetic targets during in utero exposure in this model.

The agouti model has also recently been utilized to examine whether diet has transgenerational effects on phenotype. Interestingly, passing the A$^{vy}$ allele through 3 maternal generations resulted in amplification of obesity in the offspring of the mice fed a standard diet in each successive generation (Waterland et al. 2008). By the third generation, however, offspring of methyl-supplemented animals had a significant decrease in body weight relative to the unsupplemented group. These results suggest a preventive effect of methyl-supplementation on transgenerational amplification of obesity in adulthood. Interestingly, the transgenerational effect was not explained by progressive A$^{vy}$ methylation in the methyl-supplemented group, and the authors speculate that rather than occurring at A$^{vy}$ locus, the transgenerational effect of methyl supplementation on body weight among isogenic A$^{vy}$/a mice may involve epigenetic mechanisms operating at other loci. This study is important because it examines the effect of maternal obesity among three generations of genetically identical mice which had an inclination to overeat due to the A$^{vy}$ allele. The authors mention the intriguing hypothesis that the supplements investigated affected body weight by interfering with the area of the brain that regulates appetite.

It is important to clarify how observations in the agouti model may relate to DNA methylation and other epigenetic alterations, phenotype and disease risk in humans. Although the A$^{vy}$ locus (a retrovirus-like transposon sequence) is not found in the human genome, there is the possibility that metastable epialleles ("metastable" refers to the labile nature of the epigenetic state of these alleles; "epiallele" refers to the capacity to maintain epigenetic marks transgenerationally) (Rakyan et al. 2002) associated with other transposable elements could likewise be influenced by methylation or another epigenetic regulating process by in utero exposure to dietary factors. An intriguing hypothesis is that transposable elements in the mammalian genome may cause significant phenotypic variability, making each individual mammal a "compound epigenetic mosaic" (Whitelaw and Martin 2001). Whether or not such an epigenetic mosaic exists in humans, whether it can be modulated by (early) diet, whether there is a critical time period for exposure, and whether such phenotypes alter susceptibility to chronic disease in adulthood requires further study. Additionally, regions of the genome that may be susceptible to epigenetic variation need to be identified and characterized in human tissues.

## MECHANISTIC INSIGHTS FROM METHYL-DEFICIENT DIET-INDUCED HEPATOCARCINOGENESIS

A classic example for the influence of diet in DNA methylation and cancer development is the finding that dietary methyl-deficiency has been shown to alter hepatic DNA methylation patterns and induce hepatocarcinogenesis in rodents (Poirier 1994). The methyl-deficient model of endogenous hepatocarcinogenesis in rodents is distinctive in that dietary omission rather than the addition of chemical carcinogens can lead to tumor development (Pogribny et al. 2007). In particular, deficiency of the major dietary sources of methyl groups—methionine, choline, folic acid, and vitamin B-12—is sufficient to induce liver tumor formation in male rats and certain mouse strains (Newberne 1986; Poirier 1994; Denda et al. 2002; Christman 2003). The methyl-deficiency generated in these experimental animals has been associated with a number of defects, including increased genome-wide and gene-specific hypomethylation (Christman 2003; Wainfan and Poirier 1992; Christman et al. 1993; Pogribny et al. 2004). Importantly, the aberrant epigenetic alterations imposed by this diet have been thought to be the primary mechanism responsible for malignant transformation of rat liver cells in these animals (Christman 2003; Wainfan and Poirier 1992; Pogribny et al. 2004, 2006a,b). In further experiments, the effects of feeding a methyl-deficient diet (for 9, 18, 24, and 36 weeks) followed by a methyl-adequate diet (for a total of 54 weeks) on liver DNA methylation and on the development and progression of altered hepatic foci were investigated (Pogribny et al. 2006b). Three major effects were observed: a rapid decrease in hepatic SAM levels, which was increased to control levels upon refeeding the methyl-adequate diet; refeeding the complete diet restored DNA methylation to normal levels in the group that had been fed the methyl-deficient diet for 9 weeks; and the formation of persistent altered hepatic foci (as placental isoform of glutathione-S-transferase (GSTpi)-positive lesions) at all investigated time points. The results suggest that biological and biochemical changes associated with the early stages of hepatocarcinogenesis occurred between 9 and 18 weeks of methyl deprivation. The persistence of altered hepatic foci in liver even after providing the methyl-adequate diet suggests the carcinogenic potential of a methyl-deficient diet. Furthermore, the association between DNA hypomethylation and expansion of foci suggests that stable DNA hypomethylation is a promoting factor for clonal expansion of initiated cells. These results provide experimental evidence and a mechanistic basis by which epigenetic alterations may contribute to the initiation and promotion steps of carcinogenesis and also point to the importance of timing of an environmental impact in carcinogenesis. Significant changes in other aspects of the epigenetic machinery have been observed in early stages of hepatocarcinogenesis induced by methyl deficiency in rats, including aberrant expression of DNA methyltransferases and methyl CpG binding proteins (Ghoshal et al. 2006), defects in histone methyltransferase protein expression and histone post-translational modifications (Pogribny et al. 2006a). Evidence also suggests that early dysregulation of microRNAs in the process of hepatocarcinogenesis could be reversed by restoring methyl donors (Kutay et al. 2006). The epigenetic changes observed in the methyl-deficiency induced hepatocarcinogenesis model are thought to be important contributing factors in the formation of epigenetically reprogrammed cells that become neoplastic. How these epigenetic mechanisms are interrelated and which mechanism is the earliest trigger toward fixing the neoplastic state requires further study. The sequence of pathological and molecular events in the methyl-deficient model of liver carcinogenesis is astonishingly similar to the development of human hepatocellular carcinoma associated with viral hepatitis B and C infections, alcohol exposure, and metabolic liver diseases (Powell et al. 2005). The biochemical and molecular events predisposing to cancer in the methyl-deficient-induced hepatocarcinogenesis are the result of chronic metabolic stress and the model provides an ideal system to study progressive epigenetic alterations that might occur during human carcinogenesis and may also provide insight for understanding prevention and intervention strategies for this cancer.

## EPIGENETIC EPIGEMIOLOGY: DNA METHYLATION AS A MOLECULAR TARGET

Waterland and Michels (2007) define epigenetic epidemiology as the study of the associations between epigenetic variation and risk of disease. Investigators are beginning to use population-based epidemiological studies for characterizing the relationship between epigenetic variability and cancer, as well as for identifying associations between dietary exposure, epigenetic marks, and cancer risk. It has been suggested that DNA methylation may be the most suitable epigenetic mark for large-scale epidemiological studies, since methyl groups are covalently bound to CpG dinucleotides and are not lost during routine DNA extraction, unlike histone modifications (Talens et al. 2010). A recent study indicated that there are fairly good prospects for the use of biobanks for such epigenetic studies (Talens et al. 2010). It was found that loci suitable for testing in epigenetic studies demonstrate interindividual variation in DNA methylation, stability of this variation in DNA methylation over time, and a correlation between DNA methylation measured in blood to the tissue of interest. The investigators stressed, however, that meeting these criteria was found to be locus-dependent.

In an observational setting, investigators explored the relationships of lifestyle, demographic, dietary, and genetic factors with genomic methylation, using a LINE-1 (long interspersed nucleotide elements) pyrosequencing assay, in normal mucosal biopsies from individuals in a clinical trial of aspirin and folate for the prevention of large bowel adenomas (Figueiredo et al. 2009). Overall, there was relatively little variability in LINE methylation. Samples from the right side of the normal colon had significantly lower mean LINE-1 methylation levels than those on the left. Otherwise, no significant associations were found between LINE-1 methylation and folate treatment, age, sex, body mass index, smoking status, alcohol use, dietary intake, or circulating levels of B vitamins, homocysteine, or selected genotypes. Because these investigators did not find that LINE-1 methylation is influenced by folic acid supplementation, they suggested that if diet is associated with colorectal cancer it may not operate through genomic methylation. It may also be that other epigenetic loci besides LINE-1 methylation are targeted in colorectal cancer development. Technologies are rapidly being developed to scan the whole genome for locus-specific variation in DNA methylation. Once available, such epigenomic tools will accelerate the discovery of human loci at which epigenetic regulation is correlated with dietary and other environmental exposures.

Gene promoter methylation has putative utility for use as a biomarker for early lung cancer detection and to identify persons who would benefit most from chemoprevention. In addition, epidemiologic studies suggest that dietary fruits and vegetables and the dietary constituents they contain may reduce risk of lung cancer. For these reasons, researchers recently evaluated whether diet and multivitamin use influenced the prevalence of gene promoter methylation in cells exfoliated from the aerodigestive tract of current and former smokers (Stidley et al. 2010). In this study, subjects in the Lovelace Smokers Cohort completed the Harvard Food Frequency Questionnaire and provided a sputum sample that was assessed for promoter methylation of eight genes commonly silenced in lung cancer and associated with risk for this disease. Methylation status was categorized as low (fewer than two genes methylated) or high (two or more genes methylated). Significant protection against methylation was observed for consumers of leafy green vegetables and folate as well as with current users of multivitamins. These findings support the notion that novel interventions to prevent lung cancer could be explored based on the ability of diet and dietary supplements to affect reprogramming of the epigenome. Future prospective investigations are required to determine whether subjects with specific dietary patterns develop epigenetic alterations over time and, in turn, whether such alterations increase or decrease the risk of cancer.

## DIETARY POLYPHENOLS REACTIVATE SILENCED GENES VIA EPIGENETIC MECHANISMS

Dietary components have been shown to reactivate inappropriately silenced genes (e.g., tumor suppressor genes) by enhancing demethylation or reversing hypermethylation-induced inactivation of key tumor suppression or DNA repair genes in cell culture systems. For example, dietary components have been shown to modulate DNA methylation by interfering with DNMT activity, which may have implications for cancer prevention. Epigallocatechin 3-gallate (EGCG) (5-50 microM) from green tea and genistein (2-20 μmol/L) from soybean have been found to restore methylation patterns and gene expression of tumor suppressor genes in neoplastic cells in culture (Fang et al. 2003, 2005, 2007). Both dietary components have been shown to inhibit DNA methyltransferase activity by binding to the enzyme. This observation was associated with demethylation of CpG islands in the gene promoters and the reactivation of methylation-silenced genes such as *p16INK4a*, retinoic acid receptor beta, O6-methylguanine methyltransferase, human mutL homolog 1, and glutathione S-transferase-pi (Fang et al. 2003, 2005, 2007). Changes in the expression of these genes have been associated with growth inhibition in human esophageal, colon, prostate, and mammary cancer cell lines. The ability of these dietary constituents to directly inhibit DNMT activity needs to be confirmed in other settings.

Epigenetic silencing of gluthathione-S-transferase pi (*GSTP1*) is a common observation in human prostate cancer. Given the previous findings for EGCG (Fang et al. 2003, 2005, 2007), a recent study examined the effects of green tea polyphenols (GTPs) on *GSTP1* re-expression (Pandey et al. 2010). Exposure of human prostate cancer LNCaP cells to 1–10 lg/ml of GTP for 1–7 days resulted in a concentration- and time-dependent re-expression of *GSTP1*, which correlated with DNMT1 inhibition. Methyl-specific-polymerase chain reaction and sequencing uncovered extensive demethylation in the proximal *GSTP1* promoter and regions distal to the transcription factor binding sites. These investigators also examined changes in the epigenetic silencing machinery following GTP exposure with intriguing results. GTP exposure diminished the mRNA and protein levels of the methyl binding proteins MBD1, MBD4, and MeCP2 as well as HDAC 1–3 and increased the levels of acetylated histone H3 (LysH9/18) and H4. Chromatin immunoprecipitation assays also established that cells treated with GTP had reduced MBD2 association with accessible Sp1 binding sites leading to increased binding and transcriptional activation of the *GSTP1* gene. These interesting results demonstrate that GTPs have the potential to alter both DNA methylation and chromatin modeling and the authors also suggest that their lack of toxicity makes them excellent candidates for the chemoprevention of prostate cancer.

Another group tested whether oral consumption of GTPs affects normal or cancer-specific DNA methylation *in vivo*, using wild-type and transgenic adenocarcinoma of mouse prostate (TRAMP) mice (Morey Kinney et al. 2009). Mice were given 0.3% GTPs in drinking water beginning at 4 weeks of age. The methylation parameters monitored were 5-methyl-deoxycytidine (5mdC) levels, methylation of the B1 repetitive element, and methylation of the *Mage-a8* gene. In wild-type mice a decrease of 5mdC in the liver at 12 weeks was the only parameter to change. In GTP-treated TRAMP mice, 5mdC levels and the methylation status of four loci hypermethylated during tumor progression were unaltered in TRAMP prostates at 12 or 24 weeks. Furthermore, no alterations in DNA methylation status were observed following a dose-response experiment in the TRAMP model. Genome-wide DNA methylation profiling also revealed no significant hypomethylating effect of GTP. Unlike the reports by others using the TRAMP model, GTP treatment also did not inhibit tumor progression in these mice. These data indicate that oral administration of GTPs does not affect normal or cancer-specific DNA methylation in the murine prostate. It is necessary for potential DNA methylation inhibitors to be active *in vivo* in order to use them to prevent or treat human cancer. This study does not support the contention that green tea is a DNA methylation inhibitor, and emphasizes the importance of combining *in vitro* analyses with *in vivo* validation studies. That these investigators did not find GTP treatment to inhibit tumor progression in these mice may also provide clues as to why methylation changes were not observed.

Polyphenols from other food sources have also recently been examined for their ability to reverse hypermethylation-induced inactivation of key tumor suppression genes. The Annurca apple, a variety of southern Italy, is rich in cancer protective polyphenols. Interestingly, it has been reported that populations in southern Italy who consume these apples have lower incidences of colorectal cancer than elsewhere in the Western world. Annurca polyphenol extract (APE) was evaluated for cancer protective effects in RKO, SW48, and SW480 human colorectal cancer cells (Fini et al. 2007). Because some sporadic colorectal cancers display CIMP, DNA methylation of selected tumor suppressor genes were also evaluated in these cells after treatment with APE and compared with the synthetic demethylating agent 5-aza-2′deoxycytidine (5-aza-2dC). APE administration (a polyphenol dose comparable to that from dietary consumption of 1 apple) decreased cell viability and enhanced apoptosis in the RKO and SW48 cell lines, both *in vitro* models for CIMP. A similar dose of APE reduced DNA methylation in the promoters of *hMLH1*, *p14(ARF)*, and *p16(INK4a)* genes with subsequent restoration of normal mRNA expression in RKO cells. These effects were qualitatively comparable with those obtained with 5-aza-2dC. In addition, a significant reduction in expression of DNMT-1 and -3b proteins without changes in mRNA was also observed after treatment with APE. Thus, APE appears to decrease gene-specific DNA methylation through the inhibition of DNMT proteins. The ability of polyphenols to reverse hypermethylation-induced inactivation of key tumor suppression genes is an intriguing anticancer activity that requires confirmation in human cells as well as *in vivo*.

Although the interactions between bioactive food components and DNA methylation are among the earliest studies of the relationship between diet and epigenetics in cancer prevention, there continues to be a growing body of literature encompassing more and more dietary factors that may impact DNA methylation at various times of vulnerability, including during cancer development and prevention. How these single observations will be united to reflect the complexity of human dietary patterns as well as how the many epigenetic mechanisms and marks will be integrated to provide an understanding of specific and global gene regulation are the hopeful outcomes of future endeavors.

## HISTONE ACETYLATION/DEACETYLATION, BIOACTIVE FOOD COMPONENTS, AND CANCER PREVENTION

Several dietary factors that have been associated with cancer prevention have also been shown to inhibit HDAC enzymes (Garfinkel and Ruden 2004; Myzak and Dashwood 2006), and to enhance histone acetylation (Myzak and Dashwood 2006). These include butyrate (formed in the colon from the fermentation of dietary fiber), diallyl disulfide (present in garlic and other *Allium* vegetables), and isothiocyanates (found in cruciferous vegetables). Moreover, these dietary components have all been shown to inhibit cell proliferation and stimulate apoptosis in a manner analogous to other nondietary HDAC inhibitors. For example, sulforaphane (SFN) (3-15 µM) has been shown to inhibit HDAC activity and in tandem increase acetylated histones in several cell systems, including human embryonic kidney 293 cells, HCT116 human colorectal cancer cells, and various prostate epithelial cells lines (BPH-1, LnCaP, and PC-3) (Myzak and Dashwood 2006). This response by sulforaphane was linked with increased apoptosis and cell cycle arrest at the $G_2/M$ phase (Myzak and Dashwood 2006). Moreover, these observations were associated with increased acetylated histone H4 in the *p21* promoter and concomittant increased *p21* protein expression in a dose dependent fashion (Myzak et al. 2004, 2006b). Using the $Apc^{min}$ mouse model, sulforaphane (443 mg/kg diet) suppressed tumor development and increased acetylated histones in gastrointestinal polyps, including acetylated histones specifically associated with the promoter region of the *p21* and *bax* genes (Myzak et al. 2006a). Another intriguing finding concerned healthy human subjects (N = 3) wherein a single ingestion of 68 g (1 cup) of broccoli sprouts rich in SFN inhibited HDAC activity in circulating peripheral blood mononuclear cells 3–6 hours after consumption, with coincident induction of histone H3 and H4 acetylation (Myzak et al. 2007). The biological consequences of reduced HDAC activity and enhanced histone acetylation in normal cells (as compared to cancer cells) requires further study. Although SFN has been shown to selectively induce apoptosis and growth inhibition in cancer cells but not in normal cells, additional research is warranted to determine beneficial versus harmful responses to bioactive dietary components during vulnerable periods.

Another isothiocyanate present in cruciferous vegetables, namely phenethyl isothiocyanate (PEITC, 1 µM), was also found to inhibit the level and activity of HDACs in both androgen-dependent and androgen-independent prostate cancer cells, induce selective histone acetylation and histone methylation for chromatin unfolding (Wang et al. 2007). Chromatin immunoprecipitation revealed that the *p21* gene was associated with PEITC-induced hyperactylated histones (Wang et al. 2008). As a result, the chromatin opening/unfolding permitted the transcriptional activation of the *p21* gene. PEITC also significantly reduced the expression of *c-Myc* which represses *p21*. These results demonstrate that cruciferous compounds can inhibit HDAC activity *in vitro* and *in vivo*, which appears to contribute to their cancer preventive activity.

The mechanism of action for selenium in cancer prevention has been an active area of research since the observation that selenium supplementation decreased prostate cancer incidence in the Nutritional Prevention of Cancer Trial (Clark et al. 1996). In fact, a recent question in selenium research concerns why selenium-enriched yeast was protective against prostate cancer in the

Nutritional Prevention of Cancer trial, yet selenomethionine was not protective in the Selenium and Vitamin E Cancer Prevention Trial (SELECT) (Lippman et al. 2009). Recent observations suggest that the answer may involve selenium metabolites influencing HDAC activity (Lee et al. 2009; Nian et al. 2009). While the majority of selenium in selenium-enriched yeast is selenomethionine, other selenium metatbolites, such as Se-methyl-selenocysteine (MSC) are also present. Colon cancer cells treated with MSC but not selenomethionine had increased acetylation of histone H3 and dose dependent inhibition of HDAC activity. Many mammalian cells contain the enzyme glutamine transaminase K, which can metabolize MSC but not selenomethionine to its $\alpha$-keto acid metabolite. The $\alpha$-keto acid metabolite of MSC is $\beta$-methylselenopyruvate (MSP) which is a competitive inhibitor of HDAC8. Moreover, MSP increased the activity of a *P21WAF1* luciferase reporter in HT29 cells and *p21* mRNA and protein levels were markedly elevated. These results suggest that MSC, and possibly other organoselenium compounds, can generate $\alpha$-keto acid metabolites which are HDAC inhibitors, with the potential to modulate histone status and chromatin remodeling, leading to expression of aberrantly silenced tumor suppressor genes, including *p21*.

A recent report revealed that genistein (at both 10 and 25 mol/L) induced the expression of the tumor suppressor genes *p21* and *p16* (INK4a) with a concomitant decrease in cyclins in prostate cancer cells (Majid et al. 2008). These investigators found that genistein increased acetylation of histones H3, H4, and H3 lysine 4 (H3K4) at the *p21* and *p16* transcription start sites with concomitant increased expression of HATs. Interestingly, DNA methylation analysis revealed the absence of *p21* promoter methylation prior to genistein exposure. Furthermore, these same investigators found that genistein (50 µM) activated expression of several aberrantly silenced tumor suppressor genes that have unmethylated promoters such as *PTEN*, *CYLD*, *p53*, and *FOXO3a* in prostate cancer cells (Kikuno et al. 2008). Instead of turning on tumor suppressor genes through promoter demethylation, these investigators found that genistein influenced remodeling of the heterochromatic domains at promoters by reducing/modulating histone H3-Lysine 9 (H3K9) methylation and deacetylation. Other studies have suggested that DNA methylation and histone deacetylation may cooperate to repress gene expression. Recent findings revealed that the tumor suppressor gene *BTG3*, which is epigenetically silenced in prostate cancer cells, was reactivated by genistein induced promoter demethylation and active histone modification (Majid et al. 2010). Overall, these findings suggest that genistein may be protective against cancers with various epigenetic profiles. Furthermore, the relationship between genistein, histone and DNA methylation modifications in gene reactivation, which may be gene specific, is not entirely clear and requires further investigation.

The story is further complicated for soy consumption because of the activity of other compounds found in soy that may modify chromatin in a somewhat opposing fashion. For example, lunasin, a unique 43-amino acid soybean peptide that also has cancer prevention properties, has been found to inhibit acetylation of core histones in mammalian cells and selectively kills cells that are in the process of transformation (e.g., E1A-transfected mouse fibroblast NIH 3T3 cells), but does not affect the growth rate of normal and established cancer cell lines at 10 µM concentrations (Lam et al. 2003). An epigenetic mechanism of action has been postulated whereby lunasin selectively kills cells being transformed or newly transformed cells by binding to deacetylated core histones exposed by the transformation event, thereby disrupting the dynamics of histone acetylation-deacetylation. These results point to the importance of understanding the timing of cellular vulnerability to epigenetic modulation. Recently, investigators observed the histone H3- and H4-acetylation inhibitory properties of lunasin from different Korean soybean varieties used for various food purposes (Jeong et al. 2007). They found that amounts of lunasin varied in the soybean varieties (4.40– 70.49 ng of lunasin per µg of protein), and that the amount was directly correlated with the extent of inhibition of core histone acetylation. Furthermore, the blood from rats fed lunasin-enriched soy protein, but not the blood from control fed rats, was found to inhibit histone acetylation activity. Defining the epigenetic activity of various soy constituents and products in different cellular contexts is another potential area for future research.

Maternal diet was shown to alter histone acetylation and gene expression profiles in the developing offspring of primates (Aagaard-Tillery et al. 2008). In this study, chronic consumption of a maternal high-fat diet (35% fat versus 13% fat for control animals) resulted in a threefold increase in fetal liver triglycerides and histologic correlates of fatty liver disease, which was accompanied by hyperacetylation of fetal hepatic tissue at histone 3 lysine 14 (H3K14) and decreased HDAC1 mRNA, protein, and activity. Gene expression changes were also observed, including increased glutamic pyruvate transaminase (alanine aminotransferase) 2 *(GPT2)*, *DNAJA2* (a heat shock protein 70 cochaperone), and *Rdh12* (an all-trans and 9-cis retinol dehydrogenase responsive to oxidative stress) in fetal hepatic tissue from maternally high-fat-fed animals when compared with controls. Furthermore, the gene *Npas2*, a peripheral circadian regulator, was significantly downregulated in the offspring of animals fed the high-fat diet. Definitive conclusions regarding the role of H3K14 acetylation with respect to the observed altered gene expression requires additional study. These results, however, suggest that a caloric-dense or high fat maternal diet leading to obesity epigenetically alters fetal chromatin structure in primates via covalent modifications of histones and hence offers a molecular basis to the fetal origins of adult-onset of disease hypothesis.

Caloric restriction has been shown to decrease cancer susceptibility in many different experimental models. One of the molecular targets of this effect is thought to be the NAD-dependent, type III HDACs. In yeast and worms, caloric restriction increases longevity by modulating the expression of the NAD-dependent HDAC SIR2 (Vaquero et al. 2006). Evidence to support the caloric-restriction reduced cancer hypothesis comes from studies utilizing the glucose analogue 2-deoxyglucose, which blocks glucose metabolism and results in decreased energy metabolism as well as decreased carcinogen-induced rat mammary tumor incidence, multiplicity, and prolonged tumor latency (Zhu et al. 2005). That the expression of Sirt-1, a mammalian homologue of SIR2, was induced in a dose- and time-dependent manner by treatment with 2-deoxyglucose in these animals, suggests that the sirtuin family of genes may be involved in the cancer protective effects of caloric restriction in mammals (Zhu et al. 2005). Caloric restriction has also been shown to increase the carbonylation of histones (Sharma et al. 2006). However, the functional significance of this observation is not known. A number of plant polyphenols have been shown to activate Sirt1 including quercetin and resveratrol (Allard et al. 2009). Resveratrol, a polyphenolic compound found in grapes, wine, and peanuts, has been shown to increase SIR2 as well as to mimic the caloric restriction pathways for longevity in both *Caenorhabditis elegans* and *Drosophila melanogaster* models (Wood et al. 2004). A recent report indicates that resveratrol treatment produces beneficial effects similar to the effects of caloric restriction in mice (Lagouge et al. 2006). Resveratrol has also demonstrated cancer prevention activity in animal models (Sengottuvelan et al. 2009). This activity appears to be mediated by the pleiotropic interactions of resveratrol with different enzyme targets, including Sirt1 (Calamini et al. 2010). Uncovering the specific molecular targets for the cancer protective activity of resveratrol requires further study. Overall, the data above suggest that dietary components can either stimulate or inhibit various HDAC enzymes depending on the cellular context.

## OTHER HISTONE POST-TRANSLATIONAL MODIFICATIONS IMPACTED BY DIET

As described in the *Histone Acetylation/Deacetylation, Bioactive Food Components, and Cancer Prevention* section, the majority of the evidence for dietary induced histone post-translational modifications concerns the effects of dietary HDAC inhibitors and the effects of diet on histone acetylation. Additional histone post-translational modifications and their enzymatic associates have been shown to be influenced by dietary factors that have also been implicated in cancer prevention pathways. These include interactions between a methyl deficient diet and histone methylation in early hepatocellular carcinogenesis (Pogribny et al. 2006a) and the nutrient biotin and histone biotinylation in repression of transposable elements in cancer cells (Chew et al. 2008). Biotinylation of histone 4 (at lysine 8 and 12), which is mediated by holocarboxylase synthetase, has been associated with heterochromatin structures, gene silencing, mitotic condensation of chromatin, and DNA repair. Like many of the histone modifications, biotinylation appears to be a reversible process, although debiotinylases have not been characterized. Histone biotinylation depends on dietary biotin and it has been hypothesized that some effects of biotin deficiency can be attributed to abnormal chromatin structures (Hassan and Zempleni 2006). These reports provide evidence for the impact of dietary factors on histone modification and in determining chromatin structures, including whether the chromatin is in the open (euchromatin, active) or closed (heterochromatin, inactive) state. Research on the identification and characterization of dietary triggers of histone modifications and associated effects such as gene silencing or activation is emerging. Some of these efforts will likely examine specificity of bioactive food factors for particular histone modifying enzymes and perhaps will utilize epigenomic approaches to map the numerous histone post-translational marks in normal and cancer cells following dietary exposure(s).

## BIOACTIVE FOOD COMPONENTS MODULATE CHROMATIN PROTEINS

The metastasis-associated protein 1 (MTA1) is part of the nucleosome remodeling deacetylation (NuRD) corepressor complex that mediates posttranslational modifications of histones and nonhistone proteins resulting in transcriptional repression. Importantly, MTA1 over-expression in prostate cancer has been shown to be associated with tumor aggressiveness and metastasis. Recent evidence suggests that resveratrol can impact chromatin remodeling anticancer activity mediated through the MTA1/NuRD corepressor complex (Kai et al. 2010). In this study, resveratrol (50 μM) was found to downregulate the MTA1 protein, leading to destabilization of MTA1/NuRD complex, which resulted in acetylation/activation of *p53* and subsequent activation of proapoptotic genes in prostate cancer cells. This study defines a novel epigenetic pathway for resveratrol that requires further investigation to inform future prostate cancer prevention interventions.

Investigators are also beginning to examine the interplay between bioactive food components, PcG complexes and their impact on cancer prevention. One example involves the protein B-cell-specific Moloney murine leukemia virus integration site 1 (BMI1), which is a component of the PcG complex 1, and is over-expressed in some human cancers, including colorectal cancer (Kim et al. 2004) and human non-small-cell lung cancer (Vonlanthen et al. 2001), as well as markedly elevated in epidermal squamous cell carcinoma cells (Lee et al. 2008). The polyphenol EGCG (40 μM) was found to suppress BMI1 levels and reduce BMI1 phosphorylation, resulting in displacement of the BMI1 polycomb protein complex from chromatin and reducing survival of skin cancer cells (Balasubramanian et al. 2008). More recently, these investigators examined the effect of EGCG on BMI1 and the enhancer of zeste homolog 2 (EZH2), which is a component of PcG complex 2 (Balasubramanian et al. 2010). EGCG treatment of SCC-13 cells reduced BMI1 and EZH2 level, which was associated with reduced cell survival and global reduction in histone H3 lysine 27 trimethylation, a hallmark of PcG complex 2 action. These changes in PcG protein expression (BMI1 and EZH2) were also associated with reduced expression of key proteins that enhance progression through the cell cycle, increased expression of proteins that inhibit cell cycle progression, and impacted expression of proteins involved in apoptosis. This study is important because it suggests that GTPs reduce skin tumor cell survival by influencing PcG-mediated epigenetic regulatory mechanisms.

Future research is needed to clarify the role of dietary regulation of chromatin remodeling, including the NuRD and PcG complexes, during the carcinogenic process. The importance of the polycomb repressive complexes in the development of cancer is currently an active research enterprise with much of the work focused on PcG complex involvement in acquiring and maintaining cancer stem and adult stem cell identify. Investigators will likely gain an understanding of how the gene silencing process goes awry by dissecting the factors that control PcG targeting. How diet participates in this process is an emerging area of research.

## FUTURE DIRECTIONS AND CONCLUSIONS

The impact of epigenetic mechanisms in the etiology of cancer has been increasingly recognized in recent years. Evidence also indicates that diet and other environmental factors may be significant regulators of epigenetic events. Several observations suggest that many different bioactive food components exert cancer protective effects through modulation of epigenetic mechanisms, such as DNA methylation of CpG islands in promoters and other regions of the genome, chromatin silencing complexes, and post-translational modifications of histone tail domains (Figure 1.1). Furthermore, dietary alteration of these epigenetic marks have been associated with modulation of several cellular processes associated with carcinogenesis, including differentiation, inflammation, apoptosis, cell cycle control/proliferation, carcinogen metabolism, and angiogenesis, among others. Although the science of epigenetics has advanced, much remains to be revealed especially with respect to potential modification by bioactive dietary components. Issues remain about the quantity of dietary components and the optimal dietary pattern needed to bring about a biological effect, as well as the timing of exposure and other variables (chemical form, duration of exposure, interactions between dietary exposures) that can influence the response. Additionally, the interaction between diet and other environmental stimuli that may potentially function as epigenetic modifiers, such as exposures to heavy metals and aromatic hydrocarbons, make the scenario even more complicated. Understanding the relationships between genetics and epigenetics may provide further insights about transcriptional regulation during carcinogenesis and how dietary factors participate in these interactions. Moreover,

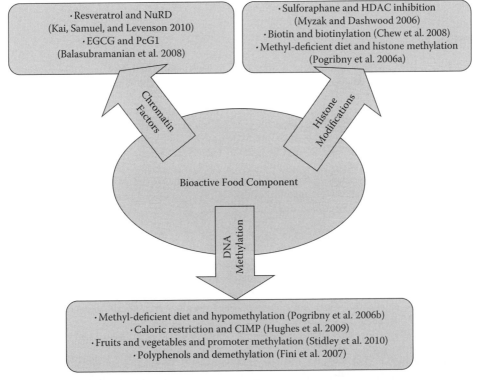

**FIGURE 1.1** Relationships between bioactive food components and epigenetic mechanisms. Several observations suggest that many different bioactive food components exert cancer protective effects through modulation of epigenetic mechanisms, such as DNA methylation of CpG islands in promoters and other regions of the genome, chromatin silencing complexes, and post-translational modifications of histone tail domains. CIMP, (promoter) CpG island methylation phenotype; EGCG, epigallocatechin 3-gallate; HDAC, histone deacetylases; NuRD, nucleosome remodeling deacetylation corepressor complex; PcG1, polycomb group complex 1.

the identification and characterization of novel epigenetic marks and mechanisms with the capacity to regulate gene expression are likely to surface over the next few years.

The epigenome has been described as the study of epigenetic changes at the level of the genome. Recent advances in genomics technology, in particular high-throughput sequencing, have enabled genome-wide analysis of histone modifications and analysis of DNA methylation at nucleotide resolution. Such epigenomic approaches are likely to assist in characterizing genome-wide epigenetic marks that are targets for dietary regulation and that impact phenotype. The ability to characterize cell and context specific epigenomes (be it profiles of DNA methylation or histone modifications) will greatly impact the ability to determine how diet impacts differential epigenetic effects on normal versus cancer cells. This information should also provide the tools to illuminate epigenetic changes resulting from dietary exposures during critical periods of prenatal and postnatal development, adolescence, and senescence, as well as examine the potential impact of diet on transgenerational transmission of epigenetic changes. Epigenetic analyses to reveal function or significance of target gene regions or loci identified through epigenomic approaches or from other endeavors will continue to be an active research activity. Emerging evidence indicates that epigenomics and epigenetics approaches hold substantial potential for developing biological markers to predict which exposures put subjects at risk or benefit and which individuals will be more susceptible to developing or averting disease. Investigations are required to determine whether exposed subjects develop epigenetic alterations over time and, in turn, whether such alterations increase the risk of disease. Importantly, for the future of nutrigenomics and personalized nutrition, epigenetic marks may be useful as biomarkers of cancer prevention, early disease, or nutritional status, as well as function as potential molecular targets that are modulated by dietary interventions.

## REFERENCES

Aagaard-Tillery, K. M., K. Grove, J. Bishop, X. Ke, Q. Fu, R. McKnight, and R. H. Lane. 2008. Developmental origins of disease and determinants of chromatin structure: Maternal diet modifies the primate fetal epigenome. *J Mol Endocrinol* 41(2):91–102.

Adams, L. S., S. Phung, X. Wu , L. Ki, and S. Chen. 2008. White button mushroom (*Agaricus bisporus*) exhibits antiproliferative and proapoptotic properties and inhibits prostate tumor growth in athymic mice. *Nutr Cancer* 60(6):744–56.

Allard, J. S., E. Perez, S. Zou, and R. de Cabo. 2009. Dietary activators of Sirt1. *Molec Cell Endocrin* 299(1):58–63.

Atsumi, A., A. Tomita, H. Kiyoi, and T. Naoe. 2006. Histone deactylase 3 (HDAC3) is recruited to target promoters by PML-RARalpha as a component of the N-CoR co-repressor complex to repress transcription *in vivo*. *Biochem Biophys Res Commun* 345(4):1471–80.

Balasubramanian, S., G. Adhikary, and R. L. Eckert. 2010. The Bmi-1 polycomb protein antagonizes the (-)-epigallocatechin-3-gallate-dependent suppression of skin cancer cell survival. *Carcinogenesis* 31(3):496–503.

Balasubramanian, S., K. Lee, G. Adhikary, R. Gopalakrishnan, E. A. Rorke, and R. L. Eckert. 2008. The bmi-1 polycomb group gene in skin cancer: Regulation of function by (-)-epigallocatechin-3-gallate (EGCG). *Nutrition Rev* 66(Suppl 1):S65–8.

Boland, C. R., S. K. Shin, and A. Goel. 2009. Promoter methylation in the genesis of gastrointestinal cancer. *Yonsei Med J* 50(3):309–21.

Bracken, A. P., and K. Helin. 2009. Polycomb group proteins: Navigators of lineage pathways led astray in cancer. *Nat Rev Cancer* 9(11):773–84.

Calamini, B., K. Ratia, M. Malkowski, M. Cuendet, J. M. Pezzuto, B. D. Santarsiero, and A. D. Mesecar. 2010. Pleiotropic mechanisms facilitated by resveratrol and its metabolites. *Biochem J* 429(2):273–82.

Chew, Y. C., J. T. West, S. J. Kratzer, A. M. Ilvarsonn, J. C. Eissenberg, B. J. Dave, D. Klinkebiel, J. K. Christman, and J. Zempleni. 2008. Biotinylation of histones represses transposable elements in human and mouse cells and cell lines and in Drosophila melanogaster. *J Nutr* 138(12):2316–22.

Christman, J. K. 2003. Diet, DNA methylation and cancer. In H. Daniel, and J. Zempleni, eds., *Molecular Nutrition*, 237–65. Oxon, UK: CABI Publishing.

Christman, J. K., G. Sheikhnejad, M. Dizik, S. Abileah, and E. Wainfan. 1993. Reversibility of changes in nucleic acid methylation and gene expression in rat liver by severe dietary methyl deficiency. *Carcinogenesis* 14(4):551–7.

Clark, L. C., G. F. Combs Jr., B. W. Turbull, E. H. Slate, D. K. Chalker, J. Chow, L. S. Davis, R. A. Glover, G. F. Graham, E. G. Gross, A. Krongrad, J. L. Lesher Jr., H. K. Park, B. B. Sanders Jr., C. L. Smith and J. R. Taylor. 1996. Effects of selenium supplementation for cancer prevention in patients with carcinoma of the skin: A randomized controlled trial. Nutritional Prevention of Cancer Study Group. *J Am Medic Assoc* 276(24):1957–63.

Cooney, C. A., A. A. Dave, and G. L. Wolff. 2002. Maternal methyl supplements in mice affect epigenetic variation and DNA methylation of offspring. *J Nutr* 132(8 Suppl):2393S–400S.

Cropley, J. E., C. M. Suter, K. B. Beckman, and D. I. Martin. 2006. Germ-line epigenetic modification of the murine A^vy allele by nutritional supplementation. *Proc Natl Acad Sci USA* 103(46):17308–12.

Cropley, J. E., C. M. Suter, K. B. Beckman, and D. I. Martin. 2010. CpG methylation of a silent controlling element in the murine A^vy allele is incomplete and unresponsive to methyl donor supplementation. *PLoS One* 5(2):e9055.

Davis, C. D., and J. A. Milner. 2007. Molecular targets for nutritional preemption of cancer. *Curr Cancer Drug Targets* 7(5):410–5.

Denda, A., W. Kitayama, H. Kishida, N. Murata, M. Tsutsumi, T. Tsujiuchi, D. Nakae, and Y. Konishi. 2002. Development of hepatocellular adenomas and carcinomas associated with fibrosis in C57BL/6J male mice given a choline-deficient, L-amino acid-defined diet. *Jpn J Cancer Res* 93(2):125–32.

Dirx, M. J., P. A. van den Brandt, R. A. Goldbohm, and L. H. Lumey. 2003. Energy restriction early in life and colon carcinoma risk: Results of The Netherlands Cohort Study after 7.3 years of follow-up. *Cancer* 97(1):46–55.

Dolinoy, D. C., J. R. Weidman, R. A. Waterland, and R. L. Jirtle. 2006. Maternal genistein alters coat color and protects A^vy mouse offspring from obesity by modifying the fetal epigenome. *Environ Health Perspect* 114(4):567–72.

Duhl, D. M., H. Vrieling, K. A. Miller, G. L. Wolff, and G. S. Barsh. 1994. Neomorphic agouti mutations in obese yellow mice. *Nat Genet* 8(1):59–65.

Ehrlich, M. 2002. DNA methylation in cancer: too much, but also too little. *Oncogene* 21(35):5400–13.

Esteller, M. 2005. Aberrant DNA methylation as a cancer-inducing mechanism. *Annu Rev Pharmacol Toxicol* 45(Feb):629–56.

Esteller, M. 2007. Cancer epigenomics: DNA methylomes and histone-modification maps. *Nat Rev Genet* 8(4):286–98.

Fabbri, M., C. M. Croce, and G. A. Calin. 2008. MicroRNAs. *Cancer J* 14(1):1–6.

Fang, M., D. Chen, and C. S. Yang. 2007. Dietary polyphenols may affect DNA methylation. *J Nutr* 137(1 Suppl):223S–8S.

Fang, M. Z., Y. Wang, N. Ai, Z. Hou, Y. Sun, H. Lu, W. Welsh, and C. S. Yang. 2003. Tea polyphenol (-)-epigallocatechin-3-gallate inhibits DNA methyltransferase and reactivates methylation-silenced genes in cancer cell lines. *Cancer Res* 63(22):7563–70.

Fang, M. Z., D. Chen, Y. Sun, Z. Jin, J. K. Christman, and C. S. Yang. 2005. Reversal of hypermethylation and reactivation of p16INK4a, RARbeta, and MGMT genes by genistein and other isoflavones from soy. *Clin Cancer Res* 11(19 Pt 1):7033–41.

Fearon, E. A., and B. Vogelstein. 1990. A genetic model for colorectal tumorigenesis. *Cell* 61(5):759–67.

Figueiredo, J. C., M. V. Grau, K. Wallace, A. J. Levine, L. Shen, R. Hamdan, X. Chen, R. S. Bresalier, G. McKeown-Eyssen, R. W. Haile, J. A. Baron, and J. P. Issa. 2009. Global DNA hypomethylation (LINE-1) in the normal colon and lifestyle characteristics and dietary and genetic factors. *Cancer Epidemiol Biomarkers Prev* 18(4):1041–9.

Fini, L., M. Selgrad, V. Fogliano, G. Graziani, M. Romano, E. Hotchkiss, Y. A. Daoud, E. B. De Vol, C. R. Boland, and L. Ricciardiello. 2007. Annurca apple polyphenols have potent demethylating activity and can reactivate silenced tumor suppressor genes in colorectal cancer cells. *J Nutr* 137(12):2622–8.

Fraga, M. F., E. Ballestar, A. Villar-Garea, M. Boix-Chornet, J. Espada, G. Schotta, T. Bonaldi, C. Haydon, S. Ropero, K. Petrie, N. G. Iyer, A. Pérez-Rosado, E. Calvo, J. A. Lopez, A. Cano, M. J. Calasanz, D. Colomer, M. A. Piris, N. Ahn, A. Imhof, C. Caldas, T. Jenuwein, and M. Esteller. 2005. Loss of acetylation at Lys16 and trimethylation at Lys20 of histone H4 is a common hallmark of human cancer. *Nat Genet* 37(4):391–400.

Garfinkel, M. D., and D. M. Ruden. 2004. Chromatin effects in nutrition, cancer and obesity. *Nutrition* 20(1):56–62.

Gayther, S. A., S. J. Batley, L. Linger, A. Bannister, K. Thorpe, S. F. Chin, Y. Daigo, P. Russell, A. Wilson, H. M. Sowter, J. D. Delhanty, B. A. Ponder, T. Kouzarides, C. Caldas. 2000. Mutations truncating the EP300 acetylase in human cancers. *Nat Genet* 24(3):300–3.

Geier, M. S., R. N. Butler, and G. S. Howarth. 2006. Probiotics, prebiotics and synbiotics: a role in chemoprevention for colorectal cancer? *Cancer Biol Ther* 5(10):1265–9.

Ghoshal, K., X. Li, J. Datta, S. Bai, I. Pogribny, M. Pogribny, Y. Huang, D. Young, and S. T. Jacob. 2006. A folate- and methyl-deficient diet alters the expression of DNA methyltransferases and methyl CpG binding proteins involved in epigenetic gene silencing in livers of F344 rats. *J Nutr* 136(6):1522–7.

Gibbons, R. J. 2005. Histone modifying and chromatin remodeling enzymes in cancer and dysplastic syndromes. *Hum Mol Genet* 14(Review issue 1):R85–R92.

Gieni, R. S., and M. J. Hendzel. 2009. Polycomb group protein gene silencing, non-coding RNA, stem cells, and cancer. *Biochem Cell Biol* 87(5):711–46.

Hake, S. B., A. Xiao, and C. D. Allis. 2004. Linking the epigenetic "language" of covalent histone modifications to cancer. *Br J Cancer* 90(4):761–9.

Hanahan, D., and R. A. Weinberg. 2000. The hallmarks of cancer. *Cell* 100(1):57–70.

Hassan, Y. I., J. and Zempleni. 2006. Epigenetic regulation of chromatin structure and gene function by biotin. *J Nutr* 136(7):1763–5.

Hawkins, P. G., and K. V. Morris. 2008. RNA and transcriptional modulation of gene expression. *Cell Cycle* 7(5):602–7.

Heijmans, B. T., E. W. Tobi, A. D. Stein, H. Putter, G. J. Blauw, E. S. Susser, P. E. Slagboom, and L. H. Lumey. 2008. Persistent epigenetic differences associated with prenatal exposure to famine in humans. *Proc Natl Acad Sci USA* 105(44):17046–9.

Hermann, A., H. Gowher, and A. Jeltsch. 2004. Biochemistry and biology of mammalian DNA methyltransferases. *Cell Mol Life Sci* 61(19-20):2571–87.

Holliday, R. 1987. The inheritance of epigenetic defects. *Science* 238(4824):163–70.

Hughes, L. A., P. A. van den Brandt, A. P. de Bruïne, K. A. Wouters, S. Hulsmans, A. Spiertz, R. A. Goldbohm, A. F. de Goeij, J. G. Herman, M. P. Weijenberg, and M. van Engeland. 2009. Early life exposure to famine and colorectal cancer risk: A role for epigenetic mechanisms. *PLoS One* 4(11):e7951.

Hughes, L. A., P. A. van den Brandt, R. A. Goldbohm, A. F. de Goeij, A. P. de Bruïne, M. van Engeland, and M. P. Weijenberg. 2010. Childhood and adolescent energy restriction and subsequent colorectal cancer risk: Results from the Netherlands Cohort Study. *Int J Epidemiol* 39(5):1333–44.

Jenuwein, T. 2001. Re-SET-ting heterochromatin by histone methyltransferases. *Trends Cell Biol* 11(6):266–73.

Jenuwein, T., and C. D. Allis. 2001. Translating the histone code. *Science* 293(5532):1074–80.

Jeong, H. J., J. B. Jeong, D. S. Kim, and B. O. de Lumen. 2007. Inhibition of core histone acetylation by the cancer preventive peptide lunasin. *J Agri Food Chem* 55(3):632–37.

Jones, P. A., and S. B. Baylin. 2002. The fundamental role of epigenetic events in cancer. *Nat Rev Genet* 3(6):415–28.

Kai, L., S. K. Samuel, and A. S. Levenson. 2010. Resveratrol enhances p53 acetylation and apoptosis in prostate cancer by inhibiting MTA1/NuRD complex. *Int J Cancer* 126(7):1538–48.

Kautiainen, T. L., and P. A. Jones. 1986. DNA methyltransferase levels in tumorigenic and nontumorigenic cells in culture. *J Biol Chem* 261(4):1594–8.

Kelley, N. S., N. E. Hubbard, and K. L. Erickson. 2007. Conjugated linoleic acid isomers and cancer. *J Nutr* 137(12):2599–607.

Kikuno, N., H. Shiina, S. Urakami, K. Kawamoto, H. Hirata, Y. Tanaka, S. Majid, M. Igawa, and R. Dahiya. 2008. Genistein mediated histone acetylation and demethylation activates tumor suppressor genes in prostate cancer cells. *Int J Cancer* 123(3):552–60.

Kim, D. H., P. Saetrom, O. Snøve, Jr., and J. J. Rossi. 2008. MicroRNA-directed transcriptional gene silencing in mammalian cells. *Proc Natl Acad Sci USA* 105(42):16230–5.

Kim, J. H., S. Y. Yoon, C. N. Kim, J. H. Joo, S. K. Moon, I. S. Choe, Y. K. Choe, and J. W. Kim. 2004. The Bmi-1 oncoprotein is overexpressed in human colorectal cancer and correlates with the reduced p16INK4a/p14ARF proteins. *Cancer Lett* 203(2):217–24.

Kim, Y., and J. Mason. 1995. Folate, epithelial dysplasia and colon cancer. *Proc Assoc Am Physicians* 107:218–27.

Kothapalli, N., G. Camporeale, A. Kueh, Y. C. Chew, A. M. Oommen, J. B. Griffin, and J. Zempleni. 2005. Biological functions of biotinylated histones. *J Nutr Biochem* 16(7):446–8.

Kutay, H., S. Bai, J. Datta, T. Motiwala, I. Pogribny, W. Frankel, S. T. Jacob, and K. Ghoshal. 2006. Downregulation of miR-122 in the rodent and human hepatocellular carcinomas. *J Cell Biochem* 99(3):671–8.

Lagouge, M., C. Argmann, Z. Gerhart-Hines, H. Meziane, C. Lerin, F. Daussin, N. Messadeq, J. Milne, P. Lambert, P. Elliott, B. Geny, M. Laakso, P. Puigserver, and J. Auwerx. 2006. Resveratrol improves mitochondrial function and protects against metabolic disease by activating SIRT1 and PGC-1alpha. *Cell* 127(6):1109–22.

Lam, Y., A. Galvez, and B. O. de Lumen. 2003. Lunasin suppresses E1A-mediated transformation of mammalian cells but does not inhibit growth of immortalized and established cancer cell lines. *Nutr Cancer* 47(1):88–94.

Lee, J. I., H. Nian, A. J. L. Cooper, R. Sinha, J. Dai, W. H. Bisson, R. H. Dashwood, and J. T. Pinto. 2009. ∀-Keto acid metabolites of naturally occurring organoselenium compounds as inhibitors of histone deacetylase in human prostate cancer cells. *Cancer Prev Res* 2(7):683–93.

Lee, K., G. Adhikary, S. Balasubramanian, R. Gopalakrishna, T. McCormick, G. P. Dimri, R. L. Eckert, and E. A. Rorke. 2008. Expression of Bmi-1 in epidermis enhances cell survival by altering cell cycle regulatory protein expression and inhibiting apoptosis. *J Invest Dermatol* 128(1):9–17.

Li, E. 2002. Chromatin modification and epigenetic reprogramming in mammalian development. *Nat Rev Genet* 3(9):662–73.

Lin, R. J., L. Nagy, S. Inooue, W. Shao, W. H. Miller, Jr., and R. M. Evans. 1998. Role of the histone deacetylase complex in acute promylelocytic leukaemia. *Nature* 391(6669):811–4.

Lippman, S. M., E. A. Klein, P. J. Goodman, M. S. Lucia, I. M. Thompson, L. G. Ford, H. L. Parnes, L. M. Minasian, J. M. Gaziano, J. A. Hartline, J. K. Parsons, J. D. Bearden 3rd, E. D. Crawford, G. E. Goodman, J. Claudio, E. Winquist, E. D. Cook, D. D. Karp, P Walther, M. M. Lieber, A. R. Kristal, A. K. Darke, K. B. Arnold, P. A. Ganz, R. M. Santella, D. Albanes, P. R. Taylor, J. L. Probstfield, T. J. Jagpal, J. J. Crowley, F. L. Meyskens Jr, L. H. Baker, and C. A. Coltman, Jr. 2009. Effect of selenium and vitamin E on risk of prostate cancer and other cancers: The Selenium and Vitamin E Cancer Prevention Trial (SELECT). *J Am Med Assoc* 301(1):39–51.

Mahlknecht, U., and D. Hoelver. 2000. Histone acetylation modifiers in the pathogenesis of malignant disease. *Mol Med* 6(8):623–44.

Majid, S., A. A. Dar, V. Shahryari, H. Hirata, A. Ahmad, S. Saini, Y. Tanaka, A. V. Dahiya, And R. Dahiya. 2010. Genistein reverses hypermethylation and induces active histone modifications in tumor suppressor gene B-Cell translocation gene 3 in prostate cancer. *Cancer* 116(1):66–76.

Majid, S., N. Kikuno, J. Nelles, E. Noonan, Y. Tanaka, K. Kawamoto, H. Hirata, L. C. Li, H. Zhao, S. T. Okino, R. F. Place, D. Pookot, and R. Dahiya. 2008. Genistein induces the p21WAF1/CIP1 and p16INK4a tumor suppressor genes in prostate cancer cells by epigenetic mechanisms involving active chromatin modification. Cancer Res 68(8):2736–44.

Malecová, B., and K. V. Morris. 2010. Transcriptional gene silencing through epigenetic changes mediated by non-coding RNAs. *Curr Opin Mol Ther* 12(2):214–22.

Manson, M. M., B. E. Foreman, L. M. Howells, and E. P. Moiseeva. 2007. Determining the efficacy of dietary phytochemicals in cancer prevention. *Biochem Soc Trans* 35(Pt 5):1358–63.

Marks, P. A., and M. Dokmonovic. 2005. Histone deacetylase inhibitors: Discovery and development as anticancer agents. *Expert Opin Investig Drugs* 14(12):1497–511.

McGowan, P. O., M. J. Meaney, and M. Szyf. 2008. Diet and the epigenetic (re)programming of phenotypic differences in behavior. *Brain Res* 1237(Oct):12–24.

Mohammad, H. P., Y. Cai, K. M. McGarvey, H. Easwaran, L. Van Neste, J. E. Ohm, H. M. O'Hagan, and S. B. Baylin. 2009. Polycomb CBX7 promotes initiation of heritable repression of genes frequently silenced with cancer-specific DNA hypermethylation. *Cancer Res* 69(15):6322–30.

Morey, L., C. Brenner, F. Fazi, R. Villa, A. Gutierrez, M. Buschbeck, C. Nervi, S. Minucci, F. Fuks, and L. Di Croce. 2008. MBD3, a component of the NuRD complex, facilitates chromatin alteration and deposition of epigenetic marks. *Mol Cell Biol* 28(19):5912–23.

Morey Kinney, S. R., W. Zhang, M. Pascual, J. M. Greally, B. M. Gillard, E. Karasik, B. A. Foster, and A. R. Karpf. 2009. Lack of evidence for green tea polyphenols as DNA methylation inhibitors in murine prostate. *Cancer Prev Res (Phila Pa)* 2(12):1065–75.

Myzak, M. C., and R. H. Dashwood. 2006. Histone deacetylases as targets for dietary cancer preventive agents: Lessons learned with butyrate, diallyl disulfide and sulforaphane. *Current Drug Targets* 7(4):443–52.

Myzak, M. C., W. M. Dashwood, G. A. Orner, E. Ho, and R. H. Dashwood. 2006a. Sulforaphane inhibits histone deacetylase *in vivo* and suppresses tumorigenesis in APC[min] mice. *FASEB J* 20(3):506–8.

Myzak, M. C., K. Hardin, R. Wang, R. H. Dashwood, and E. Ho. 2006b. Sulforaphane inhibits histone deacetylase activity in BPH-1, LnCaP and PC-3 prostate epithelial cells. *Carcinogenesis* 27(4):811–9.

Myzak, M. C., A. Karplus, F-L. Chung, and R. H. Dashwood. 2004. A novel mechanism of chemoprotection by sulforaphane: Inhibition of histone deactylase. *Cancer Res* 64(16):5767–74.

Myzak, M. C., P. Tong, W. M. Dashwood, R. H. Dashwood, and E. Ho. 2007. Sulforaphane retards the growth of human PC-3 xenografts and inhibits HDAC activity in human subjects. *Exp Biol Med (Maywood)* 232(2):227–34.

Newberne, P. M. 1986. Lipotropic factors and oncogenesis. *Adv Exp Med Biol* 206:223–51.

Nian, H., W. H. Bisson, W-M. Dashwood, J. T. Pinto, and R. H. Dashwood. 2009. ∀-Keto acid metabolites of organoselenium compounds inhibit histone deacetylase activity in human colon cancer cells. *Carcinogenesis* 30(8):1416–23.

Oki, M., H. Aihara, and T. Ito. 2007. Role of histone phosphorylation in chromatin dynamics and its implications in diseases. *Subcell Biochem* 41:319–36.

Pandey, M., S. Shukla, and S. Gupta. 2010. Promoter demethylation and chromatin remodeling by green tea polyphenols leads to re-expression of GSTP1 in human prostate cancer cells. *Int J Cancer* 126(11):2520–33.

Pogribny, I. P., S. J. James, S. Jernigan, and M. Pogribna. 2004. Genomic hypomethylation is specific for preneoplastic liver in folate/methyl deficient rats and does not occur in non-target tissues. *Mutat Res* 548(1-2):53–9.

Pogribny, I. P., S. A. Ross, V. P. Tryndyak, M. Pogribna, L. A. Poirier, and T. V. Karpinets. 2006a. Histone H3 lysine 9 and H4 lysine 20 trimethylation and the expression of Suv-20h2 and Suv-39h1 histone methyltransferases in hepatocarcinogenesis induced by methyl deficiency in rats. *Carcinogenesis* 27(6):1180–6.

Pogribny, I. P., S. A. Ross, C. Wise, M. Pogribna, E. A. Jones, V. P. Tryndyak, S. J. James, Y. P. Dragan, and L. A. Poirier. 2006b. Irreversible global DNA hypomethylation as a key step in hepatocarcinogenesis induced by dietary methyl deficiency. *Mutat Res* 593(1-2):80–7.

Pogribny, I. P., V. P. Tryndyak, L. Muskhelishvili, I. Rusyn, and S. A. Ross. 2007. Methyl deficiency, alterations in global histone modifications, and carcinogenesis. *J Nutr* 137(1 Suppl):216S–22S.

Poirier, L. A. 1994. Methyl group deficiency in hepatocarcinogenesis. *Drug Metab Rev* 26(1-2):185–99.

Powell, C. L., O. Kosyk, B. U. Bradford, J. S. Parker, E. K. Lobenhofer, A. Denda, F. Uematsu, D. Nakae, and I. Rusyn. 2005. Temporal correlation of pathology and DNA damage with gene expression in a choline-deficient model of rat liver injury. *Hepatology* 42(5):1137–47.

Rakyan, V. K., M. E. Blewitt, R. Druker, J. I. Preis, and E. Whitelaw. 2002. Metastable epialleles in mammals. *Trends Genet* 18(7):348–51.

Rodenhiser, D., and M. Mann. 2006. Epigenetics and human disease: Translating basic biology into clinical applications. *Can Med Assoc J* 174(3):341–8.

Rosato, R. R., and S. Grant. 2003. Histone deacetylase inhibitors in cancer therapy. *Cancer Biol Ther* 2(1):30–7.

Ross, S. A. 2003. Diet and DNA methylation interactions in cancer prevention. *Ann NY Acad Sci* 983(Mar):197–207.

Sengottuvelan, M., K. Deeptha, and N. Nalini. 2009. Influence of dietary resveratrol on early and late molecular markers of 1,2-dimethylhydrazine-induced colon carcinogenesis. *Nutrition* 25(11-12):1169–76.

Sharma, R., A. Nakamura, R. Takahashi, H. Nakamoto, and S. Goto. 2006. Carbonyl modification in rat liver histones: Decrease with age and increase by dietary restriction. *Free Radical Biol Med* 40(7):1179–84.

Shiio, Y., and R. N. Eisenman. 2003. Histone sumoylation is associated with transcriptional repression. *Proc Natl Acad Sci USA* 100(23):13225–30.

Shilatifard, A. 2006. Chromatin modifications by methylation and ubiquitination: Implications in the regulation of gene expression. *Annu Rev Biochem* 75(July):243–69.

Shukla, V., T. Vaissière, and Z. Herceg. 2008. Histone acetylation and chromatin signature in stem cell identity and cancer. *Mutation Research* 637(1-2):1–15.

Simon, J. A., and C. A. Lange. 2008. Roles of the EZH2 histone methyltransferase in cancer epigenetics. *Mutat Res* 647(1-2):21–9.

Sparmann, A., and M. van Lohuizen. 2006. Polycomb silencers control cell fate, development and cancer. *Nature Reviews Cancer* 6(11):846–56.

Stidley, C. A., M. A. Picchi, S. Leng, R. Willink, R. E. Crowell, K. G. Flores, H. Kang, T. Byers, F. D. Gilliland, and S. A. Belinsky. 2010. Multivitamins, folate, and green vegetables protect against gene promoter methylation in the aerodigestive tract of smokers. *Cancer Res* 70(2):568–74.

Taft, R. J., K. C. Pang, T. R. Mercer, M. Dinger, J. S. Mattick. 2010. Non-coding RNAs: Regulators of disease. *J Pathol* 220(2):126–39.

Takihara, Y. 2008. Role of polycomb-group genes in sustaining activities of normal and malignant stem cells. *Int J Hematol* 87(1):25–34.

Talens, R. P., D. I. Boomsma, E. W. Tobi, D. Kremer, J. W. Jukema, G. Willemsen, H. Putter, P. E. Slagboom, and B.T. Heijmans. 2010. Variation, patterns, and temporal stability of DNA methylation: Considerations for epigenetic epidemiology. *FASEB J* Apr 16. [Epub ahead of print]

Tobi, E. W., L. H. Lumey, R. P. Talens, D. Kremer, H. Putter, A. D. Stein, E. Slagboom, and B. T. Heijmans. 2009. DNA methylation differences after exposure to prenatal famine are common and timing- and sex-specific. *Hum Mol Genet* 18(21):4046–53.

Trujillo, E., C. Davis, and J. Milner. 2006. Nutrigenomics, proteomics, metabolomics, and the practice of dietetics. *J Am Diet Assoc* 106(3):403–13.

Tsou, J. A., J. A. Hagen, C. L. Carpenter, and I. A. Laird-Offringa. 2002. DNA methylation analysis: a powerful new tool for lung cancer diagnosis. *Oncogene* 21(35):5450–61.

Unterberger, A., M. Szyf, P. W. Nathanielsz, and L. A. Cox. 2009. Organ and gestational age effects of maternal nutrient restriction on global methylation in fetal baboons. *J Med Primatol* 38(4):219–27.

Urnov, F. D., and A. P. Wolffe. 2001. Above and within the genome: Epigenetics past and present. *J Mammary Gland Biol Neoplasia* 6(2):153–67.

Vaquero, A., M. B. Scher, D. H. Lee, A. Sutton, H. L. Cheng, F. W. Alt, L. Serrano, R. Sternglanz, and D. Reinberg. 2006. SirT2 is a histone deacetylase with preference for histone H4 Lys16 during mitosis. *Genes and Develop* 20(10):1256–61.

Varga-Weisz, P. D., and P. B. Becker. 2006. Regulation of higher-order chromatin structures by nucleosome-remodelling factors. *Curr Opin Genet Dev* 16(2):151–6.

Vire, E., C. Brenner, R. Deplus, L. Blanchon, M. Fraga, C. Didelot, L. Morey, A. Van Eynde, D. Bernard, J. M. Vanderwinden, M. Bollen, M. Esteller, L. Di Croce, Y. De Launoit, and F. Fuks. 2006. The polycomb group protein EZH2 directly controls DNA methylation. *Nature* 439(7078):871–4.

Vonlanthen, S., J. Heighway, H. J. Altermatt, M. Gugger, A. Kappeler, M. M. Borner, M. van Lohuizen, and D. C. Betticher. 2001. The bmi-1 oncoprotein is differentially expressed in non-small cell lung cancer and correlates with INK4A-ARF locus expression. *Br J Cancer* 84(10):1372–6.

Wade, P. A., D. Pruss, and A. P. Wolffe. 1997. Histone acetylation: chromatin in action. *Trends Biochem Sci* 22(4):128–32.

Wainfan, E., and L. A. Poirier. 1992. Methyl groups in carcinogenesis: effects on DNA methylation and gene expression. *Cancer Res* 52(7 Suppl):2071S–7S.

Wang, L. G., A. Belkemisheva, X. M. Liu, A. C. Ferrari, J. Feng, and J. W. Chiao. 2007. Dual action on promoter demethylation and chromatin by an isothiocyanate restored GSTP1 silenced in prostate cancer. *Mol Carcinog* 46(1):24–31.

Wang, L. G., X. M. Liu, Y. Fang, W. Dai, F. B. Chiao, G. M. Puccio, J. Feng, D. Liu, and J. W. Chiao. 2008. De-repression of the p21 promoter in prostate cancer cells by an iothiocyanate via inhibition of HDACs and c-Myc. *Int J Oncol* 33(2):375–80.

Waterland, R. A., D. C. Dolinoy, J. R. Lin, C. A. Smith, X. Shi, and K. G. Tahiliani. 2006a. Maternal methyl supplements increase offspring DNA methylation at axin fused. *Genesis* 44(9):401–6.

Waterland, R. A., and R. L. Jirtle. 2003. Transposable elements: Targets for early nutritional effects on epigenetic gene regulation. *Mol Cell Biol* 23(15):5293–300.

Waterland, R. A., J. R. Lin, C. A. Smith, and R. L. Jirtle. 2006b. Post-weaning diet affects genomic imprinting at the insulin-like growth factor 2 (IGF2) locus. *Hum Mol Genet* 15(5):705–16.

Waterland, R. A., and K. B. Michels. 2007. Epigenetic epidemiology of the developmental origins hypothesis. *Annu Rev Nutr* 27(August):363–88.

Waterland, R. A., M. Travisano, K. G. Tahiliani, M. T. Rached, and S. Mirza. 2008. Methyl donor supplementation prevents transgenerational amplification of obesity. *Int J Obes (Lond)* 32(9):1373–9.

Wendel, M., and A. R. Heller. 2009. Anticancer actions of omega-3 fatty acids—current state and future perspectives. *Anticancer Agents Med Chem* 9(4):457–70.

Whitelaw, E., and D. I. Martin. 2001. Retrotransposons as epigenetic mediators of phenotypic variation in mammals. *Nat Genet* 27(4):361–5.

Widshwendter, M., H. Fiegl, D. Egle, E. Mueller-Holzner, G. Spizzo, C. Marth, D. J. Weisenberger, M. Campan, J. Young, I. Jacobs, and P. W. Laird. 2007. Epigenetic stem cell signature in cancer. *Nat Genet* 39(2):157–8.

Wolff, G. L., R. L. Kodell, S. R. Moore, and C. A. Cooney. 1998. Maternal epigenetics and methyl supplements affect agouti gene expression in Avy/a mice. *FASEB J* 12(11):949–57.

Wolffe, A. P. 1994. Inheritance of chromatin states. *Dev Genet* 15(6):463–70.

Wood, J. G., B. Rogina, S. Lavu, K. Howitz, S. L. Helfand, M. Tatar, and D. Sinclair. 2004. Sirtuin activators mimic caloric restriction and delay ageing in metazoans. *Nature* 430(7000):686–9.

WCRF/AICR. 2007. Food, Nutrition, Physical Activity, and the Prevention of Cancer: a Global Perspective. Washington DC: AICR. http://www.aicr.org/site/PageServer?pagename=research_science_expert_report, August 4, 2011.

Xue, Y., J. Wong, G. T. Moreno, M. K. Young, J. Côté, and W. Wang. 1998. NURD, a novel complex with both ATP-dependent chromatin-remodeling and histone deacetylase activities. *Mol Cell* 2(6):851–61.

Zhang, K., and S. Y. R. Dent. 2005. Histone modifying enzymes and cancer: Going beyond histones. *J Cell Biochem* 96(6):1137–48.

Zhu, P., E. Martin, J. Mengwasser, P. Schlag, K. P. Janssen, and M. Göttlicher. 2004. Induction of HDAC2 expression upon loss of APC in colorectal tumorigenesis. *Cancer Cell* 5(5):455–63.

Zhu, Z., W. Jiang, J. N. McGinley, and H. J. Thompson. 2005. 2-Deoxyglucose as an energy restriction mimetic agent: Effects on mammary carcinogenesis and on mammary tumor cell growth *in vitro*. *Cancer Res* 65(15):7023–30.

# 2 Diet-Influenced Chromatin Modification and Expression of Chemopreventive Genes by the Soy Peptide Lunasin

*Alfredo F. Galvez, Liping Huang, Mark J. M. Magbanua,*
*Kevin Dawson, Somen Nandi, and Raymond L. Rodriguez*

## CONTENTS

## DIET, CANCER, AND MECHANISMS REGULATING CHEMOPREVENTIVE GENE EXPRESSION

It has been estimated that up to 33% of cancer cases can be prevented by diet and associated factors (Kant et al. 2009). Although many reports conclude that higher consumption of plant-based foods can probably protect against cancers of various types, examples of specific bioactive food components responsible for cancer risk reduction are few in number and their mechanisms of action are, for the most part, not well understood (Ma and Chapman 2009). Other epidemiological studies, animal experiments, and *in vitro* studies show that increased consumption of soy products can also reduce cancer risks (Greenwald et al. 2002). These findings could account for the reduced risk of breast and prostate cancer in some Asian countries where soy consumption is eight to twelve times higher than in Western nations (Messina et al. 1994). Unfortunately, our understanding of how soy and its components influence long-term health and disease outcomes at the molecular level is limited. The principal reason for this is that most dietary signals (e.g., primary and secondary metabolites, vitamins, minerals, peptides, and lipids) are present in the cell at low concentrations, have low affinities and specificities for their receptors, and have poor pharmacokinetics and pharmacodynamics. Consequently, molecular mechanisms that employ the binding of dietary signals to receptors for different signal transduction pathways (e.g., EGCG) or various transcription factors (e.g., genistein) cannot fully account for how diet can alter the expression of thousand of genes in hundreds of pathways in a highly coordinated fashion. For example, in one study androgen-dependent human prostate (LNCaP) cells were treated with increasing concentrations of resveratrol (3,5,40-trihydroxy-*trans*-stilbene), a phytoalexin found in grapes, peanuts, and mulberries with antioxidant, anti-inflammatory, and anticancer properties (Jones et al. 2005). Resveratrol is believed to be responsible for the inverse correlation between red wine consumption and the incidence of cardiovascular diseases known as French paradox (Baur and Sinclair 2006). Of the 42,000 transcripts examined, over 1,600 genes showed consistent dose-dependent changes in response to resveratrol treatment in the 10 to 150μM/l range in as little as 6 hours. Of the 412 genes induced by androgen, 50% were downregulated by resveratrol and of the 105 genes normally repressed by androgen, 92% were reversed by resveratrol treatment. Several genes associated with prostate cancer, such as PSA and TMEPAI, were downregulated by resveratrol. In another study involving mice fed a standard diet (SD) or a highly caloric diet (HC), the latter mice showed a reduced rate of survival (Baur et al. 2006). When a separate group of mice were fed the HC diet with resveratrol (HCR), their risk of death was reduced by 31% and their life span was nearly the same as the SD-fed mice. Microarray analysis of 41,534 transcripts showed that only 782 (<2%) genes were dysregulated in the HCR mice compared to the SD mice. Moreover, resveratrol was found to oppose the affects of the HC diet in 144 of 153 significantly altered pathways, suggesting that this dietary signal does not act randomly but rather in a targeted fashion to re-regulate those gene expression patterns required for a normal life expectancy.

## DIET, EPIGENETICS, AND GENE EXPRESSION

The significance of these and related studies is not that resveratrol alters gene expression in human cell cultures or mice but rather that gene expression patterns are altered in a coordinated and apparently targeted fashion. That the authors of the mouse study identified the class III histone deacetylase, Sir2 (SIRT1 in humans), as the likely target of resveratrol raises the possibility that dietary signals like resveratrol may be altering gene expression indirectly by deacetylating powerful transcription coactivators like PGC1-α (Rodgers et al. 2005) and/or by deacetylating lysine residues in histones. Because PGC1-α is associated with a histone acetyl transferase (HAT) complex in addition to SIRT1, it has the ability to epigenetically initiate a regulatory cascade of gene expression starting with various transcription factor genes and ending with the activation of thousands of genes involved in a variety of metabolic functions in the brain, liver, skeletal muscle, and brown adipose tissue (Lin et al. 2005). Another epigenetic modification, the addition of an acetyl group to a lysine 16 on the amino-terminus of histone 4 (H4K16), has been found to play a critical role in making chromatin accessible to the basal transcription apparatus and related coregulators (Shogren-Knaak et al. 2006). The prevailing view for how this particular modification alters chromatin accessibility is based on the disruption of electrostatic attractions between positive and negative charged amino acid residues in adjacent nucleosomes by HAT-dependent neutralization of H4K16 (Shogren-Knaak and Peterson 2006). Therefore, a simple chemical modification involving the addition (acetyalation) or removal (deacetylation) of an acetyl group on specific amino acids on the H4 tail can affect gene expression patterns by altering chromatin accessibility. Moreover, the balance between condensed chromatin (inaccessible) and its decondensed chromatin (accessible) utilizes an abundant byproduct of metabolism, acetyl-CoA, the immediate substrate for the HAT reaction. Such a mechanism links a key hierarchical control point for gene expression (i.e., chromatin accessibility) with the metabolism of common dietary substrates like carbohydrates, proteins, and lipids. Likewise, the activation of the histone deacetylase (HDAC), SIRT1, by resveratrol involves $NAD^+$, another common metabolite that can be derived intracellularly from amino acid metabolism or from dietary sources such as niacin (vitamin $B_3$) (Brody 1999). On the other hand, the sulfuraphanes from broccoli are known to be weak inhibitors of HDACs (Dashwood et al. 2005). These examples underscore the tight linkage between food, its metabolism, and the dietary control of gene expression in networks involved in important physiological processes. Indeed, there is an increasing appreciation for the role the epigenetic modification of chromatin plays in the dietary control of mammalian gene expression (Gallou-Kabani et al. 2007).

## EPIGENETICS AND CANCER

A growing body of evidence supports the role of epigenetic modifications in the dysregulation of gene expression in cancer formation, especially in the early stages of carcinogenesis (Timp et al. 2009; Bonifer and Bowen 2010). In the case of the oncogene for the small early region 1a protein of adenovirus (e1a), significant and coordinated reprogramming of gene expression was observed in human fibroblast cells 24 hours post-infection (Ferrari et al. 2009). Microarray analysis revealed three distinct phases in gene expression patterns over this 24-hour period. The first phase (2 to 6 hours) was enriched for genes involved in inflammation and pathogen response, while the second phase (6 to 12 hours) involved genes related to cell division, growth, and DNA synthesis. The third phase (12 to 26 hours) was characterized by gene expression changes involving development and differentiation, including homeobox domain–containing genes and cell–cell signaling. Most importantly, these researchers found a global relocalization of the Rb (retinoblastoma) proteins (Rb, p130, and p107) and p300/CBP (CREB cyclic AMP response element binding protein) HATs on the promoters for these genes. A follow-up study by this group showed that the e1a protein drove these cells into S phase, overriding cell cycle control genes (Howitz et al. 2008). The e1aoncoprotein was shown to directly bind to the Rb proteins as well as to the closely related HATs, CBP, and p300 and it is believed that it is this binding that is responsible for reversing the repression of cell cycle genes and the concomitant threefold reduction in total cellular histone H3 lysine 18 acetylation (H3K18ac). Gene knockdown experiments have shown that both CBP and p300 are required for acetylation of H3K18 and that global hypoacetylation at this site is associated with poor prognosis in prostate carcinomas (Seligson et al. 2009). Fraga and coworkers (2005) examined global epigenetic modifications of histone H4 in 36 fresh human tumors and 25 different tumorigenic cell lines, and two epigenetic modifications emerge as common hallmarks for all these samples: hypoacetylation of H4K16 and loss of trimethylation at H4K20. Histone hypoacetylation coupled with DNA hypermethylation of CpG islands (Kim et al. 2010) is frequently found in promoters of tumor suppressor and proapoptotic genes and is associated with aberrant gene silencing during carcinogenesis (Fahrner et al. 2002; Jones and Baylin 2007; Widschwendter et al. 2007). While some investigators believe that global epigenetic modifications to the genome (i.e., modifications in 5′, 3′, intronic, exonic, and intergenic regions) can serve as informative markers for tumor heterogeneity and broad variations in clinical behaviors in cancer patients (Seligson et al. 2009), other believe that these modifications may be specific to cancer-related genes or regions and that these site- or regio-specific epigenetic marks will serve as potential targets for next-generation drug therapies and dietary interventions.

## CHEMOPREVENTIVE PROPERTIES OF THE SOY PEPTIDE, LUNASIN

Lunasin is a 43 amino acid, bioactive peptide found in soy protein that is derived from the cotyledon-specific 2S albumin (Galvez et al. 1997). It is also found in significant amounts in a variety of soy-based foods (De Mejia et al. 2004). Below is a brief summary of our current understanding of the cellular and molecular properties of the lunasin peptide. When applied exogenously to mammalian cells in culture (Galvez et al. 2001), the lunasin peptide

- Enters mammalian cells through its internalizing RGD (arginine/glycine/aspartic acid) cell-adhesion motif
- Colocalizes with hypoacetylated chromatin in telomeric regions at prometaphase
- Binds preferentially to deacetylated histones H3 and H4, which is facilitated by its conserved chromobinding domain
- Inhibits histone H3 di-acetylation and H4 tetra-acetylation

- Upregulates the expression of genes involved in apoptosis and cell cycle control (Magbanua et al. 2006)
- Induces apoptosis in e1a-transfected cells (Lam et al. 2003)
- Suppresses transformation of normal cells to cancerous foci induced by the e1a oncogene

When constitutively expressed as a transgene in mammalian cells, the lunasin peptide is also known to disrupt kinetochore formation and block mitosis, leading to cell death (Galvez and de Lumen 1999).

As discussed in the *Epigenetics and Cancer* section, there is growing evidence that cancer formation and malignant transformation involve epigenetically mediated changes (e.g., histone acetylation, deacetylation, and DNA methylation) in gene expression. The objective of recent studies on lunasin has been to integrate its intracellular properties into a molecular mechanism that can explain its role in reducing cancer risk. Based on recent studies, a chemopreventive mechanism is emerging by which lunasin enters the nucleus, binds specifically to H4 amino termini in a way that masks a deacetylated lysine residue at position 8 (H4K8), and promotes the acetylation of H4K16. The latter modification triggers decondensation of chromatin and the concomitant expression of genes like those controlled by the *Rb* tumor suppressor and *ras* oncogene. Such a mechanism could explain how dietary constituents like soy can modulate gene networks and physiological systems to lower risk for cancer. In the next section is a description of some of the experimental data supporting this novel chemopreventive model.

## Lunasin Inhibits Carcinogen- and Oncogenes-Induced Tumor Formation

To investigate lunasin's tumor suppressing activity, an *in vitro* transformation assay was performed on normal, nontumorigenic C3H 10 T1/2 cells (C3H), exposed for 24 hours to synthetic lunasin peptide. For these assays, polycyclic hydrocarbon chemical carcinogens, 7,12-dimethylbenz-[a]anthracene (DMBA) and 3-methylcholanthrene (MCA), were used to induce foci formation. The soybean Bowman-Birk protease inhibitor (BBI), previously shown to inhibit foci formation with this assay (Kennedy 1993), was used as a positive control. BBI has been shown to be cancer preventive in several *in vitro* and animal model studies and is currently in phase III clinical trial for prevention of oral and neck cancer (Kennedy 1993, 1995). As shown in Figure 2.1, there was a 75% reduction of tumorous foci formation in both DMBA and MCA-treated cells when 125 nM of lunasin was added

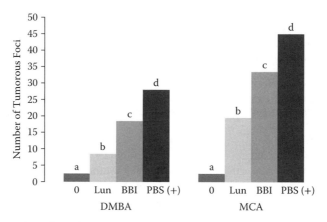

**FIGURE 2.1** Suppression of carcinogen-induced transformation of C3H cells by the lunasin peptide. Chemical carcinogens, 7,12-dimethylbenz[a]anthracene (DMBA) and 3-methylcholanthrene (MCA), were used to induce transformation of C3H cells in 24-well plates, treated with equimolar amounts (125 nM) of lunasin peptide and the Bowman-Birk protease inhibitor (BBI). Negative controls (O) were not treated with carcinogen, while the positive controls (PBS+) were treated with carcinogens without lunasin. Treatment means were compared using Duncan's multiple range test (DMRT) and means with similar letters are not significantly different from each other.

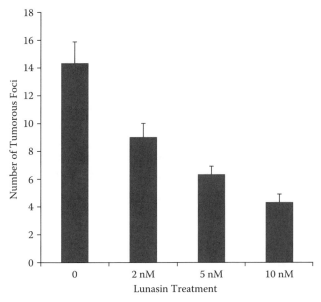

**FIGURE 2.2**   Effect of lunasin treatment on *Ki-ras* induced foci formation in NIH3T3 cells. Approximately $8 \times 10^5$ cells were grown in each well of a 12-well plate and then treated with different concentrations of luna-sin peptide for 24 hours before cells were transfected with pCMV-Ki-ras construct. The number of foci from three replicates was determined after 5 weeks and compared using DMRT. The mean and standard deviation from each treatment are shown.

to the culture media for 24 hours. The antitransformation effect of lunasin was found to be significantly higher than that of BBI, which was similarly applied in equimolar amounts to the growth media and produced a 33% reduction in foci formation in both DMBA and MCA-treated C3H cells (Figure 2.1).

To examine the chemopreventive effects of lunasin on oncogene-induced carcinogenesis, NIH 3T3 cells were transiently transfected with the *Ki-ras* oncogene and then incubated with media containing increasing amounts of the synthetic lunasin peptide. The number of foci was counted and the mean number of foci for each treatment was compared. Figure 2.2 shows that the number of *Ki-ras*-induced foci decreased in transfected NIH 3T3 cells treated with lunasin in a dose-dependent fashion. The maximum suppression (40%) of foci formation was observed at 10 μM lunasin concentration.

## UPREGULATION OF CHEMOPREVENTIVE GENES IN NONTUMORIGENIC PROSTATE EPITHELIAL CELLS

Transformation experiments showed that the 24-hour incubation of nontumorigenic mammalian cells with the lunasin peptide was sufficient to reduce tumor formation induced by chemical carcinogens and oncogenes. This is consistent with the finding that adenovirus e1a-induced transformation and global gene expression changes occur within 24 hours post-infection of the virus (Ferrari et al. 2008). To explore the global changes in gene expression induced by 24-hour lunasin treatment in prostate epithelial cells, microarray analysis was conducted on nontumorigenic (RWPE1) and tumorigenic (RWPE2) prostate epithelial cells. RWPE2 differs from RWPE1 by the presence of a stably transfected *Ki-ras* oncogene (Bello et al. 1997). Prostate epithelial cells were chosen for this study because prostate cancer is a common and slow-growing cancer, and thus a good candidate for dietary intervention with food bioactives like lunasin.

Microarray analysis showed that of the 14,500 genes interrogated, 114 genes had a greater than twofold change in expression in the cells exposed to 2 μM lunasin for 24 hours (Magbanua et al. 2006). Of these genes, 112 were upregulated in RWPE1 cells and only two genes were upregulated in RWPE2 cells. Surprisingly, no genes were downregulated twofold or more in either cells line treated with lunasin. Genes that were upregulated in RWPE1 cells include genes that are involved in the control of cell division, tumor suppression, and cell death (Table 2.1). At the transcriptional level, microarray analysis revealed that four representative genes (the proapoptotic genes, THBS1 and HIF1, and the antiproliferative genes, PRKAR1A and TOB1) were upregulated in RWPE1 with PRKAR1A showing the highest fold induction of 2.5 (Figure 2.3a). The microarray results were validated using quantitative RT-PCR (qRT-PCR) with HIF1A showing the highest induction (2.3-fold) in gene expression and THBS1 the lowest (1.4-fold) induction in gene expression, after lunasin treatment (Figure 2.3b). Taken together, these results suggest that lunasin acts as an indirect, positive regulator of chemopreventive genes to protect nontumorigenic cells from transformation. These findings are seemingly at odds with previous mechanistic models suggesting that lunasin

## TABLE 2.1
## Partial List of Genes Upregulated in Normal Epithelial Cells (RWPE1) after 24-Hour Exposure to 2 μM Lunasin

|  | Symbol | Fold-change |
|---|---|---|
| **Tumor uppressive (anticell proliferation) genes** | | |
| Protein kinase, cAMP-dependent, regulatory, type I, alpha | PRKAR1A, PKA | 2.45 |
| Transducer of ERBB2, 1 | TOB1 | 2.32 |
| ERBB2 interacting protein | ERBB2IP, ERBIN | 2.23 |
| **Genes involved in apoptosis** | | |
| Protein kinase C-like 2 | PRKCL2, PRK2 | 2.33 |
| BCL2/adenovirus E1B 19kDa interacting protein 3 | BNIP3, NIP3 | 2.05 |
| Thrombospondin 1 | THBS1, TSP1 | 2.05 |
| Pro-oncosis receptor inducing membrane injury gene | PORIMIN | 2.26 |
| Serine palmitoyltransferase, long chain base subunit 1 | SPTLC1 | 2.12 |
| **Mitotic checkpoint control genes** | | |
| BUB1 budding uninhibited by benzimidazoles 1 homolog beta (yeast) | BUB1B, BUBR1 | 2.23 |
| TTK protein kinase | TTK | 2.08 |
| MAD2 mitotic arrest deficient-like 1 (yeast) | MAD2L1 | 2.12 |
| **Protein degradation genes** | | |
| Proteasome (prosome, macropain) 26S subunit, ATPase, 6 | PSMC6 | 2.58 |
| RAN binding protein 2 | RANBP2 | 2.36 |
| E3 ubiquitin ligase SMURF2 | SMURF2 | 2.29 |
| Ubiquitin specific protease 1 | USP1 | 2.23 |
| Ubiquitin-activating enzyme E1C (UBA3 homolog, yeast) | UBE1C | 2.15 |
| **Cellular communication** | | |
| Gap junctionprotein, alpha 1, 43kDa (connexin 43) | GJA1 | 2.58 |

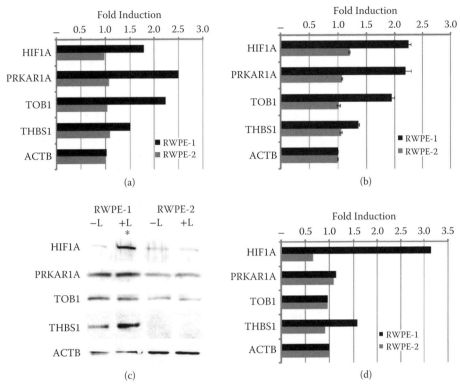

**FIGURE 2.3** Expression of H1F1A, PRKAR1A, TOB1, and THBS1 in RWPE-1 and RWPE-2 cells treated with lunasin. (a) mRNA expression of *H1F1A*, *PRKAR1A*, *TOB1*, and *THBS1* in RWPE-1 and RWPE-2 cells treated with lunasin. Lunasin-induced gene expression was detected by Affymetrix microarrays. (b) Confirmation of lunasin-induced gene expression of *H1F1A*, *PRKAR1A*, *TOB1*, and *THBS1* using qRT-PCR. The expression of each sample was normalized to the housekeeping gene, *ACTB*. All measurements of fold induction were based on 1xPBS-treated samples of each cell type. (c) Western blot analysis of *H1F1A*, *PRKAR1A*, *TOB1*, and *THBS1* expression in RWPE1 and RWPE2 cells. RWPE1 and RWPE2 cells were treated with 2 μM lunasin (+L) or 1xPBS, pH 7.4(−L), for 24 h before Western blot analysis. (d) Densitometry data of Western blot analyses. Each bar represents the average fold induction of four blots generated from two independent experiments. The expression of each sample was normalized to the expression of the housekeeping gene, ACTB.

silences tumorigenic gene expression by blocking acetylation of hypoacetylated histones H3 and H4 (Galvez et al. 2001). Presumably, by blocking the acetylation of these histones, chromatin in and around tumorigenic genes would remain condensed and thus inaccessible to the basal transcription apparatus, even in the absence or inactivation of tumor suppressors such as Rb.

To confirm that transcriptional upregulation of the four representative genes, THBS1, HIF1, PRKAR1A, and TOB1 (as revealed by microarray analysis and qRT-PCR), could be detected at the protein level, Western blot analysis was performed on proteins from lunasin-treated RWPE1 and RWPE2 cells (Figure 2.3c and d). The protein levels of HIF1A, PRKAR1A, THBS, and TOB were overall higher in RWPE1 than RWPE2 (Figure 2.3c) with or without lunasin treatment. On a fold-induction basis however, HIF1A protein in lunasin-treated RWPE1 cells showed a threefold induction over the lunasin-treated RWPE2, followed by THBS1, which showed a more than 1.5-fold increase. Upregulation of both HIF1A (Chen et al. 2009) and THBS1 (Komuro et al. 2009) has been linked to the suppression of prostate tumors and gastric carcinomas, respectively. PRKAR1A showed a slight induction by lunasin in RWPE1, while TOB1 showed no response to lunasin treatment, at least at the protein level (Figure 2.3d). The results from microarray, qRT-PCR, and Western blot analyses showed consistent upregulation of three of the four chemopreventive genes in lunasin-treated RWPE1, while remaining unchanged in lunasin-treated RWPE2. Only TOB1 failed to show the differential expression between lunasin-treated RWPE1 and RWPE2 cells, but this was observed only at the protein level.

## LUNASIN INHIBITS H4-LYSINE (K) 8 AND INCREASES H4-LYSINE (K) 16 ACETYLATION

Lunasin is known to bind deacetylated histones and inhibit histone acetylation in mammalian cells (Galvez et al. 2001), suggesting an epigenetic mechanism for lunasin in preventing carcinogenesis. Since it was discovered that lunasin upregulates chemopreventive gene expression, instead of downregulating the expression of tumorigenic genes, at least in RWPE1 cells, this raises the question of how a chromatin-binding peptide can both block histone acetylation *and* upregulate gene expression. The answer to this question may lie in lunasin's specificity for amino acid residues in the H4 tail.

Lunasin has been shown to significantly inhibit tetra-acetylation of H4 (at K5, K8, K12, and K16) in C3H 10T1/2 and MCF-7 cells treated with the HDAC inhibitor, Na butyrate (Galvez et al. 2001). Lunasin is known to have a chromobinding motif that binds more tightly to the deacetylated H4 than its negatively charged poly-aspartyl tail (Galvez et al. 2001), suggesting that lunasin may mask a specific H4 lysine residue via its chromobinding domain instead of masking all deacetylated H4 lysines through nonspecific interactions with its poly-aspartyl tail. To investigate whether lunasin differentially affected the acetylation of the four lysine residues in H4, *in vitro* HAT assays were performed to measure the acetylation of K5, K8, K12, and K16 in the H4 N-terminus as a function of lunasin treatment.

Densitometric quantitation of acetylation levels, normalized relative to the acid-extracted histone template for H4K5, K8, K12, and K16, and tetra-acetylated H4, are shown in Figure 2.4. In HAT assays without lunasin, HAT1 preferentially acetylated K8 and to a lesser extent, K12 (Figure 2.4c). This is consistent with HAT1's role as a cytoplasmic enzyme involved in acetylating newly synthesized, cytoplasmic histonesat H4K5 and K12 prior to nuclear transport and deposition in nucleosomes during DNA replication (Sobel et al. 1995). Conversely, the nucleosomal HATs, PCAF and p300, are involved in regulating gene expression. The p300 enzyme is a global coactivator of gene expression and PCAF is a secondary coactivator that requires association with p300 or CBP to locally acetylate chromatin (Schiltz et al. 1999). Without lunasin, both p300 and PCAF acetylated H4K8, consistent with the previous report of their H4 lysine specificity (Schiltz et al. 1999). p300 also acetylated H4K16, but to a slightly lesser degree than PCAF, which showed preferential acetylation of H4K16, similar to the acetylation specificity of GCN5, a PCAF orthologue from yeast (Kuo et al. 1996; Roth and Allis 1996). In the presence of lunasin, p300 and PCAF, and to a lesser extent HAT1, showed significant decrease in K8 acetylation and an increase in K16 acetylation (Figure 2.4a through c). PCAF showed the most significant increase in K16 acetylation, followed by p300 and HAT1.

Tetra-acetylation of H4 was significantly inhibited by lunasin in p300 and PCAF HAT reactions but not in HAT1 reactions. Both p300 and PCAF acetylated H4K8, while HAT1 preferentially acetylated K5 and K12. The inhibition of H4 tetra-acetylation by p300 and PCAF but not by HAT1 revealed in this study reflects the specific masking property of lunasin at H4K8 as opposed to H4K5, K12, and K16. This specific masking of H4K8 by lunasin also suggests that the inhibition of H4 acetylation is not due to nonspecific binding between the acidic poly-aspartyl tail of lunasin and the positively charged deacetylated H4 lysine residues but rather to a specific affinity of lunasin's chromobinding motif to the deacetylated H4K8.

An unexpected finding from our HAT assays is the ability of lunasin to increase H4K16 acetylation, a principal chromatin modification involved in chromatin decondensation and subsequent gene expression (Robinson et al. 2008; Shogren-Knaak et al. 2006; Shogren-Knaak and Peterson 2006) H4K16 acetylation in chromatin has been shown in structural studies to destabilize interactions between

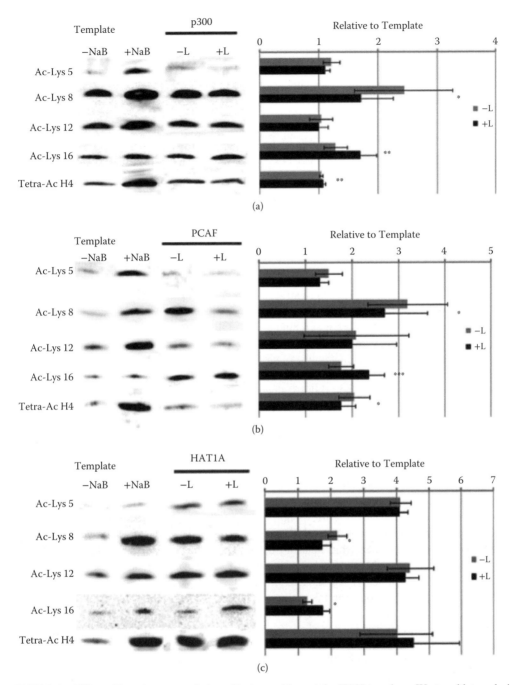

**FIGURE 2.4**  Effect of lunasin on acetylation of lysine residues at the H4 N-terminus. Western blot analysis was conducted on products of HAT assays using acid-extracted histones from HeLa cells. HAT reactions were conducted using HAT enzymes p300 (a), PCAF (b), andHAT1A (c) in response to lunasin treatment (+L) or 1xPBS (−L). Blots were immunostained with antibodies specific to acetylated H4K5, K8, K12, or K16. The tetra-acetylated H4 antibody only reacts with H4 acetylated at all four lysine residues. Negative template control (−NaB) and positive control (+NaB) correspond to acid-extracted histones from Na butyrate untreated and treated HeLa cells, respectively. Histograms represent densitometric scans of −L and +L bands normalized to their corresponding −NaB template controls. Averages of at least 3–5 replicates were plotted as −L (gray bars) and +L (black bars). Error bars represent S.E.M. values (n = 3–5). Statistical significance was calculated using paired Student t-test. *: p<0.05, **: p<0.01, ***: p<0.001.

adjacent nucleosomes, thus increasing accessibility of transcriptional complexes to regulatory DNA sequences at the 5′ end of the gene (Luger et al. 1997; Dorigo et al. 2003, 2004). The ability of lunasin to bind the deacetylated H4K8 while promoting the acetylation of H4K16 provides a credible and testable mechanism for how lunasin can coordinately upregulate expression of chemopreventive genes in nontumorigenic cells. We hypothesize that the binding of lunasin to the deacetylated H4K8 induces a conformational change in the N-terminus of H4 resulting in increased accessibility of H4K16 to various HATs.

## Lunasin Increases H4K16 Acetylation by PCAF but Not P300 in Hypoacetylated Recombinant H4 Template

The question remains why lunasin does not upregulate chemopreventive genes in the tumorigenic RWPE2. In HAT assays using acid-extracted histones, all three HAT enzymes were able to acetylate H4K16 in the presence of lunasin. In HAT assays using nonacetylated recombinant H4 (rH4) as template, H4K8 acetylation was significantly decreased in all three HATs, but only PCAF, a secondary transcriptional coactivator, was able to acetylate H4K16 (Figure 2.5a through c). This suggests that p300, the ubiquitous and global coactivator HAT, is unable to acetylate H4K16 residues when H4 is completely unacetylated, such as those found in hypoacetylated chromatin. The inability of p300 to acetylate rH4K16 regardless of lunasin implies that the bromodomain of this HAT (Marmorstein and Berger 2001; Zenget et al. 2008) requires preexisting acetylated H4 lysine residues at K5, K8, or K12 before it can bind and acetylate K16. Chromatin in or around chemopreventive genes in RWPE2 may be hypoacetylated at these residues as a result of *Ki-ras*-induced transformation. This could explain the loss of p300-mediated acetylation of H4K16 and the inability of lunasin to upregulate chemopreventive genes in RWPE2. In support of this hypothesis, it has been shown that the loss of acetylation at H4K16, albeit in regions of repetitive DNA sequences, is a common epigenetic hallmark of many tumors and transformed cells (Fraga et al. 2005).

## Chromatin Immunoprecipitation (ChIP) with Acetylated H4K16

Lunasin treatment has been shown to increase H4K16 acetylation in both RWPE1 and RWPE2 relative to the untreated controls (Figure 2.6a). If lunasin upregulates chemopreventive genes in RWPE1 by promoting acetylation of H4K16, chromatin in or around these genes should be enriched in acetylated H4K16. Conversely, corresponding chromatin regions in tumorigenic RWPE2 should contain predominantly hypoacetylated H4K16. To investigate these two possibilities, ChIP was performed on chromatin isolated from lunasin-treated or untreated RWPE1 and RWPE2 cells using antibody against acetylated H4K16. The resulting ChIP DNA was then analyzed by PCR to detect gene regions enriched in acetylated H4K16. Among the chemopreventive genes interrogated using primers derived from 5′, exonic, and 3′ regions, THBS1 produced a PCR amplicon with the ChIP DNA (Figure 2.6c and d), whereas HIF1A did not (data not shown). As shown in Figure 2.6c, primers that flanked a CpG island located at the 5′ end of THBS1 produced more PCR amplicons with the ChIP DNA from lunasin-treated than untreated RWPE1. However, the same primer pair generated little or no amplicon with ChIP DNA from lunasin-treated or untreated RWPE2. It should be noted that H4K16 acetylation could also be detected in exon 11 of THBS1, as evidenced by a strong amplicon signal with primers flanking this region. Interestingly, this primer pair produced strong amplicon signals with ChIP DNA from both lunasin-treated and untreated RWPE2 (Figure 2.6c).

## CpG Island Promoter of THBS1 Is Hypermethylated

Histone hypoacetylation coupled with DNA hypermethylation of CpG islands is frequently found in promoters of tumor suppressor and proapoptotic genes during carcinogenesis and is associated with aberrant gene silencing (Fahrneret al. 2002; Jones and Baylin 2007; Widswendter et al. 2007; Chen and Blumenthal, 2010). Bioinformatic analysis revealed that about one-third of the lunasin-upregulated genes

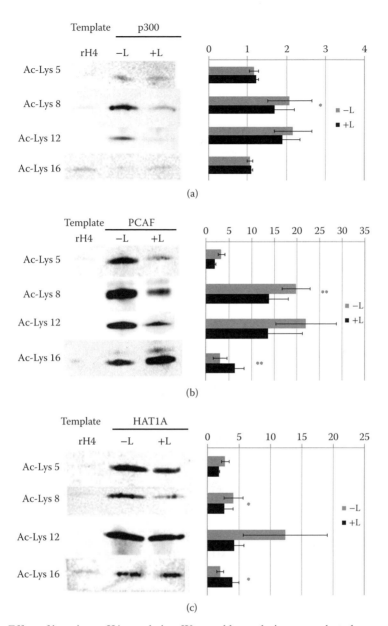

**FIGURE 2.5**    Effect of lunasin on rH4 acetylation. Western blot analysis was conducted on products of HAT assays using rH4 as the template. HAT reactions were performed using HAT enzymes *p300* (a), PCAF (b), and HAT1A (c) in response to lunasin treatment (+L) or 1xPBS (−L). Blots were immunostained with antibodies specific to acetylated H4K5, K8, K12, or K16. Histograms represent densitometric scans of −L and +L bands normalized to their corresponding rH4 template controls. Averages of 3–8 replicates were plotted as −L (gray bars) and +L (black bars). Error bars represent S.E.M. values (n = 3–8). Statistical significance was calculated using paired Student t-test. *: $p<0.05$, **: $p<0.01$, ***: $p<0.001$.

in RWPE1, including THBS1, have CpG islands located within 1000 bp (or approximately 5 nucleosomes) from their transcription start sites (Table 2.2). The absence of THBS1 expression and failure of lunasin to upregulate this gene in RWPE2 suggested that gene silencing could be due to the combination of reduced H4K16 acetylation (Figure 2.6c) associated with chromatin hypoacetylation and CpG island methylation at the 5′ end of THBS1. To assess the methylation status of the THBS1 CpG island, bisulfite-specific

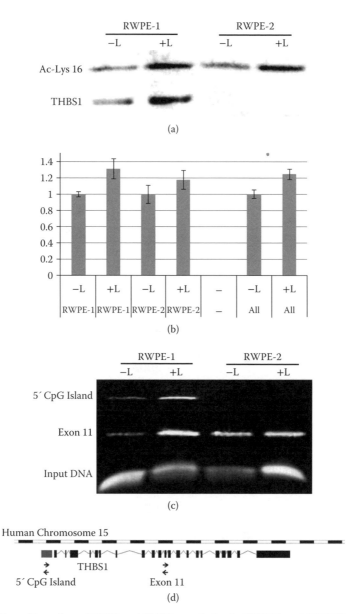

**FIGURE 2.6**   Effect of lunasin on H4K16 and H4K8 acetylation in RWPE1 and RWPE2 cells. (a) Western blot analysis of acetylated H4K16 and H4K8 protein levels in RWPE-1 and RWPE-2 cells treated with 2 μM lunasin (+L) or 1xPBS (−L). (b) Densitometric quantification of H4K16 acetylation in lunasin treated (+L) or 1xPBS treated (−L) RWPE-1, RWPE-2 cells, and both cells combined (All). * indicates p<0.05 with a Student t-test based on data normalized to L controls. (c) Upregulation of THBS1 was associated with increased H4K16 acetylation. PCR amplicons using template DNA from chromatin immunoprecipitation (ChIP) with an acetylated H4K16 antibody and primers derived from the 5′ CpG island and exon 11 of THBS1. Input DNA control represents PCR amplicons using template DNA from unprecipitated chromatin and 5′ CpG island primers. (d) The THBS1 locus with locations of PCR primers in the 5′ CpG island and exon 11.

PCR (BSP) was performed on genomic DNA isolated from RWPE1 and RWPE2. Using unmethylated primers specific to the THBS1 CpG island, more THBS1 amplicons were found in RWPE1 than in RWPE2 (Figure 2.7a), indicating that the THBS1 CpG island in RWPE1 was hypomethylated. On the other hand, using methylation-specific primers within the same region, RWPE2 showed significantly

**TABLE 2.2**
**List of Genes Upregulated by Lunasin in RWPE1 with a Known CpG Island and the Number of Base Pairs between the CpG Island and the Transcription Start Site**

| Gene Name | Number of bp of CpG Island from TSS |
|---|---|
| Thrombospondin 1 | 0 |
| Heterogeneous nuclear ribonucleoprotein H2 (H′) | 0 |
| NADH dehydrogenase (ubiquinone) 1 alpha subcomplex, 5, 13kDa | 0 |
| Golgiautoantigen, golgin subfamily a, 4 | 0 |
| Likely ortholog of mouse immediate early response, erythropoietin 4 | 0 |
| Tumor protein D52 (1) | 0 |
| Tumor protein D52 (2) | 0 |
| Translocated promoter region (to activated MET oncogene) | 0 |
| Leucine rich repeat (in FLII) interacting protein 1 | 0 |
| Serine palmitoyltransferase, long chain base subunit 1 | 0 |
| Cysteine-rich motor neuron 1 | 0 |
| Sema domain, immunoglobulin domain (Ig), short basic domain, secreted, (semaphorin) | 0 |
| SMC2 structural maintenance of chromosomes 2-like 1 (yeast) | 0 |
| Bullous pemphigoid antigen 1, 230/240kDa | 0 |
| TTK protein kinase | 0 |
| High-mobility group box 2 | 0 |
| Ubiquitin-activating enzyme E1C (UBA3 homolog, yeast) | 0 |
| Chondroitin sulfate proteoglycan 6 (bamacan) | 0 |
| Thioredoxin domain containing | 0 |
| NP220 nuclear protein | 0 |
| KIAA0143 protein | 0 |
| LYRIC/3D3 (1) | 0 |
| LYRIC/3D3 (2) | 0 |
| Lamina-associated polypeptide 1B | 0 |
| LPS-responsive vesicle trafficking, beach and anchor containing | 0 |
| Jagged 1 (Alagille syndrome) | 0 |
| CD44 antigen (homing function and Indian blood group) CELL ADHESION | 0 |
| Asp (abnormal spindle)-like, microcephaly associated, cell cycle, | 138 |
| A kinase (PRKA) anchor protein 2 | 230 |
| RNA-binding region (RNP1, RRM) containing 2 | 387 |
| Matrin 3 MATR3 | 440 |
| Matrin 3 | 440 |
| E3 ubiquitin ligase SMURF2 | 509 |
| ERBB2 interacting protein | 907 |

more THBS1 amplicons than RWPE1, indicating that the THBS1 CpG island in RWPE2 was hypermethylated (Figure 2.7a). Bisulfite sequencing of the THBS1 CpG island in RWPE1 and RWPE2 showed no methylation in RWPE1 whereas the same region was highly methylated in RWPE2 (Figure 2.7b). We believe that the hypermethylation of the 5′ CpG island of THBS1 in RWPE2 leads to hypoacetylation and a loss of H4K16 acetylation, thus making THBS1 unresponsive to lunasin treatment.

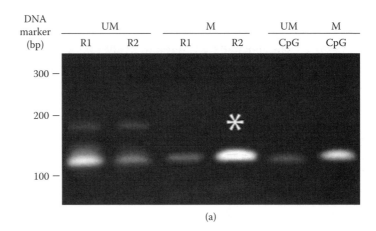

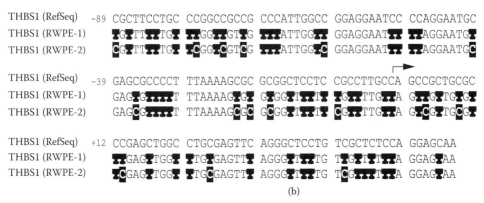

**FIGURE 2.7**  Methylation status of the THBS1 CpG island in RWPE1 and RWPE2 cells. (a) PCR of bisulfite-treated RWPE1 (R1) and RWPE2 (R2) genomic DNA with unmethylated and methylated primers. Methylated genomic DNA (CpG) was used as the template control in PCR using unmethylated (UM) and methylated (M) primers. * indicates CpG methylation at the 5′ end of THBS1 in RWPE2 cells. (b) DNA methylation mapping of the THBS1 CpG island by bisulfite genomic sequencing. Genomic DNA reference sequence (RefSeq) of the THBS1 CpG island is shown together with bisulfite sequence of the same region in RWPE1 and RWPE2 cells. Arrow indicates transcription start site (TSS). Negative and positive numbers correspond to base pairs upstream and downstream of TSS, respectively. Bold T letters indicate conversion of unmethylated cytosines in genomic DNA by bisulfite treatment. White **C** letters on black background correspond to methylated cytosines present in RWPE2 cells.

## CONCLUSION

Epigenetic silencing of tumor suppressors and proapoptosis genes in cancer cells, unlike genetic mutations, can potentially be reversed by the use of DNA demethylating agents (to remove methylation marks on the DNA) and HDAC inhibitors (to increase histone acetylation). It is now well established that plant-based foods are a natural source of modifiers of the epigenetic machinery. We believe that the ability of lunasin to upregulate expression of chemopreventive genes is associated with, and possibly due to, a lunasin-dependent increase in H4K16 acetylation in nontumorigenic RWPE1 cells. Furthermore, the inability of lunasin to upregulate chemopreventive genes in the tumorigenic RWPE2 cell line suggests that epigenetic changes (i.e., histone hypoacetylation coupled with DNA hypermethylation) within the promoters of these genes may have occurred during the course of Ki-ras oncogenic transformation of RWPE1 into RWPE2. As a dietary peptide capable of upregulation of chemopreventive gene expression by specific epigenetic modifications of the human genome, we believe the lunasin represents a novel regulatory motif and another food bioactive with the potential to reduce cancer risk. This could explain the long-standing inverse correlation between increased soy consumption and risk of various cancers. Future studies will be directed toward investigating the synergistic affects of lunasin in therapeutic combination with other natural chromatin remodeling bioactive agents.

## REFERENCES

Baur, J. A., and D. Sinclair. 2006. Therapeutic potential of resveratrol: The *in vivo* evidence. *Nature Rev Drug Disc* 5:493–506.

Baur, J. A., K. Pearson, N. L. Price, H. A. Jamieson, C. Lerin, A. Kalra, V. V. Prabhu, J. S. Allard, G. Lopez-Lluch, K. Lewis, P. J. Pistell, S. Poosala, K. G. Becker, O. Boss, D. Gwinn, M. Wang, S. Ramaswamy, K. W. Fishbein, R. G. Spencer, E. G. Lakatta, R. Le Couteur, R. J. Shaw, P. Navas, P. Puigserver, D. K. Ingram, R. de Cabo, and D. A. Sinclair. 2006. Resveratrol improves health and survival of mice on a high-calorie diet. *Nature* 444:337–42.

Bello, D., M. M. Webber, H. K. Kleinman, D. D. Wartinger, and J. S. Rhim. 1997. Androgen responsive adult human prostatic epithelial cell lines immortalized by human papillomavirus 18. *Carcinogenesis* 18:1215–23.

Bonifer, C., and D. T. Bowen. 2010. Epigenetic mechanisms regulating normal and malignant haematopoiesis: New therapeutic targets for clinical medicine. *Expert Reviews in Mol Med* 12:1–21.

Brody, T. 1999. *Nutritional Biochemistry*. 2nd ed. San Diego: Academic Press.

Chen, N., X. Chen, R. Huang, H. Zeng J. Gong, W. Meng, Y. Lu, F. Zhao, L. Wang, and Q. Zhou. 2009. BCL-xL is a target gene regulated by hypoxia-inducible factor-1{alpha}. *J Biol Chem* 284:10004–12.

Chen, X., and R. M. Blumenthal. 2010. Coordinated chromatin control: Structural and functional linkage of DNA and histone methylation. *Biochemistry* 49:2999–3008.

Dashwood, R. H., M. C. Myzak, and E. Ho. 2005. Dietary HDAC inhibitors: Time to rethink weak ligands in cancer prevention. *Carcinogenesis* 27:344–9.

De Mejia, E. G., M. Vasconez, B. O. de Lumen, and R. Nelson. 2004. Lunasin concentration in different soybean genotypes, commercial soy protein and isoflavone products. *J Agric Food Chem* 52:5882–7.

Dorigo, B., T. Schalch, A. Kulangara, S. Duda, R. R. Schroeder, and T. J. Richmond. 2004. Nucleosome arrays reveal the two-start organization of the chromatin fiber. *Science* 306:1571–3.

Dorigo, B., T. Schalch, K. Bystricky, and T. J. Richmond. 2003. Chromatin fiber folding: Requirement for the histone H4 N-terminal tail. *J Mol Biol* 327:85–96.

Fahrner, J. A., S. Eguchi, J. G. Herman, and S. B Baylin. 2002. Dependence of histone modifications and gene expression on DNA hypermethylation in cancer. *Cancer Res* 62:7213–8.

Ferrari, R., M. Pellegrini, G. A. Horwitz, W. Xie, A. J. Berk, and S. K. Kurdistani. 2008. Epigenetic reprogramming by adenovirus e1a. *Science* 321:1086–8.

Ferrari, R., A. J. Berk, and S. K. Kurdistani. 2009. Viral manipulation of the host epigenome for oncogenic transformation. *Nat Rev Genet* 10:290–4.

Fraga, M. F., E. Ballestar, A. Villar-Garea, M. Boix-Chornet, J. Espada, G. Schotta, T. Bonaldi, C. Haydon, S. Ropero, K. Petrie, N. G. Iyer, A. Pérez-Rosado, E. Calvo, J. A. Lopez, A. Cano, M. J. Calasanz, D. Colomer, M. A. Piris, N. Ahn, A. Imhof, C. Caldas, T. Jenuwein, and M. Esteller. 2005. Loss of acetylation at K16 and trimethylation at Lys20 of histone H4 is a common hallmark of human cancer. *Nat Genet* 37:391–400.

Gallou-Kabani, C., A. Vige, M. S. Gross, and C. Junien. 2007. Nutri-epigenomics: Lifelong remodelling of our epigenomes by nutritional and metabolic factors and beyond. *Clin Chem Lab Med* 45:321–7.

Galvez, A. F., M. J. R. Revilleza, and B. O. de Lumen. 1997. A novel methionine-rich protein from soybean cotyledon: Cloning and characterization of cDNA. *Plant Physiol* 114:1567.

Galvez, A. F., and B. O. de Lumen. 1999. A soybean cDNA encoding a chromatin-binding peptide inhibits mitosis of mammalian cells. *Nature Biotech* 17:495–500.

Galvez, A. F., N. Chen, J. Macasieb, and B. O. de Lumen. 2001. Chemopreventive property of a soybean peptide. *Cancer Res* 61:7473–8.

Greenwald, P., C. K. Clifford, and J. A. Milner. 2002. Diet and cancer prevention. *Eur J Cancer* 37:948–65.

Horwitz, G. A., K. Zhang, M. A. McBrian, M. Grustein, S. K. Kurdistani, and A. J. Berk. 2008. Adenovirus small e1a alters global patterns of histone modification. *Science* 321:1084–5.

Jones, S. B., S. E. DePrimo, M. L. Whitfield, and J. D. Brooks 2005. Resveratrol-induced gene expression profiles in human prostate cancer cells. *Cancer Epidemiol Biomakers Prev* 14:596–604.

Jones, P. A., and S. B. Baylin. 2009. The epigenomics of cancer. *Cell* 128:683–92.

Kant, A. K., M. F. Leitzmann, Y. Park, A. Hollenbeck, and A. Schatzkin. 2009. Patterns of recommended dietary behaviors predict subsequent risk of mortality in a large cohort of men and women in the United States. *J Nutr* 139:1374–80.

Kennedy, A. R. 1993. Cancer prevention by protease inhibitors. *Preventive Med* 22:796–811.

Kennedy, A. R. 1995. The evidence for soybean products as cancer preventive agents. *J Nutr* 125:733S–743S.

Kim, M. S., J. Lee, and D. Sidransky. 2010. DNA methylation markers in colorectal cancer. *Cancer Metastasis Rev* 29:181–206.

Komuro, A., M. Yashiro, C. Iwata, Y. Morishita, E. Johansson, Y. Matsumoto, A. Watanabe, H. Aburatani, H. Miyoshi, K. Kiyono, Y. T. Shirai, H. I. Suzuki, K. Hirakawa, M. R. Kano, and K. Miyazono. 2009. Diffuse-type gastric carcinoma: Progression, angiogenesis, and transforming growth factor beta signaling. *J Natl Cancer Inst* 101:592–604.

Kuo M. H., J. E. Brownell, R. E. Sobel, T. A. Ranalli, R. G. Cook, D. G. Edmondson, S. Y. Roth, and C. D. Allis. 1996. Transcription-linked acetylation by Gcn5p of histones H3 and H4 at specific lysines. *Nature* 383:269–72.

Lam, Y., A. F. Galvez, and B. O. de Lumen. 2003. Lunasin suppresses E1A-mediated transformation of mammalian cells but does not inhibit growth of immortalized and established cancer cell lines. *Nutrition & Cancer* 47:88–94.

Lin, J., C. Handschin, and B. M. Spiegelman. 2005. Metabolic control through the PGC-1 family of transcriptional coactivators. *Cell Metab* 1:361–71.

Luger, K., A. W. Mader, R. K. Richmond, D. F. Sargent, and T. J. Richmond. 1997. Crystal structure of the nucleosome core particle at 2.8 A resolution. *Nature* 389:251–60.

Ma, R. W., and K. Chapman. 2009. A systematic review of the effect of diet in prostate cancer prevention and treatment. *J Hum Nutr Diet* 22:187–99.

Magbanua, M., K. Dawson, L. Huang, W. Malyj, J. Gregg, A. Galvez, and R. L. Rodriguez. 2006. Nutrient–gene interactions involving soy peptide and chemopreventive genes in prostate epithelial cells. In J. Kaput and R. L. Rodriguez, eds., *Nutritional Genomics—Discovering the Path to Personalized Nutrition*. Hoboken, NJ: John Wiley & Sons.

Marmorstein, R., and S. L. Berger. 2001. Structure and function of bromodomains in chromatin-regulating complexes. *Gene* 272:1–9.

Messina, M., V. Persky, K. D. R. Setchell, and S. Barnes. 1994. Soy intake and cancer risk: A review of *in vitro* and *in vivo* data. *Nutr Cancer* 21:113–31.

Robinson, P. J., W. An, A. Routh, F. Martino, L. Chapman, R. G. Roeder, and D. Rhodes. 2008. 30 nm chromatin fibredecompaction requires both H4K16 acetylation and linker histone eviction. *J Mol Biol* 381:816–25.

Rodgers, J. T., C. Lerin, W. Haas, S. P. Gygi, B. M. Spiegelman, and P. Pulgserver. 2005. Nutrient control of glucose homeostasis through complex of PGC-1α and SIRT1. *Nature* 434:113–8.

Roth, S. Y., and C. D. Allis. 1996. Histone acetylation and chromatin assembly: A single escort, multiple dances? *Cell* 87:5–8.

Schiltz, R. L., C. A. Mizzen, A. Vassilev, R. G. Cook, C. D. Allis, and Y. Nakatani. 1999. Overlapping but distinct patterns of histone acetylation by the human coactivators p300 and PCAF within nucleosomal substrates. *J BiolChem* 274:1189–92.

Seligson, D. B., S. Horvath, M. A. McBrian, V. Mah, H. Yu, S. Tze, Q. Wang, D. Chia, L. Goodglick, and S. K. Kurdistani. 2009. Global levels of histone modifications predict prognosis in different cancers. *Am J Pathol* 174:1619–28.

Shogren-Knaak, M., H. Ishii, J. M. Sun, M. J. Pazin, J. R. Davie, and C. L. Peterson. 2006. Histone H4K16 acetylation controls chromatin structure and protein interactions. *Science* 311:844–7.

Shogren-Knaak, M., and C. L. Peterson. 2006. Switching on chromatin: Mechanistic role of histone H4K16 acetylation. *Cell Cycle* 5:1361–5.

Sobel, R. E., R. G. Cook, C. A. Perry, A. T. Annunziato, and C. D. Allis. 1995. Conservation of deposition-related acetylation sites in newly synthesized histones H3 and H4. *Proc NatlAcadSci USA* 92:1237–41.

Timp, W., A. Levchenko, and A. P. Feinberg. 2009. A new link between epigenetic progenitor lesions in cancer and the dynamics of signal transduction. *Cell Cycle* 8:383–90.

Widschwendter, M., H. Fiegl, D. Egle, E. Mueller-Holzner, G. Spizzo, C. Marth, D. J. Weisenberger, M. Campan, J. Young, I. Jacobs, and P. W. Laird. 2007. Epigenetic stem cell signature in cancer. *Nat Genet* 39:157–8.

Zeng, L., Q. Zhang, G. Gerona-Navarro, N. Moshkina, and M. M. Zhou. 2008. Structural basis of site-specific histone recognition by the bromodomains of human coactivators PCAF and CBP/p300. *Structure* 16:643–52.

# 3 Role of Epigenetics in the Complications Associated with Diabetes and Related Metabolic Disorders

*Louisa M. Villeneuve and Rama Natarajan*

## CONTENTS

## DIABETES AND ITS CONSEQUENCES

The incidence of diabetes, obesity, and related metabolic disorders is growing rapidly and has become a major healthcare issue worldwide in both developed and developing countries. Changes in the environment and nutrition have been implicated as major players. While diabetes itself is treatable at least to a large extent, it is associated with significantly increased incidence of numerous complications, including cardiovascular and kidney diseases. It is also one of the leading causes of blindness and lower limb amputations. Since diabetes and its complications would require lifelong or long-term treatment, diabetic patients face significantly higher medical costs. With the number of diabetic patients predicted to double in the next 20 years, access to affordable health care is expected to become an even greater challenge. Current treatments have added focus on preventative approaches so as to avoid future diabetic complications. Unfortunately, diabetic patients quite often continue to develop various complications even after achieving glucose control. This has been suggested to be attributable to a mysterious phenomenon termed "metabolic memory" of the prior glycemic state. Recent evidence suggests that this metabolic or hyperglycemic memory may be due to epigenetic changes in specific target cells and tissues that can alter gene expression without changes in the genetic code itself. While the genetics of diabetes has been the focus of scientific research for many years, much less is known about the role of epigenetics and the related molecular pathways that could lead to diabetes and the accelerated complications associated with such disorders. Further studies of epigenetic mechanisms are therefore timely and could provide valuable new insights to help clarify the crosstalk between genes and the environment and also uncover new therapeutic targets for diabetes and its complications.

## TYPES OF DIABETES

Diabetes is a complex disease with key subtypes, all of which manifest in elevated blood glucose levels resulting from either insufficient insulin production or insulin resistance or both. Type 1 diabetes (T1D) is an autoimmune disorder characterized by the destruction and loss of pancreatic beta cell function and loss of insulin production. Type 2 diabetes (T2D) involves progressive insulin resistance and beta cell dysfunction. Gestational diabetes results from insulin resistance during pregnancy and can lead to increased risk for the future development of diabetes for both the mother and child. Metabolic syndrome (MetS) is a condition that usually comprises three or more disorders including obesity, insulin resistance, glucose intolerance, dyslipidemia, and hypertension (Hotamisligil and Erbay 2008; de Luca and Olefsky 2008; Reaven 2006). While there is still some controversy as to whether MetS qualifies as disease by itself, it is associated with increased risk for the future development of T2D and cardiovascular disease (Liu et al. 2008; Reaven 2006). Other monogenic forms of diabetes are associated with heritable genetic mutations, such as maturity-onset diabetes of the young (MODY), which results in disruption of insulin production in most cases due to mutations in transcription factors involved in insulin signaling pathways (Vaxillaire and Froguel 2008). Notably, several of these transcription factors are known to play a role in the epigenetic regulation of gene expression (Ling and Groop 2009).

Interestingly, in most cases, the development of diabetes is not simply based on genetics, and not all people with diabetes-associated risk factors appear to succumb to this disease. Increasing evidence seems to implicate a key role for other factors, including the environment, in disease pathogenesis (Figure 3.1). Since the environment can alter epigenetic states, current research has focused on the potential role of the chromatin and associated epigenetic factors in the development of various common diseases such as diabetes, as this could help clarify connections between genes and the environment and how cells with the same genetic code and DNA sequence can express different genes which in some cases are associated with pathological conditions.

## OBESITY, EXERCISE, NUTRITION, AGING, AND THE ENVIRONMENT

The prevalence of obesity has increased dramatically in both adults and children over the last 20–30 years, with teenage obesity becoming a major concern as obesity is associated with increased risk for the development of T2D, cardiovascular disease, cancer, and mortality (Ogden et al. 2007). Fat and inflammatory cells in the adipose tissue can greatly affect metabolism through the release of free fatty acids, various hormones, and inflammatory cytokines and chemokines which can contribute to insulin resistance (Hotamisligil and Erbay 2008; Jin and Patti 2009; Ferrante 2007; Olefsky and Glass 2010). While genetics may factor into the susceptibility to obesity, the surprising and dramatic increase in obesity is more likely related to environmental changes, with increased access to calorie-rich foods coupled with decreased physical activity (Ogden et al. 2007). Exercise clearly affects metabolic states and protects against diabetes, while decreased physical activity can lead to insulin resistance and mitochondrial dysfunction (Ling and Groop 2009).

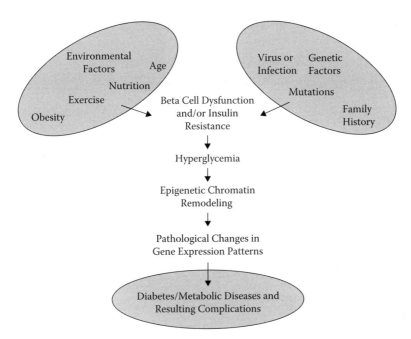

**FIGURE 3.1**   Factors involved in the development and progression of diabetes and metabolic diseases. Various environmental and genetic factors are associated with pancreatic beta cell dysfunction and/or insulin resistance, leading to loss of insulin action or production and hyperglycemia. Current evidence suggests hyperglycemia can result in altered epigenetic chromatin remodeling and aberrant gene expression patterns in various cells and tissues, which over time can contribute to the development of diabetes as well as other metabolic diseases and their associated complications.

Improper nutrition can not only contribute to obesity, it can also play a key role from the earliest stages of life with both pre- and postnatal nutritional environment being linked to T2D susceptibility. Maternal malnutrition and low birth weights are associated with increased susceptibility to diabetes and obesity later in life, the mechanisms for which might include epigenetic regulation of developmental genes (Jin and Patti 2009; Devaskar and Thamotharan 2007). Aging is also associated with increased risk for T2D thought to be related to genetic, epigenetic, and environmental factors (Ling and Groop 2009), and one factor that likely contributes to the increased prevalence of T2D is the increased life expectancy in developed countries over the last hundred years (Jin and Patti 2009) (Figure 3.1).

## MECHANISTIC BASIS FOR DIABETIC COMPLICATIONS

Micro- and macrovascular diseases are some of the most prevalent complications in diabetes. These include retinopathy, nephropathy, and neuropathy, as well as atherosclerosis, hypertension, and other cardiovascular diseases (Orasanu and Plutzky 2009). Great progress has been made into understanding many of the cellular biochemical signaling mechanisms associated with diabetic complications, and hyperglycemia has been identified as one of the key upstream players involved. Elevated glucose levels can result in long-term detrimental changes to various cells and tissues involved in diabetic complications.

Several signaling pathways have been found to be activated by hyperglycemia, such as increased flux in the hexosamine and polyol pathways (Brownlee 2001), increased levels of oxidant stress (Baynes 1991; Brownlee 2001; Clempus and Griendling 2006), increased formation of advanced glycation end products (AGEs) (Kim et al. 2005; Brownlee et al. 1988), and activation of protein kinase C (PKC) (Ishii et al. 1998) and transforming growth factor-beta-Smad-mitogen activated protein kinase

(MAPK) signaling pathways (Kanwar et al. 2008; Sharma and Ziyadeh 1995). Hyperglycemic activation of these pathways has been linked to increased mitochondrial superoxide anion formation, nicotinamide adenine dinucleotide phosphate (NADPH) oxidase activation, and associated oxidant stress (Devaraj et al. 2010; Brownlee 2005). Therefore, while reactive oxygen species such as superoxide anions may be short-lived, their effects can be long lasting due to subsequent activation of multiple downstream signaling pathways. Hyperglycemia and consequent increases in oxidant stress, AGEs, PKC, and MAPKs can also lead to increased activation of NF-κB proinflammatory transcription factor which can promote the expression of key inflammatory genes and tissue inflammation (Devaraj et al. 2006; Schmidt et al. 1994; Shanmugam et al. 2004). Inflammation plays a major role in several diabetic complications and also contributes to beta cell failure in T1D as well as insulin resistance in T2D. Overall, the consequences of hyperglycemia result in deleterious changes to the expression patterns of various inflammatory and other pathologic genes and proteins, therefore contributing to the development of diabetic micro- and macrovascular complications (King 2008). While these biochemical mechanisms have provided important information, much less is known about the molecular and nuclear chromatin based mechanisms involved in the development of diabetic complications and how they might be reversed.

## EPIGENETICS AND THE REGULATION OF GENE EXPRESSION

Regulation of gene expression is a tightly controlled process that occurs at multiple levels and layers. Epigenetics refers to heritable changes in gene expression that occur independent of changes to the DNA sequence. DNA is wrapped around the H2A, H2B, H3, and H4 core histone proteins to make up the nucleosome, the basic subunit of chromatin (Luger et al. 1997). Repeating units of nucleosomes assemble in a highly organized fashion to form higher-order chromatin, which can regulate transcription depending on the degree of accessibility. A more open and accessible chromatin, termed euchromatin, will allow for recruitment and binding of various transcription factors and RNA polymerase II (RNA pol II), while a closed condensed chromatin, termed heterochromatin, is inaccessible to these factors. The N-terminal tails of histones in chromatin can also be subjected to various post-translational modifications such as acetylation, methylation, and phosphorylation, forming a "histone code" (Jenuwein and Allis 2001; Strahl and Allis 2000) (Figure 3.2). Different combinations of these modifications will affect accessibility of transcription and other chromatin factors and play a large role in transcriptional regulation by controlling gene activation or silencing. Therefore these epigenetic modifications form an additional layer of genetic information capable of regulating gene expression while leaving the DNA sequence itself unchanged (Jenuwein and Allis 2001; Strahl and Allis 2000).

Epigenetic histone modifications allow for dynamic flexibility of the chromatin via local recruitment of various nonhistone proteins and complexes that recognize specific combinations of chromatin marks. Changes in epigenetic histone modifications can result from cellular or environmental stimuli and disruption in nucleosome–nucleosome interactions to enable gene expression or form a compact silent or repressed conformation, thereby regulating chromatin accessibility and transcriptional outcomes (Kouzarides 2007).

### HISTONE LYSINE ACETYLATION AND DEACETYLATION

Histone lysine acetylation is one of the more transient histone modifications. It is mediated by histone acetyltransferases (HATs) and removed by histone deacetylases (HDACs). Acetylation can occur on a number of lysine residues and is generally associated with gene activation. HATs are capable of acetylating more than one lysine residue on both histone and nonhistone proteins. This allows for a more accessible chromatin structure associated with transcription factor and RNA Pol II recruitment permissible for transcription. HDACs in turn are often found in repressor complexes or to be involved in various signaling pathways associated with transcriptional silencing (Roth et al. 2001; Kouzarides 2007).

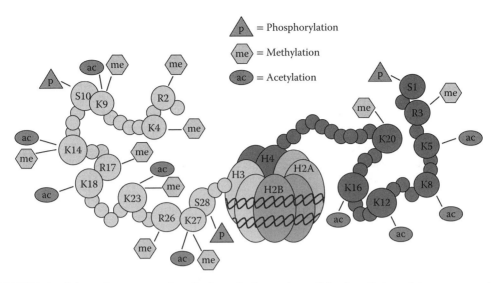

**FIGURE 3.2** Schematic representation of epigenetic chromatin modifications that regulate gene expression. DNA wrapped around a histone octamer which consists of two copies each of histones H2A, H2B, H3, and H4 comprise the nucleosome. The N-terminal histone tails of each of the histones are free and accessible to various post-translational modifications such as phosphorylation, methylation, and acetylation. H3 and H4 histone tail modifications are the best characterized and have been found to play key roles in regulating gene transcription. Lysine (K) acetylation, arginine (R) methylation, and serine (S) phosphorylation are typically associated with gene activation. Lysine methylation has proven more complex and is associated with both activation and repression depending on the lysine residue modified as well as the location within the gene (promoter versus coding region). Depending on the combination of histone modifications, the chromatin will either be accessible for transcription or condensed and silent. Therefore these epigenetic modifications form an additional layer of genetic information or a histone code capable of regulating gene expression without changing the DNA code itself.

## HISTONE METHYLATION AND DEMETHYLATION

Histone methylation occurs on either arginine or lysine residues and is usually thought to be more stable than lysine acetylation. Arginine residues can be either mono- or di-methylated by protein arginine methyltransferases (PRMTs) and are usually associated with gene activation (Lee et al. 2005). Histone lysine methylation is mediated by histone lysine methyltransferases (HMTs) (Lee et al. 2005), and lysine residues can be mono- or di- or tri-methylated. Unlike HATs and HDACs, HMTs are typically specific for one particular lysine (Kouzarides 2007; Zhang and Reinberg 2001). The SET domain containing family of HMTs, which is named for a conserved sequence motif found in three *Drosophila* proteins, suppressor of position effect variegation 3-9 (S), enhancer of zeste (E), and trithorax (T), are responsible for mediating histone lysine methylation (Zhang and Reinberg 2001; Feng et al. 2002). Lysine methylation can result in either gene activation or repression depending on the lysine residue modified. Several HMTs such as MLL1-4 (mixed lineage leukemia 1-4), SET1, SET7/9, and SMYD3 (SET and MYND domain containing 3) are known to mediate histone H3 lysine 4 methylation (H3K4me), which is typically associated with gene activation (Ruthenburg et al. 2007; Shi and Whetstine 2007), while G9a, SUV39H1 (Suppressor of variegation 3-9 homolog 1), and SETDB1/ESET (SET domain, bifurcated 1/ERG-associated protein with SET domain) mediate histone H3 lysine 9 methylation (H3K9me), generally associated with gene repression (Shi and Whetstine 2007). In addition, methylation at various other lysine residues can result in either activation or repression. Therefore, the transcriptional outcome is not always easy to predict. Having multiple HMTs capable of methylating the same residue suggests a degree of fine tuning allowing for other chromatin factors to play additional roles in recruiting specific HMTs, as well as redundancy in the case of loss of a crucial HMT.

Histone lysine methylation, while considered one of the more durable and stable epigenetic marks next to DNA methylation, can also undergo dynamic regulation through the concerted and balanced actions of various HMTs and lysine demethylases (KDMs). The first KDM was discovered only in 2004 by Shi et al. (2004) who demonstrated that lysine demethylase 1 (LSD1) could specifically remove H3K4me. It was not long before other KDMs were also discovered with varying specificities for different histone lysine residues (Shi and Whetstine 2007; Whetstine et al. 2006; Tsukada et al. 2006; Metzger et al. 2005). Currently, there is much interest in further characterizing these demethylases and their potential roles in various diseases (Shi 2007; Trojer and Reinberg 2006). Interestingly, a recent report demonstrated that the histone demethylase JHDM2A is associated with obesity and affects genes related to metabolism in rodent models (Tateishi et al. 2009). Despite the dynamic flexibility of histone lysine methylation, it is still a relatively stable epigenetic modification with some histone lysine methylation states demonstrating very low turnover rates, and hence could be a key factor in metabolic memory.

## EPIGENETICS AND DNA METHYLATION

DNA methylation at CpG islands is one of the best-characterized epigenetic modifications, and a few studies have demonstrated changes in DNA methylation in diabetes. The agouti mouse is a prominent example where agouti gene expression levels and DNA methylation have an effect on the development of obesity and diabetes (Morgan et al. 1999). Another study showed increased DNA methylation at the PPAR-gamma coactivator 1α gene (*PPARGC1A*) promoter in diabetic islets which was found to play a key role in regulating mitochondrial genes and diabetes (Ling et al. 2008), while in skeletal muscle, diabetic conditions led to DNA hypermethylation at the PPARGC1A locus at non-CpG regions (Barres et al. 2009). Intrauterine growth retardation was associated with T2D linked to epigenetic silencing possibly through both DNA methylation and histone modifications of a key transcription factor, Pdx1, involved in insulin expression and beta cell differentiation (Park et al. 2008). Interestingly, DNA methylation was found at the insulin promoter in mouse embryonic stem cells, only becoming demethylated as the cells differentiated into insulin expressing cells. In addition, both human and mouse insulin promoters were specifically demethylated in adult pancreatic beta cells, suggesting epigenetic regulation of insulin expression (Kuroda et al. 2009). While a specific role for DNA methylation in diabetes has remained elusive, several studies have demonstrated a strong association between aberrant DNA methylation and inflammation as well as hyperlipidemia (Maier and Olek 2002; Stenvinkel et al. 2007; Zaina et al. 2005).

Since the environment can affect epigenetic states, in more recent years, the focus has shifted from heritability to a more inclusive and unifying definition of epigenetics focused more on "the structural adaptation of chromosomal regions so as to register, signal or perpetuate altered activity states" (Bird 2007, p. 398). Histone modifications are now widely accepted to play a role in epigenetics; however, several questions remain regarding their specific role, whether they modulate or are modulated by DNA methylation, and whether they initiate or simply maintain a transcriptional memory (Berger et al. 2009). Much progress has been made in recent years in elucidating the basic biology of epigenetic states and potential roles of epigenetics in the regulation of tumor suppressor genes and other cancer related genes (Sharma et al. 2010). However, much less is known in the field of diabetes, obesity, nutritional disorders, and related common metabolic diseases and are hence of great current interest.

## MICRORNAS AS EPIGENETIC REGULATORS

MicroRNAs (miRs) are recently discovered small RNAs that act as negative regulators of gene expression. They are short (21–25 nucleotides in length), noncoding RNAs that typically bind to the 3′UTR of target mRNA transcripts leading to post-transcriptional silencing and translational repression or mRNA degradation (Bartel 2009; Kim et al. 2009). miRs provide another novel means for the cell/tissue/organism to respond to environmental stimuli without changing the DNA sequence. They can target and downregulate various signaling pathways, transcription factors, and numerous other proteins and enzymes. miRs also have the potential to target other chromatin modifying enzymes, leading to further epigenetic modifications affecting gene expression levels. In addition, there is in turn the potential for histone modifications and chromatin structural changes to affect the transcription and expression of miRs themselves. Therefore, miRs may also be epigenetically regulated.

Increasing evidence suggests that miRs play a significant role in numerous diseases. They have been identified as tumor suppressors and oncogenes in various cancers (Esquela-Kerscher and Slack 2006) and are also involved in development and differentiation (Bartel 2009; Kim et al. 2009). Several key miRs have been associated with insulin secretion, fat metabolism, cholesterol biosynthesis, and adipogenesis, crucial pathways in the pathogenesis of diabetes (Poy et al. 2004, 2007; Heneghan et al. 2009). Key miRs have also been associated with TGF-β signaling and diabetic kidney disease (Kato et al. 2009). The role of miRs in diabetic complications is a fast-paced area of research both as a means to understanding the molecular pathways leading to complications and to discover potentially new therapeutic targets.

## METABOLIC MEMORY

Several clinical trials have investigated the long-term effects of hyperglycemia and their relationships to diabetic complications. Results suggest that, even after glucose normalization, quite often there is a sustained metabolic memory of the prior glycemic state wherein diabetic patients continue to develop debilitating complications. Interestingly, the Diabetes Control and Complications Trial (DCCT) demonstrated that T1D patients treated with intensive insulin therapy versus conventional therapy clearly exhibited a decreased progression of key diabetic complications, including retinopathy, nephropathy, and neuropathy (DCCT Research Group 1993). Because of such pronounced benefits, the DCCT trial was terminated, and all participating patients were placed on intensive therapy for long term follow-up in the Epidemiology of Diabetes Interventions and Complications (EDIC) study.

The results from the EDIC trial further demonstrated that there was significantly decreased progression of microvascular complications in patients who had previously been on intensive insulin therapy during the DCCT and continued with intensive glycemic control for the EDIC trial relative to those who were on conventional therapy during the DCCT and only later placed on intensive control during EDIC (Pop-Busui et al. 2009; EDIC study 2003; Writing Team DCCT/EDIC Research Group 2002). Patients under continued intensive therapy were also found to have better outcomes for macrovascular complications such as nonfatal heart attack, stroke, or death from cardiovascular disease (Nathan et al. 2005), as well as decreased progression of intima-media thickness and coronary artery calcification associated with atherosclerosis (Cleary et al. 2006; Nathan et al. 2003). Overall, patients under intensive glycemic therapy during the DCCT trial exhibited long-term beneficial effects of prior tight control relative to patients who were on prior conventional therapy, a phenomenon termed "metabolic memory" (Nathan et al. 2005) (Figure 3.3).

Studies with T2D patients have also identified similar long-lasting beneficial effects of early glucose control. The United Kingdom Prospective Diabetes Study (UKPDS) demonstrated a decrease in cardiovascular risk in people with lower fasting plasma glucose at the time of diagnosis (Colagiuri et al. 2002; Holman et al. 2008), and the Action in Diabetes and Vascular Disease: Preterax and Diamicron Modified Release Controlled Evaluation (ADVANCE) trial showed that intensive glycemic control led to a decrease in both macro- and microvascular disease, primarily due to a reduction in nephropathy (Patel et al. 2008). The Steno-2 Study, found a decrease in cardiovascular disease as well as end-stage renal disease following intensive multifactor therapy including, but not limited to, strict glycemic control (Gaede et al. 2008).

The overall findings from these major clinical trials confirm the detrimental effects of hyperglycemia and demonstrate the importance of early glycemic control to reduce long-term complications.

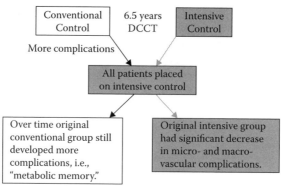

17 years DCCT/EDIC

**FIGURE 3.3** The metabolic memory phenomena. The Diabetes Control and Complications Trial/ Epidemiology of Diabetes Interventions and Complications (DCCT/EDIC) trial found that T1D on conventional glycemic control continued to develop diabetic complications at a faster rate compared to those on intensive glycemic control. Therefore the DCCT portion of the trial ended after 6.5 years, and all patients were placed on intensive control and enrolled in the EDIC trial for additional follow up. While those on intensive control for both the DCCT and EDIC trials exhibited a decreased risk for vascular complications, T1D who were originally on conventional control for the DCCT and only later placed on intensive therapy continued to develop diabetic complications at a faster rate demonstrating a "metabolic memory" of the prior glycemic state.

However, the mechanisms responsible for the continued development of diabetic complications resulting from the prior hyperglycemic state are not fully understood.

## MODELS OF METABOLIC MEMORY

Numerous animal and cell culture models have been used to examine the molecular mechanisms responsible for "metabolic memory" in the hopes of designing better therapeutic treatments for diabetic patients. Early studies in diabetic dogs found that even after reversing hyperglycemia, there was a continuation of retinal complications (Engerman and Kern 1987). Studies in diabetic rats demonstrated beneficial effects of islet transplantation after a short 6-week period of diabetes resulting in slower progression of retinopathy, relative to rats with longer prior exposure to diabetes (Hammes et al. 1993). Studies in streptozotocin (STZ)-induced diabetic rats showed that reinstitution of glycemic control after a short period of hyperglycemia had protective effects on retinal nitric oxide levels and lipid peroxides as well as other pathological indices of diabetic retinopathy. However, reinstitution of normal glucose after prolonged diabetes and hyperglycemia in the rats failed to reverse many of these pathological factors including increases in nitrative and oxidative stress, NF-κB activity, and inflammation, and this was attributed to metabolic memory (Chan et al. 2010; Kowluru 2003; Kowluru et al. 2004b, 2007). Some similar results were seen in kidneys from STZ-injected rats (Kowluru et al. 2004a).

Using cell culture models, early *in vitro* studies with endothelial cells treated with high glucose showed sustained expression of fibronectin and collagen extracellular matrix proteins even after normalization of glucose levels (Roy et al. 1990). More recent studies demonstrated the persistence of oxidant stress for up to 1 week after glucose normalization and that these high glucose effects could be partially blocked by antioxidants or NADPH oxidase inhibitors (Ihnat et al. 2007a,b). In another cell model, endothelial cells were treated for a short period of 16 hours with high glucose followed by glucose normalization and demonstrated a sustained increase in NF-κB p65 subunit expression, inflammatory genes, and oxidant stress that persisted for several days despite a return to normoglycemia (Brasacchio et al. 2009; El-Osta et al. 2008).

In yet another cell culture model, vascular smooth muscle cells (VSMC) derived from aortas of T2D, insulin resistant, obese *db/db* mice were found to exhibit a preactivated phenotype and metabolic memory even after culturing *in vitro* for several passages outside the diabetic animal. The diabetic *db/db* VSMC expressed increased levels of inflammatory cytokines and chemokines as well as increased migration, oxidant stress, activation of NF-κB and CREB transcription factors, and key signaling pathways associated with growth and migration, relative to VSMC derived from control nondiabetic *db/+* littermates (Li et al. 2006). In addition, *db/db* VSMC exhibited increased monocyte adhesion relative to *db/+* cells likely due to the increase in inflammatory chemokine production (Li et al. 2006). Overall the results from these various cell culture models suggest a metabolic memory of vascular dysfunction due to acute hyperglycemic spikes or prior chronic exposure to hyperglycemic conditions.

These findings in both animal and cell culture models further substantiate the results from clinical trials demonstrating that early tight control of glucose levels is essential to slowing down the progression of diabetic complications. They also implicate oxidant stress as a key factor in perpetuating this metabolic memory by potentially modifying or damaging essential lipids, proteins, and/or DNA (Ceriello et al. 2009; Ihnat et al. 2007a). Hyperglycemia, oxidant stress, and increased activity in the polyol pathway can also result in accumulation of AGEs which can lead to irreversible glycation of various proteins and lipids changing the structure and function of the vasculature, further perpetuating and amplifying local inflammation and oxidant stress resulting in long-term vascular damage. Thus AGEs, acting through receptors such as RAGE, can also contribute to hyperglycemic or metabolic memory (Calcutt et al. 2009; Yan et al. 2004; Meerwaldt et al. 2008). Similarly, excess food intake or improper nutrition might also be a key contributor as they can also increase the levels of AGEs. We are just beginning to understand some of the biochemical and signaling aspects of metabolic memory and how they may adversely affect target tissues and organs susceptible to diabetic complications. However, the subtle molecular, nuclear, and epigenetic mechanisms responsible for the sustained "memory" over time through multiple cell divisions at the transcriptional and epigenetic level need more attention and have evolved as an exciting area of research.

## ROLE OF EPIGENETICS IN DIABETIC COMPLICATIONS

Recent interesting reports have demonstrated key roles for epigenetic histone modifications in diabetes and its complications. A review by Gray and De Meyts (2005) demonstrates how HATs and HDACs play important roles in the activation and regulation of several key genes linked to diabetes. Additionally, the sirtuin (SIRT) family of HDACs has recently been discovered to modulate several factors related to diabetes such as insulin secretion, metabolism, and adipogenesis (Wellen et al. 2009). NF-κB transcriptional activity is also known to be regulated by several HATs and HDACs, resulting in altered downstream inflammatory gene expression (Ashburner et al. 2001; Gerritsen et al. 1997; Ito et al. 2004; Vanden Berghe et al. 1999).

Interestingly, in cultured monocytes, high glucose itself was shown to increase recruitment of HATs CPB and p/CAF, leading to increased expression of inflammatory genes cyclooxygenase-2 (COX-2) and tumor necrosis factor-alpha (TNF-α) corresponding to increased histone lysine acetylation at their promoters (Miao et al. 2004). Increased histone lysine acetylation at these inflammatory gene promoters was also seen in monocytes from both T1D and T2D patients relative to healthy control volunteers demonstrating *in vivo* relevance (Miao et al. 2004). Another more recent study showed that oxidized lipids can also lead to increased inflammatory gene promoter histone acetylation in a CREB/p300 HAT-dependent manner with a corresponding increase in gene expression (Reddy et al. 2009). p300 HAT was also shown to play a role in poly (ADP-ribose) polymerase (PARP) signaling pathways in diabetic kidney and retinal tissues due to increased oxidant stress leading to the expression of extracellular matrix (ECM) proteins (Xu et al. 2008). These studies implicate a role for chromatin histone acetylation in promoting inflammatory gene expression under diabetic conditions. However, acetylation is thought to be a more transient and potentially less gene-specific histone modification. Therefore a role for acetylation/deacetylation in metabolic memory remains to be determined.

Several *in vitro* studies have demonstrated characteristic histone epigenetic modifications at the insulin promoter associated with insulin production. The insulin promoter exhibited increased histone acetylation with a corresponding increase in p300 HAT recruitment along with increased H3K4me and recruitment of the corresponding HMT SET7/9, while H3K9me was undetectable. This pattern of active epigenetic histone modifications characterized the insulin gene promoter only in cells that were associated with insulin production as compared to other non-insulin-producing cell types (Chakrabarti et al. 2003; Mutskov et al. 2007). Additional studies in rat pancreatic cells using HDAC inhibitors further confirmed the essential role of histone modifications in pancreatic development (Haumaitre et al. 2008).

Histone methylation and its potential role in diabetic complications is another area of great interest. One method used to study histone methylation patterns genome-wide is the chromatin immunoprecipitation (ChIP) coupled to DNA microarray (ChIP-on-chip) technique. ChIP-chip studies in monocytes revealed dynamic changes in both H3K9me2 repressive marks and H3K4me2 activation marks at key genes in response to high glucose, and relevant changes were also seen in blood monocytes obtained from diabetic patients (Miao et al. 2007). Interestingly, primary human blood monocytes and lymphocytes demonstrated distinct cell-type specific histone methylation patterns which proved to be relatively stable within the cell type regardless of age or gender (Miao, Wu et al. 2008). Additional ChIP-chip studies comparing blood monocyte and lymphocyte H3K9me2 profiles from T1D relative to healthy volunteers found an increase in H3K9me2 in a subset of genes in diabetic lymphocytes linked to inflammatory and immune pathways associated with T1D and its complications (Miao, Smith et al. 2008). Being relatively stable, histone methylation patterns are likely maintained in healthy individuals over time in a cell-type-specific setting, and observation of key variations in these genome-wide studies suggest that this profile can be disrupted by changes in the environment or in a disease state. In addition, they also provide a glimpse of the inflammatory cell epigenome in the diabetic state and suggest that new information about diabetes, its complications, and metabolic memory can be obtained by such profiling approaches.

The molecular mechanisms responsible for these diabetes-associated changes in the epigenetic machinery are currently of great interest. Recent studies in monocytes have demonstrated an essential role for SET7/9, a H3K4 HMT, in NF-κB regulation of key inflammatory genes. SET7/9 knockdown led to decreased NF-κB p65 subunit as well as p300 HAT occupancies at monocyte chemoattractant protein-1 (MCP-1) and TNF-α promoters, which corresponded to reduced expression of these inflammatory genes (Li, Reddy et al. 2008). A role for SET7/9 in regulating NF-κB expression and inflammatory gene regulation was also shown in endothelial cells (Brasacchio et al. 2009; El-Osta et al. 2008).

## POTENTIAL EPIGENETIC MECHANISMS FOR METABOLIC MEMORY

Exciting recent studies have begun to unravel potential chromatin-based epigenetic mechanisms responsible for metabolic memory. Relative to VSMC derived from control non-diabetic *db/+* mice, VSMC derived from the aortas of diabetic *db/db* mice were found to express increased levels of key inflammatory genes associated with diabetes and diabetic complications even after culturing *ex vivo* for several passages (Li et al. 2006; Villeneuve et al. 2008). Additional studies demonstrated a reciprocal decrease in the repressive H3K9me3 mark at the promoters of interleukin-6 (IL-6), macrophage colony stimulation factor (MCSF), and MCP-1 genes in these cultured diabetic *db/db* VSMC relative to the control *db/+* suggesting a sustained loss of repression in the diabetic state. The loss of H3K9me3 paralleled a decrease in protein levels of the H3K9me3 HMT Suv39h1. Overexpression of Suv39h1 partially reversed the diabetic phenotype, thus verifying a role for Suv39h1 in inflammatory gene regulation. Interestingly, TNF-α induced expression of these inflammatory genes was greater in *db/db* VSMC relative to *db/+* VSMC, and this corresponded to a sustained reciprocal decrease in Suv39h1 occupancy as well as promoter H3K9me3 at these gene promoters. Furthermore, normal human VSMC treated with high glucose demonstrated similar histone

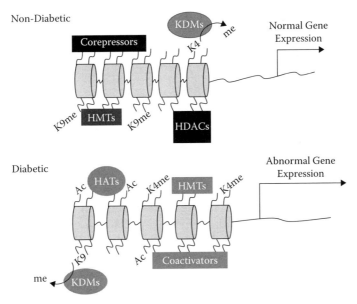

**FIGURE 3.4**   Proposed model for diabetes associated epigenetic regulation of gene expression. The regulation of post-translational modifications occurs through the actions of various chromatin modifiers such as HMTs and KDMs, which regulate lysine methylation (Kme), and HATs and HDACs, which mediate lysine acetylation (Ac). The proposed model suggests that under nondiabetic conditions normal gene expression levels are maintained through strict balanced control of repressive and active histone marks. However, in the diabetic milieu there may be a loss of repressive and a corresponding increase in active histone modifications leading to relaxation or accessibility of the chromatin surrounding pathological genes leading to abnormal gene expression levels. Other histone modifications and DNA methylation may also be involved.

lysine methylation changes, suggesting that the sustained alteration of these epigenetic marks could be due to the prior chronic exposure to a hyperglycemic environment in the diabetic *db/db* mice (Villeneuve et al. 2008).

Short-term hyperglycemia was also shown to induce long-lasting epigenetic changes in endothelial cells, suggesting that even short hyperglycemic spikes may be a risk for complications independent of HbA1c levels. Endothelial cells exposed to 16 hours of high glucose followed by a return to normal glucose for up to 6 days demonstrated sustained increases in NF-κB p65 subunit expression and inflammatory genes MCP-1 and VCAM-1. This corresponded to increased Set7 occupancy and H3K4me1 activation marks at the p65 promoter (Brasacchio et al. 2009; El-Osta et al. 2008). Further mechanistic studies showed that reducing mitochondrial superoxide production or the highly reactive methylglyoxal which accumulates in response to high glucose could prevent these epigenetic changes (El-Osta et al. 2008). Transient high glucose followed by glucose normalization in mice demonstrated similar changes in p65 expression and promoter H3K4me1 in the endothelial cells derived from the aortas of these mice (El-Osta et al. 2008). Overall, these results support the idea that prior exposure to chronic hyperglycemia and even periods of transient high glucose can alter epigenetic patterns and chromatin structure in target cells and thereby have long-lasting effects on the expression of genes associated with the pathology of diabetic vascular complications (Figure 3.4).

## EPIGENETIC TRANSMISSION

Clinical trials have demonstrated the progression of diabetic complications in spite of subsequent glycemic control suggesting a metabolic memory of the prior hyperglycemic state. Current research has provided insight into the possible mechanisms responsible for this phenomenon. Hyperglycemia can have pathological effects by increasing oxidant stress, AGEs, PKC, and inflammation, which recently have been linked to epigenetic chromatin changes. While it is becoming apparent that histone lysine methylation could play a key role in metabolic memory, it is a challenge to determine how these histone modifications are transmitted through multiple cell cycles and modulate diabetic complications.

## TRANSMISSION OF HISTONE LYSINE METHYLATION

Several studies have demonstrated the transmission of histone lysine methyl marks through DNA replication. SETDB1, H3K9 HMT, can be recruited to chromatin during replication by methyl-CpG binding protein (MBD1) allowing for methylation of newly deposited histones thus coupling the transmission of histone K9 methylation with DNA methylation (Sarraf and Stancheva 2004). Polycomb complex family of proteins is associated with long-term silencing, components of which have been found to play a role in epigenetic regulation of pancreatic beta-cell regeneration associated with diabetes and aging (Chen et al. 2009; Dhawan et al. 2009). Interestingly, Polycomb complexes have recently been found to remain bound to chromatin through replication (Francis et al. 2009), and polycomb repressive complex 2 (PRC2), containing the H3K27 HMT EZH2, can bind H3K27me3 modifications at ongoing sites of replication, which could then allow transmission or copying of the parental H3K27 methyl marks to the new histones deposited on the daughter strand (Hansen et al. 2008). However, it is yet to be determined how hyperglycemia-induced changes in epigenetic histone modifications are transmitted through cell replication.

## HISTONE VARIANTS AND CHROMATIN REMODELING COMPLEXES

Some potential mechanisms for the stable transmission of epigenetic modifications include histone variants, chromatin remodeling complexes as well as replication independent nucleosome assembly pathways (Hake and Allis 2006; Henikoff et al. 2004; Ng and Gurdon 2008). Functions of histone variants are distinct from canonical histones with differences in the basic properties of the histones as well as their localization to distinct regions of the genome. Histone variants can result in changes to nucleosome–nucleosome and nucleosome–protein interactions as well as to chromatin structure, which can lead to changes in gene transcription (Abbott et al. 2004; Fan et al. 2004; Hake and Allis 2006). Histone variants have also been proposed to play a role in establishing and maintaining epigenetic memory (Hake and Allis 2006; Henikoff et al. 2004; Ng and Gurdon 2008) and are susceptible to post-transcriptional modifications, similar to canonical histones.

Several ATP-dependent chromatin remodeling enzyme complexes have been identified that play varying roles in chromatin remodeling. Chromatin remodeling can occur through disruption of DNA-histone interactions either by altering, relocating, or replacing the nucleosomes (Korber and Horz 2004; Martens and Winston 2003). Chromatin remodeling enzymes have been found in complexes with various histone modifying proteins including HATs, HDACs, and HMTs, as well as DNA methyltransferases (Geiman et al. 2004; Martens and Winston 2003; Hassan et al. 2001; Nielsen et al. 2002). A role for chromatin remodeling enzymes has also been shown in diabetes where the SWI/SNF remodeling enzyme was reported in the induction of peroxisome proliferator-activated receptor gamma (PPARgamma) during adipogenesis (Salma et al. 2004). Overall, these results suggest the need for further investigation into a potential role for histone variants and nucleosome remodeling in epigenetics and metabolic memory of diabetic complications.

# POTENTIAL NUTRITIONAL EFFECTORS OF EPIGENETIC CHANGES IN DIABETES

Numerous nutritional supplements have been used extensively for the treatment of diabetes or its complications with mixed results. Some of these include soy, magnesium, curcumin, and several antioxidants such as lipoic acid, vitamins C and E, and zinc (Bartlett and Eperjesi 2008). Curcumin has shown some promising results. It has anti-inflammatory properties and is a well-known inhibitor of NF-κB activation and downstream consequences including increased expression of various inflammatory genes. Curcumin's therapeutic potential has been demonstrated in various inflammatory diseases including cardiovascular disease, obesity, and diabetes (Aggarwal and Harikumar 2009). *In vitro* studies have shown that curcumin can decrease cholesterol as well as VSMC proliferation and migration. A potential role in epigenetic regulation of gene expression is of great interest since curcumin can also inhibit HAT activity and HAT-dependent signaling pathways associated with gene activation (Li, Liu et al. 2008; Morimoto et al. 2008). Curcumin appears to decrease blood glucose levels, probably due to its antioxidant and anti-inflammatory properties. The drawback is that curcumin has low bioavailability due to poor absorption and rapid metabolism and elimination. Increasing curcumin bioavailability is a promising area of current research (Aggarwal and Harikumar 2009).

Folic acid as well as other diet-derived cofactors play key roles in the availability of methyl groups for various cellular methylation reactions, including methylation of histones, DNA methylation, and numerous other proteins and enzymes essential to cellular function. Methyl group metabolism is altered in both T1D and T2D, thus linking dietary cofactors with epigenetic changes in chromatin (MacFarlane et al. 2009). However, the downstream effects of altered epigenetic methylation states due to dietary factors and how they affect specific gene expression patterns related to diabetic complications are not well understood and require further research.

In addition to nutritional supplements, other environmental factors such as diet and exercise also play significant roles in the pathology of diabetes, obesity, and metabolic syndrome, and the resulting complications. However, there are still several unanswered questions. While some studies suggest that exercise might affect histone modifications (Ling and Groop 2009), very little is known in this area, and particularly whether exercise and dietary changes can reverse key pathologic chromatin marks. Since hyperglycemia-induced oxidant stress has been implicated in epigenetic modifications, expression of pathologic genes, and even metabolic memory, further studies into the beneficial effects of antioxidant therapies as well as therapies aimed at blocking specific related factors and pathways including AGE/RAGE, growth factors, and lipids on epigenetic mechanisms would be worthy of further investigation.

## FUTURE DIRECTIONS AND CONCLUSIONS

Fine tuning of gene expression and cellular homeostasis can occur through epigenetic modifications to the chromatin structure in response to environmental stimuli or altered nutrition. In turn, changes in epigenetic modifications play essential roles in altering gene expression patterns associated with disease states (Liu et al. 2008). Both clinical and *in vitro* studies have clearly demonstrated the importance of maintaining good diet and glycemic control to prevent or slow the progression of diabetic complications. Furthermore, these studies have shown that high glucose can induce epigenetic modifications in target cells resulting in activation of various signaling pathways and genes associated with these complications. While research to date in this area has focused mainly on a few chromatin marks and enzymes, cells, and gene types, it is likely that in the near future, additional chromatin marks, HMTs, and KDMs at other relevant genes will be uncovered in many target cells and tissues associated with hyperglycemic damage and diabetic complications. However, diabetes is a complex disease, and apart from hyperglycemia, there are several other risk factors. Particularly T2D involves insulin resistance, obesity, dyslipidemia, genetics, and environmental stimuli in addition to hyperglycemia. In conjunction with elevated glucose, each of these risk factors have the potential to induce epigenetic changes to the chromatin structure, ultimately affecting gene expression patterns in various target tissues and cells. Further exploration into how these risk factors affect the epigenetic mechanisms responsible for metabolic memory will enhance our understanding of diabetic complications and how they might be prevented.

An ambitious plan to characterize the human epigenome is currently ongoing (Qiu 2006; Human Epigenome Task Force 2008). The human epigenome project will provide more detailed genome-wide information on DNA methylation and other specific epigenetic modifications associated with various cells, tissues, and different developmental stages. Therefore, this project will broaden our view of the chromatin landscape ultimately providing a high resolution reference map to compare with the disease state.

The incidence of diabetes, metabolic syndrome, and related complications as well as the associated healthcare costs are expected to increase significantly over the next several years. New strategies are therefore greatly needed to curb these trends. Epigenetic research has advanced rapidly in recent years mainly due to marked advances in basic mechanisms, technologies, and bioinformatics; however, a more comprehensive understanding of the epigenetic code under normal versus diabetic states will provide better therapeutic opportunities beyond those currently available (Calcutt et al. 2009). Inhibitors of DNA methylation, HMTs, and HDACs are already being evaluated as epigenetic drugs for cancer and other diseases (Gray and De Meyts 2005; Shi 2007). Currently available therapies and drugs for the treatment of diabetic complications could be tested for their ability to alter epigenetic marks. Greater understanding of the epigenetic aspects of the disease state would enable the discovery of new therapeutic targets and treatment options for diabetes and its complications as well as numerous other diseases.

## ACKNOWLEDGMENTS

We gratefully acknowledge funding from the National Institutes of Health (NIDDK and NHLBI) and the Juvenile Diabetes Research Foundation.

## REFERENCES

Abbott, D. W., M. Laszczak, J. D. Lewis, H. Su, S. C. Moore, M. Hills, S. Dimitrov, and J. Ausio. 2004. Structural characterization of macroH2A containing chromatin. *Biochemistry* 43(5):1352–9.

Aggarwal, B. B., and K. B. Harikumar. 2009. Potential therapeutic effects of curcumin, the anti-inflammatory agent, against neurodegenerative, cardiovascular, pulmonary, metabolic, autoimmune and neoplastic diseases. *Int J Biochem Cell Biol* 41(1):40–59.

Ashburner, B. P., S. D. Westerheide, and A. S. Baldwin, Jr. 2001. The p65 (RelA) subunit of NF-kappaB interacts with the histone deacetylase (HDAC) corepressors HDAC1 and HDAC2 to negatively regulate gene expression. *Mol Cell Biol* 21(20):7065–77.

Barres, R., M. E. Osler, J. Yan, A. Rune, T. Fritz, K. Caidahl, A. Krook, and J. R. Zierath. 2009. Non-CpG methylation of the PGC-1alpha promoter through DNMT3B controls mitochondrial density. *Cell Metab* 10(3):189–98.

Bartel, D. P. 2009. MicroRNAs: Target recognition and regulatory functions. *Cell* 136(2):215–33.

Bartlett, H. E., and F. Eperjesi. 2008. Nutritional supplementation for type 2 diabetes: A systematic review. *Ophthalmic Physiol Opt* 28(6):503–23.

Baynes, J. W. 1991. Role of oxidative stress in development of complications in diabetes. *Diabetes* 40(4):405–12.

Berger, S. L., T. Kouzarides, R. Shiekhattar, and A. Shilatifard. 2009. An operational definition of epigenetics. *Genes Dev* 23(7):781–3.

Bird, A. 2007. Perceptions of epigenetics. *Nature* 447(7143):396–8.

Brasacchio, D., J. Okabe, C. Tikellis, A. Balcerczyk, P. George, E. K. Baker, A. C. Calkin, M. Brownlee, M. E. Cooper, and A. El-Osta. 2009. Hyperglycemia induces a dynamic cooperativity of histone methylase and demethylase enzymes associated with gene-activating epigenetic marks that coexist on the lysine tail. *Diabetes* 58(5):1229–36.

Brownlee, M. 2001. Biochemistry and molecular cell biology of diabetic complications. *Nature* 414(6865): 813–20.

———. 2005. The pathobiology of diabetic complications: A unifying mechanism. *Diabetes* 54(6):1615–25.

Brownlee, M., A. Cerami, and H. Vlassara. 1988. Advanced glycosylation end products in tissue and the biochemical basis of diabetic complications. *N Engl J Med* 318(20):1315–21.

Calcutt, N. A., M. E. Cooper, T. S. Kern, and A. M. Schmidt. 2009. Therapies for hyperglycaemia-induced diabetic complications: From animal models to clinical trials. *Nat Rev Drug Discov* 8(5):417–29.

Ceriello, A., M. A. Ihnat, and J. E. Thorpe. 2009. Clinical review 2: The "metabolic memory": Is more than just tight glucose control necessary to prevent diabetic complications? *J Clin Endocrinol Metab* 94(2):410–5.

Chakrabarti, S. K., J. Francis, S. M. Ziesmann, J. C. Garmey, and R. G. Mirmira. 2003. Covalent histone modifications underlie the developmental regulation of insulin gene transcription in pancreatic beta cells. *J Biol Chem* 278(26):23617–23.

Chan, P. S., M. Kanwar, and R. A. Kowluru. 2010. Resistance of retinal inflammatory mediators to suppress after reinstitution of good glycemic control: Novel mechanism for metabolic memory. *J Diabetes Complications* 24(1):55–63.

Chen, H., X. Gu, I. H. Su, R. Bottino, J. L. Contreras, A. Tarakhovsky, and S. K. Kim. 2009. Polycomb protein Ezh2 regulates pancreatic beta-cell Ink4a/Arf expression and regeneration in diabetes mellitus. *Genes Dev* 23(8):975–85.

Cleary, P. A., T. J. Orchard, S. Genuth, N. D. Wong, R. Detrano, J. Y. Backlund, B. Zinman, A. Jacobson, W. Sun, J. M. Lachin, and D. M. Nathan. 2006. The effect of intensive glycemic treatment on coronary artery calcification in type 1 diabetic participants of the Diabetes Control and Complications Trial/Epidemiology of Diabetes Interventions and Complications (DCCT/EDIC) study. *Diabetes* 55(12):3556–65.

Clempus, R. E., and K. K. Griendling. 2006. Reactive oxygen species signaling in vascular smooth muscle cells. *Cardiovasc Res* 71(2):216–25.

Colagiuri, S., C. A. Cull, and R. R. Holman. 2002. Are lower fasting plasma glucose levels at diagnosis of type 2 diabetes associated with improved outcomes? U.K. prospective diabetes study 61. *Diabetes Care* 25(8):1410–7.

DCCT Research Group. 1993. The effect of intensive treatment of diabetes on the development and progression of long-term complications in insulin-dependent diabetes mellitus: The Diabetes Control and Complications Trial Research Group. *N Engl J Med* 329(14):977–86.

de Luca, C., and J. M. Olefsky. 2008. Inflammation and insulin resistance. *FEBS Lett* 582(1):97–105.

Devaraj, S., M. R. Dasu, and I. Jialal. 2010. Diabetes is a proinflammatory state: A translational perspective. *Expert Rev. Endocrinol. Metab.* 5(1):19–28.

Devaraj, S., N. Glaser, S. Griffen, J. Wang-Polagruto, E. Miguelino, and I. Jialal. 2006. Increased monocytic activity and biomarkers of inflammation in patients with type 1 diabetes. *Diabetes* 55(3):774–9.

Devaskar, S. U., and M. Thamotharan. 2007. Metabolic programming in the pathogenesis of insulin resistance. *Rev Endocr Metab Disord* 8(2):105–13.

Dhawan, S., S. I. Tschen, and A. Bhushan. 2009. Bmi-1 regulates the Ink4a/Arf locus to control pancreatic beta-cell proliferation. *Genes Dev* 23(8):906–11.

EDIC study. 2003. Sustained effect of intensive treatment of type 1 diabetes mellitus on development and progression of diabetic nephropathy: The Epidemiology of Diabetes Interventions and Complications (EDIC) study. *JAMA* 290(16):2159–67.

El-Osta, A., D. Brasacchio, D. Yao, A. Pocai, P. L. Jones, R. G. Roeder, M. E. Cooper, and M. Brownlee. 2008. Transient high glucose causes persistent epigenetic changes and altered gene expression during subsequent normoglycemia. *J Exp Med* 205(10):2409–17.

Engerman, R. L., and T. S. Kern. 1987. Progression of incipient diabetic retinopathy during good glycemic control. *Diabetes* 36(7):808–12.

Esquela-Kerscher, A., and F. J. Slack. 2006. Oncomirs—microRNAs with a role in cancer. *Nat Rev Cancer* 6(4):259–69.

Fan, J. Y., D. Rangasamy, K. Luger, and D. J. Tremethick. 2004. H2A.Z alters the nucleosome surface to promote HP1alpha-mediated chromatin fiber folding. *Mol Cell* 16(4):655–61.

Feng, Q., H. Wang, H. H. Ng, H. Erdjument-Bromage, P. Tempst, K. Struhl, and Y. Zhang. 2002. Methylation of H3-lysine 79 is mediated by a new family of HMTases without a SET domain. *Curr Biol* 12(12):1052–8.

Ferrante, A. W., Jr. 2007. Obesity-induced inflammation: A metabolic dialogue in the language of inflammation. *J Intern Med* 262(4):408–14.

Francis, N. J., N. E. Follmer, M. D. Simon, G. Aghia, and J. D. Butler. 2009. Polycomb proteins remain bound to chromatin and DNA during DNA replication *in vitro*. *Cell* 137(1):110–22.

Gaede, P., H. Lund-Andersen, H. H. Parving, and O. Pedersen. 2008. Effect of a multifactorial intervention on mortality in type 2 diabetes. *N Engl J Med* 358(6):580–91.

Geiman, T. M., U. T. Sankpal, A. K. Robertson, Y. Zhao, Y. Zhao, and K. D. Robertson. 2004. DNMT3B interacts with hSNF2H chromatin remodeling enzyme, HDACs 1 and 2, and components of the histone methylation system. *Biochem Biophys Res Commun* 318(2):544–55.

Gerritsen, M. E., A. J. Williams, A. S. Neish, S. Moore, Y. Shi, and T. Collins. 1997. CREB-binding protein/p300 are transcriptional coactivators of p65. *Proc Natl Acad Sci USA* 94(7):2927–32.

Gray, S. G., and P. De Meyts. 2005. Role of histone and transcription factor acetylation in diabetes pathogenesis. *Diabetes Metab Res Rev* 21(5):416–33.

Hake, S. B., and C. D. Allis. 2006. Histone H3 variants and their potential role in indexing mammalian genomes: The "H3 barcode hypothesis." *Proc Natl Acad Sci USA* 103(17):6428–35.

Hammes, H. P., I. Klinzing, S. Wiegand, R. G. Bretzel, A. M. Cohen, and K. Federlin. 1993. Islet transplantation inhibits diabetic retinopathy in the sucrose-fed diabetic Cohen rat. *Invest Ophthalmol Vis Sci* 34(6):2092–6.

Hansen, K. H., A. P. Bracken, D. Pasini, N. Dietrich, S. S. Gehani, A. Monrad, J. Rappsilber, M. Lerdrup, and K. Helin. 2008. A model for transmission of the H3K27me3 epigenetic mark. *Nat Cell Biol* 10(11):1291–300.

Hassan, A. H., K. E. Neely, and J. L. Workman. 2001. Histone acetyltransferase complexes stabilize swi/snf binding to promoter nucleosomes. *Cell* 104(6):817–27.

Haumaitre, C., O. Lenoir, and R. Scharfmann. 2008. Histone deacetylase inhibitors modify pancreatic cell fate determination and amplify endocrine progenitors. *Mol Cell Biol* 28(20):6373–83.

Heneghan, H. M., N. Miller, and M. J. Kerin. 2009. Role of microRNAs in obesity and the metabolic syndrome. *Obes Rev.*

Henikoff, S., T. Furuyama, and K. Ahmad. 2004. Histone variants, nucleosome assembly and epigenetic inheritance. *Trends Genet* 20(7):320–6.

Holman, R. R., S. K. Paul, M. A. Bethel, D. R. Matthews, and H. A. Neil. 2008. 10-year follow-up of intensive glucose control in type 2 diabetes. *N Engl J Med* 359(15):1577–89.

Hotamisligil, G. S., and E. Erbay. 2008. Nutrient sensing and inflammation in metabolic diseases. *Nat Rev Immunol* 8(12):923–34.

Human Epigenome Task Force. 2008. Moving AHEAD with an international human epigenome project. *Nature* 454(7205):711–5.

Ihnat, M. A., J. E. Thorpe, and A. Ceriello. 2007a. Hypothesis: The "metabolic memory," the new challenge of diabetes. *Diabet Med* 24(6):582–6.

Ihnat, M. A., J. E. Thorpe, C. D. Kamat, C. Szabo, D. E. Green, L. A. Warnke, Z. Lacza, A. Cselenyak, K. Ross, S. Shakir, L. Piconi, R. C. Kaltreider, and A. Ceriello. 2007b. Reactive oxygen species mediate a cellular "memory" of high glucose stress signalling. *Diabetologia* 50(7):1523–31.

Ishii, H., D. Koya, and G. L. King. 1998. Protein kinase C activation and its role in the development of vascular complications in diabetes mellitus. *J Mol Med* 76(1):21–31.

Ito, K., T. Hanazawa, K. Tomita, P. J. Barnes, and I. M. Adcock. 2004. Oxidative stress reduces histone deacetylase 2 activity and enhances IL-8 gene expression: Role of tyrosine nitration. *Biochem Biophys Res Commun* 315(1):240–5.

Jenuwein, T., and C. D. Allis. 2001. Translating the histone code. *Science* 293(5532):1074–80.

Jin, W., and M. E. Patti. 2009. Genetic determinants and molecular pathways in the pathogenesis of type 2 diabetes. *Clin Sci (Lond)* 116(2):99–111.

Kanwar, Y. S., J. Wada, L. Sun, P. Xie, E. I. Wallner, S. Chen, S. Chugh, and F. R. Danesh. 2008. Diabetic nephropathy: Mechanisms of renal disease progression. *Exp Biol Med (Maywood)* 233(1):4–11.

Kato, M., L. Arce, and R. Natarajan. 2009. MicroRNAs and their role in progressive kidney diseases. *Clin J Am Soc Nephrol* 4(7):1255–66.

Kim, V. N., J. Han, and M. C. Siomi. 2009. Biogenesis of small RNAs in animals. *Nat Rev Mol Cell Biol* 10(2):126–39.

Kim, W., B. I. Hudson, B. Moser, J. Guo, L. L. Rong, Y. Lu, W. Qu, E. Lalla, S. Lerner, Y. Chen, S. S. Yan, V. D'Agati, Y. Naka, R. Ramasamy, K. Herold, S. F. Yan, and A. M. Schmidt. 2005. Receptor for advanced glycation end products and its ligands: a journey from the complications of diabetes to its pathogenesis. *Ann N Y Acad Sci* 1043:553–61.

King, G. L. 2008. The role of inflammatory cytokines in diabetes and its complications. *J Periodontol* 79(8 Suppl):1527–34.

Korber, P., and W. Horz. 2004. SWRred not shaken: Mixing the histones. *Cell* 117(1):5–7.

Kouzarides, T. 2007. Chromatin modifications and their function. *Cell* 128(4):693–705.

Kowluru, R. A. 2003. Effect of reinstitution of good glycemic control on retinal oxidative stress and nitrative stress in diabetic rats. *Diabetes* 52(3):818–23.

Kowluru, R. A., S. N. Abbas, and S. Odenbach. 2004a. Reversal of hyperglycemia and diabetic nephropathy: Effect of reinstitution of good metabolic control on oxidative stress in the kidney of diabetic rats. *J Diabetes Complications* 18(5):282–8.

Kowluru, R. A., S. Chakrabarti, and S. Chen. 2004b. Re-institution of good metabolic control in diabetic rats and activation of caspase-3 and nuclear transcriptional factor (NF-kappaB) in the retina. *Acta Diabetol* 41(4):194–9.

Kowluru, R. A., M. Kanwar, and A. Kennedy. 2007. Metabolic memory phenomenon and accumulation of peroxynitrite in retinal capillaries. *Exp Diabetes Res* 2007:21976.

Kuroda, A., T. A. Rauch, I. Todorov, H. T. Ku, I. H. Al-Abdullah, F. Kandeel, Y. Mullen, G. P. Pfeifer, and K. Ferreri. 2009. Insulin gene expression is regulated by DNA methylation. *PLoS One* 4(9):e6953.

Lee, D. Y., C. Teyssier, B. D. Strahl, and M. R. Stallcup. 2005. Role of protein methylation in regulation of transcription. *Endocr Rev* 26(2):147–70.

Li, H. L., C. Liu, G. de Couto, M. Ouzounian, M. Sun, A. B. Wang, Y. Huang, C. W. He, Y. Shi, X. Chen, M. P. Nghiem, Y. Liu, M. Chen, F. Dawood, M. Fukuoka, Y. Maekawa, L. Zhang, A. Leask, A. K. Ghosh, L. A. Kirshenbaum, and P. P. Liu. 2008. Curcumin prevents and reverses murine cardiac hypertrophy. *J Clin Invest* 118(3):879–93.

Li, S. L., M. A. Reddy, Q. Cai, L. Meng, H. Yuan, L. Lanting, and R. Natarajan. 2006. Enhanced proatherogenic responses in macrophages and vascular smooth muscle cells derived from diabetic db/db mice. *Diabetes* 55(9):2611–9.

Li, Y., M. A. Reddy, F. Miao, N. Shanmugam, J. K. Yee, D. Hawkins, B. Ren, and R. Natarajan. 2008. Role of the histone H3 lysine 4 methyltransferase, SET7/9, in the regulation of NF-kappaB-dependent inflammatory genes: Relevance to diabetes and inflammation. *J Biol Chem* 283(39):26771–81.

Ling, C., S. Del Guerra, R. Lupi, T. Ronn, C. Granhall, H. Luthman, P. Masiello, P. Marchetti, L. Groop, and S. Del Prato. 2008. Epigenetic regulation of PPARGC1A in human type 2 diabetic islets and effect on insulin secretion. *Diabetologia* 51(4):615–22.

Ling, C., and L. Groop. 2009. Epigenetics: A molecular link between environmental factors and type 2 diabetes. *Diabetes* 58(12):2718–25.

Liu, L., Y. Li, and T. O. Tollefsbol. 2008. Gene-environment interactions and epigenetic basis of human diseases. *Curr Issues Mol Biol* 10(1-2):25–36.

Luger, K., A. W. Mader, R. K. Richmond, D. F. Sargent, and T. J. Richmond. 1997. Crystal structure of the nucleosome core particle at 2.8 A resolution. *Nature* 389(6648):251–60.

MacFarlane, A. J., A. Strom, and F. W. Scott. 2009. Epigenetics: Deciphering how environmental factors may modify autoimmune type 1 diabetes. *Mamm Genome* 20(9-10):624–32.

Maier, S., and A. Olek. 2002. Diabetes: a candidate disease for efficient DNA methylation profiling. *J Nutr* 132(8 Suppl):2440S–2443S.

Martens, J. A., and F. Winston. 2003. Recent advances in understanding chromatin remodeling by Swi/Snf complexes. *Curr Opin Genet Dev* 13(2):136–42.

Meerwaldt, R., T. Links, C. Zeebregts, R. Tio, J. L. Hillebrands, and A. Smit. 2008. The clinical relevance of assessing advanced glycation endproducts accumulation in diabetes. *Cardiovasc Diabetol* 7:29.

Metzger, E., M. Wissmann, N. Yin, J. M. Muller, R. Schneider, A. H. Peters, T. Gunther, R. Buettner, and R. Schule. 2005. LSD1 demethylates repressive histone marks to promote androgen-receptor-dependent transcription. *Nature* 437(7057):436–9.

Miao, F., I. G. Gonzalo, L. Lanting, and R. Natarajan. 2004. *In vivo* chromatin remodeling events leading to inflammatory gene transcription under diabetic conditions. *J Biol Chem* 279(17):18091–7.

Miao, F., D. D. Smith, L. Zhang, A. Min, W. Feng, and R. Natarajan. 2008. Lymphocytes from patients with type 1 diabetes display a distinct profile of chromatin histone H3 lysine 9 dimethylation: An epigenetic study in diabetes. *Diabetes* 57(12):3189–98.

Miao, F., X. Wu, L. Zhang, A. D. Riggs, and R. Natarajan. 2008. Histone methylation patterns are cell-type specific in human monocytes and lymphocytes and well maintained at core genes. *J Immunol* 180(4):2264–9.

Miao, F., X. Wu, L. Zhang, Y. C. Yuan, A. D. Riggs, and R. Natarajan. 2007. Genome-wide analysis of histone lysine methylation variations caused by diabetic conditions in human monocytes. *J Biol Chem* 282(18):13854–63.

Morgan, H. D., H. G. Sutherland, D. I. Martin, and E. Whitelaw. 1999. Epigenetic inheritance at the agouti locus in the mouse. *Nat Genet* 23(3):314–8.

Morimoto, T., Y. Sunagawa, T. Kawamura, T. Takaya, H. Wada, A. Nagasawa, M. Komeda, M. Fujita, A. Shimatsu, T. Kita, and K. Hasegawa. 2008. The dietary compound curcumin inhibits p300 histone acetyltransferase activity and prevents heart failure in rats. *J Clin Invest* 118(3):868–78.

Mutskov, V., B. M. Raaka, G. Felsenfeld, and M. C. Gershengorn. 2007. The human insulin gene displays transcriptionally active epigenetic marks in islet-derived mesenchymal precursor cells in the absence of insulin expression. *Stem Cells* 25(12):3223–33.

Nathan, D. M., P. A. Cleary, J. Y. Backlund, S. M. Genuth, J. M. Lachin, T. J. Orchard, P. Raskin, and B. Zinman. 2005. Intensive diabetes treatment and cardiovascular disease in patients with type 1 diabetes. *N Engl J Med* 353(25):2643–53.

Nathan, D. M., J. Lachin, P. Cleary, T. Orchard, D. J. Brillon, J. Y. Backlund, D. H. O'Leary, and S. Genuth. 2003. Intensive diabetes therapy and carotid intima-media thickness in type 1 diabetes mellitus. *N Engl J Med* 348(23):2294–303.

Ng, R. K., and J. B. Gurdon. 2008. Epigenetic inheritance of cell differentiation status. *Cell Cycle* 7(9):1173–7.

Nielsen, A. L., C. Sanchez, H. Ichinose, M. Cervino, T. Lerouge, P. Chambon, and R. Losson. 2002. Selective interaction between the chromatin-remodeling factor BRG1 and the heterochromatin-associated protein HP1alpha. *Embo J* 21(21):5797–806.

Ogden, C. L., S. Z. Yanovski, M. D. Carroll, and K. M. Flegal. 2007. The epidemiology of obesity. *Gastroenterology* 132(6):2087–102.

Olefsky, J. M., and C. K. Glass. 2010. Macrophages, inflammation, and insulin resistance. *Annu Rev Physiol* 72:219–46.

Orasanu, G., and J. Plutzky. 2009. The pathologic continuum of diabetic vascular disease. *J Am Coll Cardiol* 53(5 Suppl):S35–42.

Park, J. H., D. A. Stoffers, R. D. Nicholls, and R. A. Simmons. 2008. Development of type 2 diabetes following intrauterine growth retardation in rats is associated with progressive epigenetic silencing of Pdx1. *J Clin Invest* 118(6):2316–24.

Patel, A., S. MacMahon, J. Chalmers, B. Neal, L. Billot, M. Woodward, M. Marre, M. Cooper, P. Glasziou, D. Grobbee, P. Hamet, S. Harrap, S. Heller, L. Liu, G. Mancia, C. E. Mogensen, C. Pan, N. Poulter, A. Rodgers, B. Williams, S. Bompoint, B. E. de Galan, R. Joshi, and F. Travert. 2008. Intensive blood glucose control and vascular outcomes in patients with type 2 diabetes. *N Engl J Med* 358(24):2560–72.

Pop-Busui, R., P. A. Low, B. H. Waberski, C. L. Martin, J. W. Albers, E. L. Feldman, C. Sommer, P. A. Cleary, J. M. Lachin, and W. H. Herman. 2009. Effects of prior intensive insulin therapy on cardiac autonomic nervous system function in type 1 diabetes mellitus: The Diabetes Control and Complications Trial/Epidemiology of Diabetes Interventions and Complications study (DCCT/EDIC). *Circulation* 119(22):2886–93.

Poy, M. N., L. Eliasson, J. Krutzfeldt, S. Kuwajima, X. Ma, P. E. Macdonald, T. Pfeffer, T. Tuschl, N. Rajewsky, P. Rorsman, and M. Stoffel. 2004. A pancreatic islet-specific microRNA regulates insulin secretion. *Nature* 432 (7014):226–30.

Poy, M. N., M. Spranger, and M. Stoffel. 2007. microRNAs and the regulation of glucose and lipid metabolism. *Diabetes Obes Metab* 9(Suppl 2):67–73.

Qiu, J. 2006. Epigenetics: Unfinished symphony. *Nature* 441(7090):143–5.

Reaven, G. M. 2006. The metabolic syndrome: is this diagnosis necessary? *Am J Clin Nutr* 83(6):1237–47.

Reddy, M. A., S. Sahar, L. M. Villeneuve, L. Lanting, and R. Natarajan. 2009. Role of Src tyrosine kinase in the atherogenic effects of the 12/15-lipoxygenase pathway in vascular smooth muscle cells. *Arterioscler Thromb Vasc Biol* 29(3):387–93.

Roth, S. Y., J. M. Denu, and C. D. Allis. 2001. Histone acetyltransferases. *Annu Rev Biochem* 70:81–120.

Roy, S., R. Sala, E. Cagliero, and M. Lorenzi. 1990. Overexpression of fibronectin induced by diabetes or high glucose: Phenomenon with a memory. *Proc Natl Acad Sci USA* 87(1):404–8.

Ruthenburg, A. J., C. D. Allis, and J. Wysocka. 2007. Methylation of lysine 4 on histone H3: intricacy of writing and reading a single epigenetic mark. *Mol Cell* 25(1):15–30.

Salma, N., H. Xiao, E. Mueller, and A. N. Imbalzano. 2004. Temporal recruitment of transcription factors and SWI/SNF chromatin-remodeling enzymes during adipogenic induction of the peroxisome proliferator-activated receptor gamma nuclear hormone receptor. *Mol Cell Biol* 24(11):4651–63.

Sarraf, S. A., and I. Stancheva. 2004. Methyl-CpG binding protein MBD1 couples histone H3 methylation at lysine 9 by SETDB1 to DNA replication and chromatin assembly. *Mol Cell* 15(4):595–605.

Schmidt, A. M., O. Hori, J. Brett, S. D. Yan, J. L. Wautier, and D. Stern. 1994. Cellular receptors for advanced glycation end products: Implications for induction of oxidant stress and cellular dysfunction in the pathogenesis of vascular lesions. *Arterioscler Thromb* 14(10):1521–8.

Shanmugam, N., I. T. Gaw Gonzalo, and R. Natarajan. 2004. Molecular mechanisms of high glucose-induced cyclooxygenase-2 expression in monocytes. *Diabetes* 53(3):795–802.

Sharma, K., and F. N. Ziyadeh. 1995. Hyperglycemia and diabetic kidney disease: The case for transforming growth factor-beta as a key mediator. *Diabetes* 44(10):1139–46.

Sharma, S., T. K. Kelly, and P. A. Jones. 2010. Epigenetics in cancer. *Carcinogenesis* 31(1):27–36.

Shi, Y. 2007. Histone lysine demethylases: Emerging roles in development, physiology and disease. *Nat Rev Genet* 8(11):829–33.

Shi, Y., F. Lan, C. Matson, P. Mulligan, J. R. Whetstine, P. A. Cole, R. A. Casero, and Y. Shi. 2004. Histone demethylation mediated by the nuclear amine oxidase homolog LSD1. *Cell* 119(7):941–53.

Shi, Y., and J. R. Whetstine. 2007. Dynamic regulation of histone lysine methylation by demethylases. *Mol Cell* 25(1):1–14.

Stenvinkel, P., M. Karimi, S. Johansson, J. Axelsson, M. Suliman, B. Lindholm, O. Heimburger, P. Barany, A. Alvestrand, L. Nordfors, A. R. Qureshi, T. J. Ekstrom, and M. Schalling. 2007. Impact of inflammation on epigenetic DNA methylation—a novel risk factor for cardiovascular disease? *J Intern Med* 261(5):488–99.

Strahl, B. D., and C. D. Allis. 2000. The language of covalent histone modifications. *Nature* 403(6765):41–5.

Tateishi, K., Y. Okada, E. M. Kallin, and Y. Zhang. 2009. Role of Jhdm2a in regulating metabolic gene expression and obesity resistance. *Nature* 458(7239):757–61.

Trojer, P., and D. Reinberg. 2006. Histone lysine demethylases and their impact on epigenetics. *Cell* 125(2):213–7.

Tsukada, Y., J. Fang, H. Erdjument-Bromage, M. E. Warren, C. H. Borchers, P. Tempst, and Y. Zhang. 2006. Histone demethylation by a family of JmjC domain-containing proteins. *Nature* 439(7078):811–6.

Van den Berghe, W., K. De Bosscher, E. Boone, S. Plaisance, and G. Haegeman. 1999. The nuclear factor-kappaB engages CBP/p300 and histone acetyltransferase activity for transcriptional activation of the interleukin-6 gene promoter. *J Biol Chem* 274(45):32091–8.

Vaxillaire, M., and P. Froguel. 2008. Monogenic diabetes in the young, pharmacogenetics and relevance to multifactorial forms of type 2 diabetes. *Endocr Rev* 29(3):254–64.

Villeneuve, L. M., M. A. Reddy, L. L. Lanting, M. Wang, L. Meng, and R. Natarajan. 2008. Epigenetic histone H3 lysine 9 methylation in metabolic memory and inflammatory phenotype of vascular smooth muscle cells in diabetes. *Proc Natl Acad Sci USA* 105(26):9047–52.

Wellen, K. E., G. Hatzivassiliou, U. M. Sachdeva, T. V. Bui, J. R. Cross, and C. B. Thompson. 2009. ATP-citrate lyase links cellular metabolism to histone acetylation. *Science* 324(5930):1076–80.

Whetstine, J. R., A. Nottke, F. Lan, M. Huarte, S. Smolikov, Z. Chen, E. Spooner, E. Li, G. Zhang, M. Colaiacovo, and Y. Shi. 2006. Reversal of histone lysine trimethylation by the JMJD2 family of histone demethylases. *Cell* 125(3):467–81.

Writing Team DCCT/EDIC Research Group. 2002. Effect of intensive therapy on the microvascular complications of type 1 diabetes mellitus. *Jama* 287(19):2563–9.

Xu, B., J. Chiu, B. Feng, S. Chen, and S. Chakrabarti. 2008. PARP activation and the alteration of vasoactive factors and extracellular matrix protein in retina and kidney in diabetes. *Diabetes Metab Res Rev* 24(5):404–12.

Yan, S. F., R. Ramasamy, L. G. Bucciarelli, T. Wendt, L. K. Lee, B. I. Hudson, D. M. Stern, E. Lalla, D. U. Yan S, L. L. Rong, Y. Naka, and A. M. Schmidt. 2004. RAGE and its ligands: A lasting memory in diabetic complications? *Diab Vasc Dis Res* 1(1):10–20.

Zaina, S., K. B. Dossing, M. W. Lindholm, and G. Lund. 2005. Chromatin modification by lipids and lipoprotein components: An initiating event in atherogenesis? *Curr Opin Lipidol* 16(5):549–53.

Zhang, Y., and D. Reinberg. 2001. Transcription regulation by histone methylation: interplay between different covalent modifications of the core histone tails. *Genes Dev* 15(18):2343–60.

# Section I

*Transforming Dietary Signals
into Gene Expression*

*Systems Biology—Cell Signaling Regulation of
Gene Expression*

# 4 Systems Biology Approaches to Studying Diet x Genome Interactions

*Xia Yang, Zhidong Tu, and Jun Zhu*

## CONTENTS

## INTRODUCTION

Common complex human diseases result from a combination of genetics, environment, and the interactions between the two. Although genome-wide association studies (GWAS) have produced many clues regarding the genetic contributions to the pathogenesis of complex human disease, the genetic risk factors identified so far only explain a small proportion of disease risks (Altshuler, Daly, and Lander 2008; Lusis, Attie, and Reue 2008). Environmental factors are also a significant contributor for human diseases. Air pollution, for example, has been shown to increase risks of lung, heart, and possibly liver disease, and many studies have shown that stress contributes to heart disease. It is also well accepted that diet is one of the main factors contributing to the current obesity pandemic and its associated health consequences, such as cardiovascular diseases and diabetes. High-fat diets increase the risk of breast, colon, prostate, and possibly pancreatic and ovarian cancers, and 30%–40% of cancers have been directly linked to dietary factors (Glade 1999). Diet not only contributes to diseases directly, but also interacts with other risk factors by influencing metabolic pathways, homeostasis, and the flow of genetic information from gene expression to protein synthesis and degradation. A systems biology approach, which gives a comprehensive and holistic view of how diet contributes to complex human diseases and disease prevention, is needed to understand these complex interactions.

There are three components in systems biology as defined by Ideker et al. (2001): (1) systematic perturbations, (2) systematic observations, and (3) integration of high-throughput data into mathematic models that can describe complex biological systems. Recent advances in high-throughput technologies for genetic, genomic, transcriptomic, proteomic, and metabolomic analyses have made comprehensive and systematic studies of biological systems possible. Integration of this diverse data into systems, network views, and mathematic models has enormous potential to further our understanding of the molecular basis of human disease and the environmental factors that trigger it.

When analyzing and interpreting high-throughput "-omics" data, nutrient status is one of the major covariance factors to be considered. Diet can have both short-term effects on human gene expression levels observable immediately after a meal, and profound long-term effects on DNA methylation and expression, which can be passed from generation to generation (Pembrey 2002). In the next section, we review some systems biology approaches for studying dietary effects on biological systems which help elucidate the mechanisms of complex human diseases.

## DIET SHORT-TERM EFFECTS

Human blood is one of the most accessible tissues for monitoring molecular phenotypes associated with disease and drug response. The patterns of gene expression from human blood have been used to diagnose, assess status, and predict disease risk (Borovecki et al. 2005; Bull et al. 2004; Rus et al. 2004; Tanaka et al. 2005; Wu et al. 2005; Yu et al. 2005) as well as determine an individual's responsiveness to various treatment regimens (Choi et al. 2005). The primary aim of many of these studies is to identify predictive biomarkers for disease and drug response: genes whose activities can be robustly measured and used to shed light on cellular processes occurring in a given physiological state. These studies have shown that gene expression traits from human peripheral blood tissue vary significantly by age, sex, time of day, and disease status (Borovecki et al. 2005; Bull et al. 2004; Radich et al. 2004; Whitney et al. 2003; Emilsson et al. 2008). We have recently shown that fasting/feeding status also has a profound impact on gene expression in human blood (Leonardson et al. 2010). By employing a standard two-arm crossover design, we were able to distinguish expression changes due to individual and time-specific effects from those due to food intake, shown in Figure 4.1.

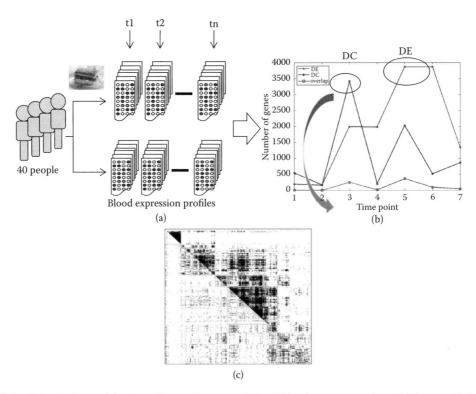

(a)

(b)

(c)

**FIGURE 4.1**  Fasting and feeding effect on human peripheral blood gene expression. (a) The overall design of the study. (b) Gene expression differences at multiple time points as measured by differential expression and differential connectivity. (c) Topological overlap map of connectivity at the time point 3 when the fasted and fed states have the largest number of genes with significant differential connectivity. (Modified from Leonardson, A. S. et al. 2010. *Hum Mol Genet* 19(1):159–69.)

Coexpression networks are frequently used to study mechanisms and biological processes involved in regulating gene expression changes (Lum et al. 2006; Gargalovic et al. 2006; Ghazalpour et al. 2006; Chen et al. 2008; Emilsson et al. 2008). When comparing coexpressed modules between fasted and fed states, we found that multiple gene ontology (GO) biological processes—including RNA processing, ribonucleoprotein complex biogenesis and assembly, and ubiquitin cycle—are enriched in both fasted and fed modules. RNA processing, especially RNA splicing and RNA localization, has been previously identified as a key responder to fasting and feeding cycles (Amir-Ahmady and Salati 2001). Genes involved splicing and RNA processing pathways are significantly enriched in a module that correlates with diabetes and obesity related end points when a whole brain expression profile data set was analyzed. This suggests that similar biological processes are regulated in different tissues responding to the same perturbations (Lum et al. 2006). Fasting and feeding are also known regulators of ribosome biogenesis and ubiquitin cycle processes (Conde and Franze-Fernandez 1980; Yoshizawa et al. 1997). In general, researchers look for genes that are differentially expressed (DE) between conditions, where $de = \log 10 \left( \dfrac{expression_A}{expression_B} \right)$. Here we defined a measurement of connectivity changes among gene expression traits in the context of coexpression networks as differential connectivity (DC) $dc = \log 10 \left( \dfrac{connectivity_A + \alpha}{connectivity_B + \alpha} \right)$ and found clear connectivity changes over time and between the fasted and fed states. Genes responding to food intake at the expression level (DE genes) and at the connectivity structure (DC genes) do not overlap well, shown in Figure 4.1. DC genes provide insights into the pathways and mechanisms involved in the dynamic response to food intake. On the other hand, DE genes reflect more on the accumulated end effects of the dynamic changes, so that differences in gene expression become more significant over time, shown in Figure 4.1. We also demonstrated that the transcriptional response to food intake is robust. A classifier from the gene expression traits was constructed and achieved greater than 90% accuracy in distinguishing the fasted and fed states. The gene expression traits that most accurately discriminated between the fasting and fed states were also the more heritable, gave rise to stronger linkage signals, and exhibited greater coherence with respect to pathways associated with metabolic traits. These findings suggest that feeding-induced signaling in the transcriptional network is an important state to consider in studies of clinical traits related to metabolic diseases like obesity.

## CONSTRUCTING CAUSAL NETWORKS FOR FASTED AND FED STATES

Coexpression networks are based on pair-wise gene–gene correlations of expression data, revealing functional modules in the network that elucidate pathways driving core biological processes (Ghazalpour et al. 2006; Lum et al. 2006) or pathways underlying complex human disease (Chen et al. 2008; Emilsson et al. 2008). Although coexpression networks provide global views of network structures, by themselves they cannot yield causal relationships between genes or between genes and clinical traits. To maximally leverage the time series data described in Figure 4.1, we applied the Granger causality test to derive causal blood transcriptional networks for fasted and fed states.

The time series–based causality test was proposed by Wiener (1956) based on the idea that if one time series prediction could be improved by incorporating the knowledge of a second, then the second series has a causal influence on the first. Granger (1969) was the first to formalize the idea in the context of a linear regression model, so the time series based causality test is generally referred to as the Granger causality test. A variety of models exist to test for Granger causality, including the multivariate autoregressive model (MVAR) and bivariate autoregressive model (BVAR). Typically, the Granger causality test is applied to long time series, however it is difficult and expensive to collect long time course gene expression data from human samples. Our data consists of many short time series from multiple individuals. Assuming response slopes for individuals are similar, these short time series can be combined together. To reduce individual specific variation, which could affect the response slope, an individual's gene expression data was normalized according to its own expression data at the first time point.

For an individual $n$, the test of Granger causality between two genes $X$ and $Y$ under BVAR model can be carried out by comparing the full model $y_{n,t} = \alpha_n y_{n,t-1} + \beta_n x_{n,t-1} + \mu_{n,t}$ with the autoregressive model $y_{n,t} = \alpha_n y_{n,t-1} + \delta_{n,t}$. The significance of the Granger causality test $X \to Y$ (the full model explains more variance than the autoregressive model) is then measured by F-test statistics

$$S_n = \frac{(RSS0_n - RSS1_n)}{RSS1_n/(T-3)} \sim F_{1,T-3}, \text{ where } RSS1_n = \sum_{t=2}^{T} \mu_{n,t}^2 \text{ and } RSS0_n = \sum_{t=2}^{T} \delta_{n,t}^2 \text{ are the sum of squared}$$

residuals of full model and autoregressive model respectively, and $T$ is the length of the time series. By assuming $\alpha_n$ and $\beta_n$ are similar, these short series can be combined together to infer Granger causality, and the F-test statistics can be modified as $S = \dfrac{(RSS0 - RSS1)}{RSS1/(N*(T-1)-2)} \sim F_{1,N*(T-1)-2}$

$(RSS1 = \sum_{n=1}^{N}\sum_{t=2}^{T} \mu_{n,t}^2$ and $RSS0 = \sum_{n=1}^{N}\sum_{t=2}^{T} \delta_{n,t}^2$ are the sum of squared residuals of full model and autoregression model).

Granger causal networks constructed for the fasted and fed states are comprised of 2010 and 967 causal links, respectively (Zhu et al. 2010). The fasted network consists of many small subnetworks and the fed network consists of mainly two subnetworks, as shown in Figure 4.2. The top causal gene in the fed network is *PER1*, a transcription factor regulating the circadian clock, cell growth, and apoptosis. The genes under *PER1* regulation are enriched for genes correlated to plasma concentration of triglyceride (p-value = 0.00045) in the Icelandic family blood (IFB) cohort (Emilsson et al. 2008). *PER1*'s downstream genes are involved in diverse biological processes including *CREB5* in circadian rhythm, *PTEN* and *P53INP2* in apoptosis, *IL1R1*, *IL1RAP*, and *TLR2* involved in inflammation response, *FASN* and *ACSL1* in fatty acid metabolism, and *MVK* in cholesterol biosynthesis. These results suggest that food intake interacts with circadian rhythm, reinforcing previous studies associating circadian rhythm and many biological

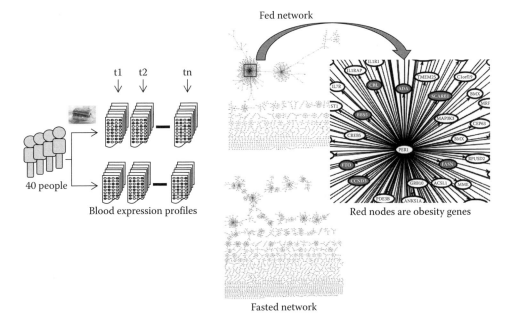

**FIGURE 4.2**  Granger causal networks for fasted and fed networks for human peripheral blood. There are clear differences between the fasted and fed Granger causal networks. In the fed causal network, *PER1* is the key regulator whose downstream genes include many obesity causal genes. (Modified from Zhu, J. et al. 2010. *PLoS Comput Biol* 6(2):e1000671.)

processes (Kohsaka et al. 2007; Dobrin et al. 2009). Furthermore, previous research has linked circadian gene (*PER1*, *PER2*, *PER3*, etc.) mRNA expression rhythm in human peripheral blood cells to individual circadian phenotypes (Kusanagi et al. 2008; Burioka et al. 2007). Our blood causal network where *PER1* is a top causal gene suggests a potential mechanism for how central nervous system (CNS) control and environmental influences (e.g., sunlight) can affect circadian rhythm gene expression, in turn regulating a host of other biological functions. More specifically, circadian rhythm genes (*PER1* in particular) play important roles in cell cycle regulation and cancer processes (Gery et al. 2006; Sheng-Fung et al. 2004). These reports support our observations that in the fed network, several genes under *PER1* control are involved in apoptosis and cell cycle regulation (e.g., *PTEN* and *P53INP2*). Recent GWAS have identified seven genes (*ADA*, *BBS5*, *CBL*, *CCND3*, *FASN*, *FTO*, and *SCARB1*) downstream from *PER1* as human obesity genes, and it has been shown that circadian rhythm is linked to metabolic processes in mice (Yang et al. 2006; Zvonic et al. 2006). For example, mutations in mouse genes involving circadian rhythm regulation, such as *Clock*, can lead to obesity (Turek et al. 2005). Our results provide evidence that human obesity causal genes are under circadian rhythm control in peripheral tissues like blood.

## DIET x GENOME INTERACTIONS

Genetic factors predispose some populations to risks of certain human diseases, whereas dietary choices can increase or decrease disease risks (Willett et al. 1993; Dwyer et al. 2004; Shin et al. 2006). Diet x genome interactions have been studied in many model organisms, as summarized in Figure 4.3. Using quantitative trait locus (QTL) analysis, Ron Korstanje et al. showed that several HDL QTLs are dependent on diet in an F2 mouse study (Korstanje et al. 2004), shown

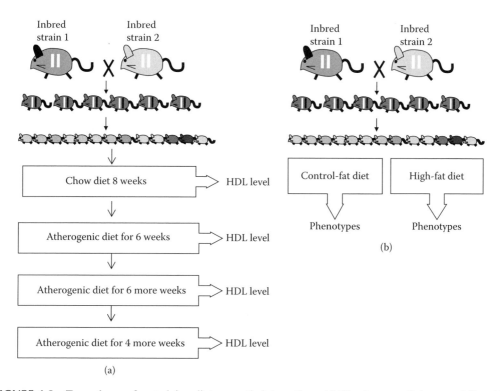

**FIGURE 4.3** Two schemes for studying diet x genetic interactions. (a) F2 mice were fed sequentially with different diets, then phenotypes were measured at the ends of each diet regiment. (b) F2 mice were randomly assigned to different diet regiments and phenotypes were measured at the end of the diet regiment.

in Figure 4.3a. Using a similar design, Mashimo et al. (2007) demonstrated that the cholesterol QTLs for normal diet and high-fat diet did not overlap in F2 rat study, which could be explained by diet x genetic interactions. There is a clear drawback in the design where the diet effect cofounds with age effect; a better design would be to randomly assign mice to different diet groups then map QTLs for traits of interest, as shown in Figure 4.3b. Using this design, Gordon et al. (2008) showed that there were QTL x diet interactions for many body weight and obesity related traits in an F2 mouse cross, indicating that such interactions play a significant role in studying complex human diseases such as obesity. Similarly, Ehrich et al. (2005) showed that there are locus-by-diet interactions for all previously identified adiposity QTLs using a mouse advanced intercross line. Even through mice are randomly assigned to two diet groups, there could still be hidden genetic differences between mice in the two diet groups that could give rise to different QTLs. An ideal design would be to test mice with the same genetic compositions under different conditions. Smith and Kruglyak (2008) showed that gene expression QTLs (eQTLs) also shift under different nutrition conditions using an identical F2 yeast cross, shown in Figure 4.4. All these results imply that there are considerable differences in metabolic pathways under different diet conditions. Such differences may be due to different underlying network structures for different diet conditions, or different states of the same underlying network structure for different diet conditions. When comparing the yeast coexpression networks under glucose and ethanol conditions, we have shown that the coexpression networks are similar (Narayanan et al. 2010). Thus, we can construct one common causal network for each biological system, and then refine the causal network under different conditions.

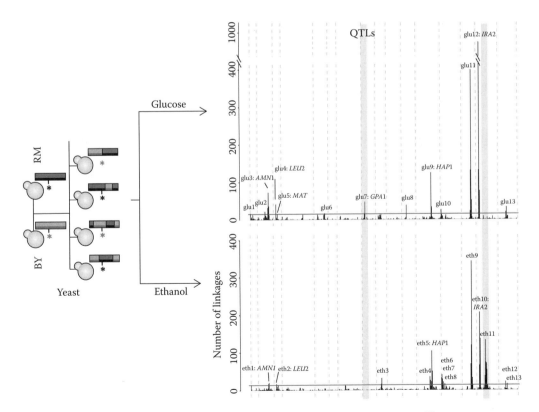

**FIGURE 4.4** Studying diet x genetic interactions using the same F2 yeast segregants. Yeast segregants were grown under glucose and ethanol conditions. Gene eQTL analysis showed that there were nutrition x genetic interactions at gene transcription regulation. (Modified from Smith, E. N., and L. Kruglyak. 2008. *PLoS Biol* 6(4):e83.)

# BAYESIAN NETWORK AS A FRAMEWORK FOR INTEGRATING DIVERSE INFORMATION

Bayesian networks (Jensen 1996) consist of a set of nodes, directed edges between pairs of nodes, and a set of conditional probability tables (for discrete nodes) or conditional probability functions (for continuous nodes). In general, a probabilistic causal network (a graph $G$ over a set of variables $X$) can be described by a joint probability distribution $p(X_1,...,X_n)$. If the graph $G$ is a directed acyclic graph (DAG), then the distribution can be decomposed into the product $p(X_1,...,X_n) = \prod_{i=1}^{n} p(X_i | Pa_i^G)$, where $Pa_i^G$ is the set of parents of $X_i$ in $G$. Learning Bayesian network structure is a process to find the structure with the maximum probability given the data. Bayes formula allows us to determine the probability of a network model $M$ given observed data $D$ as a function of our prior belief that the model is correct and the probability of the observed data given the model $p(M|D) - p(D|M) * p(M)$, where $p(M) = \prod_{(X_i \to Y_j)^G} p(X_i \to Y_j)$ is the structure prior. Constructing different structure priors for Bayesian networks is an efficient way to integrate diverse information, shown in Figure 4.5. Structure priors not only alter the probability of a specific relationship in the final structure, but they can also enhance true signals. One way to break Markov equivalent classes is by assigning different structure priors $p(X_i \to Y_j)$ and $p(X_j \to Y_i)$. The genetic data is a good source for constructing structure priors. Systematic simulations show that integration of genetics and gene expression increases the accuracy of reconstructed Bayesian networks (Zhu et al. 2007), shown in Figure 4.6. The similar improvement has also been shown using empirical data (Zhu et al. 2004). In addition to genetic data, there are many types of data complementary to gene expression data, including protein–protein interaction data, protein phosphorylation state data, protein–DNA binding data, RNA–DNA binding data, and so on. Gene expression is regulated by transcription factors, and many gene–gene expression correlations can be explained by coregulation via the same transcription factors. When high-confidence protein–protein interactions are identified, the corresponding gene pair expression levels tend to correlate as well. When the Bayesian network model was applied to a yeast F2 cross, we showed that integration of protein–protein interactions and transcription factor binding information with genetic and gene expression data further improved the accuracy of Bayesian networks (Zhu et al. 2008).

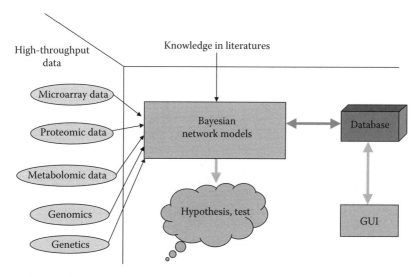

**FIGURE 4.5** A general scheme for data integration using Bayesian network as a framework.

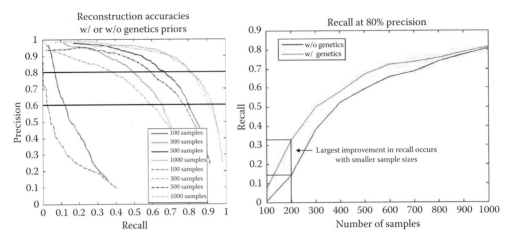

**FIGURE 4.6**  Assessment of Bayesian network reconstruction accuracies with and without genetic data as a prior. (Modified from Zhu, J. et al. 2007. *PLoS Comput Biol* 3(4):e69.)

## DYNAMIC BAYESIAN NETWORK

A Bayesian network structure is a direct acyclic graph, which cannot accommodate feedback loops, although feedback loops are common in biological systems. A dynamic Bayesian network (DBN), which contains some general features of a Bayesian network but allows feedback loops, is a useful model of dynamic changes in biological systems. However, constructing a DBN requires a large amount of data, which may be impossible to collect for human studies. As discussed in the section *Diet x Genome Interactions*, even though diet x genome interactions affect ultimate phenotypes, the underlying networks could be similar. This suggests that a global static network can be constructed and used to study system responses under different conditions. When a static structure is known, then the amount of data required for constructing a DBN is dramatically reduced. To construct a DBN for fasted and fed states, we combined two large data sets, shown in Figure 4.7.

A large dataset of profiled peripheral blood samples was obtained from the IFB study and it is already described and available (Emilsson et al. 2008). The fasting/feeding study group and the IFB cohort are derived from the same population in terms of geological location and genetic background, therefore the static networks based on these two studies are presumably similar. We have previously demonstrated the high quality of Bayesian networks constructed by integrating gene expression data and genotype data (Zhu et al. 2004, 2007, 2008). The IFB data set consisting of both fasting state gene expression and genotype data is used to construct a static Bayesian network (SBN) following the described procedures (Zhu et al. 2004, 2007, 2008).

When comparing the SBN and the DBN based on it, some interesting biological insights emerge. The largest change between the fed DBN and SBN is at the subnetwork center at *CDCA7*, a transcription regulator for the cell cycle. In the DBN for fed state, *CDCA7* connects a subnetwork consisting of genes involved in lipid metabolism such as *NPC1*, *FABP5*, and *APOE*, as well as to a subnetwork of inflammatory response genes such as *STAT3*, *STAT5*, *GPR109A*, *TNF*, *NTSR1*, *ORM1*, and *IL1RN*. This connection suggests that gene expression involved in either inflammatory response or changes in lipid metabolism responds to food intake. Circadian rhythm regulator *PER1* is also in the same subnetwork, which includes many genes involved in inflammatory response pathways. Inflammation has been identified as a key factor driving metabolic diseases such as diabetes and obesity (Chen et al. 2008; Emilsson et al. 2008). The results from DBN not only validate the connection between inflammation response and lipid metabolism processes, but also suggest that cell cycle regulation and lipid metabolism are linked to circadian rhythms, consistent with our Granger causal network results (Zhu et al. 2010).

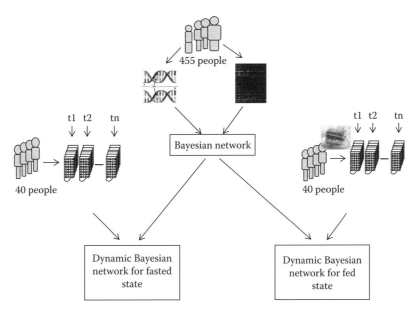

**FIGURE 4.7**    The scheme for integrating multiple data sets to construct dynamic Bayesian networks (DBNs).

## DIET LONG-TERM EFFECTS

Diet x genome interactions can be both short-term and long-term. An increasing body of scientific evidence suggests nutrition during gestation and lactation—among other environmental factors occurring later in life—predisposes infants to metabolic syndrome in adulthood (Gallou-Kabani and Junien 2005; Jirtle and Skinner 2007; Simmons 2005). The Dutch Famine Birth Cohort Study (http://www.dutchfamine.nl/) showed that pregnant women exposed to famine and malnutrition produced offspring who were more susceptible to diabetes, obesity, cardiovascular disease, micro-albinuria, and other metabolic symptoms. Studies done on animal models have also shown that maternal protein restriction can lead to hypertension, hyperglycemia, and obesity (Armitage et al. 2004; Bieswal et al. 2006; Brennan, Olson, and Symonds 2006; Burns et al. 1997), and that a high-fat maternal diet can result in obesity and diabetic phenotypes in adult offspring (Srinivasan et al. 2006). These results suggest that maternal nutritional imbalance (e.g., high fat or low protein) contributes to metabolic syndrome in offspring. Interestingly, Waterland et al. (2008) found that providing promethylation dietary supplements prevents transgenerational amplification of obesity, suggesting that epigenetics may play a role in the process.

The past few years have seen growing interest in identifying gene expression, epigenetic changes, and pathways affected by environmental factors in order to understand the molecular mechanisms underlying the phenotypic effects of maternal diets. Lillycrop et al. (2005, 2007) found that maternal protein restriction in rats resulted in decreased methylation in the promoter regions of the glucocorticoid receptor (GR) and the peroxisomal proliferator-activated receptor alpha (*PPARα*) genes and increased expression of these genes, possibly via decreased expression of DNA methyltransferase 1 (*DNMT1*). A maternal high-fat diet was found to induce hyperacetylation of fetal hepatic tissue at *H3K14*, *H3K9*, and *H3K18*, as well as alterations in the expression of *HDAC1* and other candidate genes in primates (Aagaard-Tillery et al. 2008). In humans, individuals who experienced the Dutch famine were shown to have persistent hypomethylation of *IGF2* gene (Heijmans et al. 2008). These lines of evidence support the presence of transcriptional and/or epigenetic changes associated with maternal nutrition.

In order to fully characterize the diet-induced alterations in the transcriptome, we studied mouse pups exposed in utero to different maternal diets (chow, high fat, and low protein) and different

**Study Design**

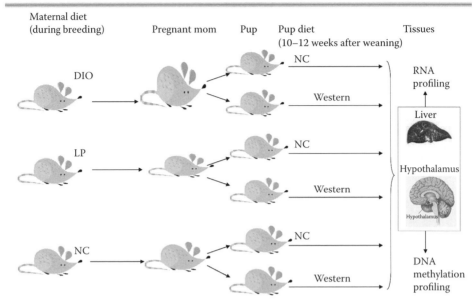

DIO: Diet-induced obesity (high fat, high sugar); LP: low protein; NC: normal chow; Western: high fat

**FIGURE 4.8**  The experimental design to study long-term effects of diet.

diets after weaning (chow and high fat), shown in Figure 4.8. At the end of the diet period, liver and hypothalamus tissues were collected for RNA profiling. We observed striking gene expression differences due to both maternal and pup diets, represented by thousands of genes with altered expression in both tissues. Many signature genes identified are not only involved in biological processes relevant to metabolic syndrome (e.g., lipid metabolism, immune response, and angiogenesis), but are also enriched for genes supported as causal for various metabolic traits including adiposity, aortic lesions, hypertension, lipids, glucose, and insulin in several previously studied mouse crosses (Chen et al. 2008; Schadt et al. 2005, 2008). What are the main changes that drive this striking differential expression? To address this question, we applied a key driver analysis, shown in Figure 4.9, which generalizes approaches we published earlier (Zhu et al. 2008; Argmann et al. 2009). Given a set of signatures, key drivers are central nodes in the Bayesian subnetworks that overlap significantly with the input signature. When we applied the key driver analysis to liver genes affected by diets using a liver Bayesian network constructed from mouse F2 crosses (Schadt et al. 2008), multiple key drivers including *Hmgcr* and *Igfbp1* were identified. It is encouraging that *Hmgcr* is one of the central nodes controlling large sets of liver maternal diet signature genes in the Bayesian network, as statins are *Hmgcr* inhibitors that dramatically mitigate cardiovascular risks (Davignon 2004). *Igfbp1* is insulin-like growth factor–binding protein 1, which plays a role in regulating glucose metabolism and inducing apoptosis, and has also been associated with type II diabetes (18, 19). The Bayesian subnetwork for *Igfbp1* is highly enriched for genes affected by maternal diet in both tissues and also for many metabolic pathways including pyruvate metabolism, polysaccharide biosynthesis, glucose metabolism, gluconeogenesis. The subnetwork also includes pathways such as cell cycle control, MAPK signaling, and circadian rhythm. Therefore, the network analysis suggests some key drivers and relevant mechanisms on how changes of these key drivers are amplified into large gene expression changes. However, many questions demand further investigation, particularly why these key drivers change in the first place and whether methylation patterns in these key drivers' regulatory regions change under different diet conditions. Methylation profiles in conjunction with gene expression profiles may answer some of these questions.

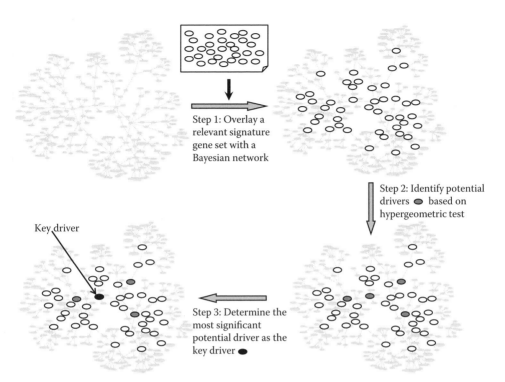

**FIGURE 4.9** The schematic diagram of key driver analysis using a Bayesian network.

## DISCUSSION

In summary, diet has profound short-term and long-term effects on biological systems at different levels. To outline a coherent and comprehensive summary of these effects, we need to integrate all different layers of information into networks. Beyond the multiple layers of information involved for a single tissue (genetic, epigenetic, epigenomic, transcriptomic data, etc.), how multiple tissues interact and synchronize with each other is also important. Many different tissues and organs are involved in digestion; the stomach and intestine in nutrient absorption, the liver, adipose and muscle tissues in fat and sugar metabolism, and the CNS in regulation of appetite are but a few examples. When gene expression profiles from multiple tissues were compared, we showed that genes involved in circadian rhythm are at the center of the cross tissue network (Dobrin et al. 2009). Diet, circadian rhythm, and obesity are thus interconnected (Turek 2008), showing that interactions across multiple tissues should be considered when studying diet x genome interactions.

## REFERENCES

Aagaard-Tillery, K. M., K. Grove, J. Bishop, X. Ke, Q. Fu, R. McKnight, and R. H. Lane. 2008. Developmental origins of disease and determinants of chromatin structure: Maternal diet modifies the primate fetal epigenome. *J Mol Endocrinol* 41(2):91–102.

Altshuler, D., M. J. Daly, and E. S. Lander. 2008. Genetic mapping in human disease. *Science* 322(5903):881–8.

Amir-Ahmady, B., and L. M. Salati. 2001. Regulation of the processing of glucose-6-phosphate dehydrogenase mRNA by nutritional status. *J Biol Chem* 276(13):10514–23.

Argmann, C., R. Dobrin, S. Heikkinen, A. Auburtin, L. Pouilly, T. A. Cock, H. Koutnikova, J. Zhu, E. E. Schadt, and J. Auwerx. 2009. Ppargamma2 is a key driver of longevity in the mouse. *PLoS Genet* 5(12):e1000752.

Armitage, J. A., I. Y. Khan, P. D. Taylor, P. W. Nathanielsz, and L. Poston. 2004. Developmental programming of the metabolic syndrome by maternal nutritional imbalance: How strong is the evidence from experimental models in mammals? *J Physiol* 561(Pt 2):355–77.

Bieswal, F., M. T. Ahn, B. Reusens, P. Holvoet, M. Raes, W. D. Rees, and C. Remacle. 2006. The importance of catch-up growth after early malnutrition for the programming of obesity in male rat. *Obesity (Silver Spring)* 14(8):1330–43.

Borovecki, F., L. Lovrecic, J. Zhou, H. Jeong, F. Then, H. D. Rosas, S. M. Hersch, P. Hogarth, B. Bouzou, R. V. Jensen, and D. Krainc. 2005. Genome-wide expression profiling of human blood reveals biomarkers for Huntington's disease. *Proc Natl Acad Sci USA* 102(31):11023–8.

Brennan, K. A., D. M. Olson, and M. E. Symonds. 2006. Maternal nutrient restriction alters renal development and blood pressure regulation of the offspring. *Proc Nutr Soc* 65(1):116–24.

Bull, T. M., C. D. Coldren, M. Moore, S. M. Sotto-Santiago, D. V. Pham, S. P. Nana-Sinkam, N. F. Voelkel, and M. W. Geraci. 2004. Gene microarray analysis of peripheral blood cells in pulmonary arterial hypertension. *Am J Respir Crit Care Med* 170(8):911–9.

Burioka, N., M. Takata, M. Endo, M. Miyata, K. Takeda, H. Chikumi, K. Tomita, Y. Fukuoka, H. Nakazaki, H. Sano, and E. Shimizu. 2007. Treatment with beta2-adrenoceptor agonist *in vivo* induces human clock gene, Per1, mRNA expression in peripheral blood. *Chronobiol Int* 24(1):183–9.

Burns, S. P., M. Desai, R. D. Cohen, C. N. Hales, R. A. Iles, J. P. Germain, T. C. Going, and R. A. Bailey. 1997. Gluconeogenesis, glucose handling, and structural changes in livers of the adult offspring of rats partially deprived of protein during pregnancy and lactation. *J Clin Invest* 100(7):1768–74.

Chen, Y., J. Zhu, P. Y. Lum, X. Yang, S. Pinto, D. J. MacNeil, C. Zhang, J. Lamb, S. Edwards, S. K. Sieberts, A. Leonardson, L. W. Castellini, S. Wang, M. F. Champy, B. Zhang, V. Emilsson, S. Doss, A. Ghazalpour, S. Horvath, T. A. Drake, A. J. Lusis, and E. E. Schadt. 2008. Variations in DNA elucidate molecular networks that cause disease. *Nature* 452(7186):429–35.

Choi, S., H. S. Park, M. S. Cheon, and K. Lee. 2005. Expression profile analysis of human peripheral blood mononuclear cells in response to aspirin. *Arch Immunol Ther Exp (Warsz)* 53(2):151–8.

Conde, R. D., and M. T. Franze-Fernandez. 1980. Increased transcription and decreased degradation control and recovery of liver ribosomes after a period of protein starvation. *Biochem J* 192(3):935–40.

Davignon, J. 2004. Beneficial cardiovascular pleiotropic effects of statins. *Circulation* 109(23 Suppl 1): III39–43.

Dobrin, R., J. Zhu, C. Molony, C. Argman, M. L. Parrish, S. Carlson, M. F. Allan, D. Pomp, and E. E. Schadt. 2009. Multi-tissue coexpression networks reveal unexpected subnetworks associated with disease. *Genome Biol* 10(5):R55.

Dwyer, J. H., H. Allayee, K. M. Dwyer, J. Fan, H. Wu, R. Mar, A. J. Lusis, and M. Mehrabian. 2004. Arachidonate 5-lipoxygenase promoter genotype, dietary arachidonic acid, and atherosclerosis. *N Engl J Med* 350(1):29–37.

Ehrich, T. H., T. Hrbek, J. P. Kenney-Hunt, L. S. Pletscher, B. Wang, C. F. Semenkovich, and J. M. Cheverud. 2005. Fine-mapping gene-by-diet interactions on chromosome 13 in a LG/J x SM/J murine model of obesity. *Diabetes* 54(6):1863–72.

Emilsson, V., G. Thorleifsson, B. Zhang, A. S. Leonardson, F. Zink, J. Zhu, S. Carlson, A. Helgason, G. B. Walters, S. Gunnarsdottir, M. Mouy, V. Steinthorsdottir, G. H. Eiriksdottir, G. Bjornsdottir, I. Reynisdottir, D. Gudbjartsson, A. Helgadottir, A. Jonasdottir, A. Jonasdottir, U. Styrkarsdottir, S. Gretarsdottir, K. P. Magnusson, H. Stefansson, R. Fossdal, K. Kristjansson, H. G. Gislason, T. Stefansson, B. G. Leifsson, U. Thorsteinsdottir, J. R. Lamb, J. R. Gulcher, M. L. Reitman, A. Kong, E. E. Schadt, and K. Stefansson. 2008. Genetics of gene expression and its effect on disease. *Nature* 452(7186):423–8.

Gallou-Kabani, C., and C. Junien. 2005. Nutritional epigenomics of metabolic syndrome: New perspective against the epidemic. *Diabetes* 54(7):1899–906.

Gargalovic, P. S., M. Imura, B. Zhang, N. M. Gharavi, M. J. Clark, J. Pagnon, W. P. Yang, A. He, A. Truong, S. Patel, S. F. Nelson, S. Horvath, J. A. Berliner, T. G. Kirchgessner, and A. J. Lusis. 2006. Identification of inflammatory gene modules based on variations of human endothelial cell responses to oxidized lipids. *Proc Natl Acad Sci USA* 103(34):12741–6.

Gery, S., N. Komatsu, L. Baldjyan, A. Yu, D. Koo, and H. P. Koeffler. 2006. The circadian gene per1 plays an important role in cell growth and DNA damage control in human cancer cells. *Mol Cell* 22(3):375–82.

Ghazalpour, A., S. Doss, B. Zhang, S. Wang, C. Plaisier, R. Castellanos, A. Brozell, E. E. Schadt, T. A. Drake, A. J. Lusis, and S. Horvath. 2006. Integrating genetic and network analysis to characterize genes related to mouse weight. *PLoS Genet* 2(18):e130. doi:10.1371/journal.pyen.0020130.

Glade, M. J. 1999. Food, nutrition, and the prevention of cancer: a global perspective. American Institute for Cancer Research/World Cancer Research Fund, American Institute for Cancer Research, 1997. *Nutrition* 15(6):523–6.

Gordon, R. R., K. W. Hunter, P. Sorensen, and D. Pomp. 2008. Genotype X diet interactions in mice predisposed to mammary cancer. I. Body weight and fat. *Mamm Genome* 19(3):163–78.

Granger, C. W. J. 1969. Investigating causal relations by econometric models and cross-spectral methods. *Econometrica* 37:424–438.

Heijmans, B. T., E. W. Tobi, A. D. Stein, H. Putter, G. J. Blauw, E. S. Susser, P. E. Slagboom, and L. H. Lumey. 2008. Persistent epigenetic differences associated with prenatal exposure to famine in humans. *Proc Natl Acad Sci USA* 105(44):17046–9.

Ideker, T., T. Galitski, and L. Hood. 2001. A new approach to decoding life: Systems biology. *Annu Rev Genomics Hum Genet* 2:343–72.

Jensen, Finn V. 1996. *An Introduction to Bayesian Networks*. London: UCL Press Limited.

Jirtle, R. L., and M. K. Skinner. 2007. Environmental epigenomics and disease susceptibility. *Nat Rev Genet* 8(4):253–62.

Kohsaka, A., A. D. Laposky, K. M. Ramsey, C. Estrada, C. Joshu, Y. Kobayashi, F. W. Turek, and J. Bass. 2007. High-fat diet disrupts behavioral and molecular circadian rhythms in mice. *Cell Metab* 6(5):414–21.

Korstanje, R., R. Li, T. Howard, P. Kelmenson, J. Marshall, B. Paigen, and G. Churchill. 2004. Influence of sex and diet on quantitative trait loci for HDL cholesterol levels in an SM/J by NZB/BlNJ intercross population. *J Lipid Res* 45(5):881–8.

Kusanagi, H., A. Hida, K. Satoh, M. Echizenya, T. Shimizu, J. S. Pendergast, S. Yamazaki, and K. Mishima. 2008. Expression profiles of 10 circadian clock genes in human peripheral blood mononuclear cells. *Neurosci Res* 61(2):136–42.

Leonardson, A. S., J. Zhu, Y. Chen, K. Wang, J. R. Lamb, M. Reitman, V. Emilsson, and E. E. Schadt. 2010. The effect of food intake on gene expression in human peripheral blood. *Hum Mol Genet* 19(1):159–69.

Lillycrop, K. A., E. S. Phillips, A. A. Jackson, M. A. Hanson, and G. C. Burdge. 2005. Dietary protein restriction of pregnant rats induces and folic acid supplementation prevents epigenetic modification of hepatic gene expression in the offspring. *J Nutr* 135(6):1382–6.

Lillycrop, K. A., J. L. Slater-Jefferies, M. A. Hanson, K. M. Godfrey, A. A. Jackson, and G. C. Burdge. 2007. Induction of altered epigenetic regulation of the hepatic glucocorticoid receptor in the offspring of rats fed a protein-restricted diet during pregnancy suggests that reduced DNA methyltransferase-1 expression is involved in impaired DNA methylation and changes in histone modifications. *Br J Nutr* 97(6):1064–73.

Lum, P. Y., Y. Chen, J. Zhu, J. Lamb, S. Melmed, S. Wang, T. A. Drake, A. J. Lusis, and E. E. Schadt. 2006. Elucidating the murine brain transcriptional network in a segregating mouse population to identify core functional modules for obesity and diabetes. *J Neurochem* 97(Suppl 1):50–62.

Lusis, A. J., A. D. Attie, and K. Reue. 2008. Metabolic syndrome: From epidemiology to systems biology. *Nat Rev Genet* 9(11):819–30.

Mashimo, T., H. Ogawa, Z. H. Cui, Y. Harada, K. Kawakami, J. Masuda, Y. Yamori, and T. Nabika. 2007. Comprehensive QTL analysis of serum cholesterol levels before and after a high-cholesterol diet in SHRSP. *Physiol Genomics* 30(2):95–101.

Narayanan, M., A Vetta, E. E. Schadt, and J. Zhu. 2010. Simultaneous clustering of multiple gene expression and physical interaction datasets. *PLoS Comput Biol* 6(4):e1000742. doi:101371/journal.pcbi.1000742 (accessed April 15, 2010).

Pembrey, M. E. 2002. Time to take epigenetic inheritance seriously. *Eur J Hum Genet* 10(11):669–71.

Radich, J. P., M. Mao, S. Stepaniants, M. Biery, J. Castle, T. Ward, G. Schimmack, S. Kobayashi, M. Carleton, J. Lampe, and P. S. Linsley. 2004. Individual-specific variation of gene expression in peripheral blood leukocytes. *Genomics* 83(6):980–8.

Rus, V., H. Chen, V. Zernetkina, L. S. Magder, S. Mathai, M. C. Hochberg, and C. S. Via. 2004. Gene expression profiling in peripheral blood mononuclear cells from lupus patients with active and inactive disease. *Clin Immunol* 112(3):231–4.

Schadt, E. E., J. Lamb, X. Yang, J. Zhu, S. Edwards, D. Guhathakurta, S. K. Sieberts, S. Monks, M. Reitman, C. Zhang, P. Y. Lum, A. Leonardson, R. Thieringer, J. M. Metzger, L. Yang, J. Castle, H. Zhu, S. F. Kash, T. A. Drake, A. Sachs, and A. J. Lusis. 2005. An integrative genomics approach to infer causal associations between gene expression and disease. *Nat Genet* 37(7):710–7.

Schadt, E. E., C. Molony, E. Chudin, K. Hao, X. Yang, P. Y. Lum, A. Kasarskis, B. Zhang, S. Wang, C. Suver, J. Zhu, J. Millstein, S. Sieberts, J. Lamb, D. GuhaThakurta, J. Derry, J. D. Storey, I. Avila-Campillo, M. J. Kruger, J. M. Johnson, C. A. Rohl, A. van Nas, M. Mehrabian, T. A. Drake, A. J. Lusis, R. C. Smith,

F. P. Guengerich, S. C. Strom, E. Schuetz, T. H. Rushmore, and R. Ulrich. 2008. Mapping the genetic architecture of gene expression in human liver. *PLoS Biol* 6(5):e107.

Sheng-Fung, L., M.-Y. Yang, J.-G. Chang, H.-H. Hsiao, Y.-C. Liu, S.-B. Tseng, H.-J. Tsai, T.-C. Liu, and T.-P. Chen. 2004. Downregulation of circadian genes, PER1, PER2, and PER3, in chronic myeloid leukemia. *ASH Annual Meeting Abstracts* 104(11):4317–7.

Shin, M. J., Y. Jang, S. J. Koh, J. S. Chae, O. Y. Kim, J. E. Lee, J. M. Ordovas, and J. H. Lee. 2006. The association of SNP276G>T at adiponectin gene with circulating adiponectin and insulin resistance in response to mild weight loss. *Int J Obes (Lond)* 30(12):1702–8.

Simmons, R. 2005. Developmental origins of adult metabolic disease: Concepts and controversies. *Trends Endocrinol Metab* 16(8):390–4.

Smith, E. N., and L. Kruglyak. 2008. Gene-environment interaction in yeast gene expression. *PLoS Biol* 6(4):e83.

Srinivasan, M., S. D. Katewa, A. Palaniyappan, J. D. Pandya, and M. S. Patel. 2006. Maternal high-fat diet consumption results in fetal malprogramming predisposing to the onset of metabolic syndrome-like phenotype in adulthood. *Am J Physiol Endocrinol Metab* 291(4):E792–9.

Tanaka, T., H. Takada, A. Nomura, S. Ohga, R. Shibata, and T. Hara. 2005. Distinct gene expression patterns of peripheral blood cells in hyper-IgE syndrome. *Clin Exp Immunol* 140(3):524–31.

Turek, F. W. 2008. Circadian clocks: Tips from the tip of the iceberg. *Nature* 456(7224):881–3.

Turek, F. W., C. Joshu, A. Kohsaka, E. Lin, G. Ivanova, E. McDearmon, A. Laposky, S. Losee-Olson, A. Easton, D. R. Jensen, R. H. Eckel, J. S. Takahashi, and J. Bass. 2005. Obesity and metabolic syndrome in circadian Clock mutant mice. *Science* 308(5724):1043–5.

Waterland, R. A., M. Travisano, K. G. Tahiliani, M. T. Rached, and S. Mirza. 2008. Methyl donor supplementation prevents transgenerational amplification of obesity. *Int J Obes (Lond)* 32(9):1373–9.

Whitney, A. R., M. Diehn, S. J. Popper, A. A. Alizadeh, J. C. Boldrick, D. A. Relman, and P. O. Brown. 2003. Individuality and variation in gene expression patterns in human blood. *Proc Natl Acad Sci USA* 100(4):1896–901.

Wiener, N. 1956. The theory of prediction. In E. F. Beckenback, ed., *Modern Mathematics of Engineers,* Chapter 8: pp. 165–90. New York: McGraw-Hill.

Willett, W. C., M. J. Stampfer, J. E. Manson, G. A. Colditz, F. E. Speizer, B. A. Rosner, L. A. Sampson, and C. H. Hennekens. 1993. Intake of trans fatty acids and risk of coronary heart disease among women. *Lancet* 341(8845):581–5.

Wu, Y. S., R. Q. Fan, D. C. Chen, and G. W. Xuan. 2005. Gene expression profiling of peripheral leukocytes from patients with systemic lupus erythematosus using oligonucleotide DNA microarray. *Di Yi Jun Yi Da Xue Xue Bao (Academic Journal First Medical College of PLA)* 25(8):929–34.

Yang, X., M. Downes, R. T. Yu, A. L. Bookout, W. He, M. Straume, D. J. Mangelsdorf, and R. M. Evans. 2006. Nuclear receptor expression links the circadian clock to metabolism. *Cell* 126(4):801–10.

Yoshizawa, F., T. Nagasawa, N. Nishizawa, and R. Funabiki. 1997. Protein synthesis and degradation change rapidly in response to food intake in muscle of food-deprived mice. *J Nutr* 127(6):1156–9.

Yu, S. Y., Y. W. Hu, X. Y. Liu, W. Xiong, Z. T. Zhou, and Z. H. Yuan. 2005. Gene expression profiles in peripheral blood mononuclear cells of SARS patients. *World J Gastroenterol* 11(32):5037–43.

Zhu, J., Y. Chen, A. S. Leonardson, K. Wang, J. R. Lamb, V. Emilsson, and E. E. Schadt. 2010. Characterizing dynamic changes in the human blood transcriptional network. *PLoS Comput Biol* 6(2):e1000671.

Zhu, J., P. Y. Lum, J. Lamb, D. GuhaThakurta, S. W. Edwards, R. Thieringer, J. P. Berger, M. S. Wu, J. Thompson, A. B. Sachs, and E. E. Schadt. 2004. An integrative genomics approach to the reconstruction of gene networks in segregating populations. *Cytogenet Genome Res* 105(2-4):363–74.

Zhu, J., M. C. Wiener, C. Zhang, A. Fridman, E. Minch, P. Y. Lum, J. R. Sachs, and E. E. Schadt. 2007. Increasing the power to detect causal associations by combining genotypic and expression data in segregating populations. *PLoS Comput Biol* 3(4):e69.

Zhu, J., B. Zhang, E. N. Smith, B. Drees, R. B. Brem, L. Kruglyak, R. E. Bumgarner, and E. E. Schadt. 2008. Integrating large-scale functional genomic data to dissect the complexity of yeast regulatory networks. *Nat Genet* 40(7):854–61.

Zvonic, S., A. A. Ptitsyn, S. A. Conrad, L. K. Scott, Z. E. Floyd, X. Wu, B. C. Goh, R. L. Mynatt, and J. M. Gimble. 2006. Characterization of peripheral circadian clocks in adipose tissues. *Diabetes* 55(4):962–70.

# 5 Nutritional Genomics
## *Cellular Signaling and Molecular Targets*

*Wayne R. Bidlack*

## CONTENTS

## INTRODUCTION TO CELLULAR COMMUNICATION

Human cells, individually or in the aggregate, are exquisitely sensitive to changes in the extracellular environment. These environmental changes can be physical, chemical or biological and they are typically communicated to the cell and its genome by chemical signals. Cellular metabolites, peptides and proteins are formed to activate specific cells to regulate requisite physiologic effects. In some instances, chemical signals are provided exogenously by dietary constituents such as nutrients and bioactive non-nutrient components. Regardless of their origin, chemical signals can be distributed to other cells, tissues and organs via the circulatory system and interstitial fluids. These signaling molecules help initiate and coordinate cellular activity in response to environmental change and can have a dramatic effect on metabolism, physiology and long-term health. In this chapter, signaling molecules are defined as any chemical entity capable of traversing the space between cells to deliver biologically relevant information from the sending cell to the receiving cell. How this information is detected, amplified and transmitted to specific genes is the subject of this chapter.

Internal cellular signals originate from three primary sources: endocrine signals (hormones) generated in distant endocrine tissues that release their signals into the blood to reach the target cells; paracrine signals are produced from nearby cells and diffuse to and affect metabolism of target cells; and autocrine signals are produced by the same cell that also receives them. In addition, cell-surface imbedded glycoproteins provide surface to surface contact and serve as signal proteins binding to surface receptors in adjacent cells, coordinating common cellular activities (Berridge 2005).

Although cells are exposed to a multitude of signals, they respond specifically to only a select few. Each signal can affect only those cells containing the ligand-specific receptor that specifically recognizes and accepts that signal molecule. The binding affinity of the signal to its receptor needs to fall within the concentration range of the circulating signal to trigger a physiologic response. The minimum concentration to stimulate a response may occur at the lower end of the binding concentration, approximately 0.1 nM.

The receptor may be a transmembrane receptor with an extracellular domain that binds large molecular weight or polar molecules that cannot diffuse across the plasma membrane, or the receptor may be located intracellularly to receive small (lipophilic) molecules that diffuse into the cell. Both types of protein receptors undergo conformation changes that trigger cell signaling and stimulates transduction pathways, transferring the signal to the nucleus and regulating gene expression.

Direct transduction occurs as a function of the receptor itself at the plasma membrane. The interaction between the signal ligand (primary messenger) and the receptor results in a conformation change that triggers the required cellular response (Pawson and Nash 2000 a,b).

The conformation change may affect a transmembrane gated ion channel, allowing increased or decreased ion ($Na^+$ or $Ca^{+2}$) movement into or out of the cell, or it may be more complex, transmitting the message through the transmembrane receptor to activate protein kinases to relay the signal through phosphorylation of other proteins, altering their conformation and enzymatic activities (Adjei and Hidalgo 2005). The eventual outcome of these interactions is an alteration in cellular metabolism and changes in the expression of specific genes within the responding cells.

Indirect transductance occurs when a second messenger is activated by the transmembrane receptor. Relay proteins are activated (phosphorylated) to transmit the signal to a second messenger or the messenger itself diffuses through the cytoplasm and initiates additional steps in the transduction pathway. In either case, the signal can regulate one or more activities. The initial signal can be amplified and further distributed to cause varying metabolic responses in different target cells. Signals initiated from unrelated receptors, with each having affinity for distinctly different ligands, can converge to activate a common effector pathway. Conversely, signals from the same ligand can diverge to activate different effectors, leading to diverse cellular responses. The signal pathways may be linked together by scaffolding proteins, which act as relay proteins or may work with transcription factors (TFs) to enter the nucleus to regulate expression of specific genes (Pawson and Scott 1997).

Multistep pathways offer greater opportunities to coordinate and regulate the required response by producing a cascade of protein interactions that amplify the cellular impact of weak signals like dietary metabolites. Importantly, each activation step is countered by a related inactivation step, for example, phosphorylation and dephosphorylation reactions returning the cell to a state of homeostasis. It is the rapidity and duration of each activation/deactivation step in the pathway that determines the balance between cellular response and homeostasis. The original extracellular signal molecule may not enter the cell; yet the signal directs a series of protein–protein communications, affecting the regulation of the expression of specific genes.

Signal transmission can occur directly if small lipophilic molecules diffuse across the plasma membrane to nonmembrane cytoplasmic receptors located inside the cell. Signal molecules may migrate across the cytoplasm bound to cytosolic carrier proteins, or chaperone, which prevents degradation of the molecule and restricts the entrance of the lipophilic ligand directly into the nucleus. When the signal molecule binds to the receptor, the chaperone is released, and the receptor changes its shape, enabling the ligand-receptor to enter the nucleus. The ligand-receptor then participates as a TF affecting gene expression. These lipophilic receptors are designated nuclear receptors (NRs), and include the large family of steroid and thyroid hormone receptors, vitamin D receptors, RXR (retinoid receptor), and others. Each receptor has a specific lipophilic ligand binding domain, a DNA binding domain, and a transcriptional activator domain that attracts the factors that regulate gene expression.

Transmembrane receptors are proteins that penetrate through the lipid bilayer of the plasma membrane, have an extracellular ligand binding domain serving as the binding site for the signal molecule, a hydrophobic domain that spans the membrane matrix one or more times, and an intracellular domain that activates proteins in the cytoplasm to relay the signal across the cytoplasm and eventually regulate gene transcription in the nucleus. Some transmembrane receptors, such as the growth factor receptors and insulin receptors (IR), have intrinsic tyrosine kinase activity that is capable of autophosphorylation, as well as phosphorylation of related relay proteins

(Plowman et al. 1994). In addition, membrane receptors include other intrinsic activities like protein tyrosine phosphatases (PTPs), such as the CD45 protein of T cells and macrophages, the guanylate cyclases in natriuretic peptide receptors, and serine/threonine kinases (RSTK), such as activin and TGF-β receptors (Silvestri et al. 2010). Other membrane receptors lack an intrinsic enzyme activity, but attract intracellular protein kinases, such as tyrosine kinases (TK) and serine/threonine kinase (STK). These kinases bind by direct protein–protein interactions with the receptor triggering phosphorylation and signal activation (Plowman et al. 1994).

## SIGNAL TRANSDUCTION RECEPTORS

### RECEPTOR PROTEIN KINASE

The protein kinases are key regulators of cellular function and play a prominent role in signal transduction. They constitute one of the largest and most functionally diverse gene families (Schlessinger 2000). The majority belong to a single super family whose catalytic domains are related in their amino acid sequence. Protein kinases include two specific classes of kinases, the protein tyrosine kinases (PTK) and protein STK. The phosphorylation of tyrosine residues occurs less often than that of the serines/threonine residues (Heldin 2010; Hubbard 2010).

Per their name, the protein kinases transfer the gamma phosphate group from adenosine triphosphate (ATP) to hydroxyl groups on tyrosine, serine, and threonine residues in the receptor proteins. Phosphorylation alters protein conformation, directs protein and enzyme activities, enhances protein–protein interactions, and directs the activity of most cellular processes. Through signal transduction these enzymes modulate cell growth, metabolism, differentiation, and proliferation.

The protein kinases have both regulatory and catalytic domains (Kennedy 2010). The regulatory domain may be part of the same polypeptide that also contains the active site, such as protein kinase C (Ca$^{+2}$/calmodulin-dependent protein kinases) or it may be contained on separate polypeptide subunits, such as protein kinase A (cAMP-dependent [cyclic adenosine monophosphate] protein kinase). The binding of a second messenger changes the receptor conformation, releasing the regulatory domain, unmasking the active binding site for ATP and the target protein, and enabling activation of the kinase activity at the cytoplasmic domain.

The protein kinases catalyse the phosphorylation of target proteins including the activation of other protein kinases resulting in a cascade effect, which amplifies the original signal hundreds of times. The specificity of the signal is produced by the relay proteins involved in the transduction pathways that eventually lead to activation of TFs that bind to DNA responsive element sequence involved in specific gene expression. When the extracellular signal is absent, phosphatase enzymes become active to dephosphorylate the tyrosine, serine, or threonine residues in the activation site, thereby decreasing the signaling pathway and the cellular response (Barford et al. 1998; Barford 2001). There are as many tyrosine phosphatases as there are tyrosine kinases, suggesting a close regulation serving as "on"/"off" switches for tyrosine phosphorylation. There are fewer serine/threonine phosphatases suggesting serine and threonine phosphates may remain activated for longer periods of time. Abnormal protein kinase activity (increase or decrease in phosphorylation) can alter cell growth and contribute to the development of cancer.

### RECEPTOR TYROSINE KINASES

Of the tyrosine kinases identified in the human genome, two-thirds are receptor tyrosine kinases (RTKs) and one-third are nonreceptor, or cytoplasmic, tyrosine kinases (NRTKs). The RTK system is especially effective when the cell needs to regulate and coordinate a variety of activities and trigger several signal pathways at once (Fantl et al. 1993). The RTK receptors include various growth factors—platelet derived growth factor (PGDF), epidermal growth factor (EGF), fibroblast growth factor (FGF), and insulin like growth factor (IGF) receptors—and insulin receptors (IR) (Plowman et al. 1994).

Representative examples of transmembrane signaling and intracellular signal communication are presented in Figure 5.1a through d. These figures are provided as examples of the transmembrane signaling to the intracellular signaling pathways discussed throughout the sections describing the signal transduction receptor, and non-receptor, protein kinases. Many of these protein linked pathways vary between cells and serve to cross-talk between pathways within cells, clearly indicating the complexity and sophistication of these proteins in communication of the initial signal to regulate specific nuclear gene expression.

The amino acid sequences of the tyrosine kinase domains of RTKs are highly conserved within the ATP binding and substrate binding regions. They consist of extracellular signal binding domains, a single alpha helix that spans the membrane, and an intracellular tail with several tyrosines forming a regulatory domain (Heldin 2010). The cytoplasmic side of the receptor contains the TK activity, for example epidermal growth factor (EGF) binding to the Epidermal Growth Factor Receptor (EGFR) (Figure 5.1). When a signal-ligand binds to the external domain, two receptor polypeptides aggregate, forming a homo-dimeric receptor (Zwick et al. 2002). The dimeric conformation change activates the TK section of both polypeptides, and autophosphorylation occurs, adding phosphates to the exposed tyrosine residues of the neighboring polypeptide tails. Phosphorylation activity is extended to other targeted proteins as well. The activated receptor proteins attract a variety of specific relay proteins that bind to the phosphorylated tyrosine molecules. A signaling cascade is promoted by each RTK dimer, simultaneously activating ten or more different intracellular proteins. These activated relay proteins then trigger many different transduction pathways and responses. Addition or removal of the phosphate groups regulates the active or inactive state of these receptor proteins (Figure 5.1a).

The EGF receptor, has three closely related receptors (ErbB2, ErbB3, ErbB4), and the Ras-Raf-ERK signal transduction cascade [Ras-(a guanosine nucleotide binding protein; GDP (guanosine diphosphate) is exchanged for GTP to become active). Raf-(serine-threonine specific protein kinase). Erk-(extracellular signal regulated kinase)], defined by the extracellular signal regulated kinase-1 (ERK-1) and ERK-2 (Seger and Krebs 1995). Upon binding, the receptor undergoes autophosphorylation on the cytosolic side of the receptor. The phosphotyrosines that are formed then serve to attract SHC (Src homology) domain and GRB2 (growth factor receptor bound 2) which binds SOS (son of sevenless). SOS acts as a GDP (guanosine diphosphate) exchange factor, releasing GDP and inserting guanosine triphosphate (GTP) into RAS (a G protein), forming the active protein (Figure 5.1a).

The relay proteins contain a specific SH2 domain (Src homology domain 2) of amino acid sequences, which have high-affinity binding for phosphotyrosine residues. Interaction of the SH2-domain proteins with RTK activity leads to phosphorylation of tyrosine in the SH2 domains. The resulting phosphorylation of the SH2 domain may alter enzyme activity (either positively or negatively), change the conformation of the SH2 protein, allowing it to bind to another protein, or cause the SH2 protein to be translocated to another part of the cell (Pawson and Nash 2010). There are several SH2 proteins that have intrinsic enzyme activity, including phospholipase C (PLC), the proto-oncogene Ras-associated GTPase activating protein (Ras-GAP), phosphatidylinositol 3′-kinase (PI3K), protein phosphatase-1C (PTP1C), and members of the Src family of PTKs (Pawson et al. 2004).

Many of the relay proteins are also tyrosine kinases and together they enable a cascade of expanding phosphorylations that amplifies the ligand signal within the cytoplasm. After entering the nucleus, some of these tyrosine kinases phosphorylate TFs, enhancing the binding of coactivators and activation of their regulatory function affecting gene transcription. A cell exists in a state of homeostasis that can be activated when increased activity is needed, but must also be able to return to homeostasis in the absence of the signal.

## Receptor Serine/Threonine Kinase

Two types of single-pass transmembrane proteins form receptors that, when bound by their ligands become serine/threonine kinases attaching phosphate groups to serine and/or threonine residues of their target proteins. Ligands for these receptors include: transforming growth factor-β (TGF-β),

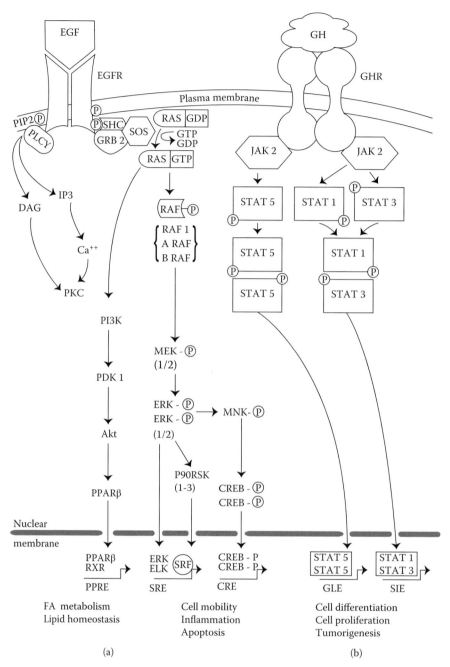

**FIGURE 5.1** Transmembrane signaling and cellular communication. (a) Epidermal growth factor. (b) Growth hormone. (c) Tumor necrosis factor α. (d) Interleukin 1. ([a] based on Qi and Elion 2005; and Kolch et al. 2002; [b] based on Yamaoka et al. 2004; and Qiagen and Protein Lounge, SABiosciences.com/pathway.php?sn=Growth_Hormone_Signaling 2011; [c] based on Wu and Zhou 2010; Qi and Elion 2005; and Qiagen and protein Lounge, SABiosciences.com/pathway.php?sn=MAPK_Signaling and SABiosciences.com/pathway.php?sn=TNF_Signaling 2011; [d] based on Qi and Elion 2005; and Qiagen and Protein Lounge, SABiosciences.com/pathway.php?sn=NFKappaB_Family_Pathway 2011.)

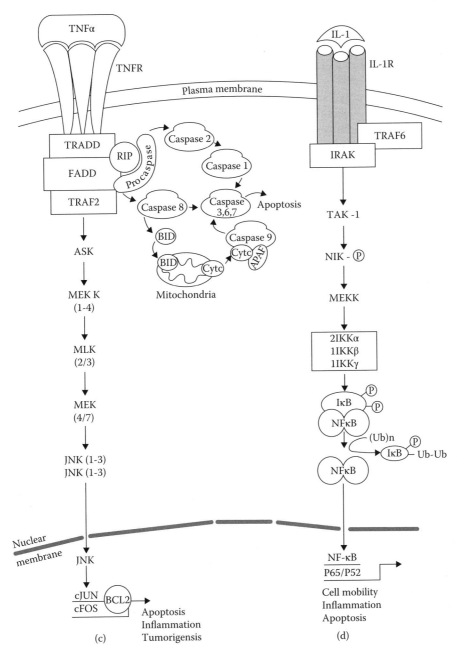

**FIGURE 5.1** (*Continued*)

activins (regulate FSH synthesis and degradation in menses), bone morphogenic proteins (BMPs), and myostatin, an inhibitor of skeletal muscle growth (Silvestri et al. 2010). The TGF-β signaling pathway normally suppresses the cell cycle in several ways. Defects in the pathway and mutations in the genes of the TGF-β receptors are associated with increased incidence of cancer, specifically pancreatic and colon cancer. Thus, the normal pathway activities appear to inhibit cancer processes by serving as tumor suppressor genes.

The serine/threonine kinases include a variety of enzymes, such as cAMP-dependent protein kinases (PKA), glycogen phosphorylase kinase (GPK), cGMP-dependent protein kinases (PKG),

Ca$^{+2}$/calmodulin-dependent protein kinases (PKC), mitogen-activated protein kinases (MAPK), potential ongogenes (Raf), and the TGF-β superfamily (Hubbard 2010). The receptors for the TGF-β superfamily of ligands have intrinsic RSTK activity. TGF-β can induce and/or inhibit cellular proliferation or differentiation and regulate migration and adhesion of various cell types. The signaling pathways utilized by the TGF-β, activin, and BMP receptors differ from those stimulated by RTK activity or that are associated with intracellular NRTK activities.

The RSTKs have been identified as type I (55 kDa) and type II (75 kDa) receptors, with differing electromobility due to a betaglycan attachment on the type II receptor (Attisano et al. 1994). Ligands first bind to the type II receptors, which interact with the type I receptors. Once the complex forms between ligand and the two receptor subtypes, the type II receptor phosphorylates the type I receptor, leading to initiation of the signaling cascade. This dual receptor link has been identified with activins and BMPs (Attisano et al. 1994; Wrana et al. 1994).

One predominant effect of TGF-β is regulation of progression through the cell cycle. The proto-oncogene, MYC, is a nuclear protein that directly affects the expression of genes harboring MYC-binding elements. The MYC gene encodes a transcription factor (TF) that regulates 15% of all genes through binding to enhancer box sequences. In response to TGF-β, MYC recruits histone acetyl-transferases (HATs) that prepare chromatin to receive the basal transcription apparatus. A mutated form of MYC leads to the dysregulation of MYC-dependent gene expression, subsequent increase in cell proliferation, and increased incidence of cancer. (Gurel et al. 2008).

## NONRECEPTOR PROTEIN KINASES

### NONRECEPTOR TYROSINE KINASES

There are numerous intracellular NRTKs that are responsible for phosphorylation of tyrosine residues on intracellular proteins that vary depending on the signal being transmitted. RTK, G protein coupled receptors (GPCRs), and receptors that lack intrinsic TK activity recruit NRTK (Hubbard and Till 2000). Other receptors include the cytokine receptors (e.g., the interleukin-2 receptor, IL2R) as well as the CD4 and CD8 cell surface glycoproteins of T cells and the T-cell antigen receptor (TCR).

Although the insulin receptor (IR) has intrinsic tyrosine kinase activity, it does not directly interact with enzymatically active proteins containing SH2 domains. Instead, the IR interacts specifically with IRS-1, which contains several SH2 binding domains. IRS-1 acts as a docking or relay protein to couple the IR to SH2 containing signaling proteins (He et al. 1996). This mode of coupling receptors to intracellular NRTKs is distinctly different from that of RTKs. Two types of NRTKs include the PTK family related to Src and Src-related kinases, and another family related to the Janus kinases (JAKs). These cytosolic receptor proteins have similar amino acid sequences and structure.

### SRC KINASE

The Src protein is a non-receptor protein tyrosine kinase identified as the transformation protein in the Rous sarcoma virus. Src plays a multitude of roles in cell signalling, including stimulation of cell adhesion, growth, movement, and differentiation. Src is widely expressed in many cell types, and can have different locations within a cell. The IgM receptors in B cells and the T cell receptors contain transmembrane subunits containing immunoreceptor tyrosine-based activation motifs (ITAMs). When an antigen is present, Src kinases are dephosphorylated in part by CD45, a surface-level tyrosine phosphatase.

The subcellular location of the Src proteins can affect their function, as they associate with cellular membranes. At the plasma membrane, Src can transduce signals from a variety of receptors to internal signalling pathways that convey these signals to the nucleus, cytoskeleton, and other cellular components. Src can act through the growth factor receptors to affect cell growth and proliferation. Within the nucleus, the Src protein may help regulate the cell cycle and cell division by its interactions with other proteins (Frame 2002).

## Janus Kinase (JAK)

The JAK proteins are coupled to cellular receptors that lack intrinsic enzyme activity (Figure 5.1b). They include all of the cytokine receptors. The JAKs are cytoplasmic kinases—constitutively associated with the cytoplasmic region of the cytokine receptors. JAKs are essential for cytokine transduction (Behrmann et al. 2004). The JAK family is composed of four NRTKs (JAK1, 2, 3, and tyrosine kinase 2 [TYK2]). JAKs are widely distributed among different types of cells and are essential to cell function.

JAK1, JAK2, and Tyk2 are predominately located at the cell membranes. JAK1 has a typical tyrosine kinase domain, identified by a highly conserved eight-residue sequence within the catalytic domain that has only been identified in the JAKs family (Behtmann et al. 2004). A second kinase-like domain exists on the N-terminal side of the kinase domain, but it lacks the residue sequence needed for either the tyrosine kinases or the serine/threonine kinases (Behrmann et al. 2004). Thus, this domain lacks intrinsic catalytic activity and may only serve as a binding site.

A JAK1 mutant (L80A/Y81A) has been identified and found to be unable to associate with cytokine receptors; due to a lack of the kinase activity. Thus, the cytokine receptor for JAK complexes can be regarded essential to establish the RTK activity (Behrmann et al. 2004). JAK2 interacts with the erythropoietin (Epo) receptor whose signals lead to the growth, differentiation, and survival of red blood cell precursors. JAK2 is assembled with newly synthesized erythropoietin receptors in the endoplasmic reticulum (ER), processed, and transferred to the cell surface (Bonafacino 2002). The receptors that bind JAKs consist of two identical single-pass transmembrane proteins (homodimers) embedded in the plasma membrane. Each of their cytoplasmic ends binds a molecule of JAK (Murray 2007). When the receptor is activated, the JAKs are rapidly phosphorylated on specific tyrosine residues. The mechanism of action occurs when the ligand occupied receptors aggregate and form multimeric complexes with JAKs, which then catalyse phosphorylation of the receptors themselves and the STAT (signal transducers and activators of transcription) proteins (Yamaoka et al. 2004).

## Signal Transducers and Activators of Transcription

Six members of the STAT (signal transducers and activators of transcription) family have been identified. All STAT molecules have SH2 and SH3 domains and require activation of specific tyrosine residues by phosphorylation. Dimerization results when the phosphorylated binding site of one STAT molecule binds with the SH2 site of a second STAT molecule. The activated STAT dimers are transported into the nucleus where they bind to specific DNA nucleotide sequences in the promoter regions of genes to initiate transcription. These nucleotide sequences are common to all STAT dimers (Murray 2007). The JAK-STAT pathways are much shorter and simpler than the pathways triggered by the RTKs (Figure 5.1b), increasing the rate at which the cells respond to these signals (Rawlings et al. 2004).

Many signal-ligands trigger JAK-STAT pathways (Shuai 2000), including interferons (IFN), most of the interleukins (IL 2-7, IL 11-13), growth hormone (GH), leptin, prolactin (Prl), Epo, thrombopoietin, and granulocyte-macrophage colony-stimulating factor (GM-CSF). IFN, Epo, and GH receptors self associate and each has been found to bind JAK2 and initiate phosphorylation of JAK2 when the ligand binds. The Epo receptor may bind JAKs in the absence of the ligand, while the GH receptor is ligand dependent (Song and Schindler 2010).

When growth hormone (GH) binds to the GHR the conformation change initiates autophosphorylation. The phosphorylated tyrosines attract JAK to bind and activate STAT (signal transducer and activator of transcription) proteins by phosphorylation. The STAT proteins form either a homodimeric (eg P-ATAT 5) or a heterodimeric (eg P-STAT 1 and P-STAT 3). The dimers then act as nuclear receptors entering the nucleus and bind to response elements, in this case GLE and SIE respectively to initiate gene expression.

## CYTOKINE RECEPTOR

Cytokines are small proteins and peptides that are secreted by specific cells of the immune system and the glial cells. Cytokines participate in several essential roles in the development, differentiation, and function of myeloid and lymphoid cells. Cytokines are separated by functional categories—type I includes IL-2, IL-12, IFN-γ, and TNF-β, and type II includes IL-4, IL-5, IL-6, IL-10, and IL-13. Dozens of cytokine receptors have been discovered, but they all signal through the JAK-STAT pathway and other pathways that usually involve the activation of the MAP kinase cascade (Yoshimura 2009). Cytokine binding to the cytokine receptor induces oligomerization, initiating a signaling response. This signal brings associated JAK kinases (JAK1-3, Tyk2) into close contact and allows cross phosphorylation and activation (Ward et al. 2000). The cytoplasmic domains of the cytokine receptors associate with the JAKs binding sites located close to the membrane to stimulate the kinase activity of JAK (Haan et al. 2006). Phosphorylation of the receptor, cytoplasmic domains creates docking sites for SH2-containing relay proteins. Among the substrates of tyrosine phosphorylation are members of the STAT family of proteins. STAT recruitment is followed by tyrosine, and in some cases, serine, phosphorylation on key residues, which enables the STAT-STAT complex to move into the nucleus and activate specific genes.

PTPs are associated with many cytokine receptors and participate in downregulation of ligand-induced signaling by dephosphorylation of the activated JAK receptors. In addition, dephosphorylation of activated STATS occurs by other PTPs in the nucleus, essentially terminating the signaling process (Haque et al. 1997). Ligands that affect this pathway include the cytokines, growth factors, and hormonal factors that also activate JAK and/or STAT proteins. Two examples include IL-6, which binds to the IL-6 receptor and to the gp130 (cytokine glycoprotein receptor) both of which activate JAK1 and STAT3, and IFNγ, which can activate JAK1 and JAK2 but primarily activates STAT1 (Pixley and Stanley 2010). Other cytokines activate the Ras-ERK pathways, which may be more important than the STAT pathway for the proliferation of the hematopoietic cells (Smithgall 1998; Yoshimura 2009).

Cytokine signaling is affected by two regulator protein families. The suppressor of cytokine signaling (SOCS) proteins bind to JAK and to certain receptors, thereby inhibiting binding, phosphorylation, and further signaling events (Yoshimura 2009). The sprouty related EVH1 domain (SPRED) protein family interacts with Ras and Raf, thereby suppressing ERK activation (Wakioka et al. 2001). These proteins are key physiologic regulators of immunity, hematopoiesis, and angiogenesis and may play a role in human diseases involving allergy, inflammation, and tumorigenesis (Yoshimura 2009).

## TUMOR NECROSIS FACTOR-ALPHA (TNF-α) RECEPTORS AND NF-κB PATHWAY

The Tumor Necrosis Factor (TNF) is a cytokine involved in systemic inflammation. TNF-α is produced by macrophages and other cells of the immune system. TNF is able to induce apoptotic cell death, induce inflammation, and to inhibit tumorigenesis and viral replication (Figure 5.1c).

TNF-α and TNF-β form homotrimeric complexes as they bind to the TNF- receptors (TNFR1 and TNFR2). TNF-R1 is activated by both the membrane-bound and soluble trimeric forms of TNF. Most information regarding TNF signaling has been derived from TNF-R1. The role of TNF-R2 is less understood. TNF-R2 is found only in the cells of the immune system and responds only to the membrane-bound form of the TNF homotrimer (Chan 2007).

Within their extracellular sequences and the N-terminal pre-ligand assembly domain (PLAD), TNFRs have four common structural units identified as the cysteine rich domains (CRD). The CRDs contain about 40 amino acids with three disulfide bridges, and the interacting loops provide binding sites for corresponding ligands (Chan et al. 2000). Uniquely, the TNFR family is also distinguished by an α-helical protein–protein interaction domain, called the death domain (DD).

A bundle of six antiparallel α-helices form an essential platform for formation of the death-inducing signaling complex (DISC), which is a multiprotein complex required for processing and activation of the caspases-8 and -10 (Andera 2009).

The TNF-related apoptosis-inducing ligand (TRAIL) leads to cell death. Upon binding to the receptors, TNFR1 (CD120a, p55/60) and TNFR2 (CD120b, p75/80), both soluble and membrane bound TNFα's induce cell specific prosurvival and proinflammatory signaling or cell death (Chan 2007). Other noted death receptors include Fas/CD95, and the TRAIL receptors, TRAIL-R1/DR4 and TRAIL-R2/DR5 (Andera 2009). The death receptors contain the PLAD domain, which enhances ligand mediated activation (Chan 2007). It is the activation of the TNF-α receptor that leads to apoptosis of the cell.

TNFR1 forms a large multi-protein complex (Complex 1) including the adapter proteins TRADD (TNFR associated death domain), FADD (Fas associated death domain), TRAF2 (TNF receptor associated factor), DD-containing RIP1 kinase (receptor interacting protein kinase).

The procaspase binds to the complex and independently activates caspase 2 and 8, which promotes the proteolytic cascade of events toward apoptosis. Caspase 2 activates caspase 1 which activates caspase 3, 6, and 7. Caspase 8 may interact with the caspase 3, 7 complex or it may activate BID (BH3 interacting domain death agonist) which stimulates mitochondria to release cytochrome c. The cytochrome becomes part of a complex, including APAF (apoptotic protease activating factor), cytochrome c, and caspase 9, which also activates the caspase 3, 6, 7 complex, leading to apoptosis.

Complex I also participates in the activation of MAP kinases and TNF-RI DD-interacting riboflavin kinase and enhances NADPH oxidase (Nox1/2)-activated production of reactive oxygen species (ROS). Complex IIA and IIB can be formed, following internalization of the activated TNFRI, with the participation of the adapter protein FADD and the cysteine protein, procaspase-8. Complex IIA is composed of RIP1/FADD/pro-caspase-8, while complex IIB contains TRADD/FADD/pro-caspase-8. Over-expression of the CI FLIP protein suppresses activation of complex IIB by inhibiting the death receptor–mediated apoptosis. With inhibition of the caspases, secondary complexes are formed with RIP-3 containing necroptosis-activating complexes. The default pathway, activated by Fas/CD95 or TRAIL receptors, is caspase-8 activation in DISC, which induces apoptosis. Cell internalization enhances cell death (Chan et al. 2000).

Signal binding causes a conformational change in the receptor resulting in the dissociation of the inhibitory protein, suppressor of death domain (SODD), from the intracellular death domain. The dissociation enables the adaptor protein, TRADD, to bind to the death domain, and provides a platform for subsequent protein binding (Wajant et al. 2003). Another adaptor protein, TRAF2, binds and reduces the recruitment of inhibitor-of-apoptosis proteins (IAPs) and suppresses TRAF2 mediated apoptosis. TRAF2 is required for TNF-alpha-mediated activation of MAPK8/JNK and NF-κB. Other proteins such as UBC13 (DNA damage inducible gene post-replication repair) and cIAP1/2 (inhibitor of apoptosis proteins) participate in regulation of apoptosis. TRAF2 E3 ligase ubiquinates RIP1 directly and TAKI kinase (TGFβK-1, MAP3K) indirectly activates the phosphorylation by the IKK kinase complex causing subsequent degradation of IκB releasing the NF-κB signal as a transcription factor (Gaur and Aggarwal 2003).

As expected, an extensive cross-talk between multiple pathways exists. For example, NF-κB enhances the transcription of inhibitory proteins, C-FLIP, Bcl-2, and cIAP1/cIAP2, that interfere with the death signal. While activated, the caspases cleave several components of the NF-κB pathway, including the kinases, RIP and IKK, and the subunits of NF-κB itself (Wajant et al. 2003). Alternatively, TRAF initiates signaling through ASK (activator S-phase kinase), MEKK (1-4) (MAPK kinase kinase), MLK (2/3) (mixed lineage kinase), MEK (4/7), and JNK (1-3). JNK enters the nucleus attracting cJUN and cFOS (AP1TF) to bind to the TPA response element attracting BCL2 as a coactivator to initiate gene expression. Other factors, such as cell type, concurrent stimulation of other cytokines, or the amount of ROS can also shift the balance in favor of one pathway over another. Such complex signaling ensures that, whenever TNF is released, a variety of cells with vastly diverse functions and conditions can all respond appropriately to inflammation.

NF-κB was first identified as a TF bound to the enhancer of the kappa light chain antibody gene (Figure 5.1d). NK-κB turns on the genes encoding IL-1 and other cytokines that promote inflammation, as well as genes needed for cell proliferation, cell adhesion, and angiogenesis. The immunosuppressive and anti-inflammatory effects of glucocorticoids are caused by their enhanced production of IκB, an inhibitory protein.

Interleukin-1 (IL-1) binds to the IL-1 receptor, attracting TRAF6 and IRAK (Interleukin receptor associated kinase). IRAK activates TAK-1 (TGFβ activated kinase-1), which phosphorylates NIK (nuclear factor inducing kinase). Activated MEKK phosphorylates IKK (I-kappa kinase), which is composed of three isoforms, α, β and γ. They stimulate formation of IκB-P (I kappa B) binding to NF-κB (nuclear factor kappa B). Ubiquitin conjugates the IκB-P form with multiple UB molecules causing release of NF-κB and leading IκB degradation. NF-κB enters the nucleus, interacts with p65/p52 (or other p proteins), and binds to the respective response element.

IκB-α normally binds to NF-κB and inhibits its translocation to the nucleus. IKK phosphorylates IκBα releasing of NF-κB. The phosphorylated IκB then becomes ubiquinated initiating degradation by the proteasomes. The NF-κB then becomes a heterodimeric TF that translocates to the nucleus and binds to the promoter DNA region. It enhances the expression of more than 60 genes involved in cell survival and proliferation, inflammatory response, and antiapoptosis (Wajant et al. 2003).

Although little is known about the tocotrienols, and they differ from the tocopherols by having an unsaturated isoprenoid side chain, Ahn et al. (2007) evaluated the effects of γ-tocotrienol on the NF-κB pathway on tumorigenesis in a variety of cancer cell lines—squamous cell carcinoma (SCC4), multiple myeloma (U266), lung adenocarcinoma (H1299), and breast cancer cells (MCF-7). γ-Tocotrienol (25–50μM) completely abolished TNF-α induced NF-κB activation, while α-tocopherol had no effect. Reducing agents had no effect on the down-regulation of NF-κB. γ-Tocotrienol also blocked TNF-phosphorylation and degradation of IκBα through the inhibition of IκBα kinase activation, leading to suppression of the phosphorylation and nuclear translocation of p65. NF-κB regulated gene products associated with antiapoptosis, proliferation, and angiogenesis were down regulated by γ-tocotrienol (Ahn et al. 2007).

## NONRECEPTOR SERINE/THREONINE KINASES

There are several serine/threonine kinases that function in signal transduction pathways. The most commonly known are cAMP-dependent protein kinase (PKA), protein kinase C (PKC), Ca$^{+2}$/calmodulin-dependent protein kinase (CaMK), and cGMP-dependent protein kinases (PKG). Additional serine/threonine kinases important for signal transduction are the MAPKs, several oncogenes (RAF), and the TGF-β superfamily, which contains intrinsic RSTK activity.

PKC was originally identified as a serine/threonine kinase that was maximally active in the presence of diacylglycerols (DAG) and calcium ion. There are at least ten proteins comprising the PKC family. Each of these enzymes exhibits specific patterns of tissue expression and activation by lipids and calcium. PKCs are involved in the signal transduction pathways initiated by certain hormones, growth factors, and neurotransmitters. The phosphorylation of various proteins by PKC can lead to either increased or decreased activity. Of particular importance is the phosphorylation of the EGF receptor by PKC, which downregulates the tyrosine kinase activity of the receptor. This limits the length of the cellular responses initiated through the EGF receptor (Skarpen et al. 2005).

MAPKs were identified by their activation in response to growth factor stimulation of cells in culture, hence the name "mitogen activated protein kinases." The MAPKs control a variety of cellular processes including proliferation, differentiation, and apoptosis. Although MAP kinases are not direct substrates for the RTKs or the receptor associated tyrosine kinases, they are in fact activated by an additional class of kinases (Figure 5.1a and 5.1c). Four discrete groups of MAPK have been identified in mammalian cells (Chang and Karin 2001). They include extracellular signal regulated kinases (ERK1 and ERK2), the c-Jun amino terminal kinases (JNK1, JNK2 and JNK3), the p38 kinases (p38 α, β, γ, δ), and ERK5/Big MAP kinase 1 (BMK1). The MAP kinases are part of a three-tiered phosphorylation cascade consisting of an MAPK, an MAPK kinase (MAPKK, or MEK), and an MAPK kinase kinase (MAPKKK or MEKK) (Theodosiou and Ashworth 2002; Qi and Elion 2005). The MAPKs progressively phosphorylate each other and generate a cascade effect on signal amplification. Phosphorylation of the threonine and tyrosine residues located in the T-loop (phosphorylation motif Thr-Xaa-Tyr) of the MAPK results in activation. The duration and magnitude of the activation determines the physiologic outcome (Qi and Elion 2005).

MAPKs are attenuated in a site-specific manner by dual-specificity MAPK phosphatases (MKPs), tyrosine phosphatases, and serine/threonine phosphatases (Keyse 2000). There are ten MAPK phosphatases specific for dephosphorylation of the phosphorylated MAPKs. Frequently, active MAPKs are translocated from the cytoplasm to the nucleus to phosphorylate nuclear targets. MAPKKs can shuttle in and out of the nucleus in association with the MAPK and retain them until needed in the cytoplasm. The MAPK scaffold proteins function similarly (Elion 2001). MAPKs are localized with numerous subcellular structures including microtubules, endosomes, the ER, and the actin cytoskeleton. Based on the *in vitro* substrates and the tissue locations, the MAPKs have been identified as microtubule associated protein-2 kinase (MAP-2 kinase), myelin basic protein kinase (MBP kinase), ribosomal S6 protein kinase (RSK-kinase), and EGF receptor threonine kinase (ERT kinase). All of these proteins have similar immunocross reactivities, amino acid sequence, and the ability to *in vitro* phosphorylate similar substrates. The activated MAPKs bind tightly to MAPKKs and scaffold proteins through multiple docking domains, distinct from their active sites, and that can recognize homologous sites on different target domains (Tanoue and Nishida 2003). MAPKs can also inhibit signaling by binding substrates while in their inactive form.

Maximal MAPK activity requires that both tyrosine and threonine residues be phosphorylated, suggesting that MAPKs can act as switch kinases, changing from increased intracellular tyrosine phosphorylation to that of serine/threonine phosphorylation. Although MAPK activation was first observed in response to activation by the growth factors (EGF, PDGF, NGF, and insulin receptors), other cellular stimuli such as T cell activation, phorbol ester activation of PKC, thrombin, bombesin, and bradykinin through G proteins, and electrical stimulation rapidly induce tyrosine phosphorylation of MAPKs.

Importantly, there are a limited number of MAPK cascades; yet the various pathway interactions provide the requisite specificity for the cascades. The association of the MAPKs with scaffold, activator, or effector proteins produces differences in kinetics and affect compartmentalization (Caunt et al. 2006). The various pathway interactions provide the requisite specificity for the specific MAPK cascades. Two of the major MAPK cascades are initiated by TNF and RAF. TNF induces a strong activation of the stress-related JNKs that bind and phosphorylate c-Jun at Ser63 and Ser73 (Li et al. 2004). Within its transcriptional activation domain, MAPKs respond to stress stimuli, such as the cytokines, and are involved in T cell differentiation and the apoptosis process. TNF also evokes a moderate response of the p38-MAPK, and is responsible for minimal activation of the classical ERKs.

Raf, is a proto-oncogenic serine/threonine kinase (Dhillon and Koch 2004; Johannessen et al. 2010). When Raf becomes an active protein kinase, it initiates an orderly phosphorylation cascade from Raf to MEK to MAPK (Figure 5.1c). TRAF2 activates the kinases of MEKK1 and ASK1 which phosphorylates MEK, which activates JNK (Johnson and Lapadat 2002). The JNK translocates to the nucleus and activates TFs such as c-Jun and ATF2 (activating TF 2). Once the MAPK enters the nucleus, it phosphorylates and activates specific TFs. They bind to promoter regions of the genes, c-FOS and c-Jun, and form a heterodimeric TF called AP-1, which activates genes involved in cell proliferation. The JNK pathway is involved in cell differentiation, proliferation, and is generally proapoptotic (Caunt et al. 2006; Johnson and Lapadat 2002). The targets of the MAP kinases include several transcriptional regulators, including serum response factor (SRF) and the proto-oncogenes FOS, MYC, and JUN, as well as members of the steroid/thyroid hormone receptor family of proteins.

## Phosphatidylinositol-3-Kinase

Various RTKs and receptor-associated PTKs activate PI3K by phosphorylation of tyrosines in the SH2 domain. PI3K is responsible for the production of phosphatidylinositol 3-phosphate (PI3P), PI 3,4 bisphosphate, and PI 3,4,5 triphosphate (Cantley 2002; Foster et al. 2003).

PI3K forms a heterodimeric protein containing two subunits. The p85 subunit contains SH2 domains that interact with activated receptors or other receptor-associated PTKs and is itself subsequently phosphorylated and activated. Even though this subunit contains a domain homologous to GTPase activating proteins (GAP), it lacks catalytic activity. The second subunit is enzymatically active (Fruman et al. 1998).

Several signals—PDGF, EGF, insulin, IGF-1, HGF, and NGF—bind to their respective receptors, which trigger MAPK to attract and activate PI3K. PI3K phosphorylates the 3 position on the inositol ring of membrane phosphatidylinositol forming PIP2 (Figure 5.1a). This activity generates substrates for PLC, which can then hydrolyse PIP2 producing both DAG and IP3. DAG then activates PKC and IP3 activates $Ca^{++}$ release from ER stores, which further activates PKC, calmodulin, and other cellular enzyme activities (Rhee 2001; Rhee and Bae 1997).

Both the insulin and IGF-1 receptors activate PIK3 indirectly through phosphorylation of the insulin receptor substrate (IRS) family of adaptor molecules (Saltiel and Kahn 2001). PI3K initiates metabolic signaling downstream of the IR and IGF-1 receptors, and mediates cell differentiation, survival, and proliferation.

These receptors phosphorylate tyrosine residues on IRS to create p85 binding sites. The monomeric form of p85 acts as a negative regulator of insulin signaling *in vivo* by controlling the extent of PI3K activation downstream of the receptor. P85 and IRS-1 assemble into large complexes in the cytosol and suggests a novel mechanism for limiting IRS-1/PI3K signaling, suggesting why insulin/IGF-1 signaling is sensitive to p85 levels (Luo et al. 2005).

## G Protein/G Protein Coupled Receptors

The most widely used signaling pathway in mammalian systems involves GGPCRs, the largest family of transmembrane proteins in vertebrates. There are 400–450 genes encoding GPCRs, which conserve a 7-transmembrane domain. The activation of G proteins and the GPCRs occurs in response to diversified extracellular stimuli to modulate cellular function (Wettschureck and Offermanns 2005).

Any individual cell of the body may express numerous GPCRs, which are activated by phosphorylation through GPCR kinases (GRKs). Traditionally, the activated receptors couple to G proteins, transduce downstream signals using second messengers, adenylyl cyclase, phospholipases, and membrane channels (Cervantes et al. 2010; Ritter and Hall 2009). When the GPCR transmembrane receptor is activated by an extracellular stimulus, an allosteric conformation change occurs, changing the shape of the cytoplasmic region and exposes a binding site for the mobile membrane bound G protein (Figure 5.2a). The G protein is partially inserted into the cytosolic side of the plasma membrane lipid bilayer. In addition, many G proteins have 3-polypeptide subunits (Gα, Gβ, Gγ) that can bind three different molecules including the receptor protein, the GDP or GTP nucleotide, and an effector protein.

When GPCRs are activated, they bind the Gα protein. The guanine nucleotide exchange factors (GEFs) bind to the inactive G protein and stimulate the disassociation of GDP. GTP then binds, activating the Gα protein and both the GTP-bound α-subunit as well as the GβGγ-subunits may activate downstream target effector proteins. This structural change reverses the receptor conformation and releases the extracellular ligand from the receptor ending the initial signal transduction (Kazio et al. 1991).

The active GTP bound Gα-subunit separates from the GβGγ subunits which remain linked as a dimer (Neer 1994). Gα diffuses along the lipid bilayer until it reaches the effector protein to which it can bind. Gα has GTPase activating protein function, so when it binds to the effector, conformation change initiates its cellular effect and quickly converts GTP to GDP and inactivates Gα The inactive Gα protein separates from the effector and diffuses along the bilayer until it rebinds to the

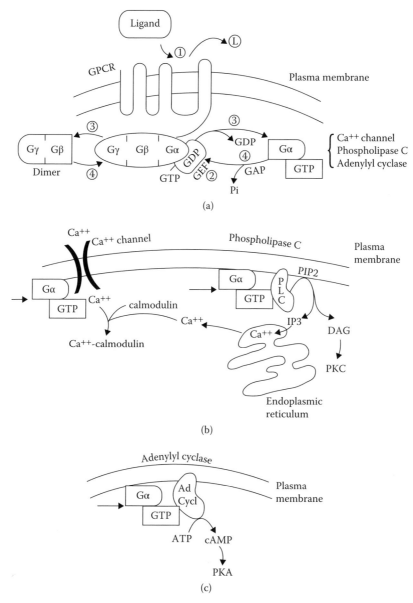

**FIGURE 5.2** Transmembrane regulation of GPCR and G protein activation. (a) Regulation of G protein activation. (b) Activation of calcium channel and PLCγ triggering influx of calcium. (c) Activation of adenylyl cyclase results in conversion of ATP to cAMP, which can activate PKA. (Based on New, D. C., and Y. H. Wong, *J Molecular Signaling* 2:2 doi:10.1186/1750-2187-2-2, 2007.)

reassembling the GβGγ subunits, returning the G protein to its inactive state. Each of the three subunits (Gα, Gβ, Gγ) can direct the G protein to different effector molecules. The G protein can again bind to the receptor and the cycle repeats.

The GTPase cycle of G protein activation and deactivation is subject to regulation by the regulator of G protein signaling (RGS) family of proteins that activate the GTPase function, thereby maintaining the GDP Gα form and preventing G signaling (Wieland et al. 2007). The paradigm of G protein activation occurs most often through activated GPCRs (Blumer et al. 2007). However, a class of G protein activator proteins (AGS-proteins) can directly interact with Gα-subunits and βγ dimers, and are capable of initiating G protein dependent signaling without GPCR-activation.

Nucleoside-diphosphate kinase (NDPK) acts as a direct activator of G protein signaling, serving as a histidine kinase (Wieland 2007). The B isoform of NDPK phosphorylates Gβ at His 266 using ATP. The high-energy form of the His-phosphate allows transfer to GDP reforming GTP in close proximity to the Gα subunit and thus mediates local activation of G proteins (Kimura and Shimada 1988; Kimura et al. 2003).

The G-protein activation by NDPK occurs in the mammalian heart and contributes significantly to the basal formation of cAMP (Wieland et al. 2007). Importantly, the expression of a mutant βγ-complex, specifically lacking His 266 and affecting the NDPK-mediated Gα activation, confirmed this role as the basal cAMP formation decreased by half. The β-adrenergic receptor mediated cAMP formation through adenylyl cyclase and muscle contractility were unaffected by disruption of NDPK-mediated Gβ phosphorylation, suggesting the two G protein activation pathways have separate functions.

Interactions between the GPCRs occur as individual GPCRs form heterodimers between pairs of GPCR (Milligan et al. 2006). GRKs are a family of seven related serine/threonine protein kinases that have differential distribution across tissues and different binding affinities to certain receptors (Bridges and Lindsley 2008). GRKs phosphorylate the GPCR receptors, providing another mechanism of regulation to activate the G protein (Bunemann and Hosey 1999). When the GRK phosphorylates an activated GPCR receptor, the signaling by that receptor is terminated, initiating the uncoupling of the receptor from heterotrimeric G proteins (Premont et al. 1995; Reiter and Lefkowitz 2006).

The GRK1 is involved in the vision process as a rhodopsin kinase, which requires a farnesyl isoprenoid to be added following translation to enable binding to the light activated rhodopsin. The β-adrenergic receptor kinases, GRK2 and GRK3, associate with heterotrimeric G protein GβGγ subunits, which are released upon receptor activation of G proteins. Phosphorylation decreases the interaction between GPCR and G proteins and increases interaction with arrestins. The arrestins further inhibit the GPCR activity by signaling their removal (Premont and Gainetdinov 2007). The arrestins directly link active receptors to clathrin-coated pits to facilitate receptor endocytosis and degradation, decreasing the GPCR in the membrane (Lefkowitz and Shenoy 2005; Mills 2007).

There are several GPCRs that relay signal transduction through G proteins to affect second messenger activities specific for a variety of ligands (Figure 5.2b and 5.2c), including adenylate cyclase, forming cAMP; phospholipase C, which forms IP3 and DAG; and phosphodiesterase (PDE), which decreases cGMP (Soderling and Beano 2000). GPCR modulation of adenylate cyclase activity leads to the production of cAMP as the second messenger. Receptors include the β-adrenergic receptors and glucagon receptor. As the ligand binds, the receptor stimulates the increased production of cAMP, which stimulates an increase in phosphorylation of the PKA activity in both β-adrenergic and glucagon receptors and results in the activation of ion channels. Conversely, the $\alpha_1$-type adrenergic receptors and the muscarinic $M_2$ receptor are coupled to inhibitory G proteins (Gi subunit) that repress adenylate cyclase activity upon receptor activation (Taussig et al. 1993).

In the case of cAMP, the enzymatic changes can activate the TF CREB, cAMP response element binding protein (Cha-Molstad et al. 2004). When CREB binds to its response element, 5′ TGACGTCA 3′, in the promoter regions of genes that are able to respond to the ligand activated CREB, gene transcription is initiated (Quinn 2002). The cell begins to produce the appropriate gene products in response to the signal it received at the cell surface.

Other GPCRs can activate PLCγ (Figure 5.2b) to hydrolyze phosphoinositides (PIP2) in the membrane generating the second messengers, diacylglycerol (DAG) and inositol 1,4,5 trisphosphate (IP3). The GPCR include the $\alpha_2$-adrenergic receptors, angiotensin receptors, bradykinin receptors, and vasopressin receptors. The DAG formed stimulates PKC, which initiates phosphorylation of a broad spectrum of protein kinases resulting in a rapid cascade response. The IP3 diffuses through the cytoplasm to the ER, stimulating a ligand gated calcium channel increasing intracellular $Ca^{++}$ levels, which regulates numerous effector proteins, such as the $\alpha_1$-adrenergic receptor and muscarinic acetylcholine receptors.

GPCR may also activate phospholipase A2, which releases arachidonic acid, initiating formation of prostaglandins (PG) and eicosinoid mediators to open/close ion channels as well (Jenkins et al. 2009). During pathologic states such as ischaemia or congestive heart failure, eicosanoids contribute to multiple maladaptive changes including inflammation, alterations of cellular growth programmes, and activation of multiple transcriptional events leading to the deleterious sequelae of these pathologic states. The central roles of myocardial $PLA_2$ in eicosanoid signalling in the heart, the major COX, LOX, and CYP pathways of eicosanoid generation in the myocardium, and the effects of important eicosanoids on receptor, ion channel, and transcription mediated processes that facilitate cardiac hypertrophy, mediate ischaemic preconditioning, and precipitate arrhythmogenesis in response to pathologic stimuli.

Another novel class of GPCRs includes the photoreceptors expressed in rods and cones associated with vitamin A and vision. This class is coupled to the heterotrimeric G protein, transducin ($\alpha$, $\beta$, $\gamma$ subunits), which responds to a light-induced structural change of rhodopsin. Activated transducin causes cGMP phosphodiesterase to increase, decreasing the level of cGMP. The decrease in cGMP results in the closing of a $Na^+/Ca^{2+}$ ion channel leading to hyperpolarization of the cell membrane (Downs et al. 2006). Termination of the transducin activity occurs when the bound GTP is hydrozyed to the transducin GDP form.

## G Protein Regulators

Several accessory regulatory proteins can alter regulation of G proteins activity, including the GRKs that phosphorylate the GPCR and block association with the G protein; the guanine nucleotide disassociation inhibitors (GDIs), which inhibit the release of GDP from the G protein, keeping the G protein inactive; and the GAPs that regulate GTP hydrolysis (Ross and Wilkie 2000).

G protein activity is regulated by a GDP/GTP cycle determined by a family of GTPase activating proteins. Protein kinases start with receptor autophosphorylation to enable the binding of adaptor proteins to phosphotyrosine groups on the receptor, activating the guanine nucleotide release factor (GRF), which then removes GDP from Ras, and allows Ras to bind GTP. Ras becomes active when GTP is bound, which is the constituitive form of the oncogenic Ras protein. Regulation occurs by a GAP protein which hydrolyses GTP to GDP and by removal of the GDP by a guanine nucleotide exchange protein, SOS, which displaces the GDP (Tian and Feig 2001).

GTP-bound Ras protein recruits and activates Raf, initiating a phosphorylation cascade of MEK, which phosphorylates ERK, which then moves into the nucleus (Figure 5.1a) Activated ERK-P then phosphorylates several different TFs, including Elk-1 associated with apoptosis. The TFs bind to a promoter region of the gene to be expressed and initiates transcription. The proto-oncogenic protein Ras is a G protein involved in carcinogenesis. Ras was found first in the Src oncogene (a viral oncogene). It is a small, single unit G protein regulated by several GTPase activating proteins that activate the Ras GTPase 100-fold, returning the Ras protein to its inactive state. The Src codes for a mutated form of normal tyrosine kinase associated within the inner side of the plasma membrane.

RTK activates Ras by phosphorylation, which then activates Raf. During cell division, this pathway promotes mitosis. The protein is susceptible to numerous mutations. Over activation of the RAS gene occurs with higher frequency than any other gene in the development of colorectal cancers. Mutations in the Ras or Raf genes results in increased risk of many types of cancer. Mutations in the Raf gene occur in 15% of all human tumors, and 66% of all melanomas (Dhillon and Koch 2004; Johannessen et al. 2010).

The Ras proteins, and their downstream pathways, play pivotal roles in cell proliferation, differentiation, survival, and cell death. Their roles in several autosomal dominant multiple congenital anomaly syndromes (Noonan syndrome, Costello syndrome, and cardio-facio-cutaneous syndrome) have been characterized from mutations in genes coding for proteins in the Ras/MAPK pathway (Aoki et al. 2008).

## Ion Channel Receptors

Ligand gated ion channels ($Na^+$ or $Ca^{+2}$) are transmembrane protein receptors in the plasma membrane. Binding of a ligand to the receptor changes the protein conformation and opens the channel. Each type of ion channel has its own signal to open or close in response to that signal. The ion flow rapidly changes the concentration inside the cell. The channel recloses when the ligand disassociates (Gouaux and Mackkinou 2005).

For example, when 2-acetylcholine binds to the sodium channel receptor, the receptor only opens for 1/1000th of a second. $Na^+$ from the outside (high concentration) rushes in, due to both the concentration and the electrical potential gradient of the $Na^+$ ions. Although the actual change in concentration inside the cell is low, the existing homeostatic level is also very low, so the relative change is sufficient to alter a series of cellular events.

## REGULATORY FACTORS

### Phosphatases in Signal Transduction

Substantial evidence links both tyrosine and serine/threonine phosphorylation with increased cellular growth, proliferation, and differentiation. Removal of the incorporated phosphates is essential to turn off the activation of signal pathways. This suggests that the phosphatase genes may function as growth suppressor genes and anti-oncogenes. Proteins are phosphorylated predominantly on Ser (86%), Thr (12%), and Tyr (2%) residues of the phosphoproteome in mammals. Protein phosphatases (PPs) are the primary effectors of dephosphorylation that occur in three main classes based on sequence, structure, and catalytic function. The largest class of PPs is the phosphoprotein phosphatase (PPP) family, comprising PP1, PP2A, PP2B, PP4, PP5, PP6, and PP7, and the protein phosphatase $Mg^{2+}$- or $Mn^{2+}$-dependent (PPM) family, composed primarily of PP2C. The PTP superfamily forms the second group, and the aspartate-based protein phosphatases the third (Cohen 2009).

### Protein Tyrosine Phosphatases

The PTPs include transmembrane enzymes, containing the phosphatase activity domain in the intracellular portion of the membrane protein. PTPs include both the receptor class of PTPs (PTPR) and the intracellularly localized nontransmembrane PTP (PTPN) class (Armad et al. 1993). The regulation of the activities of the transmembrane PTPs and of the intracellular PTPs is not clearly defined. Examinations of two intracellular PTPs (PTP1C and PTP1D) have indicated the presence of SH2 domains. These domains allow the PTPs to directly interact with RTKs and PTKs to stimulate tyrosine dephosphorylation and terminate the ligand signal.

The C-terminal residues of most intracellular PTPs are very hydrophobic, suggesting they serve as membrane attachment domains. A link between insulin function and PTP1B has been established. PTP1B interacts directly with the insulin receptor and removes the tyrosine phosphates added during response to insulin binding, turning off the insulin signal. Interestingly, when the PTP1B gene was deleted in mice, the outcome indicated an increase in insulin sensitivity as well as an increased resistance to obesity induced by a high-fat diet (Delibegovic et al. 2009).

The leukocyte common antigen protein, CD45, has homology to the intracellular PTP1B. The CD45 is involved in the regulation of the tyrosine kinase activity of leukocyte-specific protein tyrosine kinase (LCK) in T cells. LCK is also involved in T-cell activation following tyrosine phosphorylation of T-cell antigens CD4 and CD8. The CD45 dephosphorylates a specific regulatory tyrosine–phosphate site located in the C-terminus of LCK, increasing the LCK activity towards its substrate(s) (Mustelin et al. 2004; Mustelin and Tasken 2003; Nakahira et al. 2007).

## Protein Serine/Threonine Phosphatases

Another class of phosphatases recognizes serine and/or threonine phosphorylated proteins in cells. At least 15 distinct protein serine phosphatases (PSPs) have been identified. Primarily, type 2A PSPs exhibit selective substrate specificity towards PKC phosphorylated proteins (serine and threonine phosphorylated receptors). These protein kinases are involved in signaling cascades and catalyzing phosphorylation of ribosomal S6 protein. Type 2A PSPs are more effective than other PSPs in dephosphorylation of receptor serine kinases (RSKs), while type 1 PSP is required for dephosphorylation of ribosomal S6 (Kuang et al. 2009; Belandia et al. 1994).

The loss of a functional phosphatase involved in regulation of growth promoting signals could lead to neoplasia. However, examples have also been shown where dephosphorylation may be required for promotion of cell growth. Specifically, this is true of kinases that are directly involved in the regulation of cell cycle progression. Phosphatases do not generally function as tumor suppressor genes; yet one-third of the tyrosine phosphatases have been determined to be mutated or over-expressed in cancer cells (Ostman et al. 2006).

The type 2A PSPs have a regulatory and a catalytic subunit, both of which can associate with one of the polyoma, DNA tumor antigens. Transformation by DNA tumor viruses, such as polyoma, appears to be mediated by the formation of a signal transduction unit containing a viral encoded T antigen and multiple host encoded Src of tyrosine kinase proteins. The polyoma middle T antigen can also bind to PI3K (Ichaso and Dilworth 2001). Thus, the role of type 2A PSPs in these complexes may be to dephosphorylate the phosphoserine/phosphothreonine sites, increasing signal transduction and subsequent cellular proliferation.

## Second Messengers

Certain signal molecules and ions are required components of signaling pathways as second messengers. Second messengers participate in pathways initiated by both G protein linked receptors and tyrosine kinase receptors.

## Adenylyl Cylase and cAMP

Adenylyl cyclase catalyzes the conversion of ATP to 3′,5′-cyclic AMP and pyrophosphate. cAMP is an important molecule in eukaryotic signal transduction as a second messenger. Hormones or other stimuli can bind to membrane receptors, and initiate a G protein response to activate or inhibit adenylyl cyclase. Following activation of adenylyl cyclase, the formation of cAMP acts as a second messenger, which interacts with and regulates other proteins such as protein kinase A (PKA) and cyclic nucleotide-gated ion channels. Cyclic adenylyl nucleotide phosphodiesterase, nucleoside-2′,3′-cyclic-phosphate 3′-nucleotide hydrolase (PDE), decreases the level of cAMP available forming nucleoside 3′-phosphate, regulating the cAMP stimulatory effect (Furman and Pyne 2006).

## Guanylyl Cyclase and cGMP

Cyclic guanosine monophosphate (cGMP) is a cyclic nucleotide derived from GTP by guanylyl cyclase (GC). cGMP acts as a second messenger similar to cAMP, most notably activating intracellular protein kinases in response to the binding of membrane-impermeable peptide hormones to the membrane receptors guanylyl cyclase exists in membrane bound and soluble forms. The membrane-bound GC is activated by peptide hormones, while the soluble GC is typically activated by nitric oxide, an external signal that diffuses into the cytoplasm, to stimulate cGMP synthesis. cGMP is produced as a key regulator to cell proliferation, differentiation, and apoptosis, and many pathophysiologic events (Pilz and Broderick 2005).

cGMP modulates metabolic responses via the activation of its downstream effectors, protein kinase G (PKG) and cyclic nucleotide-gated ion channels. Cyclic guanylyl nucleotide phosphodiesterase, nucleoside-3′,5′-cyclic-phosphate 5′-nucleotidohydrolase (PDE), decreases the level of cGMP available forming nucleoside 5′-phosphate.

Cross-talk between cAMP and cGMP signaling pathways involves substrate regulation by the cyclic PDEs. The PDEs regulate cAMP mediated signaling by controlling the level of cAMP available. cGMP regulates the cAMP PDE2 through an allosteric enhancement of cAMP hydrolysis, whereas cGMP acts as a competitive inhibitor of PDE3, regulating cAMP hydrolysis (Zaccolo and Movesian 2007).

In response to cGMP regulation, at least 60 genes are expressed, including those leading to inflammation (iNOS, COX-2, TNFa), cell proliferation (MAPK-phosphatase-1), apoptosis (pro- and antiapoptotic effects), and angiogenesis and extracellular matrix (vascular endothelial growth factor [VEGF], Thrombspondin-1). Some of these effects are indirect, occurring through cGMP modulation of other signaling pathways, such as the MAPK pathway and modulation of TFs (Erk1 and 2, p38 and JNK). The direct regulation effects on transcription have been noted by cGMP-dependent phosphorylation (CREB/ATF-1, TFII-1, NF-κB), cGMP regulation of genes encoding TF (AP-1, regulation of c-fos and junB, growth arrest-specific homobox gene [GAX], Egr-1, PPARγ and its coactivator PGC-1) (Pilz and Broderick 2005).

## PHOSPHOLIPASES AND PHOSPHOLIPIDS

Membrane phospholipids are converted to second messengers by activated phospholipases. Phospholipases and phospholipids are involved in the processes of transmitting ligand-receptor induced signals from the plasma membrane to intracellular proteins. The primary protein affected by the activation of phospholipases is PKC which becomes maximally active in the presence of calcium ion and DAG (Blobe et al. 1994).

The principal mediators of PKC activity are receptors coupled to G proteins, forming heterodimers, and activation of phospholipase C (PLC). PLC contains SH2 domains that enable it to interact with the phosphorylated RTKs (Figures 5.1a and 5.2). PLC becomes closely linked with the signal transduction complexes of the membrane as well as membrane phospholipids, which serve as substrates. Activation of PLC leads to the hydrolysis of membrane phosphatidylinositol bisphosphate (PIP2), increasing intracellular diacylglycerides (DAG) and inositol 1,4,5 trisphosphate (IP3) (Foskett et al. 2007).

## IP3

The IP3 diffuses into the cytosol, binds to the IP3 receptor which regulates a gated, $Ca^{+2}$ channel in the smooth ER, thereby releasing $Ca^{+2}$ into the cytoplasm (Patterson et al. 2004). The IP3 effects are transient, because it is rapidly degraded by a phosphatase. Together, the increased DAG and intracellular free calcium ion concentrations further increase the PKC activity.

## CALCIUM

Many signal molecules induce responses in target cells via signal transduction pathways that increase intracellular concentraions of $Ca^{+2}$. Calcium acts as a second messenger in both G protein and tyrosine kinase pathways. The $Ca^{+2}$ concentration in the cytosol (100 nM) is normally much lower (1/10,000) than the concentration outside of the cell (1–2 mM). Various membrane protein pumps maintain this ratio of calcium. Because of the low intacellular concentration, only small changes in the absolute $Ca^{+2}$ level can produce a large percentage increase in $Ca^{+2}$ concentration, altering sensitive calcium regulated steps in the transduction pathways. The transduction pathways trigger the release of $Ca^{+2}$ bound to the ER. Calcium and Ca-ATPase support active transport pumps in the plasma membrane and the ER (Berridge et al. 2003). $Ca^{+2}$ released by IP3 further activates the PKC, increasing PIP2 hydrolysis and stimulating plasma membrane ion channels (Taylor and Laude 2002).

In the cytoplasm, a $Ca^{+2}$ binding protein, calmodulin, binds $Ca^{+2}$, and the resulting complex binds to enzymes and other effectors, regulating their activity (Lewis 2007; Means 2010). Calmodulin regulates numerous proteins in the cellular system, such as myosin protein kinase, phosphorylase kinase, protein kinase I-IV, NO synthase, regulation of ion flux, and plasma membrane $Ca^{+2}$ ATPase (Eldik and Watterson 1998).

## OTHER PHOSPHOLIPASE ACTIVITIES

Phospholipase D (PLD) and phospholipase A2 (PLA2) are also involved in the sustained activation of PKC through their hydrolysis of membrane phosphatidylcholine (PC). The enzymatic action of PLD on PC leads to the release of phosphatidic acid and choline. Then the phosphatidic acid is hydolyzed to DAG by lipid phosphate phosphohydrolase, removing the phosphate group. PLA2 removes the sn2 fatty acid, yielding free fatty acids and lysoPC, both of which potentiate the DAG mediated activation of PKC.

## LYSOPHOSPHOLIPIDS

Lysophospholipids (LPLs) are minor lipid components compared to the major membrane phospholipids such as phosphatidylcholine (PC), phosphatidylethanolamine (PE), and sphingomyelin. LPLs were originally presumed to be simple metabolic intermediates in the de novo biosynthesis of phospholipids. However, subsequent studies demonstrated that LPLs exhibited biological properties resembling those of extracellular growth factors or signaling molecules. The most biologically significant LPLs are lysophosphatidic acid (LPA), lysophosphatidylcholine (LPC), sphingosine 1-phosphate (s9P), and sphingosylphosphorylcholine (SPC). Each of these LPLs functions by interaction with specific GPCRs, which then leads to autocrine or paracrine effects.

Currently there are fifteen characterized LPL receptors (LPR). Because several of the LPRs were independently identified in unrelated assays, several different names exist for some members. However, Choi et al. (2010) have provided an excellent review on the LDR classification. In particular, there is a group of genes that were originally identified as GPCRs and called endothelial differentiation genes (EDGs) that were later found to be identical to several of the LP receptors (LPAR1-3) and Sphingosine-1-P receptors (s9PR1-5).

Activation of the LPARs triggers several different downstream signaling cascades, including activation of MAP kinase (MAPK), activation of PLCγ, Akt/PKB activation, calcium mobilization, release of arachidonic acid, inhibition or activation of adenylate cyclase, and activation of several small GTPases, such as Ras, Rho, and Rac (Tang et al. 1999). The LPLs exert a wide range of biochemical and physiological responses, including platelet activation, smooth muscle contraction, cell growth, and fibroblast proliferation. LPA is produced by activated platelets, activated adipocytes, neuronal cells, and several other cell types. LPA is produced in the serum through the action of several different enzymes including monoacylglycerol kinase, phospholipase A1 (PLA1), secretory phospholipase A2 (sPLA2), and lysophospholipase D (lysoPLD). Degradation of LPA occurs via lysophospholipase, lipid phosphate phosphatase, or LPA acyl transferase.

# NUCLEAR RECEPTORS AND TRANSCRIPTION FACTORS

## NUCLEAR RECEPTORS

NRs belong to a superfamily of proteins, serving as ligand-regulated TFs that transduce signals from steroid hormones and other lipophilic ligands to the nucleus. The steroid hormones are small, hydrophobic molecules that can diffuse across the plasma membrane and move through the cytoplasm. The NRs provide the last step in the signaling process. These signaling molecules can then enter the nucleus and become activated as TFs (Evans 1988, 2004).

The NR/TF complexes regulate metabolic processes by binding to specific DNA sequences in the promoter region. Coregulators are attracted to the NR/TF forming a complex that includes histone acetyltransferases, methyltransferases, or ubiquitin ligases (Lonard and O'Mally 2006, 2007). Thus, the transcriptional machinery influences transcription of genes, including initiation of transcription, elongation, splicing, and translation of the mRNA introns. Translation of the mRNA into proteins provide key enzymes, transporters, and other proteins involved in metabolic homeostasis (Nagy and Schwabe 2004; Evans 2004). If dysregulation of the NRs or TFs occurs, they may initiate metabolic disease processes related to type 2 diabetes, dyslipidemia, hypercholesterolemia, obesity, and cancer.

Importantly, coregulators are both targets and propagators of post-translational modification (PTM) codes. This underlies a sophisticated epigenetic regulatory scheme from which a complex and dynamic mammalian phenotype emanates (Lonard and O'Mally 2006, 2007). They also regulate a variety of biological processes involved in development, reproduction, and metabolism of lipids and energy through expression of selective genes (Gronemeyer et al. 2004).

The family of NRs (Table 5.1) are high-affinity receptors for fat-soluble hormones and vitamins, various steroid hormones (SHR), TR, vitamin A (RAR), and vitamin D (VDR), which are essential for homeostatic control of the endocrine system (Nuclear Receptor Nomenclature Committee 1999; Germain et al. 2006). The steroid receptors—estrogen receptor (ER), mineralcorticoid receptor (MR), and progesterone receptor (PR), androgen receptor (AR)—function as homodimers, while testicular receptor (TR), vitamin D receptor (VDR), and retinoic acid receptor (RAR) form heterodimers with retinoid X receptor (RXR) (Li et al. 2003; Baxter et al. 2004). The steroid receptors form dimers with zinc-finger proteins that bind to specific HRE nucleotide sequences in the DNA.

NRs are characterized by their structural domains that reflect their structural function (Sonoda et al. 2008). The C-terminal end contains the ligand binding domain (LBD), which is functionally unique to the NRs for ligand recognition and binding, and provides the ability for receptor dimerization. The LBDs consist of about 12 helices, the last one, AF-2 (activation factor-2), is structurally active. The binding of the ligand to the LBD induces conformational changes in the AF-2 region, which facilitates the release of corepressors and histone deacetylase (HDAC) complexes and stimulates the recruitment of coactivators and histone acetyltransferase (HAT) complexes to activate gene transcription (Li et al. 2003).

The receptors also contain a ligand-dependent activation domain, AF-1, and a central DNA binding domain (DBD) consisting of two highly conserved zinc fingers that direct the receptors to bind to the HRE (hormone responsive element), a specific DNA-nucleotide sequence, in the promoter region of the gene. The NRs contain two hexa-nucleotide motifs AGGTCA or its variants, separated by a sequence of several nucleotides. The hinge region provides structural flexibility in the receptor dimers allowing a single receptor dimer to interact with multiple HRE nucleotide sequences (Sonoda et al. 2008). Some apo-steroid receptors within the nucleus associate with HDACs, preventing gene expression in those regions of the chromosome, until the ligand attaches to the apo-receptor and interacts with other regulatory proteins involved in basal transcription (Collingwood et al. 1999). Both activation and repression require participation of other chromatin remodeling factors (Figure 5.3).

Chromatin remodeling enzymes contribute to the dynamic changes that occur in chromatin structure during transcription, recombination, repair, and replication (Marfella and Imbalzano 2007). In addition, several chromatin proteins have been implicated in human disease. The chromatin remodeling factors are large, multiprotein complexes that use ATP hydrolysis to mobilize and restructure the nucleosome. These proteins have unique roles in altering histone–DNA interactions and target gene promoters to open the chromatin to provide access for the TFs (Aranda and Pascual 2001). The presence of the bound HDAC blocks access to the specific regions of DNA by removing acetatyl markers from the histone tails. The DNA strand is tightly coiled (1.8 loops) around eight histone proteins to compress the storage form of the chromosome. This enables compact packaging of the chromosomes within the nucleus, blocking access to specific regions of DNA. In

**TABLE 5.1**

**Human Nuclear Receptors**

| NR Nomenclature | Name | Trivial Name | NR Gene | Ligand |
|---|---|---|---|---|
| NR1A1 | Thyroid Hormone Receptor | TRα | THRA | Thyroid Hormone |
| NR1A2 | −α, −β | TRβ | THRB | Thyroid Hormone |
| NR1B1 | Retinoic Acid Receptor | RARα | RARA | Retinoic Acid |
| NR1B2 | −α, −β, −γ | RARβ | RARB | Retinoic Acid |
| NR1B3 | | RARγ | RARG | Retinoic Acid |
| NR1C1 | Peroxisome Proliferator Activated | PPARα | PPARA | Fatty Acids, Leukotriene B4 |
| NR1C2 | Receptor −α, −β, −γ | PPARβ | PPARD | Fatty Acids |
| NR1C3 | | PPARγ | PPARG | Fatty Acids, Prostaglandin J2 |
| NR1D1 | Rev-ErbAα | RevErbAα | NR1D1 | Orphan |
| NR1D2 | Rev-ErbAβ | RevErbAβ | NR1D2 | Orphan |
| NR1F1 | RAR-Related Orphan Receptor | RORα | RORA | Cholesterol, Cholesteryl sulfate |
| NR1F2 | −α, −β, −γ | RORβ | RORB | Retinoic Acid |
| NR1F3 | | RORγ | RORC | Orphan |
| NR1H3 | Liver X Receptor −α, −β | LXRα | NR1H3 | Oxysterols |
| NR1H2 | | LXRβ | NR1H2 | Oxysterols |
| NR1H4 | Farnesoid X Receptor | FXRα | NR1H4 | Bile Acids |
| NR1I1 | Vitamin D Receptor | VDR | VDR | Vitamin D, 1,25(OH)$_2$ D$_3$ |
| NR1I2 | Pregnane X Receptor | PXR | NR1I2 | 16α-cyanopregnenolone, Xenobiotics |
| NR1I3 | Constitutive Androstane Receptor | CAR | NR1I3 | Xenobiotics, Phenobarbital |
| NR2A1 | Hepatocyte Nuclear Factor 4 −α, −γ | HNF4α | HNF4A | Fatty Acids |
| NR2A2 | | HNF4γ | HNF4G | Fatty Acids |
| NR2B1 | Retinoid X Receptor −α, −β, −γ | RXRα | RXRA | Retinoic Acid |
| NR2B2 | | RXRβ | RXRB | Retinoic Acid |
| NR2B3 | | RXRγ | RXRG | Retinoic Acid |
| NR2C1 | Testicular Receptor −α(2), −β(4) | TR2 | NR2C1 | Orphan |
| NR2C2 | | TR4 | NR2C2 | Orphan |
| NR2E3 | Photoreceptor-specific Nuclear R | PNR | NR2E3 | Orphan |
| NR2F1 | Coup Transcription Factor I, II | Coup-TFI | NR2F1 | Orphan |
| NR2F2 | | Coup-TFII | NR2F2 | Orphan |
| NR3A1 | Estrogen Receptor −α, −β | ERα | ESR1 | 17β-Estradiol, Tamoxifen, Genistein |
| NR3A2 | | ERβ | ESR2 | 17β-Estradiol, Phytoestrogens |
| NR3B1 | Estrogen Related Receptor −α, −β, −γ | ERRα | ESRRA | Orphan |
| NR3B2 | | ERRβ | ESRRB | 4-OH Tamoxifen |
| NR3B3 | | ERRγ | ESRRG | 4-OH Tamoxifen |
| NR3C1 | Glucocorticoid Receptor | GR | NR3C1 | Cortisol, Dexamethasone |
| NR3C2 | Mineralcorticoid Receptor | MR | NR3C2 | Aldosterone, Spirolactone |
| NR3C3 | Progesterone Receptor | PR | PGR | Progesterone |
| NR3C4 | Androgen Receptor | AR | AR | Testosterone, Flutamide |
| NR4A1 | Nerve Growth Factor Induced | NGFI-B | NR4A1 | Orphan |
| NR4A2 | Nuclear Receptor Related | NURR1 | NR4A2 | Orphan |
| NR4A3 | Neuron-derived Orphan Receptor 1 | NOR1 | NR4A3 | Orphan |
| NR5A1 | Steroidogenic Factor 1 | SF1 | NR5A1 | Orphan |
| NR5A2 | Liver Receptor Homolog 1 | LRH1 | NR5A2 | Orphan |
| NR6A1 | Germ Cell Nuclear Factor | GCNF | NR6A1 | Orphan |
| NR0B1 | DSS-AHC critical region on X chromosome protein 1 | DAX1 | NR0B1 | Orphan |
| NR0B2 | Small Heterodimer Partner | SHP | NR0B2 | Orphan |

*Source:* Adapted from Germain, P. et al. (2006) "Overview of Nomenclature of Nuclear Receptors" Table 1. *Pharmacological Review* 58:685–704 with some modifications. With permission.

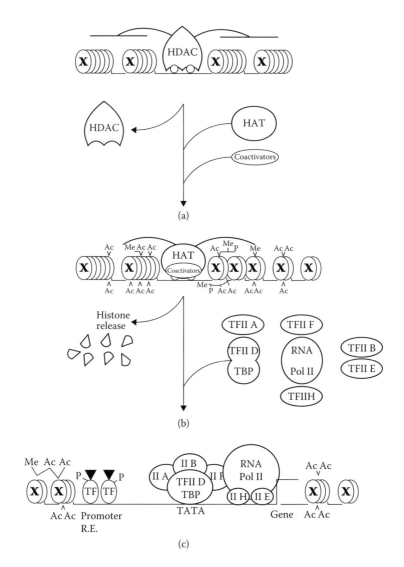

**FIGURE 5.3**  Nucleosome remodeling to enable transcription. (a) HDAC maintains chromatin condensation. (b) HAT and chromosome remodeling. (c) Transcription regulatory complex. (Based on Li, B., M. Carey, and J. Workman, *Cell* 128:707–719, 2007; and Gronemeyer, H., J. A. Gustafsson, and V. Laudet, *Nat Rev Drug Disc* 3:950–64, 2004.)

the presence of the bound HDAC, access to the specific region of the DNA, leading to initiation of gene transcription is blocked. The complex releases the HDACs and recruits HATs. As HAT and coactivators replace HDAC, and with acetylation of the histones, the chromosome begins to open and various histones are released providing access to segments of the DNA (Figure 5.3a). A series of coactivators, TFs, TATA binding protein (TBP) and TFIID, and the RNA polymerase II binds to the specific DNA regions upstream from the gene to be expressed. Each critical protein has DNA binding domains that interact with specific regions of DNA, such as the promoter region responsive element and the TATA binding site. NRs or activated TFs can bind to the promoter RE region and initiate the transcription process (Figure 5.3).

Critical proteins have DNA binding domains which attach to specific regions of DNA, such as the promoter region responsive elements and TATA, and protein binding sites to form a platform for assembly of the transcription complex (Figure 5.3c). The assembly of the RNA polymerase II preinitiation complex includes the TATA binding protein TF IID (reorganizes the corepressors and recruits TFIIB), and a group of TFs: TF IID (assists transcription activators and promoter recognition), TF IIA (stabilizes TF IID and promoter binding), TF IIB (recruits RNA Polymerase II and TF IIF), TF IIF (assists RNA polymerase II in reaching promoter), TF IIE (recruits and modulates TF IIH helicase, ATPase and kinase activities), and TF IIH (helicase activity and DNA repair). TFs can bind to the promoter RE region. Transcription of the target gene is initiated to form the mRNA needed for translation into the required protein.

In addition to contacting the basal transcription machinery directly, NRs alter, inhibit, or enhance the transcription process by recruiting an array of coactivator or corepressor proteins to the transcription complex (Horwitz 1996; Naar et al. 2001). As the ligand binds to the receptor, a conformation change occurs that allows recruitment of multiple coactivator complexes to regulate expression. In the absence of a ligand, the apoRAR or apoTR can interact with corepressors when they are associated with RXR forming heterodimeric partners. The corepressor factors include HDAC activity and maintain chromatin compaction (closed state) repressing transcription.

The apo-NRs are either present in the cytoplasm awaiting the ligand for activation, or they may form a complex in the nucleus constituitively bound to the HRE, or they may form a complex with a corepressor, such as a NR corepressor 1 (NCOR1), which is a transcriptional coregulatory protein containing multiple NR interacting domains, a silencing mediator of retinoid (RAR) and thyroid (TR) receptor (SMRT) proteins, or the HDAC complexes (Rosenfeld et al. 2006; Cote et al. 2004; Hong and Privalsky 2000; Privalsky 2004). The corepressor box is a region between the receptor LBD and the DBD, and is essential for the association of corepressors. The transcription intermediary factors (TIF1α and β) bind to the receptor. The TIF1α interacts with the ligand-binding domain of the RAR in the presence of the hormone-ligand, while the TIF1β binds to a region in the zinc fingers and with the heterochromatin protein (HP1).

Except for the glucocorticoid receptor, which resides in the cytoplasm until the glucocorticoid ligand binds to it, most other NRs reside within the nucleus. The glucocorticoid hormone receptor (GR) complex forms in the cytoplasm, translocates into the nucleus and binds to the hormone responsive element (HRE) attracting other TFs that alter the transcription rates of the associated genes. Each of the receptor functions depends upon a particular region of the protein. If a mutation occurs in any one region, the function of that region may be disrupted, but other functions related to the other regions of the protein may not be altered.

As noted previously, each NR contains a ligand binding domain (LBD) and a DNA binding domain (DBD). Some of these NR/TFs bind to DNA as homodimers, such as the ER, MR, PR, AR and GR, while others such as TR, VDR, and RAR, bind lipophilic ligands and form heterodimers through interactions with RXR, peroxisome proliferator activate receptor (PPAR), liver X receptor-like (LXR), farnesoid X receptor (FXR), and pregname X receptor (PXR).

The RXRs represent a class of receptors that bind the retinoid, 9-cis-retinoic acid, and have three isforms, RXRα, RXRβ, and RXRγ. The receptors have a high affinity for its natural ligand, 9-cis RA, and a lower affinity for ligands delivered from the diet, such as lipids to form heterodimers activated by RXR (Shulman and Mangelsdorf 2005). The RXRs serve as obligatory heterodimeric partners for other members of the NR family as well, such as the PPARs.

The PPAR family is composed of three members, PPARα, PPARβ/δ, and PPARγ. Each of these receptors forms a heterodimer with the RXRs (Evans et al. 2004; Semple et al. 2006). The PPARs serve as sensors for fatty acids and fatty acid derivatives, and control important metabolic pathways involved in the maintenance of energy balance. The PPAR-FA receptor complex regulates lipid homeostasis and influences insulin action (Berry and Noy 2009). PPARα was the first member identified as the endogenous receptor for polyunsaturated fatty acids in liver, skeletal muscle, heart, and kidney. Its function in the liver is to induce peroxisomal fatty acid oxidation during periods of fasting, while the expression of PPARα in macrophage foam cells and vascular endothelium is considered to be activation of protective anti-inflammatory and antiatherogenic effects.

PPARγ has a wide influence on metabolism. It is a master regulator of adipogenesis and is expressed most in adipose tissue (Barish et al. 2006). Low levels of expression are also observed in liver and skeletal muscle. The activation of the PPARγ activity, and the consequent activation of adipocytes, leads to increased fat storage and secretion of insulin-sensitizing adipocytokines, such as adiponectin (Li et al. 2007). In addition, the activation of PPARγ contributes to the control of glycemia, lipemia, adipogenesis, endothelium function, and reduction of insulin resistance (Luan et al. 2001). The most frequent gene mutations of the PPARγ proteins, Pro467Leu and Val290Met, are connected with severe insulin resistance; Pro115Gln is connected with obesity; and Pro12Ala can affect the development of diabetes (Gacka and Adamiec 2004). The Pro12Ala is associated with a decreaed risk of T2DM (Groop and Lander 2000). Thus, PPARγ has become one of the most important gene regulators associated with the metabolic syndrome.

PPARδ is broadly expressed in most tissues and is involved in the promotion of mitochondrial fatty acid oxidation, energy consumption, and thermogenesis. PPARδ serves as the receptor for polyunsaturated fatty acids and VLDLs. PPARδ activates fat metabolism to prevent obesity (Wang et al. 2003).

Continuing research indicates the physiological and morphological changes that occur during obesity and obesity-related disorders are related to adipocyte hypertrophy and inflammation in the white adipose tissue (Li et al. 2007). The body has developed a regulatory system designed to mediate the systemic response to obesity, utilizing free fatty acids and their metabolites as nutrient messengers to signal adaptations in peripheral tissues. These messages are predominantly transmitted through the PPARs, which serve as a network of lipid sensors throughout the body (Guri et al. 2006).

Understanding how and why nutrients and their metabolites exert their physiological effects are the key goals in the study of molecular nutrition. By understanding the mechanisms and tissue-specific affects of endogenous PPAR ligands and expanding our knowledge of the body's integrated homeostatic system, we may increase the successful design of safe and effective preventive and therapeutic interventions to diminish obesity-related diseases.

NRs regulate the expression of a prescribed set of target genes. In the nucleus, PPARs form heterodimers with retinoid X receptor-α (RXRα) which bind to DNA with corepressor molecules. Upon ligand activation, PPARs undergo conformational changes that facilitate the dissociation of the corepressor molecules and stimulate the recruitment of TFs, including coactivators and coactivator-associated proteins (Yu and Reddy 2007). Coactivators may influence the function of many regulators and thereby affect the transcription of many genes (Viswakarma et al. 2010). Some of the coactivators, such as PPAR-binding protein (PPARBP), thyroid hormone receptor-associated protein 220 (TRAP220), and mediator complex subunit 1 (MED1), may exert a broader influence on the functions of several NRs and their target genes.

LXR exists in two isoforms, LXRα and LXRβ, both of which serve as sensors for oxysterol (hydroxysterol) and are essential for whole body cholesterol homeostasis (Tontonoz and Mangelsdorf 2003; Kalanny and Mangelsdorf 2006). The LXRs form heterodimers with the RXRs and as such can regulate gene expression upon binding of hydroxysterols, for example, 22R-hydroxycholesterol, becoming important in the regulation of whole body cholesterol levels.

LXRs and FXRs also work together to balance the metabolic pathways for sterols and bile acids, respectively. They induce transcriptional responses that regulate cholesterol and bile acid metabolism (Kalanny and Mangelsdorf 2006). LXRs also permit the efficient storage of carbohydrate- and fat-derived energy, whereas FXR activation results in an overall decrease in triglyceride levels and modulation of glucose metabolism. In the liver, LXRs mediate cholesterol metabolism by inducing the expression of several genes involved in cholesterol synthesis, such as the sterol responsive element binding protein-1c (SREBP-1c) (Cha and Repa 2007). LXRs also act as sensors for glucose levels, and after binding glucose they activate the gene expression of the carbohydrate-response element-binding protein, ChREBP (Mitro et al. 2007; Commerford et al. 2007). In addition, the

LXR receptors provide reciprocal regulation of inflammation and lipid metabolism (Joseph et al. 2003). The immune response depends on LXR-dependent gene expression (Joseph et al. 2004). LXR activation extends their regulatory function to participate in macrophage survival as well as enabling apoptotic cells to control their own clearance and immune tolerance (A-Gonzales et al. 2009).

The FXRs (FXRα and FXRβ) form heterodimers with members of the RXR family as well. Once the FXR-RXR heterodimers form, the complex binds to specific RE nucleotide sequences in the promoter region, regulating the expression of genes involved in bile acid, cholesterol, lipid, and glucose metabolism (Wang et al. 2008). FXRs serve as receptors for bile acids providing the primary mechanism by which bile acids negatively regulate their own expression. FXRs also bind polyunsaturated fatty acids, such as the ω-3 polyunsaturated fatty acids docosahexaenoic acid (DHA) and α-linolenic acid (ALA). Most recently, FXR has been shown to bind androsterone, the androgen hormone derived via testosterone metabolism. In addition to the NRs discussed here, additional members are being identified all the time, such as the estrogen-related receptors (ERRβ and ERRγ) (Chang et al. 2008), the retinoid-related orphan receptor (RORα), and the constitutive androstane receptor (CAR).

One major target of FXR is the small heterodimer partner (SHP) protein, which functions to compete with coactivators and diminish their transcriptional repressor function (Wang et al. 2006). The protein has been shown to interact with retinoid and thyroid hormone receptors, inhibiting their ligand-dependent transcriptional activation. SHP represses the expression of the cholesterol 7-hydroxylase gene (CYP7A1), which is the rate-limiting enzyme in the synthesis of bile acids from cholesterol. Uniquely, SHP lacks a DNA binding domain. Thus, it is neither a TF nor an NR, but due to a high sequence homology with other NR family members it remains classified as an NR. The main function of SHP appears to be repression of other NRs through a protein–protein binding association that produces a non-functional heterodimer.

## BIOACTIVES AND MOLECULAR TARGETS

### FUNCTIONAL FOODS

The amount and composition of food consumed at various stages of life may impact the expression of certain diseases. As a complex mixture of chemicals, food has been selected and cultivated to provide essential nutrients and requisite calories, and only by fortuitous events has it provided other physiologically active constituents needed for health and enhanced longevity.

Food is more plentiful now than at any other time in history, and the nutritional value of food has been elevated dramatically by agriculture selection and food processing. Our food supply is diverse, nutritious, safe, and available out of season, and today fresh foods are available from around the world (Gravani 2010). The food industry has met these challenges with consistent success through technology, and can use emerging technologies to further develop foods to meet more specialized nutritional demands through individual diets.

A new health paradigm has been discussed for almost 35 years now in which the goal is to optimize nutrition, inclusive of the positive health benefits of food components beyond the basic nutrients. Epidemiologic evidence has repeatedly correlated the positive effects of fruits and vegetables and grains with a lower incidence of cancer and coronary heart disease and other diseases; however, the correlations do not always agree solely with nutrient content. Non-nutrient constituents have been identified as having beneficial, physiological effects that may retard or prevent disease. Foods that combine these qualities have been called functional foods (Clydesdale 2004; Milner 2000). Once the mechanisms by which nutrients, and non-nutrient constituents, function physiologically are better understood, food scientists will be able to design food products for a healthier diet. Clydesdale (1997) proposed scientific criteria to establish health claims that required identification of function and possible efficacy.

Functional foods contain physiologically active components which may improve health and prevent disease beyond the intake of essential daily nutrients (Thomas and Earl 1994). Active components of plants include soy phytoestrogens (genistein, daitzen), tea and cocoa catechins (epigallocatechins, epigallocatechin-3-gallate [EGCG]), vegetable oil tocopherols and tocotrienols, phenolic antimicrobials, organosulfur antitumor agents (diallyl disulfide, sulforaphane), carotenoid (lycopene, lutein, β-carotene) ROS quenchers, and fiber and prebiotic (fructo-oligosaccharides) enhancers of the intestinal microbiome and immune function. Fish oil omega-3 fatty acids and dairy product conjugated linoleic acids alter triglyceride and cholesterol metabolism (Bidlack and Wang 1999; Clydesdale 2004).

The acceptance of functional foods in the marketplace supports the widespread belief that healthy eating is a good way to reduce the risk of chronic illness. Development of functional foods and their popularity has occurred in part due to the aging population and increasing health care costs. Functional foods contain nutrients and other healthful components that have been part of the diet of large populations for many centuries, and are for the most part GRAS (generally regarded as safe). It remains important that functional foods remain safe for everyone and under all conditions regardless of the amount consumed. In a whole food, bioactive food components act with other dietary elements, enhancing each other's digestion, absorption, distribution, and activity to support overall health (Bidlack and Wang 1999; Land 2010).

Better health through improved diet and nutrition can increase the quality of life, enhance productivity, maximize learning potential, and enhance the state of mind, in addition to providing its nutritional value for each individual. In so doing, functional foods can contribute to reduction of health care costs by preventing or delaying the onset of chronic disease (Clydesdale et al. 2005). Perhaps the food industry will be given better recognition for its contributions to a healthier, safer diet for the mutual benefit of everyone.

## DIETARY BIOACTIVE COMPONENTS

By definition, functional foods benefit health beyond the effect of nutrients alone; yet few of these bioactive agents have been established for specific health benefits when evaluated by population studies (Ferguson 2009). Thousands of natural polyphenolic compounds have been identified in plant foods (Shahidi and Naczk 1995). Polyphenols are almost ubiquitous in the plant foods and beverages consumed daily (Bravo 1998). The content of polyphenolic compounds contribute to the sensory qualities, astringency and bitterness, characteristic of many foods and beverages.

One-third of these polyphenolics are phenolic acids while the remaining two-thirds are flavonoids, composed primarily of flavanols such as the catechins, proanthocyanidins, anthocyanins, and their oxidation products (Table 5.2, Figure 5.4). Many of these compounds have been characterized for bioactivity in cellular processes leading to improved health. The amount of these compounds varies greatly between cultivars of the same species (Herrmann 1988). In addition, the highest concentration of flavones and flavonol glycosides was dependent on the amount of light exposure on the plants, indicating the concentration of these compounds was greatest in the leaves (Herrmann 1988). The amount and type of polyphenols depends on plant genetics, environmental conditions, as well as degree of ripeness, variety, processing, and storage (Porter 1989; Peleg et al. 1991). Thus, the polyphenolic content can vary by several orders of magnitude (Peleg et al. 1991; Hertog et al. 1992).

Dietary sources include fruits and vegetables, juices, wine, tea, coffee, chocolate, and beer (Packer 2000; Daayf and Lattanzio 2008). The total polyphenolic content of selected foods (mg/100 g dry matter) and beverages (mg/L) include the following selected examples from Shahidi and Naczk (1995) and Bravo (1998): barley (1200 mg/100 g dm), onions (100–2025 mg/100 g dm) and celery (94 mg/100 g dm), apples (27–298 mg/100 g dm), blueberries (135–280 mg/100 g dm), cranberries (77–247 mg/100 g dm), grapes (50–490 mg/100 g dm), raspberries (37–429 mg/100 g dm), strawberries (38–218 mg/100 g dm), and tomatoes (85–130 mg/100 g dm). Beverages include orange juice (370–7100 mg/L), tea (750–1000 mg/L), coffee (1260–3300 mg/L) and wine, white (200–300 mg/L) and red (1000–4000 mg/L).

**TABLE 5.2**

**Bioactive Food Components: Flavonoids and Phenolic Derivatives (Partial List)**

| Category[1] | Name Aglycone[1,2,3] (Acylated Derivative) | Food Sources[2,3,4,5] | Food Content Range[2] (mg/100 g or mg/200 ml) |
|---|---|---|---|
| | | **I. Flavonoids** | |
| A. Flavonols | | Onions | 35–120 mg/100 g |
| | | Broccoli | 4–10 mg/100 g |
| | | Blueberries | 3–16 mg/100 g |
| | | Apricots | 2.5–5 mg/100 g |
| | | Apples | 2–4 mg/100 g |
| | | Green tea | 4–7 mg/200 ml |
| | | Black tea | 6–9 mg/200 ml |
| | Quercetin (quercetin-3-O-glucoside; quercetin-4-O-glucoside) | Red and yellow onions, beans, broccoli, cabbage, cauliflower, green and black tea, wine, apples, blueberries, cranberries | |
| | Kaempferol | Strawberries, cranberries, apples, peas, grapefruit, tomatoes, brassica, chives | |
| | Myricetin | Grapes, walnuts | |
| B. Flavanones | | Orange juice | 40–140 mg/200 ml |
| | | Grapefruit juice | 20–130 mg/200 ml |
| | | Lemon juice | 10–60 mg/200 ml |
| | Naringenin (naringin; naringenin-7-O-neohesperidoside) | Citrus fruits Grapefruit Tomatoes | |
| | Hesperetin (hesperidin; hesperetin-7-O-rutinoside) | Citrus fruits Oranges (rind) | |
| | Eriodictyol | Citrus fruits (lemon) | |
| C. Flavones | | Parsley | 24–185 mg/100 g |
| | | Celery | 2–14 mg/100 g |
| | | Capsicum pepper | 0.5–1 mg/100 g |
| | Luteolin | Celery, carrots, green peppers, broccoli, legumes, olives, olive oil, rosemary, oregano, thyme, cherries | |
| | Apigenin | Celery, parsley, vegetables, plant seeds, chamomile | |
| | Tangeritin (polymethoxylated flavone) | Tangerines, citrus skins | |
| | Nobiletin (polymethoxylated flavone) | Oranges Citrus skins | |
| D. Isoflavones | | Soy flour | 80–180 mg/100 g |
| | | Soybeans | 20–90 mg/100 g |
| | | Miso | 25–90 mg/100 g |
| | | Tofu | 8–70 mg/100 g |
| | | Soy milk | 6–35 mg/200 ml |
| | Genistein | Soy beans, soy foods, alfalfa sprouts, red clover, chickpeas, legumes | |

*(Continued)*

**TABLE 5.2** (*Continued*)
**Bioactive Food Components: Flavonoids and Phenolic Derivatives (Partial List)**

| Category[1] | Name Aglycone[1,2,3] (Acylated Derivative) | Food Sources[2,3,4,5] | Food Content Range[2] (mg/100 g or mg/200 ml) |
|---|---|---|---|
| | Daidzein | Soy, alfalfa sprouts, red clover, chickpeas, legumes | |
| | Glycitein | Soy | |
| *E. Flavan-3-ols* | | Apples | 2–12 mg/100 g |
| | | Cherries | 5–22 mg/100 g |
| | | Chocolate | 46–61 mg/100 g |
| | | Red wine | 16–60 mg/200 ml |
| | | Green tea | 0–160 mg/200 ml |
| | | Black tea | 12–100 mg/200 ml |
| | (+)Catechin (−)Epicatechin (EGC, EGCG, EC3G, GC) | White tea, green tea, black tea, red wine, beer, apple juice, grapes, strawberries, cocoa, dark chocolate, black eyed peas | |
| | Proanthocyanidins | Grapes, peaches, apples, pears, berries, and dark chocolate | |
| *F. Anthocyanins* | | Blackberries | 100–400 mg/100 g |
| | | Blueberries | 25–500 mg/100 g |
| | | Cherries | 35–450 mg/100 g |
| | | Strawberries | 15–75 mg/100 g |
| | | Plums | 2–25 mg/100 g |
| | | Red cabbage | 25 mg/100 g |
| | Anthocyanidins | Red wine, red, purple, or blue fruits and vegetables | |
| | Pelargonidin (Pelargonidin-3-O-glucoside) | Bilberries, raspberries, strawberries | |
| | Cyanidin (cyanidin-3-rutinoside-glucoside-galactoside) | Red apples, pears, bilberries, blackberries, blueberries, cherries, cranberries, peaches, plums, loganberries, purple cabbage, purple/black carrots, cocoa | |
| | Peonidin (peonidin-3-caffeoysophoroside-5-glucoside) | Bilberries, blueberry, cherries, cranberries, peaches, purple sweet potatoes | |
| | Delphinidin (delphinidin-3-O-β-D-glucopyranoside) | Bilberries, blueberries | |
| | Nasunin (delphinidin-3-(p-courmaroyl rutinoside)-5-glucoside) | Eggplant (skin) | |

**II. Phenolic Acids**

| | | | |
|---|---|---|---|
| | | Blueberries | 200 mg/100 g |
| | | Kiwis | 60–100 mg/100 g |
| | | Cherries | 18–115 mg/100 g |
| | | Plums | 14–115 mg/100 g |
| | | Coffee | 70–350 mg/200 ml |

**TABLE 5.2    (Continued)**

**Bioactive Food Components: Flavonoids and Phenolic Derivatives (Partial List)**

| Category[1] | Name Aglycone[1,2,3] (Acylated Derivative) | Food Sources[2,3,4,5] | Food Content Range[2] (mg/100 g or mg/200 ml) |
|---|---|---|---|
| **A.** *Hydroxycinnamic Acid* | Caffeic acid | Artichokes, beans, thyme, oregano, basil, pears, apples, coffee | |
| | p-Coumeric acid | Plums | |
| | Ferulic acid (curcumin: dimeric ferulic acid) | Turmeric, mustard, curry | |
| | Chlorogenic acid (Caffeoylquinic acid) | Strawberries, pineapples, blueberries, sunflower seeds, pumpkin seeds, poppy seeds | |
| **B.** *Hydroxybenozic Acid* | | Purple carrots | 54 mg/100 g |
| | | Blackberries | 8–27 mg/100 g |
| | | Raspberries | 6–10 mg/100 g |
| | | Strawberries | 2–9 mg/100 g |
| | Vanillic acid | Vanilla beans, cloves, fruits, vegetables, cereals, spices, oil seeds, tree nuts, coffee | |
| **III. Tannins** | | | |
| **A. Derived Tannins** | Theaflavins (Theaflavin-3-gallate) | Black tea | |
| **B. Hydrolysable Tannins** | Gallic acid esters | Strawberries, rhubarb, grapes, wine, mangoes, guavas, green tea, black tea | |
| | Ellagic acid esters | Walnuts, raspberries, strawberries, cranberries, blackberries, guavas, pomegranates, persimmons, avocados | |
| | Punicalagin Ellagitannin | Pomegranates | |
| **IV. Stilbenes** | | | |
| | Resveratrol | Wine, grape skins, seeds, nuts, peanuts | |
| **V. Other** | | | |
| | Coumestrol | Legumes, soybeans, spinach, brussel sprouts, citrus, fruits, maize | |
| | Capsaicin | Red chili peppers | |
| | Gingerol | Ginger | |

[1] Liu, R. H. 2004. *J Nutr* 134:3479S–3485S; [2]Manach, C. et al. 2004. *Am J Clin Nutr* 79(5):727–47; [3]Crozier, A., I. B. Jaganath, and M. N. Clifford. 2009. *Nat Prod Rep* 26:1001–1043; [4]Holst, B. and G. Williamson. 2008. *Current Opin Biotechnol* 19(2):73–82; [5]Daayf, F. and V. Lattanzio. 2008. *Recent Advances in Polyphenol Research*. Wiley-Blackwell.

[2] The data for the food content range is adapted from Manach, C. et al. 2004. *Am J Clin Nutr* 79(5):727–47. (With permission.)

Most of the polyphenols identified in these samples were phenolic acids and flavonoids including anthocyanins, procyanidins, flavanones, flavonols, and so on, and to lesser extent the tannins. The most recent estimates of daily prothroanthocyanins from major food sources in the US diet was reported by Wang et al. (2011).

Dietary consumption of polyphenols through food intake has been estimated only a few times with limited agreement. Kuhnau (1976) estimated the average daily intake of dietary flavonoids in the United States to be about 1 g/day, while the specific intake of the flavones and flavonols was determined to be only 115 mg/day. Hertog et al. (1993) also estimated the intake of flavones and flavonols from the Dutch diet and determined only 23 mg/day, and Leh and Justesen (1997) in the diet of the Danes determined an intake of 28 mg/day. The earlier values may have been overestimated due to the analytical methods used (Ho et al. 1992). In addition, the more recent results specifically

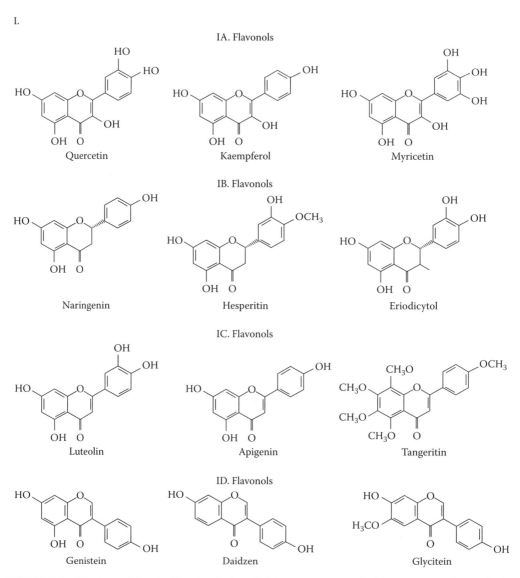

**FIGURE 5.4** Structure of the significant polyphenolic bioactive compounds. The structures of the polyphenolic bioactive compounds are grouped under their appropriate classification. I. Flavonoids, II. phenolic acids, III. tannins, IV. stilbenes, and V. other.

IE. Flavan-3-ols

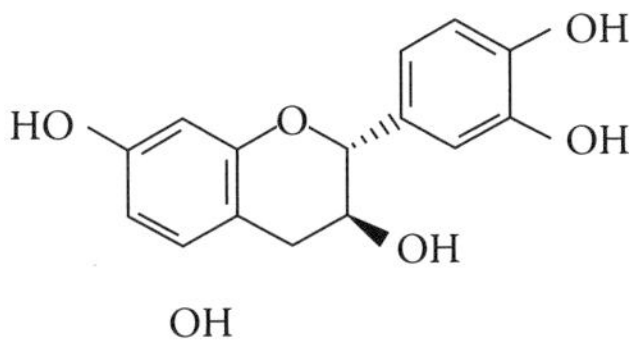

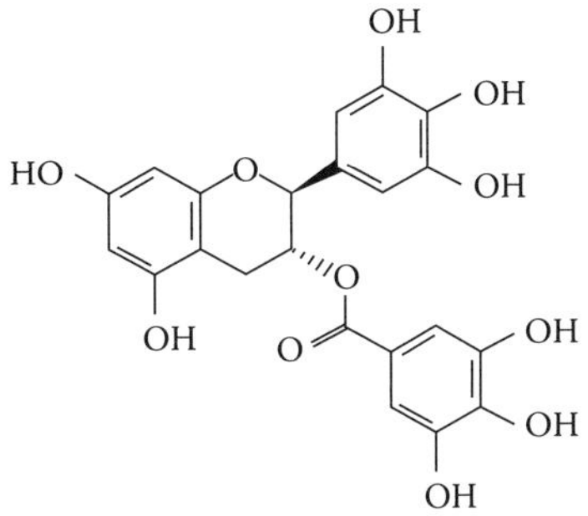

(+)Catechins

IF. Anthocyanins

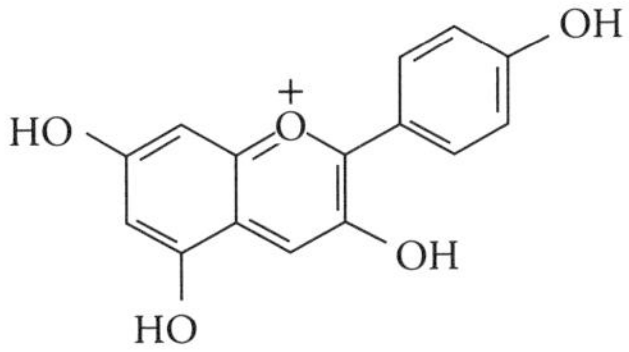

Pelargonidin 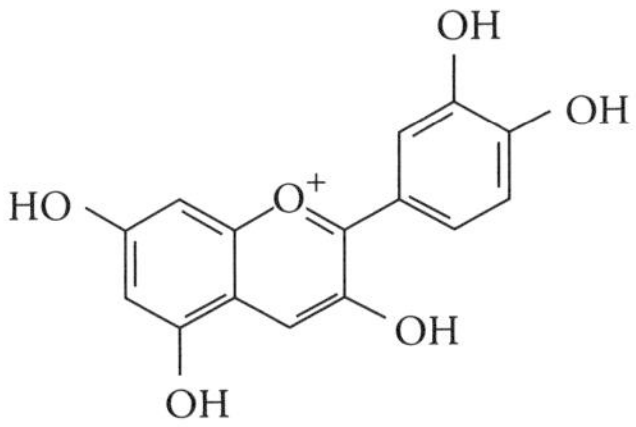          Cyanidin 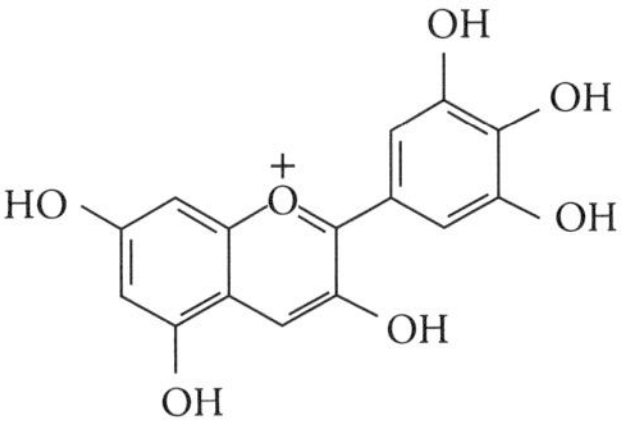          Delphinidin

II.

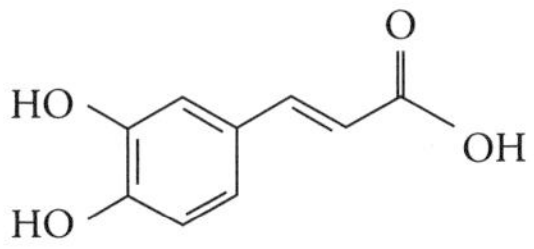

Caffeic acid 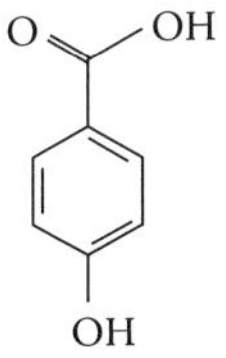          p-Coumeric 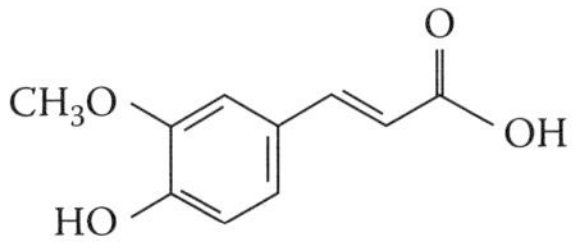          Ferulic acid

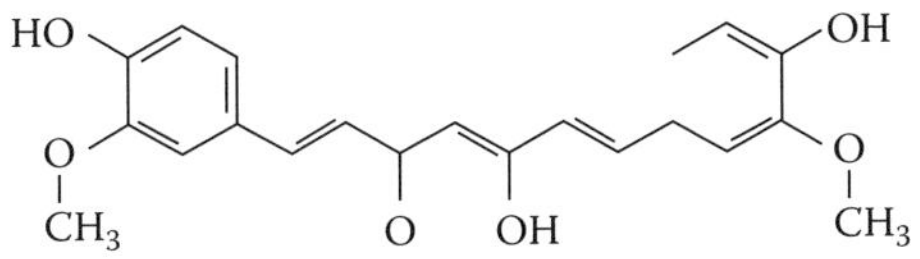

Curcumin, dimeric ferulic acid 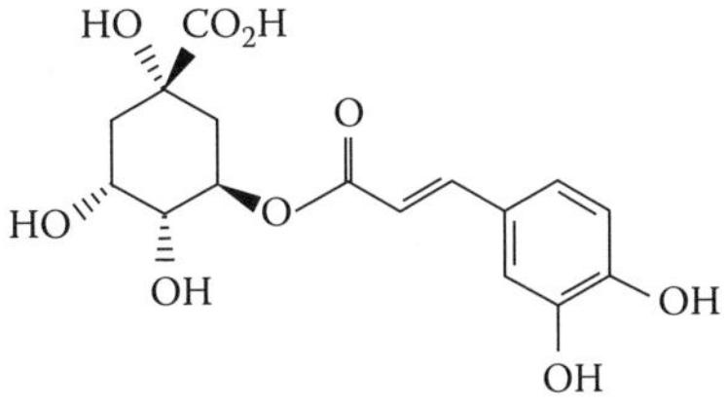          Chlorogenic acid

**FIGURE 5.4    (Continued)**

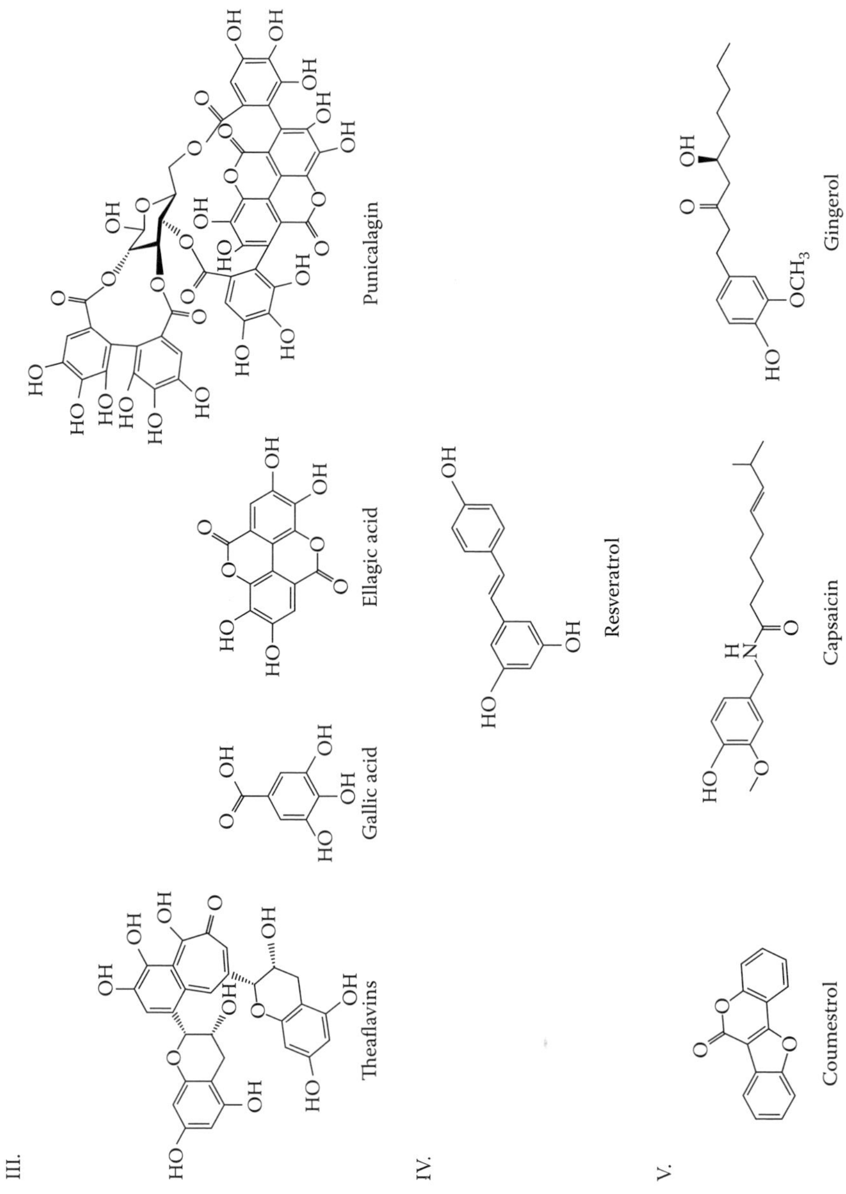

FIGURE 5.4 (Continued)

analyzed only selected types of flavonoids, e.g. Wang et al (2011). The lack of agreement on the appropriate method to analyze the different types or families of polyphenolics hinders comparison of published data which has lead to contradictory findings (Shahidi and Naczk 1995; Leh and Justesen 1998).

## BIOAVAILABILITY

Bioavailability has numerous variables that determine the final concentration of the polyphenolics available to the target cells. Factors that affect bioavailability include the structure of the food matrix, the processing or cooking method used, the chemical structure of the bioactive compound, the amount that is present, and the interaction between coingested compounds including nutrients and non-nutrient food components (Clevidence et al. 2004).

Bioavailability depends on the intestinal absorption, the influence of chemical structure (glycosylation, esterification, and polymerization), and enterocyte metabolism. In addition, chemicals released from the enterocyte might be further metabolized in the liver. Other factors include plasma kinetics, the nature of circulating metabolites, binding to albumin, cellular uptake, intracellular metabolism, accumulation in tissues, and biliary and urinary excretion. The challenge is to integrate the related variables to provide health effects, since each compound may be affected differentially at each step. The bioavailability of the bioactive food components consumed in our diet determines the amount of each agent that enters the circulation and thereby actually reaches the cell surface in the body (Crozier et al. 2009). This is the point at which the bioactive agent either binds to a surface receptor to initiate cell signal transduction or it penetrates the cell membrane to reach an NR, which triggers regulation of gene expression.

Variability of intake has to do with the dose of the exposure and the type of polyphenols that are absorbed. The nutritional significance and potential systemic effects will depend on the behavior of the flavonoids and their polyphenolic glycosides in the digestive tract. There is very little known about the absorption of polyphenols in the gastrointestinal tract (GI) in animals or humans, whether they are metabolized or retained in the body, and of what significance the compounds have at the circulating concentrations (Bravo 1998).

Intestinal absorption of monomer flavonoids has been estimated to be 10–100 mg for each compound consumed from food, which results in plasma concentrations that rarely exceed 1μM, but if consumed in a supplement form, the plasma level could be greater, reaching 5uM (Kuhnau 1976; Aziz et al. 1998). In the colon, the aglycones are absorbed by the gut epithelium and methylated or conjugated with either glucuronic acid or sulfate in the liver. The metabolic derivatives of catechin, quercetin, and genistein have been detected in the plasma and they have been excreted into the urine or bile. In the latter case these conjugates can not be hydrolyzed in the small intestine but could be by the colonic bacteria, allowing the aglycones to reenter the body by the enterohepatic circulation (Hollman et al. 1995; Lee et al. 1995). Less is known about the efficiency of the absorption or their conjugate derivatives while in the body.

Absorption of polyphenolics is determined by the basic chemical structure of the bioactive compound, and by the presence of specific glycosylation/acylation groups, conjugation with other phenolics, molecular size, and their solubility. Both *in vitro* and *in vivo* studies have examined the effects of different chemical structures and solubility to determine susceptibility to digestion and absorption from the GI. Absorption has been evaluated based on extractable low and intermediate molecular mass. Phenolics that can be extracted with water, methanol, or aqueous acetone include hydrolysable tannins and proanthocyanidins, whereas nonextractable polyphenols are high molecular weight compounds or phenolics that are bound to dietary fiber or proteins that remain insoluble in the usual solvents (Bravo et al. 1994). Nonextractable polyphenols were extensively recovered in the feces, indicating a lack of digestion and absorption (Saura-Calixto and Bravo 1996). Similarly, in rat studies, animals were fed apple pulp or grape pomace. The results indicated that 95%–97% of the extractable polyphenols were absorbed, while only 10%–30% of the nonextractble polyphenols were absorbed. The latter compounds are metabolized in the large intestine, enabling release and absorption of smaller phenolic components.

Aglycones and free simple phenolic compounds, flavonoids (quercetin, genestein), and phenolic compounds, can be absorbed through the small intestinal mucosa (Manach et al. 1997; King et al. 1996). Free phenolics (cinnamic acid and derivatives using p-coumaric, ferrulic, caffeic acids) are absorbed through the intestinal tract in both *in vivo* and *in vitro* rat experiments (Wolffram et al. 1995). Glycosides must be hydrolyzed to their aglycone forms before being absorbed; yet mammals lack the appropriate β-glycosidases and rely on partial hydrolysis by bacteria in the large intestine and absorbed as the aglycones. In a model study, quercetin glycosides are absorbed in the intestine near the terminal ileum, but most pass into the large intestine to be hydrolyzed by cecal microflora yielding free aglycones. Indeed, the flavonoid glycosides were excreted intact in the feces of germ-free rats (Griffiths and Barrow 1972; Hollman et al. 1995).

In the colon, the aglycones are absorbed by the gut epithelium and methylated, then carried to the liver and conjugated with glucuronic acid or sulfate. Conjugated and methylated derivatives of catechin, quercetin, and genistein are detected in the plasma and are excreted into the urine or bile. In the later case these conjugates may be hydrolyzed by colonic bacteria, again allowing the aglycones to reenter the enterohepatic circulation (Hollman et al. 1995; Lee et al. 1995). Less is known about the efficiency of the absorption or the retention of the compounds or their conjugate derivatives in the body. Individual variability may result from differences in intestinal transit time, the integrity of the mucosal lining, the rate of gastric emptying, and the metabolic degradation of the conjugates, albumin, or other protein binding in blood or tissues.

Absorption studies indicate that dimeric or trimeric forms of flavonols can cross the epithelium membrane of the small intestine (Deprez et al. 2001). Strawberry anthocyanins are absorbed and are excreted almost linearly with an increasing dose (Carkeet et al. 2008). For purple cabbage and purple carrots, the acylated anthocyanins were absorbed much less than the nonacylated forms, as noted by lower plasma and urine levels (Charron et al. 2007; Kurilich et al. 2005). The oligomeric forms of proanthocyanidins are least likely to be absorbed. Thus, the bioactivity of proanthocyanidin may be restricted to the intestinal cells directly.

Hydroxycinnamic acids are found in a wide variety of foods, often at high concentrations and in esterified forms, for example caffeic acids, a type of phenolic acids, are esterified with sugar, organic acids and lipids. The ester form decreases their intestinal absorption. With a lack of an efficient esterase, they are poorly hydrolyzed until they reach the colon bacteria. Ellagitannins are a family of bioactive polyphenols found in fruits and nuts such as raspberries, strawberries, and pomegranates, and almonds and walnuts (Heber 2008). Punigcalagin is the largest molecular weight polyphenol known. They are not absorbed directly, but can be hydrolysed in the colon to release ellagic acid, which is absorbed. The gut flora further metabolize ellagic acid into urolithins, which are also bioactive metabolites. They are conjugated in the liver and eliminated in the urine.

The inflammatory pathways in the prostate and other cancer cells elicit multiple targets for cancer development. The metabolites of pomegranate juice inhibit prostate cancer cell growth by inhition of NF-κB activation. In clinical studies following therapeutic radiation and chemotherapy, prostate cancer patients were given pomegranate juice over a two-year period. These patients saw an 85% decrease in the elevation of prostate cancer specific antigen during the study (Heber 2008).

Human studies have similarly indicated partial absorption of polyphenols. In ileostomy volunteers, quercetin absorption varied between 24%–52% of the ingested aglycone and glycoside respectively (Hollman et al. 1995). Soybean isoflavones were absorbed in healthy subjects varying from 9%–21%, depending on the isoflavone (Xu et al. 1994). Plasma concentrations of total catechins following consumption of black tea and green tea were determined to be 0.17 and 0.55 umol/L, respectively, which was similar to that reported by Lee et al. (1995). The peak level of quercetin consumed in onions was reached after 3.3 hours and the half-life for elimination was 16.8 hours (Hollman et al. 1996).

Recent experimental and clinical studies have linked ROS to major pathologies related to diseases. Reduction of oxidative stress by flavonoids has been-described as the protective mechanism of these compounds (Molavi and Mehta 2004). In addition, ROS enhance activation of nuclear factor-κB (NF-κB), activator protein 1 (AP-1) and nuclear histone acetylation and deacetylation (Biesalski 2007). The most common activity of polyphenols is their antioxidative action and modulating effects -on NF-κB chromatin structure, glutathione synthesis, and activation of nuclear redox factor (Nrf2) and glutathione peroxidase. Inflammatory genes in macrophages and different target tissues are regulated (Biesalski 2007). They may also modulate actions of protein kinase and lipid kinase signaling pathways (Angeloni et al. 2008). Among the flavonoids tested, quercetin, kaempferol, malvidin, peonidin, daidzein, and genistein all demonstrated an inverse relationship to the serum C-reactive protein concentrations (Chun et al. 2008). Thus, a close association appears to exist with flavonoid rich foods, decreasing inflammation mediated chronic disease.

## Lumen Perfusion

Using a more focused experimental design, lumen perfusion was used to examine the fate of six different flavonoids (quercetin, kaempferol, luteolin, eriodictyol, genestein, catechin) in the splanchic area of intestinal absorption (Crespy et al. 2003). Following individual aglycone perfusion in the rat jejunum, net transfer across the brush border was used to determine net absorption. The percent recovery was analyzed and established for the nonabsorbed aglycone in the lumen. The uptake of the aglycones by the enterocyte indicated some were metabolized and the conjugated forms were partially secreted back into the intestine, and the remaining aglycone passed through to the mesenteric vein. The circulation delivered a signifiant portion of the absorbed polyphenol to the peripheral tissues (9%–49%), including the hepatocyte which secretes conjugated metabolites back to the intestine through the biliary system. Because of these clearing mechanisms, the intestine was determined to eliminate 51%–91% of the dose given.

Of the six polyphenolics examined, three clear differences in absorption and metabolism can be seen by comparing three of the phenolic compounds, quercetin, kaempferol, and catechin. First, one-third of the quercetin was not absorbed, and 52% was conjugated by the enterocyte and returned to the intestine. Of the 15% that was absorbed, only 9% went to peripheral tissues while 6% returned through hepatocyte conjugation to the intestine. Second, one-fifth of the Kaempferol was not absorbed, and another 20% was conjugated in the enterocyte and released into the intestine. Thus, 58% was absorbed, of which 49% went to the peripheral tissues with only 9% conjugated through the hepatocyte and delivered back to the lumen. Third, the catechins were poorly absorbed. Only 35% was absorbed, none was conjugated in the enterocyte and only 1% was conjugated in the hepatocyte, leaving 34% distributed in the peripheral tissues. Thus, for the examples provided, the absorption and distribution to the peripheral tissues ranged from a low of 9% for quercetin, to a high of 49% for kaempferol. The other three phenolics were similar at 20%–34%. Although the results of these perfusion experiments may still differ from results obtained from absorption of phenols from food sources, since they could be enhanced or diminished by other food components, this research study is an impressive assessment of polyphenol absorption and retention that should continue to serve as a model for evaluation of bioavailability of other bioactive components (Crespy et al. 2003).

## Absorption and Distribution

Numerous articles on the bioavailability of various dietary polyphenolic compounds have been published recently by Williamson and colleagues (Holst and Williamson 2004, 2008; Scholz and Willliamson 2007; Williamson and Manach 2005; Manach et al. 2004, 2005a,b). The relative bioavailability was determined for quercetin and the quercetin-4′-O-glucoside from various foods (Hollman et al. 1997; Aziz et al. 1998). The quercetin partition coefficient (log octanol/water concentration ratio) was 1.2 indicating a hydrophobic nature that allows membrane passage, while the glycoside partition coefficient decreased to 0.37 indicating a greater hydrophilicity and less membrane penetration (Nakagawa et al. 1997).

The plasma level of ingested quercetin reached maximum peak levels at 9 hours. High-affinity binding to albumin may have slowed elimination and allowed quercetin to accumulate. Due to the low dietary intake at any time (<100 mg/serving) and enterocyte and liver conjugation, the polyphenols were mostly conjugated and very little free aglycones exist (Scalbert and Williamson 2000). Plasma quercetin was recovered as conjugated products, indicating rapid conjugation forming quercetin glucuronides, sulfates, and methyl groups. The formation of the conjugates can dramatically alter the biological properties of the circulating metabolites. To date other authors have simply reported total phenols, providing an incomplete picture.

The catechins are acylated with gallic acid to form (-) epigallocatechin-3-gallate and (-) epigallocatechin. The acylated compounds remain hydrophobic and pass through cellular membranes, yet the plasma level reflects only 0.2%–2.0% of the ingested dose of the catechins (Nakagawa et al. 1997; Lee et al. 1995). Assessment of ingested polyphenols have indicated that the majority (75%–99%) was not excreted in the urine (only 1%–25% was excreted); however, individual polyphenol variability was high (5%–57%).

Engler et al. (2004) identified dark chocolate as a rich source of the flavonoids, epicatechins, and procyanidins. The authors observed cardioprotective effects including antioxidant properties, inhibition of platelet aggregation, and activation of endothelial nitric oxide synthetase. Similarly, Rein et al. (2000) and Wang et al. (2000) evaluated the consumption of chocolate and the absorption of the polyphenols. The plasma epicatechin levels rose and the oxidative status (lipid oxidation and platelet activity) of plasma decreased proportionately to the increase in catechin levels.

A random double-blind, placebo-controlled design was conducted over a 2-week period in 21 healthy adult subjects, randomly assigned daily intake of high flavonoid (213 mg procyanidins, 46 mg epicatechin) chocolate bars or low-flavonoid (trace only of proanthocyanidins and epicatechins) dark chocolate bars (46 g of chocolate for each group). At 2 weeks the high-flavonoid group had elevated plasma epicatechin concentrations (204 ± 18.5 nmol/L) but the low-flavonoid group did not (17.5 ± 9 nmol/L). Improved endothelial function was associated with an increase in epicatechin concentrations in healthy adults. Thus, a small repetitive dose of phenolics indicated an increased delivery and an increase in circulating plasma levels of the polyphenolic agents (Engler et al. 2004).

## FLAVONOID POLYPHENOLS

The flavonoids are a group of several thousand naturally occurring plant phenolic compounds. They are not produced in the human body and must therefore be acquired from foods in our diet or with use of dietary supplements. The flavonoids are polyphenoic compounds categorized by structure and function as the flavonols, flavones, isoflavones, flavanones, flavanonols, flavanols (flavan-3-ols), and anthocyanidins (Table 5.2, Figure 5.4). The flavonoid compounds are found in plants either as free or as conjugates forming glycosidic linkages to sugar molecules (glucose, galactose, arabinose, glucuronic acid, and rhamnose). The basic flavonoid skeleton can have numerous substituents, for example, hydroxyl groups and sugar attachments are common. The majority of the flavonoids exist as natural glycosides. The sugars and hydroxyl groups make the flavonoids more water soluble, while substituents such as methyl groups and isopentyl groups make them more lipophilic (Crozier et al. 2009).

### Flavonols

Flavonols are the most widespread flavonoids found in foods. The three main flavonols are quercetin, kaempferol, and myricetin, and most exist as their O-glycosides (Table 5.2, part IA). The different aglycones are limited, but there are more than 200 different sugar conjugates for kaempferol alone. The amounts of the flavonols found in fruits, vegetables, and beverages varies depending on the varieties, seasonal differences, and effects of processing (Crozier et al. 1997).

These compounds are normally present at low concentrations (approximately 15–30 mg/kg fresh wt), mostly in their glycosylated forms with glucose or rhamnose and with lesser levels of other sugars (Manach et al. 2004). The richest dietary sources are found in onions, broccoli, tomato and kale, blueberries, cherries, and apples, red wine and tea (45 mg/L), and spices like cumin and caraway. Fruit sources frequently contain 5–10 different flavonol glucosides (Macheix et al. 1990).

Quercetin is a very potent polyphenolic antioxidant. Quercetin is found in numerous foods such as the brassica vegetables (e.g., broccoli, cauliflower, cabbage, brussel sprouts, bok choy), apples, berries, red onions, citrus fruits, and green tea as well as many seeds, nuts, leaves, barks, flowers, and capers on the order of 1–2 mg/g of plant material. Onions provide a range of flavonols that are comparatively restricted to quercetin-4'-O-glucoside and quercetin-3,4-O-diglucoside (Mullen et al. 2004). Quercetin is available in highly purified extracts for sale as a dietary supplement, which enables consumption of 500 mg per day (equivalent to eating 5 kilograms of apples each day).

The glycosides of quercetin compounds were thought to be poorly absorbed from the GI tract following consumption. However, evidence now indicates that about 50% of quercetin glycosides are absorbed, whereas only 25% of the aglycone form is absorbed (Hollman et al. 1995). The biochemical basis for this absorption difference is believed to be due to the intestinal uptake process that involves a carrier-mediated transport or a coupled deglycosylation transport mechanism. Following uptake by a carrier-mediated process, the quercetin glycosides were hydrolyzed by intracellular glycosidases, releasing the aglycone. The preferential uptake of the glycoside was confirmed in absorption studies using human ileostomy subjects, which eliminated the potential hydrolysis of the glycosides by the microbiota in the colon.

Following absorption, quercetin is metabolized by the small intestine, colon, liver, and kidney. In animal models of quercetin absorption and tissue distribution, the highest concentrations were found in the lung, liver, and kidney. Because the half-life of quercetin in the plasma and tissues is long (ca 28 hours), repeated intake with high-level food sources, such as red onions (2 g/kg) or fruits (10–200 mg/kg) or supplements can lead to accumulation and elevated plasma levels of the flavonols (Hollman et al. 1996). Quercetin's chemical structure provides a strong antioxidant effect and increases the activity of superoxide dismutase, catalase, and glutathione peroxidase, which enhances the celluar oxidative defense mechanisms. One major result is a significant decrease in the oxidation and peroxidation of membrane lipids, thereby preventing cell damage (Boots et al. 2008).

The role of quercetin in the cardiovascular system may result from these protective effects. In animal models, quercetin has also been shown to lower blood pressure and ameliorate hyperglycemia and conditions caused by hyperglycemia. In a trial involving prehypertensive and stage 1 hypertensive patients, the consumption of 730 mg/day of quercetin for 4 weeks led to a reduction in blood pressure, but did not affect the parameters of oxidative stress (Bischoff 2008). Quercetin is active in many metabolic processes related to diseases of aging, such as cardiovascular and neurodegenerative disorders. In addition, quercetin also exhibits anti-inflammatory, antiproliferative, and apoptotic effects both on cells in culture as well as when ingested *in vivo*.

Changes in gene expression were examined by microarray analysis, using cultured rat primary cardiomyocytes treated with quercetin (Angeloni et al. 2008). A strong upregulation of phase two enzymes was noted. Numerous studies have examined the effects of quercetin in the treatment of breast, ovarian, and colon cancers and leukemias. The antitumor properties of quercetin are diverse and include modulation of carcinogen metabolism through inhibition or induction of xenobiotic enzymes, induction of cell cycle arrest, and DNA strand breakage resulting in apoptosis.

The p53 gene produces a protein that regulates the progression of cells through the cell cycle. Although quercetin has little effect on the expression of the normal p53 gene, the mutated form of p53 gene is found in numerous types of cancer cells. In this case, quercetin appears to down regulate the expression of the mutant p53 gene in breast cancer cells to nearly undetectable levels (Kuo et al. 2004). The downregulation arrests the cell cycle prior to cell division.

Additional activities attributed to quercetin include regulation of caspase-3 (apoptosis), telomerase and DNA replication, lymphocyte tyrosine kinase, and other tyrosine kinases and serine/threonine kinases. Quercetin blocks JNK and p38 MAPK related signaling triggered by the oxidant and may regulate expression of apoptotic downstream genes, preventing apoptosis and promoting cell survival (Choi et al. 2005).

## Flavanones

Flavanones are found in tomatoes, mint, and in large amounts in citrus fruit. Naringenin is the most abundant citrus-derived flavanone, while the naringenin chalcone is found in tomato. The chalcone structure lacks the oxygen linkage (-O-) in the second aromatic ring. The main aglycones are naringenin in grapefruit, hesperitin in oranges, and eriodictyol in lemons. The main glycosidic forms include naringin and hesperidin formed from naringen and hesperitin, respectively (Table 5.2, part IB).

The white fleshy pulp on the inner surface of the citrus skin and the membranes that separate the segments have a high flavanone content. Interestingly, only small amounts of these components are transferred to the juice during processing, but are primarily consumed when the whole fruit is eaten without over removal of the inner white pulp (pith) remaining bound to the fruit after peeling (Manach et al. 2004). The flavanones are present in high concentrations in citrus fruits, juice, and fruit segments. The most common glycosides are hesperitin-7-O-rutinoside from bitter orange and naringenin-7-O-neohesperidoside from grapefruit peel, both of which are very bitter (Tomas-Barberian and Clifford 2000; Morand et al. 2011).

Absorption of the flavonoids only occurred in cultured cells having β-glucosidase activity. Two β-glucosidases having activity towards flavonoid glycosides were isolated from human small intestine mucosa (lactase-phlorizin hydrolase, LPH) localized to the apical membrane of small intestinal epithelial cells and cytosolic β-glucosidase (CBG). Thus, the absorption of dietary flavonoid glycosides in humans involves a critical deglycosylation step that is mediated by epithelial β-glucosidases (LPH and CBG). The significant variation in β-glucosidase activity between individuals may be a factor in the variation in flavonoid bioavailability that has been reported (Nemeth et al. 2003).

The consumption of citrus fruit has been associated with lower risk of acute coronary events and strokes (Johnsen et al. 2003; Grassi et al. 2009). Citrus juice consumption has been reported to reduce oxidative DNA damage and improves plasma concentrations of markers of inflammation and oxidative stress. The major flavonoid in oranges, hesperidin, was evaluated in overweight men. Four weeks consuming 500 mL of orange juice per day resulted in a decrease in the diastolic blood pressure (DBP) as well as postprandial increases of endothelium-dependent microvascular reactivity and cardiovascular risk biomarkers (Morand et al. 2011).

Naringenin exerts antitumor, antioxidant, and anti-inflammatory effects. Naringenin is a phyoestrogen and expresses estrogenic activity in cell culture, activating the ERα and ERβ estrogen receptors. In addition, the anti-inflammatory effects of naringenin chalone are produced by the inhibition of the synthesis and release of the proinflammatory cytokines from macrophages (Bodet et al. 2008). Although commercially available dietary supplements containing naringenin are on the market, the evaluation of their bioavailability has yet to be completed.

## Flavones

Flavones occur less frequently than the flavonols in fruits and vegetables. The most common flavones are luteolin, apigenin, and chryslin and their respective glycosides (Table 5.2, part IC). Flavones such as apigenin and luteolin include a wide range of hydroxylation, methylation, O- and C-alkylation, and glycosylation metabolites. Polymethoxylated flavones, such as tangeretin (tangerines) and nobelitin (oranges), occur in the skin of citrus fruit. Most flavones occur as the 7-O-glycosides. Flavones are found in broccoli, celery, legumes, cherries, tea, olives, thyme, parsley, and some herbs (Manach et al. 2004).

## Isoflavones

The flavones and the isoflavones are considered the most bioavailable flavonoids, especially from soy foods and fruit. Genistein is synthesized in plants from naringenin, which is a common precursor in the biosynthesis of phytoalexins and phytoanticipins in legumes, which act as natural antimicrobials synthesized in plants (Naidu 2000). The most common isoflavones include genestein, daidzein, and glycitein (Table 5.2, part ID).

Isoflavones are flavonoids similar in structure to estrogens, but do not have steroidal hormonal activity. They have hydroxyl groups in a configuration similar to the hydroxyl groups on the estradiol molecule so they can bind to estrogen receptors (ERα and ERβ), although at only 0.01% and 7.4% of 17β estradiol binding, respectively. Since they confer pseudohormonal properties they are classified as phytoestrogens. Isoflavones are found almost exclusively in leguminous plants, with the highest levels in soy beans and processed soy foods. The soy isoflavones include three compounds, genistein, daidzein, and glycitein, and their glycosides, as well as coumesterol, which have low estrogenic activity. These compounds can block estrogenic effects on ovulation (Martin et al. 1978). Many processing steps used in soyfood manufacture, soy milk heating, and miso fermentation result in hydrolysis of the glycosides and formation of the aglycones. Soybeans contain 500–3800 mg isoflavones/kg fresh wt, and processed soy milk has 30–178 mg/L (Wang and Murphy 1994; Franke et al. 1994; Setchell et al. 2002).

Soy isoflavones were absorbed in healthy subjects varying from 9%–21% depending on the isoflavone. Daidzein was determined to be the most bioavailable (Xu et al. 1994). Concluded daidzen and genestein were absorbed to the same extent, but daidzein was excreted faster through the urinary system (King and Bursill 1998). Slow absorption, that is, peak plasma level reached in 7.4 hours, was determined to be the result of hydrolysis of the glycosides enabling absorption of the aglycones in the large intestine.

The soy aglycone forms were absorbed more efficiently to a much greater extent than the isoflavone glycosides in human subjects (Kano et al. 2006). Isoflavones from soymilk were absorbed slowly with appearance in the plasma of isoflavone and daidzein metabolites at about 1 μmol/L in 6 hours, but when the soymilk was pretreated with β-glycosidase the rate of appearance occurred much faster (1.75 umol/L in 1 hour). Fermented soymilk enhanced the rate to about the same level.

In rat studies, isoflavones (genestein) were absorbed and excreted through the biliary system (Sfakianos et al. 1997). Genistein has been shown to possess a wide variety of activities using *in vitro* animal cells and *in vivo* animal models, including tyrosine kinase inhibition, chemoprevention of breast and prostate cancers, prevention of cardiovascular disease, and amelioration of postmenopausal ailments.

Dietary consumption of genistein and daidzein from soy products is thought to reduce the incidence of prostrate and breast cancers in humans. The phytoestrogens block the synthesis of the androgen testosterone, which stimulates prostate cancer cell formation. The phytoestrogens lower androgen levels inhibiting tumour growth. Genistein has been shown to promote the growth of estrogen-responsive breast and endometrial cancer cells in culture. Similar to the known effect of 17β-estradiol, treatment of breast (T47D and MCF-7) and endometrial (ECC-1) cancer cells with phytoestrogens, such as genistein, induces cell proliferation, cell-cycle progression, and transactivation of the estrogen response element (ERE). Of interest, the effects of genistein on these types of cancer cells can be reversed by treatment with carotenoids such as lycopene, phytoene, and phytofluene. Importantly, the cancer-promoting effect of genistein on estrogen-responsive cancers is not seen in estrogen receptor-negative breast cancer, and in fact, in these cell types, genistein inhibits cell growth. The inhibitory action of genistein is effected through its ability to inhibit tyrosine kinase activity, such as that associated with many cell surface growth factor receptors and activation of cell signaling transduction.

Hong et al. (2004) evaluated genistein, daidzein, and equol (daidzein metabolite) using the ERα-mediated HAT assay. Estrogenic ligands exerted their effects by elevating histone acetylation and coactivator activity of ER, and suggested genistein may reduce estrogen related diseases. Similar effects were established during the evaluation of coactivator recruitment by ERα and ERβ (Routledge et al. 2000).

## Flavanols (Flavan-3-ols)

Flavanols, specifically flavan-3-ols, are the most complex structures, including (+)-catechins and the isomer (–)-epicatechin (Table 5.2, part IE), which can be hydroxylated and acylated by gallic acid to form gallocatechin, and formation of oligomeric and polymeric proanthocyanidins (dimers). Gallocatechin and epigallocatechin and epigallocatechin3-gallate are found in certain seeds of leguminous plants, grapes, tea, and cocoa (Cassidy et al. 2000). The amounts of the flavan-3-ols found in foods include fruits such as apricots (250 mg/kg fresh wt), red wine (300 mg/L), green tea (4572 mg/L), and dark chocolate (100 mg/20 g). These are the most studied sources of flavanols (Crozier et al. 2006).

Interestingly, the partition coefficient of the catechins, and the gallic acid derivatives, indicate a strong hydrophobic character enabling passage through the cellular membranes. The catechin gallates are absorbed without hydrolysis or deconjugation, yet the plasma level reflects only 0.2%–2.0% of the ingested dose (Nakagawa et al. 1997; Scalbert and Williamson 2000). The plasma concentration of total catechins following consumption of black tea and green tea was noted to be 0.17 and 0.55 umole/L, respectively (Lee et al. 1995). Further studies on absorption, metabolism, and excretion for Choladi green tea flavon-3-ols have been reported (Stalmach et al. 2009).

Similarly, consumption of chocolate indicated proportional increases in the plasma catechin levels. In two studies, a 65-fold increase was observed with 53 g of chocolate (Wang et al. 2000) and a 12-fold increase was determined with 80 g of chocolate (Rein et al. 2000). However, no increase in antioxidant properties (no change in oxidative stress measures), lipid profiles, blood pressure, body weight, or body mass index (BMI) was observed. The specific function of the catechins on transduction pathways may provide a more accurate biomarker, since repeated doses would establish adjustment in homeostasis not found in the broader physiologic markers chosen above.

Proanthocyanidins are dimers or oligomers of catechin and epicatechin and their gallic acid esters. Proanthocyanidins provide the characteristic astringent taste in grapes, peaches, apples, pears, berries, and in beverages such as wine, tea, beer and cider, and for the bitterness in chocolate (Santos-Buelga and Scalbert 2000). The glycosidic cyanins are the naturally occurring forms of the cyanidins.

In beer, the phenolics and polyphenolics are derived from the barley (70%) and the hops (30%). A variety of monmers, (+)-catechins and (–)-epicatechins, and dimmers, procyanidin B3 and prodelphinidin B3 were determined. The malt contributes most of the simple phenolics, including 3, 4-dihydroxybenzoic acid, caffeic acid, and ferulic acid with small amounts also found in the hops. The hops contain quercetin conjugates (Gu et al. 2003).

The polymeric proanthocyanidin dimmers are not well absorbed as such in the small intestine. However, absorption studies indicate that dimeric or trimeric forms of flavonols can cross the epithelium membrane (Deprez et al. 2001). Detection of proanthocyanidin dimers B1 and B2 in human plasma have been reported. The absorption of these dimers was about 1% of the monomeric flavonols (Manach et al. 2005 a,b; Sano et al. 2003). However, these compounds were determined to have direct effects on the intestinal mucosa and protect it against oxidative stress and the actions of carcinogens. In addition, consumption of proanthocyanidin foods (cocoa, red wine, grape seed extracts) were shown to increase the plasma antioxidant capacity, to have positive effects on vascular function, and to reduce platelet activity in humans (Rios et al. 2002).

Apples provide a major source of proanthocyanidins in the diet. Apple and apple products are consumed in large amounts, and are good sources of flavonoids and phenolic compounds (2310–4880 mg/kg). The principal ingredients include 5-O-caffeoylquinic acid and quercetin-3-O-glycosides (glucoside, galactoside, rhamnoside, xyloside, or rutinoside). They also contain flavan-3-ols, including (-)-epicatechin and its procyanidin dimmers B1 and B2 (Clifford et al. 2003). In addition, proanthocyanidins are present in high concentrations in nuts, such as hazelnuts, pecans (5 g/kg), almonds, and pistachios, (1.8–2.4 g/kg) (Clifford 2000).

### Anthocyanidins

Anthocyanidins are also widely dispersed in the plant kingdom. The most common anthocyanidins are pelargonidin, cyanidin, and delphinidin (Table 5.2, part IF). They usually occur as anthocyanins, the sugar conjugates. The conjugates may also include hydroxycinnamates and organic acids. Strawberry anthocyanins, such pelargonidin-3-glucoside, are absorbed and are excreted in the urine almost linearly with an increasing dose (Carkeet et al. 2008). Five anthocyanin metabolic products were identified, including three monoglucuronides (80%), one sulfo-conjugate of pelargonidin, and the non-conjugated pelagonidin itself (Felgines et al. 2003), suggesting liver conjugation.

For purple cabbage and black/purple carrots, lower doses of the acylated anthocyanins, for example, cyanidin-3-rutinoside-glucoside-galactoside acylated with one hydroxylcinnamic acid derivative (p-coumaric, ferulic, or sinapic) were absorbed to a greater extent than the nonacylated forms, as reflected by lower plasma and urine levels (Charron et al. 2007, 2009; Kurilich et al. 2005). Using a rat model, the absorption of the caffeic acid acylated anthocyanins in purple-fleshed sweet potatoes was also studied. The peonidin 3-caffeoylsophoroside-5-glucoside present in the purple-fleshed sweet potato was directly absorbed and determined as an intact acylated form in the plasma. Oral administration of a purple-fleshed sweet potato anthocyanin concentrate also resulted in the anthocyanin being detected in the plasma, reaching a maximum of 50 nmol/L at its peak in 30 minutes. Furthermore, the plasma antioxidant capacity was significantly elevated by 50% in the same period (Suda et al. 2002).

Nasunin, the coumaroyl acylated anthocyanin in eggplant peel, contains two isomers, cis-nasunin and trans-nasunin. In rats, gastrointestinal absorption of the cis- and trans-nasunins indicated rapid absorption of their original acylated forms appearing maximally in the plasma within 15 minutes with no significant difference in the uptake efficiency between the two isomers (Ichiyanagi et al. 2006). Deacylated and glycolytic products of nasunins, such as delphinidin 3-O-β-D-glucopyranoside or delphinidin (aglycone), were not detected in blood plasma or urine even after 8 hours. Thus the nasunins were absorbed best in their original acylated forms (Ichiyanagi et al. 2006). Conversely, Matsumoto et al. (2006) reported that about 2.67% of the delphinidin glycosides were absorbed and excreted as the glycosides in the urine and as the methylated form in the bile.

Absorption of cyanidin-based anthocyanins is not well understood with respect to dose or the anthocyanin structure. Using whole foods, nonacylated anthocyanins are more bioavailable than their acylated counterparts. Thus, to evaluate the effect of the food matrix on availability for absorption, purple carrot juice containing acylated (75%) and nonacylated (25%) anthocyanins was used to assess differences in their absorption. Ten healthy adults were used to evaluate the relative absorption over 8 hours. The nonacylated anthocyanins were absorbed four times more efficiently than the acylated anthocyanins tested over a three- to fivefold concentration range (Charron et al. 2009), but it was concluded that the presence of the plant matrix did not alter absorption.

## NONFLAVONOID POLYPHENOLS

The main nonflavonoids are a major class of plant-derived phenolic acids, gallic acid, which is the precursor to formation of hydrolysable tannins, the hydroxycinnamic acids, the hydroxybenzoic acids and their conjugate derivatives, and the polyphenolic stilbenes.

## PHENOLIC ACIDS

The phenolic acids represent a major class of plant-derived phenolic compounds, predominantly the hydroxycinnamic acids and the hydroxybenzoic acids. The most common hydroxycinnamates are p-coumaric acid, caffeic acid, ferulic acid, chlorogenic acid, and sinapic acid, are frequently found in their acylated form (Table 5.1, part II).

## HYDROXYCINNAMIC ACID

Caffeic acid occurs most often esterified with sugars, organic acids, and lipids. They occur as conjugates with tartaric acid or quinic acid, forming chlorogenic acid, and are found in coffee in high amounts, 300 mg/L (Hammerstone et al. 1999). Thus, typical consumers of the beverage may have an intake in excess of 1 g/day. Chlorogenic acid and its conjugates are also found in orange, purple, and white carrots. The level of 5-O-caffeoylquinic acid in purple carrots is 540 mg/kg, about 10 times higher than the other varieties. The hydroxylcinnamic acid compounds occur mostly as simple esters of carboxylic acids or glucose, or amides. The esterified structures are less absorbed in the intestine due to a lack of the requisite esterase. Once they reach the colon, the bacterial hydrolysis occurs and the aglycone can be absorbed.

### Curcumin (Ferulic Acid)

The curcuminoid compounds are derivatives of ferulic acid, having two molecules of ferulic acid linked together. There are three curcuminoids: curcumin, demethoxycurcumin, and bisdemethoxycurcumin, which are yellow and impart their color and flavor to spices such as turmeric and mustard. The curcuminoid compounds have been shown to possess antioxidant, anti-inflammatory, antithrombotic, antifibrosis, antimicrobial, antiparasitic, antiviral, anticarcinogenic, antimutagenic, and hepatoprotective properties (Bisht et al. 2010). Curcumin is the most well studied of this class of compounds.

When curcumin is eaten, very little appears to be absorbed from the GI tract. In studies where from 2 to 10 grams of curcumin were eaten without other foods present, there were undetectable to very low levels of the compound detected in the serum. While in the gut, curcumin is unstable and the traces that are absorbed are taken up by the liver and rapidly degraded or conjugated with glucuronic acid and subsequently excreted in the bile.

Curcumin has been reported to suppress tumor promotion and proliferation, inflammatory signaling, and angiogenesis. Solid tumors cannot grow unless they can develop new blood vessels to bring oxygen-rich blood to the cells within the cancerous tissue. Therefore, the antiangiogenic properties of curcumin may contribute to its anticancer activity. Curcumin has been determined to be effective on an extremely wide range of cancer cell lines. It appears to be particularly useful in the treatment of mantle cell lymphoma, which has a greater mortality rate than do other non-Hodgkin lymphomas (Shishodia et al. 2005). The anti-inflammatory activity of curcumin is, in part, due to its ability to inhibit enzymes that are necessary for the synthesis of lipid mediators of inflammation (Bisht et al. 2010). Curcumin inhibits cyclooxygenase-2 (COX-2) and lipoxygenase. Curcumin also inhibits inflammatory responses initiated by stimuli that activate macrophages and T-cells, both of which mediate inflammatory response.

In studies on the effects of curcumin using human cells in culture, it has been shown that curcumin blocks the release of inducible nitric oxide synthase (iNOS) and COX-2 from airway epithelial cells, prevents COX-2 expression in mammary epithelial cells, inhibits cytokine secretion from macrophages, and blocks the release of cytokines and ROS from arterial cells. Curcumin also exerts cytoprotective effects that enhance cellular survival, due to its antioxidant properties.

Curcumin may exert anti-inflammatory effects by interfering with the signaling cascade of IL-1 by blocking the earliest event, the recruitment of IRAK to the IL-1RI. Recruitment of adapter molecules and kinases leads to the IL-1RI complex formation, which establishes the signaling to further downstream events (Bol et al. 2003; Jurrmann et al. 2005). A variety of events may be inhibited by curcumin, including inhibition of IL-1 mediated chemokine production and matrix metaloproteinase-3 upregulation in human colonic subepithelial myofibroblasts. Each of these effects was attributed to inhibition of NK-κB activation, produced by the inactivation of the signal leading to IKK activity (Jobin et al. 1999). IRAK also serves as a conserved component in signal cascades activated by IL-18 and TNFα (Kojima et al. 1998). After recruitment and phosphorylation, IRAK disassociates from the receptor complex and translocates to the nucleus (Bol et al. 2000). Although not clearly established, the phosphorylation of STAT3 by IRAK in the nucleus has been indicated in mononuclear cells of atherosclerotic patients. Curcumin does not prevent IL-1 interaction with the IL1RI (receptor), rather, it modifies protein thiols especifically associated with IRAK, not by oxidation but rather by alkylation with an α, β-unsaturated carbonyl component.

*In vivo* studies have demonstrated that administration of curcumin can lead to decreases in the level of cholesterol in the blood. The first consideration was that curcumin might enhance upregulation of the LDL receptors. However, it was determined that curcumin exerts its cholesterol lowering effects via inhibition of the gut specific cholesterol transporter, the Niemann-Pick C1-like 1 protein (Jia et al. 2011). Using intestinal cell culture system (Caco-2 cells), treatment with curcumin resulted in a downregulation of the expression of the NPC1L1 gene, resulting in reduced levels of the protein present in the membrane of Caco-2 cells (Jia et al. 2011).

The NPC1L1 protein is also highly expressed in human liver. The hepatic function of NPC1L1 participates in limiting excessive biliary cholesterol loss. The cholesterol uptake by NPC1L1 is regulated by cellular cholesterol content. Inhibition of NPC1L1 may result in healthful benefits on chronic disease components of the metabolic syndrome, such as obesity, insulin resistance, and fatty liver, as well as atherosclerosis. Therefore, consumption of curcumin may have clinical benefits in the mangement of these chronic diseases. Patients suffering from inflammatory bowel disease were given 550 mg curcumin twice daily, which resulted in significant amelioration of inflammatory symptoms (Hanai and Sugimoto 2009).

Structurally related polyphenols that are considered members of the phenolic acid analog family include rosmarinic acid, tyrosol, hydroxytyrosol, gossypol, ellagic acid, cynarin, paradol, and salvianolic acid B. Each of the naturally occurring phenolic acids are found free or conjugated to a sugar molecule.

## Hydroxylbenzoic Acids

The hydoxybenzoic acids occur as esters of glucosides. Related carboxylic acid structures include vanillic acid, p-hydroxybenzoic acid, syringic acid, and protocatechuic acid. These compounds are found in fruits, vegetables, cereals, spices, coffee, oil seeds, and tree nuts (Hermann 1989).

### Tannins (Hydrolysable)

The most common phenolic acid is gallic acid. The nonsugar galloyl esters found in grapes, wine, mangoes, green tea, and black tea are the major sources of gallic acid in the human diet. The related compounds, ellagic acid and ellagitannin (ellagic glucoside), are found in raspberries and strawberries are also present in other fruits like pomegranate, persimmons, and blackberries (Cerda et al. 2004, 2005).

The pomegranate has a high antioxidant content created by gallagic acid, containing four gallic acid residues, and punicalagin, a large polyphenol ellagitannin in which ellagic and gallic acid are linked to the glucose sugar (Gil et al. 2000). The punicalagin is water soluble and has high bioavailability. Punicalagin is the largest molecule absorbed intact to be found in the plasma of rats. They are partially hydrolyzed into smaller components of ellagic acid.

## Stilbenes

Stilbenes are phytoallexins produced by plants in response to injury and stress. The stilbenes are compsed of two aromatic rings linked together and are found as monomers, oligomers, and conjugated to sugars. They exhibit antioxidant, anti-inflammatory, anticancer, antibacterial, and antiviral activities (Bisht et al. 2010). The main dietary source of stilbenes is resveratrol from red wine and peanuts. Lesser amounts occur in berries, red cabbage, spinach, and some herbs.

The most well-known stilbene is trans resveratrol and t-resveratrol-3-0-glucoside (Mateus et al. 2002). The protective effects of red wine are ascribed to resveratrol, but it is unlikely since the levels of resveratrol in red wines are low (1–18 mg/L), and the quantity needed by humans to develop a protective effect would be in excess of 100 L of red wine/day (Corder et al. 2003; Kaeberlein and Rabinovitch 2006).

Resveratrol is found in grape skins and is currently one of the most extensively studied phytochemicals at this time. Resveratrol has a broad range of chemopreventive capabilities. Its antiproliferative effect occurs by the inhibition of matrix metallopeptidase-9 (MMP-9), which is associated with the downregulation of the MAPK signaling pathway. Resveratrol was shown to suppress the phosphorylation of extracellular signal regulated kinases 1 and 2 (ERK ½), which are associated with the invasion of breast cancer cells. In carcinoma cells, the diminished expression of E2F TFs leads to cell cycle arrest near the G1 to S phase transition (Tang et al. 2008). Resveratrol also promotes cancer cell destruction by upregulation of the p53 apoptotic pathway.

Three polyphenols, genistein from soy, resveratrol from grapes and peanuts, and epigallocatechin 3-gallate from green tea and chocolate, have protective effects reducing cancer (Lamartiniere 2002). Feeding of significant amounts of resveratrol (1000 ppm) to rats from birth to weaning and then continuing from 100 days of age produced a 50% reduction in the number of mammary tumors induced by carcinogens (Whitsett et al. 2006). In contrast, epigallocatechin3-gallate administered in the drinking water had no effect in this model.

## Other

Structurally-varied bioactive compounds that contribute unique effects include gingerol, the spicy flavor of ginger, and capsaicin, the hot spicy flavor of red chili peppers.

### *Capsaicin*

Capsaicin is the well-recognized active compound in chili peppers. The compound is an irritant and causes a burning sensation when contacting any mucous membrane or tissue, such as the skin. Current medicinal uses for capsaicin include its use in topical creams for the relief of the itching and inflammation associated with psoriasis or relief of joint pain resulting from sports injuries. Topical ointments with capsaicin are also used to treat the pain associated with peripheral neuropathy such as that experienced by patients suffering from shingles.

## POLYPHENOLS, TRANSDUCTION, AND GENE EXPRESSION

In closing this discussion of bioactive polyphenolics, a few examples are provided indicating some of the mechanisms being assessed by which specific phenolics interact with the key proteins involved in signal transduction pathways and gene expression. Interestingly, many bioactives have been evaluated as an extract or partially purified, such as cinnamon, cinnamon extract (CE), or CE polyphenols; green tea and purified catechins and gallo-catechins; dark chocolate and cocoa, catechins; pomegranate juice and ellagitannins; and red wine and resveratrol. The physiologic changes are noted early, and then researchers begin assessing the specific effects of purified components in cell culture or animal models.

## Cinnamon Polyphenolic Extract

People having normal and impaired glucose tolerance, type 2 diabetes, insulin resistance, and the metabolic syndrome have been reported to benefit from consumption of cinnamon and CE

(Cao et al. 2010). Using a cell culture system of 3T3-L1 adipocytes, a variety of regulators and pathways have been assessed. TNFα treated adipocytes initiated an inflammatory response, but when 10 ug/ml of CE was added to the medium, the expression of the inflammatory genes (IL-1β, IL6, and TNFα) was reduced, while the expression of the insulin signaling pathway mRNA (IR, IRS1, IRS2, PI3K, and Akt) was improved, as was the expression of the impaired SREBP-1c (Qin et al. 2009). The effects of CE (100 ug/ml) on expression of genes coding for adipokines, glucose transporter family, and insulin signaling components were increased for GLUT1 mRNA levels up to sevenfold of the control and maintained for 16 hours, while expression of genes encoding insulin signaling pathway proteins (GSK3B, IGF1R, IGF2R and PIK3R) was decreased (Sheng et al. 2008). The tristetraprolin (TTP) family proteins have anti-inflammatory effects by decreasing proinflammatory RNAs. The expression of TTP is reduced in the adipose of subjects that have metabolic syndrome. The effect of CPE (100 ug/ml) on the regulation of TTP, VEGF, and related gene expression in cell culture increased the expression of TTP mRNA levels up to tenfold, which were sustained for 16 hours. The level of VEGF mRNA, a putative target for TTP, was decreased by 40%–50% (Cao and Anderson 2011). From these preliminary cell experiments, CE decreased inflammation gene expression, decreased dyslipidemia, enhanced TTP, increased GLUT1 and diminished insulin signaling factors, and VEGF. A consistently positive response between experimental outcomes was acheived using this system. In a series of similar experiments, using two diabetic mice models DIO (early T2 diabetes) and a db/db mouse (severe diabetes) fed a high-caloric diet to induce obesity. The inclusion of CE provided an improved insulin resistance, lower fasting glucose, lower free fatty acid, and lower LDL-C levels (Sheng et al. 2008).

*In vitro* studies indicated that CE increased the expression of both PPARγ and PPARα, and their target genes: LPL, CD36, GLUT 4, and ACO in the 3T3-L1 adipocytes cell model. The reporter assay confirmed the transactivities of both full length and ligand binding domains of PPARγ and PPAR α were increased by cinnamon (Sheng et al. 2008). Thus, the active participation of PPARγ and α on the regulation of insulin resistance and adipogenesis was enhanced in the presence of CE.

CE containing procyanidin Type A polymers display insulin like activity. CE was incorporated into the media at three concentrations (0, 10, and 100 ug/ml). At the highest concentration, CE increased the TTP mRNA sixfold at 30 and 60 minutes and decreased the activity by half at 90 and 120 minutes. The IRβ protein and IR mRNA levels were constant at 60 minutes but decreased 60% at 90 minutes only to return in 120 minutes. The GLUT 4 mRNA levels exhibited a decreased effect at 100 ug/ml through 90 minutes, but increased at both EC concentrations at 2 hours (Cao et al. 2007).

## Genestein and Daidzein

Genistein and daizein have been reported to interact with the estrogen NR, ER α, and β. Hong et al. (2004) examined the effect of genistein, daidzein and equol on the ER mediated HAT assay. Estrogenic ligands exert their effects through elevation of histone acetylation and coactivator activity of ER. 17B-estradiol-dependent HAT activity of steroid receptor coactivators 2 (SRC2) and p300 mediated by ERβ was weaker than that mediated by ERα. As ligands, genistein and equol, with ERα mediated histone acetylation was significant, but equol, genestein, and daidzein all markedly stimulated ERβ mediated histone acetylation.

## Myricetin and Quercetin

Flavonoids may exert chemopreventive effects acting at protein kinase signaling pathways (Hou and Kumamoto 2010). Flavonoids can bind directly to some protein kinases, such as Akt protein kinase B (Akt/PKB), JAK1, MAPK 1 (MEK1, phosphoinositide 3-kinase (PI3K), and MAP kinase 4 (MKK4), and alter their phosphorylation state to reglate multiple cell signaling pathways in normal and carcinogenic processes.

Direct binding of the phytochemical can alter the protein kinase function and inhibit its action. Myrecetin binds directly with and inhibits JAK1, Akt, MEK1, MKK4, and PI3K. Quercetin similarly binds to MEK1 and PI3k, and procyanidins bind to MEK1 and MEK4. Thus, the pathways that are affected include PI3K-Akt signaling, Raf-MEK1-MAPK signaling, and JAK-STAT3 signaling pathways. Importantly, the binding sites are selective for specific flavonoids due to their attachment to either the ATP binding sites, the activation loop or the allosteric regulatory site, and of course the affinity of the flavonoid to its binding site (Hou and Kumamoto 2010). These naturally occurring compounds are able to bind to the ATP binding site of several of the kinases. Currently, it appears that the chromenone moiety of the flavonoid mimics the adenine moiety of the nucleotide cofactor, ATP, and enables flavonoids to bind to the kinases (Teillet et al. 2008).

Myrecetin had the strongest inhibitory effect on cell transformation. EGF binds to the EGFR causing a conformation change enabling it to undergo phosphorylation. EGF also induces phosphorylation of JAK1, but not of JAK2. In the presence of myricetin, the flavonoid did bind to JAK1 and inhibited the phosphorylation of JAK1. Myricetin was determined not to bind to the EGFR and did not alter an increase in EGFR autophosphorylation. Myricetin also inhibited STAT3 binding to DNA and prevented the transcription activity, perhaps by inhibiting the phosphorylation of STAT3 and preventing formation of the active homodimer. Myricetin bound to both JAK1 and STAT3, but the binding affinity was greater for JAK1 than STAT3. The authors concluded that myricetin might target JAK1 to block cell transformation in mouse JB6 cells (Kumamoto et al. 2009).

## Epigallocatechin 3-Gallate

An increase in apoptosis as determined by increased caspase 3, 8, 9, and cyt c, was noted for epigallocatechin 3-gallate (EGCG), theaflavin, ellagic acid, cucumin, quercetin, and luteolin. Each of these agents inhibit carcinogenesis through arresting of the cell cycle and inhibition of cyclins, induction of apoptosis through cytochrome c release, activation of the caspases and up-down regulation of Bcl2 family members, inhibition of survival/proliferation signals (Akt, MAPK, NF-κB), inflammation, and suppression of key proteins involved in angiogenesis and metastasis. Cancer cells seem to be more sensitive/more specific to the actions of the phytochemicals than their control cells (Ramos 2008; Surh 2003). The gallolyl and hydroxyl groups at the 3′ position on EGCG correlate most with the anti-inflammatory properties. EGCG, evaluated in both *in vitro* and *in vivo* experiments, can affect multiple pathways (Khan et al. 2006). NF-κB signaling pathway is inactive with IκB bound to it, but phosphorylation by IκB Kinase (activated by NF-κB kinase) releases IκB which is conjugated with ubiquitin and degraded. The NF-κB is released and enters the nucleus. EGCG has been determined to inhibit NF-κB activity in human cancer cells (Shimizu et al. 2005).

There are three types of MAPKs in mammalian cells, ERKs, p38 MAPKs, and the JNKs. The major pathways are downstream of the membrane associated RTKs. In this cascade Ras interacts with and activates MAP/ERK kinase ½ (MEK1/2). Activated MEK1/2 then phosphorylates and activates ERK1/2. The JNK 1/2/3 and p38a/B/g pathways are parallel MAPK cascades in mammalian cells. Once activated, MAPKs (ERK, JNK, and p38) activate ELK and c-Jun. PI3K is activated by the RTKs and it then synthesizes the second messenger. Inositol-3, 4, 5-triphosphate, which is necessary for the phosphorylation of Akt and Akt initiates activation of apoptosis (Gupta et al. 2004).

EGCG (5–20 umol/L) inhibits the MAPK pathway in the mouse epidermal cell line (Dong et al. 1997). Recently, EGCG at 10–20 ug/ml was determined to inhibit MAPK pathway and AP-1 activity in human colon cancer cells (Shimizu et al. 2005). These results suggest a potential to reverse the deregulation of MAPK pathway observed in numerous human cancers. EGCG has been found to inhibit 12-O-tetradecanoylphorbal-13 acetate or the EGF induced transformation of mouse epidermal cells (JB6 cell line), which was closely linked to the inhibition of AP-1 (Dong et al. 1997). AP-1 is a TF composed of the dimer c-Fos and cJun with ATF present. Elevated AP-1 appears to be involved in tumor promotion and progression.

EGFR is a plasma membrane receptor with an extracellular ligand binding domain, a single transmembrane region, and an intracellular domain that exhibits intrinsic TK activity. Overexpression of EGFR produces a neoplastic phenotype in tumor cells. EGCG at 10–20 ug/ml has been determined to inhibit the activation of EGFR and other signaling pathways in colon cancer cell lines (Shimizu et al. 2005).

Inhibition of VEGF, a mitogen for endothelial cells, has been associated with tumor induced angiogenesis. VEGF binds to VEGFR 1 and 2. VEGFR2 is most related to mitogenic and chemotactic effects. EGCG has been reported to decrease VEGFR phosphorylation and induced apoptosis in chronic lymphocytic leukemia B cells (Lee et al. 2005). Treatment of HUAEC (human umbilical arterial endothelial cells) with ECGC (5-50 umol/L) inhibited mitogenesis. EGCG also inhibited phosphorylation of EGFR½, ERK 1/2, and mRNA expression of the early growth factor-1 in EGCG pretreated cells (Neuhaus et al. 2004). Thus, VEGF binding and activation of its receptor EGFR may contribute to the antiangiogenisis and cancer chemopreventive effects of EGCG.

EGCG has potent anti-inflammatory properties and inhibits TNF-α mediated activation of the NF-κB pathway, in part by inhibition of IκB kinase (IKK) (Chen et al. 2002; Wheeler et al. 2004), The NF-κB pathway may also be activated by interleukin-1β (IL-1β) stimulation through a specific transduction pathway. EGCG markedly inhibited IL-1β mediated IL-1β receptor associated kinase (IRAK) degradation and the related IRAK dependent signaling events downstream from IRAK degradation: IKK activation, IκBα degradation and NF-κB activation. In addition, EGCG inhibited phosphorylation of the p65 subunit of NF-κB (Wheeler et al. 2004).

## Pomegranate

NF-κB expression was increased in advanced lesions of prostate cancer collected following prostatectomy (Heber 2008). NF-κB leads to immune activity, inflammation and cell proliferation in most cell types. NK-κB also upregulates genes that produce collagenase, cell adhesion molecules (CAM), and inflammatory cytokines (TNFa, IL-1, 2, 6, and 8).

Pomegranate polyphenols, ellagitannin-rich extract, exhibit strong antioxidant activity *in vitro* and *in vivo* and inhibit the growth of prostate cancer cells (Pantuck et al. 2006; Seeram et al. 2007). They were determined to downregulate the expression of androgen synthesizing genes and the AR, using the LNCaP-AR cell line, which overexpresses AR similar to androgen-independent prostate cancer. Thus, the specific polyphenolic inhibition of gene expression related to the androgen-synthesizing enzymes and the AR may indicate potential treatment in androgen-independent prostate cancer cells, specifically in those cases where the human prostate cancers where the AR is upregulated (Hong et al. 2008). As a follow up, 8 oz/day of pomegranate juice was given to prostate cancer patients following their primary treatment, radiation, or surgery, and the results indicated a significant prolonging of the time for the level of the PSA antigen to double, from 15–54 months (Pantuck et al. 2006), indicating a slow rate prostate cancer growth. Ellagitannin also suppressed prostate tumor formation in a severe combined Immunodeficient (SCID) mouse xenograft model (Seeram et al. 2007).

As a brief assessment, the phytochemicals can be seen to have cellular effects that are very promising. When translated into animal model systems, the same outcomes are observed but perhaps not to the same extent. Several reasons explain the need for further research. Experimental conditions need to be standardized (dose, cell type, culture conditions, duration, animal models, and diet composition), so that the results between laboratories can be more easily compared. Researchers must restrict their exhuberance using cellular data alone. The differences between cell culture experiments and animal studies, where digestion, absorption, and clearance mechanisms alter the concentration of the bioactive form of the polyphenolic that reaches the tissue cells will greatly alter the physiologic response.

## CONCLUSION

The wholesomeness of any diet depends on the balance of nutritional food components. However, it has become apparent that food delivers thousands of chemicals, most of which remain uncharacterized for nutritional or health benefits. The bioactive phytochemicals prove to be of interest because of their wide distribution in our food supply, and the defensive role they play in plant survival. Indeed, the few that have been examined indicate strong potential for providing health benefits.

Although not all encompassing, this chapter has provided a detailed overview of the intricate mechanisms that enable cellular signaling from the environment to be transmitted through a variety of transmembrane receptors, signal transduction by a variety of membrane and nonmembrane receptor kinases, a myriad of related protein pathways that also use phosphorylation and dephosphorlations to regulate the signals and their duration, and in the end to transfer NRs with lipophilic signal ligands bound to them or TFs that interact by specific amino acid-phosphate markers, to bind to regulatory sites on the DNA. Many bind to DNA sequences related to the promoter region of DNA related to specific target genes, establishing the assembly of the TATA-binding protein, RNA polymerase II, and the requisite coactivators/repressors and participating TFs that initiate and carry out gene expression. To keep the myriad of related reactions occurring throughout the cell and arrive at the nucleus as highly ordered, specific signals, requires interactive cross-talk to precisely modulate the genetic response to a variety of ligands reaching the cell at the same time.

The characterization of many polyphenolic flavonoids and related compounds, includes establishing their chemical characteristics, their acylated and aglycone structures, and their bioavailability following digestion and absorption. To date the caution has been to not overextrapolate results obtained using purified components in cell culture experiments and concluding that the same effects would be noted consuming the same agents in whole foods. On the other hand, the ability of dietary metabolites such as phenolics, carotenoids, fatty acids and peptides to produce large and coordinated changes in gene expression patterns in cultured human cells is well documented. These studies underscore the important role of dietary metabolites play in controlling the flow of extracellular information through the complex array of interaction networks and pathways required for cellular homeostasis. As more of these studies are performed on laboratory animals and human subjects, a clearer picture of how these dietary ligands work will continue to emerge.

Previous chapters have examined individual agents in great detail, identifying specific regulatory events that affect metabolic outcomes and control physiologic expression of phenotype. Others have clearly identified that these metabolic processes are involved in the major chronic diseases we develop with age and questionable dietary selections. As the role of bioactive ingredients are further characterized to control an individual's genetic potential, enhancing physiologic performance and delaying or blocking disease expression, we must move cautiously from outright rejection to overenthusiastic exhuberance and assure the facts being considered are science based.

## REFERENCES

A-Gonzalez, N., S. J. Bensinger, C. Hong, S. Beceiro, M. N. Bradley, N. Zelcer, J. Deniz, C. Ramirez, M. Díaz, G. Gallardo, C. R. de Galarreta, J. Salazar, F. Lopez, P. Edwards, J. Parks, M. Andujar, P. Tontonoz, and A. Castrillo. 2009. Apoptotic cells promote their own clearance and immune tolerance through activation of the nuclear receptor LXR. *Immunity* 31(2):245–58.

Adjei, A., and M. Hidalgo. 2005. Intracellular signal transduction pathway proteins as targets for cancer therapy. *J Clin Oncol* 23:5386–403.

Ahn, K. S., G. Sethi, K. Krishnaw, and B. B. Aggarwal. 2007. γ-Tocotrienol inhibits nuclear factor-κB signaling pathway through inhibition of receptor-interacting protein TAK1 leading to suppression of antiapoptotic gene products and potentiation of apoptosis. *J Biol Chem* 282(1):809–20.

Andera, L. 2009. Signaling activated by the death receptors of the TNFR family. *Biomed Pap Med Fac Univ Palacky Olomouc Czech Repub* 153:173–80.

Angeloni, C., E. Leoncini, M. Malaguti, S. Angelini, P. Hrelia, and S. Hrelia. 2008. Role of quercetin in modulating rat cardiomyocye gene expression profile. *Am J Physiol Heart Circ Physiol* 294: H1233–43.

Aoki, Y., T. Niihori, Y. Narumi, S. Kure, and Y. Matsubara. 2008. The RAS/MAPK syndrome: Novel roles of the RAS pathway in human genetic disorders. *Hum Mutat* 29(8):992–1006.

Aranda, A., and A. Pascual. 2001. Nuclear hormone receptors and gene expression. *Physiol Rev* 8(3):1269–304.

Armad, S., D. Banville, Z. Zhao, E. H. Fisher, and S. H. Shen. 1993. A widely expressed human protein-tyrosine phosphatase containing Src homology 2 domains. *Proc Nat Acad Sci* 90:2197–2201.

Attisano, L., J. L. Wrana, F. López-Casillas, and J. Massagué. 1994. TGF-.beta receptors and actions. *J Biochim Biophys Acta* 1222:71–80.

Aziz, A. A., C. A. Edwards, M. E. Lean, and A. Crozier. 1998. Absorption and excretion of conjugated flavonols, including quercetin-4′-O-β-glucoside and isorhamnetin-4′-O-β glucoside by human volunteers
after the consumption of onions. *Free Radic Res* 29(3):257–69.

Barford, D. 2001. The mechanisms of protein kinase regulation by protein phosphatases. *Biochem Soc Trans*
29:385–91.

Barford, D., A. K. Das, and M. P. Egloff. 1998. The structure and mechanism of protein phosphatases: Insights
into catalysis and regulation. *Annu Rev Biophys Biomol Struct* 27:133–64.

Barish, G. D., V. A. Narkar, and R. M. Evans. 2006. PPAR delta: A dagger in the heart of the metabolic syndrome. *J Clin Invest* 116:590–7.

Baxter, J. D., P. Webb, G. Grover, and T. S. Scanlan. 2004. Selective activation of thyroid hormone signaling
pathways by GC-1: A new approach to controlling cholesterol and body weight. *Trends Endocrinol
Metab* 15:154–7.

Behrmann, I., T. Smyczek, P. C. Heinrich, H. Schmitz-Van de Leur, W. Komyod, B. Giese, G. Muller-Newen,
S. Haan, and C. Haan. 2004. Janus kinase (Jak) subcellular localization revisited: The exclusive membrane localization of endogenous Janus kinase 1 by cytokine receptor interaction uncovers the Jak receptor complex to be equivalent to a receptor tyrosine kinase. *J Biol Chem* 279:35486–493.

Belandia, B., D. Brautigan, and J. Martin-Perez. 1994. Attenuation of ribosomal protein S6 phosphatase activity in chicken embryo fibroblasts transformed by Rous Sarcoma virus. *Mol Cell Biol* 14(1):200–206.

Berger, J., and D. E. Moller. 2002. The mechanisms of action of PPARs. *Annu. Rev. Med.* 53:409–435.

Berridge, M. J. 2005. Unlocking the secrets of cell signaling. *Ann Rev Physiol* 67:1–21.

Berridge, M. J., M. D. Bootman, and H. L. Roderick. 2003. Calcium signaling dynamics, homeostasis and
remodeling. *Nat Rev Mol Biol* 4(7):517–29

Berry, D. C., and N. Noy. 2009. All trans retinoic acid represses obesity and insulin resistance by activating both peroxisome proliferation-activated receptor beta/delta and retinoic acid receptor. *Mol Cell Biol*
29:3286–96.

Bidlack, W. R., and W. Wang. 1999. Designing functional foods. In M. E. Shils, J .A. Olson, M. Shike and
A. C. Ross, eds., *Modern Nutrition in Health and Disease*. Philadelphia: Williams and Wilkins, 1823–33.

Biesalski, H. K. 2007. Polyphenols and inflammation: Basic interactions. *Curr Opin Clin Nutr Metab Care*
10:724–8.

Bischoff, S. C. 2008. Quercetin: Potentials in the prevention and therapy of disease. *Curr Opin Clin Nutr Metab
Care* 11:733–40.

Bisht, K., K. H. Wagner, and A. C. Bulmer. 2010. Curcumin, resveratrol and flavonoids as anti-inflammatory,
cyto- and DNA-protective dietary compounds. *Toxicology* 278(1):88–100.

Blobe, G. C., L. M. Obeid, and Y. A. Hannun. 1994. Regulation of protein kinase C and role in cancer biology.
*Cancer Metastasis Rev* 13:411–31.

Blumer, J. B., A. V. Smrcka, and S. M. Lanier 2007. Mechanistic pathways and biological roles for receptor
independent activators of G-protein signaling. *Pharmacol Ther* 113:488–506.

Bodet C., V. D. Le, F. Epifano, and D. Grenier 2008. Naringenin has anti-inflammatory properties in macrophage and *ex vivo* human whole-blood models. *J Periodontal Res* 43(4):400–7.

Bol, G. F., N. Jurrmann, and R. Brigelius-Flohe. 2003. Recruitment of the interleukin-1 receptor (IL-1RI)-
associated kinase IRAK to the IL-1RI is redox regulated. *Biol Chem* 384:609–17.

Bol, G. F., O. J. Kreuzer, and R. Brigelius-Flohe. 2000. Translocation of the interleukin-1 receptor associated
kinase-1 (IRAK-1) into the nucleus. *FEBS Lett* 477:73–78.

Bonifacino, J. S. 2002. Quality control of receptor kinase signaling complexes. *Dev Cell* 2:1–2.

Boots, A. W., G. R. Haenen, and A.Bast. 2008. Health effects of quercetin: From antioxidant to nutraceutical.
*Eur J Pharmacol* 585:325–37.

Bravo, L. 1998. Polyphenols: Chemistry, dietary sources, metabolism, and nutritional significance. *Nutr Rev*
56(11):317–33.

Bravo, L., R. Abia, and F. Saura-Calixto 1994. Polyphenols as dietary fiber associated compounds: comparative
study on *in vivo* and *in vitro* properties. *J Agric Food Chem* 42:1481–87.

Bridges, T. M., and C. W. Lindsley. 2008. G-protein coupled receptors from classical modes of modulation to
allosteric mechanisms. *ACS Chem Biol* 3:530–41.

Brindle, P. K., and M. R. Montminy. 1992. The CREB family of transcription activators. *Curr. Opin. Genet.
Dev.* 2:199–204.

Bunemann, M., and M. M. Hosey. 1999. G-protein coupled receptor kinases as modulators of G-protein signaling. *J Physiol* 517(pt 1):5–23.

Cantley, L. C. 2002. The phosphoinositide 3-kinase pathway. *Science* 296(5573):1655–7.

Cao, H. and R. A. Anderson. 2011. Cinnamon polyphenol extract regulates tristetrprolin and relted gene expression in mouse adipocytes. *J Agric food Chem* 59(6):2739–44.

Cao, H., D. J. Graves, and R. A. Anderson. 2010. Cinnamon extract regulates glucose transporter and insulin-signaling gene expression in mouse adipocytes. *Phytomedicine* 17(13):1027–32.

Cao, H., M. M. Polansky, and R. A. Anderson. 2007. Cinnamon extract and polyphenols affect the expression of tristetraprolin, insulin receptor and glucose transporter 4 in mouse 3T3-L1 adipocytes. *Arch Biochem Biophys* 459:214–22.

Carkeet, C., B. A. Clevidence, and J. Novotny. 2008. Anthocyanin excretion by humans increases linearly with increasing strawberry dose. *J Nutr* 138(5):897–902.

Cassidy, A., B. Hansley, and R. M. Lamuela-Raventos. 2000. Isoflavones, lignans, and stillbenes—origins, metabolism, and potential importance to human health. *J Sci Food Agric* 80:1044–62.

Caunt, C. J., A. R. Finch, K. R. Sedgley, and C. A. McArdle. 2006. Seven transmembrane receptor signaling and ERK compartmentalization. *Trends Endocrinol Metab* 17:276–83.

Cerda, B., J. C. Espin, S. Parra, P. Martinez, and F. A. Tomas-Barberan. 2004. The potent *in vitro* antioxidant ellagitannins from pomegranate juice are metabolised into bioavailable but poor antioxidant hydroxy-6H-dibenzopyran-6-one derivatives by the colonic microflora of healthy humans. *Eur J Nutr* 43:205–20.

Cerda, B., F. A. Tomas-Barberan, and J. C. Espin. 2005. Metabolism of antioxidant and chemopreventive ellagitannins from strawberries, raspberries, walnuts, and oak-aged wine in humans: Identification of biomarkers and individual variability. *J Agric Food Chem* 53:227–35.

Cervantes, D., C. Crosby, and Y. Xiang. 2010. Arrestin orchestrates crosswalk between G protein coupled receptors to modulate the spatiotemporal activation of ERK MAPK. *Circ Res* 106:79–88.

Cha, J. Y., and J. J. Repa. 2007. The LXR and hepatic lipogenesis. *J Biol Chem* 282:743–51.

Cha-Molstad, H., D. M. Keller, G. S. Yochum, S. Impey, and R. H. Goodman. 2004. Cell type specific binding of the transcription factor CREB to the cAMP response element. *Proc Natl Acad Sci USA* 101(37):13572–77.

Chan, F. K. 2007. Three is better than one: pre-ligand-independent receptor assembly in regulation of TNF receptor signaling. *Cytokine* 37:101–7

Chan, F. K., H. J. Chun, L. Zheng, R. M. Siegel, V. L. Bui, and M. J. Leonard. 2000. A domain in the TNF receptors that mediate lipid independent receptor assembly and signaling. *Science* 288:2351–54.

Chang, L., and M. Karin. 2001. Mammalian map kinase signaling cascades. *Nature* 410:37–40.

Chang, E. C., T. H. Charn, S. H. Park, W. G. Helferich, B. Komm, J. A. Katzenellenbogen, and B. S. Katzenellenbogen. 2008. Estrogen receptors α and β as determinants of gene expression: Influence of ligand, dose, and chromatin binding. *Mol Endocrinol* 22 (5):1032–43.

Charron, C. S., B. A. Clevidence, S. J. Britz, and J. A. Novotny. 2007. The effect of dose size on bioavailability of acylated and nonacylated anthocyanins from red cabbage (*Brassica oleracea* L. var. capitata). *J Agric Food Chem* 55(13): 5354–62.

Charron, C. S., A. C. Kurilich, B. A. Clevidence, P. W. Simon, D. J. Harrison, S. J. Britz, D. A. Baker, and J. A. Novotny. 2009. Bioavailability of anthocyanidins from purple carrot juice: Effects of acylation and plant matrix. *J Agric Food Chem* 57:1226–30.

Chen, P. C., D. S. Wheeler, V. Malhorta, K. Odoms, A. G. Denenberg, and H. R. Wong. 2002. A green tea-derived polyphenol.epigallocatechin-3 gallate, inhibits IκB kinase activation and IL-8 gene expression. *Inflammation* 26:233–41.

Chin, D., and A. R. Means. 2000. Calmodulin: A prototypical calcium sensor. *Trends Cell Biol* 10(8):322–8.

Choi, J. W., D. R. Herr, K. Noguchi, Y. C. Yung, G. W. Lee, T. Mutch, M. E. Lin, S. T. Tee, K. E. Park, A. N. Mosely, and J. Chun. 2010. LPA Receptors: Subtypes and biochemical actions. *Ann Rev Pharm Toxicol* 50:157–86.

Choi Y. J., Y. J. Jeong, Y. J. Lee, H. M. Kwon, and Y. H. Kang. 2005. (-) Epigallocatechin gallate and quercetin enhance survival signaling in response to oxidant-induced human endothelial apoptosis. *J Nutr* 135(4):707–13.

Chun, O. K., S. J. Chung, K. J. Claycombe, and W. O. Song. 2008. Serum C-Reacyive protein concentrations are inversely associated with dietary flavonoid intake in U.S. adults. *J Nutr* 138:763–760.

Clevidence, B. A., A. C. Kurilich, S. Britz, P. W. Simon, and J. Novotny-Dura. 2004. Bioavailability of anthocyanins from raw and cooked purple carrots. *UJNR* 33:171–4

Clifford, M. N. 2000. Miscellaneous phenols in foods and beverages—nature, occurrence and dietary burden. *J Sci Food Agric* 80:1126–37.

Clifford, M. N., K. L. Johnston, S. Knight, and N. Kuhnert. 2003. Hierarchical scheme for LC-MSn identification of chlorogenic acids. *J Agric Food Chem* 51:2900–11.

Clydesdale, F. 1997. A proposal for the establishment of scientific criteria for health claims for functional foods. *Nutr Rev* 55:413–22.

Clydesdale, F. 2004. Functional foods: Opportunities and challenges. *Food Tech* 58:35–40.

Clydesdale F., W. R. Bidlack, D. F. Birt, B. R. Bistrain, J. F. Borzelleca, R. A. Clemens, M. L. Dreher, J. W. Erdman, N. Fogg-Johnson, L. Israelsen, M. Leahy, G. A. Leveille, D. B. McColl, S. H. McNamara, K. C. Mercurio, J. A. Milner, S. K. Sathe, and J. E. Vanderveen. 2005. Functional foods: Opportunities and challenges. *IFT Expert Report*, 1–66.

Cohen, P. T. W. 2009. Phospahatase families dephosphorylating serine and threonine residues in proteins. In R. A. Bradshaw and E. A. Dennis, eds., *Handbook of Cell Signaling*, pp. 659–75. New York: Elsevier.

Collingwood, T. N., F. D. Urnov, and A. P. Wolffe. 1999. Nuclear receptors: Coactivators, corepressors and chromatin remodeling in the control of transcription. *J Mol Endocrinol* 23:255–75.

Commerford, S. R., L. Vargas, S. E. Dorfman, N. Mitro, E. C. Rocheford, P. A. Mak, X. Li, P. Kennedy, T. L. Mullarkey, and E. Saez. 2007. Dissection of the insulin-sensitizing effect of liver X receptor ligands. *Mol Endocrinol* 21(12):3002–12.

Corder, R., A. Crozier, and P. A. Kroon. 2003. Drinking your health? It's too early to say. *Nature* 426:119–6.

Cote, S., S. McNamara, D. Brambilla, A. Bianchini, G. Rizzo, S. V. del Rincon, F. Grignani, C. Nervi, and W. H. Miller Jr. 2004. Expression of SMRT (beta) promotes ligand induced activation of mutated and wild type retinoid receptors. *Blood* 104(13):4226–35.

Crespy, V., C. Morand, C. Besson, N. Cotelle, H. Vezin, C. Demigne, and C. Remesy. 2003. The splanchic metabolism of flavonoids highly differed according to the nature of the compound. *Am J Physiol Gastrointest Liver Physiol* 284:G980–88.

Crozier, A., I. B. Jaganath, and M. N. Clifford. 2009. Dietary phenolics: Chemistry, bioavailability and effects on health. *Nat Prod Rep* 26:1001–43.

Crozier, A., M. E. Lean, M. S. McDonald and C. Black. 1997. Quantitative analysis of the flavonoid content of commercial tomatoes, onions, lettuce, and celery. *J Agric Food Chem* 45:590–95.

Crozier, A., M. N. Clifford and H. Ashihara. 2006. *Plant Secondary Metabolites, Occurrence, Structure and Role in the Human Diet*. Oxford, UK: Blackwell Publishing.

Daayf, F., and V. Lattanzio. 2008. *Recent Advances in Polyphenol Research*. Hoboken, NJ: Wiley-Blackwell.

Delibegovic, M., D. Zimmer, C. Kauffman, K. Rak, E. G. Hong, Y. R. Cho, J. K. Kim, B. B. Kahn, B. G. Neel, and K. K. Bence. 2009. Liver-specific deletion of protein-tyrosine phosphatase 1B (PTP1B) improves metabolic syndrome and attenuates diet-induced endoplasmic reticulum stress. *Diabetes* 58(3):590–99.

Deprez, S., I. Mila, J. F. Huncau, D. Tome, and A. Scalbert. 2001. Transport of proanthocyanidin dimmer, trimer and polymer across monolayers of human intestinal epithelial Caco-2 cells. *Antioxid Redox Signal* 3:957–67.

Dhillon, A. S. and W. Koch. 2004. Oncogenic B-Raf mutations: Crystal clear at last. *Cancer Cell* 5:303–4.

Dong, Z., W. Ma, C. Huang, and C. S. Yang. 1997. Inhibition of tumor promoter induced activator protein-1 activation and cell transformation by tea polyphenols, (-)-epicgallocatechin gallatate and the aflavins. *Cancer Res* 57:4414–9.

Downs, M. A., R. Arimoto, G. R. Marshall, and O. G. Kisselev. 2006. G-protein alpha and beta-gamma subunits interact with conformationally distinct signaling states of rhodopsin. *Vision Res* 46(27):4442–48.

Eldik, I. and D. Watterson. 1998. *Calmodulin and Calcium Signal Transduction: An Introduction*. New York: Academic Press.

Elion, E. A. 2001. The Ste5p scaffold. *J Cell Sci* 114:3967–78.

Engler, M. B., M. M. Engler, C. Y. Chen, M. J. Malloy, A. Browne, E. Y. Chiu, H. K. Kwak, S. M. Paul, J. Blumberg, and M. L. Mietus-Snyder. 2004. Flavonoid-rich dark chocolate improves endothelial function and increases plasma epicatechin concentrations in healthy adults. *J Am Coll Nutr* 23:197–204.

Evans, R. M. 1988. The steroid and thyroid hormone receptor superfamily. *Science* 240:889–95.

Evans, R. M. 2004. A transcriptional basis for physiology. *Nat Med* 10:1022–26.

Evans, R. M., G. D. Barish, and Y. X. Wang. 2004. PPARs and the complex journey to obesity. *Nat Med* 10:355–61.

Fantl, W. J., D. E. Johnson, and L. T. Williams. 1993. Signaling by receptor tyrosine kinases. *Annu Rev Biochem* 62:453–81.

Felgines, C., S. Talavéra, M. P. Gonthier, O. Texier, A. Scalbert, J. L. Lamaison, and C. Rémésy. 2003. Strawberry anthocyanins are recovered in urine as glucuro- and sulfoconjugates in humans. *J Nutr* 133:1296–1301.

Ferguson, L. R. 2009. Nutrigenomics approaches to functional foods. *J Am Diet Assoc* 109(3):452–58.

Foskett, J. K., C. White, K. H. Cheung, and D. O. Mak. 2007. Inositol triphosphate receptor Ca++ release channels. *Physiol Rev* 87(2):593–658.

Foster, F. M., C. J. Ttraer, S. M. Abraham, and M. J. Fry. 2003. The phosphoinositide (PI) 3-kinase family. *J Cell Sci* 116:3037–40.

Frame, M. 2002. Src in cancer: Deregulation and consequences for cell behaviour. *Biochim Biophys Acta* 1602:114–30.

Franke, A. A., L. J. Custer, C. M. Cerna, and K. Narala. 1994. Quantification of phytoestrogens in legumes by HPLC. *J Agri Food Chem* 42:1905–13.

Fruman, D. A., R. E. Meyers, and L. C. Cantley. 1998. Phosphoinositide kinases. *Annu Rev Biochem* 67:481–507.

Furman, B., and N. Pyne. 2006. Modulation of cyclic nucleotides and cyclic nucleotide phosphodiesterases in pancreatic islet beta-cells and intestinal L-cells as targets for treating diabetes mellitus. *Curr Opin Investig Drugs* 7:898–905.

Gacka, M., and R. Adamiec. 2004. Mutations of peroxisome proliferaqtor-activated receptor gamma (PPARgamma): Clinical implications. *Postepy Hig Med Dosw* 58:483–89.

Gaur, U., and B. B. Aggarwal. 2003. Regulation of proliferation, survival and apoptosis by members of the TNF superfamily. *Biochem Pharmacol* 66(8):1403–8.

Germain, P., B. Staels, C. Dacquet, M. Spedding, and V. Laudet. 2006. Overview of nomenclature of nuclear receptors. *Pharmacol Rev* 58:685–704.

Gil, M. I., F. A. Tomas, B. Hess-Pierce, D. M. Holcroft, and A. A. Kader. 2000. Antioxidant activity of pomegranate juice and its relationship with phenolic composition and processing. *J Agric Food Chem* 48:4581–89.

Gouaux, E. and R. Mackkinnon. 2005. Principles of selective ion transport in channels and pumps. *Science* 310:1461–5.

Grassi, D., G. Desideri, G. Croce, S. Tiberti, A. Aggio, and C. Ferri. 2009. Flavonoids, vascular function and cardiovascular protection. *Curr Pharm Des* 15:1072–84.

Gravani, R. B. 2010. A marvel of modern times. *Food Tech* 64(10):11.

Griffiths, L. A. and A. Barrow. 1972. Metabolism of flavonoid compounds in germ-free rats. *Biochem J* 130:1161–2.

Gronemeyer, H., J. A. Gustafsson, and V. Laudet. 2004. Principles for Modulation of the Nuclear receptor Superfamily. *Nat Rev Drug Disc* 3:950–64.

Groop, L., and E. S. Lander. 2000. The common PPAR gamma Pro12Ala polymorphism is associated with decreased risk of type 2 diabetes. *Nat Genet* 26:76–80.

Gu, L., M. A. Kelm, J. F. Hammerstone, G. Beecher, J. Holden, D. Haytowitz, and R. L. Prior. 2003. Screening of foods containing proanthocyanidins and their structural characterization. *J Agric Food Chem* 51:7513–21.

Gupta, S., K. Hastak, F. Afaq, N. Ahmad, and H. Mukhtar. 2004. Essential role of caspases in eligallocatechin-3-gallate mediated inhibition of nuclear factor kB and induction of apoptosis. *Oncogene* 23:2507–22.

Gurel, B., T. Iwata, C. M. Koh, R. B. Jenkins, F. Lan, C. V. Dang, J. L. Hicks, J. Morgan, T. C. Cornish, S. Sutcliffe, W. B. Isaacs, J. Luo, and A. M. De Marzo. 2008. Nuclear MYC protein overexpression is an early alteration in human prostate carcinogenesis. *Mod Pathol* 21:1156–67.

Guri, A. J., R. Hontecillas, and J. Bassaganya-Riera. 2006. Peroxisome proliferator-activated receptors: Bridging metabolic syndrome with molecular nutrition. *Clin Nutr* 25:871–85

Haan, C., S. Kreis, C. Margue, and I. Behrmann. 2006. JAKs and cytokine receptors—an intimate relationship. *Biochem Pharmacol* 72:1538–46.

Hammerstone, J. F., S. A. Lazarus, A. E. Mitchell, R. Rucker, and H. H. Schmitz. 1999. Identification of procyanidins in cocoa (*Theobroma cacao*) and chocolate using high pressure liquid chromatography/mass spectrometry. *J Agric Food Chem* 47:490–96.

Hanai, H., and K. Sugimoto. 2009. Curcumin has bright prospects for the treatment of inflammatory bowel disease. *Curr Pharm Des* 15(18):2087–94.

Haque, S. J., Q. Wu, W. Kammer, K. Friedrich, J. M. Smith, I. M. Kerr, G. R. Stark, and B. R. Williams. 1997. Receptor associated constitutive protein tyrosine phosphatase activity controls the kinase function of JAK1. *Proc Natl Acad Sci USA* 94:8563–68.

He, W., A. Comparo, Y. Zhu, T. J. Oneil, L. M. Wang, J. Pierces, and T. A. Gustafson. 1996. Interaction of insulin receptor substrate (IRS-2) with insulin and insulin like growth factor I receptors. *J Biol Chem* 271:31855–62.

Heber, D. 2008. Multitargeted therapy of cancer by ellagitannins. *Cancer Lett* 269:262–68.

Heldin, C. H. 2010. Protein tyrosine kinase receptor signaling overview. In R. A. Bradshaw and E. A. Dennis, eds., *Handbook of Cell Signaling*, pp. 419–26. New York: Elsevier.

Herrmann, H. 1988. On the occurrence of flavonol and flavones glycosides in vegetables. *Z Lebensm Unters Forsch* 186:1–5.

Hermann, K. 1989. Occurrence and content of hydroxycinnamic and hydroxybenzoic acid compounds in foods. *Crit Rev Food Sci Nutr* 28(4):315–47.

Herrington, J., L. S. Smith, J. Schwartz, and C. Carter-Su 2000. The Role of STAT Proteins in Growth Hormone Signaling. *Oncogene* 19(21):2585–97.

Hertog, M. G. L., P. C. H. Hollman, and M. B. Katan. 1992. Content of potentially anticarcinogenic flavonoids of 28 vegetables and 9 fruits commonly consumed in the Netherlands. *J Agic Food Chem* 40:2379–83.

Hertog, M. G. L., P. C. H. Hollman, M. B. Katan, and D. Kromhout 1993. Intake of potentially anticarcinogenic flavonoids and their determinants in adults in the Netherlands. *Nutr Cancer* 20:21–9.

Ho, C. T., C. Y. Lee, and M. T. Huang. 1992. Phenolic compounds in food and their effects on health. I. Analysis, occurrence and chemistry. Washington, DC: American Chemical Society.

Hollman, P. C, J. H. deVries, S. D. van Leeuwen, M. J. Mengelers, and M. B. Katan. 1995. Absorption of dietary quercetin glycosides and quercetin in health ileostomy volunteers. *Am J Clin Nutr* 62:1276–82.

Hollman, P. C., J. M. van Trijp, M. N. Buysman, M. S. van der Gaag, M. J. Mengelers, J. H. de Vries, and M. B. Katan. 1997. Relative bioavailability of the antioxidant flavonoid quercetin from various foods in man. *FEBS Lett* 418:152–6.

Hollman, P. C., M. vd Gaag, M. J. Mengelers, J. M. van Trijp, J. H. de Vries, and M. B. Katan. 1996. Absorption and disposition kinetics of the dietary antioxidant quercetin in man. *Free Radic Biol Med* 21:703–7.

Holst, B., and G. Williamson. 2004. A critical review of the bioavailability of glucosinolates and related compounds. *Nat Prod Rep* 21:425–47.

Holst, B., and G. Williamson. 2008. Nutrients and phytochemicals: From bioavailability to bioefficiency beyond antioxidants. *Current Opin Biotechnol* 19(2):73–82.

Hong, M. Y., N. P. Seeram, and D. Heber. 2008. Pomegranate polyphenols down regulate expression of androgen-synthesizing genes in human prostate cancer cells overexpressing the androgen receptor. *J Nutr Biochem* 19:848–55.

Hong, S. H., and M. L. Privalsky. 2000. The SMRT corepressor is regulated by a MEK-1 kinase pathway: Inhibition of corepressor function is associated with SMRT phosphorylation and nuclear export. *Mol Cell Biol* 20(17): 6612–25.

Hong, T., T. Nakagawa, W. Pan, M. Y. Kim, W. L. Kraus, T. Ikehara, K. Yasui, H. Aihara, M. Takebe, M. Muramatsu, and T. Ito. 2004. Isoflavones stimulate estrogen receptor-mediated core histone acetylation. *Biochem Biophys Res Commun* 317:259–64.

Horwitz, K. B., T. A. Jackson, D. L. Bain, J. K. Richer, G. S. Takimoto, and L. Tung. 1996. Nuclear receptor coactivators and corepressors. *Molec Endocrin* 10:1167–77.

Hou, D. X., and T. Kumamoto. 2010. Flavonoids as protein kinase inhibitors for cancer chemoprevention: Direct binding and molecular modeling. *Antioxid Redox Signal* 13:691–719.

Hubbard, S. R. 2010. Structure of serine/threonine and tyrosine kinases. In R. A. Bradshaw and E. A. Dennis, eds., *Handbook of Cell Signaling*, pp. 413–17. New York: Elsevier.

Hubbard S. R., and J. H. Till. 2000. Protein tyrosine kinase structure and function. *Annu Rev Biochem* 69:373–98.

Ichaso, N. and S. M. Dilworth. 2001. Cell transformation by the middle T-antigen of polyoma virus. *Oncogene* 20:7908–16.

Ichiyanagi, T., N. Terahara, M. M. Rahman, and T. Konishi. 2006. Gastrointestinal uptake of nasunin, acylated anthocyanin in eggplant. *J Agric Food Chem* 54:5306–12.

Jenkins, C. M., A. Cedars, and R. W. Gross. 2009. Eicosanoid signalling pathways in the heart. *Cardiovasc Res* 82(2):240–49.

Jia, L., J. L. Betters, and L. Yu. 2011. Niemann-pick C1-like-1 (NPC1L1) protein in intestinal and hepatic cholesterol transport. *Annu Rev Physiol* 73:239–59.

Jobin, C., C. A. Bradham, M. P. Russo, B. Juma, A. S. Narula, D. A. Brenner, and R. B. Sartor. 1999. Curcumin blocks cytokine mediated NF-kappaB activation and proinflammatory gene expression by inhibiting inhibitory factor I-kappa B kinase activity. *J Immunol* 163:3474–83.

Johannessen, C. M., J. S. Boehm, S. Y. Kim, S. R. Thomas, L. Wardwell, L. A. Johnson, C. M. Emery, N. Stransky, A. P. Cogdill, J. Barretina, G. Caponigro, H. Hieronymus, R. R. Murray, K. Salehi-Ashtiani, D. E. Hill, M. Vidal, J. J. Zhao, X. Yang, O. Alkan, S. Kim, J. L. Harris, C. J. Wilson, V. E. Myer, P. M. Finan, D. E. Root, T. M. Roberts, T Golub, K. T. Flaherty, R. Dummer, B. L. Weber, W. R. Sellers, R. Schlegel, J. A. Wargo, W. C. Hahn, and L. A. Garraway. 2010. COT drives resistance to RAF inhibition through MAP kinase pathway reactivation. *Nature* 468:968–72.

Johnsen, S. P., K. Overvad, C. Stripp, A. Tjonneland, S. E. Husted, and H. T. Sorensen. 2003. Intake of fruit and vegetables and the risk of ischemic stroke in a cohort of Danish men and women. *Am J Clin Nutr* 78:57–64.

Johnson, G. L., and R. Lapadat. 2002. Mitogen activated protein kinase pathways mediated by ERK, JNK and p38 protein kinases. *Science* 298:1911–12.

Joseph, S. B., A. Castrillo, B. A. Laffitte, D. J. Mangelsdorf, and P. TonTonoz. 2003. Reciprocal regulation of inflammation and lipid metabolism by liver X receptors. *Nat Med* 9:213–9.

Joseph S. B., M. N. Bradley, A. Castrillo, K. W. Bruhn, P. A. Mak, L. Pei, J. Hogenesch, R. M. O'Connell, G. Cheng, E. Saez, J. F. Miller, and P. Tontonoz. 2004. LXR-dependent Gene Expression is important for Macrophage survival and the innate immune system. *Cell* 119:299–309.

Jurrmann, N., R. Brigelius-Flohe, and G. F. Bol. 2005. Curcumin blocks interleukin-1 (IL-1) signaling by inhibiting the recruitment of the IL-1 receptor-associated kinase IRAK in murine thymoma EL-4 Cells. *J Nutr* 135:1859–64.

Kaeberlein, M., and P. S. Rabinovitch. 2006. Medicine: Grapes versus gluttony. *Nature* 444:280–281.

Kalaany, N. Y., and D. J. Mangelsdorf. 2006. LXRS and FXR: The yin and the yang of cholesterol and fat metabolism. *Annu Rev Physiol* 68:159–91.

Kano, M., T. Takayanagi, K. Harada, S. Sawada, and F. Ishikawa 2006. Bioavailability of isoflavones after ingestion of soy beverages in healthy adults. *J Nutr* 136:2291–6.

Kazio, Y., H. Itoh, T. Kozasa, M. Nakafuku, and T. Sadoh. 1991. Structure and function of signal-transducing GTP-binding proteins. *Annu Rev Biochem* 60:349–400.

Kennedy, M. B. 2010. $Ca^{++}$/Calmodulin dependent protein kinase II. In R. A. Bradshaw and E. A. Dennis, eds., *Handbook Cell Signaling*, pp. 565–8. New York: Elsevier.

Keyse, S. M. 2000. Protein phosphatases and the regulation of mitogen-activated protein kinase signaling *Curr Opin Cell Biol* 12:186–92.

Khan, N., F. Afag, M. Saleem, N. Ahmad, and H. Mukhtar. 2006. Targeting multiple signaling pathways by green tea polyphenol. *Cancer Res* 66:2500–5.

Kimura, N., and N. Shimada. 1988. Direct interactions between membrane-associated nucleotide diphosphate kinase and GTP-binding protein (Gs), and its regulation by hormones and guanine nucleotides. *Biochem Biophys Res Commun* 151:248–56.

Kimura, N., N. Shimada, Y. Ishijima, M. Fukuda, Y. Takagi, and N. Ishikawa. 2003. Nucleoside diphosphate kinases in mammalian signal transduction systems: recent development and perspective. *J Bioenerg Biomembr* 35:41–7.

King, R. A., J. L. Broadbent, and R. J. Head. 1996. Absorption of the soy isoflavone, genistein, in rats. *J Nutr* 126:176–82.

King, R. A., and D. R. Bursill. 1998. Plasma and urinary kinetics of the isoflavones daidzein and genistein after a single soy meal in humans. *Am J Clin Nutr* 67:867–72.

Kojima, H., M. Takeuchi, T. Ohta, Y. Nishida, N. Arai, M. Ikeda, H. Ikegami and M. Kurimoto. 1998. Interleukin-18 activates the IRAK-TRAF6 pathway in mouse EL-4 cells. *Biochem Biophys Res Commun* 244:183–6.

Kolch, W., A. Kotawaliwale, K. Vass, and P. Janosch. 2002. The organization and function of the Ras Raf MEK ERK Pathway. (Fig 2) *Expert Reviews in Molecular Medicine*, www.expertreviews.org. (accessed April, 2002).

Kuang, E., F. Wu, and F. Zhu. 2009. Mechanism of sustained activation of ribosomal S6 kinase (RSK) and ERK by kaprosi sarcoma- associated herpevirus ORF45: multiprotein complexes retain active phosphorylated ERK and RSK and protect them from dephosphorylation. *J Biol Chem* 284(20):13958–68.

Kuhnau, J. 1976. The flavonoids. A class of semi-essential food components: Their role in human nutrition. *World Rev Nutr Diet* 24:117–91.

Kumamoto, T., M. Fujii, and D. X. Hou. 2009. Myrecetin directly targets JAK1 to inhibit cell transformation. *Cancer Lett* 275:17–28.

Kuo, P. C., H. F. Liu, and J. I. Chao. 2004. Survivin and p53 modulate quercetin-induced cell growth inhibition and apoptosis in human lung carcinoma cells. *J Biol Chem* 279:55875–85.

Kurilich, A. C., B. A. Clevidence, S. J. Britz, P. W. Simon, and J. A. Novotny. 2005. Plasma and urine responses are lower for acetylated vs. nonacetylated anthocyanins from raw and cooked purple carrot. *J Agric Food Chem* 53:6537–42.

Lamartiniere, C. A. 2002. Timing of exposure and mammary cancer risk. *J Mammary Gland Biol Neoplasia* 7:67–76.

Land, D. 2010. Optimizing bioactive ingredients. *Food Tech* 64(10):50–6.

Lee, M. J., Z. Y. Wang, H. Li, L. Chen, Y. Sun, S. Gobbo, D. A. Balentine, and C. S. Yang. 1995. Analysis of plasma and urinary tea polyphenols in human subjects. *Cancer Epidemiol Biomarkers Prev* 4:393–9.

Lee, Y. K., T. D. Shanafelt, N. D. Bone, A. K. Strege, D. F. Jelenik, and N. E. Kay. 2005. VEGF receptors on chronic lymphocytic leukemia (CLL) B cells interact with STAT 1 and 3: Implication for apoptosis resistance. *Leukemia* 19:513–23.

Lefkowitz, R. J. and S. K. Shenoy. 2005. Transduction of receptor signals by beta-arrestins. *Science* 308:512–17.

Leh, T., and U. Justesen. 1997. Analysis of flavonoids in fruits, vegetable, and beverages by HPLC-UV and LC-MS and estimation of the total daily flavonoid intake in Denmark. *Proceedings of the 1st Workshop on Polyphenols in Food: COST 916*, p. 416. European Commission, Brussels.

Lewis, R. S. 2007. The molecular choreography of store operated calcium channel. *Nature* 446:247–84.

Li, B., M. Carey, and J. Workman. 2007. The Role of Chromatin During Transcription. *Cell* 128:707–719.

Li, D., Q. Kang, and D. M. Wang. 2007. Constituitive coactivator of peroxisome proliferator activated receptor (PPAR gamma), a novel coactivator of PPAR gamma that promotes adipogenesis. *Mol Endocrinol* 21(10):2320–33.

Li, Y., M. H. Lambert, and H. E. Xu. 2003. Activation of nuclear receptors: A perspective from structural genomics. *Structure* 11:741–6.

Li, L., Z. Feng, and A. G. Porter. 2004. JNK-dependent phosphorylation of c-Jun on serine 63 mediates nitric oxide-induced apoptosis of neuroblastoma cells. *J Biol Chem* 279(6):4058–65.

Liu, R. H. 2004. Potential synergy of phytochemicals in cancer prevention: Mechanism of action. *J Nutr* 134:3479S–85S.

Lonard, D. M., and B. W. O'Malley. 2006. Expanding functional diversity of the coactivators. *Trends Biochem Sci* 30:126–32.

Lonard, D. M., and B. W. O'Malley. 2007. Nuclear receptor coregulators: Judges, juries and executioners of cellular regulation. *Molecular Cell* 27:691–700.

Luan, J., P. O. Browne, A. H. Harding, D. J. Halsall, S. O'Rehilly, V. K. Chatterjee, and N. J. Wareham. 2001. Evidence for gene-nutrient interaction at the PPAR gamma locus. *Diabetes* 50:686–89.

Luo, J., S. J. Feld, J. Y. Lee, J. A. Engleman, and L. C. Cantley. 2005. The p85 regulatory subunit of phosphoinositide 3-kinase down regulates IRS-1 signaling via formation of a sequestration complex. *J Cell Biol* 170(3):455–64.

Machiex, J. J., A. Fleuriet, and J. Billot. 1990. *Fruit Phenolics*. Boca Raton, FL: CRC Press.

Manach, C., C. Morand, C. Demigne, O. Texier, F. Regerat, and C. Remesy. 1997. Bioavailability of rutin and quercetin in rats. *FEBS Let* 409:12–16.

Manach, C., A. Scalbert, C. Morand, C. Remesy, and L. Jimenez. 2004. Polyphenols: Food sources and bioavailability. *Am J Clin Nutr* 79(5):727–47.

Manach, C., G. Williamson, C. Morand, A. Scalbert and C. Remsey. 2005a. Bioavailability and bioefficiency of polyphenols in humans. I. Review of 97 bioavailability studies. *Am J Clin Nutr* 81:230S–242S.

Manach, C., G. Williamson, C. Morand, A. Scalbert and C. Remsey. 2005b. Bioavailability and bioefficiency of polyphenols *in vivo*. *Int J Vit Nutr Res* 77:224–35.

Marfella, C. G and A. N. Imbalzano. 2007. The Chd family of chromatin remodelers. *Mutat Res* 618:30–40.

Martin P. M., K. B. Horwitz, D. S. Ryan, and W. L. McGuire. 1978. Phytoestrogen interaction with estrogen receptors in human breast cancer cells. *Endocrinology* 103:1860–7.

Mateus, N., A. M. S. Silva, C. Santos-Buelga, J. C. Rivas-Gonzalo, and V. de Freitas. 2002. Isolation and structural characterization of new acylated anthocyanins-vinyl flavanol pigments occurring in red wine. *J Agric Food Chem* 51:277–82.

Matsumoto, H., T. Ichiyanagi, H. Iida, K. Ito, T. Tsuda, M. Hirayama, and T. Konishi. 2006. Ingested delphinidin-3-rutinoside is primarily excreted to urine as the intact form and to bile as the methylated form in rats. *J Agric Food Chem* 54(2):578–82.

Means, A. R. 2010. Calmodulin-mediated signaling. In R. A. Bradshaw and E.A. Dennis, eds., *Handbook of Cell Signaling*, 2nd ed., pp. 978–81. New York: Elsevier.

Milligan, G., M. Canals, J. D. Pediani, J. Ellis, and J. F. Lopez-Gimenez. 2006. The Role of GPCR dimerization/oligomerization in receptor signaling. *Ernst Schering Foundation Symposium Proc* 2:145–61.

Mills, I. G. 2007. The interplay between clathrin coated vesicles and cell signaling. *Semin Cell Dev Biol* 18:459–70.

Milner, J. 2000. Functional Foods: the US perspective. *Am J Clin Nutr* 71:1654S–59S.

Mitro, N., P. A. Mak, L. Vargas, C. Godio, E. Hampton, V. Molteni, A. Kreusch, and E. Saez. 2007. The nuclear receptor LXR is a glucose sensor. *Nature* 445:219–23.

Molavi, B. and J. L. Mehta. 2004. Oxidative stress in cardiovascular disease: Molecular basis of its deleterious effects, its detection, and therapeutic considerations. *Curr Opin Cardiol* 19:488–93.

Morand, C., C. Dubray, D. Milenkovic, D. Lioger, J. F. Martin, A. Scalbert, and A. Mazur. 2011. Hesperin contributes to the vascular protective effects of orange juice: A randomized crossover study in healthy volunteers. *Am J Clin Nutr* 93:73–80.

Mullen, W., A. Boitier, A. J. Stewart, and A. Crozier. 2004. Flavonoid metabolites in human plasma and urine after the consumption of red onions: Analysis by liquid chromatography with photodiode array and full scan tandem mass spectrometric detection. *J Chromatogr A* 1058:163–8.

Murray, P. J. 2007. The JAK-STAT signaling pathway input and output integration. *J Immunology* 178:2623–29.

Mustelin, T., A. Alonso, N. Bottini, H. Huynh, S. Rahmouni, K. Nika, C. Louis-dit-Sully, L. Tautz, S. H. Togo, S. Bruckner, A. V. Mena-Duran, and A. M. al-Khouri. 2004. Protein tyrosine phosphatases in T cell physiology. *Mol Immunol* 41(6-7):687–700.

Mustelin, T., and K. Tasken. 2003. Positive and negative regulation of T-cell activation through kinases and phosphatases. *Biochem J* 371:15–27.

Naar, A. M., B. D. Lemon, and R Tjian. 2001. Transcriptional coactivator complexes. *Annu Rev Biochem* 70:475–501.

Nagy, L., and J. W. Schwabe. 2004. Mechanism of the nuclear receptor molecular switch. *Trends Biochem Sci* 29:317–24.

Naidu, A. S. 2000. *Natural Food Antimicrobial Systems*. Boca Raton, FL: CRC Press.

Nakagawa, K., S. Okuda, and T. Miyazawa. 1997. Dose dependent incorporation of tea catechins, (-)-epigallocatechin-3-gallate and (-)-epigaloocatechin, into human plasma. *Biosci Biotechnol Biochem* 61:1981–5.

Nakahira, M., T. Tanaka, B. E. Robson, J. P. Mizgerd, and M. J. Grusby. 2007. Regulation of signal transduction and activation of transcription signaling by the tyrosine phosphatase PTP-BL. *Immunity* 26(2):163–176.

Nandakumar, V, T Singh and SK Katiyar (2008) Multi-targeted prevention and therapy of cancer by proantho-cyanidins. *Cancer Letters* 269(2):378–87.

Neer, E. J. 1994. G proteins: Critical control points for transmembrane signals. *Protein Sci* 3(1):3–14.

Németh, K., G. W. Plumb, J. G. Berrin, N. Juge, R. Jacob, H. Y. Naim, G. Williamson, D. M. Swallow, and P. A. Kroon. 2003. Deglycosylation by small intestinal epithelial cell β-glucosidases is a critical step in the absorption and metabolism of dietary flavonoid glycosides in humans. *Eur J Nutr* 42:29–42.

Neuhaus, T., S. Pabst, S. Stier, A. A. Weber, K. Schrör, A. Sachinidis, H. Vetter, and Y. D. Ko. 2004. Inhibition of the vascular endothelial growth factor-induced intracellular signaling and mitogenesis of human endothelial cells by epigallocatechin-3 gallate. *Eur J Pharmacol* 483:223–7.

New, D. C., and Y. H. Wong. 2007. Molecular mechanisms mediating the G protein-coupled receptor regulation of cell cycle progression. *J Molecular Signaling* 2:2 doi:10.1186/1750-2187-2-2.

Nuclear Receptor Nomenclature Committee. 1999. A unified nomenclature system for the nuclear receptor superfamily. *Cell* 97(2):161–3.

Ostman, A., C. Hellberg, and F. D. Bohmer. 2006. Protein-tyrosine phosphatases and cancer. *Nat Rev Cancer* 6:307–20.

Packer, L. 2000. *Bioflavonoids and Polyphenols*. Orlando, FL: Academic Press.

Pantuck, A. J., J. T. Leppert, N. Zomorodian, W. Aronson, J. Hong, R. J. Barnard, N. Seeram , H. Liker, H. Wang, R. Elashoff, D. Heber, M. Aviram, L. Ignarro, and A. Belldegrun. 2006. Phase II study of pomegranate juice for men with rising prostate-specific antigen following surgery or radiation for prostate cancer. *Clin Cancer Res* 12:4018–26.

Patterson, R. L., D. Boehning, and S. H. Snyder. 2004. Inositol 1,4,5-trisphosphate receptors as signal integrations. *Annu Rev Biochem* 73:437–65.

Pawson, T., G. D. Gish, and P. Nash. 2004. The SH2 domain: A prototype for protein interaction domains. In *Modular Protein Domains*, ed. G. Cesareni, pp. 5–36. Weinheim, Germany: Wiley VCH Press.

Pawson, T., and P. Nash. 2000a. Protein-protein interactions define specificity in signal transduction. *Genes Dev* 14:1027–47.

Pawson, T., and P. Nash. 2000b. Assembly of cell regulatory systems through protein interaction domains. *Science* 300:445–52.

Pawson, T., and P. Nash. 2010. Modular protein interaction domain in cellular communication. In R. A. Bradshaw and E. A. Dennis, eds., *Handbook of Cell Signaling*, pp. 399–416. New York: Elsevier.

Pawson, T., and J. D. Scott. 1997. Signaling through scaffold, anchoring and adaptor proteins. *Science* 278:2075–80.

Peleg, H., M. Naim, R. L. Rouseff, and U. Zehavi. 1991. Distribution of bound and free phenolic acid in oranges (*Citrus sinensis*) and grapefruits (*Citrus paradise*). *J Sci Food Agric* 57:417–26.

Pilz, R. B., and K. E. Broderick. 2005. Role of cGMP in gene regulation. *Front Biosci* 10:1239–68.

Pixley, F. J. and E. R. Stanley. 2010. Cytokines and cytokine receptors regulation cell survival, proliferation and differentiation in hematopoiesis. In R. A. Bradshaw and E. A. Dennis, eds., *Handbook of Cell Signaling*, pp. 2733–42. New York: Elsevier.

Plowman, G. D., A. Ullrich, and L. K. Shawver. 1994. Receptor tyrosine kinases as targets for drug intervention. *Drug News and Perspectives* 7(6):334–9.

Porter, L. W. 1989. Tannins. In J. B. Harborne, ed., *Methods in Plant Biochemistry. I. Plant Phenolics*, pp. 389–419. London: Academic Press.

Premont, R. T., J. Inglese, and R. J. Lefkowitz. 1995. Protein kinases that phosphorylate activated G protein-coupled receptors. *FASEB J* 9:175–82.

Premont, R. T., and R. R. Gainetdinov. 2007. Physiological roles of G-protein-coupled receptor kinases and arrestins. *Annu Rev Physiol* 69:511–34.

Privalsky, M. L. 2004. The role of corepressors in transcriptional regulation by nuclear hormone receptors. *Ann Rev Physiol* 66:315–60.

Qi, M., and E. A. Elion. 2005. MAP kinase pathways. *J Cell Sci* 118:3569–72.

Qiagen and Protein Lounge. 2011. General pathway source. www.SABioscience.com/pathwaycentral.php. (accessed August 26, 2011).

Qin, B., H. Dawson, M.M. Polansky, and R.A. Anderson. 2009. Cinnamon extract attenuates TNFalpha induced intestinal lipoprotein ApoB48 overproduction by regulating inflammatory, insulin and lipoprotein pathways in enterocytes. *Horm Metab Res* 41(7):516–22.

Quinn, P. G. 2002. Mechanisms of basal and kinase inducible transcription activation by CREB. *Prog Nucleic Acid Res Mol Biol* 72:269–305.

Ramos. S. 2008. Cancer chemoprevention and chemotherapy: Dietary polyphenls and signaling pathways. *Mol Nutr Food Res* 52:507–26.

Rawlings, J. S., K. M. Rosler, and D. A. 2004. The JAK/STAT signaling pathway. *J Cell Sci* 117:1281–3.

Rein, D., S. Lotito, R. R. Holt, C. L. Keen, H. H. Schmitz, and C. G. Fraga. 2000. Epicatechin in human plasma: *in vivo* determination and effect of chocolate consumption on plasma oxidation status. *J Nutr* 130:2109S–14S.

Reiter, E., and R. J. Lefkowitz. 2006. GRKs and B-arrestin: Roles in receptor silencing, trafficking and signaling. *Trends Endocrinol Metab* 17:159–65.

Rhee, S. G. 2001. Regulation of phosphoinositide-specific phospholipase C. *Annual Rev Biochem* 70:281–312.

Rhee, S. G., and Y. S. Bae. 1997. Regulation of phosphoinositide-specific phospholipase C isozymes. *J Biol Chem* 272(24):15045–48.

Rios, L. Y., R. N. Bennett, S. A. Lazarus, C. Remesy, A. Scalbert, and G. Williamson. 2002. Cocoa procyanidins are stable during gastric transit in humans. *Am J Clin Nutr* 76:1106–10.

Ritter, S., and R. A. Hall. 2009. Fine tuning of GPCR activity by receptor-interacting proteins. *Nat Rev Mol Cell Biol* 10:819–30.

Rosenfeld, M. G., V. V. Lunyak, and C. K. Glass. 2006. Sensors and signals: A coactivator/corepressor/epigenetic code for integrating signal-dependent programs of transcriptional response. *Genes Dev* 20:1405–28.

Ross, E. M., and T. M. Wilkie. 2000. GTPase-activating proteins for heterotrimeric G proteins: regulators of G protein signaling (RGS) and RGS-like proteins. *Annu Rev Biochem* 69:795–827.

Routledge, E. J., R. White, M. G. Parker, and J. P. Sumpter. 2000. Differential effects of xenoestrogens on coactivator recruitment by estrogen receptor (ER) alpha and ERbeta. *J Biol Chem* 275(46):35986–93.

Saltiel, A. R., and C. R. Kahn. 2001. Insulin signalling and the regulation of glucose and lipid metabolism. *Nature* 414(6865):799–806.

Sano, A., J. Yamakoshi, S. Takutake, K. Tobe, Y. Kubota, and M. Kibuchi. 2003. Procyanidin B1 is detected in human serum after intake of proanthocyanidin-rich grape seed extract. *Biosci Biotechnol Biochem* 67(5):1140–3.

Santos-Buelga, C., and A. Scalbert. 2000. Proanthocyanidins and tannin-like compounds: Nature, occurrence, dietary intake and effects on nutrition and health. *J Sci Food Agric* 80:1094–117.

Saura-Calixto, F., and L. Bravo. 1996. Intestinal degradation of polyphenols. In Y. Malkki and J. H. Cummings, eds., *Dietary Fibre and Fermentation in the Colon*, pp. 87–92. Finland: Office for Official Publications of the European Communities.

Scalbert, A., and G. Williamson. 2000. Dietary intake and bioavailability of polyphenols. *J Nutr* 130:2073S–2085S.

Schlessinger, J. 2000. Cell signaling by receptor tyrosine kinase. *Cell* 103:211–25.

Scholz, S., and G. Williamson. 2007. Interactions affecting the bioavailability and bioefficiency of polyphenols *in vivo*. *Int J Vit Nutr Res* 77:224–35.

Seeram N. P., W. J. Aronson, Y. Zhang, S. M. Henning, A. Moro, R. P. Lee, M. Sartippour, D. M. Harris, M. Rettig, M. A. Suchard, A. J. Pantuck, A. Belldegrun, and D. Heber. 2007. Pomegranate ellagitannin-derived metabolites inhibit prostate cancer growth and localize to the mouse prostate gland. *J Agric Food Chem* 55:7732–7.

Seger, R., and E. G. Krebs. 1995. The MAPK signaling cascade. *FASEB J* 9(9):726–35.

Semple, R. K., V. K. Chatterjee, and S. O'Rahilly. 2006. PPAR gamma and human metabolic disease. *J Clin Invest* 116:581–9.

Setchell, K. D., N. M. Brown, L. Zimmer-Nechemias, W. T. Brashear, B. E. Wolfe, A. S. Kirschner, and J. E. Heubi. 2002. Evidence for lack of absorption of soy isoflavone glycosides in humans, supporting the crucial role of intestinal metabolism for bioavailability. *Am J Clin Nutr* 76:447–53.

Sfakianos, J., L. Coward, M. Kirk, and S. Barnes. 1997. Intestinal uptake and biliary excretion of the isoflavone genestein in rats. *J Nutr* 127:1260–8.

Shahidi, F., and M. Naczk. 1995. *Food Phenolics: Sources, Chemistry, Effects, Applications.* Lancaster: Technomic Publishing.

Sheng, X., Y. Zhang, Z. Gong, C. Huang, and Y. Q. Zhang. 2008. Improved insulin resistance and lipid metabolism by cinnamon extract through activation of peroxisome proliferator-activated receptors. *PPAR Res* 2008:581348.

Shimizu, M., A. Deguchi, J. T. Lim, H. Moriwaki, L.Kopelovich, and I. B. Weinstein. 2005. (-)-Epigallocatechin gallate and polyphenon E inhibit growth and activation of the epidermal growth factor receptor and human epidermal growth factor receptor-2 signaling pathways in human colon cancer cells. *Clin Cancer Res* 11:2735–46.

Shishodia, S., H. M. Amin, R. Lai, and B. B. Aggarwal. 2005. Curcumin (diferuloylmethane) inhibits constitutive NF kappaB activation, induces G1/S arrest, suppresses proliferation, and induces apoptosis in mantle cell lymphoma. *Biochem Pharmacol* 70(5):700–13.

Shuai, K. 2000. Modulation of STAT signaling by STAT interacting proteins. *Oncogene* 19(21):2638–2644.

Shulman, A. I. and D. F. Mangelsdorf. 2005. Retinoid X receptor heterodimers in the metabolic syndrome. *N Engl J Med* 353:604–15.

Silvestri, C., R. Bose, L. Attisano, and J. L. Wrana. 2010. TGFB Signal Transduction. In R. A. Bradshaw and E. A. Dennis, eds., *Handbook of Cell Signaling*, pp. 21–532. New York: Elsevier.

Skarpen, E., M. P. Oksvold, H. Grasvik, C. Widnes, and H. S. Huitfeldt. 2005. Altered regulation of EGF receptor signaling following a partial hepatectomy. *J Cell Physiol* 202:707–716.

Smithgall, T. E. 1998. Signal transduction pathways regulating hematopoietic differentiation. *Pharmacol Rev* 50:1–19.

Soderling, S. H., and J. A. Beano. 2000. Regulation of cAMP and cGMP signaling: New phospho-diesterases and new functions. *Curr Opin Cell Biol* 12(2):174–79.

Song, L., and C. Schindler. 2010. JAK-STAT signaling. In R. A. Bradshaw and E. A. Dennis, eds., *Handbook of Cell Signaling*, pp. 2041–2048. New York: Elsevier.

Sonoda, J., L. Pei, and R. M. Evans. 2008. Nuclear receptors: Decoding metabolic disease. *FEBS Lett* 582(1):2–9.

Stalmach, A., S. Troufflard, M. Serafini, and A. Crozier. 2009. Absorption, metabolism and excretion of choladi green tea flavan-3-ols by humans. *Mol Nutr Food Res* 53:S44–S53.

Suda, I., T. Oki, M. Masuda, Y. Nishiba, S. Furuta, K. Matsugano, K. Sugita, and N. Terahara. 2002. Direct absorption of acylated anthocyanin in purple-fleshed sweet potato into rats. *J Agric Food Chem* 50(6):1672–76.

Surh, Y. J. 2003. Cancer chemoprevention with dietary phytochemicals. *Nat Rev Cancer* 3:768–80.

Tang, F., E. I. Chiang, and Y. Sun. 2008. Resveratrol inhibits heregulin-B1-mediated matrix metalloproteinase-9 expression and cell invasion in human breast cancer cells. *J Nutr Biochem* 19(5):287–94.

Tang, Y., J. Yu, and J. Field. 1999. Signals from the Ras, Rac, and Rho GTPases converge on the Pak protein kinase in rat-1 fibroblasts. *Mol Cell Biol* 19(3):1881–91.

Tanoue, T., and E. Nishida. 2003. Molecular recognitions in the MAP kinase cascades. *Cell Signal* 15:455–462.

Taussig, R., J. A. Iniguez-Lluhi, and A. G. Gilman. 1993. Inhibition of adenylyl cyclase by Gi alpha. *Science* 261:218–21.

Taylor, C. W. and A. J. Laude. 2002. IP$_3$ receptors and their regulation by calmodulin and cytosolic Ca$^2$+. *Cell Calcium* 32:321–34.

Teillet, F., A. Boumendjel, J. Boutonnat, and X. Ronot. 2008. Flavonoids as RTK inhibitors and potential anticancer agents. *Med Res Rev* 28(5):715–45.

Theodosiou, A., and A. Ashworth. 2002. MAP kinase phosphatases. *Genome Biol* 3(7): 3009.1–3009.10.

Thomas, P. R. and R. Earl. 1994. Committee on Opportunities in the Nutrition and Food Science, Institute of Medicine. *Opportunities in the Nutrition and Food Sciences: Research Challenge and the Next Generation of Investigators.* Washington, DC: National Academy Press.

Tian, X., and L. A. Feig. 2001. Basis for signaling specificity difference between Sos and Ras-GRF guanine nucleotide exchange factors. *J Biol Chem* 276:47248–56.

Tomas-Barberian, F. A., and M. N. Clifford. 2000. Flavanones, chalcones and dihydrochalcones—nature, occurrence and dietary burden. *J Sci Food Agric* 80:1073–80

Tontonoz, P., and D. J. Mangelsdorf. 2003. Liver X receptor signaling pathways in cardiovascular disease. *Mol Endocrinol* 17:985–93.

Viswakarma, N., Y. Jia, L. Bai, A. Vluggens, J. Borensztajn, J. Xu, and J. K. Reddy. 2010. Coactivators in PPAR-regulated gene expression. *PPAR Res* Epub DOI:10.1155/2010/250126.

Wajant H., K. Pfizenmaier, and P. Scheurich. 2003. Tumor necrosis factor signaling. *Cell Death Differ* 10(1):45–65.

Wakioka, T., A. Sasaki, R. Kato, T. Shouda, A. Matsumoto, K. Miyoshi, M. Tsuneoka, S. Komiya, R. Baron, and A. Yoshimura. 2001. Spred is a sprouty-related suppressor of ras signalling. *Nature* 412(6847):647–51.

Wang, H. J., and P. A. Murphy. 1994. Isoflavone content in commercial soybean foods. *J Agric Food Chem* 42:1666–73.

Wang, J. F., D. D. Schramm, R. R. Holt, J. L. Ensunsa, C. G. Fraga, H. H. Schmitz, and C. L. Keen. 2000. A dose response effect from chocolate consumption on plasma epicatechin and oxidative damage. *J Nutr* 130:2115S–19S.

Wang, S., K. D. Lai, F. J. Moy, A. Bhat, H. B. Hartman, and M. J. Evans. 2006. The nuclear receptor farnesoid X receptor (FXR) is activated by androsterone. *Endocrinology* 147:4025–33.

Wang, Y., S. J. Chung, W. O. Song, and O. K. Chun. 2011. Estimation of daily proanthocyanidin intake and major food sources in the U.S. diet. *J Nutr* 141(3):447–52.

Wang, Y. D., W. D. Chen, D. D. Moore, and W. Huang. 2008. FXR: A metabolic regulator and cell protector. *Cell Res* 18:1087–95.

Wang, Y. X., C. H. Lee, S. Tiep, R. T. Yu, J. Ham, H. Kang, and R. M. Evans. 2003. Peroxisome proliferator-activated receptor delta activates fat metabolism to prevent obesity. *Cell* 113:159–170.

Ward, A. C., I. Touw, and A. Yoshimura. 2000. The Jak-Stat pathway in normal and perturbed hematopoiesis. *Blood* 95:19–29.

Wettschureck, N., and S. Offermanns. 2005. Mammalian G proteins and their cell type specific functions. *Physiol Rev* 85:1159–1204.

Wheeler, D. S., J. D. Catravas, K. Odoms, A. Denenberg, V. Malhorta, and H. R. Wong. 2004. Epigallocatechin-gallate, a Green tea-derived polyphenol, inhibits IL-1B dependent proinflammatory signal transduction in cultured respiratory epithelial cells. *J Nutr* 134:1039–44.

Whitsett, T., M. Carpenter, and C. A. Lamartiniere. 2006. Resveratrol, but not EGCG, in the diet suppresses DMBA-induced mammary cancer in rats. *J Carcinog* 5:15.

Wieland, T. 2007. Interaction of nucleoside diphosphate kinase B with heterotrimeric G protein BG dimmers: Consequences on G protein activation and stability. *Naunyn Schmiedebergs Arch Pharmacol* 374:373–83.

Wieland, T, S. Lutz, and P. Chidiac. 2007. Regulators of G protein signaling: A spotlight on emerging functions in the cardiovascular system. *Curr Opin Pharmacol* 7:201–7.

Williamson, G., and C. Manach. 2005. Bioavailability and bioefficiency of polyphenols in humans. II. Review of 93 intervention studies. *Am J Clin Nutr* 81:243S–55S.

Wolffram, S., T. Weber, B. Grenacher, and E. Scharrer. 1995. A $Na^+$-dependent mechanism is involved in mucosal uptake of cinnamic acid across the jejuna brush border in rats. *J Nutr* 125:1300–8.

Workman, J. L., and R. E. Kingston. 1998. Alteration of Nucleosome Structure as a Mechanism of Transcriptional Regulation. *Annu. Rev. Biochem.* 67:545–79.

Wrana, J. L., L. Attisano, R. Wieser, F. Ventura, and J. Massague. 1994. Mechanism of activation of the TGF-β receptor. *Nature* 370:341–7.

Wu, Y. and B. P. Zhou. 2010. TNF-a/NFκB/Snail pathway in cancer cell migration and invasion. *British J of Cancer* 102:639–644. doi:10.1038/sj.bjc.6605530 www.bjcancer.com.

Xu, X., H. J. Wang, P. A. Murphy, L. Cook, and S. Hendrich. 1994. Daidzein is a more avavailable isoflavone than is genistein in adult women. *J Nutr* 124:825–32.

Yamaoka K., P. Saharinen, M. Pesu, V. E. Holt III, O. Silvennoinen, and J. J. O'Shea. 2004. The Janus kinases (Jaks). *Genome Biol* 5(12):253.

Yoshimura, A. 2009. Regulation of cytokine signaling by the SOCS and spred family proteins. *Keio J Med* 58(2):73–83.

Yu, S., and J. K. Reddy. 2007. Transcription coactivators for peroxisome proliferator-activated receptors. *Biochim Biophys Acta* 1771(8):936–51.

Zaccolo, M., and M. A. Movesian. 2007. cAMP and cGMP signaling cross-talk: Role of phosphodiesterases and implications for cardiac pathophysiology. *Circ Res* 100:1569–78.

Zwick, E., J. Bange, and A. Ullrich. 2002. Receptor tyrosine kinases as targets for anticancer drugs. *Trends Mol Med* 8:17–23.

# 6 Modulation of Atherosclerosis by N-3 Polyunsaturated Fatty Acids

*Darshan S. Kelley, Yuriko Adkins, Sumeet Sharma, and Dawn M. Fedor*

## CONTENTS

## INTRODUCTION

Atherosclerosis comes from the Greek words *athero* (meaning gruel or paste) and *sclerosis* (hardness). It is a common disorder of the arteries which occurs when fat, cholesterol, and other substances build up in the walls of arteries and form hard structures called plaques (Rader and Daugherty 2008). The plaques can make the artery narrow and less flexible, making it harder for blood to flow. If the coronary arteries become narrow, blood flow to the heart can slow down or stop. This can cause chest pain (stable angina), shortness of breath, heart attack, and other symptoms. Pieces of plaque can break off and move through the bloodstream (embolization). This is a common cause of heart attack and stroke. Blood clots can also form around a tear (fissure) in the plaque and block blood flow. If the clot moves into an artery in the heart, lungs, or brain, it can cause a stroke, heart attack, or pulmonary embolism (Rader and Daugherty 2008). Eighty million American adults (approximately 1 in 3) have cardiovascular disease (CVD) (Lloyd-Jones et al. 2009). CVD is the number 1 cause of deaths in United States, accounting for 34% of all deaths or 2400 deaths each day (AHA 2005). It is a major public health problem with annual economic loss of $400 billion (Lloyd-Jones et al. 2009).

The arterial wall is comprised of three layers. Adventitia, the outermost layer, carries blood and nerve supply to the artery itself. Media, the middle layer, is comprised of smooth muscle cells and controls vascular tone. Intima, the innermost or basement membrane, is covered by a single layer of endothelial cells which have adhesion molecules (intracellular adhesion molecule-1, or ICAM-1, and vascular cell adhesion molecule-1, or VCAM-1) on their surface to which the immune cells can stick. It is the site of atherosclerosis and it regulates haemostasis, thrombosis, vascular tone, and permeability. Atherosclerosis process starts within the walls of the artery and not in the lumen (Sanz and Fayad 2008).

Atherosclerosis develops from low-density lipoprotein molecules (LDL) becoming oxidized by free radicals, particularly oxygen free radicals (ROS). When oxidized LDL comes in contact with an artery wall, it alters the permeability of the arterial wall and penetrates under the endothelial cells. The body's immune system responds by sending specialized white blood cells (WBC; monocytes and T-lymphocytes) to repair the damage to the arterial wall; the WBC attach to the adherence molecules on the endothelial cells. Next the white blood cells pass under the endothelial cell layer, where monocytes absorb the oxidized LDL and are transformed into specialized foam cells. Unfortunately, these WBC are not able to process the oxidized LDL, and ultimately grow then rupture, depositing a greater amount of oxidized cholesterol into the artery wall. This triggers more WBC, continuing the cycle. Eventually, the artery becomes inflamed. The growth factors released from monocytes and T cells cause the muscle cells to enlarge and form a hard cover over the affected area. This hard cover is what causes a narrowing of the artery, reduces the blood flow, and increases blood pressure (BP).

Atherosclerosis starts at a very early age and progresses with age. There are six stages in the development of atherosclerosis (Sanz and Fayad 2008). The first four stages (foam cells, fatty streak, extracellular fatty streak, and lipid core) are asymptomatic for decades and are compensated by artery enlargement. The last two stages (lipid core embedded in fibrosis, plaque rupture leading to thrombosis, and hemorrhage) cause angina, heart attack, and stroke. The atheromatous plaques, though long compensated for by artery enlargement, eventually lead to plaque ruptures and clot formation inside the artery lumen over the ruptures. The clots heal and usually shrink but leave behind stenosis (narrowing) of the artery, or worse, complete closure, and therefore an insufficient blood supply to the tissues and organ it feeds. If the compensating artery enlargement process is excessive, it results in aneurysm.

Multiple factors including dyslipidemia, increased oxidative stress, inflammation, endothelial dysfunction, plaque rupture, age, sex, and smoking contribute to the pathogenesis of atherosclerosis (Rader and Daugherty 2008). Increased LDL cholesterol, triglycerides (TG), and the number of small dense LDL and chylomicron particles and low high density lipoprotein (HDL) cholesterol are atherogenic. Endothelial dysfunction induced by dyslipidemia is an initial step in atherosclerosis. Increased oxidative stress increases lipid peroxidation and inflammation. Macrophages and endothelial cells that have been modified by the oxidized LDL release a variety of inflammatory substances, cytokines, and growth factors. Plasma concentration of markers of systemic inflammation are directly associated with the risk of atheroscleosis (Danesh et al. 2004; Burdge and Calder 2005).

## N-3 POLYUNSATURATED FATTY ACIDS AND CVD

First indications suggesting the improvement of cardiovascular health by n-3 polyunsaturated fatty acids (PUFA) came from epidemiological studies. Sinclair (1956) noted low incidence of coronary heart disease (CHD) mortality rate in Greenland Eskimos, who consumed a high-fat diet, but one rich in n-3 PUFA. Similar conclusions were drawn based on the comparisons of CVD between the Greenland Inuits and Danish population or the Japanese and North American populations (Dyerberg et al. 1975; Menotti et al. 1999; Holub 2002; Lee and Lip 2003). Other studies have reported an inverse association between dietary intake of n-3 PUFA or whole blood n-3 PUFA or adipose tissue PUFA and CVD (Dolecek 1992; Albert et al. 2002; Hu et al. 2002; Breslow 2006; Harris 2007).

N-3 PUFA can regulate atherosclerosis by modulating plasma concentrations of blood lipids, inflammation and adhesion molecules, lipid peroxidation, plaque formation and stability, platelet aggregation, thrombosis, BP, and heart rate (HR). Because of the large number of studies with fish oils, the results of which can be found in other recent reviews (Psota et al. 2006; Marik and Varon 2009; Russo 2009; Vrablik et al. 2009), this chapter will discuss the effects of purified eicosapentaenoic acid (EPA; 20:5 n-3) and docosahexaenoic acid (DHA; 22:6 n-3). Alpha linolenic acid (ALA; 18:3 n-3) from flaxseed oil and walnuts has been reported to mimic several effects of EPA and DHA, but has been determined to be less potent than the long chain (LC) n-3 PUFA; therefore, the effects of ALA on atherosclerosis will not be discussed.

DHA and not EPA is the major LC n-3 PUFA in human tissues. DHA concentration in brain, retina, and reproductive organs is greater than 20 wt% of total fatty acids; its concentration in human heart is approximately 5.1 wt% and EPA is 0.5 wt% (Rocquelin et al. 1985). Mice heart contains about 13 wt% DHA, while EPA is not detectable (Kelley et al. 2006). Human red blood cell (RBC) membranes phospholipids (PL) contain 3 wt% DHA and less than 0.5 wt% EPA when not supplemented with fish oils (Cao et al. 2006). Thus, most human tissues contain several fold more DHA than EPA. When an animal is put on n-3 PUFA free diet, the body specifically retains DHA not EPA. Human beings can readily retroconvert DHA to EPA, but the elongation of EPA to DHA is minimal (Arterburn et al. 2006). In men and postmenopausal women, conversion of ALA to EPA is less than 5% and only negligent to DHA (Brenna et al. 2009). While both EPA and DHA have important roles, the above facts indicate that adequate supply of DHA may be able to meet the needs for both EPA and DHA, while EPA can not meet the need for DHA. Hence, the effects of DHA will be emphasized here, while the effects of EPA will be discussed only to highlight the differences between the effects of these two fatty acids. Effects of fish oils will be discussed only for those response variables where studies with EPA and DHA individually have not been conducted.

## EFFECT OF DHA ON BLOOD LIPIDS

The effects of DHA on blood lipids and other risk factors for CVD have recently been reviewed (Holub 2009; Ryan et al. 2009). As an example of the effects of n-3 PUFA on blood lipids, the results from a recent DHA study with hypertriglyceridemic men are reported here (Kelley et al. 2007, 2008, 2009). Hypertriglyceridemic men were selected for this study because they have a preponderance of small dense LDL particles and are at increased risk for CVD (Rizzo and Berneis 2006). The study group of hypertriglyceridemic men, aged 39–66 years participated in a double-blind, randomized, placebo-controlled, parallel 98-day study (17/group). They received no supplements for the first 8 days and received either 7.5 g/day DHA Algal oil (3 g DHA/d) or olive oil (placebo) for the last 90 days (n = 17/group). Fasting and postprandial blood samples were drawn twice before the start of supplements, once in the middle, and twice at the end of supplements. The results from two blood draws, at the start and end of the study, were each averaged (Kelley et al. 2007).

DHA decreased the fasting TG by 25% within 45 days of its supplementation and this decrease was maintained for the next 45 days. It did not alter fasting total cholesterol, but increased LDL-C by 15% and HDL-C by 8% (Table 6.1). Postprandial plasma triglyceride (PPTG) and cholesterol concentrations were determined every 2 hours after a test meal. Curves for the PPTG concentrations at the start, middle, and end of the study overlapped in the placebo group, but in the DHA group the PPTG concentrations were blunted by approximately 25% at the middle and end of the study when compared to the corresponding values at the start of the study (Figure 6.1).

The area under curve (AUC) for postprandial concentrations of TG, total-, LDL-, and HDL-C mimicked the corresponding changes in their fasting concentrations in both dietary groups (Table 6.1 and Figure 6.2). Increases in LDL- and HDL-C following DHA supplement were significant within the DHA group, but the interaction between day and diet were not significant. An increase in the LDL-C could be viewed as an increased risk for CVD. However, the risk is not determined by the total LDL-C but by its distribution and number of total and small dense LDL "particles," which were both decreased following DHA supplementation (Mora et al. 2007). DHA supplementation decreased the number of total LDL particles by 11%, those of small dense LDL particles by 22%, and those of intermediate density lipoproteins (IDL) particles by 53%, while it increased the number of large LDL particles by 127% (Table 6.2 and Figure 6.3). A decrease in the number of small dense LDL particles caused by DHA indicates a reduction in the risk for atherosclerosis, because these particles can penetrate through the endothelial lining of arterial wall. The increase in the LDL-C was due to an increase in the number of large LDL particles which cannot

## TABLE 6.1
### Effect of DHA Supplementation on Fasting Plasma Lipids, Lipoproteins, and Apoproteins

| Variable | DHA Group (n = 17) Baseline | Mid Intervention | End Intervention | Placebo Group (n = 17) Baseline | Mid Intervention | End Intervention | Day X Diet P Value |
|---|---|---|---|---|---|---|---|
| TGs, mmol/L | 2.79 ± 0.27[a] | 2.09 ± 0.15[b] | 2.11 ± 0.15[b] | 2.90 ± 0.25 | 2.63 ± 0.24 | 2.67 ± 0.23 | 0.03 |
| Cholesterol, mmol/L | 5.43 ± 0.21 | 5.56 ± 0.24 | 5.69 ± 0.21 | 5.39 ± 0.24 | 5.31 ± 0.28 | 5.30 ± 0.21 | 0.22 |
| LDL-C, mmol/L | 3.10 ± 0.20[a] | 3.49 ± 0.22[b] | 3.58 ± 0.19[b] | 3.11 ± 0.23 | 3.13 ± 0.23 | 3.10 ± 0.21 | 0.05 |
| HDL-C, mmol/L | 1.07 ± 0.06[a] | 1.15 ± 0.08[b] | 1.15 ± 0.07[b] | 0.95 ± 0.03 | 0.97 ± 0.05 | 0.98 ± 0.04 | 0.46 |
| Total Chol: HDL-C | 5.09 ± 0.33 | 4.83 ± 0.30 | 4.95 ± 0.30 | 5.67 ± 0.31 | 5.47 ± 0.25 | 5.5 ± 0.3 | 0.74 |
| TGs: HDL-C | 6.78 ± 1.22[a] | 4.51 ± 0.59[b] | 4.57 ± 0.50[b] | 7.32 ± 0.87 | 6.57 ± 0.76 | 6.65 ± 0.73 | 0.02 |
| LDL-C: HDL-C | 2.90 ± 0.2[a] | 3.03 ± 0.23[b] | 3.11 ± 0.21[b] | 3.27 ± 0.24 | 3.23 ± 0.20 | 3.16 ± 0.23 | 0.03 |
| Apo A1, μmol/L | 45.41 ± 1.46 | 45.95 ± 1.57 | 45.02 ± 1.36 | 41.34 ± 1.36 | 41.8 ± 1.54 | 41.06 ± 1.46 | 0.53 |
| HDL-C:Apo A1 | 0.32 ± 0.01[a] | 0.34 ± 0.02[b] | 0.35 ± 0.01[b] | 0.32 ± 0.01 | 0.32 ± 0.01 | 0.33 ± 0.01 | 0.06 |
| Apo B μmol/L | 3.94 ± 0.20 | 4.05 ± 0.21 | 4.15 ± 0.19 | 4.06 ± 0.24 | 4.05 ± 0.22 | 3.98 ± 0.19 | 0.45 |
| LDL-C:Apo B | 1.15 ± 0.04[a] | 1.26 ± 0.03[b] | 1.26 ± 0.03[b] | 1.11 ± 0.03 | 1.12 ± 0.04 | 1.13 ± 0.03 | 0.03 |
| Apo CIII, μmol/L | 20.29 ± 1.25[a] | 17.56 ± 0.8[b] | 17.67 ± 0.8[b] | 21.31 ± 1.48 | 20.86 ± 1.37 | 20.55 ± 1.25 | 0.01 |
| Apo E, μmol/L | 2.29 ± 0.41 | 1.99 ± 0.26 | 2.05 ± 0.29 | 2.11 ± 0.23 | 1.99 ± 0.18 | 1.93 ± 0.15 | 0.2 |
| Lp-a, μmol/L | 0.63 ± 0.16 | 0.61 ± 0.14 | 0.60 ± 0.13 | 0.63 ± 0.16 | 0.69 ± 0.19 | 0.67 ± 0.17 | 0.62 |
| Systolic BP, mm Hg | 123.6 ± 2.98[a] | 116.6 ± 3.85[b] | 120.8 ± 2.66[ab] | 120.8 ±3.19 | 123.4 ± 4.17 | 120.4 ± 2.57 | 0.02 |
| Diastolic BP, mm Hg | 72.3 ± 2.08[a] | 69.35 ± 2.3[b] | 71.9 ± 2.63[ab] | 70.8 ± 1.79 | 72.9 ± 2.12 | 72.3 ± 2.37 | 0.06 |
| HR, BPM | 69.2 ± 2.33[a] | 63.47 ± 1.76[b] | 65.8 ± 2.73[ab] | 66.7 ± 3.39 | 70.47 ± 1.91 | 69.5 ± 2.31 | 0.02 |

*Source:* Kelley, D. S. et al. 2007. *Am J Clin Nutr* 86(2):324–33. With permission.

*Notes:* All values are mean ± SEM (n = 17). Data were analyzed by repeated measures ANOVA with single degree of freedom contrasts between baseline, mid-, and end of intervention periods. Variables within each row and treatment group, with different superscripts are significantly different (P < 0.05). For HDL–C, the P value for the day effect was 0.35. For the baseline period, none of the response variables significantly differed between the two groups (nonpaired t test). Abbreviations: TGs, triacylglycerols; LDL–C, LDL–cholesterol; HDL–C, HDL–cholesterol; BP, blood pressure; BPM, beats per minute.

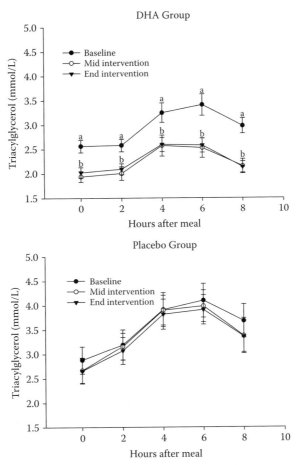

**FIGURE 6.1**  Postprandial plasma triacylglycerol concentrations for baseline (mean d -7 and 0), mid intervention (d 45), and end intervention (mean d 84 and 91). Each curve represents one of the study periods showing triacylglycerol concentrations at 0, 2, 4, 6, and 8 hours after the test meal. All values are mean ± SEM, n = 14. Data were analyzed by repeated measures analysis of variance (ANOVA) with single degree of freedom contrasts between baseline, mid-, and end of intervention periods. Different letters on the curves indicate significant (P < 0.05) differences among different study periods at a give time point. Interaction between diet and day was significant (P < 0.05). Triacylglycerol concentrations at baseline did not differ between the two groups. (From Kelley, D. S. et al. 2007. *Am J Clin Nutr* 86(2):324–33. With permission.)

penetrate through the endothelial linings and are not atherogenic. Similar to the changes in the distribution of LDL-C particles into groups of different sizes, the concentration of large HDL-C particles increased and those of small HDL-C particles decreased (Table 6.2 and Figure 6.4). In the placebo group none of the response variables changed during the course of the study.

DHA supplementation did not alter the number of total and medium VLDL particles, but it decreased the number of atherogenic large VLDL particles by 87% and increased the number of small VLDL particles by 133% (Table 6.2 and Figure 6.5). A change in the VLDL particle size indicates an increased clearance of VLDL by the DHA treatment. Again, placebo oil did not have any effect on the number of VLDL particles of all sizes.

The mean particle size for the VLDL particles decreased by 11 nm (22%), for LDL increased by 0.6 nm (3%), and it did not change for HDL (Table 6.2). DHA caused a large and clinically relevant significant decrease in the mean VLDL particle size. An increase of 0.6 nm in the mean size of LDL particles may seem trivial compared to the change in mean size of VLDL particles, but it is highly

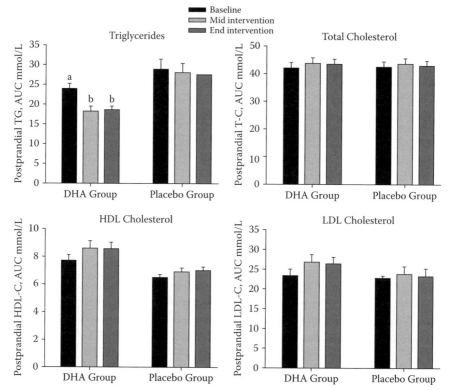

**FIGURE 6.2**    Area under curve (AUC) for postprandial (0, 2, 4, 6, and 8 hours) plasma triacylglycerol, total-, HDL-, and LDL-cholesterol concentrations. Baseline, mean of d -7 and 0; mid intervention, d 45; end intervention, mean d 84 and 91. All values are mean ± SEM, n = 14. Data were analyzed by repeated measures ANOVA with single degree of freedom contrasts between baseline, mid-, and end of intervention periods. Within each treatment group, bars with different superscripts are significantly different (P < 0.05). Interaction between day and diet was significant (P < 0.05) for triacylglycerol area only. For the baseline period, none of the response variables significantly differed between the two groups (nonpaired t test). (From Kelley, D. S. et al. 2007. *Am J Clin Nutr* 86(2):324–33. With permission.)

significant and also physiologically relevant. It is important because there is a difference of only 1 nm in the mean size of LDL particles found in healthy middle-aged men and men with metabolic syndrome; DHA produced 60% of this change in diameter (0.6 nm) within 90 days of its supplementation.

Another independent marker used to assess the risk for CVD is the n-3 index, which is the sum of EPA and DHA as a percentage of the total fatty acids in RBC lipids. This index ranges from 3 to 9 in most people; an n-3 index < 4 compared with an index of 8 is associated with a tenfold increase in the risk of sudden cardiac death (Harris 2007). DHA supplementation resulted in an increase in the n-3 index from less than 4 to greater than 8 (Figure 6.6a). The increase in n-3 index resulted from increases in the concentrations of both EPA (2X) and of DHA (3X). Since there was no EPA in the supplement, an increase in RBC-EPA occurred by the retroconversion of DHA. DHA supplementation also caused a greater than a 100% increase in the EPA:arachidonic acid (AA; 20-:4 n-6) ratio in RBC lipids (Figure 6.6b) which is used to evaluate inflammation status.

Plasma RLP-C or remnants like particle-C or remnant lipoprotein-C are considered a novel risk factor for CVD. RLP are produced from VLDL and are the major atherogenic lipoproteins that can be taken up by macrophages without oxidative modification to form foam cells (Nakajima et al. 2006). DHA supplementation decreased the fasting RLP-C by 35% and 25% at the middle and end of the study, respectively, when compared to the corresponding values prior to the supplement (Figure 6.6c). Since RLP-C are formed from the chylomicrons and VLDL particles by the

**TABLE 6.2**

**Treatment Effects on Fasting Lipoprotein Particle Concentrations and Sizes as Determined by NMR**

| Plasma Lipid Variables | DHA Group (n = 17) | | | Placebo Group (n = 17) | | | Day X Diet P Value |
|---|---|---|---|---|---|---|---|
| | Baseline | Mid Intervention | End Intervention | Baseline | Mid Intervention | End Intervention | |
| **Very-Low-Density Lipoprotein Particles (nmol/L)** | | | | | | | |
| Total | $135.5 \pm 7.0^a$ | $149.1 \pm 9.6^b$ | $146.8 \pm 9.1^{ab}$ | $136.9 \pm 12.6$ | $134.7 \pm 11.0$ | $137.4 \pm 11.9$ | 0.59 |
| Large | $10.8 \pm 2.8^a$ | $0.9 \pm 0.4^b$ | $1.4 \pm 0.6^b$ | $13.1 \pm 2.6$ | $11.1 \pm 1.9$ | $12.0 \pm 1.5$ | 0.0001 |
| Medium | $97.0 \pm 7.5^a$ | $84.8 \pm 10.1^{ab}$ | $82.0 \pm 9.0^b$ | $94.0 \pm 12.7$ | $86.8 \pm 12.9$ | $85.4 \pm 10.9$ | 0.70 |
| Small | $27.7 \pm 4.5^a$ | $64.5 \pm 6.8^b$ | $63.5 \pm 8.0^b$ | $29.8 \pm 4.7$ | $36.9 \pm 4.9$ | $40 \pm 3.4$ | 0.02 |
| **Low-Density Lipoprotein Particles (nmol/L)** | | | | | | | |
| Total | $1693 \pm 133^a$ | $1507 \pm 120^b$ | $1567 \pm 110^b$ | $1753 \pm 126$ | $1685 \pm 125$ | $1668 \pm 111$ | 0.75 |
| Intermediate | $75.5 \pm 14.0^a$ | $35.9 \pm 10.8^b$ | $33.9 \pm 6.0^b$ | $69.2 \pm 13.6$ | $67.4 \pm 12.9$ | $60.4 \pm 10.8$ | 0.02 |
| Large | $179.2 \pm 40.2^a$ | $393.3 \pm 61.6^b$ | $406.7 \pm 60.5^b$ | $128.8 \pm 31.5$ | $144.7 \pm 41.5$ | $189.8 \pm 43.5$ | 0.007 |
| Small | $1439 \pm 131^a$ | $1079 \pm 122^b$ | $1127 \pm 113^b$ | $1555 \pm 118$ | $1473 \pm 116$ | $1417 \pm 110$ | 0.19 |
| **High-Density Lipoprotein Particles (µmol/L)** | | | | | | | |
| Total | $29.1 \pm 0.9^a$ | $27.5 \pm 1.0^b$ | $27.5 \pm 0.8^b$ | $27.5 \pm 1.1$ | $28.5 \pm 1.1$ | $28.0 \pm 0.9$ | 0.04 |
| Large | $3.2 \pm 0.5^a$ | $5.2 \pm 0.7^b$ | $4.5 \pm 0.7^c$ | $2.0 \pm 0.3$ | $2.0 \pm 0.3$ | $2.4 \pm 0.3$ | 0.0002 |
| Medium | $0.9 \pm 0.5$ | $0.6 \pm 0.3$ | $0.6 \pm 0.4$ | $0.4 \pm 0.2$ | $0.9 \pm 0.4$ | $0.6 \pm 0.2$ | 0.13 |
| Small | $25.0 \pm 0.8^a$ | $21.6 \pm 0.7^b$ | $22.4 \pm 0.7^b$ | $25.1 \pm 0.8$ | $25.6 \pm 0.8$ | $25.0 \pm 0.8$ | 0.0003 |
| **Mean Particle Size (nm)** | | | | | | | |
| VLDL | $51.8 \pm 1.6^a$ | $40.7 \pm 0.9^b$ | $41.3 \pm 1.1^b$ | $53.6 \pm 1.5$ | $52.8 \pm 1.6$ | $53.4 \pm 1$ | 0.0001 |
| LDL | $19.9 \pm 0.1^a$ | $20.5 \pm 0.2^b$ | $20.5 \pm 0.2^b$ | $19.6 \pm 0.1$ | $19.7 \pm 0.1$ | $19.8 \pm 0.1$ | 0.05 |
| HDL | $8.5 \pm 0.1$ | $8.6 \pm 0.07$ | $8.5 \pm 0.1$ | $8.3 \pm 0.1$ | $8.3 \pm 0.04$ | $8.3 \pm 0.03$ | 0.11 |

*Source:* Kelley, D. S. et al. 2007. *Am J Clin Nutr* 86(2):324–33. With permission.

*Notes:* All values are mean ± SEM. Data were analyzed by repeated measures ANOVA with single degree of freedom contrasts between baseline, mid-, and end of intervention periods. Within each treatment group, numbers with different superscripts for a given variable are significantly different ($P < 0.05$). P values for the individual effects of diet and day effects were 0.66 and 0.56 (medium VLDL particles), 0.51, and 0.06 (total LDL particles), 0.12 and 0.005 (small LDL particles), respectively. For the baseline period, none of the response variables significantly differed between the two groups (nonpaired t test).

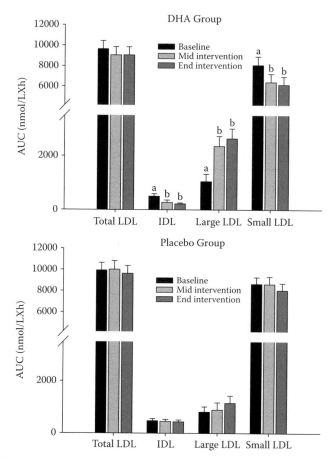

**FIGURE 6.3** AUC for postprandial (0, 4, and 6 hours) LDL subclasses determined by NMR. Baseline, mean of d -7 and 0; mid intervention, d 45; end intervention, mean of d 84 and 91. All values are mean ± SEM, n = 14. Data were analyzed by repeated measures ANOVA with single degree of freedom contrasts between baseline, mid-, and end of intervention periods. Within each dietary group and subfractions, bars with different superscripts are significantly different (P < 0.05). Day x diet p < 0.05 for IDL, large LDL, and small LDL. For the baseline period, none of the response variables significantly differed between the two groups (nonpaired t test). (From Kelley, D. S. et al. 2007. *Am J Clin Nutr* 86(2):324–33. With permission.)

removal of TG, the observed decrease in the concentration of RLP-C are in good agreement with the decrease in the concentration of plasma TG discussed earlier in this section.

DHA supplementation decreased plasma TG in more than a dozen human studies. Only a few studies examined its effects on HDL and LDL particle sizes and found an increase in both large LDL and HDL particles and a decrease in total and small dense LDL particles (Mori et al. 2000; Kelley et al. 2007; Ryan et al. 2009; Vrablik et al. 2009). Purified EPA also decreased the plasma TG but in none of those studies it increased the number of large LDL or HDL particles or decreased those of small dense LDL particles (Ryan et al. 2009; Vrablik et al. 2009). These findings suggest that the overall lipid profile in subjects taking DHA supplements may be healthier than those taking EPA supplements.

## PRO- AND ANTI-INFLAMMATORY METABOLITES OF FATTY ACIDS

Human beings do not have the delta 12 and delta 15 desaturases needed for the synthesis of n-6 and n-3 fatty acids, respectively. Hence, linoleic acid (LA; 18:2 n-6) and ALA (18:3 n-3) must be supplied in food as essential fatty acids. Humans can desaturate and elongate both LA and ALA to

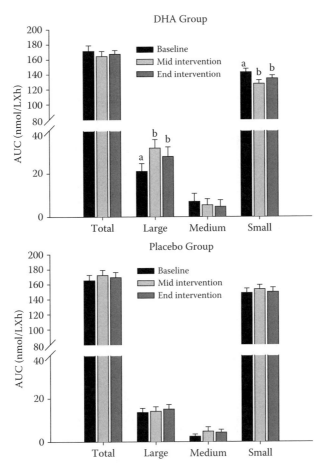

**FIGURE 6.4** AUC for postprandial (0, 4, and 6 hours) HDL subclasses determined by NMR. Baseline, mean of d -7 and 0; mid intervention, d 45; end intervention, mean of d 84 and 91. All values are mean ± SEM, n = 14. Data were analyzed by repeated measures ANOVA with single degree of freedom contrasts between baseline, mid-, and end of intervention periods. Within each treatment group and subfraction, bars with different superscripts are significantly different (P < 0.05). Day x diet p < 0.05 for large and small HDL. For the baseline period, none of the response variables significantly differed between the two groups (nonpaired t test). (From Kelley, D. S. et al. 2007. *Am J Clin Nutr* 86(2):324–33. With permission.)

long-chain polyunsaturated fatty acids (LC-PUFA) of both n-6 (AA) and n-3 (EPA and DHA) series (Figure 6.7). However, because of the large amounts of LA consumption and competition between LA and ALA for the elongation and desaturation enzymes, conversion of ALA to EPA in humans is limited and to DHA is negligent (Brenna et al. 2009). Since EPA and DHA have metabolic effects distinct from those of ALA, these fatty acids also need to be provided in the diet or alternatively the consumption of LA needs to be decreased with a concomitant increase in the consumption of ALA. Because of the abundance of LA in the diet, there is no shortage of AA in the tissues.

In the body, the LC-PUFA are used to produce hormone-like substances that regulate a wide range of functions, including BP, blood clotting, blood lipid levels, and the inflammation response to injury or infection. Free AA released from membrane phospholipids by phospholipase $A_2$ serves as a substrate for the enzymes cyclooxygensases and lipooxygenases (Figure 6.7). AA is metabolized to prostaglandins and thromboxanes by the cyclooxygenases and to leukotrienes by 5-lipooxygenase (5-LOX), lipoxins by 15-LOX, and hydroxy eicosatetraenoic acids (HETES) by 12-LOX. PGE2 and $LTB_4$ have been known as the two most proinflammatory eicosanoids, while lipoxins have been

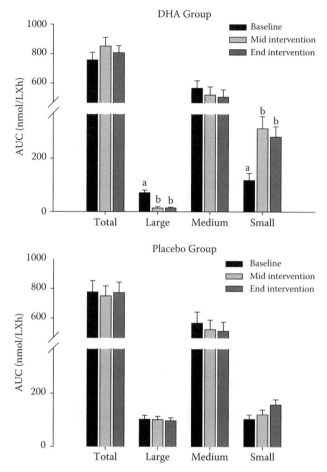

**FIGURE 6.5** AUC for postprandial (0, 4, and 6 hours) VLDL subclasses determined by NMR. Baseline, mean of d -7 and 0; mid intervention, d 45; end intervention, mean of d 84 and 91. All values are mean ± SEM, n = 14. Data were analyzed by repeated measures ANOVA with single degree of freedom contrasts between baseline, mid-, and end of intervention periods. Within each group and subfraction, bars with different superscripts are significantly different (P < 0.05). Day x diet p < 0.05 for large and small VLDL. For the baseline period, none of the response variables significantly differed between the two groups (nonpaired t test). (From Kelley, D. S. et al. 2007. *Am J Clin Nutr* 86(2):324–33. With permission.)

known as the anti-inflammatory eicosanoids (Kelley and Rudolph 2000). However, we now know that PGE$_2$ provides a proinflammatory effect at low concentrations, and an anti-inflammatory effect at high concentrations, because at high concentrations PGE$_2$ inhibits 5-LOX and decreases the production of proinflammatory LTB$_4$, and increases the formation of anti-inflammatory lipoxins by stimulating 15-LOX (Calder 2006).

EPA competes with AA for the cycloxygenase and lipoxygenase enzymes and decreases the production of PGE$_2$ and LTB$_4$. The eicosanoids synthesized from EPA (PGE$_3$ and LTB$_5$) are much less inflammatory than the corresponding metabolites formed from AA (Stulnig 2003). Furthermore, EPA is a precursor of the resolvins of the E series (RvE1 and RvE2) which have anti-inflammatory and protective properties (Serhan 2007; Calder 2009). DHA shares some of the beneficial effects of EPA, however, some of its effects are distinct from those of EPA (Woodman et al. 2002; Yusufi et al. 2003; Rahman et al. 2008). It can inhibit the release of AA from membrane phospholipids by decreasing the activity of phospholipase A$_2$ and can also affect the activity of cell surface receptors (Calder 2006). DHA has a higher potency than EPA in activating (PPARs, RXR) or inhibiting

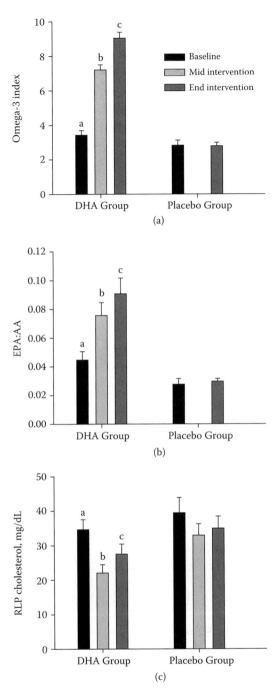

**FIGURE 6.6** Effect of DHA supplementation on (a) RBC (n-3) index, (b) plasma EPA:AA, and (c) RLP-C in hypertriglyceridemic men. Data shown are mean ± SEM, DHA group n = 14 for all variables; for placebo group n = (a) 6, (b) 10, and (c) 14. None of these variables differed between the two groups at baseline. Bars with different superscripts are significantly different ($p < 0.05$, repeated measures ANOVA). Interactions between day and treatments were significant ($P < 0.05$) for all three response variables. (n-3) index on d 45 for the placebo group was not analyzed because this variable did not change at the end of the study. (From Kelley, D. S. et al. 2008. *J Nutr* 138(1):30–5. With permission.)

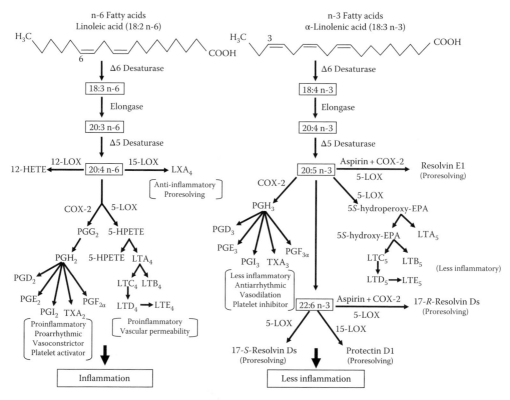

**FIGURE 6.7** The metabolism of n-3 and n-6 PUFA and the biosynthesis of their respective eicosanoid and proresolving mediators. N-3 PUFA are generally less inflammatory than the n-6 PUFA. However, PGE$_2$ derived from n-6 PUFA can have an anti-inflammatory effect by decreasing LTB$_4$ production by the inhibition of 5-LOX and increasing production of LXA$_4$ by stimulating 15-LOX. N-3 PUFA-derived eicosanoids have different physiological potencies than n-6 PUFA-derived eicosanoids. Abbreviations: COX, cyclyoxygenase; EPA, eicosapentaenoic acid; HETE, hydroxyeicosatetraenoic acid; HPETE, hydroperoxyeicosatetraenoic acid; LTA$_4$, leukotriene A$_4$; LXA$_4$, lipoxin A$_4$; LOX, lipoxygenase; PG, prostaglandin; TXA$_2$, thromboxane A$_2$. (From Adkins, Y. and D. S. Kelley. 2010. *J Nutr Biochem* 21(9):781–92. With permission.)

(NFKB, SREBP1c) nuclear receptors (Fedor and Kelley 2009). These nuclear receptors modulate the expression of many genes involved in inflammation. In addition, DHA is a precursor for the synthesis of resolvins of the D series (RvD1, RvD2, RvD3, and RvD4) and docosatrienes, such as 17S-hydroxy-docosahexaenoic acid (17-HDHA) and protectin D1 (PD1), which belongs to the family of protectins (Chiang et al. 2006; Serhan 2006). These compounds have strong anti-inflammatory and inflammation resolving effects. These DHA metabolites are made through pathways distinct from the biosynthesis of prostaglandins, leukotrines, and lipoxins, metabolites of AA (Hong et al. 2003; Mukherjee et al. 2004).

## EFFECTS OF DHA ON THE CONCENTRATIONS OF INFLAMMATORY MARKERS

Inflammation is the normal response of an organism's immune system to the damage caused to its cells and vascularized tissues by viruses, bacteria, injurious chemicals, or physical insults. Blood flow to infected site is increased so WBC can neutralize and remove the damage causing agents. Although painful, inflammation is usually a healing response. Inadequate inflammatory response leads to immunodeficiency, cancer, and infections; however, in some instances inflammation proceeds to a chronic state and is associated with debilitating diseases such as diabetes and CVD. Progression of inflammatory diseases is associated with an increase in the plasma concentrations of one or more markers of inflammation (Calder 2006; Calder 2009).

Plasma markers of inflammation include an increase in the number of circulating WBC, acute phase proteins (C-reactive protein [CRP], serum amyloid A [SAA], fibrinogen), cytokines and their soluble receptors (TNFα, interleukins 1, 6, 7, 8, and 18, interferon γ), adhesion molecules (ICAM-1, VCAM-1, E and P selectin), and plasminogen activator inhibitor 1 (PAI1). An increase in the concentration of insulin and decrease in the concentrations of leptin and adiponectin are also associated with inflammation. Plasma concentrations of several of these markers are increased in a number of inflammatory diseases and are used to evaluate the disease status. Plasma CRP is one of the most commonly used markers of inflammation. There is a 4.4-fold increase in the relative risk (RR) for CVD comparing the highest and lowest quartiles of CRP, while this increase is only 2.4 fold comparing the quartiles of cholesterol (Ridker 2003; Danesh et al. 2004). CRP stimulates mononuclear cells to release tissue factors which are central to initiation of coagulation reactions, complement activation, and neutralization of PAF. Together these factors promote thrombotic response.

Several lines of evidence support the claim that n-3 PUFA have anti-inflammatory effects. First, epidemiological studies showed lesser incidence of inflammatory diseases such as CVD and arthritis in populations consuming more fish than those populations consuming less or no fish (Daviglus et al. 1997; Albert et al. 1998; Zhang et al. 1999; Hu et al. 2002). Second, several studies indicated inverse associations between the estimated n-3 PUFA consumption and the plasma concentrations of inflammatory markers (Zampelas et al. 2005; Niu et al. 2006; Murakami et al. 2008). Third, other studies demonstrated an inverse association between tissue concentrations of n-3 PUFA and of inflammatory markers (Madsen et al. 2001; Pischon et al. 2003; Lopez-Garcia et al. 2004; Klein-Platat et al. 2005; Ferrucci et al. 2006). Fourth, a number of intervention studies with fish oil showed a decrease in the symptoms for inflammatory diseases and a decrease in the *in vivo* and *ex vivo* secretion of inflammatory markers. There have been dozens of studies regarding the effects of fish oil and of individual n-3 PUFA on the *ex vivo* production of inflammatory cytokines and eicosanoids. Those have been summarized in several recent reviews (Calder 2006; Calder 2009). In general, the results indicate anti-inflammatory effects of fish oil, but there have been a number of inconsistencies due to health and age of the participants, amount, duration, and fatty acid composition of the supplements, diets, and the methods used to evaluate the inflammatory status.

Only a limited number of studies have investigated the anti-inflammatory effects of individual n-3 PUFA in human subjects and we found only six studies that examined the effects of DHA alone. Three of these studies found decrease in markers of inflammation following DHA supplementation (Kelley et al. 1998, 2009; Lopez-Garcia et al. 2004) the other three did not (Thies et al. 2001; Mori et al. 2003; Kew et al. 2004). In one study, healthy men were given a supplement of DHA (6 g/day for 90 days) producing a 10% decrease in the number of circulating WBC which resulted from a 20% reduction in the number of circulating neutrophils (Kelley et al. 1998). The absolute number of other types of WBC did not change, but the percentage of lymphocytes was increased because of a reduction in the number of granulocytes. The change in the number of circulating neutrophils was detectable within 50 days of DHA supplementation. In the same study, DHA supplementation caused a 60%–75% decrease in the *ex vivo* secretion of $PGE_2$ and $LTB_4$, and a 30%–40% decrease in the secretion of IL-1β and TNFα within 12 weeks (Kelley et al. 1999). Production of both IL-1β and TNFα were decreased within 8 weeks of DHA supplementation, but those were not significant. None of these variables changed in the placebo group.

The second DHA study was of the same duration as the one discussed above (91 days), but it was conducted in hypertriglyceridemic men and the DHA supplement was one-half (3 g/day) of the amount served in the first study (Kelley et al. 2009). In this study, the number of circulating neutrophils decreased by 11% within 45 days of DHA supplement, and this reduction was maintained until the end of study (Kelley et al. 2009). This change in neutrophil numbers is about half of what was found in the first study, most likely due to the dose of DHA used. The reduction in the

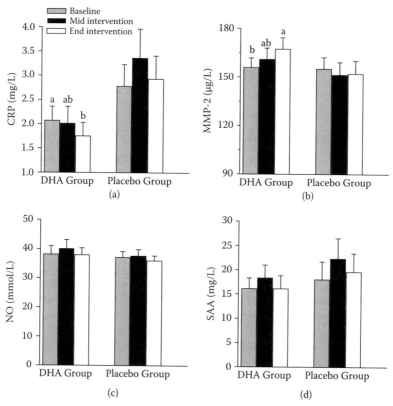

**FIGURE 6.8** Effect of DHA supplementation on circulating concentrations of (a) CRP, (b) MMP-2, (c) NO, and (d) SAA in hypertriglyceridemic men. Baseline represents the mean from study days -7 and 0, middle intervention represents study day 45, and end intervention represents the mean from study days 84 and 91. Data are mean ± SEM (n = 17). Means without a common letter differ (P < 0.05) (ANOVA). (From Kelley, D. S. et al. 2009. *J Nutr* 139(3):495–501. With permission.)

number of circulating neutrophils caused by DHA may be clinically important in conditions like acute respiratory distress syndrome, which result from an increase in the activity and number of circulating neutrophils. The number of circulating neutrophils did not change in the placebo group. The concentration of other markers of inflammation did not significantly change within 45 days, but by 91 days, CRP decreased by 15%, IL-6 by 23% and GM-CSF by 21%, and the anti-inflammatory marker, MMP-2, increased by 7% (Figures 6.8 and 6.9). Plasma concentrations of other cytokines (IL-1β, IL-2, IL-8, IL-10, and TNFα) and adhesion molecules (ICAM-1, VCAM-1, and E-selectin) did not change in both the DHA and placebo groups.

The third human study examined the effects of doubling the dose of DHA (200, 400, 800, 1,600 mg/day, over 2-week intervals each) supplementation to healthy men on the *ex vivo* secretion of $LTB_4$ and $LTB_5$ by the cultured peripheral blood mononuclear cells stimulated with the calcium ionophore A23187 (Stanke-Labesque et al. 2008). The ratio of *ex vivo* secreted $LTB_4$ and $LTB_5$ was significantly increased at DHA concentrations of 800 and 1600 mg/day, and they returned back to normal within 2 weeks of discontinuation of DHA. These results show that DHA supplementation of 800 mg/day or higher reduced the inflammatory response.

The lack of an effect of DHA on the concentration of inflammatory markers in the other three studies was most likely to the low dose of DHA (0.7 g/day, 12 weeks) (Thies et al. 2001) or short duration (4 or 6 weeks) (Mori et al. 2003; Kew et al. 2004) of its supplementation. EPA supplementation also decreased the concentration of inflammatory markers in several studies. There have been a limited number of studies which compared the anti-inflammatory effects of equal amounts of EPA

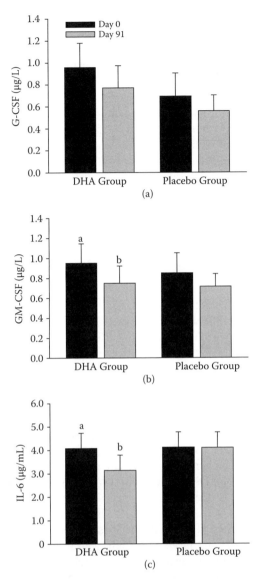

**FIGURE 6.9** Effect of DHA supplementation on plasma concentrations of (a) G-CSF, (b) GM-CSF, and (c) IL-6 in hypertriglyceridemic men. Data are mean ± SEM (n = 17 for all except IL-6 where n = 11 for the DHA group). Means without a common letter differ ($P < 0.05$) (ANOVA). (From Kelley, D. S. et al. 2009. *J Nutr* 139(3):495–501. With permission.)

and DHA; results varied, some suggesting more potent effects of DHA (Mori et al. 2003; Yusof et al. 2008), while others suggested greater effects of EPA (Sierra et al. 2008). Further studies are needed to distinguish the anti-inflammatory effects of EPA and DHA.

## EFFECTS OF N-3 PUFA ON PLASMA CONCENTRATIONS OF ADHESION MOLECULES

There are only a few studies that examined the effects of purified EPA and/or DHA on the plasma concentrations adhesion molecules (VCAM1, ICAM1, and E and P selectins). In one study, Omacor 4 g/day (EPA-EE 465 + DHA-EE 375 mg/g) supplementation given to hypertriglyceridemic subjects for seven months significantly decreased the plasma concentrations of sCAM-1 and sE selectin, but did not change sVCAM-1 concentration (Abe et al. 1998). In another study with type 2 diabetic patients, EPA (1.8 g/day, 4 weeks) significantly decreased plasma concentrations of E selectin, and those of platelet and monocyte-derived micro particles that have procoagulant activities; concentration of P-selectin was also decreased but did not attain significance (Nomura et al. 2003). In contrast to the results from above studies with hypertriglyceridemic or diabetic subjects supplementing 2 or 6.6 g/day of a 1:1mixture of EPA and DHA to healthy subjects for 12 weeks did not decrease the plasma concentrations of VCAM and ICAM; only the higher concentration of n-3 PUFA decreased that of E-selectin (Eschen et al. 2004). In another study with healthy subjects, DHA + EPA but not DHA alone decreased plasma concentrations of sVCAM1 and E-selectin (Thies et al. 2003). In another DHA study with hypertriglyceridemic men circulating concentrations of none of the adhesion molecules changed (Kelley et al. 2009). Thus, the effects of n-3 PUFA on plasma concentrations of adhesion molecules have been variable. Further studies are needed to determine the effects of individual n-3 PUFA on the plasma concentration of these molecules.

## EFFECTS OF n-3 PUFA ON BLOOD PRESSURE AND HEART RATE

Initial indications for the BP-lowering effects of n-3 PUFA came from the studies of Dyerberg et al. with Greenland Inuits, when they found a negative association between plasma concentration of n-3 PUFA and diastolic BP (Dyerberg et al. 1975). Preliminary studies with normotensive and hypertensive subjects support the BP-lowering effects of n-3 PUFA, although the results have been variable (Cicero et al. 2009; Holub 2009; Ryan et al. 2009). Two different meta-analyses concluded that the effect of n-3 PUFA on BP is dose dependent with a minimal efficacious dose of 3 g/day, a BP decrease of −0.66/−0.35 mm Hg/g n-3 PIFA (Mori and Woodman 2006; Cicero et al. 2009).

These conclusions were based on results from studies that used a mixture of EPA and DHA. More recent studies have been conducted with EPA and DHA individually. In a study with mildly hypercholesterolemic individuals, DHA but not EPA significantly reduced BP and HR (Mori et al. 1999). However, results from another study by the same investigators in treated hypertensive type 2 diabetic patients, neither EPA nor DHA reduced BP (Woodman et al. 2002). This discrepancy may be because of hyperglycemia or the use of pharmacologic treatments. In two other studies with healthy normotensive subjects DHA did not decrease BP (Conquer and Holub 1998; Geppert et al. 2006). DHA decreased both BP and HR in two studies with hypertriglyceridemic subjects (Keller et al. 2007; Kelley et al. 2007) and in another study with healthy middle-aged men and women (Theobald et al. 2007). These discrepancies seem to be related to the dose and duration of DHA treatment and the health status of the study participants. Similarly, in a recent study, DHA supplementation (3 g/day) for 45 days decreased HR by 8%, systolic BP by 6% and diastolic BP by 4% (Kelley et al. 2007). With the continued supplementation of DHA for 91 days, HR and BP were still reduced compared to the corresponding values prior to the start of DHA supplementation, but the decreases were not statistically significant. At the end of the study HR was decreased by only 5%, systolic BP by 2.3%, and diastolic BP by 0.5%. It is possible that the BP-lowering effect of DHA is transient. Further studies are needed to compare the effects of EPA and DHA on BP and HR.

## EFFECT OF n-3 PUFA ON LIPID PEROXIDATION

Increased lipid peroxidation has been associated with the development and progression of a number of chronic human diseases including cardiovascular disease and diabetes (Kaikkonen et al. 2004; Spiteller 2005, 2007; Yoshida and Niki 2006). It damages biological membranes which changes membrane fluidity and functions including receptor activity, and nutrient and ion transport. Peroxidation of LDL lipids renders the lipoprotein pro-atherogenic. Many lipid peroxidation products exert cytotoxic effects and alter cell signaling (Leonarduzzi et al. 2000; Tang et al. 2002;

Niki et al. 2005; Yoshida et al. 2007). Thus, control of lipid peroxidation plays a critical role in health maintenance and disease prevention.

Lipid peroxidation in biological systems is believed to increase in proportion to an increase in the number of double bonds in the fatty-acid chain (PUFA) and inversely to the antioxidant levels within the cells. Long-chain PUFA, particularly of the n-3 type reduce the risk of a number of chronic diseases, yet there remains a concern that they may also increase the risk for chronic diseases by increasing lipid peroxidation.

Human dietary studies with n-3 PUFA regarding their effects on lipid peroxidation have had variable results (Harats et al. 1991; Meydani et al. 1991; Suzukawa et al. 1995; Palozza et al. 1996; Wu et al. 2006; Egert et al. 2007; Parra et al. 2007). The effect of DHA ranged from protection to increased lipid peroxidation depending upon its dose (Calzada et al. 2010; Guillot et al. 2009). These inconsistencies result from the different methods used (Cu [2+] catalyzed lipid peroxidation *ex vivo*, F-2 isoprostanes, MDA, TBARS, ORAC, etc.) and their limitations. Future studies using more sensitive and reproducible methods are needed to establish if there is any risk of lipid peroxidation with increased consumption of n-3 PUFA and how to minimize it by increasing the intake of anti-oxidant nutrients.

## EFFECTS OF N-3 PUFA ON INTIMA-MEDIA THICKNESS, FLOW-MEDIATED DILATION AND PLAQUE STABILITY

Carotid intima-media thickness (CIMT) has been used as a surrogate endpoint for evaluating the regression and/or progression of atherosclerotic CVD. In general, wall thickening may be in the intimal layer or in the muscular, medial layer. One study compared the intima-media thickness (IMT) and coronary artery calcification (CAC) among the Japanese men (300 subjects, 40–49 years old) from Kusatsu, Shiga, Japan and 306 U.S. white men from Allegheny County, Pennsylvania (Sekikawa et al. 2008). The Japanese men had twice the plasma concentrations of n-3 PUFA and significantly less IMT and CAC than the U.S. men. Once they corrected for the serum n-3 PUFA, the differences between the Japanese and American men became smaller, though still significant. Part of these differences between subject groups was due to n-3 PUFA and the remainder due to genetic, environmental, and lifestyle factors.

When blood flow increases through a vessel, the vessel dilates which is called flow-mediated dilatation (FMD). This property is an effective means to evaluate endothelial dysfunction non-invasively; it is useful for early detection of atherosclerosis and for determining the efficacy of treatment. In a study with hypercholesterolemic subjects, supplementing a mixture of EPA and DHA, 4 g/day for 120 days, significantly increased FMD while an equivalent amount of corn oil did not (Goodfellow et al. 2000). N–3 PUFA supplementation in this study did not alter the endothelial independent dilation induced by glyceryl trinitrate.

Carotid endarterectomy is an operation during which a vascular surgeon removes the plaque from the carotid artery to restore blood flow. In a randomized study with patients who were waiting for carotid endarterectomy (7–189 days, 52–57/group), they took a supplement of either fish oil or sunflower oil, or a placebo made of soybean and palm oils (4 g/day) (Thies et al. 2003). On the day of surgery carotid plaques were collected, rinsed, fixed, and evaluated according to the American Heart Association (AHA) criterion for the plaque stages and for cap thickness. The group receiving the fish oil had significantly higher number of stage 4, and less of the stage 5 plaques than those in the sunflower oil group (stage 5 is more advanced than 4). There were also fewer plaques with thinner caps in the fish oil group than those in sunflower oil group; thinner caps rupture easier than those with thicker caps. Since n-3 PUFA from fish oil had been incorporated into the caps, the authors concluded that n-3 PUFA may enhance the stability of the caps.

Coronary artery diameters and stenosis were monitored at the start and end of a study in 228 postmenopausal women with coronary artery disease (CAD), who were supplemented either with or without estrogen therapy for 3 years (Erkkila et al. 2006). There was no effect of hormone therapy on either response tested. The subjects were then divided into groups with those below and above the median concentrations of ALA, EPA, and DHA in plasma PL. The differences between the initial and final readings for mean diameter and stenosis were not affected by the plasma PL ALA and EPA concentrations; both response variables were significantly lower in the subjects who had DHA concentrations above the median than those who had it below the median. The number of new lesions that appeared was also lower in the high than in the low DHA group. Authors concluded that plasma DHA significantly reduced progression of coronary atherosclerosis over the 3-year follow up in postmenopausal women with established CAD. Together the results from these studies suggest that n-3 PUFA affect the development of new atherosclerotic lesions, calcification, FMD of blood vessels, and the stability of the plaques.

## CONCLUSIONS AND FUTURE DIRECTIONS

Dietary intervention with long chain n-3 PUFA decreased the risk factors for atherosclerosis in most human studies discussed in this chapter. These benefits resulted from decreases in fasting and post-prandial plasma triglycerides, large VLDL particles, total and small dense LDL particles, RLP-C, and an increase in large LDL particles. The results from most studies document the anti-inflammatory and BP-lowering effects of n-3 PUFA, although there have been some exceptions. n-3 PUFA decreased plasma concentrations of adhesion molecules in some, but not an equal number of other studies. Limited numbers of studies have demonstrated the effects of n-3 PUFA on blood vessel calcification, FMD, and plaque stability. In general, both EPA and DHA provided health benefits, but some effects, such as the reduction in BP and the number of small dense LDL particles, seem to be limited to DHA. Future studies are needed to determine the optimal doses of EPA and DHA individually and to determine their synergistic, additive, or antagonistic effects. These doses will vary with the health status of the subjects, response tested, and composition of the basal diets. Studies are also needed to understand the underlying mechanisms of each of these physiologic responses.

## REFERENCES

Abe, Y., B. El-Masri, K. T. Kimball, H. Pownall, C. F. Reilly, K. Osmundsen, C. W. Smith, and C. M. Ballantyne. 1998. Soluble cell adhesion molecules in hypertriglyceridemia and potential significance on monocyte adhesion. *Arterioscler Thromb Vasc Biol* 18(5):723–31.

Adkins, Y. and D. S. Kelley. 2010. Mechanisms underlying the cardioprotective effects of omega-3 polyunsaturated fatty acids. *J Nutr Biochem* 21(9):781–92.

Albert, C. M., C. H. Hennekens, C. J. O'Donnell, U. A. Ajani, V. J. Carey, W. C. Willett, J. N. Ruskin, and J. E. Manson. 1998. Fish consumption and risk of sudden cardiac death. *JAMA* 279(1):23–8.

Albert, C. M., H. Campos, M. J. Stampfer, P. M. Ridker, J. E. Manson, W. C. Willett, and J. Ma. 2002. Blood levels of long-chain n-3 fatty acids and the risk of sudden death. *N Engl J Med* 346(15):1113–8.

Arterburn, L. M., E. B. Hall, and H. Oken. 2006. Distribution, interconversion, and dose response of n-3 fatty acids in humans. *Am J Clin Nutr* 83(6 Suppl):1467S–76S.

Brenna, J. T., N. Salem, Jr., A. J. Sinclair, and S. C. Cunnane. 2009. Alpha-linolenic acid supplementation and conversion to n-3 long-chain polyunsaturated fatty acids in humans. *Prostaglandins Leukot Essent Fatty Acids* 80(2-3):85–91.

Breslow, J. L. 2006. n-3 fatty acids and cardiovascular disease. *Am J Clin Nutr* 83(6 Suppl):1477S–82S.

Burdge, G. C. and P. C. Calder. 2005. Plasma cytokine response during the postprandial period: A potential causal process in vascular disease? *Br J Nutr* 93(1):3–9.

Calder, P. C. 2006. n-3 polyunsaturated fatty acids, inflammation, and inflammatory diseases. *Am J Clin Nutr* 83(6 Suppl):1505S–19S.

Calder, P. C. 2009. Polyunsaturated fatty acids and inflammatory processes: New twists in an old tale. *Biochimie* 91(6):791–5.

Calzada, C., R. Colas, N. Guillot, M. Guichardant, M. Laville, E. Vericel, and M. Lagarde. 2010. Subgram daily supplementation with docosahexaenoic acid protects low-density lipoproteins from oxidation in healthy men. *Atherosclerosis* 208:467–472.

Cao, J., K. A. Schwichtenberg, N. Q. Hanson, and M. Y. Tsai. 2006. Incorporation and clearance of omega-3 fatty acids in erythrocyte membranes and plasma phospholipids. *Clin Chem* 52(12):2265–72.

Chiang, N., S. Hurwitz, P. M. Ridker, and C. N. Serhan. 2006. Aspirin has a gender-dependent impact on anti-inflammatory 15-epi-lipoxin A4 formation: A randomized human trial. *Arterioscler Thromb Vasc Biol* 26(2):e14–7.

Cicero, A. F., S. Ertek, and C. Borghi. 2009. Omega-3 polyunsaturated fatty acids: their potential role in blood pressure prevention and management. *Curr Vasc Pharmacol* 7(3):330–7.

Conquer, J. A., and B. J. Holub. 1998. Effect of supplementation with different doses of DHA on the levels of circulating DHA as non-esterified fatty acid in subjects of Asian Indian background. *J Lipid Res* 39(2):286–92.

Danesh, J., J. G. Wheeler, G. M. Hirschfield, S. Eda, G. Eiriksdottir, A. Rumley, G. D. Lowe, M. B. Pepys, and V. Gudnason. 2004. C-reactive protein and other circulating markers of inflammation in the prediction of coronary heart disease. *N Engl J Med* 350(14):1387–97.

Daviglus, M. L., J. Stamler, A. J. Orencia, A. R. Dyer, K. Liu, P. Greenland, M. K. Walsh, D. Morris, and R. B. Shekelle. 1997. Fish consumption and the 30-year risk of fatal myocardial infarction. *N Engl J Med* 336(15):1046–53.

Dolecek, T. A. 1992. Epidemiological evidence of relationships between dietary polyunsaturated fatty acids and mortality in the multiple risk factor intervention trial. *Proc Soc Exp Biol Med* 200(2):177–82.

Dyerberg, J., H. O. Bang, and N. Hjorne. 1975. Fatty acid composition of the plasma lipids in Greenland Eskimos. *Am J Clin Nutr* 28(9):958–66.

Egert, S., V. Somoza, F. Kannenberg, M. Fobker, K. Krome, H. F. Erbersdobler, and U. Wahrburg. 2007. Influence of three rapeseed oil-rich diets, fortified with alpha-linolenic acid, eicosapentaenoic acid or docosahexaenoic acid on the composition and oxidizability of low-density lipoproteins: Results of a controlled study in healthy volunteers. *Eur J Clin Nutr* 61(3):314–25.

Erkkila, A. T., N. R. Matthan, D. M. Herrington, and A. H. Lichtenstein. 2006. Higher plasma docosahexaenoic acid is associated with reduced progression of coronary atherosclerosis in women with CAD. *J Lipid Res* 47(12):2814–9.

Eschen, O., J. H. Christensen, R. De Caterina, and E. B. Schmidt. 2004. Soluble adhesion molecules in healthy subjects: A dose-response study using n-3 fatty acids. *Nutr Metab Cardiovasc Dis* 14(4):180–5.

Fedor, D. and D. S. Kelley. 2009. Prevention of insulin resistance by n-3 polyunsaturated fatty acids. *Curr Opin Clin Nutr Metab Care* 12(2):138–46.

Ferrucci, L., A. Cherubini, S. Bandinelli, B. Bartali, A. Corsi, F. Lauretani, A. Martin, C. Andres-Lacueva, U. Senin, and J. M. Guralnik. 2006. Relationship of plasma polyunsaturated fatty acids to circulating inflammatory markers. *J Clin Endocrinol Metab* 91(2):439–46.

Geppert, J., V. Kraft, H. Demmelmair, and B. Koletzko. 2006. Microalgal docosahexaenoic acid decreases plasma triacylglycerol in normolipidaemic vegetarians: A randomised trial. *Br J Nutr* 95(4):779–86.

Goodfellow, J., M. F. Bellamy, M. W. Ramsey, C. J. Jones, and M. J. Lewis. 2000. Dietary supplementation with marine omega-3 fatty acids improve systemic large artery endothelial function in subjects with hypercholesterolemia. *J Am Coll Cardiol* 35(2):265–70.

Guillot, N., E. Caillet, M. Laville, C. Calzada, M. Lagarde, and E. Vericel. 2009. Increasing intakes of the long-chain omega-3 docosahexaenoic acid: effects on platelet functions and redox status in healthy men. *Faseb J* 23(9):2909–16.

Harats, D., Y. Dabach, G. Hollander, M. Ben-Naim, R. Schwartz, E. M. Berry, O. Stein, and Y. Stein. 1991. Fish oil ingestion in smokers and nonsmokers enhances peroxidation of plasma lipoproteins. *Atherosclerosis* 90(2–3):127–39.

Harris, W. S. 2007. Omega-3 fatty acids and cardiovascular disease: A case for omega-3 index as a new risk factor. *Pharmacol Res* 55(3):217–23.

Holub, B. J. 2002. Clinical nutrition: 4. Omega-3 fatty acids in cardiovascular care. *CMAJ* 166(5):608–15.

Holub, B. J. 2009. Docosahexaenoic acid (DHA) and cardiovascular disease risk factors. *Prostaglandins Leukot Essent Fatty Acids* 81(2–3):199–204.

Hong, S., K. Gronert, P. R. Devchand, R. L. Moussignac, and C. N. Serhan. 2003. Novel docosatrienes and 17S-resolvins generated from docosahexaenoic acid in murine brain, human blood, and glial cells: Autacoids in anti-inflammation. *J Biol Chem* 278(17):14677–87.

Hu, F. B., L. Bronner, W.C. Willett, M. J. Stampfer, K. M. Rexrode, C. M. Albert, D. Hunter, and J. E. Manson. 2002. Fish and omega-3 fatty acid intake and risk of coronary heart disease in women. *JAMA* 287(14):1815–21.

Kaikkonen, J., T. P. Tuomainen, K. Nyyssonen, J. D. Morrow, and J. T. Salonen. 2004. C18 hydroxy fatty acids as markers of lipid peroxidation *ex vivo* and *in vivo*. *Scand J Clin Lab Invest* 64(5):457–68.

Keller, D. D., S. Jurgilas, and B. Perry. 2007. Docosahexaenoic acid (DHA) lowers triglyceride levels and improves low density lipoprotein particle size in a statin-treated cardiac risk population. *J Clin Lipidol* 1:151.

Kelley, D. S., P. C. Taylor, G. J. Nelson, and B. E. Mackey. 1998. Dietary docosahexaenoic acid and immuno-competence in young healthy men. *Lipids* 33(6):559–66.

Kelley, D. S., P. C. Taylor, G. J. Nelson, P. C. Schmidt, A. Ferretti, K. L. Erickson, R. Yu, R. K. Chandra, and B. E. Mackey. 1999. Docosahexaenoic acid ingestion inhibits natural killer cell activity and production of inflammatory mediators in young healthy men. *Lipids* 34(4):317–24.

Kelley, D. S. and I. L. Rudolph. 2000. Effect of individual fatty acids of omega-6 and omega-3 type on human immune status and role of eicosanoids. *Nutrition* 16(2):143–5.

Kelley, D. S., G. L. Bartolini, J. W. Newman, M. Vemuri, and B. E. Mackey. 2006. Fatty acid composition of liver, adipose tissue, spleen, and heart of mice fed diets containing t10, c12-, and c9, t11-conjugated linoleic acid. *Prostaglandins Leukot Essent Fatty Acids* 74(5):331–8.

Kelley, D. S., D. Siegel, M. Vemuri, and B. E. Mackey. 2007. Docosahexaenoic acid supplementation improves fasting and postprandial lipid profiles in hypertriglyceridemic men. *Am J Clin Nutr* 86(2):324–33.

Kelley, D. S., D. Siegel, M. Vemuri, G. H. Chung, and B. E. Mackey. 2008. Docosahexaenoic acid supplementation decreases remnant-like particle-cholesterol and increases the (n-3) index in hypertriglyceridemic men. *J Nutr* 138(1):30–5.

Kelley, D. S., D. Siegel, D. M. Fedor, Y. Adkins, and B. E. Mackey. 2009. DHA supplementation decreases serum C-reactive protein and other markers of inflammation in hypertriglyceridemic men. *J Nutr* 139(3):495–501.

Kew, S., M. D. Mesa, S. Tricon, R. Buckley, A. M. Minihane, and P. Yaqoob. 2004. Effects of oils rich in eicosapentaenoic and docosahexaenoic acids on immune cell composition and function in healthy humans. *Am J Clin Nutr* 79(4):674–81.

Klein-Platat, C., J. Drai, M. Oujaa, J. L. Schlienger, and C. Simon. 2005. Plasma fatty acid composition is associated with the metabolic syndrome and low-grade inflammation in overweight adolescents. *Am J Clin Nutr* 82(6):1178–84.

Lee, K. W. and G. Y. Lip. 2003. The role of omega-3 fatty acids in the secondary prevention of cardiovascular disease. *QJM* 96(7):465–80.

Leonarduzzi, G., M. C. Arkan, H. Basaga, E. Chiarpotto, A. Sevanian, and G. Poli. 2000. Lipid oxidation products in cell signaling. *Free Radic Biol Med* 28(9):1370–8.

Lloyd-Jones, D., R. Adams, M. Carnethon, G. De Simone, T. B. Ferguson, K. Flegal, E. Ford, K. Furie, A. Go, K. Greenlund, N. Haase, S. Hailpern, M. Ho, V. Howard, B. Kissela, S. Kittner, D. Lackland, L. Lisabeth, A. Marelli, M. McDermott, J. Meigs, D. Mozaffarian, G. Nichol, C. O'Donnell, V. Roger, W. Rosamond, R. Sacco, P. Sorlie, R. Stafford, J. Steinberger, T. Thom, S. Wasserthiel-Smoller, N. Wong, J. Wylie-Rosett, and Y. Hong. 2009. Heart disease and stroke statistics 2009 update: A report from the American Heart Association Statistics Committee and Stroke Statistics Subcommittee. *Circulation* 119(3):e21–181.

Lopez-Garcia, E., M. B. Schulze, J. E. Manson, J. B. Meigs, C. M. Albert, N. Rifai, W. C. Willett, and F. B. Hu. 2004. Consumption of (n-3) fatty acids is related to plasma biomarkers of inflammation and endothelial activation in women. *J Nutr* 134(7):1806–11.

Madsen, T., H. A. Skou, V. E. Hansen, L. Fog, J. H. Christensen, E. Toft, and E. B. Schmidt. 2001. C-reactive protein, dietary n-3 fatty acids, and the extent of coronary artery disease. *Am J Cardiol* 88(10):1139–42.

Marik, P. E. and J. Varon. 2009. Omega-3 dietary supplements and the risk of cardiovascular events: A systematic review. *Clin Cardiol* 32(7):365–72.

Menotti, A., D. Kromhout, H. Blackburn, F. Fidanza, R. Buzina, and A. Nissinen. 1999. Food intake patterns and 25-year mortality from coronary heart disease: Cross-cultural correlations in the Seven Countries Study. The Seven Countries Study Research Group. *Eur J Epidemiol* 15(6):507–15.

Meydani, M., F. Natiello, B. Goldin, N. Free, M. Woods, E. Schaefer, J. B. Blumberg, and S. L. Gorbach. 1991. Effect of long-term fish oil supplementation on vitamin E status and lipid peroxidation in women. *J Nutr* 121(4):484–91.

Mora, S., M. Szklo, J. D. Otvos, P. Greenland, B. M. Psaty, D. C. Goff, Jr., D. H. O'Leary, M. F. Saad, M. Y. Tsai, and A. R. Sharrett. 2007. LDL particle subclasses, LDL particle size, and carotid atherosclerosis in the Multi-Ethnic Study of Atherosclerosis (MESA). *Atherosclerosis* 192(1):211–7.

Mori, T. A., D. Q. Bao, V. Burke, I. B. Puddey, and L. J. Beilin. 1999. Docosahexaenoic acid but not eicosapentaenoic acid lowers ambulatory blood pressure and heart rate in humans. *Hypertension* 34(2):253–60.

Mori, T. A., V. Burke, I. B. Puddey, G. F. Watts, D. N. O'Neal, J. D. Best, and L. J. Beilin. 2000. Purified eicosapentaenoic and docosahexaenoic acids have differential effects on serum lipids and lipoproteins, LDL particle size, glucose, and insulin in mildly hyperlipidemic men. *Am J Clin Nutr* 71(5):1085–94.

Mori, T. A., R. J. Woodman, V. Burke, I. B. Puddey, K. D. Croft, and L. J. Beilin. 2003. Effect of eicosapentaenoic acid and docosahexaenoic acid on oxidative stress and inflammatory markers in treated-hypertensive type 2 diabetic subjects. *Free Radic Biol Med* 35(7):772–81.

Mori, T. A. and R. J. Woodman. 2006. The independent effects of eicosapentaenoic acid and docosahexaenoic acid on cardiovascular risk factors in humans. *Curr Opin Clin Nutr Metab Care* 9(2):95–104.

Mukherjee, P. K., V. L. Marcheselli, C. N. Serhan, and N. G. Bazan. 2004. Neuroprotectin D1: A docosahexaenoic acid-derived docosatriene protects human retinal pigment epithelial cells from oxidative stress. *Proc Natl Acad Sci USA* 101(22):8491–6.

Murakami, K., S. Sasaki, Y. Takahashi, K. Uenishi, M. Yamasaki, H. Hayabuchi, T. Goda, J. Oka, K. Baba, K. Ohki, K. Muramatsu, and Y. Sugiyama. 2008. Total n-3 polyunsaturated fatty acid intake is inversely associated with serum C-reactive protein in young Japanese women. *Nutr Res* 28(5):309–14.

Nakajima, K., T. Nakano, and A. Tanaka. 2006. The oxidative modification hypothesis of atherosclerosis: The comparison of atherogenic effects on oxidized LDL and remnant lipoproteins in plasma. *Clin Chim Acta* 367(1–2):36–47.

Niki, E., Y. Yoshida, Y. Saito, and N. Noguchi. 2005. Lipid peroxidation: Mechanisms, inhibition, and biological effects. *Biochem Biophys Res Commun* 338(1):668–76.

Niu, K., A. Hozawa, S. Kuriyama, K. Ohmori-Matsuda, T. Shimazu, N. Nakaya, K. Fujita, I. Tsuji, and R. Nagatomi. 2006. Dietary long-chain n-3 fatty acids of marine origin and serum C-reactive protein concentrations are associated in a population with a diet rich in marine products. *Am J Clin Nutr* 84(1):223–9.

Nomura, S., S. Kanazawa, and S. Fukuhara. 2003. Effects of eicosapentaenoic acid on platelet activation markers and cell adhesion molecules in hyperlipidemic patients with type 2 diabetes mellitus. *J Diabetes Complications* 17(3):153–9.

Palozza, P., E. Sgarlata, C. Luberto, E. Piccioni, M. Anti, G. Marra, F. Armelao, P. Franceschelli, and G. M. Bartoli. 1996. n-3 fatty acids induce oxidative modifications in human erythrocytes depending on dose and duration of dietary supplementation. *Am J Clin Nutr* 64(3):297–304.

Parra, D., N. M. Bandarra, M. Kiely, I. Thorsdottir, and J. A. Martinez. 2007. Impact of fish intake on oxidative stress when included into a moderate energy-restricted program to treat obesity. *Eur J Nutr* 46(8):460–7.

Pischon, T., S. E. Hankinson, G. S. Hotamisligil, N. Rifai, W. C. Willett, and E. B. Rimm. 2003. Habitual dietary intake of n-3 and n-6 fatty acids in relation to inflammatory markers among U.S. men and women. *Circulation* 108(2):155–60.

Psota, T. L., S. K. Gebauer, and P. Kris-Etherton. 2006. Dietary omega-3 fatty acid intake and cardiovascular risk. *Am J Cardiol* 98(4A):3i–18i.

Rader, D. J., and A. Daugherty. 2008. Translating molecular discoveries into new therapies for atherosclerosis. *Nature* 451(7181):904–13.

Rahman, M. M., A. Bhattacharya, and G. Fernandes. 2008. Docosahexaenoic acid is more potent inhibitor of osteoclast differentiation in RAW 264.7 cells than eicosapentaenoic acid. *J Cell Physiol* 214(1):201–9.

Ridker, P. M. 2003. Clinical application of C-reactive protein for cardiovascular disease detection and prevention. *Circulation* 107(3):363–9.

Rizzo, M. and K. Berneis. 2006. Low-density lipoprotein size and cardiovascular risk assessment. *QJM* 99(1):1–14.

Rocquelin, G., L. Guenot, E. Justrabo, A. Grynberg, and M. David. 1985. Fatty acid composition of human heart phospholipids: data from 53 biopsy specimens. *J Mol Cell Cardiol* 17(8):769–73.

Russo, G. L. 2009. Dietary n-6 and n-3 polyunsaturated fatty acids: From biochemistry to clinical implications in cardiovascular prevention. *Biochem Pharmacol* 77(6):937–46.

Ryan, A. S., M. A. Keske, J. P. Hoffman, and E. B. Nelson. 2009. Clinical overview of algal-docosahexaenoic acid: Effects on triglyceride levels and other cardiovascular risk factors. *Am J Ther* 16(2):183–92.

Sanz, J., and Z. A. Fayad. 2008. Imaging of atherosclerotic cardiovascular disease. *Nature* 451(7181):953–7.

Sekikawa, A., J. D. Curb, H. Ueshima, A. El-Saed, T. Kadowaki, R. D. Abbott, R. W. Evans, B. L. Rodriguez, T. Okamura, K. Sutton-Tyrrell, Y. Nakamura, K. Masaki, D. Edmundowicz, A. Kashiwagi, B. J. Willcox, T. Takamiya, K. Mitsunami, T. B. Seto, K. Murata, R. L. White, and L. H. Kuller. 2008. Marine-derived n-3 fatty acids and atherosclerosis in Japanese, Japanese-American, and white men: A cross-sectional study. *J Am Coll Cardiol* 52(6):417–24.

Serhan, C. N. 2006. Novel chemical mediators in the resolution of inflammation: Resolvins and protectins. *Anesthesiol Clin* 24(2):341–64.

Serhan, C. N. 2007. Resolution phase of inflammation: Novel endogenous anti-inflammatory and proresolving lipid mediators and pathways. *Annu Rev Immunol* 25:101–37.

Sierra, S., F. Lara-Villoslada, M. Comalada, M. Olivares, and J. Xaus. 2008. Dietary eicosapentaenoic acid and docosahexaenoic acid equally incorporate as decosahexaenoic acid but differ in inflammatory effects. *Nutrition* 24(3):245–54.

Sinclair, H. M. 1956. Deficiency of essential fatty acids and atherosclerosis, etcetera. *Lancet* 270(6919):381–3.

Spiteller, G. 2005. The relation of lipid peroxidation processes with atherogenesis: A new theory on atherogenesis. *Mol Nutr Food Res* 49(11):999–1013.

Spiteller, G. 2007. The important role of lipid peroxidation processes in aging and age dependent diseases. *Mol Biotechnol* 37(1):5–12.

Stanke-Labesque, F., P. Moliere, J. Bessard, M. Laville, E. Vericel, and M. Lagarde. 2008. Effect of dietary supplementation with increasing doses of docosahexaenoic acid on neutrophil lipid composition and leukotriene production in human healthy volunteers. *Br J Nutr* 100(4):829–33.

Stulnig, T. M. 2003. Immunomodulation by polyunsaturated fatty acids: Mechanisms and effects. *Int Arch Allergy Immunol* 132(4):310–21.

Suzukawa, M., M. Abbey, P. R. Howe, and P. J. Nestel. 1995. Effects of fish oil fatty acids on low density lipoprotein size, oxidizability, and uptake by macrophages. *J Lipid Res* 36(3):473–84.

Tang, D. G., E. La, J. Kern, and J. P. Kehrer. 2002. Fatty acid oxidation and signaling in apoptosis. *Biol Chem* 383(3–4):425–42.

Theobald, H. E., A. H. Goodall, N. Sattar, D. C. Talbot, P. J. Chowienczyk, and T. A. Sanders. 2007. Low-dose docosahexaenoic acid lowers diastolic blood pressure in middle-aged men and women. *J Nutr* 137(4):973–8.

Thies, F., E. A. Miles, G. Nebe-von-Caron, J. R. Powell, T. L. Hurst, E. A. Newsholme, and P. C. Calder. 2001. Influence of dietary supplementation with long-chain n-3 or n-6 polyunsaturated fatty acids on blood inflammatory cell populations and functions and on plasma soluble adhesion molecules in healthy adults. *Lipids* 36(11):1183–93.

Thies, F., J. M. Garry, P. Yaqoob, K. Rerkasem, J. Williams, C. P. Shearman, P. J. Gallagher, P. C. Calder, and R. F. Grimble. 2003. Association of n-3 polyunsaturated fatty acids with stability of atherosclerotic plaques: A randomised controlled trial. *Lancet* 361(9356):477–85.

Vrablik, M., M. Prusikova, M. Snejdrlova, and L. Zlatohlavek. 2009. Omega-3 fatty acids and cardiovascular disease risk: Do we understand the relationship? *Physiol Res* 58(Suppl 1):S19–26.

Woodman, R. J., T. A. Mori, V. Burke, I. B. Puddey, G. F. Watts, and L. J. Beilin. 2002. Effects of purified eicosapentaenoic and docosahexaenoic acids on glycemic control, blood pressure, and serum lipids in type 2 diabetic patients with treated hypertension. *Am J Clin Nutr* 76(5):1007–15.

Wu, W. H., S. C. Lu, T. F. Wang, H. J. Jou, and T. A. Wang. 2006. Effects of docosahexaenoic acid supplementation on blood lipids, estrogen metabolism, and *in vivo* oxidative stress in postmenopausal vegetarian women. *Eur J Clin Nutr* 60(3):386–92.

Yoshida, Y., and E. Niki. 2006. Bio-markers of lipid peroxidation *in vivo*: Hydroxyoctadecadienoic acid and hydroxycholesterol. *Biofactors* 27(1-4):195–202.

Yoshida, Y., Y. Saito, M. Hayakawa, Y. Habuchi, Y. Imai, Y. Sawai, and E. Niki. 2007. Levels of lipid peroxidation in human plasma and erythrocytes: Comparison between fatty acids and cholesterol. *Lipids* 42(5):439–49.

Yusof, H. M., E. A. Miles, and P. Calder. 2008. Influence of very long-chain n-3 fatty acids on plasma markers of inflammation in middle-aged men. *Prostaglandins Leukot Essent Fatty Acids* 78(3):219–28.

Yusufi, A. N., J. Cheng, M. A. Thompson, H. J. Walker, C. E. Gray, G. M. Warner, and J. P. Grande. 2003. Differential effects of low-dose docosahexaenoic acid and eicosapentaenoic acid on the regulation of mitogenic signaling pathways in mesangial cells. *J Lab Clin Med* 141(5):318–29.

Zampelas, A., D. B. Panagiotakos, C. Pitsavos, U. N. Das, C. Chrysohoou, Y. Skoumas, and C. Stefanadis. 2005. Fish consumption among healthy adults is associated with decreased levels of inflammatory markers related to cardiovascular disease: The ATTICA study. *J Am Coll Cardiol* 46(1):120–4.

Zhang, J., S. Sasaki, K. Amano, and H. Kesteloot. 1999. Fish consumption and mortality from all causes, ischemic heart disease, and stroke: An ecological study. *Prev Med* 28(5):520–9.

# 7 Nutrigenomic Implications for Nuclear Receptor Coactivators

*David M. Lonard and Bert W. O'Malley*

## CONTENTS

## INTRODUCTION

Over the last 15 years, coactivator biology has developed as an integral part of our understanding of nuclear receptor (NR)-mediated biology. NRs are members of a large superfamily of ligand-regulated (and orphan) transcription factors that transduce steroid, retinoid, thyroid, and lipophillic endocrine hormones into distinct physiological responses. Pioneering work conducted over 40 years ago established how hormones, identified many years earlier, elicited distinct physiological responses. NRs were identified as receptor proteins for their cognate ligands and that they primarily function as ligand-activated DNA-binding transcription factors (Lonard and O'Malley 2007). Eventually, 48 NRs were identified in humans including many "orphan" NRs for which a cognate ligand has yet to be identified (Tsai and O'Malley 1994). As examples, the androgen (AR), progesterone (PR), and estrogen receptors (ERα and ERβ) play central roles in reproduction and target tissue growth; the glucocorticoid receptor (GR) regulates glucose metabolism, inflammation, and stress; thyroid hormone receptors (TR) control oxidative metabolism; and peroxisome proliferator-activated receptors (PPARs) have key roles in regulating lipid and energy metabolism. NRs have been popular targets for drug design and a large variety of synthetic ligands have been developed for clinical use. The clear relationship between many NRs and energy metabolism foreshadows the role that coactivators also have in regulating energy metabolism.

As transcription factors, NRs have a direct role in regulating the expression of hormone-response genes. This regulatory capacity of NRs occurs through their ability to recognize specific sequences in the promoters of their target genes, and their relationships with the RNA polymerase II holocomplex and the chromatin environment that surrounds these genes (Tsai and O'Malley 1994). Central to our discussion here, coactivators have broad genome-wide effects on gene expression through their ability to interact with numerous NRs and other non-NR transcription factors. Coactivators also have counterparts known as corepressors that act in an opposite manner to repress gene expression, primarily through their interaction with unliganded NRs (Torchia et al. 1998). Here, we will focus on the coactivators, as they have been more broadly studied, particularly with regard to the regulation of energy metabolism. Depending upon cell and signaling contexts, coactivators and corepressors can sometimes switch roles. Presently, more than 350 coregulators have been reported in the literature, frequently in connection with numerous physiological functions and pathological states (Lonard et al. 2007).

## MOLECULAR ASPECTS OF COACTIVATORS

It was realized early on that coactivators vary considerably in their amino acid compositions. Over time, it was found that they possess a diverse array of enzymatic and functional capabilities that control transcription, emphasizing the complex regulatory events involved in controlling transcription (Lonard and O'Malley 2007). They are not merely "bridging" agents between NRs and RNA polymerase as first thought, but possess numerous enzymatic capabilities and can act at many substeps of transcription, including initiation, RNA splicing, and mRNA transport (O'Malley et al. 2008). Initially after the identification of ERAP160, a protein that specifically interacts with agonist-bound receptors (Halachmi et al. 1994), and the cloning of the first NR coactivator, SRC-1 (Onate et al. 1995), we thought that only a handful of coactivators would be identified; instead over 300 coactivators now have been reported (Lonard et al. 2007). Given the broad scope of coactivators in transcriptional regulation, it is not surprising that they also have been found to be pervasive contributors to a wide variety of human disease states.

## COACTIVATORS FUNCTION AS MULTIPROTEIN COMPLEXES

Recent advancements in high throughput proteomic technologies are allowing us a means to understand how multiple proteins work together at a functional level (Wingren et al. 2009). Following the discovery of NR coregulators, molecular biological analyses revealed that coactivators and corepressors, like most other regulatory proteins, exist in large steady-state multiprotein complexes in mammalian cells (Jung et al. 2005; Rosenfeld et al. 2006). We now believe that gene transcription occurs as a consequence of the sequential recruitment by DNA-binding transcription factors (TFs) of a series of different coactivator complexes that are required for accurate and efficient gene expression (Metivier et al. 2006). These multisubunit complexes contain a collection of the diverse enzymes needed to direct distinct subreactions of transcription such as histone acetylation, methylation, ubiquitination, nucleosome rearrangement, transcriptional initiation and elongation, RNA splicing, and finally, degradation of the "activated" coregulators and TFs themselves (Lonard and O'Malley 2007). In short, our current understanding of transcription is quite different from the earlier theories that gave great weight to the role of a single functional protein in this process. Many examples of the compositions of these multiprotein complexes are available (NURSA.org), and the cooperative actions of different coactivators in the transcription of specific genes have been demonstrated repeatedly in cultured cells (Chen et al. 1997; McKenna et al. 1998; Feng et al. 2006).

### COACTIVATORS ARE MASTER REGULATORS OF GENE EXPRESSION PROGRAMS

The versatility of these coactivator complex enzymatic machines contributes to a great deal of regulatory flexibility in the control of NR-mediated transcription (Lonard and O'Malley 2007). In addition to functioning as histone code writing proteins that place post-translational marks (PTMs) on histones, coactivators and co-coactivators serve as recipients of PTMs themselves (Figure 7.1). Coregulator activity is determined by its phosphorylation, methylation, and acetylation status that forms a coactivator PTM code. This code then goes on to regulate the coactivator complex's transcriptional activity and preferences for different transcription factors and genes (Wu et al. 2005). This coactivator PTM code plays an essential role in regulating their function as "master genes" that control broad transcriptional programs responsible for cell growth, differentiation, and metabolic functions (Lonard et al. 2009). Our laboratory has discovered that SRC-3 is phosphorylated at distinct serine/threonine residues by upstream growth factors' signaling cascades, generating a distinct phosphorylation code on the coactivator (Figure 7.2). This upstream signaling is then transmitted

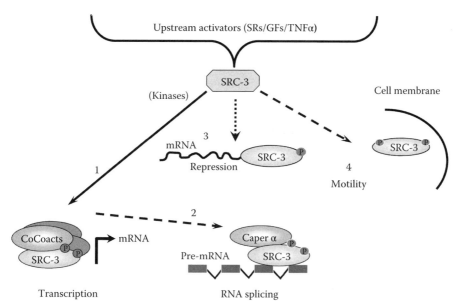

**FIGURE 7.1** SRC-3 regulates gene expression at multiple levels. Kinase targeting of SRC-3 can alter its biological activity to differentially influence its classical function as a coactivator (1). SRC-3 can also influence transcript splicing decisions in conjunction with the co-coactivator Caper α (2). In (3), SRC-3 acts as a translational repressor for cytokines and in (4), it functions at the cell membrane to regulate cell motility.

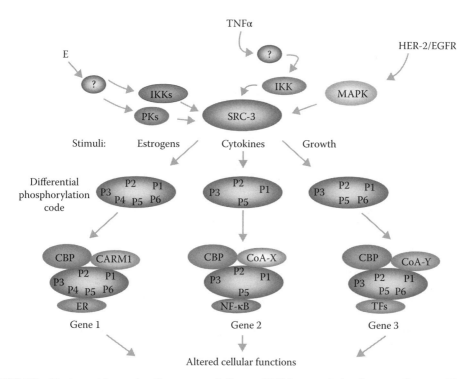

**FIGURE 7.2** Upstream kinase signaling systems influence SRC-3 transcription factor preferences. Upstream signaling systems induced by estrogens (E), tumor necrosis factor-α (TNFα), and by the Her-2 and EGFR signaling systems impinge on SRC-3, resulting in distinct post-translational modification patterns that direct the coactivator's preference for ER, NF-κB, or other TF, leading to distinct patterns of overall gene expression and altered cellular functions.

through the selective coactivation of NR and non-NR transcription factors (Wu et al. 2004). This finding is likely to underlie the observation that overexpression of both SRC-3 and the human epidermal growth factor receptor (EGFR) (her-2/neu) kinase is associated with decreased breast cancer survival and early tamoxifen resistance (Schiff et al. 2003; Shou et al. 2004). Growth factor signaling systems and coactivators thus work hand-in-hand to coordinately control gene expression programs responsible for cell growth.

Because of the extensive roles PTM events have in coactivator biology, emerging high-throughput proteomic technologies will likely contribute to a greater understanding of coactivator biology. We and other groups have shown that coactivators are subject to degradation by ubiquitin-dependent (Figure 7.3a) and -independent proteasome systems (Figure 7.3b) (Wu et al. 2007; Li et al. 2006). In normal tissues, most coactivators appear to be expressed constitutively at the mRNA level and are not subject to dynamic regulation in response to acute external stimuli (PGC-1 is an exception) (Puigserver et al. 1998); cancer may also alter the level of coactivator gene expression at the mRNA level. At the protein level, cellular coactivator concentration can be regulated by PTMs, NR ligands, and other stimuli that post-translationally affect their protein stability (Lonard et al. 2000; Wu et al. 2007).

## SRC FAMILY MEMBERS ARE KEY ONCOGENIC COACTIVATORS

Through modulating gene expression regulated by hormones, growth factors, and cytokines, coactivators play crucial roles in many biological and pathological processes including cell proliferation, differentiation, carcinogenesis, and metastasis (Figure 7.4) (Lonard et al. 2007). The combinations and concentrations of these coactivators act to determine the specificity and efficiency of gene transcription and alterations in these parameters can lead to pathologies, and cancer stands out as a prominent coactivator-related pathology (Lonard and O'Malley 2007). SRC family coactivators have well-recognized roles in hormone-dependent cancers such as breast, uterine, ovarian, and prostate cancers (Anzick et al. 1997; Zhou et al. 2005). In addition to nuclear receptors, SRC-1 and SRC-3 also interact with and coactivate other cancer-related transcription factors, including Ets-2, PEA3, and E2F1 (Yan et al. 2008) and it is becoming evident that they are playing important roles

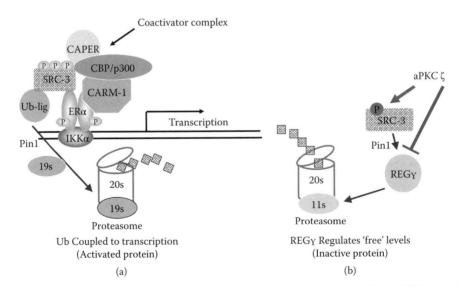

**FIGURE 7.3** Two distinct proteasome-mediated pathways are responsible for cellular SRC-3 protein degradation. (a) Transcriptionally engaged SRC-3 is targeted by ubiquitin ligases, leading to ubiquitination and degradation of the coactivator. (b) Cellular SRC-3 that is not engaged in transcription is targeted by the REGγ-proteasome for degradation in an ubiquitin-independent manner to regulate the steady-state levels of the inactive form of SRC-3.

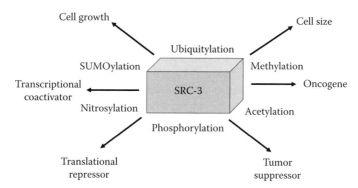

**FIGURE 7.4** Different post-translational events underlie diverse biological roles for SRC-3. Distinct post-translational events such as SUMOlyation, ubiquitination, phosphorylation, acetylation, methylation, and nitrosylation control SRC-3's participation in the regulation of its activity as a coactivator, in cell growth, cell size, as a translational repressor, as a tumor suppressor, and as an oncogene.

in other types of cancers as well. The relationship between cancer and nutrition is receiving considerable recent attention. As we shall discuss in more detail in *Coregulator Involvement in Energy Metabolism*, the fact that coactivators are involved in both cancer and metabolism should lead us to consider the possibility that coactivators play a causal role in the relationship between diet and certain types of cancer.

## COACTIVATOR BIOINFORMATICS

Nutrigenomic approaches promise to reveal a more complete picture of the role of coregulators in human metabolic disorders. Because of their overarching and pleiotropic functions, recognizing coactivator involvement in these conditions will be difficult, and will likely require genome-wide approaches to be adequately assessed. High-throughput technologies that include chromatin immunoprecipitation "ChIP-CHIP" assays, protein-interaction network maps, and genome-wide association studies (GWAS) will need to be integrated to make a holistic understanding of coactivator biology possible.

Coactivators can be seen as the "rheostats" that control the extent of gene expression from NRs and fluctuations in their expression or small changes in their biological activity will lead to significant differences in target tissue responses to hormone ligands. In human individuals, concentrations of circulating steroid hormones typically fall within a fairly narrow range. Thus, changes in the primary amino acid sequence and cellular concentration of coactivator proteins may be responsible for differences in the individual manifestation of metabolic phenotypes. As we shall discuss in *Coactivators, Diet, and Human Evolution*, coactivator gene polymorphisms exist in human population groups and could account for certain quantitative and qualitative differences to NR ligands that exist in individuals.

Many studies on human polymorphisms have investigated the biological actions of NR single nucleotide polymorphisms (SNPs) such as for ERα, PPARγ, GR, TR, and PR (Figtree et al. 2009; Thijssen 2009; van der Deure et al. 2007; Manenschijn et al. 2009). For coactivator SNPs, most are located in gene promoters, are intronic, or are synonymous noncoding variants. Nevertheless, many SNPs exist within coregulator genes and affect coregulator amino acid sequence, such as for PGC-1α. For SRC-3, a Q586H variant allele confers a protective effect toward breast cancer (Burwinkel et al. 2005; Hartmaier et al. 2009). In this population-based study of European women, a correlation between the H allele and the absence of breast cancer was seen in healthy women compared to cohorts with primary and recurrent breast cancer. Polymorphisms in p300, PGC-1α, and PGC-1β also have been shown to be low-penetrance familial breast cancer markers (Oberkofler et al. 2004; Kathiresan et al. 2009; Ek et al. 2001).

## Coregulator Involvement in Energy Metabolism

We recently performed a broad survey of the literature of reports that have identified a coactivator function for a particular protein and this number has now grown to over 350 coactivator genes (Lonard et al. 2007). Many within the list are primarily recognized as coactivators, including SRC family proteins, PGC-1α, CBP, and others. This large number of coactivator genes poses new questions about their overall scope in human biology, including energy metabolism. PPARγ coactivator-1α (PGC-1α) is a key coactivator in the regulation of metabolic function (Handschin 2009). Early work on PGC-1α revealed that this coactivator is expressed in muscle and brown adipose tissue in mice and is highly inducible by exercise, fasting, and cold exposure. It was revealed to be a coactivator for PPARγ (and for other NRs) when knocked out or ectopically expressed in mice, further reinforcing its role in metabolism (Lin et al. 2004). A polymorphism in the PGC-1α gene (G482S) and another polymorphism in the gene's promoter have been linked with an increased risk of type 2 diabetes (Andersen et al. 2005; Ek et al. 2001). A defect in the inducible expression of PGC-1α also has been linked to defects in cholesterol homeostasis. An increase in the incidence of hypertension in carriers of a G482 allele was identified, suggesting that this allele may be implicated in early-onset hypertension (Bertolotti et al. 2006). The related protein PGC-1β recently has been knocked out in mice, revealing that it also is involved in metabolic functions. PGC-1β$^{-/-}$ mice have reduced mitochondrial function and other defects in fat metabolism (Lelliott et al. 2006; Uldry et al. 2006). PGC-1β can promote the formation of oxidative type IIX muscle fiber, which is an important issue in athletic performance (Scarpulla 2008).

SRC-1 and SRC-2 have been found to play different roles in energy metabolism through mouse knockout studies. SRC-1$^{-/-}$ mice become obese due to decreased energy expenditure. On the other hand, SRC-2$^{-/-}$ mice are leaner due to the reduced transcriptional capacity of PPARγ2, an NR involved in adipoctye differentiation (Picard et al. 2002). In SRC-2$^{-/-}$ mice, a subsequent increase in PGC-1α/SRC-1 interaction occurs that promotes the thermogenic actions of PGC-1α in brown fat. SRC-3 has been shown to promote white adipose tissue development as indicated by the finding that SRC-3$^{-/-}$ mice have decreased adipose tissue mass (Louet et al. 2006). SRC-3 enhances CAAT enhancer binding protein-β (C/EBPβ)-mediated transcription of PPARγ2 needed adipoctye differentiation. Coactivation of C/EBPβ also depends upon the SWI/SNF chromatin remodeling coactivator (Caramel et al. 2008) and other studies have revealed that PPARγ2-mediated transcription is subsequently dependent upon the coactivator, TRAP220, demonstrating that coactivators are involved in multiple molecular steps involved in adipoctye differentiation (Ge et al. 2002).

We have recently uncovered a role for SRC-2 in controlling the release of glycogen stores during fasting (Chopra et al. 2008). Hepatic glucose production is essential for basal brain function and survival when dietary sources of glucose are not available. Glucose-6-phosphatase (G6Pase) is an essential, rate-limiting enzyme that functions as the gatekeeper for hepatic glucose release into the plasma. Mutations in G6Pase result in the potentially fatal disorder, Von Gierke's disease (glycogen storage disease-1a). We found that SRC-2 functions as a critical regulator responsible for releasing hepatic glucose during fasting, by stimulating the expression of G6Pase. SRC-2 coactivates G6Pase expression by acting as a coactivator for the orphan nuclear receptor RORα. SRC-2 knockout mice have a phenotype that resembles Von Gierke's disease, characterized by severe hypoglycemia during fasting. A wide variety of metabolic parameters in the SRC-2 knockout mice also clinically resemble Von Gierke's disease patients, suggesting that disruptions in SRC-2 may be identified in patients that lack disruptions in either of their G6Pase alleles.

## Coactivators, Diet, and Human Evolution

Over the course of our evolution, human beings have endured dramatic changes in their environment, placing strong selection pressures on genes involved in responding to these abrupt changes in our human existence. Differences in diet, longevity, and exposure to communicable diseases have created acute environmental challenges that have out-paced our genetic adaptation. The migrations of humans to distinctly different environments throughout the Earth (tropical, desert, arctic, temperate, etc.), agriculture, industrialization, and increases in lifespan have all placed acute selective pressures on the human genome (Cordain et al. 2005). As an example of this phenomenon, the "thrifty gene hypothesis" predicts that while energy conservation was an important issue in our recent evolutionary past, today these same energy-conserving traits are responsible in part for the emergence of obesity and diabetes since food is now easily and reliably obtainable (Neel 1962).

Prior to the Neolithic, humans were primarily hunter-gatherers whose diet came from a wide range of food sources (Cordain et al. 2005). The introduction of agriculture and animal husbandry approximately 10,000 years ago led to dramatic changes in the types of food we consumed and to seasonal variations in food supply that brought about periods of food abundance and scarcity. In contemporary Western populations, we are entering a third major transition in our diet and lifestyle that is distinct from either of the previous epochs of our human existence. During the Neolithic and the present, these changes have altered a number of key nutritional parameters from that of our ancestral hunter-gatherer diets that have existed throughout the majority of human existence. These changes have affected glycemic load, fatty acid composition, macronutrient composition, and other factors. Chronic diseases states that are now prevalent in the developed world such as obesity, diabetes, and coronary disease are products of this new dietary regime. Nevertheless, individual humans can exhibit a wide range of responses to the Western diet due to differences in our genetic makeup. Nutrigenomic approaches that incorporate genome-wide polymorphism data and diet along with information about coactivators have the potential to reveal personalized information that can guide the design of individualized healthy diet plans.

Recent evidence for positive selection of specific alleles for a variety of traits for skin color, immune response, and for specific nutritional factors such as amylase expression and lactose tolerance in adults has been identified (Gibson 2007). The evidence points to coactivator alleles being part of this process as well. Computational studies to identify alleles subject to strong selective pressure sweeps have identified a number of coactivators as agents of positive selection in different human populations (Voight et al. 2006; Lonard et al. 2007). Considerable effort is being directed towards understanding how SNPs (and other genetic variations) influence human disease susceptibility (Sebastiani et al. 2009). Of particular interest to coactivator biology, in a bioinformatic approach used to analyze positive selection pressures in HapMap project data, SRC-1 (NCOA1) was predicted to be under very strong selective pressure (the greatest for any human gene) in the Yoruba ethnic group (Voight et al. 2006). Signals for strong selective pressures exist for a number of other coactivator genes including GAC63 and CAPER (Lonard et al. 2007). Thus, many coactivators are predicted to be strong conduits for human evolutionary adaptation and may have arisen due to ethnic and migratory differences in diet or other environmental factors. However, in the context of our modern lifestyle and diet, it is easy to speculate that these changes may be maladaptive, particularly those alleles that confer for more efficient use of energy now that food is so readily available.

## Coactivators as Low Penetrance Disease Risk Alleles

Because negative selection drives out deleterious alleles from populations, genetic diseases with monogenic etiologies are relatively rare. Inherited monogenic genetic diseases are easier to understand and study, and can be traced through family pedigrees because they follow clear Mendelian rules. On the other hand, the genetic basis of common disorders such as obesity (Korner et al. 2008; Walley et al. 2006), diabetes (Ridderstrale and Groop 2009), dyslipidemia (Kathiresan et al. 2009), allergies (Goodarzi 2008; Kleeberger and Peden 2005), polycystic ovarian syndrome, hypertension (Deng 2007), and central nervous system disorders (Keller and Miller 2006) are primarily polygenic. Genome-wide association studies have had limited success in identifying the weakly penetrant alleles that underlie these diseases (Frazer et al. 2009). Already though, a single nucleotide polymorphism adjacent to SRC-1 has been identified as a significant and highly ranking risk factor for type 1 diabetes (Cooper et al. 2008). In another study, a polymorphism in SRC-3 was found to contribute to the success of chemotherapy in the treatment of acute lymphoblastic lymphoma (Yang et al. 2009). Common individual disease-associated alleles have very low phenotypic penetrance, are not subject to strong negative selective pressure, and likewise are found at much higher frequencies in human populations. For a number of reasons that were discussed throughout this chapter, and as a result of evidence obtained from mouse knockout studies, SRC family coactivator genes have relatively low phenotypic penetrance, likely standing alongside other common but weakly penetrant alleles that contribute to polygenic disease states.

Another important and somewhat counterintuitive characteristic of SRC family coactivators that also applies to other master regulators such as p53 (Donehower et al. 1992) and PGC-1α (Lin et al. 2004) is the fact that even complete loss of these genes is not lethal. We postulate that this is an important element that allows these proteins the flexibility to accommodate diverse signals from the environment. In contrast, genes that underlie core biological processes, such as RNA polymerases or histones for example, are likely to be intolerant of any change, unable to exist as weakly penetrant alleles in human populations, and ultimately unable to contribute to polygenic disease states (Lonard et al. 2009).

## CONCLUDING REMARKS

In this section, we describe some of the important ways we propose that coactivators contribute to diet, environment, and genetic background (Figure 7.5). It is evident that coactivator dysfunction is not restricted solely to rare genetic conditions or a small subset of cancers, but is instead involved in numerous human diseases, including pathologies related to energy metabolism. Mouse knockout studies attest to the physiological and pathological importance of coactivators as key regulators of energy metabolism. Future work promises an understanding of how human polymorphisms in coactivator genes relate to variations in human physiology related to energy metabolism. As seen in this chapter, existing basic and translational research efforts already have shed some light on the relationship between coactivators and our diet. Developing genomic and proteomic technologies will add greatly to our understanding of the basic roles that these master regulators play at multiple levels, from the control of gene expression up to that of system-wide regulation of the metabolic

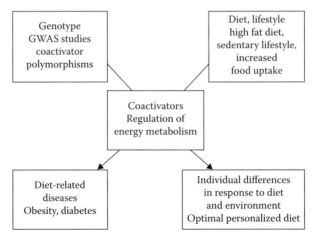

**FIGURE 7.5**  Nutrigenomic integration of genotype, environment, and coactivator biology to prevent diet-related disease and for diet optimization. By merging GWAS data with specific information on coactivator polymorphisms and our understanding of coactivator regulation of energy metabolism and an individual's diet and lifestyle, it should be possible to infer risks for diet-related disease states. This same information also should be useful for designing personalized diets for individuals prone to specific diet-related disease risks.

interaction between different tissues in the body. Overall, nutrigenomic approaches that integrate coactivator biology with information about the specific dietary, environmental, and genetic states promise to help us to understand and combat diet-related disease states that have arisen due to recent changes in our eating habits and lifestyle.

## REFERENCES

Andersen, G., L. Wegner, D. P. Jensen, C. Glumer, L. Tarnow, T. Drivsholm, P. Poulsen, S. K. Hansen, E. M. Nielsen, J. Ek, P. Mouritzen, A. Vaag, H. H. Parving, K. Borch-Johnsen, T. Jorgensen, T. Hansen, and O. Pedersen. 2005. PGC-1alpha Gly482Ser polymorphism associates with hypertension among Danish whites. *Hypertension* 45(4):565–70.

Anzick, S. L., J. Kononen, R. L. Walker, D. O. Azorsa, M. M. Tanner, X. Y. Guan, G. Sauter, O. P. Kallioniemi, J. M. Trent, and P. S. Meltzer. 1997. AIB1, a steroid receptor coactivator amplified in breast and ovarian cancer. *Science* 277(5328):965–8.

Bertolotti, M., C. Gabbi, C. Anzivino, N. Mitro, C. Godio, E. De Fabiani, M. Crestani, M. Del Puppo, M. Ricchi, L. Carulli, A. Rossi, P. Loria, and N. Carulli. 2006. Decreased hepatic expression of PPAR-gamma coactivator-1 in cholesterol cholelithiasis. *Eur J Clin Invest* 36(3):170–5.

Burwinkel, B., M. Wirtenberger, R. Klaes, R. K. Schmutzler, E. Grzybowska, A. Forsti, B. Frank, J. L. Bermejo, P. Bugert, B. Wappenschmidt, D. Butkiewicz, J. Pamula, W. Pekala, H. Zientek, D. Mielzynska, E. Siwinska, C. R. Bartram, and K. Hemminki. 2005. Association of NCOA3 polymorphisms with breast cancer risk. *Clin Cancer Res* 11(6):2169–74.

Caramel, J., S. Medjkane, F. Quignon, and O. Delattre. 2008. The requirement for SNF5/INI1 in adipocyte differentiation highlights new features of malignant rhabdoid tumors. *Oncogene* 27(14):2035–44.

Chen, H., R. J. Lin, R. L. Schiltz, D. Chakravarti, A. Nash, L. Nagy, M. L. Privalsky, Y. Nakatani, and R. M. Evans. 1997. Nuclear receptor coactivator ACTR is a novel histone acetyltransferase and forms a multimeric activation complex with P/CAF and CBP/p300. *Cell* 90(3):569–80.

Chopra, A. R., J. F. Louet, P. Saha, J. An, F. Demayo, J. Xu, B. York, S. Karpen, M. Finegold, D. Moore, L. Chan, C. B. Newgard, and B. W. O'Malley. 2008. Absence of the SRC-2 coactivator results in a glycogenopathy resembling Von Gierke's disease. *Science* 322(5906):1395–99.

Cooper, J. D., D. J. Smyth, A. M. Smiles, V. Plagnol, N. M. Walker, J. E. Allen, K. Downes, J. C. Barrett, B. C. Healy, J. C. Mychaleckyj, J. H. Warram, and J. A. Todd. 2008. Meta-analysis of genome-wide association study data identifies additional type 1 diabetes risk loci. *Nat Genet* 40(12):1399–401.

Cordain, L., S. B. Eaton, A. Sebastian, N. Mann, S. Lindeberg, B. A. Watkins, J. H. O'Keefe, and J. Brand-Miller. 2005. Origins and evolution of the Western diet: Health implications for the 21st century. *Am J Clin Nutr* 81(2):341–54.

Deng, A. Y. 2007. Genetic basis of polygenic hypertension. *Hum Mol Genet* 16(2):R195–202.

Donehower, L. A., M. Harvey, B. L. Slagle, M. J. McArthur, C. A. Montgomery, Jr., J. S. Butel, and A. Bradley. 1992. Mice deficient for p53 are developmentally normal but susceptible to spontaneous tumours. *Nature* 356(6366):215–21.

Ek, J., G. Andersen, S. A. Urhammer, P. H. Gaede, T. Drivsholm, K. Borch-Johnsen, T. Hansen, and O. Pedersen. 2001. Mutation analysis of peroxisome proliferator-activated receptor-gamma coactivator-1 (PGC-1) and relationships of identified amino acid polymorphisms to Type II diabetes mellitus. *Diabetologia* 44(12):2220–26.

Feng, Q., P. Yi, J. Wong, and B. W. O'Malley. 2006. Signaling within a coactivator complex: Methylation of SRC-3/AIB1 is a molecular switch for complex disassembly. *Mol Cell Biol* 26(21):7846–57.

Figtree, G. A., J. E. Noonan, R. Bhindi, and P. Collins. 2009. Estrogen receptor polymorphisms: Significance to human physiology, disease and therapy. *Recent Pat DNA Gene Seq* 3(3):164–71.

Frazer, K. A., S. S. Murray, N. J. Schork, and E. J. Topol. 2009. Human genetic variation and its contribution to complex traits. *Nat Rev Genet* 10(4):241–51.

Ge, K., M. Guermah, C. X. Yuan, M. Ito, A. E. Wallberg, B. M. Spiegelman, and R. G. Roeder. 2002. Transcription coactivator TRAP220 is required for PPAR gamma 2-stimulated adipogenesis. *Nature* 417(6888):563–7.

Gibson, G. 2007. Human evolution: Thrifty genes and the dairy queen. *Curr Biol* 17(8):R295–6.

Goodarzi, M. O. 2008. Looking for polycystic ovary syndrome genes: Rational and best strategy. *Semin Reprod Med* 26(1):5–13.

Halachmi, S., E. Marden, G. Martin, H. MacKay, C. Abbondanza, and M. Brown. 1994. Estrogen receptor-associated proteins: Possible mediators of hormone-induced transcription. *Science* 264(5164):1455–8.

Handschin, C. 2009. The biology of PGC-1alpha and its therapeutic potential. *Trends Pharmacol Sci* 30(6):322–9.

Hartmaier, R. J., S. Tchatchou, A. S. Richter, J. Wang, S. E. McGuire, T. C. Skaar, J. M. Rae, K. Hemminki, C. Sutter, N. Ditsch, P. Bugert, B. H. Weber, D. Niederacher, N. Arnold, R. Varon-Mateeva, B. Wappenschmidt, R. K. Schmutzler, A. Meindl, C. R. Bartram, B. Burwinkel, and S. Oesterreich. 2009. Nuclear receptor coregulator SNP discovery and impact on breast cancer risk. *BMC Cancer* 9:438.

Jung, S. Y., A. Malovannaya, J. Wei, B. W. O'Malley, and J. Qin. 2005. Proteomic analysis of steady-state nuclear hormone receptor coactivator complexes. *Mol Endocrinol* 19(10):2451–65.

Kathiresan, S., C. J. Willer, G. M. Peloso, S. Demissie, K. Musunuru, E. E. Schadt, L. Kaplan, D. Bennett, Y. Li, T. Tanaka, B. F. Voight, L. L. Bonnycastle, A. U. Jackson, G. Crawford, A. Surti, C. Guiducci, N. P. Burtt, S. Parish, R. Clarke, D. Zelenika, K. A. Kubalanza, M. A. Morken, L. J. Scott, H. M. Stringham, P. Galan, A. J. Swift, J. Kuusisto, R. N. Bergman, J. Sundvall, M. Laakso, L. Ferrucci, P. Scheet, S. Sanna, M. Uda, Q. Yang, K. L. Lunetta, J. Dupuis, P. I. de Bakker, C. J. O'Donnell, J. C. Chambers, J. S. Kooner, S. Hercberg, P. Meneton, E. G. Lakatta, A. Scuteri, D. Schlessinger, J. Tuomilehto, F. S. Collins, L. Groop, D. Altshuler, R. Collins, G. M. Lathrop, O. Melander, V. Salomaa, L. Peltonen, M. Orho-Melander, J. M. Ordovas, M. Boehnke, G. R. Abecasis, K. L. Mohlke, and L. A. Cupples. 2009. Common variants at 30 loci contribute to polygenic dyslipidemia. *Nat Genet* 41(1):56–65.

Keller, M. C., and G. Miller. 2006. Resolving the paradox of common, harmful, heritable mental disorders: Which evolutionary genetic models work best? *Behav Brain Sci* 29(4):385–404; discussion 405–352.

Kleeberger, S. R., and D. Peden. 2005. Gene-environment interactions in asthma and other respiratory diseases. *Annu Rev Med* 56:383–400.

Korner, A., W. Kiess, M. Stumvoll, and P. Kovacs. 2008. Polygenic contribution to obesity: Genome-wide strategies reveal new targets. *Front Horm Res* 36:12–36.

Lelliott, C. J., G. Medina-Gomez, N. Petrovic, A. Kis, H. M. Feldmann, M. Bjursell, N. Parker, K. Curtis, M. Campbell, P. Hu, D. Zhang, S. E. Litwin, V. G. Zaha, K. T. Fountain, S. Boudina, M. Jimenez-Linan, M. Blount, A. Lopez, A. Meirhaeghe, Y. M. Bohlooly, L. Storlien, M. Stromstedt, M. Snaith, M. Oresic, E. D. Abel, B. Cannon, and A. Vidal-Puig. 2006. Ablation of PGC-1beta results in defective mitochondrial activity, thermogenesis, hepatic function, and cardiac performance. *PLoS Biol* 4(11):e369.

Li, X., D. M. Lonard, S. Y. Jung, A. Malovannaya, Q. Feng, J. Qin, S. Y. Tsai, M. J. Tsai, and B. W. O'Malley. 2006. The SRC-3/AIB1 coactivator is degraded in a ubiquitin- and ATP-independent manner by the REGgamma proteasome. *Cell* 124(2):381–92.

Lin, J., P. H. Wu, P. T. Tarr, K. S. Lindenberg, J. St-Pierre, C. Y. Zhang, V. K. Mootha, S. Jager, C. R. Vianna, R. M. Reznick, L. Cui, M. Manieri, M. X. Donovan, Z. Wu, M. P. Cooper, M. C. Fan, L. M. Rohas, A. M. Zavacki, S. Cinti, G. I. Shulman, B. B. Lowell, D. Krainc, and B. M. Spiegelman. 2004. Defects in adaptive energy metabolism with CNS-linked hyperactivity in PGC-1alpha null mice. *Cell* 119(1):121–35.

Lonard, D. M., R. Kumar, and B. W. O'Malley. 2009. Minireview: The SRC family of coactivators: An entree to understanding a subset of polygenic diseases? *Mol Endocrinol*.

Lonard, D. M., R. B. Lanz, and B. W. O'Malley. 2007. Nuclear receptor coregulators and human disease. *Endocr Rev* 28(5):575–87.

Lonard, D. M., Z. Nawaz, C. L. Smith, and B. W. O'Malley. 2000. The 26S proteasome is required for estrogen receptor-alpha and coactivator turnover and for efficient estrogen receptor-alpha transactivation. *Mol Cell* 5(6):939–48.

Lonard, D. M., and B. W. O'Malley. 2007. Nuclear receptor coregulators: Judges, juries, and executioners of cellular regulation. *Mol Cell* 27(5):691–700.

Louet, J. F., A. Coste, L. Amazit, M. Tannour-Louet, R. C. Wu, S. Y. Tsai, M. J. Tsai, J. Auwerx, and B. W. O'Malley. 2006. Oncogenic steroid receptor coactivator-3 is a key regulator of the white adipogenic program. *Proc Natl Acad Sci USA* 103(47):17868–73.

Manenschijn, L., E. L. van den Akker, S. W. Lamberts, and E. F. van Rossum. 2009. Clinical features associated with glucocorticoid receptor polymorphisms: An overview. *Ann N Y Acad Sci* 1179:179–98.

McKenna, N. J., Z. Nawaz, S. Y. Tsai, M. J. Tsai, and B. W. O'Malley. 1998. Distinct steady-state nuclear receptor coregulator complexes exist *in vivo*. *Proc Natl Acad Sci USA* 95(20):11697–702.

Metivier, R., G. Reid, and F. Gannon. 2006. Transcription in four dimensions: Nuclear receptor-directed initiation of gene expression. *EMBO Rep* 7(2):161–7.

Neel, J. V. 1962. Diabetes mellitus: A "thrifty" genotype rendered detrimental by "progress"? *Am J Hum Genet* 14:353–62.

O'Malley, B. W., J. Qin, and R. B. Lanz. 2008. Cracking the coregulator codes. *Curr Opin Cell Biol* 20(3):310–5.

Oberkofler, H., V. Linnemayr, R. Weitgasser, K. Klein, M. Xie, B. Iglseder, F. Krempler, B. Paulweber, and W. Patsch. 2004. Complex haplotypes of the PGC-1alpha gene are associated with carbohydrate metabolism and type 2 diabetes. *Diabetes* 53(5):1385–93.

Onate, S. A., S. Y. Tsai, M. J. Tsai, and B. W. O'Malley. 1995. Sequence and characterization of a coactivator for the steroid hormone receptor superfamily. *Science* 270(5240):1354–7.

Picard, F., M. Gehin, J. Annicotte, S. Rocchi, M. F. Champy, B. W. O'Malley, P. Chambon, and J. Auwerx. 2002. SRC-1 and TIF2 control energy balance between white and brown adipose tissues. *Cell* 111(7):931–41.

Puigserver, P., Z. Wu, C. W. Park, R. Graves, M. Wright, and B. M. Spiegelman. 1998. A cold-inducible coactivator of nuclear receptors linked to adaptive thermogenesis. *Cell* 92(6):829–39.

Ridderstrale, M., and L. Groop. 2009. Genetic dissection of type 2 diabetes. *Mol Cell Endocrinol* 297(1–2):10–7.

Rosenfeld, M. G., V. V. Lunyak, and C. K. Glass. 2006. Sensors and signals: A coactivator/corepressor/epigenetic code for integrating signal-dependent programs of transcriptional response. *Genes Dev* 20(11):1405–28.

Scarpulla, R. C. 2008. Transcriptional paradigms in mammalian mitochondrial biogenesis and function. *Physiol Rev* 88(2):611–38.

Schiff, R., S. Massarweh, J. Shou, and C. K. Osborne. 2003. Breast cancer endocrine resistance: How growth factor signaling and estrogen receptor coregulators modulate response. *Clin Cancer Res* 9(1 Pt 2):447S–54S.

Sebastiani, P., N. Timofeev, D. A. Dworkis, T. T. Perls, and M. H. Steinberg. 2009. Genome-wide association studies and the genetic dissection of complex traits. *Am J Hematol* 84(8):504–15.

Shou, J., S. Massarweh, C. K. Osborne, A. E. Wakeling, S. Ali, H. Weiss, and R. Schiff. 2004. Mechanisms of tamoxifen resistance: Increased estrogen receptor-HER2/neu cross-talk in ER/HER2-positive breast cancer. *J Natl Cancer Inst* 96(12):926–35.

Thijssen, J. H. 2009. Gene polymorphisms that may influence the biological effects of progestins. *Maturitas* 62(4):366–70.

Torchia, J., C. Glass, and M. G. Rosenfeld. 1998. Co-activators and co-repressors in the integration of transcriptional responses. *Curr Opin Cell Biol* 10(3):373–83.

Tsai, M. J., and B. W. O'Malley. 1994. Molecular mechanisms of action of steroid/thyroid receptor superfamily members. *Annu Rev Biochem* 63:451–86.

Uldry, M., W. Yang, J. St-Pierre, J. Lin, P. Seale, and B. M. Spiegelman. 2006. Complementary action of the PGC-1 coactivators in mitochondrial biogenesis and brown fat differentiation. *Cell Metab* 3(5):333–41.

van der Deure, W. M., R. P. Peeters, and T. J. Visser. 2007. Genetic variation in thyroid hormone transporters. *Best Pract Res Clin Endocrinol Metab* 21(2):339–50.

Voight, B. F., S. Kudaravalli, X. Wen, and J. K. Pritchard. 2006. A map of recent positive selection in the human genome. *PLoS Biol* 4(3):e72.

Walley, A. J., A. I. Blakemore, and P. Froguel. 2006. Genetics of obesity and the prediction of risk for health. *Hum Mol Genet* 15(2):R124–30.

Wingren, C., P. James, and C. A. Borrebaeck. 2009. Strategy for surveying the proteome using affinity proteomics and mass spectrometry. *Proteomics* 9(6):1511–7.

Wu, R. C., Q. Feng, D. M. Lonard, and B. W. O'Malley. 2007. SRC-3 coactivator functional lifetime is regulated by a phospho-dependent ubiquitin time clock. *Cell* 129(6):1125–40.

Wu, R. C., J. Qin, P. Yi, J. Wong, S. Y. Tsai, M. J. Tsai, and B. W. O'Malley. 2004. Selective phosphorylations of the SRC-3/AIB1 coactivator integrate genomic reponses to multiple cellular signaling pathways. *Mol Cell* 15(6):937–49.

Wu, R. C., C. L. Smith, and B. W. O'Malley. 2005. Transcriptional regulation by steroid receptor coactivator phosphorylation. *Endocr Rev* 26(3):393–9.

Yan, J., H. Erdem, R. Li, Y. Cai, G. Ayala, M. Ittmann, L. Y. Yu-Lee, S. Y. Tsai, and M. J. Tsai. 2008. Steroid receptor coactivator-3/AIB1 promotes cell migration and invasiveness through focal adhesion turnover and matrix metalloproteinase expression. *Cancer Res* 68(13):5460–8.

Yang, J. J., C. Cheng, W. Yang, D. Pei, X. Cao, Y. Fan, S. B. Pounds, G. Neale, L. R. Trevino, D. French, D. Campana, J. R. Downing, W. E. Evans, C. H. Pui, M. Devidas, W. P. Bowman, B. M. Camitta, C. L. Willman, S. M. Davies, M. J. Borowitz, W. L. Carroll, S. P. Hunger, and M. V. Relling. 2009. Genome-wide interrogation of germline genetic variation associated with treatment response in childhood acute lymphoblastic leukemia. *JAMA* 301(4):393–403.

Zhou, H. J., J. Yan, W. Luo, G. Ayala, S. H. Lin, H. Erdem, M. Ittmann, S. Y. Tsai, and M. J. Tsai. 2005. SRC-3 is required for prostate cancer cell proliferation and survival. *Cancer Res* 65(17):7976–83.

# 8 Nutrigenomics of Fatty Acid Sensing

*Sander Kersten*

## CONTENTS

## INTRODUCTION

In the past decade, nutrigenomics has taken the field of nutritional science by storm. Nutrigenomics can be defined as "the application of high-throughput genomic tools in nutrition research" (Muller and Kersten 2003, p. 315). Through the advent of novel technologies combined with an insatiable quest for scientific progress, important nutrition-related research questions that were once considered to be beyond our technological capacity have become amenable to experimental investigation. These technologies, best represented by transcriptomics but also including proteomics, metabolomics, and high-throughput genotyping, have created a wealth of new information, a major portion of which has yet to be fully understood. Nevertheless, major advancements have been made in our knowledge of the pleiotropic effects of dietary nutrients and the underlying mechanisms. One area that has especially benefitted from the power of nutrigenomics is fatty acid sensing. Fatty acid sensing can be interpreted as the property of fatty acids to influence biological processes by serving as signaling molecules. This chapter will provide an overview of the mechanisms of fatty acid sensing, focusing on diverse mechanisms of gene regulation by dietary fatty acids.

## TRAFFICKING OF DIETARY FAT

Every day our body processes an amount of fat equivalent to almost half a cup. In the intestine, dietary triglycerides are first degraded into fatty acids and monoglycerides, and after being taken up into enterocytes, fatty acids are reconverted into triglycerides (TG) to be secreted as part of chylomicrons. After passage through the intestinal lymph vessels, the chylomicrons enter the general circulation and give rise to the postprandial peak in plasma TG. Circulating chylomicrons undergo rapid lipolytic processing via the action of lipoprotein lipase (LPL), which is anchored to the capillary endothelium, leading to the release of fatty acids and their subsequent uptake into the underlying tissue (Wang and Eckel 2009). The activity of LPL is thus a key determinant of the rate of fatty acid uptake into tissues. One of the major sinks for meal-derived fatty acids is the adipose tissue, which diverts most of the incoming fatty acids towards storage via an elevated local LPL activity. Other tissues that contribute to postprandial clearance of chylomicron TG are the muscle and the heart, the latter of which mainly uses plasma TG-derived fatty acids as its energy source (Teusink et al. 2003). The particle that remains after TG hydrolysis, the so called chylomicron remnant, delivers the remainder of the dietary fat load to the liver after being taken up via a hepatic lipase- and receptor-dependent process (Dallinga-Thie et al. 2009). Another portion of meal-derived fatty acids may reach the liver and other tissues as free fatty acids (FFA) after being released from circulating TG by LPL via so-called fatty acid spillover. Mixing with the meal derived fatty acids are FFA that are continuously released from the adipose tissue via intracellular lipolysis. This process is particularly active during fasting but even in the fed state the adipose tissue contributes large quantities of FFA to the bloodstream. Depending on the tissue, either FFA- or TG-derived fatty acids comprise the major share of fatty acids for tissue uptake (Teusink et al. 2003). An additional potential source of intracellular fatty acids is de novo lipogenesis, a postprandial process which in humans is mainly active in the liver and to a lesser extent in adipose tissue (Schutz 2004). Thus, the concentration of fatty acids and the relative contributions of the fatty acid sources can vary markedly depending on feeding status. Furthermore, circulating FFA and TG concentrations and fatty acid fluxes are often altered during obesity, type 2 diabetes, and other metabolic disturbances. What is clear though is that irrespective of disease or nutritional status, cells and tissues are continuously exposed to fatty acids. Although it is of great interest to study the mechanisms influencing cellular fatty acid uptake, this chapter will concentrate on the specific cellular detection systems enabling the cell to sense the intra- or extracellular fatty acid concentration and respond by altering gene transcription.

## PEROXISOME PROLIFERATOR-ACTIVATED RECEPTORS AS FATTY ACID SENSORS

The peroxisome proliferator-activated receptors (PPARs) perhaps comprise the best recognized sensor system for fatty acids (Figure 8.1). PPARs are transcription factors that are part of the superfamily of nuclear hormone receptors, which also include receptors for steroid hormones and fat soluble vitamins A and D (Kersten et al. 2000). Nuclear receptors function as ligand-activated transcription factors via binding of small lipophilic molecules. They share a modular structure consisting of a DNA- and ligand-binding domain and play a role in a great variety of biological processes. Three different PPARs subtypes are known to exist, each characterized by a unique tissue expression pattern. Whereas PPARα (Nr1c1) and PPARδ (Nr1c2) are found in many cell types, expression of PPARγ (Nr1c3) is more restricted, with adipocytes and macrophages expressing the highest levels (Escher et al. 2001). According to the current dogma, which likely is partially incorrect, binding of ligand triggers the physical association of PPARs to specific DNA sequences called PPAR response elements in the promoter of target genes (Feige et al. 2006). Additionally, ligand binding leads to recruitment of coactivator proteins and loss of corepressor proteins, resulting in activation of DNA transcription. Analogous to many other nuclear receptors, PPARs bind to DNA as a heterodimer with the nuclear receptor RXR, which binds the vitamin A derivative 9-cis retinoic acid.

Throughout the years, numerous compounds have been shown to be able to bind and activate PPARs. While substantial specificity toward one PPAR subtype has been achieved in the design of synthetic PPAR agonists, by contrast, endogenous ligands have shown limited subtype specificity. In several landmark papers from the 1990s it was demonstrated that all three PPARs are able to bind fatty acids with a general preference towards long-chain polyunsaturated fatty acids (PUFAs) (Forman et al. 1997; Gottlicher et al. 1992; Keller et al. 1993; Kliewer et al. 1997a; Krey et al. 1997). Subsequent studies using a variety of biochemical techniques have firmly corroborated the direct physical association between fatty acids and PPARs and have thus established fatty acids as bona fide PPAR ligands (Lin et al. 1999; Murakami et al. 1999; Sanderson et al. 2008; Xu et al. 1999a). In addition, numerous fatty acid-derived compounds and compounds showing structural resemblance to fatty acids, including acyl-CoAs, oxidized fatty acids, eicosanoids, endocannabinoids,

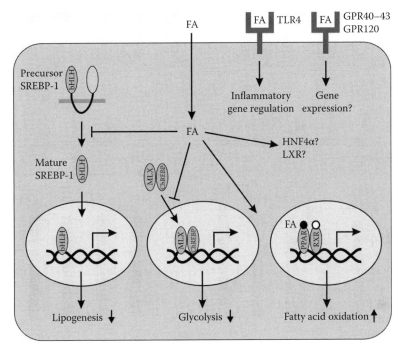

**FIGURE 8.1** General mechanisms of gene regulation by fatty acids. No differentiation is made in the behavior of the different types of fatty acids. The indicated mechanisms may be specific for a particular cell type. PUFAs reduce expression of genes involved in lipogenesis by inhibiting proteolytic processing of SREBP-1. PUFAs reduce expression of L-type pyruvate kinase (glycolysis) in liver most likely by inhibiting nuclear translocation of MLX-ChREBP. Various fatty acids but especially PUFAs act as ligand for PPARs. Activation of PPARα by PUFAs in liver leads to stimulation of fatty acids oxidation. DHA has been reported as a ligand for RXR. The GPR40 family and GPR120 serve as membrane receptors for various types of fatty acids including SCFAs. It is uncertain whether they are involved in the effects of fatty acids on gene expression. TLR4 has been proposed to be activated by saturated fatty acids. bHLH, basic helix loop helix.

and phytanic acid, have been shown to activate PPARs (Fang et al. 2006; Fu et al. 2003; Hostetler et al. 2005; Itoh et al. 2008; Li et al. 2008; Moya-Camarena et al. 1999; Zomer et al. 2000). Whereas the eicosanoid 15-deoxy-delta-12,14-prostaglandin J2 behaves as a high-affinity agonist for PPARγ, (8S)-hydroxyeicosatetraenoic acid and prostacylin PGI2 show preference for PPARα and PPARδ, respectively (Forman et al. 1995; Kliewer et al. 1995, 1997b Lim et al. 1999). Since the intracellular concentration of fatty acids (free and bound to fatty acid binding proteins) far exceeds the intracellular concentration of eicosanoids and other endogenous PPAR agonists, and since fatty acids are able to bind PPARs with high affinity, the question can be raised to what extent eicosanoids and other fatty acid-derived compounds substantially contribute to activation of PPARs *in vivo*. Rather, it can be argued that PPARs serve as general fatty acid sensors with comparatively limited ligand specificity. However, this concept is not universally embraced.

With respect to the roles of PPARs as dietary fatty acid sensor, evidence abounds indicating that especially dietary PUFAs potently activate PPARα in liver (Martin et al. 1997; Patsouris et al. 2006; Ren et al. 1997). In fact, according to a recent study in which wild type (WT) and PPARα-/- mice were provided a single bolus of synthetic TG consisting of one type of fatty acid, the effects of dietary fatty acids on hepatic gene expression are predominantly mediated by PPARα, as illustrated in Figure 8.2 for Acot3 (Sanderson et al. 2008). As discussed, dietary fatty acids mostly enter the liver as TG within chylomicron remnants, and are liberated after degradation of the remnant particles in lysosomes. In contrast and very surprisingly, circulating FFA, which primarily originate from adipose tissue lipolysis, do not seem to be able to activate PPARα. The precise mechanism

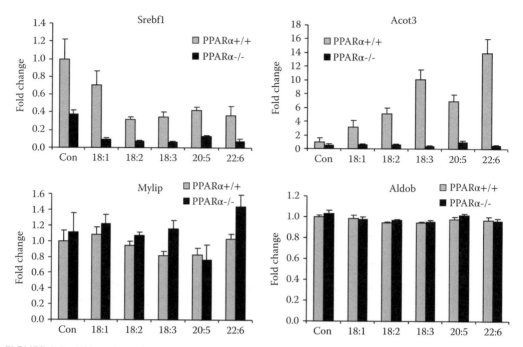

**FIGURE 8.2** Effect of oral fat load on specific receptor target genes. Mice were given an oral fat load of synthetic TG consisting of one type of fatty acid and sacrificed 6 hours thereafter. Changes in hepatic expression of selected SREBP-1 target (Srebf1, SREBP-1), PPARα target (Acot3), LXR target (Mylip, Idol), and HNF4α target (Aldob) are shown.

behind the differential effect of circulating FFA ("old fat") versus dietary and endogenously synthesized fatty acids ("new fat") on hepatic PPARα activation remains unclear but may be related to existence of distinct intracellular fatty acid pools with distinct metabolic and signaling properties (Chakravarthy et al. 2005; Sanderson et al. 2009).

PPARα can be considered the master regulator of hepatic fatty acid oxidation. In addition, activated PPARα may regulate other aspects of hepatic lipid metabolism, including fatty acid uptake, triglyceride hydrolysis, intracellular fatty acid trafficking, and fatty acid elongation (Rakhshandehroo et al. 2007, 2009). Absence of PPARα in mice leads to decreased expression of numerous enzymes involved in mitochondrial and peroxisomal fatty acid oxidation, and is associated with increased plasma FFA levels, defective ketogenesis, and hepatic steatosis (Costet et al. 1998; Kersten et al. 1999). Accordingly, activation of PPARα by fatty acids appears to be part of a feed-forward mechanism aimed at preventing the intracellular accumulation of fatty acids by stimulating their oxidation.

Similar to the situation in liver, PPARα also serves as dietary fatty acid sensor in the small intestine. Induction of numerous genes by a single oral dose of synthetic TG is abolished in PPARα-/- mice (according to our unpublished observations). The picture is less clear in other tissues that take up substantial amount of dietary fatty acids, including heart, skeletal muscle, and adipose tissue. Marine oil fatty acids have major effects on adipose tissue function and metabolism, as well as on adipose tissue gene regulation (Flachs et al. 2009). Although PUFAs are direct agonists for PPARγ, it is unclear to what extent the observed changes in adipose gene expression upon chronic PUFA feeding reflect direct ligand activation of PPARγ or other PPARs, or are secondary effects conferred by specific eicosanoids or other fatty acid-derived compounds. Compared to liver, the post-prandial effects of dietary fatty acids on gene regulation in heart are limited, likely because cardiac fatty acid uptake is relatively constant regardless of nutritional status. Nevertheless, we found that dietary fat elicited acute changes in gene expression in the heart, which were partly dependent of PPARα and PPARδ (also from our unpublished data).

## HEPATOCYTE NUCLEAR FACTOR 4α AS A FATTY ACID SENSOR

The hepatocyte nuclear factor 4α (HNF4α, Nr2a1) is a nuclear receptor that is exclusively expressed in the gastrointestinal tract, liver, and kidney (Bookout et al. 2006). Targeted disruption of HNF4α causes early embryonic lethality related to defects in the expression of visceral endoderm proteins required for maintaining gastrulation (Chen et al. 1994). Using liver-specific HNF4α-/- mice it was shown that liver HNF4α is important for hepatocyte differentiation and for governing the expression of genes involved in lipid homeostasis. In 1998 evidence was provided that saturated fatty acyl-CoA may be able to serve as agonists for HNF4α, whereas unsaturated fatty acyl-CoA were proposed to serve as antagonistic ligands (Figure 8.1) (Hertz et al. 1998). These data have been contested experimentally and have yet to gain full recognition (Bogan et al. 2000). Elucidation of the molecular structure using X-ray crystallography revealed the presence of a fatty acid, which appeared to be constitutively bound (Wisely et al. 2002; Yuan et al. 2009). Additionally, it was shown using affinity isolation/mass spectrometry that HNF4α is occupied by linoleic acid in COS-7 cells as well as in liver of fed but not fasted mice, suggesting fatty acid binding is exchangeable (Dhe-Paganon et al. 2002). However, no induction of HNF4α targets by linoleic acid was observed in a human colon cancer cell line, raising questions about the purpose of binding of linoleic acid to HNF4α. In our *in vivo* studies using dietary synthetic TG, minimal evidence was found for activation of HNF4α by various dietary unsaturated fatty acids, as judged by a lack of change in expression of known HNF4α target genes, including Aldob (Figure 8.2) (Sanderson et al. 2008). In contrast, in primary mouse hepatocytes it was observed that treatment with oleic acid, linoleic acid, or docosahexaenoic acid (DHA) increased expression of HNF4α targets Gys2, Slc2a2, and Aldob, whereas, very surprisingly, those fatty acids did not alter expression of PPAR targets (from our unpublished data). Overall, the uncertainties regarding binding and activation of HNF4α by fatty acids or acyl-CoAs are far from being resolved, and a consensus picture has yet to emerge. So far, there is limited evidence that changes in the concentration of fatty acids or acyl-CoA lead to activation of HNF4α targets.

In addition to PPARs and HNF4α, the nuclear receptors LXR, FXR, and RXR have been proposed to serve as mediators of the effects of fatty acids on gene transcription (Lengqvist et al. 2004; Ou et al. 2001; Zhao et al. 2004). Whether these findings depend on direct physical interaction between fatty acids and the nuclear receptors is controversial and requires further study. Recently, DHA was confirmed to bind directly to RXR, although with much lower affinity compared to 9cRA (Stafslien et al. 2007). In our hands, administering a single oral dose of synthetic TG consisting of different types of fatty acids did not result in any changes of LXR target genes, as illustrated for Mylip (Figure 8.2).

## STEROL REGULATORY ELEMENT BINDING PROTEIN 1 AS A FATTY ACID SENSOR

Dietary PUFAs are known to reduce the hepatic expression of genes involved in fatty acid synthesis. It is now evident that the mechanism behind the PUFA effect revolves around sterol regulatory element binding protein-1 (SREBP-1, Srebf1), a membrane-bound protein that is member of the family of basic-helix-loop-helix-leucine zipper transcription factors. In liver, SREBP-1 governs the expression of several genes involved in fatty acid and triacylglycerol synthesis. However, despite careful and prolonged investigation, the primary target of PUFAs still remains ambiguous. Studies over the last decade have indicated that PUFAs potently lower SREBP-1 mRNA levels (Figure 8.2) and inhibit proteolytic processing of SREBP-1 (Figure 8.1) (Hannah et al. 2001; Kim et al. 1999; Mater et al. 1999; Xu et al. 1999b; Yahagi et al. 1999). The latter process is required for maturation of precursor membrane-bound SREBP-1 to the mature nuclear form, which serves as the actual transcription factor. Furthermore, evidence was provided that specifically DHA, but not other PUFAs, stimulates the removal of mature nuclear SREBP-1 via a mechanism that is dependent on the 26S proteosome and ERK signaling (Botolin et al. 2006). Downregulation of SREBP-1 mRNA by PUFAs has been proposed to be mediated by stimulation of SREBP-1 mRNA decay (Xu et al. 2001), or by antagonizing the activity of the nuclear receptor LXRα, which represents a potent inducer of SREBP-1 gene transcription (Ou et al. 2001; Yoshikawa et al. 2002). Recent data argue against a role of LXR in mediating the effect of PUFAs and instead suggest that the reduction in SREBP-1 mRNA by PUFA is secondary to the inhibition of SREBP-1 maturation, which via autoregulation of SREBP-1 transcriptional activation leads to reduced SREBP-1 mRNA levels (Takeuchi et al.). In conclusion, while the intricate molecular details underlying the effects of PUFAs on SREBP-1-dependent lipogenic gene regulation are slowly filled in, the exact molecular target that physically interacts with PUFAs has yet to be identified.

PUFAs have also been shown to reduce expression of the glycolytic gene pyruvate kinase via a mechanism independent of PPARα (Figure 8.1) (Pan et al. 2000). Recently, evidence was provided that this effect is mediated by inhibiting nuclear translocation of either carbohydrate responsive element binding protein (ChREBP) or MAX-like protein X (MLX) (Dentin et al. 2005; Xu et al. 2006). ChREBP and MLX form a heterodimer that functions as glucose-responsive transcription factor and induces expression of genes involved in glycolysis and lipogenesis, including pyruvate kinase, acetyl-CoA carboxylase 1, and fatty acid synthase. However, additional data need to be collected to more precisely define how PUFAs influence ChREBP or MLX nuclear translocation and to identify the direct molecular target of PUFAs.

## TOLL-LIKE RECEPTOR 4 AS A FATTY ACID SENSOR

Numerous studies have investigated the impact of fatty acids on the inflammatory response in a great variety of cell types and tissues. These studies overwhelmingly point to a pro-inflammatory effect of saturated fatty acids, whereas n-3 PUFA exhibit mostly anti-inflammatory properties (Galli and Calder 2009). Recently, using transcriptome analysis, it was shown that a saturated fatty acid-rich diet induces a proinflammatory gene expression profile in adipose tissue of human subjects (van Dijk et al. 2009). Most of the modulatory effects of fatty acids on inflammation can probably be attributed to fatty acid metabolites, including prostaglandins, resolvins, endocannabinoids, ceramides, and diacylglycerols (Galli and Calder 2009). However, there is accumulating evidence that fatty acids may be able to directly activate or suppress inflammatory pathways.

It is believed that most of the biological activities of LPS are mediated via its lipid A moiety. Studies have shown that the fatty acids that are part of lipid A play an important role in ligand recognition and receptor activation of Toll-like receptor 4 (TLR4), leading to the suggestion that saturated fatty acids may promote inflammation by direct activation of TLR4 (Figure 8.1). Subsequent studies have provided compelling evidence that saturated fatty acids activate NF-κB and stimulate expression of NF-κB targets such as COX-2, iNOS, and IL-1α in macrophages by activating TLR4 signaling in a MyD88-, IRAK-1-, and TRAF6-dependent manner (Lee et al. 2001, 2003a,b). In contrast, unsaturated fatty acids appear to be ineffective or may even act as antagonists. Using TLR4-/- macrophages, the role of TLR4 in mediating the inflammatory effects of saturated fatty acids was convincingly demonstrated (Shi et al. 2006; Suganami et al. 2007). Loss of TLR4 was also shown to partially protect against diet-induced obesity and insulin resistance, suggesting TLR4 may be involved in mediating the detrimental effects of chronic high saturated fat consumption (Davis et al. 2008; Shi et al. 2006; Tsukumo et al. 2007). It should be emphasized that activation of TLR4 by saturated fatty acids is not universally embraced (Erridge and Samani 2009). Furthermore, experimental data showing direct physical binding of saturated fatty acids to TLR4 are still lacking (Schaeffler et al. 2009). Future studies will have to clarify to what extent the pro-inflammatory properties of saturated fatty acids can be ascribed to direct activation and how mechanistically saturated fatty acids can bring about TLR4 activation.

## NF-E2-RELATED FACTOR-2 (NRF2) AS A FATTY ACID SENSOR

An oral lipid load with PUFAs causes rapid upregulation of numerous oxidative stress genes in several organs (our unpublished data). These effects are likely mediated by the transcription factor NRF2 (NFE2L2), which governs the expression of multiple genes involved in the oxidative stress response. Compounds that activate NRF2, which range from diphenols to hydroperoxides and heavy metals, are believed to modulate the sulfhydryl group of cysteine residues within KEAP1, which serves as NRF2-specific adaptor protein for the Cullin-3 ubiquitin ligase complex (Kobayashi et al. 2004). As a result, these compounds cause the dissociation of Cullin-3 and thereby inhibit NRF2 ubiquitination, leading to stabilization and nuclear translocation of NRF2 and subsequent induction of NRF2 target genes. Studies have shown that oxidation products of linoleic acid, eicosapentanoic acid, and DHA can react with KEAP1, while the intact fatty acids cannot (Gao et al. 2007; Wang et al. 2009). Thus, the effects of (dietary) PUFA on expression of genes involved in the oxidative stress response are likely mediated by specific fatty acid oxidation products via NRF2-dependent signaling.

## G-PROTEIN-COUPLED RECEPTORS AS FATTY ACIDS SENSORS

While dietary fat has been known to influence the secretion of various gastrointestinal hormones for quite some time, only recently have we begun to learn about specific cell surface receptors that mediate the effects of fatty acids. It was shown that fatty acids act as ligands for a small family of G-protein-coupled receptors, which thus function as fatty acid sensors (Figure 8.1) (Covington et al. 2006). These fatty acid receptors, named GPR40 (FFAR1), GPR41 (FFAR3), and GPR43 (FFAR2), differ in their specificity for fatty acids, with short-chain fatty acid (SCFAs, defined by six or fewer carbon molecules) activating GPR41 and GPR43, and both saturated and unsaturated medium- to long-chain fatty acids activating GPR40. Both GPR41 and GPR43 are well expressed in the colon, which is exposed to elevated concentrations of SCFAs via bacterial fermentation. Furthermore, GPR41 is expressed in numerous immune cells and in adipose tissue, where it was shown to be involved in regulation of leptin production (Xiong et al. 2004). The relative role of GPR41 versus GPR43 as sensor for SCFAs in the enteroendocrine system is not clear. Recently, it was proposed that GPR41 mediates the effect of gut microbiota on fat mass (Samuel et al. 2008), while stimulation of GPR43 by SCFAs was shown to be necessary for the normal resolution of certain inflammatory responses (Maslowski et al. 2009). GPR40 is known as mediator of the stimulatory effect of fatty acids on glucose-stimulated insulin secretion in pancreatic β-cells (Itoh et al. 2003; Steneberg et al. 2005). The role of GPR40 in tissues other than pancreas remains unclear.

In addition to the GPR40 family members, another recently identified fatty acid receptor is GPR120, which is activated by saturated and unsaturated fatty acids with 14 or more carbons. GPR120 is most abundant in mouse large intestine, lung and adipose tissue, but is also expressed in enteroendocrine L cells, where it mediates the effect of fatty acids on release of glucagon-like peptide-1 (Hirasawa et al. 2005; Miyauchi et al. 2009). Additionally, GPR120 is involved in the induction of cholecystokinine secretion by long-chain fatty acids (Tanaka et al. 2008). Thus, it is evident that membrane fatty acid receptors play a key role in the effect of fatty acids on release of gastrointestinal hormones. Whether activation of these receptors is also linked to regulation of gene expression remains to be determined.

## CONCLUSION

While the importance of dietary fatty acids as determinants of risk for numerous chronic diseases has been well recognized, only recently have we started to gain appreciation for the vast regulatory functions of dietary fatty acids in the human body. It is now evident that fatty acids, either directly or via its metabolites, act via a great variety of signaling pathways to influence numerous metabolic, inflammatory, and other biological processes. In the past decade, nutrigenomics has provided the ideal conceptual framework and the necessary technological tools to address the global effects of dietary fatty acids, and has importantly contributed to a major advancement in our understanding of the molecular action of dietary fatty acids. So far the focus has been on the molecular characterization of specific signaling routes, coupled to the description of the whole genome effects of dietary fatty acids. In the future, greater emphasis will have to be placed on the functional consequences of specific target gene regulation in order to fully understand the functional impact of dietary fatty acids and their potentially preventive effect in specific disease conditions. It can be foreseen that nutrigenomics will continue to make a push towards a more mechanistic and genomics-driven approach within the domain of nutritional sciences and further promote the implementation of high-throughput technologies.

## REFERENCES

Bogan, A. A., Q. Dallas-Yang, M. D. Ruse, Jr., Y. Maeda, G. Jiang, L. Nepomuceno, T. S. Scanlan, F. E. Cohen, and F. M. Sladek. 2000. Analysis of protein dimerization and ligand binding of orphan receptor HNF4alpha. *J Mol Biol* 302(4):831–51.

Bookout, A. L., Y. Jeong, M. Downes, R. T. Yu, R. M. Evans, and D. J. Mangelsdorf. 2006. Anatomical profiling of nuclear receptor expression reveals a hierarchical transcriptional network. *Cell* 126(4):789–99.

Botolin, D., Y. Wang, B. Christian, and D. B. Jump. 2006. Docosahexaneoic acid (22:6,n-3) regulates rat hepatocyte SREBP-1 nuclear abundance by Erk- and 26S proteasome-dependent pathways. *J Lipid Res* 47(1):181–92.

Chakravarthy, M. V., Z. Pan, Y. Zhu, K. Tordjman, J. G. Schneider, T. Coleman, J. Turk, and C. F. Semenkovich. 2005. "New" hepatic fat activates PPARalpha to maintain glucose, lipid, and cholesterol homeostasis. *Cell Metab* 1(5):309–22.

Chen, W. S., K. Manova, D. C. Weinstein, S. A. Duncan, A. S. Plump, V. R. Prezioso, R. F. Bachvarova, and J. E. Darnell, Jr. 1994. Disruption of the HNF-4 gene, expressed in visceral endoderm, leads to cell death in embryonic ectoderm and impaired gastrulation of mouse embryos. *Genes Dev* 8(20):2466–77.

Costet, P., C. Legendre, J. More, A. Edgar, P. Galtier, and T. Pineau. 1998. Peroxisome proliferator-activated receptor alpha-isoform deficiency leads to progressive dyslipidemia with sexually dimorphic obesity and steatosis. *J Biol Chem* 273(45):29577–85.

Covington, D. K., C. A. Briscoe, A. J. Brown, and C. K. Jayawickreme. 2006. The G-protein-coupled receptor 40 family (GPR40-GPR43) and its role in nutrient sensing. *Biochem Soc Trans* 34(Pt 5):770–73.

Dallinga-Thie, G. M., R. Franssen, H. L. Mooij, M. E. Visser, H. C. Hassing, F. Peelman, J. J. Kastelein, M. Peterfy, and M. Nieuwdorp. 2009. The metabolism of triglyceride-rich lipoproteins revisited: New players, new insight. *Atherosclerosis* 211(1):1–8.

Davis, J. E., N. K. Gabler, J. Walker-Daniels, and M. E. Spurlock. 2008. Tlr-4 deficiency selectively protects against obesity induced by diets high in saturated fat. *Obesity (Silver Spring)* 16(6):1248–55.

Dentin, R., F. Benhamed, J. P. Pegorier, F. Foufelle, B. Viollet, S. Vaulont, J. Girard, and C. Postic. 2005. Polyunsaturated fatty acids suppress glycolytic and lipogenic genes through the inhibition of ChREBP nuclear protein translocation. *J Clin Invest* 115(10):2843–54.

Dhe-Paganon, S., K. Duda, M. Iwamoto, Y. I. Chi, and S. E. Shoelson. 2002. Crystal structure of the HNF4 alpha ligand binding domain in complex with endogenous fatty acid ligand. *J Biol Chem* 277(41):37973–76.

Erridge, C., and N. J. Samani. 2009. Saturated fatty acids do not directly stimulate Toll-like receptor signaling. *Arterioscler Thromb Vasc Biol* 29(11):1944–49.

Escher, P., O. Braissant, S. Basu-Modak, L. Michalik, W. Wahli, and B. Desvergne. 2001. Rat PPARs: Quantitative analysis in adult rat tissues and regulation in fasting and refeeding. *Endocrinology* 142(10):4195–202.

Fang, X., S. Hu, B. Xu, G. D. Snyder, S. Harmon, J. Yao, Y. Liu, B. Sangras, J. R. Falck, N. L. Weintraub, and A. A. Spector. 2006. 14,15-Dihydroxyeicosatrienoic acid activates peroxisome proliferator-activated receptor-alpha. *Am J Physiol Heart Circ Physiol* 290(1):H55–63.

Feige, J. N., L. Gelman, L. Michalik, B. Desvergne, and W. Wahli. 2006. From molecular action to physiological outputs: Peroxisome proliferator-activated receptors are nuclear receptors at the crossroads of key cellular functions. *Prog Lipid Res* 45(2):120–59.

Flachs, P., M. Rossmeisl, M. Bryhn, and J. Kopecky. 2009. Cellular and molecular effects of n-3 polyunsaturated fatty acids on adipose tissue biology and metabolism. *Clin Sci (Lond)* 116(1):1–16.

Forman, B. M., J. Chen, and R. M. Evans. 1997. Hypolipidemic drugs, polyunsaturated fatty acids, and eico-sanoids are ligands for peroxisome proliferator-activated receptors alpha and delta. *Proc Natl Acad Sci USA* 94(9):4312–17.

Forman, B. M., P. Tontonoz, J. Chen, R. P. Brun, B. M. Spiegelman, and R. M. Evans. 1995. 15-Deoxy-delta 12, 14-prostaglandin J2 is a ligand for the adipocyte determination factor PPAR gamma. *Cell* 83(5):803–12.

Fu, J., S. Gaetani, F. Oveisi, J. Lo Verme, A. Serrano, F. Rodriguez De Fonseca, A. Rosengarth, H. Luecke, B. Di Giacomo, G. Tarzia, and D. Piomelli. 2003. Oleylethanolamide regulates feeding and body weight through activation of the nuclear receptor PPAR-alpha. *Nature* 425(6953):90–3.

Galli, C., and P. C. Calder. 2009. Effects of fat and fatty acid intake on inflammatory and immune responses: A critical review. *Ann Nutr Metab* 55(1–3):123–39.

Gao, L., J. Wang, K. R. Sekhar, H. Yin, N. F. Yared, S. N. Schneider, S. Sasi, T. P. Dalton, M. E. Anderson, J. Y. Chan, J. D. Morrow, and M. L. Freeman. 2007. Novel n-3 fatty acid oxidation products activate Nrf2 by destabilizing the association between Keap1 and Cullin3. *J Biol Chem* 282(4):2529–37.

Gottlicher, M., E. Widmark, Q. Li, and J. A. Gustafsson. 1992. Fatty acids activate a chimera of the clofibric acid-activated receptor and the glucocorticoid receptor. *Proc Natl Acad Sci USA* 89(10):4653–57.

Hannah, V. C., J. Ou, A. Luong, J. L. Goldstein, and M. S. Brown. 2001. Unsaturated fatty acids down-regulate srebp isoforms 1a and 1c by two mechanisms in HEK-293 cells. *J Biol Chem* 276(6):4365–72.

Hertz, R., J. Magenheim, I. Berman, and J. Bar-Tana. 1998. Fatty acyl-CoA thioesters are ligands of hepatic nuclear factor-4alpha. *Nature* 392(6675):512–16.

Hirasawa, A., K. Tsumaya, T. Awaji, S. Katsuma, T. Adachi, M. Yamada, Y. Sugimoto, S. Miyazaki, and G. Tsujimoto. 2005. Free fatty acids regulate gut incretin glucagon-like peptide-1 secretion through GPR120. *Nat Med* 11(1):90–4.

Hostetler, H. A., A. D. Petrescu, A. B. Kier, and F. Schroeder. 2005. Peroxisome proliferator-activated receptor alpha interacts with high affinity and is conformationally responsive to endogenous ligands. *J Biol Chem* 280(19):18667–82.

Itoh, T., L. Fairall, K. Amin, Y. Inaba, A. Szanto, B. L. Balint, L. Nagy, K. Yamamoto, and J. W. Schwabe. 2008. Structural basis for the activation of PPARgamma by oxidized fatty acids. *Nat Struct Mol Biol* 15(9):924–31.

Itoh, Y., Y. Kawamata, M. Harada, M. Kobayashi, R. Fujii, S. Fukusumi, K. Ogi, M. Hosoya, Y. Tanaka, H. Uejima, H. Tanaka, M. Maruyama, R. Satoh, S. Okubo, H. Kizawa, H. Komatsu, F. Matsumura, Y. Noguchi, T. Shinohara, S. Hinuma, Y. Fujisawa, and M. Fujino. 2003. Free fatty acids regulate insulin secretion from pancreatic beta cells through GPR40. *Nature* 422(6928):173–6.

Keller, H., C. Dreyer, J. Medin, A. Mahfoudi, K. Ozato, and W. Wahli. 1993. Fatty acids and retinoids control lipid metabolism through activation of peroxisome proliferator-activated receptor-retinoid X receptor heterodimers. *Proc Natl Acad Sci USA* 90(6):2160–4.

Kersten, S., B. Desvergne, and W. Wahli. 2000. Roles of PPARs in health and disease. *Nature* 405(6785):421–4.

Kersten, S., J. Seydoux, J. M. Peters, F. J. Gonzalez, B. Desvergne, and W. Wahli. 1999. Peroxisome proliferator-activated receptor alpha mediates the adaptive response to fasting. *J Clin Invest* 103(11):1489–98.

Kim, H. J., M. Takahashi, and O. Ezaki. 1999. Fish oil feeding decreases mature sterol regulatory element-binding protein 1 (SREBP-1) by down-regulation of SREBP-1c mRNA in mouse liver: A possible mechanism for down-regulation of lipogenic enzyme mRNAs. *J Biol Chem* 274(36):25892–8.

Kliewer, S. A., J. M. Lenhard, T. M. Willson, I. Patel, D. C. Morris, and J. M. Lehmann. 1995. A prostaglandin J2 metabolite binds peroxisome proliferator-activated receptor gamma and promotes adipocyte differentiation. *Cell* 83(5):813–9.

Kliewer, S. A., S. S. Sundseth, S. A. Jones, P. J. Brown, G. B. Wisely, C. S. Koble, P. Devchand, W. Wahli, T. M. Willson, J. M. Lenhard, and J. M. Lehmann. 1997a. Fatty acids and eicosanoids regulate gene expression through direct interactions with peroxisome proliferator-activated receptors alpha and gamma. *Proc Natl Acad Sci USA* 94(9):4318–23.

Kobayashi, A., M. I. Kang, H. Okawa, M. Ohtsuji, Y. Zenke, T. Chiba, K. Igarashi, and M. Yamamoto. 2004. Oxidative stress sensor Keap1 functions as an adaptor for Cul3-based E3 ligase to regulate proteasomal degradation of Nrf2. *Mol Cell Biol* 24(16):7130–9.

Krey, G., O. Braissant, F. L'Horset, E. Kalkhoven, M. Perroud, M. G. Parker, and W. Wahli. 1997. Fatty acids, eicosanoids, and hypolipidemic agents identified as ligands of peroxisome proliferator-activated receptors by coactivator-dependent receptor ligand assay. *Mol Endocrinol* 11(6):779–91.

Lee, J. Y., A. Plakidas, W. H. Lee, A. Heikkinen, P. Chanmugam, G. Bray, and D. H. Hwang. 2003a. Differential modulation of Toll-like receptors by fatty acids: preferential inhibition by n-3 polyunsaturated fatty acids. *J Lipid Res* 44(3):479–86.

Lee, J. Y., K. H. Sohn, S. H. Rhee, and D. Hwang. 2001. Saturated fatty acids, but not unsaturated fatty acids, induce the expression of cyclooxygenase-2 mediated through Toll-like receptor 4. *J Biol Chem* 276(20):16683–9.

Lee, J. Y., J. Ye, Z. Gao, H. S. Youn, W. H. Lee, L. Zhao, N. Sizemore, and D. H. Hwang. 2003b. Reciprocal modulation of Toll-like receptor-4 signaling pathways involving MyD88 and phosphatidylinositol 3-kinase/AKT by saturated and polyunsaturated fatty acids. *J Biol Chem* 278(39):37041–51.

Lengqvist, J., A. Mata De Urquiza, A. C. Bergman, T. M. Willson, J. Sjovall, T. Perlmann, and W. J. Griffiths. 2004. Polyunsaturated fatty acids including docosahexaenoic and arachidonic acid bind to the retinoid X receptor alpha ligand-binding domain. *Mol Cell Proteomics* 3(7):692–703.

Li, Y., J. Zhang, F. J. Schopfer, D. Martynowski, M. T. Garcia-Barrio, A. Kovach, K. Suino-Powell, P. R. Baker, B. A. Freeman, Y. E. Chen, and H. E. Xu. 2008. Molecular recognition of nitrated fatty acids by PPAR gamma. *Nat Struct Mol Biol* 15(8):865–7.

Lim, H., R. A. Gupta, W. G. Ma, B. C. Paria, D. E. Moller, J. D. Morrow, R. N. DuBois, J. M. Trzaskos, and S. K. Dey. 1999. Cyclo-oxygenase-2-derived prostacyclin mediates embryo implantation in the mouse via PPARdelta. *Genes Dev* 13(12):1561–74.

Lin, Q., S. E. Ruuska, N. S. Shaw, D. Dong, and N. Noy. 1999. Ligand selectivity of the peroxisome proliferator-activated receptor alpha. *Biochemistry* 38(1):185–90.

Martin, G., K. Schoonjans, A. M. Lefebvre, B. Staels, and J. Auwerx. 1997. Coordinate regulation of the expression of the fatty acid transport protein and acyl-CoA synthetase genes by PPARalpha and PPARgamma activators. *J Biol Chem* 272(45):28210–7.

Maslowski, K. M., A. T. Vieira, A. Ng, J. Kranich, F. Sierro, D. Yu, H. C. Schilter, M. S. Rolph, F. Mackay, D. Artis, R. J. Xavier, M. M. Teixeira, and C. R. Mackay. 2009. Regulation of inflammatory responses by gut microbiota and chemoattractant receptor GPR43. *Nature* 461(7268):1282–6.

Mater, M. K., A. P. Thelen, D. A. Pan, and D. B. Jump. 1999. Sterol response element-binding protein 1c (SREBP1c) is involved in the polyunsaturated fatty acid suppression of hepatic S14 gene transcription. *J Biol Chem* 274(46):32725–32.

Miyauchi, S., A. Hirasawa, T. Iga, N. Liu, C. Itsubo, K. Sadakane, T. Hara, and G. Tsujimoto. 2009. Distribution and regulation of protein expression of the free fatty acid receptor GPR120. *Naunyn Schmiedebergs Arch Pharmacol* 379(4):427–34.

Moya-Camarena, S. Y., J. P. Vanden Heuvel, S. G. Blanchard, L. A. Leesnitzer, and M. A. Belury. 1999. Conjugated linoleic acid is a potent naturally occurring ligand and activator of PPARalpha. *J Lipid Res* 40(8):1426–33.

Muller, M., and S. Kersten. 2003. Nutrigenomics: Goals and strategies. *Nat Rev Genet* 4(4):315–22.

Murakami, K., T. Ide, M. Suzuki, T. Mochizuki, and T. Kadowaki. 1999. Evidence for direct binding of fatty acids and eicosanoids to human peroxisome proliferators-activated receptor alpha. *Biochem Biophys Res Commun* 260(3):609–13.

Ou, J., H. Tu, B. Shan, A. Luk, R. A. DeBose-Boyd, Y. Bashmakov, J. L. Goldstein, and M. S. Brown. 2001. Unsaturated fatty acids inhibit transcription of the sterol regulatory element-binding protein-1c (SREBP-1c) gene by antagonizing ligand-dependent activation of the LXR. *Proc Natl Acad Sci USA* 98(11):6027–32.

Pan, D. A., M. K. Mater, A. P. Thelen, J. M. Peters, F. J. Gonzalez, and D. B. Jump. 2000. Evidence against the peroxisome proliferator-activated receptor alpha (PPARalpha) as the mediator for polyunsaturated fatty acid suppression of hepatic L-pyruvate kinase gene transcription. *J Lipid Res* 41(5):742–51.

Patsouris, D., J. K. Reddy, M. Muller, and S. Kersten. 2006. Peroxisome proliferator-activated receptor alpha mediates the effects of high-fat diet on hepatic gene expression. *Endocrinology* 147(3):1508–16.

Rakhshandehroo, M., G. Hooiveld, M. Muller, and S. Kersten. 2009. Comparative analysis of gene regulation by the transcription factor PPARalpha between mouse and human. *PLoS One* 4(8):e6796.

Rakhshandehroo, M., L. M. Sanderson, M. Matilainen, R. Stienstra, C. Carlberg, P. J. de Groot, M. Muller, and S. Kersten. 2007. Comprehensive analysis of PPAR alpha-dependent regulation of hepatic lipid metabolism by expression profiling. *PPAR Res* 2007:26839.

Ren, B., A. P. Thelen, J. M. Peters, F. J. Gonzalez, and D. B. Jump. 1997. Polyunsaturated fatty acid suppression of hepatic fatty acid synthase and S14 gene expression does not require peroxisome proliferator-activated receptor alpha. *J Biol Chem* 272(43):26827–32.

Samuel, B. S., A. Shaito, T. Motoike, F. E. Rey, F. Backhed, J. K. Manchester, R. E. Hammer, S. C. Williams, J. Crowley, M. Yanagisawa, and J. I. Gordon. 2008. Effects of the gut microbiota on host adiposity are modulated by the short-chain fatty-acid binding G protein-coupled receptor, Gpr41. *Proc Natl Acad Sci USA* 105(43):16767–72.

Sanderson, L. M., P. J. de Groot, G. J. Hooiveld, A. Koppen, E. Kalkhoven, M. Muller, and S. Kersten. 2008. Effect of synthetic dietary triglycerides: A novel research paradigm for nutrigenomics. *PLoS One* 3(2):e1681.

Sanderson, L. M., T. Degenhardt, A. Koppen, E. Kalkhoven, B. Desvergne, M. Muller, and S. Kersten. 2009. Peroxisome proliferator-activated receptor beta/delta (PPARbeta/delta) but not PPARalpha serves as a plasma free fatty acid sensor in liver. *Mol Cell Biol* 29(23):6257–67.

Schaeffler, A., P. Gross, R. Buettner, C. Bollheimer, C. Buechler, M. Neumeier, A. Kopp, J. Schoelmerich, and W. Falk. 2009. Fatty acid-induced induction of Toll-like receptor-4/nuclear factor-kappaB pathway in adipocytes links nutritional signalling with innate immunity. *Immunology* 126(2):233–45.

Schutz, Y. 2004. Concept of fat balance in human obesity revisited with particular reference to de novo lipogenesis. *Int J Obes Relat Metab Disord* 28(Suppl 4):S3–S11.

Shi, H., M. V. Kokoeva, K. Inouye, I. Tzameli, H. Yin, and J. S. Flier. 2006. TLR4 links innate immunity and fatty acid-induced insulin resistance. *J Clin Invest* 116(11):301525.

Stafslien, D. K., K. L. Vedvik, T. De Rosier, and M. S. Ozers. 2007. Analysis of ligand-dependent recruitment of coactivator peptides to RXRbeta in a time-resolved fluorescence resonance energy transfer assay. *Mol Cell Endocrinol* 264(1–2):82–9.

Steneberg, P., N. Rubins, R. Bartoov-Shifman, M. D. Walker, and H. Edlund. 2005. The FFA receptor GPR40 links hyperinsulinemia, hepatic steatosis, and impaired glucose homeostasis in mouse. *Cell Metab* 1(4):245–58.

Suganami, T., K. Tanimoto-Koyama, J. Nishida, M. Itoh, X. Yuan, S. Mizuarai, H. Kotani, S. Yamaoka, K. Miyake, S. Aoe, Y. Kamei, and Y. Ogawa. 2007. Role of the Toll-like receptor 4/NF-kappaB pathway in saturated fatty acid-induced inflammatory changes in the interaction between adipocytes and macrophages. *Arterioscler Thromb Vasc Biol* 27(1):84–91.

Takeuchi, Y., N. Yahagi, Y. Izumida, M. Nishi, M. Kubota, Y. Teraoka, T. Yamamoto, T. Matsuzaka, Y. Nakagawa, M. Sekiya, Y. Iizuka, K. Ohashi, J. I. Osuga, T. Gotoda, S. Ishibashi, K. Itaka, K. Kataoka, R. Nagai, N. Yamada, T. Kadowaki, and H. Shimano. Polyunsaturated fatty acids selectively suppress sterol regulatory element-binding protein-1 through proteolytic processing and autoloop regulatory circuit. *J Biol Chem* 285(15):11681–91

Tanaka, T., S. Katsuma, T. Adachi, T. A. Koshimizu, A. Hirasawa, and G. Tsujimoto. 2008. Free fatty acids induce cholecystokinin secretion through GPR120. *Naunyn Schmiedebergs Arch Pharmacol* 377(4–6):523–7.

Teusink, B., P. J. Voshol, V. E. Dahlmans, P. C. Rensen, H. Pijl, J. A. Romijn, and L. M. Havekes. 2003. Contribution of fatty acids released from lipolysis of plasma triglycerides to total plasma fatty acid flux and tissue-specific fatty acid uptake. *Diabetes* 52(3):614–20.

Tsukumo, D. M., M. A. Carvalho-Filho, J. B. Carvalheira, P. O. Prada, S. M. Hirabara, A. A. Schenka, E. P. Araujo, J. Vassallo, R. Curi, L. A. Velloso, and M. J. Saad. 2007. Loss-of-function mutation in Toll-like receptor 4 prevents diet-induced obesity and insulin resistance. *Diabetes* 56(8):1986–98.

van Dijk, S. J., E. J. Feskens, M. B. Bos, D. W. Hoelen, R. Heijligenberg, M. G. Bromhaar, L. C. de Groot, J. H. de Vries, M. Muller, and L. A. Afman. 2009. A saturated fatty acid-rich diet induces an obesity-linked proinflammatory gene expression profile in adipose tissue of subjects at risk of metabolic syndrome. *Am J Clin Nutr* 90(6):1656–64.

Wang, H., and R. H. Eckel. 2009. Lipoprotein lipase: From gene to obesity. *Am J Physiol Endocrinol Metab* 297(2):E271–88.

Wang, R., J. T. Kern, T. L. Goodfriend, D. L. Ball, and H. Luesch. 2009. Activation of the antioxidant response element by specific oxidized metabolites of linoleic acid. *Prostaglandins Leukot Essent Fatty Acids* 81(1):53–9.

Wisely, G. B., A. B. Miller, R. G. Davis, A. D. Thornquest, Jr., R. Johnson, T. Spitzer, A. Sefler, B. Shearer, J. T. Moore, T. M. Willson, and S. P. Williams. 2002. Hepatocyte nuclear factor 4 is a transcription factor that constitutively binds fatty acids. *Structure* 10(9):1225–34.

Xiong, Y., N. Miyamoto, K. Shibata, M. A. Valasek, T. Motoike, R. M. Kedzierski, and M. Yanagisawa. 2004. Short-chain fatty acids stimulate leptin production in adipocytes through the G protein-coupled receptor GPR41. *Proc Natl Acad Sci USA* 101(4):1045–50.

Xu, H. E., M. H. Lambert, V. G. Montana, D. J. Parks, S. G. Blanchard, P. J. Brown, D. D. Sternbach, J. M. Lehmann, G. B. Wisely, T. M. Willson, S. A. Kliewer, and M. V. Milburn. 1999a. Molecular recognition of fatty acids by peroxisome proliferator-activated receptors. *Mol Cell* 3(3):397–403.

Xu, J., B. Christian, and D. B. Jump. 2006. Regulation of rat hepatic L-pyruvate kinase promoter composition and activity by glucose, n-3 polyunsaturated fatty acids, and peroxisome proliferator-activated receptor-alpha agonist. *J Biol Chem* 281(27):18351–62.

Xu, J., M. T. Nakamura, H. P. Cho, and S. D. Clarke. 1999b. Sterol regulatory element binding protein-1 expression is suppressed by dietary polyunsaturated fatty acids: A mechanism for the coordinate suppression of lipogenic genes by polyunsaturated fats. *J Biol Chem* 274(33):23577–83.

Xu, J., M. Teran-Garcia, J. H. Park, M. T. Nakamura, and S. D. Clarke. 2001. Polyunsaturated fatty acids suppress hepatic sterol regulatory element-binding protein-1 expression by accelerating transcript decay. *J Biol Chem* 276(13):9800–7.

Yahagi, N., H. Shimano, A. H. Hasty, M. Amemiya-Kudo, H. Okazaki, Y. Tamura, Y. Iizuka, F. Shionoiri, K. Ohashi, J. Osuga, K. Harada, T. Gotoda, R. Nagai, S. Ishibashi, and N. Yamada. 1999. A crucial role of sterol regulatory element-binding protein-1 in the regulation of lipogenic gene expression by polyunsaturated fatty acids. *J Biol Chem* 274(50):35840–4.

Yoshikawa, T., H. Shimano, N. Yahagi, T. Ide, M. Amemiya-Kudo, T. Matsuzaka, M. Nakakuki, S. Tomita, H. Okazaki, Y. Tamura, Y. Iizuka, K. Ohashi, A. Takahashi, H. Sone, J. Osuga Ji, T. Gotoda, S. Ishibashi, and N. Yamada. 2002. Polyunsaturated fatty acids suppress sterol regulatory element-binding protein 1c promoter activity by inhibition of liver X receptor (LXR) binding to LXR response elements. *J Biol Chem* 277(3):1705–11.

Yuan, X., T. C. Ta, M. Lin, J. R. Evans, Y. Dong, E. Bolotin, M. A. Sherman, B. M. Forman, and F. M. Sladek. 2009. Identification of an endogenous ligand bound to a native orphan nuclear receptor. *PLoS One* 4(5):e5609.

Zhao, A., J. Yu, J. L. Lew, L. Huang, S. D. Wright, and J. Cui. 2004. Polyunsaturated fatty acids are FXR ligands and differentially regulate expression of FXR targets. *DNA Cell Biol* 23(8):519–26.

Zomer, A. W., B. van Der Burg, G. A. Jansen, R. J. Wanders, B. T. Poll-The, and P. T. van Der Saag. 2000. Pristanic acid and phytanic acid: Naturally occurring ligands for the nuclear receptor peroxisome proliferator-activated receptor alpha. *J Lipid Res* 41(11):1801–7.

# 9 Polyphenol Resveratrol Alters Global Patterns of Gene Regulation and Improves Physiology through Multiple Potential Pathways

*Behzad Varamini and Joseph A. Baur*

## CONTENTS

## POLYPHENOLS AND HEALTH

Polyphenols are a diverse group of phytochemicals that include many of the molecules that give fruits and vegetables their colors. They are characterized structurally by the presence of multiple phenol rings and have numerous functions, ranging from ultraviolet (UV) shielding to pathogen resistance (Winkel-Shirley 2002). Polyphenols are also the major source of antioxidant activity in plant-derived foods, a fact that is often overlooked in favor of the more familiar antioxidant vitamins (Halliwell 2007). Recently, polyphenols have been thrust into the spotlight, due in part to the apparent health benefits of resveratrol, a relatively uncommon member of this class found in wine, berries, and traditional medicines (Figure 9.1). At least in rodents, resveratrol prevents cancer, increases endurance, and ameliorates many of the consequences of obesity, including the loss of insulin sensitivity and increase in mortality rate, albeit at doses in excess of what could be obtained from dietary sources (Jang et al. 1997; Baur et al. 2006; Lagouge et al. 2006). Although these results have garnered much attention, resveratrol is far from the only member of this class of molecules to exert beneficial effects on mammalian health. For example, many of the benefits of green tea have been attributed to the polyphenol epigallocatechin gallate (EGCG), pterostilbene in blueberries may have anticancer and anti-inflammatory effects, and a number a foods thought to promote health, including wine, contain the polyphenol quercetin, which has independent benefits and may synergize with resveratrol (Mertens-Talcott and Percival 2005). More than 8000 distinct polyphenols have been described (Bravo 1998), and in most cases almost nothing is known about these molecules. Nevertheless, a handful of specific polyphenols, including resveratrol, have been studied in some detail and may provide clues as to how these molecules interact with mammalian biology.

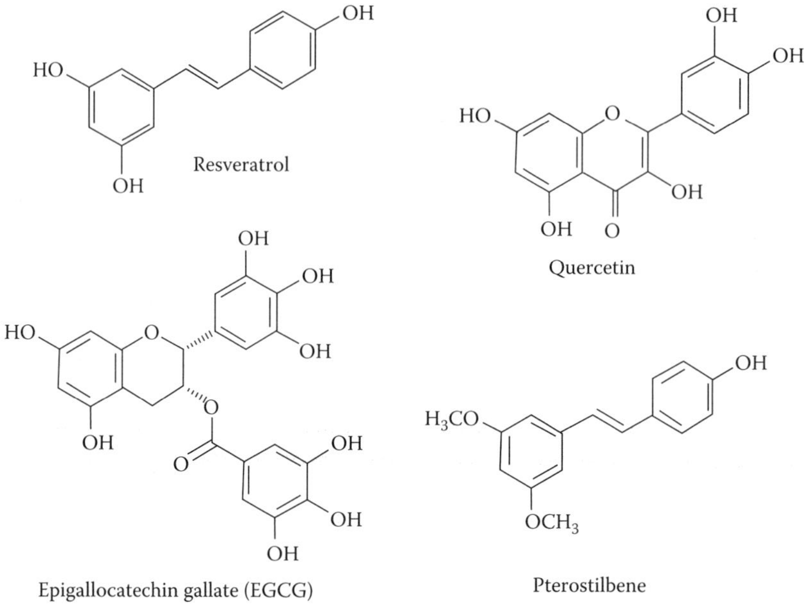

**FIGURE 9.1** Structures of four polyphenols that may be beneficial for human health. Resveratrol is found in berries and wine, quercetin is found in many foods, including apples, onions, and wine, epigallocatechin gallate (EGCG) is particularly concentrated in green tea, and pterostilbene accumulates in blueberries. All have intrinsic antioxidant capacity, and may have overlapping and discrete effects on a variety of mammalian enzymes.

## RESVERATROL, SIRT1, AND CALORIC RESTRICTION

Resveratrol owes its status as one of the few well-studied polyphenols to its identification in a number of independent screens, as well as a certain mystique that comes from being relatively specific to red wine, a substance with its own history of health benefits. In fact, a recent study justified much of the past interest in wine by showing that light consumption (averaging about half a glass per day) was associated with an approximately 5-year increase in life expectancy (Streppel et al. 2009). Anticipating this result, many of the early studies on resveratrol were directed towards recapitulating the known cardioprotective effects of red wine, without knowledge of specific targets. In 1997, however, resveratrol was identified in a screen for cyclooxygenase inhibitors, providing a direct molecular explanation for some of its effects, and predicting anticancer and anti-inflammatory actions that were promptly demonstrated (Jang et al. 1997). The rediscovery of resveratrol in a second *in vitro* screen, this time for activators of the SIRT1 enzyme, led to the more general suggestion that this small polyphenol might tap into an entire program that evolved to regulate longevity and suppress age-related disease (Howitz et al. 2003).

SIRT1 is the mammalian counterpart to the yeast Sir2 protein, and this family of enzymes (termed "sirtuins") has been proposed to mediate the beneficial effects of caloric restriction (CR), that is, the improvement in health and increase in longevity observed when energy intake is limited in the absence of malnutrition, in organisms ranging from yeast to primates (Lin et al. 2000; Colman et al. 2009). This hypothesis was originally based on the observations that *SIR2* gene dosage determines lifespan in yeast, and that the enzymatic activity of sirtuins is dependent on the cofactor nicotinamide adenine dinucleotide ($NAD^+$) (Imai et al. 2000; Smith et al. 2000). Since $NAD^+$ is intimately linked to metabolic processes within the cell, this provides a way to couple sirtuin activity to nutrient availability and thereby regulate longevity based on energy intake. Moreover, since sirtuins use $NAD^+$ to modify other proteins by deacetylation and ADP-ribosylation, this system has the potential to influence a large number of downstream processes. In support of the model, deleting the *SIR2* gene in otherwise wild-type yeast completely blocks the ability of glucose restriction (i.e., yeast CR) to extend lifespan (Lin et al. 2000). Further, lifespan extension in *eat-2* mutant worms, considered a model of CR due to a defect in pharangeal pumping, is dependent on the *SIR2* homolog in that organism (Wang and Tissenbaum 2006), and lifespan extension by CR has also been shown to require a *SIR2* homolog in flies (Rogina and Helfand 2004). It should be noted, however, that more recent studies using different experimental paradigms have shown that beneficial effects of CR can be mediated by other pathways in the absence of *SIR2* (Kaeberlein et al. 2004; Greer et al. 2007). Therefore, the view of *SIR2* as the sole mediator of CR, or CR-dependent lifespan extension, is no longer tenable. However, it remains interesting to ask whether direct stimulation of sirtuin enzymes is sufficient to recapitulate some of the beneficial effects of CR.

To test whether activation of sirtuins is sufficient to produce key benefits associated with CR, and to extend these findings to mammals, Howitz et al. (2003) designed and carried out an *in vitro* screen for activators using the mammalian enzyme SIRT1. The top hit was resveratrol, which was already suspected of mediating beneficial effects of wine, and was under investigation as a potential chemotherapeutic agent for cancer due to its ability to inhibit cyclooxygenase. Resveratrol was subsequently shown to extend yeast, worm, and fly lifespan in a sirtuin-dependent manner (Lin et al. 2000), although not all such studies have detected significant effects (Bass et al. 2007). Resveratrol has also been shown to extend lifespan in *N. furzeri*, a short-lived species of fish (Valenzano et al. 2006), however, the sirtuin dependence has not been tested in that organism.

In mammals, resveratrol has been shown to produce a variety of effects consistent with CR (Baur et al. 2006; Barger et al. 2008; Pearson et al. 2008). These results are most dramatic in obese mice, where resveratrol restores insulin sensitivity, prevents hepatic steatosis, increases endurance, and prolongs survival. Normal, nonobese rodents fed resveratrol also display improvements in cardiovascular function, bone density, and motor coordination (Barger et al. 2008; Pearson et al. 2008). Further, resveratrol has shown similarities to CR at the level of transcriptional changes. Resveratrol-fed mice on a high-calorie diet demonstrate significant transcriptional alterations in a number of pathways similar to CR mice, including downregulation of IGF-1, mTOR, and glycolysis. Transcriptional similarities have also been described between normal resveratrol-fed and CR mice. Barger et al. (2008) identified more than 700 genes whose expression was significantly altered by CR and resveratrol in the heart, over 99% of which changed in the same direction. Similar results were obtained in skeletal muscle and brain, and in a separate study, resveratrol treatment and every-other-day feeding (which produces a lifespan extension similar to conventional CR) were shown to result in similar transcriptional profiles in liver, skeletal muscle, and adipose tissue from nonobese mice (Pearson et al. 2008). In addition to general improvements in physiology and the vascular system, the anticancer effects of resveratrol in nonobese mice have been confirmed many times over (Bishayee 2009). Taken together, these results suggest that resveratrol might have a beneficial effect on health in humans. As a result, "nutraceutical" formulations containing resveratrol have been available for several years, and a number of controlled clinical trials are currently underway. However the degree to which the beneficial effects of resveratrol are mediated by SIRT1, and indeed, whether SIRT1 activation even occurs through a direct binding mechanism (Pacholec et al. 2010), remain controversial topics.

In support of the idea that activation of SIRT1 might be sufficient to mimic beneficial effects of CR, both its over-expression and activation with a novel molecule, SRT1720, have been shown to produce health benefits, including improvements in insulin sensitivity (Bordone et al. 2007; Milne et al. 2007; Feige et al. 2008; Pfluger et al. 2008). However neither of these strategies, nor resveratrol itself, has been shown to prevent spontaneous lymphomas or extend lifespan in nonobese mice, while CR accomplishes both of these, arguing that SIRT1 activation alone is not sufficient to fully recapitulate its effects (Figure 9.2). In addition, the degree to which resveratrol's effects are mediated by SIRT1 has become controversial for a number of reasons. First, resveratrol affects numerous enzymes and receptors at the same concentrations required to activate SIRT1. It has already been mentioned that resveratrol is a direct inhibitor of cyclooxygenases. It also inhibits a number of kinases, including PKCs, Akt, and S6K, other enzymes such as quinone reductase 2, riboncleotide reductase, and cytochrome p450s, and

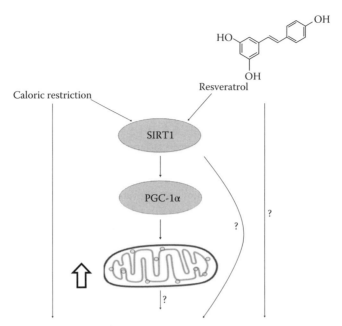

**FIGURE 9.2** Caloric restriction and resveratrol administration lead to improvements in insulin sensitivity and general health in mice. The mechanisms by which these two regimens work are thought to converge on activation of SIRT1, resulting in a number of shared effects. For example, increased SIRT1 activity leads to deacetylation and activation of its downstream target PGC-1α, which may explain the increase in mitochondrial biogenesis under either regimen, and may directly contribute to improvements in physiology. Both resveratrol and CR are likely to have critical effects that are SIRT1 independent as well, since both regimens act on a multitude of targets with potential to contribute to overall health.

interacts with the estrogen receptor and the aryl hydrocarbon receptor (Baur and Sinclair 2006; Armour et al. 2009; Shakibaei et al. 2009), making it a poor presumption that any given effect is related to SIRT1. Second, the low bioavailability of resveratrol has led many to question whether the concentrations required to activate SIRT1 *in vitro* are ever achieved *in vivo*. Pharmacokinetic studies suggest that either a higher affinity target, such as quinone reductase 2, might be mediating resveratrol's effects, or that it might be a metabolite of resveratrol that is the active form, which calls into question the interpretation of *in vitro* and cell culture studies (Goldberg et al. 2003). Third, activation of SIRT1 by resveratrol against short peptide substrates is dependent on the presence of a fluorescent tag, for reasons that remain unclear. Although these observations raise significant questions, many reports of SIRT1-dependent effects continue to support the hypothesis that resveratrol can activate the enzyme against its endogenous substrates in cells and *in vivo* (Picard et al. 2004; Nakae et al. 2006; Boily et al. 2009). Moreover, a novel and structurally unrelated SIRT1 activator, SRT1720, recapitulates many of the effects of resveratrol in mice, providing support for the activation of this enzyme as a key event (Feige et al. 2008). However, it is worth noting that, like resveratrol, SRT1720 also fails to activate SIRT1 against full-length protein substrates in the absence of a fluorophore (Pacholec et al. 2010). Clearly, further studies are necessary to fully elucidate the role of SIRT1 in the beneficial effects of resveratrol.

## ANTIOXIDANT EFFECTS OF RESVERATROL AND OTHER POLYPHENOLS

One common denominator among many phytochemicals that seem to promote health, including resveratrol, is their antioxidant capacity. Certainly direct antioxidant action has historically been considered the major mechanism contributing to the beneficial effects of a diet rich in fruits and vegetables. However, intensive study of antioxidant vitamins in recent years, particularly A, C, and E, has failed to produce convincing evidence of either disease prevention or lifespan extension, leading some to question this premise (Kamel et al. 2006; Lotito and Frei 2006). Since polyphenols, and not the more familiar vitamins, are the major source of antioxidants in natural foods, these conclusions may need to be revisited. For instance, polyphenols may have important properties, such as the ability to accumulate in specific subcellular compartments or detoxify specific radicals, which vitamin antioxidants lack. Indeed, there is good evidence that polyphenols can lower oxidative stress, at least in cells or animals exposed to a high load of damage (Mizutani et al. 2001). One argument against the relevance of direct antioxidant activity in the benefits of polyphenols, as pointed out by Halliwell (2007), is that unconjugated polyphenols rarely exceed the low micromolar range in plasma, while total antioxidant capacity is in the millimolar range. Moreover, polyphenols may even be prooxidants under some circumstances. For example, several polyphenols were reported to increase DNA damage in lymphocytes (Breinholt et al. 2003), and a preliminary study has suggested that in contrast to its effects in obese animals, resveratrol may increase oxidative stress in the livers of lean rats (Rocha et al. 2009). While a role for direct antioxidant effects in the benefits of polyphenols cannot be excluded, it is not clear that this explanation can adequately account for the improvements in physiology that have been observed.

An alternative explanation for the reduction in oxidative stress following exposure to polyphenols in many *in vivo* models is that they might induce endogenous antioxidant enzymes (Floreani et al. 2003; Robb et al. 2008; Rubiolo et al. 2008), rather than act directly on free radicals. Resveratrol, for example, induces expression of catalase and quinone reductase 1 in the cardiac tissue of guinea pigs (Floreani et al. 2003) and manganese superoxide dismutase (MnSOD) in the brains of mice (Robb et al. 2008) (Figure 9.3). Transcriptional profiling of hepatocytes has revealed a more general trend toward the induction of many antioxidant enzymes (Rubiolo et al. 2008). A prime candidate for a mechanism to explain these responses is that resveratrol and other polyphenols might induce the phase I and II xenobiotic responses (Kohle and Bock 2006). Since the phase II response in particular involves the induction of antioxidant and other protective enzymes in order to limit damage caused by foreign molecules, chronic stimulation of this pathway might be expected to have beneficial consequences. Indeed, a number of beneficial phytochemicals, including resveratrol and curcumin, as well as more complex extracts, have been shown to induce phase II enzymes and have other effects on xenobiotic metabolism (Cao and Li 2004; Kaga et al. 2005; Sharma et al. 2005; Myhrstad et al. 2006), and this most likely plays a role in their effects *in vivo*. However additional

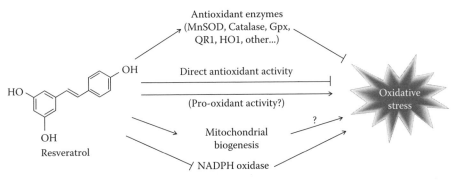

**FIGURE 9.3** Pathways by which resveratrol can influence oxidative stress in mammalian cells. Resveratrol has intrinsic antioxidant activity, and under some circumstances may also act as a pro-oxidant. However, it is likely that the influence of these direct effects on oxidative is small compared to indirect effects such as the induction of antioxidant enzymes and repression of NADPH oxidase. Resveratrol also increases the number of mitochondria, which are an important source of oxidative stress within the cell, although the net effect of this change is not clear. MnSOD, manganese superoxide dismutase; Gpx, glutathione peroxidase; QR1, quinone reductase 1 (NQO1, diaphorase); HO-1, heme oxygenase 1.

studies have demonstrated that specific polyphenols, including resveratrol, lead to changes in the activities of enzymes such as SIRT1, cyclooxygenase, PKC, and others, which are thought to be key players in mediating their beneficial effects, but are not clearly related to the phase II response or antioxidant activity. Therefore, induction of antioxidant enzymes is likely to play a role in the benefits of polyphenols, but does not provide a complete explanation, since it does not readily account for many of the specific targets that have been elucidated for resveratrol and other molecules.

Interestingly, resveratrol and other polyphenols have been shown to induce mitochondrial biogenesis, ostensibly through their effects on SIRT1 (Baur et al. 2006; Lagouge et al. 2006). This might be predicted to increase oxidative stress, since leakage from the electron transport chain during adenosine triphosphate (ATP) generation is a major source of free radicals *in vivo*. However, the available data do not seem to support this prediction, since oxidative stress is lower in resveratrol-treated animals, at least under the disease-prone or chemically induced conditions that have been studied (Mizutani et al. 2001). This may indicate that upregulation of antioxidant enzymes overcompensates for the increase in mitochondrial oxidative stress, or possibly, that spreading energy demand over a larger number of mitochondria actually decreases the rate of free radical generation per unit of ATP. Understanding the precise relationship between oxidative stress and mitochondrial biogenesis will be an interesting area for further investigation.

Finally, there is good evidence that the reduction in vascular oxidative stress following resveratrol treatment is not the result of increased antioxidant defenses at all, but rather reflects decreased production of superoxide at the level of Nicotinamide adenine dinucleotide phosphate (NADPH) oxidase (Pearson et al. 2008). This occurs through a poorly defined transcriptional mechanism, which may involve SIRT1-dependent effects. For example, SIRT1 can inhibit production of an upstream signal, tumor necrosis factor $\alpha$ (TNF-$\alpha$), by macrophages, most likely through inhibition of NF$\kappa$B (Shen et al. 2009), and overexpression of the SIRT1 target PGC-1$\alpha$ in vascular endothelial cells is sufficient to suppress NADPH oxidase expression (Kim et al. 2007). However the involvement of SIRT1 in this effect remains to be rigorously tested. Since increased expression of NADPH oxidase is associated with vascular dysfunction and obesity, restoring lower expression levels could be one of the major mechanisms by which resveratrol promotes health *in vivo*.

While further study may well support a role for direct antioxidant effects in the benefits of polyphenols, the most interesting developments in recent years have revealed that many polyphenols and, more generally, phytochemicals, exert important effects by directly interacting with and modulating specific enzymes or receptors (Singh et al. 2002; Williams et al. 2004; Baur and Sinclair 2006; Thangapazham et al. 2006). The idea that specific, individual polyphenols can act in unique ways, rather than collectively through a general property such as antioxidant activity, makes their study a daunting task, given the enormous number of molecules that have been described (Bravo 1998).

## XENOHORMESIS

An interesting question is why so many different mammalian enzymes respond to polyphenols, and in particular, why the responses are almost exclusively beneficial and at times appear coordinated. For example, resveratrol suppresses inflammation through inhibition of IL-1$\beta$ transcription (Casper et al. 1999), cyclooxygenase, and NF$\kappa$B (Baur and Sinclair 2006). The inhibition of NF-$\kappa$B appears to be due to both inhibition of the upstream activator PKC$\delta$ (Storz et al. 2004) and activation of the inhibitor SIRT1 (Yeung et al. 2004). The inhibition of cyclooxygenase appears to be due to both direct binding (Jang et al. 1997) and transcriptional repression (Subbaramaiah et al. 1998). Cyclooxygenase is also inhibited by aspirin (Vane 1971), which was derived from another phytochemical, salicylic acid, and by other polyphenols such as curcumin. These observations present a puzzling redundancy in the actions of polyphenols, and, upon consideration of the large number of targets that have been described, reveal a surprisingly consistent trend toward effects that would be considered desirable.

When considering the more limited problem of why a plant-derived molecule (resveratrol) should have any beneficial effect on yeast, (Howitz et al. 2003) proposed what they later termed the "xeno-hormesis hypothesis," "xeno-" meaning "stranger," and "hormesis" referring to the protective effect of a low intensity stress against a subsequent high intensity stress. (The most familiar examples of hormesis come from toxicology, where immunities to various poisons can be built up by ingesting small amounts). Essentially, Howitz and Sinclair suggested that evolutionarily conserved survival pathways might respond to chemical cues in the environment or the food supply that indicate stress. In this way, the hormesis response might be triggered in an animal (or fungus) by a stress that was applied to a plant food source. To a yeast cell growing on a grape, an increase in the concentration of resveratrol, which is induced in the plant after injury or infection (Langcake and Pryce 1976), might be a useful indicator that the food supply is about to become limiting. Therefore, the yeast cell might gain a selective advantage by responding to resveratrol in the same way it would to an actual deficit of calories, that is, activation of sirtuin enzymes (Lamming et al. 2004). It is more difficult to say whether xenohormesis can reasonably be postulated to explain some of the beneficial effects of polyphenols in higher organisms. For many different molecules, including resveratrol, beneficial effects on health have been demonstrated and putatively associated with specific molecular targets (Singh et al. 2002; Williams et al. 2004; Baur and Sinclair 2006; Thangapazham et al. 2006). However in most of these cases, the therapeutic dose of a single polyphenol is higher than what could be obtained from dietary sources alone (Scalbert et al. 2005), making it hard to understand how the molecules could serve as signals within the diet. Moreover, the specific profile of molecules present in the diet varies wildly depending on the plants being consumed and the stress level within those plants (Winkel-Shirley 2002), significantly decreasing the predictive value of any single molecule in determining future conditions.

Based on these observations, it is attractive to speculate that animals may have evolved to detect not just a single molecule in the diet, but more general parameters such as the total polyphenol content, common structural elements among the molecules present, or a change in the substitution or oligomerization state of such molecules. Despite the complexity of their structures and roles, there are some common elements to the behavior of polyphenols in plants. A large proportion are phyto-alexins, joining with other polyphenols, terpenoids, alkaloids, and other molecules that accumulate following UV irradiation, fungal infection, injury, possibly drought, and other stresses (Winkel-Shirley 2002). Since flavonoids contribute the bulk of the estimated 1g or approximately 14 mg/kg/day of dietary polyphenol intake (Aherne and O'Brien 2002; Scalbert et al. 2005), and many of these are greatly induced by stress in plants, for example, total flavonoids exuded from the roots of maize can be enhanced at least 15-fold (Kidd et al. 2001) and the resveratrol content of various grape varieties can be increased 3.4 to 2315-fold (Cantos et al. 2002) compared to unstressed controls, total phytoalexin accumulation could well provide a reliable cue of a dwindling food supply. In addition, polyphenols can act synergistically, rather than additively in some cases, and bioavailability may be improved by certain combinations of molecules present in the diet (Mertens-Talcott and Percival 2005; Baur and Sinclair 2006).

If the xenohormesis hypothesis is correct, target enzymes may have evolved to tolerate some variability in structure, or even to respond synergistically to a combination of molecules. One case where mammals have clearly evolved the ability to induce a highly coordinated response to a wide array of structurally diverse compounds is in the control of phase I (oxidation, reduction, and hydrolysis enzymes) and II (conjugating and antioxidant enzymes) responses to foreign molecules by the aryl hydrocarbon receptor (Ahr) and Nrf2 (Kohle and Bock 2007). Ahr directly binds small molecules with broad specificity, while Nrf2 is activated by many different reactive oxygen species and electrophiles. The dynamic induction of detoxifying and antioxidant enzymes mediated by these two transcription factors is thought to limit exposure to potentially harmful foreign molecules. These xenobiotic metabolizing enzymes, particularly phase II enzymes, may play an important role in hormetic effects. In fact, a number of carcinogens have been suggested to protect against the formation of neoplastic lesions when given at very low concentrations (Fukushima et al. 2005).

Although the Ahr/Nrf2 system evolved for reasons that are unrelated to xenohormesis, its existence illustrates the plausibility of sensing structurally diverse molecules, and it is interesting to ask whether Ahr and/or Nrf2 might have been at least partially co-opted by xenohormesis over time.

## CONCLUSIONS AND FUTURE CHALLENGES

Enormous challenges remains in the effort to understand the beneficial effects of polyphenols in general, and resveratrol in particular. There is significant evidence to support the hypothesis that resveratrol (and a subset of other polyphenols) might work at least partially through SIRT1 activation, and thereby mimic effects of caloric restriction. However, CR essentially slows aging of the entire body (and proportionally reduces oxidative stress), such that almost any beneficial effect on health is vaguely consistent with the hypothesis that a molecule is mimicking CR. Although SIRT1 over-expression produces some effects that are reminiscent of CR, and resveratrol treatment appears to activate the enzyme in cells and in mice, it remains to be proven that this mechanism is truly central to the beneficial effects on physiology that are conferred by resveratrol. Many plausible alternatives exist, and likely contribute to varying degrees.

The confusion surrounding the *in vivo* actions of one of the most intensely studied polyphenols highlights the near impossibility of understanding 8000 distinct molecules in detail. One potential ray of hope is that many of these molecules will behave in similar ways that can eventually be predicted based on their structures. In support of the idea that broad classes of polyphenols might have similar or cumulative effects, the original study that identified resveratrol as a SIRT1 activator also identified 16 other structurally related polyphenols possessing the same activity (Howitz et al. 2003). Moreover, among the few compounds that have been studied in relative detail, there is already evidence for similar effects *in vitro* and *in vivo* (Zheng and Ramirez 2000; Singh et al. 2002; Williams et al. 2004). Resveratrol and curcumin, for example, bear little structural resemblance to one another, yet both show potent anticancer and anti-inflammatory effects, block angiogenesis, activate glutathione-S-transferases, inhibit cyclooxygenase, possess intrinsic antioxidant capacity, induce apoptosis in tumor cells, and are rapidly metabolized (Sharma et al. 2005; Baur and Sinclair 2006). It will be interesting to see if future studies of additional molecules result in the emergence of larger trends that will allow a more complete understanding of the biology surrounding polyphenols.

In summary, resveratrol and other polyphenols show significant promise as therapeutic agents for human disease. Oxidative damage is an important feature of disease states, and can be alleviated by polyphenols in many cases. However, the evidence that intrinsic antioxidant capacity accounts for these effects is weak, and rather it seems more likely that in most cases oxidative stress is reduced as a downstream consequence of effects on specific enzymes and receptors. The bewildering array of polyphenols found in nature, and relatively small subset that have been studied in any detail, make it hard to draw general conclusions at this point. However, the fascinating results already obtained suggest that further study of polyphenols in relation to oxidative stress and human health will be very worthwhile.

## REFERENCES

Aherne, S. A. and N. M. O'Brien. 2002. Dietary flavonols: Chemistry, food content, and metabolism. *Nutrition* 18(1):75–81.

Armour, S. M., J. A. Baur, S. N. Hsieh, A. Land-Bracha, S. N. Thomas, and D. A. Sinclair. 2009. Inhibition of mammalian S6 kinase by resveratrol suppresses autophagy. *Aging* 1(6):515–25.

Barger, J. L., T. Kayo, J. M. Vann, E. B. Arias, J. Wang, T. A. Hacker, Y. Wang, D. Raederstorff, J. D. Morrow, C. Leeuwenburgh, D. B. Allison, K. W. Saupe, G. D. Cartee, R. Weindruch, and T. A. Prolla. 2008. A low dose of dietary resveratrol partially mimics caloric restriction and retards aging parameters in mice. *PLoS One* 3(6):e2264.

Bass, T. M., D. Weinkove, K. Houthoofd, D. Gems, and L. Partridge. 2007. Effects of resveratrol on lifespan in *Drosophila melanogaster* and *Caenorhabditis elegans*. *Mech Ageing Dev* 128(10):546–52.

Baur, J. A., K. J. Pearson, N. L. Price, H. A. Jamieson, C. Lerin, A. Kalra, V. V. Prabhu, J. S. Allard, G. Lopez-Lluch, K. Lewis, P. J. Pistell, S. Poosala, K. G. Becker, O. Boss, D. Gwinn, M. Wang, S. Ramaswamy, K. W. Fishbein, R. G. Spencer, E. G. Lakatta, D. Le Couteur, R. J. Shaw, P. Navas, P. Puigserver, D. K. Ingram, R. de Cabo, and D. A. Sinclair. 2006. Resveratrol improves health and survival of mice on a high-calorie diet. *Nature* 444(7117):337–42.

Baur, J. A. and D. A. Sinclair. 2006. Therapeutic potential of resveratrol: The *in vivo* evidence. *Nat Rev Drug Discov* 5(6):493–506.

Bishayee, A. 2009. Cancer prevention and treatment with resveratrol: From rodent studies to clinical trials. *Cancer Prev Res (Phila Pa)* 2(5):409–18.

Boily, G., X. H. He, B. Pearce, K. Jardine, and M. W. McBurney. 2009. SirT1-null mice develop tumors at normal rates but are poorly protected by resveratrol. *Oncogene* 28:2882–93.

Bordone, L., D. Cohen, A. Robinson, M. C. Motta, E. van Veen, A. Czopik, A. D. Steele, H. Crowe, S. Marmor, J. Luo, W. Gu, and L. Guarente. 2007. SIRT1 transgenic mice show phenotypes resembling calorie restriction. *Aging Cell* 6(6):759–67.

Bravo, L. 1998. Polyphenols: Chemistry, dietary sources, metabolism, and nutritional significance. *Nutr Rev* 56(11):317–33.

Breinholt, V. M., A. M. Molck, G. W. Svendsen, B. Daneshvar, A. M. Vinggaard, M. Poulsen, and L. O. Dragsted. 2003. Effects of dietary antioxidants and 2-amino-3-methylimidazo[4,5-f]- quinoline (IQ) on preneoplastic lesions and on oxidative damage, hormonal status, and detoxification capacity in the rat. *Food Chem Toxicol* 41(10):1315–23.

Cantos, E., J. C. Espin, and F. A. Tomas-Barberan. 2002. Postharvest stilbene-enrichment of red and white table grape varieties using UV-C irradiation pulses. *J Agric Food Chem* 50(22):6322–9.

Cao, Z. and Y. Li. 2004. Potent induction of cellular antioxidants and phase 2 enzymes by resveratrol in cardiomyocytes: Protection against oxidative and electrophilic injury. *Eur J Pharmacol* 489(1-2):39–48.

Casper, R. F., M. Quesne, I. M. Rogers, T. Shirota, A. Jolivet, E. Milgrom, and J. F. Savouret. 1999. Resveratrol has antagonist activity on the aryl hydrocarbon receptor: Implications for prevention of dioxin toxicity. *Mol Pharmacol* 56(4):784–90.

Colman, R. J., R. M. Anderson, S. C. Johnson, E. K. Kastman, K. J. Kosmatka, T. M. Beasley, D. B. Allison, C. Cruzen, H. A. Simmons, J. W. Kemnitz, and R. Weindruch. 2009. Caloric restriction delays disease onset and mortality in rhesus monkeys. *Science* 325(5937):201–4.

Feige, J. N., M. Lagouge, C. Canto, A. Strehle, S. M. Houten, J. C. Milne, P. D. Lambert, C. Mataki, P. J. Elliott, and J. Auwerx. 2008. Specific SIRT1 activation mimics low energy levels and protects against diet-induced metabolic disorders by enhancing fat oxidation. *Cell Metab* 8(5):347–58.

Floreani, M., E. Napoli, L. Quintieri, and P. Palatini. 2003. Oral administration of trans-resveratrol to guinea pigs increases cardiac DT-diaphorase and catalase activities, and protects isolated atria from menadione toxicity. *Life Sci* 72(24):2741–50.

Fukushima, S., A. Kinoshita, R. Puatanachokchai, M. Kushida, H. Wanibuchi, and K. Morimura. 2005. Hormesis and dose-response-mediated mechanisms in carcinogenesis: Evidence for a threshold in carcinogenicity of non-genotoxic carcinogens. *Carcinogenesis* 26(11):1835–45.

Goldberg, D. M., J. Yan, and G. J. Soleas. 2003. Absorption of three wine-related polyphenols in three different matrices by healthy subjects. *Clin Biochem* 36(1):79–87.

Greer, E. L., D. Dowlatshahi, M. R. Banko, J. Villen, K. Hoang, D. Blanchard, S. P. Gygi, and A. Brunet. 2007. An AMPK-FOXO pathway mediates longevity induced by a novel method of dietary restriction in *C. elegans*. *Curr Biol* 17(19):1646–56.

Halliwell, B. 2007. Dietary polyphenols: Good, bad, or indifferent for your health? *Cardiovasc Res* 73(2):341–7.

Howitz, K. T., K. J. Bitterman, H. Y. Cohen, D. W. Lamming, S. Lavu, J. G. Wood, R. E. Zipkin, P. Chung, A. Kisielewski, L. L. Zhang, B. Scherer, and D. A. Sinclair. 2003. Small molecule activators of sirtuins extend *Saccharomyces cerevisiae* lifespan. *Nature* 425(6954):191–6.

Imai, S., C. M. Armstrong, M. Kaeberlein, and L. Guarente. 2000. Transcriptional silencing and longevity protein Sir2 is an NAD-dependent histone deacetylase. *Nature* 403(6771):795–800.

Jang, M., L. Cai, G. O. Udeani, K. V. Slowing, C. F. Thomas, C. W. Beecher, H. H. Fong, N. R. Farnsworth, A. D. Kinghorn, R. G. Mehta, R. C. Moon, and J. M. Pezzuto. 1997. Cancer chemopreventive activity of resveratrol, a natural product derived from grapes. *Science* 275(5297):218–20.

Kaeberlein, M., K. T. Kirkland, S. Fields, and B. K. Kennedy. 2004. Sir2-independent life span extension by calorie restriction in yeast. *PLoS Biol* 2(9):E296.

Kaga, S., L. Zhan, M. Matsumoto, and N. Maulik. 2005. Resveratrol enhances neovascularization in the infarcted rat myocardium through the induction of thioredoxin-1, heme oxygenase-1 and vascular endothelial growth factor. *J Mol Cell Cardiol* 39(5):813–22.

Kamel, N. S., J. Gammack, O. Cepeda, and J. H. Flaherty. 2006. Antioxidants and hormones as antiaging therapies: High hopes, disappointing results. *Cleve Clin J Med* 73(12):1049–56, 1058.

Kidd, P. S., M. Llugany, C. Poschenrieder, B. Gunse, and J. Barcelo. 2001. The role of root exudates in aluminium resistance and silicon-induced amelioration of aluminium toxicity in three varieties of maize
(*Zea mays* L.). *J Exp Bot* 52(359):1339–52.

Kim, H. J., K. G. Park, E. K. Yoo, Y. H. Kim, Y. N. Kim, H. S. Kim, H. T. Kim, J. Y. Park, K. U. Lee, W. G. Jang,
J. G. Kim, B. W. Kim, and I. K. Lee. 2007. Effects of PGC-1alpha on TNF-alpha-induced MCP-1 and
VCAM-1 expression and NF-kappaB activation in human aortic smooth muscle and endothelial cells.
*Antioxid Redox Signal* 9(3):301–7.

Kohle, C. and K. W. Bock. 2006. Activation of coupled Ah receptor and Nrf2 gene batteries by dietary phytochemicals in relation to chemoprevention. *Biochem Pharmacol* 72(7):795–805.

Kohle, C. and K. W. Bock. 2007. Coordinate regulation of Phase I and II xenobiotic metabolisms by the Ah
receptor and Nrf2. *Biochem Pharmacol* 73(12):1853–62.

Lagouge, M., C. Argmann, Z. Gerhart-Hines, H. Meziane, C. Lerin, F. Daussin, N. Messadeq, J. Milne,
P. Lambert, P. Elliott, B. Geny, M. Laakso, P. Puigserver, and J. Auwerx. 2006. Resveratrol improves
mitochondrial function and protects against metabolic disease by activating SIRT1 and PGC-1alpha. *Cell*
127(6):1109–22.

Lamming, D. W., J. G. Wood, and D. A. Sinclair. 2004. Small molecules that regulate lifespan: Evidence for
xenohormesis. *Mol Microbiol* 53(4):1003–9.

Langcake, P. and R. J. Pryce. 1976. The production of resveratrol by *Vitis vinifera* and other members of the
Vitaceae as a response to infection or injury. *Physiol Plant Pathol* 9:77–86.

Lin, S. J., P. A. Defossez, and L. Guarente. 2000. Requirement of NAD and SIR2 for life-span extension by
calorie restriction in *Saccharomyces cerevisiae*. *Science* 289(5487): 2126–8.

Lotito, S. B. and B. Frei. 2006. Consumption of flavonoid-rich foods and increased plasma antioxidant capacity
in humans: Cause, consequence, or epiphenomenon? *Free Radic Biol Med* 41(12):1727–46.

Mertens-Talcott, S. U. and S. S. Percival. 2005. Ellagic acid and quercetin interact synergistically with resveratrol in the induction of apoptosis and cause transient cell cycle arrest in human leukemia cells. *Cancer
Lett* 218(2):141–51.

Milne, J. C., P. D. Lambert, S. Schenk, D. P. Carney, J. J. Smith, D. J. Gagne, L. Jin, O. Boss, R. B. Perni, C. B.
Vu, J. E. Bemis, R. Xie, J. S. Disch, P. Y. Ng, J. J. Nunes, A. V. Lynch, H. Yang, H. Galonek, K. Israelian,
W. Choy, A. Iffland, S. Lavu, O. Medvedik, D. A. Sinclair, J. M. Olefsky, M. R. Jirousek, P. J. Elliott,
and C. H. Westphal. 2007. Small molecule activators of SIRT1 as therapeutics for the treatment of type 2
diabetes. *Nature* 450(7170):712–6.

Mizutani, K., K. Ikeda, Y. Kawai, and Y. Yamori. 2001. Protective effect of resveratrol on oxidative damage in male and female stroke-prone spontaneously hypertensive rats. *Clin Exp Pharmacol Physiol*
28(1-2):55–9.

Myhrstad, M. C., H. Carlsen, L. I. Dahl, K. Ebihara, L. Glemmestad, K. Haffner, J. O. Moskaug, and
R. Blomhoff. 2006. Bilberry extracts induce gene expression through the electrophile response element.
*Nutr Cancer* 54(1):94–101.

Nakae, J., Y. Cao, H. Daitoku, A. Fukamizu, W. Ogawa, Y. Yano, and Y. Hayashi. 2006. The LXXLL motif of
murine forkhead transcription factor FoxO1 mediates Sirt1-dependent transcriptional activity. *J Clin
Invest* 116(9):2473–83.

Pacholec, M., B. A. Chrunyk, D. Cunningham, D. Flynn, D. A. Griffith, M. Griffor, P. Loulakis, B. Pabst,
X. Qiu, B. Stockman, V. Thanabal, A. Varghese, J. Ward, J. Withka, and K. Ahn. 2010. SRT1720,
SRT2183, SRT1460, and resveratrol are not direct activators of SIRT1. *J Biol Chem* 285(11):8340–51.

Pearson, K. J., J. A. Baur, K. N. Lewis, L. Peshkin, N. L. Price, N. Labinskyy, W. R. Swindell, D. Kamara, R. K.
Minor, E. Perez, H. A. Jamieson, Y. Zhang, S. R. Dunn, K. Sharma, N. Pleshko, L. A. Woollett, A. Csiszar,
Y. Ikeno, D. Le Couteur, P. J. Elliott, K. G. Becker, P. Navas, D. K. Ingram, N. S. Wolf, Z. Ungvari, D. A.
Sinclair, and R. de Cabo. 2008. Resveratrol delays age-related deterioration and mimics transcriptional
aspects of dietary restriction without extending life span. *Cell Metab* 8(2):157–68.

Pfluger, P. T., D. Herranz, S. Velasco-Miguel, M. Serrano, and M. H. Tschop. 2008. Sirt1 protects against high-
fat diet-induced metabolic damage. *Proc Natl Acad Sci USA* 105(28):9793–8.

Picard, F., M. Kurtev, N. Chung, A. Topark-Ngarm, T. Senawong, R. Machado De Oliveira, M. Leid, M. W.
McBurney, and L. Guarente. 2004. Sirt1 promotes fat mobilization in white adipocytes by repressing
PPAR-gamma. *Nature* 429(6993):771–6.

Robb, E. L., L. Winkelmolen, N. Visanji, J. Brotchie, and J. A. Stuart. 2008. Dietary resveratrol administration
increases MnSOD expression and activity in mouse brain. *Biochem Biophys Res Commun* 372(1):254–9.

Rocha, K. K., G. A. Souza, G. X. Ebaid, F. R. Seiva, A. C. Cataneo, and E. L. Novelli. 2009. Resveratrol toxicity: Effects on risk factors for atherosclerosis and hepatic oxidative stress in standard and high-fat diets. *Food Chem Toxicol* 47(6):1362–7.

Rogina, B. and S. L. Helfand. 2004. Sir2 mediates longevity in the fly through a pathway related to calorie restriction. *Proc Natl Acad Sci USA* 101(45):15998–6003.

Rubiolo, J. A., G. Mithieux, and F. V. Vega. 2008. Resveratrol protects primary rat hepatocytes against oxidative stress damage: Activation of the Nrf2 transcription factor and augmented activities of antioxidant enzymes. *Eur J Pharmacol* 591(1-3):66–72.

Scalbert, A., I. T. Johnson, and M. Saltmarsh. 2005. Polyphenols: Antioxidants and beyond. *Am J Clin Nutr* 81(1 Suppl):215S–217S.

Shakibaei, M., K. B. Harikumar, and B. B. Aggarwal. 2009. Resveratrol addiction: To die or not to die. *Mol Nutr Food Res* 53(1):115–28.

Sharma, R. A., A. J. Gescher, and W. P. Steward. 2005. Curcumin: The story so far. *Eur J Cancer* 41(13):1955–68.

Shen, Z., J. M. Ajmo, C. Q. Rogers, X. Liang, L. Le, M. M. Murr, Y. Peng, and M. You. 2009. Role of SIRT1 in regulation of LPS- or two ethanol metabolites-induced TNF-alpha production in cultured macrophage cell lines. *Am J Physiol Gastrointest Liver Physiol* 296(5):G1047–53.

Singh, R. P., S. Dhanalakshmi, and R. Agarwal. 2002. Phytochemicals as cell cycle modulators—a less toxic approach in halting human cancers. *Cell Cycle* 1(3):156–61.

Smith, J. S., C. B. Brachmann, I. Celic, M. A. Kenna, S. Muhammad, V. J. Starai, J. L. Avalos, J. C. Escalante-Semerena, C. Grubmeyer, C. Wolberger, and J. D. Boeke. 2000. A phylogenetically conserved NAD+-dependent protein deacetylase activity in the Sir2 protein family. *Proc Natl Acad Sci USA* 97(12):6658–63.

Storz, P., H. Doppler and A. Toker. 2004. Activation loop phosphorylation controls protein kinase D-dependent activation of nuclear factor kappaB. *Mol Pharmacol* 66(4):870–9.

Streppel, M. T., M. C. Ocke, H. C. Boshuizen, F. J. Kok, and D. Kromhout. 2009. Long-term wine consumption is related to cardiovascular mortality and life expectancy independently of moderate alcohol intake: The Zutphen Study. *J Epidemiol Community Health* 63:534–40.

Subbaramaiah, K., W. J. Chung, P. Michaluart, N. Telang, T. Tanabe, H. Inoue, M. Jang, J. M. Pezzuto, and A. J. Dannenberg. 1998. Resveratrol inhibits cyclooxygenase-2 transcription and activity in phorbol ester-treated human mammary epithelial cells. *J Biol Chem* 273(34):21875–82.

Thangapazham, R. L., A. Sharma, and R. K. Maheshwari. 2006. Multiple molecular targets in cancer chemoprevention by curcumin. *Aaps J* 8(3):E443–9.

Valenzano, D. R., E. Terzibasi, T. Genade, A. Cattaneo, L. Domenici, and A. Cellerino. 2006. Resveratrol prolongs lifespan and retards the onset of age-related markers in a short-lived vertebrate. *Curr Biol* 16(3):296–300.

Vane, J. R. 1971. Inhibition of prostaglandin synthesis as a mechanism of action for aspirin-like drugs. *Nat New Biol* 231(25):232–5.

Wang, Y. and H. A. Tissenbaum. 2006. Overlapping and distinct functions for a *Caenorhabditis elegans* SIR2 and DAF-16/FOXO. *Mech Ageing Dev* 127(1):48–56.

Williams, R. J., J. P. Spencer and C. Rice-Evans. 2004. Flavonoids: antioxidants or signalling molecules? *Free Radic Biol Med* 36(7):838–49.

Winkel-Shirley, B. 2002. Biosynthesis of flavonoids and effects of stress. *Curr Opin Plant Biol* 5(3):218–23.

Yeung, F., J. E. Hoberg, C. S. Ramsey, M. D. Keller, D. R. Jones, R. A. Frye, and M. W. Mayo. 2004. Modulation of NF-kappaB-dependent transcription and cell survival by the SIRT1 deacetylase. *Embo J* 23(12):2369–80.

Zheng, J. and V. D. Ramirez. 2000. Inhibition of mitochondrial proton F0F1-ATPase/ATP synthase by polyphenolic phytochemicals. *Br J Pharmacol* 130(5):1115–23.

# Section II

---

## Gene-Linked Networks

### Gene-Linked Chronic Disease

# 10 Mechanisms Mediating Obesity-Induced Inflammation and Insulin Resistance

*Simon Schenk, Olivia Osborn, and Jerrold M. Olefsky*

## CONTENTS

## THE COST OF OBESITY: MORE THAN JUST DOLLARS AND CENTS

The incidence and prevalence of obesity in the United States has increased alarmingly over the past 30 years. From the National Health and Nutrition Examination Survey (NHANES) II (1976–1980) to the 2003–2004 NHANES, the prevalence of individuals classified as being overweight or obese (BMI≥25) increased by about 41%, and the prevalence of individuals classified as being obese (body mass index [BMI]≥30) more than doubled. Of more concern is the fact that in this same time period, the prevalence of obesity among children and adolescents has tripled. Increasing body weight is associated with increased mortality (Lee et al. 1999) and morbidity for developing a variety of "lifestyle-related" diseases, including type 2 diabetes, high blood pressure, coronary heart disease, asthma, arthritis, and high cholesterol (Mokdad et al. 2003), and costs attributed to the treatment of obesity and obesity-related diseases are at least $93 billion per year (Finkelstein et al. 2003). Thus, it is of great importance to understand the mechanisms linking obesity to disease development, as well as interventions that may help to prevent and/or treat obesity.

In recent years, much research has focused on the role of inflammation in the etiology of obesity-related disease development. In this chapter, we will discuss possible underlying mechanisms that link obesity to inflammation, with particular emphasis on the interrelationships among obesity, inflammation, and insulin resistance. When coupled with relative insulin deficiency, insulin resistance is a key underlying cause of type 2 diabetes and disease development, and is the defining feature of syndrome X (metabolic syndrome) (Reaven 2005). Please note that many of the concepts presented herein are discussed in a recent review from our laboratory (Schenk et al. 2008).

## OBESITY IS CHARACTERIZED BY INCREASED INFLAMMATION

For many years it has been recognized that markers of inflammation are elevated in obesity. Studies in the 1950s and 1960s revealed that the plasma levels of fibrinogen and other acute phase reactants were elevated in obesity (Grace and Goldrick 1968; Fearnley et al. 1959). Since then it has been shown that a variety of other inflammatory factors are systemically or locally elevated in obesity, such as tumor necrosis factor-$\alpha$ (TNF-$\alpha$) (Hotamisligil et al. 1993; Hotamisligil et al. 1995), interleukin 1 (IL-1) (Maedler et al. 2002; Spranger et al. 2003), interleukin 6 (IL-6), (Bastard et al. 2000; Fried et al. 1998), and monocyte chemoattractant protein-1 (MCP-1) (Kamei et al. 2006; Kim et al. 2006). Nevertheless, despite evidence of increased inflammation in obesity, it wasn't until the 1980s and 1990s that investigators began to consider the novel idea that obesity-related inflammation might be an important mediator in the pathogenesis of insulin resistance and other obesity-related diseases.

## ADIPOSE TISSUE: MORE THAN JUST A STORAGE WAREHOUSE

In the past, adipose tissue was often considered a relatively passive storage depot for excess calories (which are stored as triglyceride), although more recent findings present a much different view. Firstly, adipose tissue actively releases fatty acids into the circulation through the process of lipolysis, which involves lipase-mediated breakdown of triglycerides into fatty acids and glycerol (Duncan et al. 2007). Insulin is a potent inhibitor of lipolysis, and activation of lipolysis is primarily controlled by catecholamines (epinephrine and norepinephrine) (Duncan et al. 2007). Interestingly, TNF-$\alpha$, an inflammatory cytokine secreted by the adipose tissue bed, can also activate lipolysis (Ryden et al. 2002). Understanding the control of lipolysis and the release of fatty acids is important because, as discussed in the section *A Matter of Fat*, excessive fatty acid mobilization may be an important mediator of inflammation and obesity-related metabolic derangement.

A major advance in our understanding of adipose tissue biology were the findings that adipose tissue secretes a variety of polypeptides (Figure 10.1). These peptides include hormones, cytokines, and chemokines, which can function in both an endocrine or paracrine fashion. In fact, adipose tissue is now recognized as the largest endocrine organ in the body. Adipose tissue is comprised of adipocytes or fat cells and the stromal vascular fraction, which contains preadipocytes, endothelial cells, macrophages, and fibroblasts. These different components of the adipose tissue bed are responsible for the secretion of different factors. Adipokines such as leptin and adiponectin are primarily secreted by the adipocyte. Chemokines and cytokines, such as TNF-$\alpha$, MCP-1, and plasminogen activator inhibitor-1 (PAI-1), are expressed in both the adipocyte, as well as immune cells, such as macrophages. The fact that adipose tissue secretes these factors, and that secretion of these inflammatory factors is increased in obesity, has provided the necessary impetus to understand how obesity can cause inflammation. Next we discuss some of these tissue-derived factors and how they might be involved in transducing inflammatory signals.

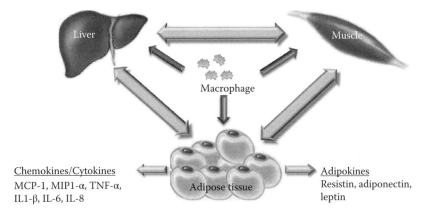

**FIGURE 10.1** Communication between insulin responsive tissues. The inflammatory response induces increased production and release of adipokines and cytokines/chemokines by both adipocytes and macrophages that play an important role in the development of systemic insulin resistance.

## TNF-α

Over 30 years ago TNF-α (also previously known as catechin and lymphotoxin) was identified as a secreted factor from monocytes and macrophages. Studies investigating the role of TNF-α in insulin resistance provided an important advance in our understanding of how obesity causes insulin resistance. Hotamisligil et al. (1993) demonstrated that TNF-α was elevated in adipose tissue of obese rodents, and that immuno-neutralization of TNF-α in obese rodents ameliorated insulin resistance. This connection between TNF-α and insulin resistance was subsequently verified in humans (Hofmann et al. 1994), and by the finding that TNF-α knockout mice are protected against obesity-related insulin resistance (Uysal et al. 1997). The mechanisms by which TNF-α induces insulin resistance are multifactorial. TNF-α activates the proinflammatory IκB kinase/nuclear factor kappa B (IKK/NF-κB) and c-Jun $NH_2$-terminal kinase (JNK) pathways, which are discussed in more detail in the *Stuck in the Middle: Propagating the Inflammatory Response* section of this chapter. JNK can negatively regulate insulin signaling by causing inhibitory serine phosphorylation of insulin receptor substrate 1 (IRS-1), which attenuates IRS tyrosine phosphorylation and downstream signaling (Hotamisligil et al. 1996). TNF-α may also propagate the inflammatory response by increasing fatty acid availability through activation of lipolysis (Ryden et al. 2002). In addition, TNF-α can induce insulin resistance by decreasing the activity of the AMP-activated protein kinase (AMPK) (Steinberg et al. 2006).

### Adiponectin

Adiponectin (also known as 30-kDa adipocyte complement-related protein, Acrp30; adipoQ; apM1; GBP28) is the most abundant, circulating adipokine and varies widely in size due to its ability to form multimeric complexes. It is categorized into three major forms: low molecular weight (LMW), middle molecular weight (MMW), and high molecular weight (HMW). The levels of adiponectin are reduced in obesity (Arita et al. 1999; Yatagai et al. 2003) and this is associated with the degree of insulin resistance. Consistent with this, administration to, or transgenic expression of, adiponectin into rodents leads to enhanced insulin sensitivity; adiponectin knockout mice are insulin resistant and have numerous adverse cardiovascular outcomes including impaired recovery from ischemia-reperfusion injury and increased risk of atherosclerosis (Maeda et al. 2002; Kubota et al. 2002; Tao et al. 2007). The beneficial effects of adiponectin on metabolic disorders are likely related to down modulation of inflammation as well as stimulation of AMPK. Adiponectin attenuates the induction of the proatherosclerotic endothelial cell adhesion factors (Ouchi et al. 1999), and reduces lipopolysaccharide- (LPS) and TNF-$\alpha$-induced activation of NF-$\kappa$B (Ajuwon and Spurlock 2005). Adiponectin also activates PPAR$\gamma$ (Ajuwon and Spurlock 2005), which itself exerts anti-inflammatory effects. Interestingly, adiponectin can sometimes be proinflammatory, an effect that appears to be specific to the multimeric complex being investigated (Haugen and Drevon 2007; Neumeier et al. 2006). While this is potentially confounding, it is possible that certain proinflammatory effects could be beneficial to the cell by creating a state of "tolerance" to subsequent inflammatory insults. Another major action of adiponectin is to activate AMPK, which then phosphorylates and inhibits acetyl coA carboxylase (ACC), leading to enhanced oxidation of fatty acids and increased glucose uptake by cells (Yamauchi et al. 2002).

## Resistin

Resistin is an inflammatory cytokine that in mice is primarily produced by the adipocyte, and in humans, appears to be produced by immune cells, such as the macrophage. Increased resistin levels are associated with obesity and increased atherosclerosis and several studies have shown that *in vivo* administration or transgenic expression of resistin causes a systemic decrease in insulin sensitivity (Steppan et al. 2001; Reilly et al. 2005). Synthesis of resistin is induced by proinflammatory stimuli such as TNF, IL-6, and fatty acids (Kaser et al. 2003). Resistin also exhibits atherogenic and insulin desensitizing effects via activation of the NF-$\kappa$B pathway, and a paralleled increase in the synthesis of proinflammatory and proatherogenic factors (Silswal et al. 2005; Verma et al. 2003). The ability of resistin to decrease insulin sensitivity could also be related to impaired AMPK activation (Palanivel and Sweeney 2005).

## MCP-1

MCP-1 (also called C-C chemokine ligand-2 [CCL2]) is a member of the CC chemokine family and is an important initiator and transducer of inflammatory responses within tissues. This occurs through its chemoattractive properties, which facilitates the recruitment of immune cells, especially macrophages, to sites of injury and inflammation. Plasma levels and adipose tissue expression of MCP-1 are elevated in obesity, in association with insulin resistance and increased atherosclerosis (Kamei et al. 2006; Kim et al. 2006; Kanda et al. 2006). An important first step of atherogenesis and atherosclerotic lesion formation is the recruitment of macrophages to lesion sites. Similarly, macrophage recruitment to adipose tissue clearly plays an important role in propagating inflammation and causing insulin resistance. To this end, mice deficient in the major receptor for MCP-1, C-C motif chemokine receptor-2 (CCR2), exhibit decreased numbers of adipose tissue resident macrophages and are protected against high-fat-diet-induced insulin resistance (Weisberg et al. 2006), although these findings are not unanimous (Chen et al. 2005; Inouye et al. 2007). In contrast, mice transgenically overexpressing MCP-1 in adipose tissue show increased macrophage recruitment and expression of inflammatory genes in adipose tissue, and become insulin resistant (Kamei et al. 2006; Kanda et al. 2006).

## Interleukin-10

Not all cytokines released by macrophages and lymphocytes are proinflammatory. Interleukin-10 (IL-10) is an anti-inflammatory cytokine that is secreted by macrophages and lymphocytes, and is negatively correlated with the degree of insulin resistance (van Exel et al. 2002). *In vitro* treatment of adipocytes with IL-10 prevents TNF-$\alpha$ induced insulin resistance, and IL-10 treated mice are protected against IL-6- and lipid-induced insulin resistance (Lumeng et al. 2007a; Kim et al. 2004). Furthermore, IL-10 limits the formation of atherosclerotic plaques, and mice deficient in IL-10 have a 30-fold higher susceptibility to develop atherosclerosis (Mallat et al. 1999).

## INTERLEUKIN 1 RECEPTOR ANTAGONIST

Interleukin 1 receptor antagonist (IL-1Ra) is another example of an endogenous anti-inflammatory cytokine produced by white adipose tissue (Juge-Aubry et al. 2003). IL-1Ra blocks the effects of the proinflammatory cytokines IL-1 α and IL-1 β by occupying the IL-1 receptor (Arend 1990; Dinarello 1991; Osborn et al. 2008b). Recent studies have shown that blockade of the IL-1 receptor with recombinant IL1Ra (Larsen et al. 2007) or sequestration of pro-inflammatory IL-1 β (Osborn et al. 2008a) results in improved glycemic control and beta-cell secretory function and reduced markers of systemic inflammation in type 2 diabetes.

## TRIGGERING INFLAMMATION IN OBESITY

While it is clear that factors that are secreted by adipose tissue can play an amplifying role in the inflammatory state of obesity, the question still remains, what are the initiating events in obesity-induced inflammation? It was the novel discovery that adipose tissue from obese mice and humans is infiltrated with bone marrow-derived macrophages that provided a major mechanistic insight into how obesity causes inflammation (Figure 10.2). Weisberg et al. (2003) demonstrated that the infiltration of adipose tissue with macrophages is positively correlated with adipose mass and adipocyte size. In fact, as much as 50% of total cell content in adipose tissue from obese human subjects was comprised of macrophages, compared with approximately 10% in lean controls (Weisberg et al. 2003). Importantly, they demonstrated that the majority of adipose tissue TNF-α and IL-6 expression was due to macrophages (Weisberg et al. 2003). The study by Xu et al. (2003) supported these findings and also demonstrated that thiazolidinediones (TZD) treatment repressed proinflammatory gene expression in adipose tissue macrophages, providing an important insight into the mechanisms by which TZDs enhance insulin sensitivity. Nevertheless, while macrophages are clearly increased in adipose tissue in obesity, and as discussed in the section *Macrophage: Fuelling Inflammation and Insulin Resistance*, are also a key contributor to inflammation and insulin resistance, it is important

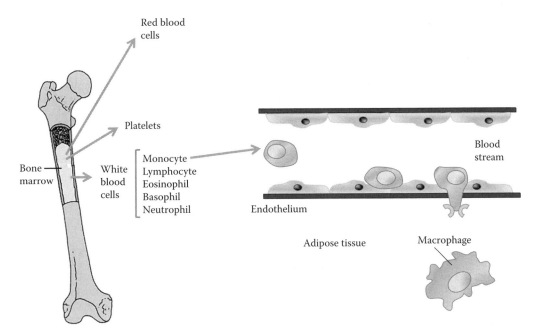

**FIGURE 10.2**  Macrophage itinerary. Monocytes are generated in the bone marrow, released into the blood and subsequently adhere to the endothelial cells and cross the endothelial layer into adipose tissue where they differentiate into macrophages.

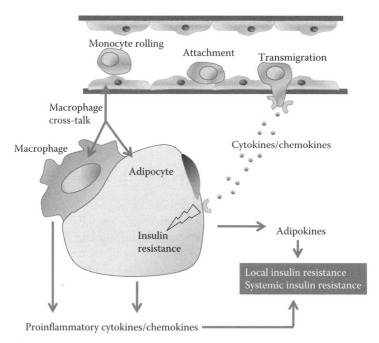

**FIGURE 10.3** The interaction between adipocytes and macrophages contributes to the development of insulin resistance in obese adipose tissues.

to address the possible mechanisms that initiate and propagate tissue inflammation and macrophage infiltration (Figure 10.3).

## STARVING FOR OXYGEN

One hypothesis linking obesity to inflammation centers on the idea that oxygen delivery to the expanding adipose tissue mass is inadequate in obesity (Trayhurn 2005). Mechanistically, restricted microcirculatory blood flow in obese adipose tissue results in a local hypoxic environment with subsequent activation of hypoxia-inducible genes. This results in activation of stress/inflammatory pathways and leads to increased gene transcription and secretion of proinflammatory adipokines/cytokines into the local environment. Along these lines, *in vitro* simulation of hypoxia increases activation of NF-κB (Ye et al. 2007). Similarly, *in vivo* studies in obese rodents demonstrate that adipose tissue beds are hypoperfused with oxygen, exhibit increased gene expression of hypoxia-related genes, and display elevated expression of genes involved in inflammation and endoplasmic reticulum (ER) stress (Hosogai et al. 2007; Ye et al. 2007). Intuitively, with increasing obesity and expansion of the adipose tissue bed, more of the adipose tissue is poorly oxygenated, exacerbating the inflamed state, resulting in further increases in the release of proinflammatory cytokines. This adipose tissue response could cause the recruitment of proinflammatory macrophages, which become resident in the adipose tissue and release additional cytokines into the local environment, thereby amplifying the inflammatory state. The majority of macrophages found in adipose tissue are localized to "crown-like" structures (CLSs; also known as multinucleate giant cells [MGCs]), around dead adipocytes. It is possible that as the size of an adipocyte increases, one mechanism causing the death of the adipocyte is an inability to provide the nutrients and oxygen required for normal function. Irrespective, it is evident that the expanding adipose tissue mass is defined by an increased influx of macrophages (Xu et al. 2003; Weisberg et al. 2003), which themselves are proinflammatory (Lumeng et al. 2007a; Lumeng et al. 2007b). Weight loss results in an increase in oxygen delivery to adipose tissue, and a reversal of inflammation and decreased expression of markers of macrophage infiltration (Ye et al. 2007).

## A MATTER OF FAT

The rate of breakdown and release of fatty acids into the systemic circulation is elevated in obesity as compared with lean individuals (Horowitz and Klein 2000), and this increased fatty acid "availability" is a major contributor to the development of obesity-related metabolic dysregulation. For example, acutely increasing plasma fatty acid flux (typically via a lipid-plus-heparin infusion) to rates found in obesity causes insulin resistance (Dresner et al. 1999) and endothelial dysfunction (Steinberg et al. 1997). Alternatively, restricting lipolysis in obesity significantly improves vascular function (de Jongh et al. 2004) and insulin sensitivity (Santomauro et al. 1999). Thus, it is evident that changes in fatty acid abundance play an important role in mediating the metabolic defects of obesity, and recent studies show an additional linkage in which fatty acids and fatty acid/nutrient flux can trigger inflammatory pathway activation.

### Fatty Acids and Toll-Like Receptors

One important mechanism by which fatty acids can induce inflammation is through activation of toll-like receptors (TLRs), in particular, TLR2 and TLR4 (Aderem and Ulevitch 2000). TLRs are part of the pattern recognition receptor system, which plays a critical role in innate immune responses by recognizing nonhost bacterial and viral components. Activation of TLR2/4 leads to stimulation of both the IKK/NF-κB and JNK/AP1 pathways, with subsequent induction of inflammatory gene expression and secretion of a variety of cytokines/chemokines (Aderem and Ulevitch 2000). Thus, fatty acid-mediated increases in inflammatory pathway activation require signals that are initiated by TLR2/4 and flow downstream through IKK/NF-κB and JNK/Activating Protein 1 (AP1). Consistent with this, the expression of TLR4 is increased in obese mice as compared to lean mice, and TLR4 knockout mice are protected from lipid- and obesity-induced insulin resistance, in association with decreased adipose tissue inflammatory markers (Shi et al. 2006; Tsukumo et al. 2007). Furthermore, hematopoietic cell-specific deletion of TLR4 ameliorates hepatic and adipose tissue insulin resistance in high-fat diet-induced mice (Saberi et al. 2009).

### Nutrient Overload and Endoplasmic Reticulum (ER) Stress

A second mechanism by which obesity may cause inflammation is via ER stress (Hotamisligil 2008). The ER is a specialized cytosolic organelle that is an essential mediator of cellular homeostasis (Gregor and Hotamisligil 2007). In particular, cellular "overnutrition," in which the cell is inundated with an excess of fatty acids and nutrients, can induce ER stress. Overburdening of the ER synthetic machinery impairs one of its key functions, which is to ensure the proper folding of newly synthesized proteins. This results in activation of the "unfolded protein response" (UPR) which leads to induction of the proinflammatory JNK and IKK pathways (Gregor and Hotamisligil 2007). Activating the UPR *in vitro* induces expression of a variety of proinflammatory genes and markers of ER stress are increased in models of obesity and insulin resistance (Ozcan et al. 2004). Moreover, genetic mouse models in which the UPR is exacerbated by deletion of key ER chaperone proteins display increased inflammation, hyperinsulinemia, glucose intolerance, and insulin resistance (Ozcan et al. 2004; Ozawa et al. 2005). Furthermore, after weight loss in obese subjects the level of ER stress in adipose tissue declines (Gregor et al. 2009).

## STUCK IN THE MIDDLE: PROPAGATING THE INFLAMMATORY RESPONSE

Fatty acids, hypoxia, ER stress, and certain cytokines can all initiate proinflammatory responses by activating two major intracellular signaling pathways involving JNK/AP1 and IKK/NF-κB, which amplify the initiating insult.

JNK is a member of the mitogen-activated protein (MAPK) kinase family and comprises three separate isoforms: *jnk1*, *jnk2*, and *jnk3*. JNK activity is increased by extracellular factors such as fatty acids and cytokines, which can stimulate the upstream activators of JNK, MAPK kinase 4 or 7 (MKK4 and MKK7). Activated JNK can phosphorylate the AP1 transcription complex (c-jun/fos), which regulates expression of numerous proinflammatory genes. JNK1 can also directly phosphorylate serine 307 of IRS-1, which inhibits IRS-1 tyrosine phosphorylation, attenuating downstream insulin signaling (Hirosumi et al. 2002). JNK is upregulated in obesity and insulin resistance (Bandyopadhyay et al. 2005; Hirosumi et al. 2002), and is a major mediator of decreased insulin sensitivity, since JNK1 knockout mice are protected against obesity-induced insulin resistance (Hirosumi et al. 2002).

NF-κB consists of a number of structurally related subunits (p50, p52, RelA [also known as p65], RelB, and RelC), which can form heterodimers or homodimers. Of these, the most characterized NF-κB heterodimer is the p50, p65/RelA heterodimer. In the basal state, this heterodimer is sequestered in the cytoplasm through its association with IκB. Inflammatory signals activate IKK, which then phosphorylates IκB causing IkB to dissociate from NF-kB and become degraded. The released NF-κB then translocates into the nucleus where it transcriptionally activates an array of inflammatory pathway genes. Interestingly, there is substantial overlap between the sets of AP1 and NFkB target genes. The IKK/NF-κB pathway is rapidly activated by numerous stimuli including ER stress and activation of TLR by fatty acids, as well as by factors such as TNF-α and resistin. Mice with heterozygous knockout of IKKβ (Kim et al. 2001; Yuan et al. 2001) or complete knockout of IKKε (Chiang et al. 2009) are protected against high-fat-diet-induced obesity and insulin resistance. Similarly, salicylic acid, which inhibits activation of NF-κB, improves insulin resistance and hyperglycemia in type 2 diabetic patients, and also attenuates lipid-induced insulin resistance (Kim et al. 2001; Yuan et al. 2001).

## MACROPHAGE: FUELING INFLAMMATION AND INSULIN RESISTANCE

The finding that obese adipose tissue is infiltrated with macrophage fueled a new series of investigations into the mechanisms linking obesity, inflammation, and metabolic dysregulation. In particular, intense research has focused on how and why macrophages are recruited to the adipose tissue bed, and if macrophage-induced inflammation can mediate metabolic diseases such as insulin resistance. Much of this research has involved genetic manipulations in mice.

### MYELOID/MACROPHAGE KNOCKOUT MODELS

As discussed above, activation of IKK/NF-κB signaling induces a proinflammatory phenotype, and inactivation of this pathway prevents the metabolic disturbances associated with inflammation. However, these studies did not identify which tissues initiate these inflammatory events. To this end, Arkan et al. (2005) studied mice with myeloid-specific knockout of IKKβ (i.e., knockout of IKKβ in macrophages) using the Cre-LoxP system. Macrophage-specific IKKβ knockout mice displayed improved glucose tolerance on chow and high-fat diet, and high-fat-diet-fed obese mice were protected from systemic insulin resistance, exhibiting enhanced insulin sensitivity in skeletal muscle and liver (Arkan et al. 2005). These studies represented the initial demonstration that disabling the macrophage inflammatory pathway could prevent obesity-induced inflammation and insulin resistance.

Studies have also shown that disrupting the other major intracellular macrophage inflammatory pathway, JNK1/AP1, prevents obesity-induced inflammation and subsequent insulin resistance (Figure 10.4) (Solinas et al. 2007). In this study, irradiated wild-type host animals were transplanted with bone marrow from JNK1 knockout animals so that, after bone marrow reconstitution, these "chimeric" animals had knockout of JNK1 in macrophages (and all bone marrow-derived myeloid

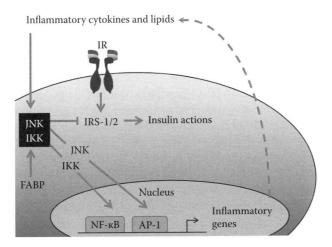

**FIGURE 10.4** Inflammatory cytokines and lipids. Inflammation induces insulin resistance through activation of *IKK*/NF-Kb or JNK/AP1 signaling pathways. (Adapted from Wellen, K. E., and G. S. Hotamisligil. 2005. *J Clin Invest* 115(5):1111–9.)

cells) but have normal JNK1 expression in all nonmyeloid tissue compartments (e.g., muscle, adipose, liver). These studies showed that deletion of myeloid/macrophage JNK1 leads to improved glucose tolerance, reduced insulinemia and protects high-fat fed mice from developing systemic insulin resistance in muscle, liver, and fat. Importantly, when placed on high-fat diets, the macrophage/myeloid JNK1 knockout animals developed the expected degree of obesity, and, therefore, these studies were able to dissociate obesity from insulin resistance, demonstrating that without the macrophage-driven inflammatory component, the simple presence of obesity does not cause insulin resistance. Interestingly, on a high-fat diet, the macrophage/myeloid JNK1 knockout animals also developed the same degree of hepatic steatosis as the wild-type animals, but did not develop hepatic insulin resistance. Kupffer cells are the bone marrow-derived macrophage cell type in liver, and in these chimeric mice, JNK1 was deleted from Kupffer cells and hepatic markers of inflammation were markedly reduced. Again, this indicates that without the inflammatory component, tissue insulin resistance does not develop, despite the expected diet-induced increase in tissue lipid accumulation.

Peroxisome proliferator-activated receptor gamma (PPARγ) is a member of the nuclear receptor super family of ligand-dependent transcription factors and controls the expression of numerous genes involved in adipogenesis, lipid metabolism, and inflammation. PPARγ is the molecular target of TZDs, a class of insulin-sensitizing drugs widely used to treat type 2 diabetes. PPARγ exerts potent anti-inflammatory effects, and recent evidence has indicated that the anti-inflammatory actions of PPARγ represent an important component of the insulin sensitizing effects of TZDs (Pascual and Glass 2006). Specifically, PPARγ broadly inhibits activation of inflammatory pathway genes through a specific transrepression mechanism, and PPARγ-mediated transrepression of inflammatory responses is particularly potent in macrophages (Pascual and Glass 2006). Based on this, it seemed possible that deletion of PPARγ from macrophages would de-repress these cells, leading to enhanced inflammation and the development of insulin resistance. In support of this, recent studies have directly demonstrated that mice with macrophage-specific deletion of PPARγ are glucose intolerant, hyperinsulinemic, and display systemic insulin resistance, even on normal-chow diet (Hevener et al. 2007; Odegaard et al. 2007). Consistent with this, adipose tissue from these mice exhibits increased proinflammatory markers, and PPARγ-depleted macrophages secrete factors that cause insulin resistance (Hevener et al. 2007; Odegaard et al. 2007). Altogether, these studies further argue that the inflammatory state of the macrophage can initiate a systemic insulin resistance response.

# FLIPPING THE INFLAMMATORY SWITCH: THE GOOD
# AND THE BAD OF MACROPHAGES

Although the aforementioned studies implicate macrophages as a central mediator of insulin resistance in obesity, it is important to appreciate that not all actions of macrophages are harmful. As key members of the innate immune system, macrophages, which are monocytic phagocytes, engulf and digest pathogens, cellular debris, and dead cells. In addition, they initiate antigen recognition by lymphocytes to respond to pathogens and instruct them to assemble at sites of tissue inflammation/damage. Interestingly, the functions of macrophages are quite heterogeneous, and macrophage "activation" has been operationally defined into two polarization states, M1 and M2 (Mantovani et al. 2004). Distinguishing between the M1 and M2 polarization states of a macrophage can be assessed by detection of different surface markers (e.g., F4/80, CD11b, CD11c) through flow cytometry. This is important because the polarization state of macrophages determines whether they are proinflammatory or anti-inflammatory, and, accordingly, determines their impact on metabolic dysregulation.

## M1 Macrophage

M1, or "classically activated," macrophages are induced by proinflammatory factors and display increased inflammatory gene expression (e.g., TNF-$\alpha$, MCP-1) and enhanced reactivity to fatty acids and LPS. As assessed using flow cytometry techniques, adipose tissue M1 macrophages are triply positive for the markers F4/80, CD11b and CD11c (F4/80+/CD11b+/CD11c+). Because of their proinflammatory properties, M1 macrophages are implicated in the pathogenesis of insulin resistance and atherosclerosis (Lumeng et al. 2007a; Bouhlel et al. 2007; Odegaard et al. 2007). Furthermore, ablation of CD11c positive cells has recently been shown to normalize insulin signaling in obese insulin-resistant animals by TZDs (Patsouris et al. 2008).

## M2 Macrophage

M2, or "alternatively activated," macrophages display low levels of inflammatory gene expression, secrete high levels of anti-inflammatory factors (e.g., IL-10) and are poorly activated by fatty acids and LPS. Similar to M1 macrophages, they are positive for F4/80 and CD11b, but in contrast to M1 macrophages, they are negative for CD11c (F4/80+/CD11b+/CD11c−). M2 macrophages are thought to regulate tissue repair, and prevent excessive inflammation and insulin resistance. Significantly, PPARγ is required for the maturation of alternatively activated macrophages, and activation of PPARγ polarizes macrophages toward the M2/anti-inflammatory state, thus providing another mechanism by which activation of PPARγ reduces inflammation and enhances insulin sensitivity (Odegaard et al. 2007; Bouhlel et al. 2007).

## Polarizing Obesity

It is evident that in the simplest terms, macrophages can be operationally separated into "good" or "bad" macrophages with respect to inflammation-induced insulin resistance. An important question, therefore, is what modulates macrophage polarization in obesity? The number of "resident" macrophages is low in lean adipose tissue, and these macrophages display "M2-like" characteristics, in that they are anti-inflammatory in nature, localize exclusively to interstitial spaces between adipocytes (Lumeng et al. 2008), and secrete high levels of IL-10, which is protective against TNF-α–induced insulin resistance (Lumeng et al. 2007a). In obesity, however, macrophage content in adipose tissue is increased. The precise mechanisms mediating increased macrophage recruitment in adipose tissue are still under investigation. The increased macrophage recruitment appears to be related to expanding adipose mass, adipose necrosis (Cinti et al. 2005), and hypoxia (Rausch et al. 2008), and recent studies have shown alterations in the composition of adipose tissue thymocytes (T cells) may also play a role (Lumeng et al. 2009). The increased macrophage content in adipose tissue in obesity is primarily due to an increased recruitment of bone marrow-derived macrophages, which display an "M1-like," F4/80+/CD11b+/CD11c+, proinflammatory phenotype.

In adipose tissue from obese mice and obese humans, more than 90% of macrophages surround dead adipocytes, where they form CLSs (Cinti et al. 2005). CLSs are a cluster of macrophages that have surrounded the adipocyte and have fused together, their main purpose being to phagocytose the necrotic adipocyte (Hernandez-Pando et al. 2000). These cells are highly proinflammatory, releasing cytokines into the local environment, which, in turn, can recruit additional macrophages into the inflamed area (Hernandez-Pando et al. 2000; Lumeng et al. 2008). They also exhibit high expression levels of scavenger receptors and endocytic activity, supporting their role in "cleaning-up" dead adipocytes (Hernandez-Pando et al. 2000; Zeyda et al. 2007).

Interestingly, weight loss in severely obese subjects decreases markers of macrophage infiltration in adipose tissue, in concert with improved insulin sensitivity (Bruun et al. 2005; Clement et al. 2004). It is also unknown if weight loss through diet and/or exercise changes the polarization state of macrophage from a proinflammatory M1 to an anti-inflammatory M2 state. It is notable, however, that exercise training and being physically active decreases TLR4 expression in blood monocytes and decreases their inflammatory response to LPS (Flynn and McFarlin 2006). Thus, one mechanism by which regular exercise and weight loss improves insulin sensitivity could be through decreasing adipose tissue inflammation and the inflammatory potential of circulating monocytes.

## SUMMARY

Elucidating the mechanisms by which obesity leads to increased morbidity and mortality has been the focus of intense research over the past three decades. Early studies demonstrated that inflammatory factors are increased in obesity, a finding that has been the focus of intense research in recent years. To this end, it is now well accepted that inflammation plays a causal role in the pathogenesis of metabolic dysregulation in obesity. The mechanisms that link obesity to inflammation and inflammation to disease development are multifactorial and involve interaction between a variety of physiological systems. From an overall perspective, the expansion of adipose tissue mass and increasing obesity are the fundamental factors that activates the inflammatory engine. At the cellular level, numerous factors provide fuel to the inflammatory engine, including poor delivery of nutrients and oxygen to the expanding adipose tissue mass, an increase in fatty acid flux, ER stress, adipocyte death, and increased cytokine/chemokine secretion. These events increase the recruitment of macrophages to the site of inflammation, which ultimately leads to insulin resistance. We view this as a "two-hit" process, in which bone marrow-derived inflammatory cells (i.e., macrophages in adipose tissue and Kupffer cells in the liver) must first become polarized toward an M1 proinflammatory, activation state (Figure 10.5). Once activated, the macrophage/Kupffer cells then release proinflammatory cytokines or other factors, which work in a paracrine fashion to activate the intracellular proinflammatory pathways in the neighboring insulin target cells (hepatocytes or adipocytes). This causes cell autonomous insulin resistance, and increases the recruitment of additional inflammatory cells, thus exacerbating the inflammatory state. Muscle cells also become insulin resistant and this may be related to circulating cytokines and/or adipokines secreted from adipose tissue and/or liver, changes in fatty acid flux, or some other mechanism. Although macrophages have been identified in skeletal muscle in obesity (Hevener et al. 2007), whether macrophages play a functional role in the generation of skeletal muscle insulin resistance is unclear (Bruun et al. 2005). In conclusion, improving our understanding of the mechanisms and importance of macrophage-mediated inflammation in the etiology of decreased insulin sensitivity could ultimately lead to new therapeutic approaches for the treatment of inflammation and insulin resistance.

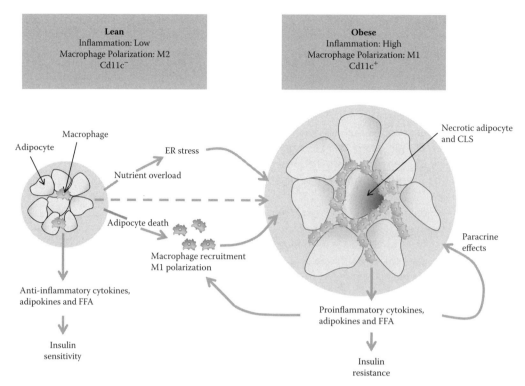

**FIGURE 10.5** Mechanisms of obesity-induced inflammation and systemic insulin resistance. In the lean state, adipocyte inflammation is low. Macrophages in lean adipose tissue secrete anti-inflammatory cyto-kines/adipokines and are F4/80+/CD11/CD11c−. Expansion of adipose tissue mass and adipocyte size during obesity results in an increase in absolute fatty acid release, adipocyte ER stress and adipocyte death. This activates localized inflammatory pathways, which initiates the recruitment of bone marrow-derived macro-phages, and polarization toward an M1, F4/80+/CD11b+/CD11c+, proinflammatory phenotype. Once activated, these macrophages then release proinflammatory cytokines/adipokines or other factors, which work in a para-crine fashion to activate the intracellular proinflammatory pathways in the neighboring cells. This causes cell autonomous insulin resistance, and increases the recruitment of additional macrophages, thus exacerbating the inflammatory state. CLS, crown-like structure; ER, endoplasmic reticulum; FFA, free fatty acids.

## REFERENCES

Aderem, A., and R. J. Ulevitch. 2000. Toll-like receptors in the induction of the innate immune response. *Nature* 406(6797):782–7.

Ajuwon, K. M., and M. E. Spurlock. 2005. Adiponectin inhibits LPS-induced NF-kappaB activation and IL-6 production and increases PPARgamma2 expression in adipocytes. *Am J Physiol Regul Integr Comp Physiol* 288(5):R1220–5.

Arend, W. P. 1990. Interleukin-1 receptor antagonist: discovery, structure and properties. *Prog Growth Factor Res* 2(4):193–205.

Arita, Y., S. Kihara, N. Ouchi, M. Takahashi, K. Maeda, J. Miyagawa, K. Hotta, I. Shimomura, T. Nakamura, K. Miyaoka, H. Kuriyama, M. Nishida, S. Yamashita, K. Okubo, K. Matsubara, M. Muraguchi, Y. Ohmoto, T. Funahashi, and Y. Matsuzawa. 1999. Paradoxical decrease of an adipose-specific protein, adiponectin, in obesity. *Biochem Biophys Res Commun* 257(1):79–83.

Arkan, M. C., A. L. Hevener, F. R. Greten, S. Maeda, Z. W. Li, J. M. Long, A. Wynshaw-Boris, G. Poli, J. Olefsky, and M. Karin. 2005. IKK-beta links inflammation to obesity-induced insulin resistance. *NatMed* 11(2):191–8.

Bandyopadhyay, G. K., J. G. Yu, J. Ofrecio, and J. M. Olefsky. 2005. Increased p85/55/50 expression and decreased phosphotidylinositol 3-kinase activity in insulin-resistant human skeletal muscle. *Diabetes* 54(8):2351–9.

Bastard, J. P., C. Jardel, E. Bruckert, P. Blondy, J. Capeau, M. Laville, H. Vidal, and B. Hainque. 2000. Elevated levels of interleukin 6 are reduced in serum and subcutaneous adipose tissue of obese women after weight loss. *J Clin Endocrinol Metab* 85(9):3338–42.

Bouhlel, M. A., B. Derudas, E. Rigamonti, R. Dievart, J. Brozek, S. Haulon, C. Zawadzki, B. Jude, G. Torpier, N. Marx, B. Staels, and G. Chinetti-Gbaguidi. 2007. PPARgamma activation primes human monocytes into alternative M2 macrophages with anti-inflammatory properties. *Cell Metab* 6(2):137–43.

Bruun, J. M., J. W. Helge, B. Richelsen, and B. Stallknecht. 2005. Diet and exercise reduce low-grade inflammation and macrophage infiltration in adipose tssue but not in skeletal muscle in severely obese subjects. *Am J Physiol Endocrinol Metab* 290(5):E961–7.

Chen, A., S. Mumick, C. Zhang, J. Lamb, H. Dai, D. Weingarth, J. Mudgett, H. Chen, D. J. MacNeil, M. L. Reitman, and S. Qian. 2005. Diet induction of monocyte chemoattractant protein-1 and its impact on obesity. *Obes Res* 13(8):1311–20.

Chiang, S. H., M. Bazuine, C. N. Lumeng, L. M. Geletka, J. Mowers, N. M. White, J. T. Ma, J. Zhou, N. Qi, D. Westcott, J. B. Delproposto, T. S. Blackwell, F. E. Yull, and A. R. Saltiel. 2009. The protein kinase IKKepsilon regulates energy balance in obese mice. *Cell* 138(5):961–75.

Cinti, S., G. Mitchell, G. Barbatelli, I. Murano, E. Ceresi, E. Faloia, S. Wang, M. Fortier, A. S. Greenberg, and M. S. Obin. 2005. Adipocyte death defines macrophage localization and function in adipose tissue of obese mice and humans. *J Lipid Res* 46(11):2347–55.

Clement, K., N. Viguerie, C. Poitou, C. Carette, V. Pelloux, C. A. Curat, A. Sicard, S. Rome, A. Benis, J. D. Zucker, H. Vidal, M. Laville, G. S. Barsh, A. Basdevant, V. Stich, R. Cancello, and D. Langin. 2004. Weight loss regulates inflammation-related genes in white adipose tissue of obese subjects. *Faseb J* 18(14):1657–69.

dc Jongh, R. T., E. H. Serne, R. G. Ijzcrman, G. de Vries, and C. D. Stehouwer. 2004. Free fatty acid levels modulate microvascular function: relevance for obesity-associated insulin resistance, hypertension, and microangiopathy. *Diabetes* 53(11):2873–82.

Dinarello, C. A. 1991. Interleukin-1 and interleukin-1 antagonism. *Blood* 77(8):1627–52.

Dresner, A., D. Laurent, M. Marcucci, M. E. Griffin, S. Dufour, G. W. Cline, L. A. Slezak, D. K. Andersen, R. S. Hundal, D. L. Rothman, K. F. Petersen, and G. I. Shulman. 1999. Effects of free fatty acids on glucose transport and IRS-1-associated phosphatidylinositol 3-kinase activity. *J Clin Invest* 103(2):253–9.

Duncan, R. E., M. Ahmadian, K. Jaworski, E. Sarkadi-Nagy, and H. S. Sul. 2007. Regulation of lipolysis in adipocytes. *Annu Rev Nutr* 27:79–101.

Fearnley, G. R., C. T. Vincent, and R. Chakrabarti. 1959. Reduction of blood fibrinolytic activity in diabetes mellitus by insulin. *Lancet* 2:1067.

Finkelstein, E. A., I. C. Fiebelkorn, and G. Wang. 2003. National medical spending attributable to overweight and obesity: How much, and who's paying? *Health Aff (Millwood)* Suppl Web Exclusives:W3-219–26.

Flynn, M. G., and B. K. McFarlin. 2006. Toll-like receptor 4: Link to the anti-inflammatory effects of exercise? *Exerc Sport Sci Rev* 34(4):176–81.

Fried, S. K., D. A. Bunkin, and A. S. Greenberg. 1998. Omental and subcutaneous adipose tissues of obese subjects release interleukin-6: Depot difference and regulation by glucocorticoid. *J Clin Endocrinol Metab* 83(3):847–50.

Grace, C. S., and R. B. Goldrick. 1968. Fibrinolysis and body bulid: Interrelationships between blood fibrinolysis, body composition and parameters of lipid and carbohydrate metabolism. *J Atheroscler Res* 8(4):705–19.

Gregor, M. F., and G. S. Hotamisligil. 2007. Thematic review series: Adipocyte biology. Adipocyte stress: the endoplasmic reticulum and metabolic disease. *J Lipid Res* 48(9):1905–14.

Gregor, M. F., L. Yang, E. Fabbrini, B. S. Mohammed, J. C. Eagon, G. S. Hotamisligil, and S. Klein. 2009. Endoplasmic reticulum stress is reduced in tissues of obese subjects after weight loss. *Diabetes* 58(3):693–700.

Haugen, F., and C. A. Drevon. 2007. Activation of nuclear factor-{kappa}B by high molecular weight and globular adiponectin. *Endocrinology* 148(11)5478–86.

Hernandez-Pando, R., Q. L. Bornstein, D. Aguilar Leon, E. H. Orozco, V. K. Madrigal, and E. Martinez Cordero. 2000. Inflammatory cytokine production by immunological and foreign body multinucleated giant cells. *Immunology* 100(3):352–8.

Hevener, A. L., J. M. Olefsky, D. Reichart, M. T. Nguyen, G. Bandyopadyhay, H. Y. Leung, M. J. Watt, C. Benner, M. A. Febbraio, A. K. Nguyen, B. Folian, S. Subramaniam, F. J. Gonzalez, C. K. Glass, and M. Ricote. 2007. Macrophage PPAR gamma is required for normal skeletal muscle and hepatic insulin sensitivity and full antidiabetic effects of thiazolidinediones. *J Clin Invest* 117(6):1658–69.

Hirosumi, J., G. Tuncman, L. Chang, C. Z. Gorgun, K. T. Uysal, K. Maeda, M. Karin, and G. S. Hotamisligil. 2002. A central role for JNK in obesity and insulin resistance. *Nature* 420(6913):333–6.

Hofmann, C., K. Lorenz, S. S. Braithwaite, J. R. Colca, B. J. Palazuk, G. S. Hotamisligil, and B. M. Spiegelman. 1994. Altered gene expression for tumor necrosis factor-alpha and its receptors during drug and dietary modulation of insulin resistance. *Endocrinology* 134(1):264–70.

Horowitz, J. F., and S. Klein. 2000. Whole body and abdominal lipolytic sensitivity to epinephrine is suppressed in upper body obese women. *Am J Physiol Endocrinol Metab* 278(6):E1144–52.

Hosogai, N., A. Fukuhara, K. Oshima, Y. Miyata, S. Tanaka, K. Segawa, S. Furukawa, Y. Tochino, R. Komuro, M. Matsuda, and I. Shimomura. 2007. Adipose tissue hypoxia in obesity and its impact on adipocytokine dysregulation. *Diabetes* 56(4):901-11.

Hotamisligil, G. S. 2008. Inflammation and endoplasmic reticulum stress in obesity and diabetes. *Int J Obes (Lond)* 32(Suppl 7):S52–4.

Hotamisligil, G. S., P. Arner, J. F. Caro, R. L. Atkinson, and B. M. Spiegelman. 1995. Increased adipose tissue expression of tumor necrosis factor-alpha in human obesity and insulin resistance. *J Clin Invest* 95(5):2409–15.

Hotamisligil, G. S., P. Peraldi, A. Budavari, R. Ellis, M. F. White, and B. M. Spiegelman. 1996. IRS-1-mediated inhibition of insulin receptor tyrosine kinase activity in TNF-alpha- and obesity-induced insulin resistance. *Science* 271(5249):665–8.

Hotamisligil, G. S., N. S. Shargill, and B. M. Spiegelman. 1993. Adipose expression of tumor necrosis factor-alpha: Direct role in obesity-linked insulin resistance. *Science* 259(5091):87–91.

Inouye, K. E., H. Shi, J. K. Howard, C. H. Daly, G. M. Lord, B. J. Rollins, and J. S. Flier. 2007. Absence of CC chemokine ligand 2 does not limit obesity-associated infiltration of macrophages into adipose tissue. *Diabetes* 56(9):2242–50.

Juge-Aubry, C. E., E. Somm, V. Giusti, A. Pernin, R. Chicheportiche, C. Verdumo, F. Rohner-Jeanrenaud, D. Burger, J. M. Dayer, and C. A. Meier. 2003. Adipose tissue is a major source of interleukin-1 receptor antagonist: Upregulation in obesity and inflammation. *Diabetes* 52(5):1104–10.

Kamei, N., K. Tobe, R. Suzuki, M. Ohsugi, T. Watanabe, N. Kubota, N. Ohtsuka-Kowatari, K. Kumagai, K. Sakamoto, M. Kobayashi, T. Yamauchi, K. Ueki, Y. Oishi, S. Nishimura, I. Manabe, H. Hashimoto, Y. Ohnishi, H. Ogata, K. Tokuyama, M. Tsunoda, T. Ide, K. Murakami, R. Nagai, and T. Kadowaki. 2006. Overexpression of monocyte chemoattractant protein-1 in adipose tissues causes macrophage recruitment and insulin resistance. *J Biol Chem* 281(36):26602–14.

Kanda, H., S. Tateya, Y. Tamori, K. Kotani, K. Hiasa, R. Kitazawa, S. Kitazawa, H. Miyachi, S. Maeda, K. Egashira, and M. Kasuga. 2006. MCP-1 contributes to macrophage infiltration into adipose tissue, insulin resistance, and hepatic steatosis in obesity. *J Clin Invest* 116(6):1494–505.

Kaser, S., A. Kaser, A. Sandhofer, C. F. Ebenbichler, H. Tilg, and J. R. Patsch. 2003. Resistin messenger-RNA expression is increased by proinflammatory cytokines *in vitro*. *Biochem Biophys Res Commun* 309(2):286–90.

Kim, C. S., H. S. Park, T. Kawada, J. H. Kim, D. Lim, N. E. Hubbard, B. S. Kwon, K. L. Erickson, and R. Yu. 2006. Circulating levels of MCP-1 and IL-8 are elevated in human obese subjects and associated with obesity-related parameters. *Int J Obes (Lond)* 30(9):1347–55.

Kim, H. J., T. Higashimori, S. Y. Park, H. Choi, J. Dong, Y. J. Kim, H. L. Noh, Y. R. Cho, G. Cline, Y. B. Kim, and J. K. Kim. 2004. Differential effects of interleukin-6 and -10 on skeletal muscle and liver insulin action *in vivo*. *Diabetes* 53(4):1060–7.

Kim, J. K., Y. J. Kim, J. J. Fillmore, Y. Chen, I. Moore, J. Lee, M. Yuan, Z. W. Li, M. Karin, P. Perret, S. E. Shoelson, and G. I. Shulman. 2001. Prevention of fat-induced insulin resistance by salicylate. *J Clin Invest* 108(3):437–46.

Kubota, N., Y. Terauchi, T. Yamauchi, T. Kubota, M. Moroi, J. Matsui, K. Eto, T. Yamashita, J. Kamon, H. Satoh, W. Yano, P. Froguel, R. Nagai, S. Kimura, T. Kadowaki, and T. Noda. 2002. Disruption of adiponectin causes insulin resistance and neointimal formation. *J Biol Chem* 277(29):25863–6.

Larsen, C. M., M. Faulenbach, A. Vaag, A. Volund, J. A. Ehses, B. Seifert, T. Mandrup-Poulsen, and M. Y. Donath. 2007. Interleukin-1-receptor antagonist in type 2 diabetes mellitus. *N Engl J Med* 356(15):1517–26.

Lee, C. D., S. N. Blair, and A. S. Jackson. 1999. Cardiorespiratory fitness, body composition, and all-cause and cardiovascular disease mortality in men. *Am J Clin Nutr* 69(3):373–80.

Lumeng, C. N., J. L. Bodzin, and A. R. Saltiel. 2007a. Obesity induces a phenotypic switch in adipose tissue macrophage polarization. *J Clin Invest* 117(1):175–84.

Lumeng, C. N., J. B. DelProposto, D. J. Westcott, and A. R. Saltiel. 2008. Phenotypic switching of adipose tissue macrophages with obesity is generated by spatiotemporal differences in macrophage subtypes. *Diabetes* 57(12):3239–46.

Lumeng, C. N., S. M. Deyoung, J. L. Bodzin, and A. R. Saltiel. 2007b. Increased inflammatory properties of adipose tissue macrophages recruited during diet-induced obesity. *Diabetes* 56(1):16–23.

Lumeng, C. N., I. Maillard, and A. R. Saltiel. 2009. T-ing up inflammation in fat. *Nat Med* 15(8):846–7.

Maeda, N., I. Shimomura, K. Kishida, H. Nishizawa, M. Matsuda, H. Nagaretani, N. Furuyama, H. Kondo, M. Takahashi, Y. Arita, R. Komuro, N. Ouchi, S. Kihara, Y. Tochino, K. Okutomi, M. Horie, S. Takeda, T. Aoyama, T. Funahashi, and Y. Matsuzawa. 2002. Diet-induced insulin resistance in mice lacking adiponectin/ACRP30. *Nat Med* 8(7):731–7.

Maedler, K., P. Sergeev, F. Ris, J. Oberholzer, H. I. Joller-Jemelka, G. A. Spinas, N. Kaiser, P. A. Halban, and M. Y. Donath. 2002. Glucose-induced beta cell production of IL-1beta contributes to glucotoxicity in human pancreatic islets. *J Clin Invest* 110(6):851–60.

Mallat, Z., S. Besnard, M. Duriez, V. Deleuze, F. Emmanuel, M. F. Bureau, F. Soubrier, B. Esposito, H. Duez, C. Fievet, B. Staels, N. Duverger, D. Scherman, and A. Tedgui. 1999. Protective role of interleukin-10 in atherosclerosis. *Circ Res* 85(8):e17–24.

Mantovani, A., A. Sica, S. Sozzani, P. Allavena, A. Vecchi, and M. Locati. 2004. The chemokine system in diverse forms of macrophage activation and polarization. *Trends Immunol* 25(12):677–86.

Mokdad, A. H., E. S. Ford, B. A. Bowman, W. H. Dietz, F. Vinicor, V. S. Bales, and J. S. Marks. 2003. Prevalence of obesity, diabetes, and obesity-related health risk factors, 2001. *JAMA* 289(1):76–9.

Neumeier, M., J. Weigert, A. Schaffler, G. Wehrwein, U. Muller-Ladner, J. Scholmerich, C. Wrede, and C. Buechler. 2006. Different effects of adiponectin isoforms in human monocytic cells. *J Leukoc Biol* 79(4):803–8.

Odegaard, J. I., R. R. Ricardo-Gonzalez, M. H. Goforth, C. R. Morel, V. Subramanian, L. Mukundan, A. R. Eagle, D. Vats, F. Brombacher, A. W. Ferrante, and A. Chawla. 2007. Macrophage-specific PPARgamma controls alternative activation and improves insulin resistance. *Nature* 447(7148):1116–20.

Osborn, O., S. E. Brownell, M. Sanchez-Alavez, D. Salomon, H. Gram, and T. Bartfai. 2008a. Treatment with an Interleukin 1 beta antibody improves glycemic control in diet-induced obesity. *Cytokine* 44(1):141–8.

Osborn, O., H. Gram, E. P. Zorrilla, B. Conti, and T. Bartfai. 2008b. Insights into the roles of the inflammatory mediators IL-1, IL-18 and PGE2 in obesity and insulin resistance. *Swiss Med Wkly* 138(45-46):665–73.

Ouchi, N., S. Kihara, Y. Arita, K. Maeda, H. Kuriyama, Y. Okamoto, K. Hotta, M. Nishida, M. Takahashi, T. Nakamura, S. Yamashita, T. Funahashi, and Y. Matsuzawa. 1999. Novel modulator for endothelial adhesion molecules: Adipocyte-derived plasma protein adiponectin. *Circulation* 100(25):2473–6.

Ozawa, K., M. Miyazaki, M. Matsuhisa, K. Takano, Y. Nakatani, M. Hatazaki, T. Tamatani, K. Yamagata, J. Miyagawa, Y. Kitao, O. Hori, Y. Yamasaki, and S. Ogawa. 2005. The endoplasmic reticulum chaperone improves insulin resistance in type 2 diabetes. *Diabetes* 54(3):657–63.

Ozcan, U., Q. Cao, E. Yilmaz, A. H. Lee, N. N. Iwakoshi, E. Ozdelen, G. Tuncman, C. Gorgun, L. H. Glimcher, and G. S. Hotamisligil. 2004. Endoplasmic reticulum stress links obesity, insulin action, and type 2 diabetes. *Science* 306(5695):457–61.

Palanivel, R., and G. Sweeney. 2005. Regulation of fatty acid uptake and metabolism in L6 skeletal muscle cells by resistin. *FEBS Lett* 579(22):5049–54.

Pascual, G., and C. K. Glass. 2006. Nuclear receptors versus inflammation: Mechanisms of transrepression. *Trends Endocrinol Metab* 17(8):321–7.

Patsouris, D., P. P. Li, D. Thapar, J. Chapman, J. M. Olefsky, and J. G. Neels. 2008. Ablation of CD11c-positive cells normalizes insulin sensitivity in obese insulin resistant animals. *Cell Metab* 8(4):301–9.

Rausch, M. E., S. Weisberg, P. Vardhana, and D. V. Tortoriello. 2008. Obesity in C57BL/6J mice is characterized by adipose tissue hypoxia and cytotoxic T-cell infiltration. *Int J Obes (Lond)* 32(3):451–63.

Reaven, G. M. 2005. The insulin resistance syndrome: Definition and dietary approaches to treatment. *Annu Rev Nutr* 25:391–406.

Reilly, M. P., M. Lehrke, M. L. Wolfe, A. Rohatgi, M. A. Lazar, and D. J. Rader. 2005. Resistin is an inflammatory marker of atherosclerosis in humans. *Circulation* 111(7):932–9.

Ryden, M., A. Dicker, V. van Harmelen, H. Hauner, M. Brunnberg, L. Perbeck, F. Lonnqvist, and P. Arner. 2002. Mapping of early signaling events in tumor necrosis factor-alpha -mediated lipolysis in human fat cells. *J Biol Chem* 277(2):1085–91.

Saberi, M., N. B. Woods, C. de Luca, S. Schenk, J. C. Lu, G. Bandyopadhyay, I. M. Verma, and J. M. Olefsky. 2009. Hematopoietic cell-specific deletion of toll-like receptor 4 ameliorates hepatic and adipose tissue insulin resistance in high-fat-fed mice. *Cell Metab* 10(5):419–29.

Santomauro, A. T., G. Boden, M. E. Silva, D. M. Rocha, R. F. Santos, M. J. Ursich, P. G. Strassmann, and B. L. Wajchenberg. 1999. Overnight lowering of free fatty acids with Acipimox improves insulin resistance and glucose tolerance in obese diabetic and nondiabetic subjects. *Diabetes* 48(9):1836–41.

Schenk, S., M. Saberi, and J. M. Olefsky. 2008. Insulin sensitivity: Modulation by nutrients and inflammation. *J Clin Invest* 118(9):2992–3002.

Shi, H., M. V. Kokoeva, K. Inouye, I. Tzameli, H. Yin, and J. S. Flier. 2006. TLR4 links innate immunity and fatty acid-induced insulin resistance. *J Clin Invest* 116(11):3015–25.

Silswal, N., A. K. Singh, B. Aruna, S. Mukhopadhyay, S. Ghosh, and N. Z. Ehtesham. 2005. Human resistin stimulates the pro-inflammatory cytokines TNF-alpha and IL-12 in macrophages by NF-kappaB-dependent pathway. *Biochem Biophys Res Commun* 334(4):1092–101.

Solinas, G., C. Vilcu, J. G. Neels, G. K. Bandyopadhyay, J-L. Luo, W. Naugler, S. Griven-nikov, A. Wynshaw-Boris, M. Scadeng, J. M. Olefsky, and M. Karin. 2007. JNK1 in hematopoietic-derived cells contributes to diet-induced inflammation and insulin resistance without affecting obesity. *Cell Metab* 6(5)386–97.

Spranger, J., A. Kroke, M. Mohlig, K. Hoffmann, M. M. Bergmann, M. Ristow, H. Boeing, and A. F. Pfeiffer. 2003. Inflammatory cytokines and the risk to develop type 2 diabetes: Results of the prospective population-based European Prospective Investigation into Cancer and Nutrition (EPIC)-Potsdam study. *Diabetes* 52(3):812–7.

Steinberg, G. R., B. J. Michell, B. J. van Denderen, M. J. Watt, A. L. Carey, B. C. Fam, S. Andrikopoulos, J. Proietto, C. Z. Gorgun, D. Carling, G. S. Hotamisligil, M. A. Febbraio, T. W. Kay, and B. E. Kemp. 2006. Tumor necrosis factor alpha-induced skeletal muscle insulin resistance involves suppression of AMP-kinase signaling. *Cell Metab* 4(6):465–74.

Steinberg, H. O., M. Tarshoby, R. Monestel, G. Hook, J. Cronin, A. Johnson, B. Bayazeed, and A. D. Baron. 1997. Elevated circulating free fatty acid levels impair endothelium-dependent vasodilation. *J Clin Invest* 100(5):1230–9.

Steppan, C. M., S. T. Bailey, S. Bhat, E. J. Brown, R. R. Banerjee, C. M. Wright, H. R. Patel, R. S. Ahima, and M. A. Lazar. 2001. The hormone resistin links obesity to diabetes. *Nature* 409(6818):307–12.

Tao, L., E. Gao, X. Jiao, Y. Yuan, S. Li, T. A. Christopher, B. L. Lopez, W. Koch, L. Chan, B. J. Goldstein, and X. L. Ma. 2007. Adiponectin cardioprotection after myocardial ischemia/reperfusion involves the reduction of oxidative/nitrative stress. *Circulation* 115(11):1408–16.

Trayhurn, P. 2005. The biology of obesity. *Proc Nutr Soc* 64(1):31–8.

Tsukumo, D. M., M. A. Carvalho-Filho, J. B. Carvalheira, P. O. Prada, S. M. Hirabara, A. A. Schenka, E. P. Araujo, J. Vassallo, R. Curi, L. A. Velloso, and M. J. Saad. 2007. Loss-of-function mutation in toll-like receptor 4 prevents diet-induced obesity and insulin resistance. *Diabetes* 56(8):1986–98.

Uysal, K. T., S. M. Wiesbrock, M. W. Marino, and G. S. Hotamisligil. 1997. Protection from obesity-induced insulin resistance in mice lacking TNF-alpha function. *Nature* 389(6651):610–4.

van Exel, E., J. Gussekloo, A. J. de Craen, M. Frolich, A. Bootsma-Van Der Wiel, and R. G. Westendorp. 2002. Low production capacity of interleukin-10 associates with the metabolic syndrome and type 2 diabetes: The Leiden 85-Plus study. *Diabetes* 51(4):1088–92.

Verma, S., S. H. Li, C. H. Wang, P. W. Fedak, R. K. Li, R. D. Weisel, and D. A. Mickle. 2003. Resistin promotes endothelial cell activation: further evidence of adipokine-endothelial interaction. *Circulation* 108(6):736–40.

Weisberg, S. P., D. Hunter, R. Huber, J. Lemieux, S. Slaymaker, K. Vaddi, I. Charo, R. L. Leibel, and A. W. Ferrante, Jr. 2006. CCR2 modulates inflammatory and metabolic effects of high-fat feeding. *J Clin Invest* 116(1):115–24.

Weisberg, S. P., D. McCann, M. Desai, M. Rosenbaum, R. L. Leibel, and A. W. Ferrante, Jr. 2003. Obesity is associated with macrophage accumulation in adipose tissue. *J Clin Invest* 112(12):1796–808.

Wellen, K. E., and G. S. Hotamisligil. 2005. Inflammation, stress, and diabetes. *J Clin Invest* 115(5):1111–9.

Xu, H., G. T. Barnes, Q. Yang, G. Tan, D. Yang, C. J. Chou, J. Sole, A. Nichols, J. S. Ross, L. A. Tartaglia, and H. Chen. 2003. Chronic inflammation in fat plays a crucial role in the development of obesity-related insulin resistance. *J Clin Invest* 112(12):1821–30.

Yamauchi, T., J. Kamon, Y. Minokoshi, Y. Ito, H. Waki, S. Uchida, S. Yamashita, M. Noda, S. Kita, K. Ueki, K. Eto, Y. Akanuma, P. Froguel, F. Foufelle, P. Ferre, D. Carling, S. Kimura, R. Nagai, B. B. Kahn, and T. Kadowaki. 2002. Adiponectin stimulates glucose utilization and fatty-acid oxidation by activating AMP-activated protein kinase. *Nat Med* 8(11):1288–95.

Yatagai, T., S. Nagasaka, A. Taniguchi, M. Fukushima, T. Nakamura, A. Kuroe, Y. Nakai, and S. Ishibashi. 2003. Hypoadiponectinemia is associated with visceral fat accumulation and insulin resistance in Japanese men with type 2 diabetes mellitus. *Metabolism* 52(10):1274–8.

Ye, J., Z. Gao, J. Yin, and Q. He. 2007. Hypoxia is a potential risk factor for chronic inflammation and adiponectin reduction in adipose tissue of ob/ob and dietary obese mice. *Am J Physiol Endocrinol Metab* 293(4)E1118–28.

Yuan, M., N. Konstantopoulos, J. Lee, L. Hansen, Z. W. Li, M. Karin, and S. E. Shoelson. 2001. Reversal of obesity- and diet-induced insulin resistance with salicylates or targeted disruption of Ikkbeta. *Science* 293(5535):1673–7.

Zeyda, M., D. Farmer, J. Todoric, O. Aszmann, M. Speiser, G. Gyori, G. J. Zlabinger, and T. M. Stulnig. 2007. Human adipose tissue macrophages are of an anti-inflammatory phenotype but capable of excessive pro-inflammatory mediator production. *Int J Obes (Lond)* 31(9):1420–8.

# 11 Nutritional Genomics of Vitamin D on Cardiovascular Disease

*Sandra F. Williams, Jorge N. Artaza, and Keith C. Norris*

## CONTENTS

## NUTRITION AND VITAMIN D

Micronutrients are substances required in small amounts, typically 5–10 micrograms daily, for proper functioning of the human body. Vitamins are one class of micronutrients, which cannot be synthesized in sufficient quantities within the body and therefore must be obtained exogenously. Of these, vitamin D, a fat-soluble vitamin can be synthesized in the human skin but is virtually absent from most natural foods (Wang 2009) with the exception of amounts found in fish liver oils, fatty fish species, whole egg, and beef liver (Ovesen et al. 2003). Vitamin D is a lipid-soluble prohormone obtained via de novo skin synthesis as a result of UV-light-induced photolytic conversion of 7-dehydrocholesterol to previtamin D3 followed by thermal isomerization to vitamin D3 or ingested through the diet. It is subsequently activated in a sequential two-step process (Holick and Adams 1998; Dusso and Brown 1998; Holick 2007) (Figure 11.1). The first step is the 25-hydroxylation in the liver to produce 25(OH) vitamin D (25D). The second and more regulated step is the 1-hydroxylation, which is classically thought to occur primarily in the kidney via the 25-hydroxyvitamin D-1-alpha hydroxylase enzyme to produce the active product $1,25(OH)_2$ vitamin D3 (1,25D) or calcitriol (Holick and Adams 1998; Dusso and Brown 1998; Holick 2007). 1,25D binds to the vitamin D receptor (VDR), a nuclear receptor and a member of the steroid/thyroid hormone superfamily receptors, which acts as a ligand-activated transcription factor. VDR functions as a heterodimer with the retinoid X receptor. Upon ligand binding the VDR undergoes a conformational change that promotes retinoid X receptor-VDR heterodimerization. The bound

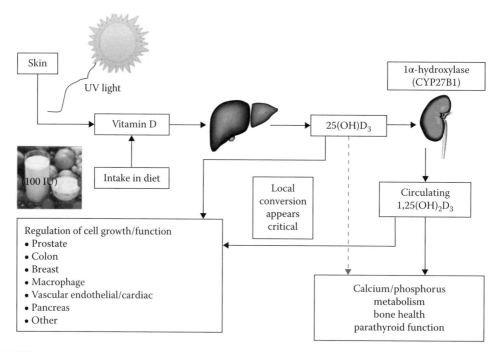

**FIGURE 11.1**  An overview of vitamin D metabolism and both traditional and nontraditional effects of vitamin D. (From Holick, M. F., and J. S. Adams. 1998. Vitamin D Metabolism and Biological Function. In L. Avioli and S. Krane, eds. *Metabolic Bone Disease and Clinically Related Disorders*. 3rd ed., pp. 124–56. San Diego, California: Academic Press. With permission.)

heterodimer translocates to the nucleus, where VDR binds to the vitamin D response element, induces chromatin-modifying enzymatic activities, and ultimately modulates gene transcription (Dusso and Brown 1998) (Figure 11.2).

Current daily requirements of vitamin D are presently under review and will likely be increased from previous standards. In 2009 the 100% daily standard used for product labels was 800 international units (IU) daily, although the Food and Nutrition Board, National Research Council, and the National Academy of Sciences have set the safe upper limit at 2000 IU daily (Food and Nutrition Board 1997).

## CLINICAL EPIDEMIOLOGY

Perhaps the greatest and most profound recent interest has been generated in the effect of vitamin D on the cardiovascular (CV) system. Cardiovascular disease (CVD), most commonly coronary artery disease and stroke, affects approximately 80 million people in the United States (1 in 3 adults) (Stephen 2009). Given this astounding statistic, even small gains in CVD prevention could translate into a significant public health impact (Lee et al. 2008). Epidemiological studies have now unequivocally linked low levels of serum vitamin D with increased incidence of hypertension, obesity, and adverse CV events (Wang et al. 2008). Although there is no consensus on optimal levels of 25D, levels at 30 ng/ml or above are generally considered sufficient, 20–29 ng/ml are insufficient and levels below 20 ng/ml are considered deficient (Holick 2007). The term severe deficiency is usually reserved for settings when serum 25D levels are less than 10 ng/ml. Recent epidemiologic studies have demonstrated statistically significant increased rates of CV disease (Martins et al. 2007; Wang et al. 2008; Artaza et al. 2009), end-stage renal disease (Melamed et al. 2009), and even death (Melamed et al. 2008; Mehrotra et al. 2009) in longitudinal cohorts

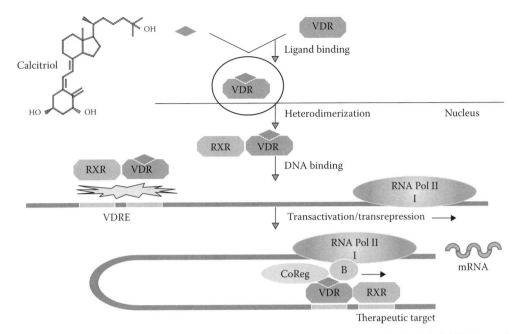

**FIGURE 11.2** Shown here is a schematic of the mechanisms involved in impaired calcitriol/VDR regulation of gene expression in chronic kidney disease. The calcitriol/VDR complex activates the VDR to interact with the RXR, causing VDR/RXR heterodimerization. The VDR/RXR complex binds to specific segments in the promoter region of vitamin D–responsive genes called vitamin D response elements. This complex recruits basal transcription factors and coregulator molecules to increase or suppress gene transcription. For example, the VDR/RXR complex suppresses PTH gene transcription by RNA pol II. mRNA = messenger ribonucleic acid; RNA pol II = ribonucleic acid polymerase; RXR = retinoid X receptor; VDR = vitamin D receptor; VDRE = vitamin D response element. (From Dusso A. S., A. J. Brown, and E. Slatopolsky. 2005. *Am J Physiol Renal Physiol* 289(1):F8–28. With permission.)

whose baseline vitamin D levels were <10–18 ng/ml. This correlation has led to the suggestion that vitamin D deficiency may well be considered an underestimated nonclassical risk factor for CVD, especially in certain high-risk populations (Levin and Li 2005; Martins et al. 2007; Wang et al. 2008). Indeed, the increased rates of CV and related diseases among African Americans may be related in part to higher rates of hypovitaminosis D and overt vitamin D deficiency (Martins et al. 2007). A combination of increased melanin reducing de novo light-induced skin synthesis of 7-dehydrocholesterol to previtamin D3 for a given degree of light exposure as well as increased rates of lactose intolerance leading to lower intake of vitamin D fortified dairy products contribute to lower vitamin D levels in African Americans. Data from the general population sampled between 1988 and 1994 as part of the third National Health and Nutrition Survey (NHANES III) revealed low rates of sufficient 25D (>30 ng/ml) in whites (50%), Hispanics (30%), and African Americans (15%), that worsened over time in the subsequent 2001–04 NHANES sampling where the prevalence of sufficient 25D was present in only 30% of whites, 10% of Hispanics, and 5% of African Americans (Martins et al. 2007; Ginde et al. 2009) (Figure 11.3). Rates of severe vitamin D deficiency (<10 ng/ml) rose between these two surveys from 1%–3% for whites, 3%–8% for Hispanics, and from 8%–30% for African Americans (Ginde et al. 2009). The high and increasing rates of hypovitaminosis D as well as severe vitamin D deficiency (<10 ng/ml), clearly below the at-risk level for CV events of 15 ng/ml, may contribute to the racial disparities in CV and related diseases. Indeed, the mean 25D level in African Americans in the NHANES 2001–04 survey was only 15 ng/ml (Ginde et al. 2009). A falling prevalence of sufficient levels of 25D over time was also noted for elderly (Figure 11.3).

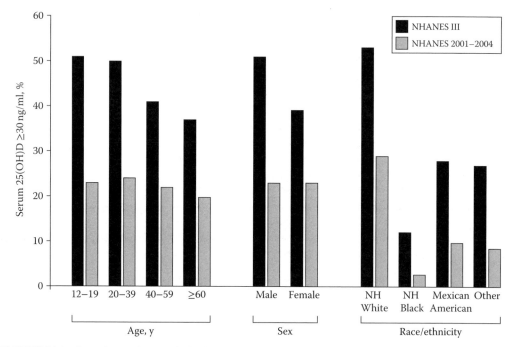

**FIGURE 11.3**   Prevalence of serum 25-hydroxyvitamin D (25[OH]D) level of 30 ng/mL or more stratified by demographic characteristics. The prevalence of sufficient levels 25(OH)D is low nationally and has fallen substantially between the Third National Health and Nutrition Examination Survey (NHANES III) (1988–1994) and the NHANES 2001–2004. (From Ginde, A. A., M. C. Liu, and C. A. Camargo Jr. 2009. *Arch Intern Med* 169(6):626–32. With permission.)

## MECHANISMS OF VITAMIN D ACTION

The mechanism via which the renal or peripherally activated forms of 25D mediates its now-recognized vast array of biological functions appears to be via primarily genomic pathways with a lesser contribution from nongenomic pathways. The genomic pathways are currently being vigorously investigated and evidence detailing their mechanisms and actions are being reported with increasing frequency. Paramount among the identified genomic pathways is the involvement of the VDR which has long been known to be involved in vitamin D's action and which appears to be ubiquitous in human tissues and cells (Holick 2007). 1,25D activates the VDR, a nuclear transcription factor of the steroid–retinoid nuclear receptor gene superfamily, ultimately promoting gene transcription (Carlberg et al. 2001; Zehnder et al. 2001), and exerting a positive or negative effect on target genes (Falkenstein et al. 2000).

In the case of vitamin D, the mechanism for its genomic action involving the VDR receptor has been well described in Figure 11.2. These receptors possess both a ligand-binding domain as well as a DNA-binding domain (Falkenstein et al. 2000). The activated form of vitamin D enters target cells and complexes with the intracellular VDR. This complex then enters the nucleus where it forms a heterodimeric complex with the retinoic acid X receptor and it is this heterodimer complex which selects for specific DNA sequences known as vitamin D response elements (Dusso and Brown 1998) (Figure 11.2). The ubiquity of the VDR supports the wide range of physiological functions of 1,25D (Dusso and Brown 1998).

The presence of at least 800 human genes for which there is a vitamin D response element has been identified (Carlberg 2003). The subsequent biologic responses which result from this genomic mechanism follow the vitamin D response element activating or inhibiting transcription of vitamin D response genes including Ca-binding protein, epithelial calcium channel, receptor activator nuclear k-B ligand (RANKL), alkaline phosphatase, prostate specific antigen, and parathyroid hormone (Carlberg 2003).

Those genomic mechanisms involving intracellular receptor binding and translocation with subsequent modulation of transcription and protein synthesis are characteristically responsible for delayed effects of hormone action (Falkenstein et al. 2000). Other mechanisms likely underlie the more rapid, acute actions of vitamin D. 1,25D has been demonstrated to within minutes both increase calcium transport across intestinal cells (Zhou et al. 1992; Norman et al. 1998) and calcium fluxes in bone-forming osteoblasts (Lieberherr 1987). Further evidence indicates that in some cells, 1,25D is able to, within seconds, increase phophorylation of both Raf kinase and MAP kinase pathways which are involved in cell growth, further substantiating that there may exist rapid response signal-transduction components of 1,25D involved in such nonclassical vitamin D functions such as cellular growth and differentiation (Marcinkowska et al. 1997; Song et al. 1998). Data from animal studies suggest that these rapid effects of vitamin D may involve a membrane-binding site for 1,25 vitamin D and implicates a mechanism distinct from the intracellular receptor genomic mechanisms responsible for the delayed effects of the hormone (Falkenstein et al. 2000).

Nonselective and selective VDR agonists (VDRAs) can induce different patterns of gene expression through differential interaction with signaling proteins or transcription factors or via differences in their activation of VDR (Takeyama et al. 1999; Issa et al. 2002; Wu-Wong et al. 2006). Studies show that VDRAs may differ in their ability to activate rapid versus delayed actions of hormone function which may reflect structural differences in their abilities to bind to postulated membrane-binding sites versus known intracellular receptors (Zhou et al. 1992; Bouillon et al. 1995; Norman et al. 1997). One role of vitamin D's rapid activation of signal transduction molecules may be to effect a process of cross-talk by which these rapidly activated pathways then influence and modify signals transmitted by the delayed intracellular-receptor pathways and therefore affect gene expression (Song et al. 1998). Finally, there is suggestion that at least some of the nonclassical paracrine effects of vitamin D are mediated by nongenomic mechanisms, which affect chloride (Zanello and Norman 1997), phosphate transport (Karsenty et al. 1985), and modulate alkaline phosphatase activity (Ben Nasr et al. 1988). Further details of the complexities of these mechanisms remain to be elucidated.

## CLASSICAL AND NONCLASSICAL FUNCTIONS OF VITAMIN D

The function of vitamin D was traditionally believed to involve 1,25D release into the circulation after activation by the renal (enzyme) 1-alpha hydroxylase, after which its effects were targeted on end organs involved in calcium and phosphorus homeostasis, namely kidney, intestine, parathyroid, and bone (Jones 2007). It is now apparent that in addition to the classical pathway for activation of 25D to 1,25D, there exists a peripheral autocrine pathway which results in calcitriol synthesis in a variety of nonrenal tissues which have now been demonstrated to possess 1-alpha hydroxylase activity (Holick and Adams 1998; Heaney 2008) (Figure 11.1). In fact, it appears that the bulk of daily metabolic utilization of 25D is via this peripheral paracrine pathway (Jones 2007; Heaney 2008). The recognition of these new pathways have now led to newly ascribed paracrine functions of vitamin D, which suggest a more expanded role of this micronutrient in biological functions outside of bone and mineral metabolism (Heaney 2008; Verstuyf et al. 2010). Multiple organ systems including, but not limited to, the CV (Wang et al. 2008), renal, and immune systems are now believed to activate vitamin D locally for regulation of cell and tissue growth and differentiation, as well as to serve as precursors of enzyme cofactors (Heaney 2008; Rostand and Warnock 2008), all vital to intact functioning of numerous metabolic processes (Jones 2007; Heaney 2008). Deficiencies in serum vitamin D levels have recently been associated with disease in these major systems (Forman et al. 2007; Wang et al. 2008; Giovannucci et al. 2008; Li et al. 2003; Holick 2007; Martins et al. 2007) and the possibility that therapeutic use of 1,25D or activated VDRAs could be effective in prevention or treatment of CVD, chronic kidney disease (CKD), or other chronic medical conditions (Mehta et al. 1997; Jones 2007; Heaney 2008; Wang et al. 2008; Li 2010) has fueled extensive investigations into the biological mechanisms which underlie vitamin D's nonclassical roles (Heaney 2008; Li 2010).

The effects of vitamin D on the CVD system are most likely mediated via its nonclassical activation by paracrine 1-alpha hydroxylase enzymatic activity and there is now extensive research underway to elucidate the molecular mechanisms responsible for these biologic effects such as inhibition of vascular smooth muscle proliferation (Carthy et al. 1989), suppression of vascular calcification (Mathew et al 2008), downregulation of proinflammatory cytokines (Levin and Li 2005; Mathieu and Adorini 2002), upregulation of anti-inflammatory cytokines (Mathieu and Adorini 2002), antifibrotic and antiapoptotic actions (Artaza and Norris 2009; Artaza et al. 2010), and the action of vitamin D as a negative endocrine regulator of the renin–angiotensin system (Li 2003; Wang et al. 2010).

## GENOMIC EFFECTS OF VITAMIN D ON CARDIOVASCULAR DISEASE

Substantial evidence has now linked hypovitaminosis D to increasing risk and incidence of CVD (Scragg et al. 1990; Lind et al. 1995; Kristal-Bonch et al. 1997; Zitterman et al. 2003, 2005; Poole et al. 2006; Wang et al 2008) and suggests the possibility that vitamin D repletion may reverse or attenuate what remains the leading cause of mortality in the United States (Zitterman et al. 2005; Wang et al. 2008). In the CV system the genomic mechanisms responsible for vitamin D's nonclassical effects appear to be mediated primarily via activated vitamin D's interaction with the intracellular VDR receptor. In keeping with their ubiquitous distribution, VDRs have been identified within vascular smooth muscle (Merke et al. 1987; Somjen et al. 2005), endothelium (Merke et al. 1989), and cardiomyocytes (Holick 2006). These mechanisms likely serve to modulate key processes involved in the pathogenesis of CVD, including vascular inflammation (Rigby et al. 1987), platelet aggregation/thrombogenesis (Aihara et al. 2004), vascular smooth muscle cell proliferation (Mitsuhashi et al. 1991), the renin–angiotensin system (Li and Glass 2002; Xiang et al. 2005), and cardiomyocyte proliferation, vascular calcification, myocardial fibrosis, and proliferation (Artaza et al. 2009, 2010). It is clear that no single gene or pathway mediates the above processes. For instance, multiple pathways are involved in the antiproliferative effects of 1,25D such as epidermal growth factor, insulin-like growth factor, transforming growth factor, prostaglandins, and Wnt-β-catenin signaling cascades (Bouillon et al. 2008; Verstuyf et al. 2010). The variety of target genes identified through these studies reflects the pleiotropic action of 1,25D in different cellular processes, including emerging roles on cell-cycle progression, apoptosis, cellular adhesion, and oxidative stress.

### INFLAMMATORY RESPONSE

The critical phases of CVD involve the initial formation of the atherosclerotic plaque followed by plaque rupture or erosion, then subsequent arterial thrombosis (Chapman 2007). Inflammation, considered a key factor in this process, is thought to drive the processes of plaque formation, progression, and rupture (Libby 2003). Studies have indicated that an inflammatory subset of monocytes and macrophages selectively concentrate in the atherosclerotic plaque and produce proinflammatory cytokines which include interleukin (IL)-1, IL-4, IL-6, interferon (INF)-γ, and tumor necrosis factor (TNF)-α (Pearson et al. 2003; Libby 2003; Levin and Li 2005). T-lymphocytes also take part in the inflammatory processes that enhance thrombosis by driving the production of collagen-degrading proteinases, such as MMPs (Timms et al. 2002), as well as the potent procoagulant, tissue factor, which is expressed by macrophages and is critical to thrombosis (Libby 2003). Macrophages and T-cells are also active participants in the formation of foam cells and in the organization of the atherosclerotic lesion within the arterial wall (Pearson et al. 2003; Levin and Li 2005).

Targeted therapeutic approaches directed at the inflammatory component of atherosclerosis are now an established part of the armamentarium in CVD prevention and treatment. Current evidence suggests that the active form of vitamin D and VDRA may in the future be effective interventions in this clinical arena. Vitamin D has long been shown to possess immunoregulatory properties (Mathieu and Adorini 2002) and cumulative data has documented vitamin D's ability to inhibit key steps in the inflammatory process (Levin and Li 2005), as summarized in Figure 11.4.

The biologic mechanism via which Vitamin D exerts its affect on CV inflammation appears to be via VDR receptors in the heart and vascular wall, as well as in immune cells (Sandgren et al. 1991; O'Connell et al. 1996; Haussler et al. 1998). In its interference with the inflammatory process, vitamin D (via its receptor, VDR) appears to inhibit antigen-presenting cell maturation (Griffin et al. 2001; Xing et al. 2002), as well as angiogenesis and smooth muscle cell proliferation (Carthy et al. 1989). It also regulates gene expression in immune cells and modulates their production of cytokines, thereby downregulating nuclear factor-kβ (NF-kβ) activity, activating IL-10 production, and inhibiting IL-6, IL-12, IFN-γ, and TNF-α alpha production (Mathieu and Adorini 2002), which results in a cytokine mixture less likely to promote inflammation. Vitamin D also regulates expression of the collagen-degrading proteinases (Timms et al. 2002; Libby 2003; Artaza and Norris 2009). Among these are matrix metalloproteinases, produced by macrophages and responsible for remodeling of the vascular wall and myocardium, which are active in the breakdown of collagen within the atherosclerotic plaque, leading to plaque rupture and thrombosis (Ross 1999; Glass and Witztum 2001; Pearson et al. 2003).

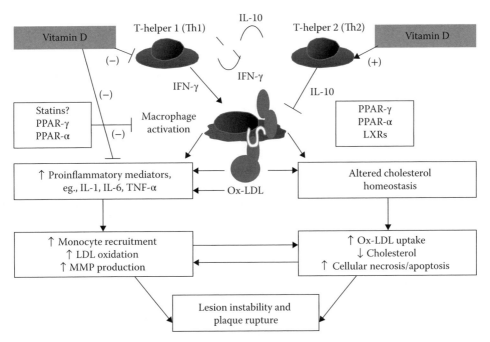

**FIGURE 11.4** Vitamin D, lymphocyte, and macrophage involvement in atherosclerosis. Recent studies suggest that lymphocytes and macrophages play the initial role in the generation of atheromas. It is hypothesized that Th1 cells start producing excess IFN-γ, which is a potent stimulator of macrophage activity. Activated macrophages secrete IL-1β, IL-6, and TNF-α. These cytokines recruit additional monocytes, increase LDL oxidation, and generate production of MMPs that can destabilize the plaque to cause rupture and thrombosis. In contrast, the Th2 lymphocyte subset is called the antiatherogenic phenotype, because these cells produce IL-10, a good cytokine that suppresses macrophage activation and Th1 proliferation. PPAR, peroxisome proliferator-activated receptor. (Adapted from Li, A. C., and C. K. Glass. 2002. *Nat Med* 8:1235–42 and Andress, D. L. 2006. *Kidney Int* 69:33–43.)

## PLATELET AGGREGATION AND THROMBOGENESIS

Parallel to the inflammatory role in plaque formation are the equally important processes of platelet aggregation and thrombogenesis. Vitamin D has been shown to influence the prevention/regression of platelet aggregation and thrombogenesis (Oshawa et al. 2000; Aihara et al. 2004). The mechanisms underlying vitamin D's role in platelet aggregation and thrombogenesis also suggest a post-translational regulatory mechanism involving the VDR. Studies in VDR knockout mice made normocalcemic through high-calcium diets showed increased platelet aggregation in comparison with wild-type mice and hypocalcemic knockout mice; these findings implicate VDR activation as the underlying mechanism which resulted in suppression of platelet aggregation (Aihara et al. 2004). Via its intracellular receptor, vitamin D also appears to modulate plasminogen-activator-inhibitor (PAI) expression in endothelial cells. Incubation of mesenchymal multipotent cells with 1,25D demonstrated decreased expression of plasminogen-activator-inhibitor (Artaza and Norris 2009), and 1,25D was shown to down-regulate expression of tissue factor, another potent coagulation factor, in monocytic cell culture (Oshawa et al. 2000).

## VASCULAR SMOOTH MUSCLE CELL PROLIFERATION

The role of vitamin D in the atherosclerotic process also appears to extend to its effect on cellular proliferation. In the process of atheroma formation, current dogma implicates proliferation of vascular smooth muscle cells during plaque formation as a key event following endothelial dysfunction (Zhang 2008). Incubation of a mesenchymal multipotent cell line with 1,25 vitamin D showed a decrease in cell numbers indicating an antiproliferative effect of vitamin D (Artaza et al. 2010). In the same study, real-time PCR microarray analysis showed a general decrease in gene expression of cyclins such as cyclin A2, B1, B2, D1 (a positive regulator of G0/G1 cell cycle progression and a well-known promoter of cell proliferation), E1, and F, and cyclin-dependent kinases such as Cdk2 and Cdk4 and genes related to check points such as Chek1 (checkpoint kinase 1). Moreover, incubation of multipotent cells with 1,25D inhibits G1 to S phase progression by downregulating the expression of the F-box protein Skp2 (p45) that induces cell growth inhibition via G1 arrest. Some of the mechanisms involved are summarized in Figure 11.5 (Verstuyf et al. 2010).

Further investigation to elucidate the underlying biologic mechanisms responsible for this effect demonstrated that in the presence of activated vitamin D there was increased VDR expression and translocation, suggesting that the mechanism of vitamin D action involved the intracellular vitamin D receptor. Confirmation of vitamin D's potent antiproliferative effect, and to assure that the observed effect was not secondary to a decrease in apoptosis, it was clearly demonstrated that there was a parallel/concomitant decrease in expression of the proliferating cell nuclear antigen (PCNA), a protein cofactor of DNA polymerase delta which increases processivity of the leading strand during DNA replication (Essers et al. 2005). 1,25D's antiproliferative effect may also be mediated through the Rho and Rhou/Wrch-1 pathways. The Wrch-1 gene is reported to be involved in regulating proliferation, and Rho, a member of the GTPase family, is also associated with biologic pathways related to cellular proliferation (Artaza et al. 2010). In the presence of 1,25D these pathways were demonstrated to be activated concomitant with demonstrated decreased cellular proliferation (Artaza et al. 2010).

## FIBROSIS

In support of the potential use of vitamin D as an effective agent against fibrosis, an important pathway contributing to the morbidity and premature mortality associated with chronic diseases. It has been demonstrated that incubating multipotent mesenchymal cells with vitamin D leads to decreased expression of different collagen isoforms (the ultimate marker of fibrosis) (Artaza and

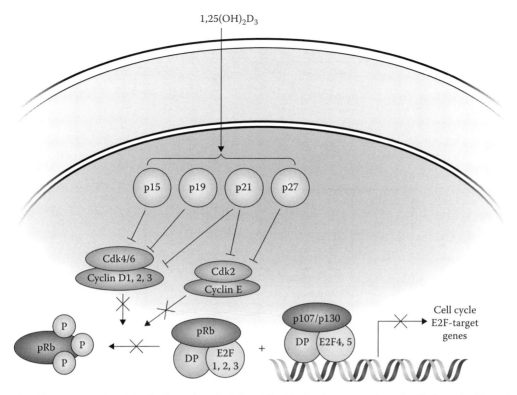

**FIGURE 11.5**   1-alpha, 25-Dihydroxyvitamin D3 (1,25D) blocks the progression of cells from the G1 to the S phase of the cell cycle. 1,25D induces the expression of different cyclin dependent kinase (Cdk) inhibitors (p15, p19, p21, and p27), which inhibit the activity of cyclin/Cdk complexes (Cyclin D1, 2, 3/Cdk4–6; Cyclin E/Cdk2). Reduced cyclin (cyclin D, cyclin E) expression after treatment with 1,25D also contributes to a reduced cyclin/Cdk activity. The pocket proteins retinoblastoma (pRb), p107, and p130 remain in an underphosphorylated state because of the reduced cyclin/Cdk activity. E2F4 and E2F5 remain in association with the hypophosphorylated p107 and p130 pocket proteins and act as transcriptional repressors in quiescent and early G1 cells. Due to the hypophosphorylated status of pRb, the transcriptional activators E2Fs1–3 remain bound and are not able to target the promoter region of cell-cycle genes to initiate the transition of G1 to S phase. As a result, cells are blocked in the G1 phase of the cell cycle after treatment with 1,25D ($\downarrow$, stimulatory effect; $\perp$, inhibitory effect). (From Verstuyf, A. et al. 2010. *Kidney Int* 78(2):140–5. With permission.)

Norris 2009; Artaza et al. 2009) (Figures 11.6 and 11.7). Concurrently, vitamin D exposure induced an antifibrotic phenotype characterized by increase expression of BMP7 (TGFβ antagonist), and Follistatin, which blocks activin signaling (Sulyok et al. 2004) and also inhibits the profibrotic factors myostatin (Artaza et al. 2008) and MMP8 (a collagen breakdown inducer) (Artaza et al. 2010) (Figure 11.6).

## Renin–Angiotensin System

Perhaps the most far-reaching and profound effect on the pathogenesis of disease within the CV system can be attributed to the renin–angiotensin system (RAS). The RAS underlies the pathophysiology of hypertension, coronary heart disease, cardiac hypertrophy, and congestive heart failure, which are all points along the continuum of CVD (Miura and Saku 2010). In this cascade there is sequential activation of angiotensin II which ultimately has deleterious effects on blood pressure, the heart and the vasculature. Both clinical and epidemiologic studies have consistently shown an inverse relationship between serum levels of 1,25D and blood pressure (Lind et al. 1995;

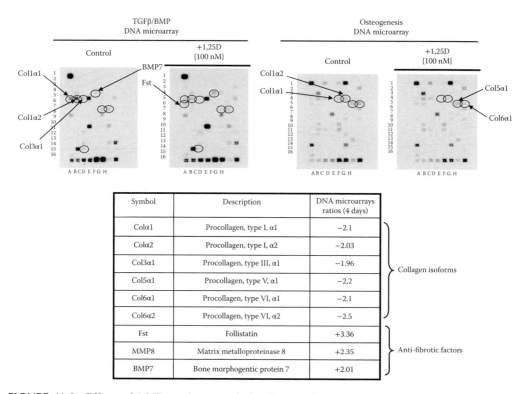

**FIGURE 11.6** Effects of 1,25D on the transcriptional expression of collagen isoforms and select fibrosis mediating genes in C3H 10T1/2 cells. Cultures of C3H 10T1/2 cells were treated with 20 μM azacytidine for 2 days to induce fibrosis and 2 days later were incubated with or without 1,25D (100 nM) for 4 days. Total RNA was isolated and subjected to DNA microarray analysis for genes related to the TGFβ/BMP signaling pathway and to the osteogenesis gene array respectively. The table shows the DNA microarray ratios of collagen isoforms and select fibrosis mediators between the 1,25D treated versus 1,25D-untreated cells corrected by GAPDH. (Adapted from Artaza, J. N., and K. C. Norris. 2009. *J Endocrinol* 200(2):207–21.)

Kristal-Bonch et al. 1997) as well as plasma renin activity (Resnick et al. 1986; Burgess et al. 1990) in both patients with elevated and normal blood pressures. Further studies have demonstrated that treatment with vitamin D (alphacalcidol) was effective in lowering systolic blood pressure in hypertensive patients with impaired glucose tolerance, in comparison to placebo (Lind et al. 1988) and 1,25D administration was also shown to result in decreased plasma renin activity, angiotensin II levels, and regresses myocardial hypertrophy (Park et al. 1999). This cumulative data suggested vitamin D suppresses plasma renin activity and blood pressure, and thereby may protect CV health through RAS modulation (Levin and Li 2005).

The mechanism via which vitamin D affects the renin–angiotensin system has been investigated utilizing genetic animal models (Li et al. 2002, 2003). Results from these studies demonstrate that vitamin D inhibits renin activity through a mechanism involving the VDR receptor (Li et al. 2002, 2003). In VDR knockout mice as well as in 25-vitamin D-1-alpha-hydroxylase knockout mice, there was an increase in renal renin production with subsequent elevations in blood pressure and cardiac hypertrophy (Li and Glass 2002; Xiang et al. 2005; Kong and Li 2003). Renin levels were also increased in normal mice made vitamin D deficient; however, when these mice were treated with activated vitamin D, renin levels were subsequently suppressed (Li and Glass 2002). Cumulative data from these studies suggests that activated vitamin D, via the VDR receptor, suppresses renin gene expression in a manner exclusive of any pathways involving calcium metabolism (Levin and Li 2005). As a result, this vitamin is emerging as an independent negative endocrine

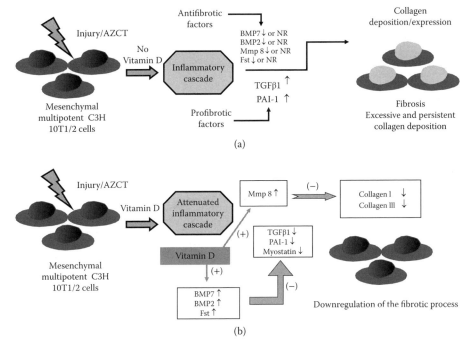

**FIGURE 11.7** Vitamin D reverts the fibrotic process induced by 5′-azacytidine (AZCT) in mesenchymal multipotent cells. (a) In the absence of vitamin D, injury activation of the inflammatory cascade can lead to a fibrotic process (progressive scarring). (b) 1,25(OH)D induces a VDR-mediated antifibrotic signaling phenotype in multipotent mesenchymal cells characterized by decreased expression of profibrotic markers (TGF-β1, plasminogen activator inhibitor 1 [PAI-1], myostatin) and increased expression of select antifibrotic markers (bone morphogenic protein 2 [BMP2] and 7 [BMP7], follistatin, MMP8), leading to reduced collagen expression, the ultimate marker of a fibrotic condition. (Adapted from Artaza, J. N., R. Mehrotra, and K. C. Norris. 2009. *Clin J Am Soc Nephrol* 4:1515–22.)

regulator of the RAS (Figure 11.8) and may provide yet another mechanism whereby therapeutic intervention may be applied to control blood pressure and improve CVD via renin inhibition (Levin and Li 2005).

## FUTURE DIRECTIONS

In view of the increasing evidence of the genomic mediated anti-inflammatory and antifibrotic actions of 1,25D as well as the ubiquity of the VDR, 1,25D or VDRA could be promising interventions to prevent and treat CVD and/or select CVD risk factors. Elucidation of the molecular mechanism underlying 1,25D related signaling factors could help identify new biomarkers for increased disease specific risk in vitamin D-deficient persons as well as targeting treatments.

## CONCLUSION

Hypovitaminosis D has emerged as an independent risk factor for all-cause and CV mortality, reinforcing its importance as a public health problem. There is a need to advance our understanding of the biologic pathways through which vitamin D affects CV health. While prospective clinical interventions to define precisely the cardioprotective effects of nutritional vitamin D repletion are still needed, the recognition of genomic mediated affects of 1,25D and VDRA within the CV system has opened new doors for strategic nutritional interventions to treat CVD.

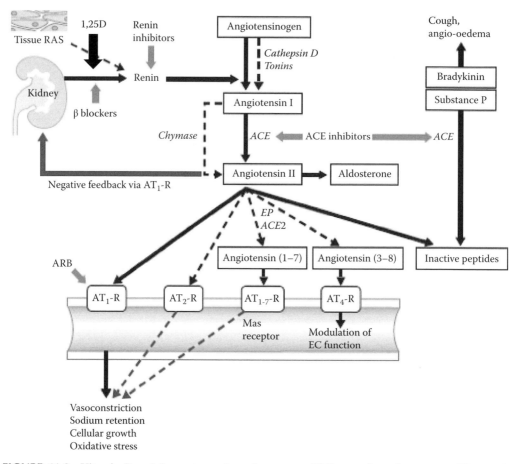

**FIGURE 11.8** Vitamin D and the renin–angiotensin system. AT-R = angiotensin receptor; EP = endopeptidases; EC = endothelial cells. (Modified from Staessen, J. A., Y. Li, and T. Richart. 2006. *Lancet* 368(9545):1449–56.)

## REFERENCES

Aihara, K., H. Azuma, M. Akaike, Y. Ikeda, M. Yaashita, T. Sudo, H. Hayashi, Y. Yamada, F. Endoh, M. Fujimura, T. Yoshida, H. Yamaguchi, S. Hashizume, M. Kato, K. Yoshimura, Y. Yamamoto, S. Kato, and T. Matsumoto. 2004. Disruption of nuclear vitamin D receptor gene causes enhanced thrombogenicity in mice. *J Biol Chem* 279:35798–802.

Andress, D. L. 2006. Vitamin D in chronic kidney disease: A systemic role for selective vitamin D receptor activation. *Kidney Int* 69:33–43.

Artaza, J. N., R. Singh, M. G. Ferrini, M. Braga, J. Tsao, and N. F. Gonzalez-Cadavid. 2008. Myostatin promotes a fibrotic phenotypic switch in multipotent C3H 10T1/2 cells without affecting their differentiation into myofibroblasts. *J Endocrinol* 196(2):235–49.

Artaza, J. N., and K. C. Norris. 2009. Vitamin D reduces the expression of collagen and key profibrotic factors by inducing an antifibrotic phenotype in mesenchymal multipotent cells. *J Endocrinol* 200(2):207–21.

Artaza, J. N., R. Mehrotra, and K. C. Norris. 2009. Vitamin D and the cardiovascular system. *Clin J Am Soc Nephrol* 4:1515–22.

Artaza, J. N., F. Sirad, M. G. Ferrini, and K. C. Norris. 2010. 1,25-(OH)$_2$Vitamin D$_3$ inhibits cell proliferation by promoting cell cycle arrest without inducing apoptosis and modifies cell morphology of mesenchymal multipotent cells. *J Steroid Biochem Mol Biol* 119(1–2):73–83.

Ben Nasr, L., J. D. Monet, and P. A. Lucas. 1988. Rapid (10-minute) stimulation of rat duodenal alkaline phosphatise activity by 1, 25-dihydroxyvitamin D3. *Endocrinology* 23:1778–82.

Bouillon, R., G. Carmeliet, L. Verlinden, E. van Etten, A. Verstuyf, H. F. Luderer, L. Lieben, C. Mathieu, and M. Demay. 2008. Vitamin D and human health: Lessons from vitamin D receptor null mice. *Endocr Rev* 29:726–76.

Bouillon, R., W. H. Okamura, and A. W. Normal. 1995. Structure-function relationships in the vitamin D endocrine system. *Endoc Rev* 16:200–57.

Burgess, E. D., R. G. Hawkins, and M. Watanabe. 1990. Interaction of 1, 25-dihydroxyvitamin D and plasma renin activity in high renin essential hypertension. *Am J Hypertens* 3:903–5.

Carlberg, C., M. Quack, M. Herdick, Y. Bury, P. Polly, and A. Toell. 2001. Central role of VDR conformations for understanding selective actions of vitamin D(3) analogues. *Steroids* 66(3–5):213–21.

Carlberg, C. 2003. Current understanding of the function of the nuclear vitamin D receptor in response to its natural and synthetic ligands. *Recent Results Cancer Res* 164:29–42.

Carthy, E. P., W. Yamashita, A. Hsu, and B. S. Ooi. 1989. 1, 25-dihydroxyvitamin D3 and rat vascular smooth muscle cell growth. *Hypertension* 13:954–9.

Chapman, M. J. 2007. From pathophysiology to targeted therapy for atherothrombosis: A role for the combination of statin and aspirin therapy in secondary prevention. *Pharmacol Therapy* 113(1):184–96.

Dusso, A. S., and A. J. Brown. 1998. Mechanism of vitamin D action and its regulation. *Am J Kidney Dis* 32(2 Suppl 2):S13–24.

Dusso A. S., A. J. Brown, and E. Slatopolsky. 2005. Vitamin D. *Am J Physiol Renal Physiol* 289(1):F8–28.

Essers, J., A. Theil, C. Baldeyron, W. A. van Capellan, A. B. Houtsonuller, R. Kanaar, and W. Vermeulen. 2005. Nuclear dynamics of PCNA in DNA replication and repair. *Mol Cell Biol* 25(21):9350–9.

Falkenstein, E., H. C. Tillman, and M. Wehling. 2000. Multiple actions of steroid hormones—a focus on rapid non-genomic effects. *Pharmacol Rev* 52(4):513–56.

Food and Nutrition Board. 1997. *Dietary Reference Intakes for Calcium, Phosphorus, Magnesium, Vitamin D, and Fluoride*. Washington, DC: National Academy Press.

Forman, J. P., E. Giovannucci, M. D. Holmes, H. A. Bischoff-Ferrari, S. S. Tworoger, W. C. Willett, and G. C. Curhan. 2007. Plasma 25-hydroxyvitamin D levels and risk of incident hypertension. *Hypertension* 49:1063–9.

Ginde, A. A., M. C. Liu, and C. A. Camargo Jr. 2009. Demographic differences and trends of vitamin D insufficiency in the U.S. population, 1988–2004. *Arch Intern Med* 169(6):626–32.

Giovannucci, E, Y. Liu, B. W. Hollis, and E. B. Rimm. 2008. 25-hydroxyvitamin D and risk of myocardial infarction in men. *Arch Intern Med* 168:1174–80.

Glass, C. K., and J. L. Witztum. 2001. Atherosclerosis. The road ahead. *Cell* 104:503–16.

Griffin, M. D., W. Lutz, V. A. Phan, L. A. Bachman, D. J. McKean, and R. Kumar. 2001. Dendritic cell modulation by 1alpha, 25 dihydroxyvitamin D3 and its analogs: A vitamin D receptor-dependent pathway that promotes a persistent state of immaturity *in vitro* and *in vivo*. *Proc Natl Acad Sci USA* 98:6800–5.

Haussler, M. R., G. K. Whitfield, C. A. Haussler, J. C. Hsieh, P. D. Thompson, S. H. Selznick, C. E. Dominguez, and P. W. Jurutka. 1998. The nuclear vitamin D receptor: Biological and molecular regulatory properties revealed. *J Bone Miner Res* 13(3):325–49.

Heaney, R. P. 2008. Vitamin D in health and disease. *Clin J Am Soc Nephrol* 3(5):1535–41.

Holick, M. F., and J. S. Adams. 1998. Vitamin D Metabolism and Biological Function. In L. Avioli and S. Krane, eds. *Metabolic Bone Disease and Clinically Related Disorders*. 3rd ed., pp. 124–56. San Diego, California: Academic Press.

Holick, M. F. 2006. High prevalence of vitamin D inadequacy and implication for health. *Mayo Clin Proc* 81:353–73.

Holick, M. F. 2007. Vitamin D deficiency. *N Engl J Med* 357:266–81.

Issa, L. L., G. M. Leong, R. L. Sutherland, and J. A. Eisman. 2002. Vitamin D analogue-specific recruitment of vitamin D receptor coactivators. *J Bone Miner Res* 17:879–90.

Jones, G. 2007. Expanding role for vitamin D in chronic kidney disease: Importance of blood 25-OH-D levels and extra-renal 1-alpha-hydroxylase in the classical and nonclassical actions of 1-alpha-dihydroxyvitamin D3. *Semin Dial* 20(4):316–24.

Karsenty, G., B. Lacour, A. Ulmann, E. Pierandrei, and T. Drueke. 1985. Early effects of vitamin D metabolites on phosphate fluxes in isolated rate enterocytes. *Am J Physiol* 248:G40–45.

Kong, J., and Y. C. Li. 2003. Effect of ANG II type I receptor antagonist and ACE inhibitor on vitamin D receptor-null mice. *Am J Physiol Regul Integr Comp Physiol* 285:R255–61.

Kristal-Bonch, E., P. Froom, G. Harari, and J. Ribak 1997. Association of calcitriol and blood pressure in normotensive men. *Hypertension* 30:1289–94.

Lee, J. H., J. H. O'Keefe, D. Bell, D. D. Hensrud, and M. F. Holick. 2008. Vitamin D deficiency an important, common, and easily treatable cardiovascular risk factor? *J Am Coll Cardiol* 52(24):1949–56.

Levin, A., and Y. C. Li. 2005. Vitamin D and its analogues: Do they protect against cardiovascular disease in patients with chronic kidney disease? *Kidney Int* 68(5):1973–81.

Li, A. C., and C. K. Glass. 2002. The macrophage foam cell as a target for therapeutic intervention. *Nat Med* 8:1235–42.

Li, Y. C. 2003. Vitamin D regulation of the renin–angiotensin system. *J Cell Biochem* 88:327–33.

Li, Y. C. 2010. Renoprotective effects of vitamin D analogs. *Kidney Int* 78(2):134–9

Li, Y. C., J. Kong, M. Wei, Z. F. Chen, S. Q. Liu, and L. P. Cao. 2002. 1,25-dihydroxyvitamin D(3) is a negative endocrine regulator of the renin–angiotensin system. *J Clin Invest* 110:229–38.

Libby, P. 2003. Vascular biology of atherosclerosis: Overview and state of the art. *Am J Card* 91(3A):3A–6A.

Lieberherr, M. 1987. Effects of vitamin D3 metabolites on cystosolic free calcium in confluent mouse osteoblasts. *J Biol Chem* Sep 25;262(27):13168–73.

Lind, L., A. Hanni, H. Lithell, A. Hvarfner, O. H. Sorensen, and S. Ljunghall. 1995. Vitamin D is related to blood pressure and other cardiovascular risk factors in middle-aged men. *Am J Hypertens* 8:894–901.

Lind, L., H. Lithell, E. Skarfors, L. Wide, and S. Ljunghall. 1988. Reduction of blood pressure by treatment with alphacalcidol: A double-blind, placebo-controlled study in subjects with impaired glucose tolerance. *Acta Med Scand* 223(3):211–17.

Marcinkowska, E., A. Wiedlocha, and C. Radzikowski. 1997. 1, 25-dihydroxyvitamin D3 induced activation and subsequent nuclear translocation of MAPK is upstream regulated by PKL in HL-60 cells. *Biochem Biophys Res Commun* 241:419–26.

Martins, D., M. Wolf, D. Pan, A. Zadshir, N. Tareen, R. Thadhani, A. Felsenfeld, B. Levine, R. Mehrotra, and K. C. Norris. 2007. Prevalence of cardiovascular risk factors and the serum levels of 25-hydroxyvitamin D in the united states: Data from the third national health and nutrition examination survey. *Arch Intern Med* 167:1159–65.

Mathew, S., R. J. Lund, L. R. Chaudhary, T. Geurs, and K. A. Hruska. 2008 Vitamin D receptor activators can protect against vascular calcification. *J Am Soc Nephrol* 19(8):1509–19.

Mathieu, C., and L. Adorini. 2002. The coming of age of 1, 25-dihydroxyvitamin D(3) analogs as immuno-modulatory agents. *Trends Mol Med* 8:174–79.

Mehrotra, R., D., Kermah, I. Salusky, M. Wolf, R. Thadhani, Y. W. Chiu, D. Martins, S. Adler, and K. C. Norris. 2009. Chronic kidney disease, hypovitaminosis D and mortality in the United States. *Kidney Int* 76(9):977–83.

Mehta, R. G., R. M. Moriarty, R. R. Mehta, R. Penmasta, G. Lazzaro, A. Constatinou, and L. Guo. 1997. Prevention of preneoplastic mammary lesion development by a novel vitamin D analogue, 1-alpha-hydroxy-vitamin D3. *J Natl Cancer Inst* 89:212–18.

Melamed, M. L., E. D. Michos, W. Post, and B. Astor. 2008. 25-hydroxyvitamin D levels and the risk of mortality in the general population. *Arch Intern Med* 168(15):1629–37.

Melamed, M. L., B. Astor, E. D. Michos, T. H. Hostetter, N. R. Powe, and P. Muntner. 2009. 25-hydroxyvitamin D levels, race, and the progression of kidney disease. *J Am Soc Nephrol* 20(12):2631–39.

Merke, J., W. Hofmann, D. Goldschmidt, and E. Ritz. 1987. Demonstration of 1, 25(OH)$_2$ vitamin D3 receptors and actions in vascular smooth muscle cells *in vitro*. *Calcif Tissue Int* 41:112–4.

Merke, J., P. Milde, S. Lewicka, U. Hugel, G. Klaus, D. J. Mangelsdorf, M. R. Haussler, E. W. Rauterberg, and E. Ritz. 1989. Identification and regulation of 1, 25-dihydroxyvitamin D3 receptor activity and biosynthesis of 1, 25-dihydroxyvitamin D3: Studies in cultured bovine aortic endothelial cells and human dermal capillaries. *J Clin Invest* 83:1903–15.

Mitsuhashi, T., R. C. Morris Jr, and H. E. Ives. 1991. 1, 25-dihydroxyvitamin D3 modulates growth of vascular smooth muscle cells. *J Clin Invest* 87:1889–95.

Miura, S., and K. Saku. 2010. Do angiotensin II type I receptor blockers have molecular effects? *Hypertens Rev* 33(2):105–6.

Norman, A. W., R. Bouillon, M. C. Farach-Carson, J. E. Bishop, L. X. Zhou, I. Nemere, J. Zhao, K. R. Muralidharan, and W. H. Okamura. 1998. Demonstration that 1-beta, 25-dihydroxyvitamin D3 is an antagonist of the nongenomic but not the genomic biological responses and biological profile of the three A-ring diastereomers of the 1alpha, 25-dihydroxyvitamin D3. *J Biol Chem* 268:20022–30.

Norman, A. W., W. H. Okamura, M. W. Hammond, J. E. Bishop, M. C. Dormanen, R. Bouillon, H. van Baelen, A. L. Ridall, E. Daane, R. Khoury, and M. C. Farach-Carson. 1997. Comparison of 6-s-cis and 6-s-trans-locked analogs of 1-alpha, 25-dihydroxyvitamin D3 indicates that the 6-s-cis conformation is preferred for rapid nongenomic biological responses and that neither 6-s-cis nor 6-s-trans-locked analogs are preferred for genomic biological responses. *Mol Endocrinol* 11:1518–31.

O'Connell, T. D., and R. U. Simpson. 1996. Immunochemical identification of the 1, 25-dihydroxyvitamin D3 receptor in the male rat. *Cell Biol Int* 20:621–24.

Ohsawa, M., T. Koyama, K. Yamamoto, S. Hirosawa, S. Kamei, and R. Kamiyama. 2000. 1alpha, 25-dihydroxyvitamin D(3) and its potent synthetic analogs downregulate tissue factor and upregulate thrombomodulin expression in monocytic cells, counteracting the effects of tumor necrosis factor and oxidized LDL. *Circulation* 102:2867–72.

Ovesen, L., C. Brot, and J. Jakobsen. 2003. Food contents and biological activity of 25-hydroxyvitamin D: A vitamin D metabolite to be reckoned with? *Ann Nutr Metab* 47(3–4):107–13.

Park, C. W., Y. S. Oh,Y. S. Shin, C. M. Kim, Y. S. Kim, S. Y. Kim, E. J. Choi, Y. S. Chang and B. K. Bang. 1999. Intravenous calcitriol regresses myocardial hypertrophy in hemodialysis patients with secondary hyperparathyroidism. *Am J Kidney Dis* 33:73–81.

Pearson, T. A., G. A. Mensah, R. W. Alexander, J. L. Anderson, R. O. Cannon 3rd, M. Criqui, Y. Y. Fadl, S. P. Fortmann, Y. Hong, G. L. Myers, N. Rifai, S. C. Smith Jr, K. Taubert, R. P. Tracy, and F. Vinicor. 2003. Markers of inflammation and cardiovascular disease: application to clinical and public health practice: A statement for healthcare professionals from the Centers for Disease Control and Prevention and the American Heart Association. *Circulation* 107(3):499–511.

Poole, K. E., N. Loveridge, P. J. Barker, D. J. Halsall, C. Rose, J. Reeve, and E. A. Warburton. 2006. Reduced vitamin D in acute stroke. *Stroke* 37:243–5.

Resnick, L. M., F. B. Muller, and J. H. Laragh. 1986. Calcium-regulating hormones in essential hypertension: Relation to plasma renin activity and sodium metabolism. *Ann Intern Med* 105:649–54.

Rigby, W. F., S. Denome, M. W. Fanger. 1987. Regulation of lymphokine production and human T lymphocyte activation by 1, 25-dihydroxyvitamin D3: Specific inhibition at the level of messenger RNA. *J Clin Invest* 79:1659–64.

Ross, R. 1999. Atherosclerosis—an inflammatory disease. *N Engl J Med* 340:115–26.

Rostand, S. G., and D. G. Warnock. 2008. Introduction to vitamin D symposium, March 14, 2008. *Clin J Am Soc Nephrol* 3(5):1534.

Sandgren, M. E., M. Bronnegard, and H. F. DeLuca. 1991. Tissue distribution of the 1,25-dihydrxyvitamin D3 receptor in the male rat. *Biochem Biophys Res Commun* 181:611–6.

Scragg, R., R. Jackson, I. Holdaway, T. Lim, and R. Beaglehole. 1990. Myocardial infarction is inversely associated with plasma 25-hydroxyvitamin D3 levels: A community-based study. *Int J Epidemiol* 19:559–63.

Somjen, D., Y. Weisman, F. Kohen, B. Gayer, R. Limor, O. Sharon, N. Jaccard, E. Knoll, and N. Stern. 2005. 25-Hydroxyvitamin D3-1alpha-hydroxylase is expressed in human vascular smooth muscle cells and is upregulated by parathyroid hormone and estrogenic compounds. *Circulation* 111:1666–71.

Song, X., J. E. Bishop, W. H. Okamura, and A. W. Norman. 1998. Stimulation of phosphorylation of mitogen-activated protein kinase by 1-alpha, 25-dihydroxyvitamin D3 I promyelocytic NB4 leukemia cells: a structure function study. *Endocrinology* 139:457–65.

Staessen, J. A., Y. Li, and T. Richart. 2006. Oral renin inhibitors. *Lancet* 368(9545):1449–56.

Stephen, M. B. 2009. Cardiac Rehabilitation. *Am Fam Physician* 80(9):955–9.

Sulyok, S., M. Wankell, C. Alzheimer, and S. Werner. 2004. Activin: An important regulator of wound repair, fibrosis, and neuroprotection. *Mol Cell Endocrinol* 225(1–2):127–32.

Takeyama, K., Y. Masuhiro, H. Fuse, H. Endoh, A. Murayama, S. Kitanaka, M. Suzawa, J. Yanagisawa, and S. Kato. 1999. Selective interaction of vitamin D receptor with transcriptional coactivators by a vitamin D analog. *Mol Cell Biol* 19:1049–55.

Timms, P. M., N. Mannan, G. A. Hitman, K. Noonan, P. G. Mills, D. Syndercombe-Court, E. Aganna, C. P. Price, and B. J. Boucher. 2002. Circulation MMP9, vitamin D and variation in the TIMP-1 response with VDR genotype: Mechanisms for inflammatory damage in chronic disorders? *QJM* 95:787–96.

Verstuyf, A., G. Carmeliet, R. Bouillon, and C. Mathieu. 2010. Vitamin D: A pleiotropic hormone. *Kidney Int* 78(2):140–5.

Wang, L., J. E. Mason, and H. D. Sesso. 2010. Systematic review: Vitamin D and calcium supplementation in prevention of cardiovascular events. *Ann Intern Med* 152(5):315–23.

Wang, S. 2009. Epidemiology of vitamin D in health and disease. *Nutr Res Rev* 2:188–203.

Wang, T. J., M. J. Pencina, S. L. Booth, P. F. Jacques, E. Ingelsson, K. Lainer, E. J. Benjamin, R. B. D'Agostino, M. Wolf, and R. S. Vasan. 2008. Vitamin D deficiency and risk of cardiovascular disease. *Circulation* 117:503–11.

Wu-Wong, J., R. M. Nakane, and J. Ma. 2006. Effects of vitamin D analogs on the expression of plasminogen activator inhibitor-1 in human vascular cells. *Thromb Res* 118(6):709–14.

Xiang, W., J. Kong, S. Chen, L. P. Cao, G. Qiao, W. Zheng, W. Liu, X. Li, D. G. Gardner, and Y. C. 2005. Cardiac hypertrophy in vitamin d receptor knockout mice: Role of the systemic and cardiac renin–angiotensin systems. *Am J Physiol Endocrinol Metab* 288(1):E125–32.

Xing, N., M.L. Maldonado, L. A. Bachman, D. J. McKean, R. Kumar, and M. D. Griffin. 2002. Distinctive dendritic cell modulation by vitamin D(3) and glucocorticoid pathways. *Biochem Biophys Res Commun* 297(3):645–52.

Zanello, L. P., and A. W. Norman. 1997. Stimulation by 1-alpha, 25(OH)2-vitamin D3 of whole cell chloride currents in osteoblastic ROS 17/2.8 cells. A structure-function study. *J Biol Chem* 272:22617–22.

Zehnder, D., R. Bland, M. C. Williams, R. W. McNinch, A. J. Howie, P. M. Stewart, and M. Hewison. 2001. Extrarenal expression of 25-hydroxyvitamin d(3)-1 alpha-hydroxylase. *J Clin Endocrinol Metab* 86(2):888–94.

Zhang, C. 2008. The role of inflammatory cytokines in endothelial dysfunction. *Basic Res Cardiol* 103(5):398–406.

Zhou, L. X., I. Nemere, and A. W. Norman. 1992. 1, 25-dihydroxyvitamin D3 analog structure-function assessment of the rapid stimulation of intestinal calcium absorption. *J Bone Miner Res* 7:457–63.

Zitterman, A., S. S. Schleithoff, and R. Koerfer. 2005. Putting cardiovascular disease and vitamin D insufficiency into perspective. *Br J Nutr* 94:483–92.

Zitterman, A., S. S. Schleithoff, G. Tenderich, H. K. Berthold, R. Korfer, and P. Stehle. 2003. Low vitamin D status: A contributing factor in the pathogenesis of congestive heart failure? *J Am Coll Cardiol* 41:105–12.

# Section II

## Gene-Linked Networks
### Gene-Linked Cancer Prevention

# 12 Network Analysis of Gene Expression Profiles in Breast Cancer Cell Lines

*T. Gregory Dewey, Katie L. Streicher,*
*and Stephen P. Ethier*

## CONTENTS

## INTRODUCTION

Systems biology has fundamentally altered the way cancer research is performed. In a few short years we have progressed from northern blot analysis of a handful of genes at a time, to probing the transcriptome, the proteome, and the metabolome of tumor cells. To date, microarray analyses of the transcriptome of breast and other cancer types have yielded novel insights that have direct relevance to our understanding of cancer progression. Gene expression analysis of primary breast cancer tissue has resulted in the definition of discrete cluster groups that have prognostic significance (Perou et al. 2000; Ross and Perou 2001; Sorlie et al. 2001; Goh et al. 2007). Specific genes have been identified that improved our understanding of both the important signaling pathways in breast cancer and the genetic basis for breast cancer progression (Bulavin et al. 2002; Hyman et al. 2002; Kauraniemi et al. 2001). To progress beyond these elegant yet descriptive analyses of the transcriptome, novel computational methods must be developed that integrates information on disease progression. Such approaches require an analysis of time series data for a more complete understanding of the system-level complexity of cancer.

In considering the full complexity of biological systems we must take into account not only the large number of interacting components, but also the complicated dynamics associated with these components. An ongoing problem in bioinformatics is to identify the relationships between the various components of a system and, specifically, how one component impacts the production of another. The molecular circuitry of gene regulation can reveal the dynamics of how one effector or agent influences the entire network. Knowledge of this circuitry implicitly allows the manipulation of gene expression and has far-reaching consequences for drug target identification. The goal of this chapter is to show how gene expression time series data can be represented in network form and how this data can be mined to develop and test hypotheses for the molecular circuitry of cancer progression.

## NETWORK MODELS AND A HYPOTHESIS ON TARGET IDENTIFICATION

Systems biology relies heavily on network analysis for representing and mining large data sets generated from high-throughput technologies. Typically, the networks generated from such analysis will consist of nodes that represent some biological component (e.g., a gene) a transcript or a protein. Links between the nodes called edges, represent some type of relationship. For instance, in a protein–protein map, the nodes are protein and the links represent a binding affinity between proteins. Most biological networks have unusual properties that differ significantly from random networks. Random networks are characterized by relative uniformity in the connectedness of individual nodes in the network. In a random network the connectivity of individual nodes follows a Poisson distribution, the peak of which is the mean connectivity per node and this value gives the network its characteristic scale. By contrast, many biological networks are scale free. Scale-free networks are characterized by the presence of a large number of nodes that have relatively few connections, and a small number of hubs, or nodes with very large numbers of connections or links. Rather than the Poisson distribution that describes the connectivity of random networks, the degree distribution of scale-free networks is described by a power-law function. Because there is no peak in the degree distribution of nodes in these networks, they are said to be scale free, which is an indication of their hub and node topology. Such highly nonuniform networks can be robust to outside perturbations (Barabasi and Oltavi 2004). Recent work has demonstrated that biological networks such as those that influence gene transcription, protein–protein interactions, and metabolism, also exhibit scale-free topology and thus can be described by the mathematics of scale-free networks (Jeong et al. 2000; Wagner and Fell 2001; Featherstone and Broadie 2002; Barabasi and Albert 1999).

There has been considerable interest in the how scale-free networks emerge, and while there are a number of mathematical models that generate scale-free networks (Barabasi and Albert 1999; Pastor-Satorras et al. 2003), it remains a challenge to understand the origins of such networks in biological systems. Nevertheless, a variety of molecular data establishes the existence of scale-free biological networks. The Ethier group has developed a working hypothesis regarding the nature of changes in network topology that occur during cancer progression. They propose that dominant genetic alterations that occur during cancer progression result in fundamental changes in network topology. Furthermore, they propose that gene amplifications can yield high levels of gene and protein expression, and can serve as the activating mechanism for many oncogenes in human cancer. The proposal is that these genes become hubs in the new cancer network. A corollary to this hypothesis is that from a network perspective, oncogenes are fundamentally different from proto-oncogenes. Whereas proto-oncogenes behave as nodes in their biological networks, Ethier proposes that oncogenes are transformed into hubs that are more highly interconnected than the proto-oncogenic form of the gene.

There are several predictions that can be made from this hypothesis. First, if oncogenes behave as hubs in a gene expression network while proto-oncogenes behave as less connected nodes, then it is possible to define oncogenes in computational terms from their network properties and not just biological terms. Since at the present time there are a relatively small number of truly validated human oncogenes and a large number of candidate oncogenes (many of which have no known function), it may be possible to understand better their oncogenic role by understanding their connectivity within a gene expression network. Second, if the hypothesis is correct, it raises the notion of cancer cell-specific hub genes. This is of potential importance because it is well known that scale-free networks are highly robust and resistant to random damage to nodes. However, it is also understood that hubs represent the points of greatest vulnerability of scale-free networks (Barabasi and Albert 1999). Thus, if cancer cell-specific hubs exist, they represent potential targets that can yield therapies with high efficacy (because of the hub nature of the target in cancers) and high therapeutic index (because the gene is a not a central hub in normal cells).

Using this hypothesis, we can generate a strategy for identifying new cancer targets. By challenging a tumor and normal cell with an external perturbation, we can examine the dynamics of the gene expression response. With an analysis of this time series data (presented in the following section), a network is determined. By comparing the networks of the normal and tumor cells, we identify candidates with an oncogenic role.

## NETWORKS FROM LINEAR MODELS

Dynamic linear models are ones in which the change in time of one of the variables is linearly related to the other variables. Here, we consider how the expression levels of a set of genes at a given time will influence the production of a given gene at a later time. The rate of production of the responding gene as measured by the mRNA level will be linearly proportional to the amount of mRNA of the influencing genes. If one considers the simple case of a transcription factor (TF) and its target genes, this model would predict that the rate of mRNA production of the target gene would be proportional to the mRNA concentration of the TF. In general, we do not know if the linear model is an accurate one and, indeed, we would anticipate that some genes would show very nonlinear responses with complex feedback loops.

Considering the complexity and inherent nonlinearity of biological phenomena, why should we even consider such linear models? First, linear models are appealing because they are simple and are computationally easy to handle. Even though they may not completely capture the underlying biological phenomena, they nevertheless are a good starting point. Before we can move to more complicated models, it is important to establish where simple models fail. This is especially true when dealing with limited data with large statistical errors, as found in most microarray studies. Secondly, the linear model may act as a first approximation to a more complicated non-linear expression. Any nonlinear function can be "linearized" through a power series expansion so a linear model can be considered the first approximation to a more complicated dependence. At this stage of sophistication and data quality, we are primarily seeking phenomenological connections, rather than quantitative mathematical connections. For such "data mining" goals, linear models can be extremely useful. Bearing in mind, the initial goal of understanding connections between genes rather than establishing a full-blown mathematical description of gene regulation, linear models provide an excellent starting point.

There are many variations on linear models and the choice of a specific model will often depend on the experimental conditions or experimental design (Chen et al. 1999; D'Haeseleer et al. 1999; Dewey and Galas 2001; Yeung et al. 2002; deHoon et al. 2003). In our work, we used a form that is perhaps the simplest one. It is a linear finite difference model and is described by Equation 12.1:

$$a_i(t) = \sum_{j=1}^{M} \lambda_{i,j} a_j(t-1) \tag{12.1}$$

In Equation 12.1, $a_i(t)$ is the experimentally measured mRNA concentration for the $i$th gene at time $t$. Equation 12.1 relates the mRNA concentration at time $t$ to the linear combination of all other mRNA concentrations at the previous time, $t - 1$. The transition coefficient $\lambda_{i,j}$ is the respective elements of the $M \times M$ transition matrix (referred to as the $\Lambda$ matrix) and is the model parameter. These coefficients are unitless and show how strongly weighted each contribution from the previous time will be to the production of the $i$th gene. This model is sometimes referred to as an autoregression model because all the variables at a later time are dependent only on the values of these same variables at an early time.

The goal of the data analysis is to determine the values for $\lambda_{i,j}$. In the following sections, we see how to calculate $\lambda_{i,j}$ and use these parameters to determine the gene expression network. Our data set consists of the measurement of $M$ gene expression levels or concentrations at $N$ different times. A single measurement is expressed as $a_i(t_k)$, indicating the mRNA level of the $i$th gene at time $t_k$

where the indices $i$ can range from 1 to $M$ and $k$ ranges from 1 to $N$. The data is ordered in a $M \times N$ matrix as follows:

$$
M \quad genes \downarrow
\overset{N \quad times \rightarrow}{
\begin{bmatrix}
a_1(t_1) & a_1(t_2) & \cdots & a_1(t_N) \\
a_2(t_1) & a_2(t_2) & \cdots & a_2(t_N) \\
\vdots & \vdots & \vdots & \vdots \\
a_M(t_1) & a_M(t_2) & \cdots & a_M(t_N)
\end{bmatrix}}
\tag{12.2}
$$

where each column is the measured values of all the genes at a given time point and each row is the value of a given gene at all time points. Our goal is to use this data set to calculate the $\lambda_{i,j}$ of the linear model in Equation 12.1. Rather than use Equation 12.1 directly, it is easier to write it as the equivalent matrix equation of the following form:

$$
\begin{bmatrix}
a_1(t_2) & a_1(t_3) & \cdots & a_1(t_N) \\
a_2(t_2) & a_2(t_3) & \cdots & a_2(t_N) \\
\vdots & \vdots & \vdots & \vdots \\
a_M(t_2) & a_M(t_2) & \cdots & a_M(t_N)
\end{bmatrix}
= \Lambda \cdot
\begin{bmatrix}
a_1(t_1) & a_1(t_2) & \cdots & a_1(t_{N-1}) \\
a_2(t_1) & a_2(t_2) & \cdots & a_2(t_{N-1}) \\
\vdots & \vdots & \vdots & \vdots \\
a_M(t_1) & a_M(t_2) & \cdots & a_M(t_{N-1})
\end{bmatrix}
\tag{12.3}
$$

where the data matrix on the left hand side of Equation 12.3 has the column with the first time point removed and is an $M \times (N-1)$ matrix called the lead matrix. The $M \times (N-1)$ matrix on the right hand side has the column with the last time point removed and is called the lag matrix. The matrix $\Lambda$ is an $M \times M$ containing the elements $\lambda_{i,j}$ as in Equation 12.1. The $\Lambda$ matrix is called transition matrix because it is used to calculate the transition from the previous time to the next time point. It can also be said to "propagate" the data matrix from the past to the present.

Equation 12.3 can be written more succinctly as

$$
\mathbf{A}(t) = \Lambda \cdot \mathbf{A}(t-1)
\tag{12.4}
$$

where $\mathbf{A}(t)$ is the time lead matrix and $\mathbf{A}(t-1)$ is the time lag matrix. The $\Lambda$ matrix can be solved by inverting $\mathbf{A}(t-1)$ and multiplying both sides of Equation 12.4 by the inverse:

$$
\Lambda = \mathbf{A}(t) \cdot \mathbf{A}(t-1)^{-1}
\tag{12.5}
$$

Computationally, all that is needed to calculate the $\Lambda$ matrix is a matrix inversion routine, a common tool in any matrix algebra package of software. However, standard methods cannot be used on this problem for two main reasons. First, most methods invert square matrices. In our case, we have a very lopsided matrix with $M$ being on the order of thousands of genes and $N$ being of the order of 10. A second problem is that $\mathbf{A}(t-1)$ is potentially a singular matrix. In any given experiment it is conceivable that two genes have identical profiles as a result of being under identical gene control (Prieto et al. 2008). The consequence of this is that two of the rows in the data matrix could be identical. Matrices with identical columns or rows are called singular and their inverse is undefined. To solve Equation 12.5, methods described as "generalized matrix inversion" techniques must be used (Dewey and Galas 2001).

We have used a matrix method known as singular value decomposition (SVD) to calculate the matrix inverse in Equation 12.5 (Dewey and Galas 2001). This is mathematically the same method used in the well-known tool of principal component analysis (PCA) used in multivariate statistics. The SVD method is derived from a theorem in matrix algebra that states that any matrix can be "decomposed" into the product of three matrices. This gives

$$
\mathbf{A}(t-1) = \mathbf{U}\mathbf{E}\mathbf{V}^{\mathrm{T}}
\tag{12.6}
$$

where $\mathbf{U}$ is an $M \times M$ orthogonal matrix of singular vectors, $\mathbf{E}$ is an $M \times (N-1)$ diagonal matrix of eigenvalues, and $\mathbf{V}$ is an $(N-1) \times (N-1)$ orthogonal matrix of singular vectors. The singular vector matrices create an abstract vector space in which the original data matrix is represented. There is often a temptation to ascribe some biological significance to the vectors derived from SVD of a data matrix. This is difficult because SVD is a mathematical device not a scientific theory and there is no a priori reason that the matrices should have significance in biological terms. An SVD can be performed on any matrix regardless of its origin or the type of data that it represents. However, one can get a very generic interpretation of the singular vectors in terms of the eigenvectors of the correlation matrices. To calculate correlations between genes averaged over the time series, one uses $\mathbf{U}$ and $\mathbf{E}$. Correlations in time points averaged over all genes can be calculated using $\mathbf{V}$ and $\mathbf{E}$. Thus, the eigenvectors in $\mathbf{U}$ capture gene correlations and the eigenvectors in $\mathbf{V}$ capture time correlations.

An SVD on the data matrix, $\mathbf{A}(t-1)$, is performed using a routine in MatLab giving $\mathbf{A}(t-1) = \mathbf{UEV}^{\mathrm{T}}$. The inverse of $\mathbf{A}(t-1)$ is now given by

$$\mathbf{A}(t-1)^{-1} = \mathbf{VE}^{-1}\mathbf{U}^{\mathrm{T}} \tag{12.7}$$

where the properties of orthogonal matrices have been used. SVD provides a way of handling the redundant data in singular matrices. When a singularity occurs, one of the eigenvalues in $\mathbf{E}$ will be zero. When taking the inverse of $\mathbf{E}$ in Equation 12.7, one truncates the eigenvalues to include only nonzero eigenvalues in $\mathbf{E}$ so that the inverse of zero is never needed. That is, all zeros in $\mathbf{E}$ are set to zero in $\mathbf{E}^{-1}$. Because of experimental error in the measurements, two rows are never exactly the same. Consequently, $\mathbf{E}$ contains very small eigenvalues that are not exactly zero. Typically, to remove the singularity, one still truncates $\mathbf{E}$ to eliminate the small eigenvalues. This procedure acts as a filter and removes noise from the data set.

Once Equation 12.5 is calculated, phenomenological networks of gene interactions can be derived from the transition matrix. The $\Lambda$ transition matrix can be viewed as a weighted graph showing the influence of one expression level on another. This is the starting point for the description of the genetic circuitry. Rather than work with these weighted graphs, we consider a simpler approach in which $\Lambda$ is converted into an adjacency matrix for digraphs, indicating the connectivity but not the strengths of the influence. We describe the operation (adj) as

$$\Gamma(\varepsilon) = adj(\Lambda) \tag{12.8}$$

where the entries in $\Lambda$ are set equal to 1 if the absolute values are above a certain threshold, $\varepsilon$, and are set equal to 0 below this threshold. For high values of the threshold, the resulting $\Gamma(\varepsilon)$ matrix will be a sparse adjacency matrix with a small network. As the value of $\varepsilon$ is lowered, we can "grow" the network to include more nodes (genes). This threshold parameter is an adjustable parameter of the model.

## EXPERIMENTAL RESULTS AND ANALYSIS

In collaboration with the Ethier group, we analyzed experiments that examined the gene expression network regulated by the epidermal growth factor receptor (EGFR) in the normal human mammary epithelial cell line MCF-10A, and in the human breast cancer cell line, SUM-149. We chose these two cell lines because of their absolute dependence on EGFR signaling for growth *in vitro* (Ethier et al. 1990, 1991; Rao et al. 2000). To identify the gene expression networks regulated by EGFR in these two cell lines, MCF-10A and SUM-149 cells were cultured to approximately 75% confluence and exposed to the EGFR inhibitor CI-1033. RNA was isolated from cells at 4 hours, 8 hours, 12 hours, 16 hours, 24 hours, and 48 hours after addition of drug, and analyzed using Affimetrix U-133a microarrays. To generate gene expression networks from the microarray data, we used the simple linear model of Equation 12.4 to analyze the time response of the system following the perturbation.

The graphs in Figure 12.1 depict small versions of the networks obtained from the two cell lines using this approach. Inspection of the genes that make up the networks reveals several important features that provide valuable insight into the biological function of the gene or protein that is targeted in the experiments. To analyze the biological functions associated with the EGFR networks in the two cell lines, we identified the gene sets that scored as hub genes in the MCF-10 network or the SUM-149 network, and the genes that were common hubs in the two networks. Not surprisingly, the genes that were identified as common hub genes in both networks are responsible for common cellular functionalities such as cell cycle progression, cell signaling, DNA repair, and DNA (metabolism). Interestingly, over 100 hub genes were identified in the MCF-10A network that were not part of the SUM-149 network, and these genes were also found to play predominant roles in cell signaling, cell growth, and cell cycle progression. The fact that these genes were lost in the SUM-149 cell EGFR-network is suggestive of fragmentation of the networks that regulate cell growth in breast cancer cells. We believe that network fragmentation will be a common feature in cancer cells, and is part of how these cells acquire autonomy of growth factors for proliferation and survival. Most

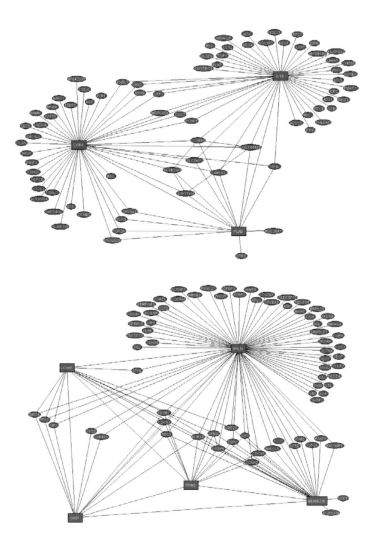

**FIGURE 12.1**  Network graphs based on gene expression time series. The graphs show differences in hubs between tumor cell line, SUM-149 (top), and a cell line from normal breast cells, MCF-10A (bottom).

interesting however, were the genes that were identified as hubs in the SUM-149 network, but which were not part of the MCF-10 network. These genes reflect the biological nature of the SUM-149 cells, which were isolated from a patient with aggressive, inflammatory breast cancer. The genes in this list include IL-1α, IL-1β, JAG1, and c20orf9. The interleukins and their receptors are known to play an important role in inflammatory breast cancer (Woodworth et al. 1995; Charaffe-Jauffret et al. 2004) and our results link these genes to EGFR signaling. The JAG1 gene is an important regulator of Notch signaling, also important in breast cancer biology, and the c20orf9 gene is a known NF-κB induced gene, and NF-κB is known to be highly active in the SUM-149 cells (Kleer et al. 2000; van Golen et al. 2000). Our network analysis identified several key features of SUM-149 cell biology and linked many of the most important genes driving the inflammatory phenotype to EGFR signaling.

A follow up annotation analysis of the gene expression networks provided novel and important biological information that provided stimulus for a new round of experiments to elucidate the inter-connectedness of biological activity of signaling molecules and oncogenes in human breast cancer. In a subsequent study, the mechanisms by which amphiregulin activation of the EGFR regulates these altered phenotypes of tumor cell lines were investigated by Streicher et al. (2007). Using the earlier network results that implicated interleukin-1α (IL-1α) and IL-1β as key mediators of amphiregulin's biological effects, new experiments were designed that showed that amphiregulin, but not epidermal growth factor, results in transcriptional upregulation of IL-1α and IL-1β. The details of these follow on results are described by the schematic of Figure 12.2. Figure 12.2 is constructed from individual experimental results and not from the gene expression analysis. Both IL-1α and IL-1β are synthesized and secreted by SUM-149 breast cancer cells, as well as MCF10A cells engineered to express amphiregulin or MCF10A cells cultured in the presence of amphiregulin. Furthermore,

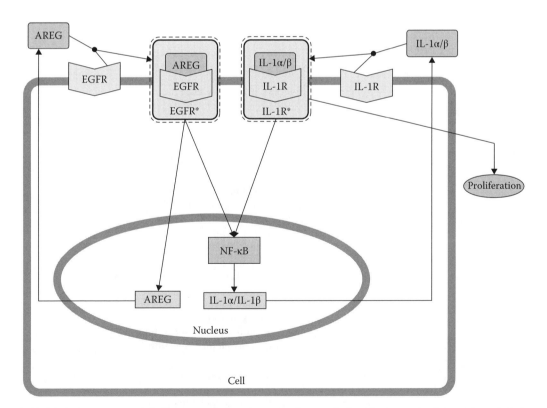

**FIGURE 12.2** Inferred signaling network based on experiments designed as a result of network analysis in Figure 12.1. New experiments lead to a model of the alteration of EGFR downstream signaling due to amphiregulin (AR). (From Streicher, K. L. et al. 2007. *Mol Cancer Res* 5:847–860. With permission.)

EGFR, activated by amphiregulin but not epidermal growth factor, results in the prompt activation of the TF nuclear factor–κB (NF-κB), which is required for transcriptional activation of IL-1. Once synthesized and secreted from the cells, IL-1 further activates NF-κB, and inhibition of IL-1 with the IL-1 receptor antagonist results in loss of NF-κB DNA binding activity and inhibition of cell proliferation. However, SUM-149 cells can proliferate in the presence of IL-1 when EGFR activity is inhibited. Thus, in aggressive breast cancer cells, such as the SUM-149 cells, or in normal human mammary epithelial cells growing in the presence of amphiregulin, EGFR signaling is integrated with NF-κB activation and IL-1 synthesis, which cooperate to regulate the growth and invasive capacity of the cells.

## CONCLUDING REMARKS

This chapter illustrates the power of network analysis to drug target identification in breast cancer. The importance of time series data is underscored as a means of determining crucial signaling and regulatory elements of a temporal response to an external perturbation. In our approach, we measure changes in the gene transcription profile of a normal and disease model system in response to a common perturbation (the pharmacological blocking of a signaling pathway). Changes over time in the transcription profiles are analyzed with a linear response model that can be used to compute a network of "influence genes." By comparing the resulting networks between tumor and healthy cell lines, we can identify new hubs that arise during cancer progression. This data mining device is used to generate new hypotheses and to influence the design of additional experiments. This interplay between modeling and experimental design is important and should be viewed as an iterative process. This method is not restricted to drug target identification but is readily adapted to problems in functional nutrigenomics.

## ACKNOWLEDGMENTS

The assistance and collaboration of Dr. Katie Streicher and Professor Stephen Ethier of the Karmanos Cancer Center at Wayne State University is acknowledged.

## REFERENCES

Barabasi, A. L., and R. Albert. 1999. Emergence of scaling in random networks. *Science* 286:509–512.
Barabasi, A. L., and Z. N. Oltvai. 2004. Network biology: Understanding the cell's functional organization. *Nature Reviews* 5:101–112.
Bulavin, D. V., O. N. Demidov, S. Saito, P. Kauraniemi, C. Phillips, S. A. Amundson, C. Ambrosino, G. Sauter, A. R. Nebreda, C. W. Anderson, A. Kallioniemi, A. J. Fornace, and E. Appella. 2002. Amplification of PPM1D in human tumors abrogates p53 tumor-suppressor activity. *Nat Genet* 31:210–215.
Charafe-Jauffret, E., C. Tarpin, V. J. Bardou, F. Bertucci, C. Ginestier, A. C. Braud, B. Puig, J. Geneix, J. Hassoun, D. Birnbaum, J. Jacquemier, and P. Viens. 2004. Immunophenotypic analysis of inflammatory breast cancers: Identification of an "inflammatory signature." *J Pathol* 202:265–273.
Chen, T., H. L. He, and G. M. Church. 1999. Modeling gene expression with differential equations. *Pac Symp Biocomput* 4:29–40.
D'Haeseleer, P., X. Wen, S. Fuhrman, and R. Somogyi. 1999. Linear modeling of mRNA expression levels during CNS development and injury. *Pac Symp Biocomput* 4:41–52.
de Hoon, M. J., S. Imoto, K. Kobayashi, N. Ogasawara, and S. Miyano. 2003. Inferring gene regulatory networks from time-ordered gene expression data of *Bacillus subtilis* using differential equations. *Pac Symp Biocomput* 8:17–28.
Dewey, T. G., and D. J. Galas. 2001. Dynamic models of gene expression and classification. *Funct Integr Genomics* 1:269–278.
Ethier, S. P., R. Moorthy, and C. A. Dilts. 1991. Secretion of an epidermal growth factor-like growth factor by epidermal growth factor-independent rat mammary carcinoma cells. *Cell Growth Differ* 2:593–602.

Ethier, S. P., R. M. Summerfelt, K. C. Cundiff, and B. B. Asch. 1990. The influence of growth factors on the proliferative potential of normal and primary breast cancer-derived human breast epithelial cells. *Breast Cancer Res Treat* 17:221–230.

Featherstone, D. E., and K. Broadie. 2002. Wrestling with pleiotropy: Genomic and topological analysis of the yeast gene expression network. *Bioessays* 24:267–274.

Goh, K., M. E. Cusick, D. Valle, B. Childs, M. Marc Vidal, and A. L. Barabási. 2007. The human disease network. *Proc Natl Acad Sci USA* 104:8685–8690.

Hyman, E., P. Kauraniemi, S. Hautaniemi, M. Wolf, S. Mousses, E. Rozenblum, E., Ringnér, G. Sauter, O. Monni, A. Elkahloun, O. P. Kallioniemi, and A. Kallioniemi. 2002. Impact of DNA amplification on gene expression patterns in breast cancer. *Cancer Res* 62:6240–6245.

Jeong, H., B. Tombor, R. Albert, Z. N. Oltvai, and A. L. Barabasi. 2000. The large-scale organization of metabolic networks. *Nature* 407:651–654.

Kauraniemi, P., M. Barlund, O. Monni, and A. Kallioniemi. 2001. New amplified and highly expressed genes discovered in the ERBB2 amplicon in breast cancer by cDNA. *Microarrays. Cancer Res* 61:8235–8240.

Kleer, C. G., K. L. van Golen, and S. D. Merajver. 2000. Molecular biology of breast cancer metastasis. Inflammatory breast cancer: Clinical syndrome and molecular determinants. *Breast Cancer Res* 2:423–429.

Pastor-Satorras, R., E. Smith, and R. V. Sole. Evolving protein interaction networks through gene duplication. *J Theor Biol* 222:199–210.

Perou, C. M., T. Sorlie, M. B. Elsen, M. van de Rijn, S. S. Jeffrey, C. A. Rees, J. R. Pollack, D. T. Ross, H. Johnsen, L. A. Akslen, O. Fluge, A. Pergamenschikov, C. Williams, S. X. Zhu, P. E. Lonning, A. I. Borresen-Dale, P. O. Brown, and D. Botstein. 2000. Molecular portraits of human breast tumours. *Nature* 406:747–752.

Prieto, C., A. Risueño, C. Fontanillo, and J. De Las Rivas. 2008. Human gene coexpression landscape: Confident network derived from tissue transcriptomic profiles. *PloS One* 3:e3911.

Rao, G. S., S. Murray, and S. P. Ethier. 2000. Radiosensitization of human breast cancer cells by a novel ErbB family receptor tyrosine kinase inhibitor. *Int J Radiat Oncol Biol Phys* 48:1519–1528.

Ross, D. T., and C. A. Perou. 2001. A comparison of gene expression signatures from breast tumors and breast tissue derived cell lines. *Dis Markers* 17:99–109.

Sorlie, T., C. M. Perou, R. Tibshirani, T. Aas, S. Geisler, H. Johnsen, T. Hastie, M. B. Eisen, M. van de Rijn, S. S. Jeffrey, T. Thorsen, H. Quist, J. C. Matese, P. O. Brown, D. Botstein, P. E. Lonning, and A. L. Borresen-Dale. 2001. Gene expression patterns of breast carcinomas distinguish tumor subclasses with clinical implications. *Proc Natl Acad Sci USA* 98:10869–10874.

Streicher, K. L., N. E. Willmarth, J. Garcia, J. L. Boerner, T. G. Dewey, and S. P. Ethier. 2007 Activation of a nuclear factor KB/interleukin-1 positive feedback loop by amphiregulin in human breast cancer cells. *Mol Cancer Res* 5:847–860.

van Golen, K. L., Z. F. Wu, X. T. Qiao, L. Bao, and S. D. Merajver. 2000. RhoC GTPase overexpression modulates induction of angiogenic factors in breast cells. *Neoplasia* 2:418–425.

Wagner, A., and D. A. Fell. 2001. The small world inside large metabolic networks. *Proc R Soc Lond B Biol Sci* 268:1803–1810.

Woodworth, C. D., E. Mcmullin, M. Iglesias, and G. D. Plowman. 1995. Interleukin 1 alpha and tumor necrosis factor alpha stimulate autocrine amphiregulin expression and proliferation of human papillomavirus-immortalized and carcinoma-derived cervical epithelial cells. *Proc Natl Acad Sci USA* 92:2840–2844.

Yeung, M. K., J. Tegner, and J. J. Collins. 2002. Reverse engineering gene networks using singular value decomposition and robust regression. *Proc Natl Acad Sci USA* 99:6163–6168.

# 13 Effects of Dietary Effectors on Signal Transduction Pathways Related to Cancer Prevention

*Ann M. Bode and Zigang Dong*

## CONTENTS

## INTRODUCTION

A major focus of much of our work has been the elucidation of molecular and cellular mechanisms and the identification of gene and protein targets in cancer development and prevention. A significant outcome of these investigations has been a clarification of signal transduction pathways induced by tumor promoters in cancer development. Signal transduction is the process by which information from a stimulus outside the cell is transmitted through the cell membrane (e.g., through a protein receptor) into the cell and along an intracellular chain of signaling proteins (e.g., protein kinases) to transcription factors, causing a cellular response (Figure 13.1). A transcription factor is comprised of one or more proteins that bind to a specific DNA sequence and act to initiate transcription that results in production of a specific protein product. DNA binding is commonly induced by protein kinases. Transcription factors acting downstream of these signaling cascades are heavily involved in numerous cellular functions that include survival, development, proliferation, apoptosis, cell cycle regulation, inflammation, and differentiation (Figure 13.1).

Protein kinase signaling, resulting in the induction of transcriptional activation of specific transcription factors, plays a critical role in carcinogenesis. Protein kinases are key regulators of

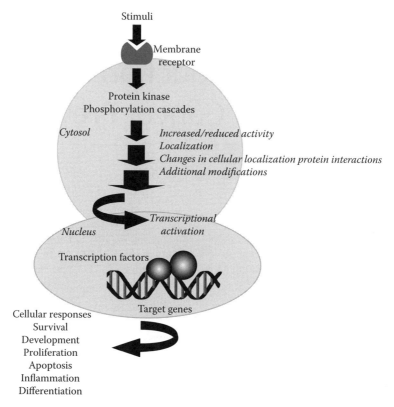

**FIGURE 13.1** Signal transduction. Extracellular stimuli are received by the cell through membrane receptors that activate several protein kinase phosphorylation cascades. Phosphorylation of protein kinases can cause a change in the kinases activity, cellular location, protein interactions, or other modifications. Protein kinases are commonly translocated to the nucleus where they activate specific transcription factors resulting in the transcription of their target genes. The net result is a variety of cellular responses that can include changes in survival, development, proliferation, apoptosis, inflammation, or differentiation.

a variety of cellular functions and comprise one of the largest and most functionally diverse gene families. Kinases are enzymes that chemically add one or more phosphate groups to other proteins (i.e., substrates), a process referred to as *phosphorylation*. Phosphorylation causes a change in the function of a target protein substrate that can include enhanced or reduced activity, alternations in cellular location, changes in associations with other proteins, and modifications of overall function (Figure 13.1). Kinases phosphorylate both serine and threonine residues and some act on tyrosine, whereas a few have dual specificity and act on all three residues. The human genome comprises about 500 protein kinase genes or about 2% of all genes; and the kinase proteins modify about 30% of all human proteins (Blume-Jensen and Hunter 2001; Manning et al. 2002; Venter et al. 2001). Protein kinases orchestrate the activity of almost every cellular process and, in particular, signal transduction, and are therefore tightly regulated. Protein kinases are themselves activated or deactivated by phosphorylation, binding of activator or inhibitory proteins, small molecules, or by localization in the cell. Aberrant regulation of kinase activity and associated transcriptional activation is closely associated with carcinogenesis and thus these molecules are suggested as ideal targets for cancer prevention and/ or therapy (Arora and Scholar 2005; Johnson 2009; Lawrence and Niu 1998; Smyth and Collins 2009).

An important group of protein kinases are the mitogen-activated protein (MAP) kinases, which are activated by translocation to the nucleus, where they phosphorylate a large number of various target transcription factors that are critical in tumor development. These transcription factors include activator protein-1 (AP-1) and nuclear factor-kappaB (NF-κB) (Angel et al. 1988; Davis 1994; Kallunki et al. 1994; Sanchez et al. 1994), p53 (Bode and Dong 2004e), nuclear factor of activated T-cells (NFAT) (Lu and Huan 2007; Medyouf and Ghysdael 2008), and cAMP response element-binding (CREB) protein (Siu and Jin 2007). The MAP kinases include the extracellular signal-regulated protein kinases (ERKs), c-Jun N-terminal kinases/stress-activated protein kinases (JNKs/SAPKs) and the p38 kinases. ERKs generally transmit signals initiated by tumor promoters such as 12-*O*-tetradecanoylphorbol-13-acetate (TPA), epidermal growth factor (EGF), and platelet-derived growth factor (PDGF) (Cowley et al. 1994). The JNKs/SAPKs and p38 kinases are strongly stimulated by stresses such as UV irradiation (Kallunki et al. 1994) and arsenic (Huang et al. 1999). The activation of these signaling cascades can result in a multitude of cellular responses (Figure 13.1).

The molecular mechanisms explaining how normal cells undergo transformation to cancer cells induced by tumor promoters are rapidly being identified and refined. A prevailing idea today is that cancer may be prevented or treated by targeting specific cancer genes, signaling proteins, and transcription factors with individual or combinations of small molecule inhibitors and especially natural or dietary factors. Many of these compounds are believed to have potent anticancer activity, low toxicity, and very few adverse side effects (reviewed in Bode and Dong 2004a,b,c,d, 2005, 2006, 2007, 2009a,b,c). In addition to being professed as generally safe, they appear to have efficacy as anticancer agents by preventing or reversing premalignant lesions and/or reducing second primary tumor incidence (Hong 2003). Because the process of cancer development is usually long, identifying the specific signal transduction pathways, gene, protein, and transcription factor targets, and mechanisms explaining the purported anticancer activity of specific dietary factors might provide effective alternatives or additions to traditional methods of cancer prevention (i.e., chemoprevention) or cancer treatment (i.e., chemotherapy).

The idea of combining agents or using individual agents that target multiple pathways is rapidly gaining acceptance. A developing belief is that multiple pathways must be targeted for a limited period of time and different cancers might require a customized combination of prevention strategies to be successful (Steele 2003; Steele and Kelloff 2005). In fact, combinations of drugs could provide multiple inhibitory mechanisms that might increase efficacy synergistically because single agents will not likely block heterogeneous cancers as has clearly been shown in clinical trials (Aggarwal et al. 2007; Aggarwal and Shishodia 2006). In addition, combinations of agents, especially natural compounds, will likely require a lower dose of each compound thereby leading to lower toxicity and fewer adverse side effects. However, powerful technologies are critically needed to accelerate the process of drug discovery especially to find compounds that can suppress multiple cellular signaling pathways. Combining supercomputer technology with protein crystal structure determination and chemical library screening to identify the multiple protein targets of natural anticancer compounds is an example of the type of technologies needed. This chapter will focus on progress made in identifying the specific protein targets of selected nutrients, including epigallocatechin gallate (EGCG), [6]-gingerol, the phytotaxin resveratrol (and analogues), the isoflavandiol equol, and various flavonols including kaempherol, quercetin, luteolin, delphinidin, procyanidins, and myricetin.

## COMPUTATIONAL APPROACH AND EXPERIMENTAL VERIFICATION FOR IDENTIFYING PROTEIN TARGETS FOR NATURAL COMPOUNDS AND SMALL MOLECULE INHIBITORS

Computer modeling and biological simulation are increasingly being used as highly effective tools in cancer research. Protein kinases and their target transcription factors are key targets for drug screening. Our overall approach comprises three general steps (Figure 13.2). The first step involves the virtual screening of various ligand databases such as the ZINC free database of commercially available compounds for virtual screening and the RCSB Protein Data Bank for available protein crystal structures. The second step involves experimental verification of the newly *in silico*-identified

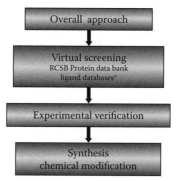

*3D coordinates of the ligands can be taken from i.e., ZINC—a free database of
commercially available compounds for virtual screening or other libraries

**FIGURE 13.2** Overall approach to the identification of novel compounds or protein targets for the development of new anticancer drugs (see text for details).

compound's effect on a predetermined target protein or the effect of a predetermined compound on a newly *in silico*-identified protein target. The final step involves the synthesis or chemical modification that would potentially enhance the verified effectiveness of the *in silico*-identified or predetermined compound on the target protein.

## COMPUTATIONAL APPROACH

The X-ray crystallographic structures of the protein of interest either alone or bound to an inhibitor or ligand is crucial in the process of identifying the protein-small molecule interaction and in designing more specific inhibitors. Once solved, the protein crystal structures are deposited within the RCSB Protein Data Bank (Bernstein et al. 1977) (http://www.rcsb.org/pdb/home). However, if a protein crystal structure with a resolution average of about 2.0 Å does not exist or the structure has not been solved, homology-modeling methods are used to create a suitable structure with which to work. Homology modeling between proteins requires that they possess a minimum protein sequence identity of 30% (Xiang 2006) and the EMBL-EBI web service provides many of the tools necessary for protein sequence searching and alignments (McWilliam et al. 2009). In addition, the National Center for Biotechnology Information provides databases on protein structures and sequences (http://www.ncbi.nlm.nih.gov/).

The crystal structure or homology modeling allows us to perform a computational technique referred to as "virtual screening." Virtual screening is used in drug discovery and involves the rapid *in silico* assessment of large libraries of compounds. From these libraries, virtual screening is used to identify the most promising structures that bind to a drug target such as a protein receptor or enzyme. Computational processes of docking and scoring are implemented using known and hypothetical drug targets on a protein, coupled with these databases of virtual chemical compounds. In docking, various computational methods (i.e., algorithms) are used to position a chemical from a virtural library into a specified target site or sites on the protein of interest. The objective of molecular docking is to determine the binding interactions between two molecules—either protein to protein or protein to ligand. Once a compound is docked, it is then scored using mathematical models. Scoring estimates the chemical interactions such as binding strength and energy state between the ligand and protein to assist in ranking the efficacy of the compound being scored. Evaluation of the docked and scored complexes leads to the selection of compounds to be moved into the laboratory for *in vitro, ex vivo*, and *in vivo* testing.

In our laboratory, we mainly use Schrödinger's suite of programs because they provide many different tools in one software package (http://www.schrodinger.com/). Glide provides a variety of docking protocols and a range of speed versus accuracy options. This program can automate calculations for large libraries of compounds and provides summaries or detailed reports of docking results. Examples of free chemical databases include the ZINC database that comprises "ready-to-dock" 3D libraries of compounds ranging from hit-to-lead candidates to FDA-approved compounds with a total number of 4.6 million searchable compounds in various formats with vendor links to order (Irwin and Shoichet 2005) (http://zinc.docking.org/). The National Cancer Institute (NCI) Database (Enhanced NCI Database Browser) contains over 250,000 compounds and the Human Metabolome Database (http://www.hmdb.ca/) comprises about 2500 metabolites linked to over 5500 protein and DNA sequences to which they correspond. The Asinex Database (http://www.asinex.com/) is an extensive library of searchable natural compounds that are also available for purchase from the site. Libraries are also available on a pay-per-use basis and include the Available Chemicals Directory (ACD; http://www.symyx.com/products/databases/sourcing/acd/index.jsp), which is a proprietary chemical database containing over 480,000 chemical compounds. More libraries are constantly being developed. In our laboratory, we have used virtual screening as described above for combining computer modeling and biological simulation as an extremely powerful technology to predict interactions between small molecule inhibitors (i.e., anticancer natural compounds) and cellular proteins (Figure 13.3). We have also used a process that we refer to as "reverse docking" (Figure 13.3). In this case, if we have a natural or synthetic compound that is an effective inhibitor of cancer growth *in vitro* or *ex vivo*, we can screen by docking that compound into every known protein crystal structure. The screening can be done in a matter of days compared to months or years of laboratory investigation. We have also used "shape-based" virtual screening of chemical libraries in which we search for chemicals in a library that are structurally similar to known inhibitors of a given protein. For shape similarity screening, we use Phase (Dixon et al. 2006), an application tool of the Schrödinger Suite, to not only search for 3D-shape complementarity but also to examine the common elements of ligands with a reference molecule. Similar chemicals are identified and then docked, enabling us to obtain detailed protein-ligand interactions.

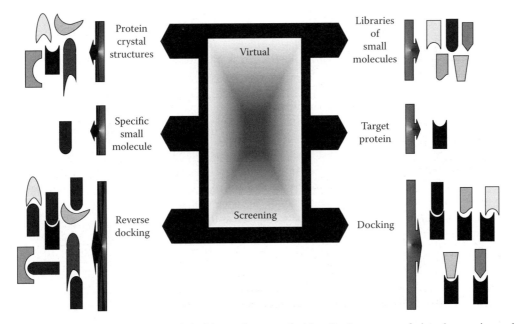

**FIGURE 13.3** Virtual screening and docking and reverse docking. In the process of virtual screening and docking, large libraries of small compounds are screened by our BlueGene/L supercomputer and docked individually into the specified protein target. Several potential compounds are identified and then verified experimentally using a variety of techniques (see text for details). In the process of reverse docking, every available protein crystal structure is docked with one specific molecule. One or more potential target proteins may be identified and then validated experimentally in the laboratory.

## Experimental Verification

Once we have narrowed drug candidates or protein targets to a small number of 10–50, we begin the process of experimental verification. Techniques include pull-down assays, affinity binding, surface plasmon resonance technology, isothermal titration calorimetry, *in vitro* kinase assays, and site-directed mutagenesis experiments.

1. *Pull-down assays*: CNBr-activated Sepharose has a long history of use in studying proteins, nucleic acids, and polysaccharides (Choi et al. 2005; David et al. 1974). In our laboratory, we have used CNBr-activated Sepharose 4B for immobilizing a variety of target molecules for affinity chromatography (Ermakova et al. 2005; Lee et al. 2007).

2. *Surface plasmon resonance (SPR) technology*: SPR is an optical technique that is used to measure biomolecular interactions in real-time in a label-free environment (reviewed in Karlsson [2004]; McDonnell [2001]; Phillips and Cheng [2007]). It can be used to identify the binding of two or more molecules with each other, to determine the affinity of the interactions, and to measure the association and dissociation rates. One of the molecules is immobilized to a sensor surface and the other is free in solution and passed over the surface. Association and dissociation are measured in arbitrary units and displayed in a graph called the sensorgram.

3. *Isothermal titration calorimetry (ITC)*: ITC is a technique for determining the thermal effects arising from molecular interactions and can quantify the interactions between molecules under physiologically relevant conditions. ITC is an excellent means for quantifying molecular interactions between two or more proteins or a protein and small molecule ligands such as a drug or inhibitor. Nearly every reaction is accompanied by the absorption or evolution of heat and therefore can be studied using calorimetry. Chromophores, fluorophores, optical transparency, chemical modification, or immobilization are not required and impurities in samples do not interfere with the measurement. ITC is commonly used to study binding interactions, including the binding affinity, binding enthalpy, and stoichiometry of the interaction.

4. *In vitro kinase assay*: The *in vitro* kinase assay is used to verify the effect of a new potential small molecule inhibitor on the kinase activity of the target protein kinase. Reactions are usually conducted with the active protein kinase, substrate protein, inhibitor, and unlabeled ATP and/or [$\gamma$-$^{32}$P] ATP. Samples are boiled and separated by SDS-PAGE and visualized by autoradiography, Western blotting, or Coomassie blue staining.

5. *Site-directed mutagenesis*: This technique is used to verify the specific amino acid sites to which the compound of interest binds. To better understand how small molecules interact with the binding site of the targeted kinse to modulate its activity, the kinase is modeled based on either its crystal structure or the crystal structure of a highly homologous protein in complex with an inhibitor if possible. Using the modeled protein, the compound(s) of interest is docked *in silico* to the proposed binding site such as the ATP-binding pocket or an allosteric site, allowing the ligand and the amino acids forming the protein binding site to be flexible, in order to achieve a more realistic view of the possible protein–ligand interactions. To validate the docking outcomes, point mutants of the predicted sites in the protein target are constructed and transfected into, for example, HEK293 cells and then an immunoprecipitation/ kinase assay is conducted with a specific antibody and protein substrate.

## THE SEARCH FOR MOLECULAR TARGETS OF SELECTED NUTRIENTS

We have successfully combined supercomputer technology with protein crystal structure determination and chemical library screening to identify a number of protein targets of natural anticancer compounds and also have identified many novel compounds that target specific protein kinases.

## PROTEIN TARGETS OF [6]-GINGEROL

One of the most popular and highly-consumed dietary substances in the world is derived from the plants of the ginger (*Zingiber officinable* Roscoe, *Zingiberaceae*) family (Surh et al. 1999). The oleoresin or oil from ginger root contains [6]-gingerol (1-[4′-hydroxy-3′-methoxyphenyl]-5-hydroxy-3-decanone) (Figure 13.4a), which is the major pharmacologically active component, and a variety of other gingerols, gingerdiols, paradols, and zingerones. The medicinal, chemical, and pharmacological properties of ginger have been extensively reviewed (Afzal et al. 2001; Bode and Dong 2004d; Grant and Lutz 2000; Langner et al. 1998). Numerous research studies have reported that components of ginger suppress cancer cell growth (Bode et al. 2001; Lee et al. 1998; Lee and Surh 1998; Surh et al. 1998). Reports suggest a variety of indirect mechanisms by which ginger compounds exert their antitumorigenic effects and several aspects of the chemopreventive effects of ginger have also been reviewed (Aggarwal et al. 2008; Bode and Dong 2004d; Shukla and Singh 2007; Surh 2002; Surh et al. 1998; Surh et al. 1999). Ginger and its constituents have been shown to inhibit tumor promotion in mouse skin (Katiyar et al. 1996). In particular,

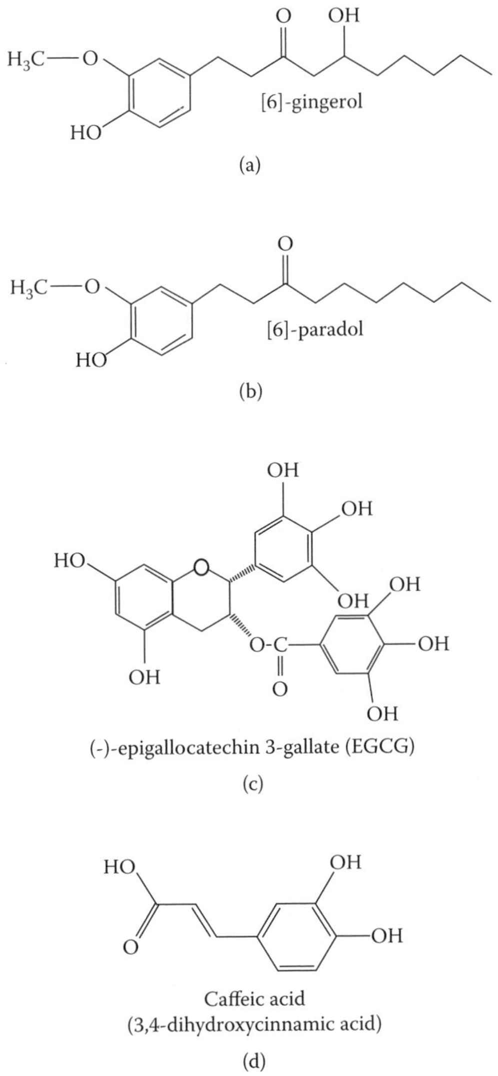

**FIGURE 13.4** Chemical structures of (a) [6]-gingerol, (b) [6]-paradol, (c) EGCG, and (d) caffeic acid.

[6]-gingerol has been reported to be highly effective as an anticancer agent against skin *in vivo* in the two-stage initiation-promotion mouse skin model. In this model, tumors are initiated by one application of 7,12-dimethylbenz[a]anthracene (DMBA), followed by repeated topical applications of TPA beginning a few days later. Topical application of [6]-gingerol onto the shaved backs of female Imprinting Control Region (IRC) mice reduced the incidence of DMBA-initiated/TPA-promoted skin papilloma formation and also suppressed TPA-induced epidermal ornithine decarboxylase activity and inflammation (Park et al. 1998). Results of a similar study indicated that in the DMBA/TPA skin tumor model, topical application of [6]-paradol or [6]-dehydroparadol prior to the application of TPA significantly reduced both the number of tumors per mouse and the number of mice exhibiting tumors (Chung et al. 2001). The effect of two structurally related compounds of the ginger family, [6]-gingerol and [6]-paradol (Figure 13.4b), on EGF-induced cell transformation and AP-1 activation was investigated (Bode et al. 2001). The results indicated that both compounds blocked EGF-induced cell transformation and although [6]-gingerol, but not paradol, inhibited AP-1 activation, both could act by inducing apoptosis (Bode et al. 2001). Cyclooxygenase-2 (COX-2) is an important enzyme in prostaglandin biosynthesis and is regarded as a promising molecular target for many anti-inflammatory as well as chemopreventive agents. Topical application of [6]-gingerol was reported to suppress TPA-induced COX-2 expression, p38 phosphorylation, and NF-κB DNA binding activity in mouse skin (Kim et al. 2004). These results were further expanded to show that pretreatment of mouse skin with [6]-gingerol resulted in decreased TPA-induced NF-κB DNA binding and transcriptional activity by suppressing both IκBα phosphorylation and degradation and p65 phosphorylation and nuclear translocation (Kim et al. 2005). The interaction of phosphorylated p65 (Ser536) with CREB (cAMP response element binding protein-binding protein), a transcriptional coactivator of NF-κB, was prevented by [6]-gingerol and the inhibitory effect of [6]-gingerol on p38 phosphorylation, an upstream mediator of COX-2 activation, was also observed (Kim et al. 2005). These results suggest that components of ginger may be effective anticancer agents, but more work is needed to determine the specific mechanism of action and cellular target molecules.

Investigators suggested that the effectiveness of ginger might be related to its ability to inhibit prostaglandin and leukotriene biosynthesis (Srivastava and Mustafa 1992). Others showed that gingerol actively inhibits arachidonate 5-lipoxygenase, an enzyme of leukotriene biosynthesis (Kiuchi et al. 1992). The leukotriene A4 hydrolase (LTA4H) protein is regarded as a relevant target for cancer therapy and our *in silico* prediction using a reverse-docking approach revealed that LTA4H might be a potential target of [6]-gingerol (Jeong et al. 2009). To find potential proteins that directly bind [6]-gingerol, the Potential Drug Target Database (PDTD; ref. 20; v. 2007) was used. The PDTD contains structural information (e.g., active site) of more than 830 known or potential protein drug targets. [6]-Gingerol was docked to each target in PDTD with the reverse docking tool TarFisDock (Li et al. 2006). The protein "hits" identified through the reverse docking method (i.e., the top 2% of ranked list) were considered as potential target candidates for further validation studies. Considering the structural similarity between [6]-gingerol and bestatin, the LTA4H crystal structure (PDB code 1HS6) was chosen for further docking studies, which were carried out using the Maestro suite of software (Maestro, version 7.5, Schrödinger). [6]-Gingerol was docked within the LTA4H binding site (Figure 13.5) using the QM-Polarized ligand docking program (Schrödinger Suite 2006; QM-Polarized Ligand Docking Protocol 2005). We supported our prediction by showing that LTA4H bound with [6]-gingerol Sepharose 4B beads and found that it also suppressed anchorage-independent cancer cell growth by binding to LTA4H and inhibiting LTA4H activity in HCT116 colorectal cancer cells. We further found that [6]-gingerol effectively suppressed colon tumor growth *in vivo* in nude mice, an effect that was mediated by inhibition of LTA4H activity. Collectively, these findings indicate a crucial role of LTA4H in cancer and also support the anticancer efficacy of [6]-gingerol targeting of LTA4H for the prevention of colorectal cancer (Jeong et al. 2009). Importantly, these are the first results to identify a direct target of [6]-gingerol to explain its anticancer activity.

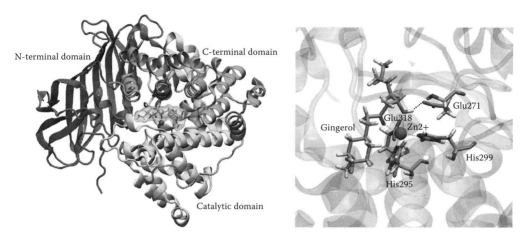

**FIGURE 13.5** [6]-Gingerol specifically binds with Glu271 of LTA4H. (a) Proposed molecular model of [6]-gingerol binding with LTA4H. The catalytic, NH2-terminal, and COOH-terminal LTA4H domains are in cartoon representation. [6]-Gingerol is depicted in stick and transparent surface area, and the zinc ion is represented as a sphere. (b) Close-up view of the interactions of [6]-gingerol within the LTA4H catalytic site. The hydrogen bond between the ligand and Glu271 (both in stick representation) is shown as a dotted black line. The amino acids coordinating the zinc ion are in stick representation and the rest of the catalytic domain is shown as a transparent cartoon. (From Jeong, C. H. et al. 2009. *Cancer Res* 69 (13):5584–91. With permission.)

## Protein Targets of EGCG

(-)-Epigallocatechin gallate (EGCG) (Figure 13.4c) is one of the most well-studied major green tea catechins (Suzuki and Isemura 2001; Tachibana et al. 2004) and is suggested to play a role as a preventive agent in cancer, obesity, diabetes, and cardiovascular disease. Searching for the EGCG "receptor" or high-affinity proteins that bind to EGCG considered to be a crucial first step in understanding the molecular and biochemical mechanisms of the anticancer effects of tea polyphenols. Several proteins that can directly bind with EGCG have been identified, including plasma proteins: fibronectin, fibrinogen, and histidine-rich glycoprotein (Sazuka et al. 1998), apoptotic Fas (Hayakawa et al. 2001), laminin and the 67-kDa laminin receptor (Hayakawa et al. 2001; Tachibana et al. 2004), fatty acid synthase (Tian 2006), HSP90 (Palermo et al. 2005; Yin et al. 2009), and T-cell receptor CD4 (Williamson et al. 2006). However, the biologic and physiologic significance for the anticancer effects of tea polyphenols is still not clear. Identification of new high affinity EGCG-binding proteins could facilitate the design of new strategies to prevent cancer. We identified the intermediate filament protein, vimentin, as a novel EGCG-binding protein (Ermakova et al. 2005). Proteins from JB6 Cl41 cell lysates were subjected to affinity chromatography using EGCG-Sepharose 4B beads and fractions containing proteins binding with EGCG were analyzed by two-dimensional electrophoresis and MALDI-TOF-MS to identify vimentin. Results were verified using a combination of a pull-down assay with EGCG-Sepharose 4B beads and an immunoprecipitation assay with a vimentin antibody (Ermakova et al. 2005). To further characterize the binding interaction, we also measured the binding affinity of GST-vimentin and [$^3$H]EGCG using a pull-down assay. Results indicated that vimentin displayed a very high affinity for binding with [$^3$H]EGCG with a $K_d$ value for the binding of EGCG to vimentin of 3.3 nM, which is certainly achievable physiologically. We also investigated the effects of EGCG on the function of glucose-regulated protein 78 (GRP78), which is associated with the multidrug resistance phenotype of many types of cancer cells. GRP78 was identified by two-dimensional gel electrophoresis and matrix-assisted laser desorption/ionization-time of flight (MALDI-TOF) analysis. EGCG was coupled to the CNBr-activated Sepharose 4B matrix, and the binding between GRP78 and EGCG was confirmed by affinity chromatography (Ermakova et al. 2006). Pull-down assays revealed that EGCG directly interacted with GRP78 at the ATP-binding site and regulated its function by competing with ATP binding, resulting in the inhibition of ATPase activity. EGCG binding caused the conversion of GRP78 from its active monomer to the inactive dimer and oligomer forms and interfered with the formation of the antiapoptotic GRP78-caspase-7 complex, which resulted in an increased etoposide-induced apoptosis in cancer cells (Ermakova et al. 2006).

EGCG was reported to directly target and inhibit a number of proteins associated with the MAP kinase/AP-1 signaling pathway. The zeta chain associated protein of 70 kDa (ZAP-70) tyrosine kinase plays a critical role in T-cell receptor-mediated signal transduction and the immune response and a high level of ZAP-70 expression is observed in leukemia. We carried out molecular docking studies using the Maestro software suite (Maestro, version 7.5, Schrödinger, New York) (Shim et al. 2008). The EGCG molecule was drawn using the builder tool in Maestro and then optimized for docking in Ligprep. The ZAP-70 crystal structure complexed with staurosporine (Protein Data Bank code 1u59) was prepared for docking following the Glide standard procedure (Friesner et al. 2004). Grids defining the protein receptor were generated considering the binding mode of staurosporine. Many protein-binding sites undergo structural rearrangements upon ligand binding, the so-called "induced fit" allows the binding site to follow the shape of the ligand with resulting better interactions. Consequently, molecular docking was performed with the induced fit docking protocol (Schrödinger Suite 2006; QM-Polarized Ligand Docking Protocol 2005). The induced fit docking is a part of the Maestro software suite that attempts to reproduce the protein conformational rearrangement upon binding. The molecular docking studies were supported by pull-down assays, kinase assays, and site-directed mutagenesis experiments and showed that EGCG could form a series of intermolecular hydrogen bonds and hydrophobic interactions within the ATP binding domain, which may contribute to the stability of the ZAP-70/EGCG complex (Shim et al. 2008). ZAP-70 and EGCG displayed high binding affinity (Kd = 0.6207 µM) and EGCG effectively suppressed ZAP-70, MAP kinase–kinase activities, and the activation of AP-1 and interleukin-2 induced by CD3 (Shim et al. 2008). Notably, EGCG induced caspase-mediated apoptosis in ZAP-70 expressing leukemia cells, but not in ZAP-70 deficient cells (Shim et al. 2008).

Insulin-like growth factor-I receptor (IGF-IR) has been implicated in cancer pathophysiology. Furthermore, impairment of IGF-IR signaling in various cancer cell lines caused inhibition of the transformed phenotype as determined by the inhibition of colony formation in soft agar and the inhibition of tumor formation in athymic nude mice. Thus, the IGF-IR might be an attractive target for cancer prevention. By using EGCG-Sepharose 4B or Sepharose 4B as affinity columns, we found that EGCG can directly bind with IGF-IR and acts as a small-molecule inhibitor of IGF-IR activity (IC$_{50}$ of 14 µM) (Li et al. 2007). To further confirm the binding results *ex vivo*, IGF-IR kinase domain mutant (Y1131F, Y1135F, and Y1136F) cells (YF3) were obtained and compared with wildtype cells for binding with EGCG. Lysates prepared from wildtype and mutant YF3 cells were incubated with EGCG-Sepharose 4B or Sepharose 4B (as a negative control), and the pulled-down proteins were analyzed by Western blot. Results indicated that the YF3 lysate, containing the mutant IGF-IR, did not interact with the EGCG-Sepharose 4B affinity column, whereas the wild-type cell lysate showed a positive signal for IGF-IR binding with EGCG Sepharose 4B (Li et al. 2007). We then determined that EGCG did not bind to other tyrosine kinases, including Abl, PDGFR, c-Src, Bmx, or Yes. Because part of the IGF-IR crystal structure is available, the docking model (Figure 13.6) of EGCG binding to the ATP binding pocket was extrapolated from computer modeling of the structure of the IGF-IR. The model predicted that several amino acid residues within the ATP binding pocket could contact EGCG, including GLN977, LYS1003, MET1052, THR1053, and ASP1123 (Li et al. 2007). These results suggested that EGCG might fit into the ATP binding pocket through a hydrogen-bonding network. This was confirmed by experiments with EGCG, ATP, and IGF-1R in which results indicated that as ATP concentration increased, binding of active IGF-1R and EGCG decreased. Furthermore, the computer modeling data also suggested that K1003 and D1123 in the ATP binding pocket of the IGF-IR are the more important amino acids for EGCG binding by supplying hydrogen bonds. Therefore, mutant (K1003A and D1123A)

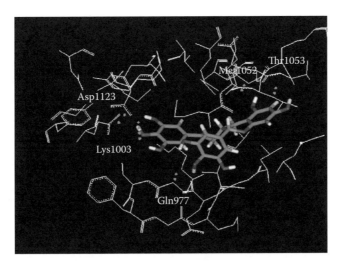

**FIGURE 13.6** Comparison of the IGF-IR binding site for EGCG and ATP. Molecular modeling of IGF-IR and docking of EGCG. The binding pocket was identified from the IGF-IR crystal structure. Hydrogen bonds between EGCG and IGF-IR were calculated (dotted lines). The participating residues include Gln977, Lys1003, Met1052, Thr1053, and Asp1123. (From Li, M. et al. 2007. *Cancer Epidemiol Biomarkers Prev* 16(3):598–605. With permission.)

IGF-IR plasmids were generated and transfected into wild-type cells. These membrane proteins were expressed and extracted from transfected cells, and incubated with an EGCG affinity column. Protein binding was analyzed by Western blot and data indicated that mutation at either Lys1003Ala or Asp1123Ala greatly decreased the binding of the IGF-IR with EGCG (Li et al. 2007).

We also found that EGCG inhibited EGF-induced cell transformation mediated by Fyn kinase activity and phosphorylation (He et al. 2008). Fyn is a member of the nonreceptor protein tyrosine kinase family. Using an *in vitro* protein-binding assay, we showed that EGCG directly bound with the GST-Fyn-SH2 domain, but not the GST-Fyn-SH3 domain, with a Kd value of 0.367 ± 0.122 μM and Bmax of 1.35 ± 0.128 nmol/mg. Knockdown of Fyn or treatment with EGCG inhibited phosphorylation of p38, ATF-2, and STAT1 and decreased the DNA binding ability of AP-1, STAT1, and ATF-2 (He et al. 2008).

Another compound, caffeic acid (3,4-dihydroxycinnamic acid) (Figure 13.4d), a well-known phenolic phytochemical present in many foods, including coffee, also targets Fyn. Recent studies suggested that caffeic acid exerts anticarcinogenic effects, but little is known about the underlying molecular mechanisms and specific target proteins. We found that Fyn was required for UVB-induced COX-2 expression, and caffeic acid suppressed UVB-induced skin carcinogenesis by directly inhibiting Fyn kinase activity (Kang et al. 2009). Caffeic acid effectively induced the downregulation of COX-2 expression at the transcriptional level mediated through the inhibition of AP-1 and NF-κB transcription activity. Pull-down assays revealed that caffeic acid directly bound with Fyn noncompetitively with ATP (Kang et al. 2009). Clearly EGCG and other compounds in tea and coffee target multiple signaling pathways exert their beneficial anticancer effects.

## MOLECULAR TARGETS OF RESVERATROL

Resveratrol (3,5,4′-trihydroxy-trans-stilbene) (Figure 13.7a) is a phytoalexin, one of a group of compounds that is produced in plants during environmental stress or pathogenic attack. This compound has been extensively studied and reviewed (Athar et al. 2009; Baxter 2008; Bishayee 2009; Bishayee and Dhir 2009; Bode and Dong 2004c; Brisdelli et al. 2009; Gatz et al. 2008; Goswami and Das 2009; Guerrero et al. 2009; Harikumar and Aggarwal 2008; Kundu and Surh 2008; Pervaiz and

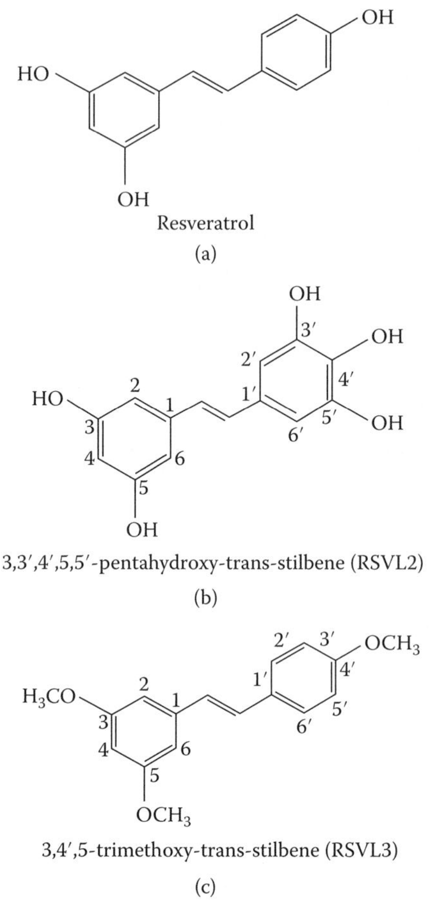

FIGURE 13.7   Chemical structures of (a) resveratrol, (b) RSVL2, and (c) RSVL3.

Holme 2009; Pirola and Frojdo 2008; Reagan-Shaw et al. 2008; Saiko et al. 2008a,b) for its potent anticarcinogenic effects.

Research findings suggest that resveratrol demonstrates nonselective COX-2 inhibition. However, we reported that resveratrol directly binds with COX-2 and this binding is absolutely required for resveratrol's inhibition of the ability of human colon adenocarcinoma HT-29 cells to form colonies in soft agar (Zykova et al. 2008). Binding of COX-2 with resveratrol was compared with two RSVL analogues, 3,3′,4′,5,5′-pentahydroxy-trans-stilbene (RSVL2) (Figure 13.7b) or 3,4′,5-trimethoxy-trans-stilbene (RSVL3) (Figure 13.7c). Notably, the results indicated that COX-2 binds with RSVL2 more strongly than with resveratrol, but did not bind with RSVL3. Resveratrol or RSVL2, but not RSVL3, inhibited COX-2-mediated $PGE_2$ production. In addition, although both resveratrol and RSVL2 were reported to bind to and inhibit COX-2 activation, RSVL2 was a stronger inhibitor (Zykova et al. 2008). HT-29 human colon adenocarcinoma cells express high levels of COX-2 and either resveratrol or RSVL2 suppressed anchorage independent growth of these cells in soft agar. Resveratrol or RSVL2 also suppressed growth of COX-2 wildtype cells but had no effect on COX-2 deficient cells (Zykova et al. 2008).

We further found that RSVL2 inhibited AP-1 transactivation, c-Fos activation, and cell transformation (Lee et al. 2008b). Further, we identified MEK/ERK signaling as the direct molecular target for the anticancer effects of RSVL2 and demonstrated that RSVL2 inhibited MEK1, but not Raf1 or ERK2 kinase activity (Lee et al. 2008b). RSVL2 inhibited H-Ras-induced cell transformation much more strongly than did PD098059 or resveratrol. Both *in vitro* and *ex vivo* pull-down assays indicated that RSVL2, but not resveratrol, directly bound with MEK1, but similar to other MEK1-selective small molecule inhibitors (i.e., PD318088, PD184352, PD098059, U0126), did not compete with ATP for binding of MEK1 (Alessi et al. 1995; Duncia et al. 1998; Favata et al. 1998; Ohren et al. 2004). This suggested that the binding region of RSVL2 with MEK1 might be similar to that of other MEK1 inhibitors. Docking data indicated that the low inhibitory activity of resveratrol might be due to the lack of the hydroxyl group at the meta position of the B ring, thereby preventing resveratrol from forming a hydrogen bond with the backbone amide group of Ser212, which is the key interaction for stabilizing the inactive conformation of the activation loop (Lee et al. 2008b). These data indicate that more potent anticancer agents can be developed, which are based on a parent compound such as resveratrol. Furthermore, these analogues can act by different mechanisms thus making their combination with the parent compound for more effective anticancer activity feasible.

Other reported binding targets for resveratrol include the $\alpha V \beta 3$ integrin receptor in breast cancer cells. The binding apparently triggers the downstream activation of the ERK1/2 pathway, leading to p53 phosphorylation at Ser15, culminating in apoptosis (Lin et al. 2006). The modeling of resveratrol and the Ref-1 pockets shows that it is likely that resveratrol can also bind to Ref-1 and regulate its activity (Yang et al. 2005).

## MOLECULAR TARGETS OF KAEMPFEROL, QUERCETIN, MYRICETIN, EQUOL, AND LUTEOLIN

Flavonoids comprise a large family of polyphenol compounds that are found in numerous and various edible plants and many have been reported to exert potent anticancer and other activities *in vitro, ex vivo*, and *in vivo*.

### Kaempferol

Kaempferol (3,4′,5,7-tetrahydroxyflavone) (Figure 13.8a) is a flavonol with anti- and pro-oxidant activity and is present in various natural sources, especially in onion leaves (832.0 mg/kg) (Miean and Mohamed 2001). Ribosomal S6 kinase 2 (RSK2), a member of the $^{p90}$RSK (RSK) family of proteins, is a widely expressed serine/threonine kinase that is activated by ERKs and phosphoinositide-dependent kinase 1 in response to many growth factors and peptide hormones. Activation of its signaling enhances cell survival. We have shown that RSK2 is a key regulator for tumor promoter-induced cell transformation and ectopic expression of RSK2 in JB6 Cl41 cells caused increased proliferation as well as anchorage-independent transformation and knockdown of RSK2 by siRNA almost totally blocked foci formation in NIH3T3 cells (Cho et al. 2007). Immunohistofluorescence and Western blot results indicated that the RSK2 protein level is markedly higher in cancer cell lines as well as cancer tissues compared with nonmalignant cell lines or normal tissues. In addition, kaempferol inhibited proliferation of malignant human cancer cell lines, including A431, SK-MEL-5 and SK-MEL-28, and HCT-116 (Cho et al. 2009). We provided evidence showing that $NH_2$-terminal kinase domain (NTD) activation of RSK2 is required for the activation of the ERK-mediated COOH-terminal kinase domain (CTD). Using homology modeling and laboratory validation methods, we also found that the NTD plays a key role in substrate phosphorylation and that kaempferol binds with the NTD but not the CTD in both the active and inactive forms (Cho et al. 2009). The NTD RSK2 is comparable to protein kinase C (PKC) (Jones et al. 1988), and therefore a homology model of NTD RSK2 (residues 65-341) could be built based on the structure of PKC. The model was refined structurally and energetically with the program Prime (Prime 1.5, Schrödinger, LLC, Portland, OR) based on the sequence alignment with the crystal structure of PKC-theta complexed with staurosporine (PDB code 1xjd). The sequence identities and similarities between the template (PKC) and the target (NTD RSK2) were 41% and 57%, respectively. The final model was analyzed for structural inconsistencies with the Molprobity web service (Davis et al. 2007). Ramachandran's analysis (Ramachandran et al. 1963) showed that 92.6% of all residues were in favored (98%) regions and 97.8% in allowed regions (>99.8%). Notably, all the

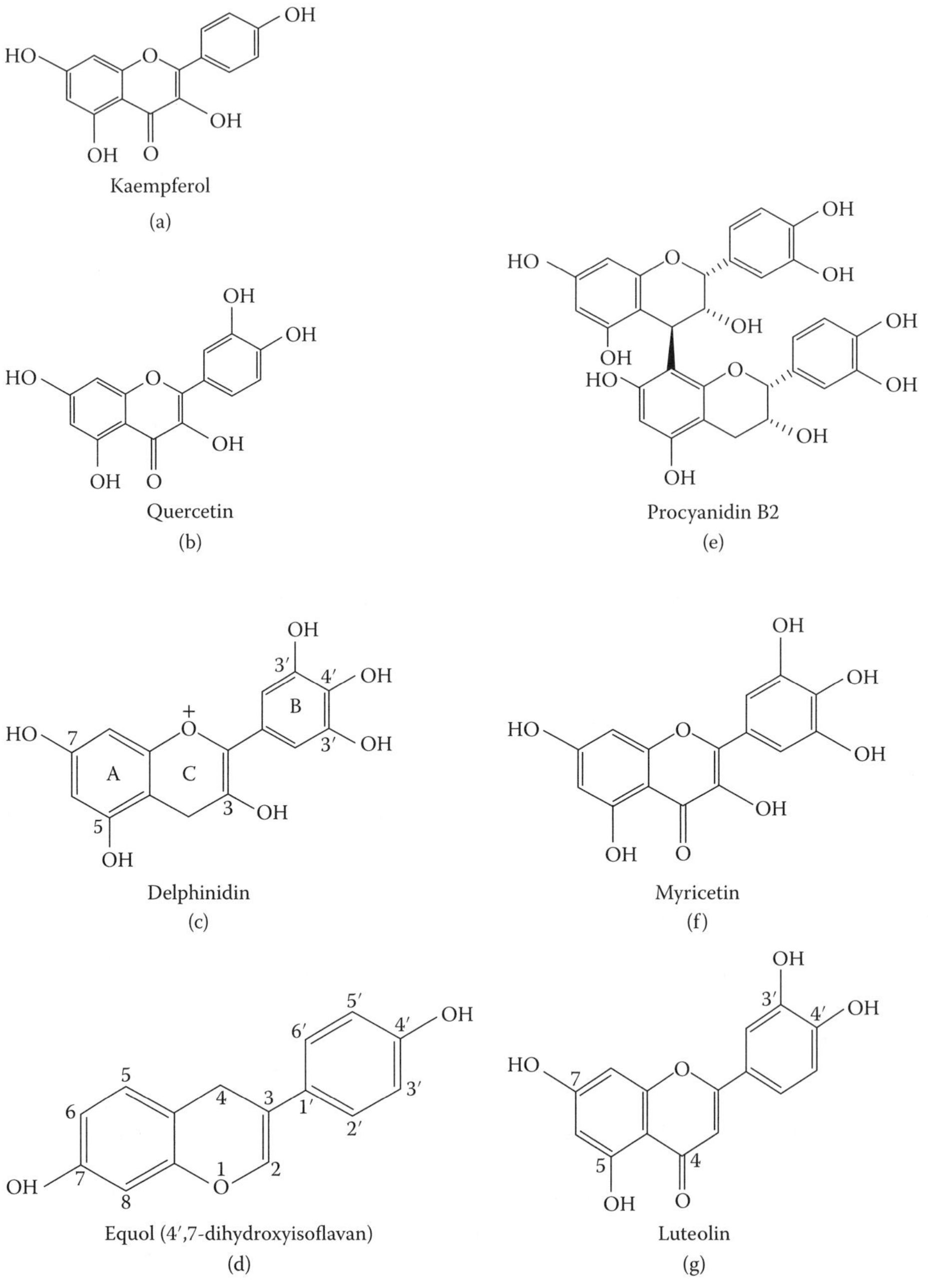

**FIGURE 13.8** Chemical structures of (a) kaempferol, (b) quercetin, (c) delphinidin, (d) equol, (e) procyanidin B2, (f) myricetin, and (g) luteolin.

residues within the ATP-binding site were in favored regions. The docking was performed similar to that described above for EGCG and ZAP-70 (Shim et al. 2008). The homology modeling of the RSK2 $NH_2$-terminal domain and small-molecule docking (Figure 13.9) was validated by mutagenesis experiments and clearly showed that Val82 and Lys100 are critical amino acids for kaempferol binding and RSK2 activity (Cho et al. 2009). These results indicate that targeting RSK2 with natural compounds, such as kaempferol, might be a good strategy for chemopreventive or chemotherapeutic application (Cho et al. 2009).

PKC-staurosporine model    RSK2-NTD structure

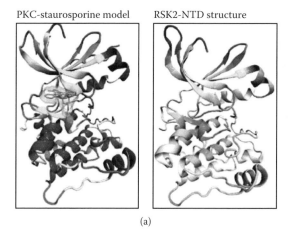

(a)

RSK2-kaempferol docking

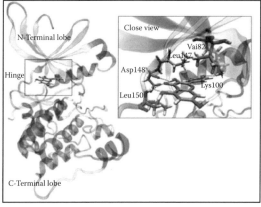

RSK2-SL0101 docking

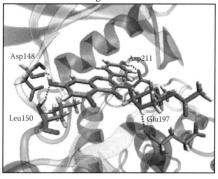

(b)

**FIGURE 13.9** Val82 and Lys100 are critical amino acids required for RSK2 activity. (a) A homology model of the RSK2-NTD (right) is compared with the crystal structure of PKC-staurosporine complex (left); both proteins are shown as cartoon, and staurosporine as atom-type plus transparent molecular surface. (b) Docking of kaempferol or SL0101 to the RSK2-NTD. The protein is in transparent cartoon representation and the ligand is depicted as an atom-type stick model. In the close views of kaempferol and SL0101 binding, the amino acids that form interactions with the ligand are represented in atom-type sticks and the hydrogen bonds as dotted lines. (From Cho, Y. et al. 2009. *Cancer Res* 69(10):4398–406. With permission.)

## Quercetin

Quercetin (Figure 13.8b) has been suggested as a potent anticarcinogenic compound because in 9,10-dimethyl-1,2-benzanthracene-initiated and TPA-promoted two-stage mouse skin cancer models, quercetin exerted strong anticarcinogenic effects (Soleas et al. 2002). Similar to kaempferol, the flavonoid quercetin is found at high levels in various foods, including red wine, and has been reported to inhibit the proliferation of cancer cells, whereas it has no effect on nonneoplastic cells. We found that red wine extract (RWE) or quercetin inhibited TPA-induced transformation of JB6 promotion-sensitive mouse skin epidermal (JB6 P+) cells, corresponding with decreases in TPA-induced activation of AP-1 and NF-κB (Lee et al. 2008a). Western blot and kinase assay data revealed that RWE or quercetin inhibited mitogen-activated protein kinase/extracellular signal-regulated kinase (ERK) kinase (MEK) 1 and Raf1 kinase activities and subsequently attenuated TPA-induced phosphorylation of ERK/p90RSK. Although either RWE or quercetin suppressed Raf1 kinase activity, they were more effective in inhibiting MEK1 activity. Importantly, quercetin exerted stronger inhibitory effects than PD098059, a well-known pharmacologic inhibitor of MEK. Pull-down assays revealed that RWE or quercetin bound with either MEK1 or Raf1. RWE or quercetin also dose-dependently suppressed JB6 P+ cell transformation induced by EGF or H-Ras, both of which are involved in the activation of MEK/ERK signaling. Docking data (Figure 13.10) suggested that quercetin formed a hydrogen bond with the backbone amide group of Ser212, which is the key interaction for stabilizing the inactive conformation of the activation loop of MEK1 (Lee et al. 2008a). We also found that RWE and quercetin inhibited the TNF-alpha-induced upregulation of MMP-9 and cell migration. The inhibitory effects of RWE and quercetin were mediated by suppression of Akt phosphorylation and AP-1 and NF-κB transactivation (Hwang et al. 2009). Aside from Akt, quercetin had no effect on the phosphorylation of other MAP kinases. RWE and quercetin inhibited phosphatidylinositol 3-kinase (PI3-K) activity and were found to bind directly to PI3-K, resulting in the inhibition of PI3-K activity (Hwang et al. 2009). Thus far, computational and experimental evidence suggests that quercetin targets Raf1, MEK1 and PI-3 kinase to exert its chemopreventive effects. This has implications for effective chemoprevention in that evidence suggests that Raf and/or Ras are constitutively activated in several tumor cell lines, and MEK generally plays a critical role in transmitting signals initiated by tumor promoters, including TPA, EGF, or PDGF. The constitutive activation of MEK1 results in cellular transformation, while a small molecular inhibitor of MEK suppresses transformation and tumor growth in both cell culture and mouse models (Cowley et al. 1994; Sebolt-Leopold et al. 1999).

Similarly, we found that delphinidin (Figure 13.8c), an anthrocyanidin found in many fruits and vegetables, inhibits EGF- or H-ras-induced transformation and cyclooxygenase-2 (COX-2) expression in JB6 P+ cells by directly targeting Raf and MEK. The activation of AP-1 and NF-κB induced by TPA was dose dependently inhibited by delphinidin treatment. Delphinidin strongly suppressed Raf1 and MEK1 kinase activities and subsequently attenuated TPA-induced phosphorylation of MEK, extracellular signal-regulated kinase (ERK), p90RSK, and MSK. Pull-down and competition assays revealed that delphinidin binds with Raf1 or MEK1 noncompetitively with ATP (Kang et al. 2008a).

## Equol

Equol (4′,7-isoflavandiol) (Figure 13.8d) is a nonsteroidal estrogen isoflavandiol metabolized from the isoflavone, daidzein, by bacterial flora in the intestines. Equol was shown to protect against UV-induced skin cancer in the hairless mouse model (Widyarini et al. 2005). The UV-induced activation of ornithine decarboxylase, a skin tumor promotion biomarker enzyme, was attenuated by equol treatment, indicating that the anticancer activity of equol may be attributed to its inhibition of the tumor promotion phase of carcinogenesis (Widyarini et al. 2005). We compared the effects of equol and daidzein on TPA-induced AP-1 activity and cell transformation in JB6 P+ cells and results indicated that equol, but not daidzein, was a potent inhibitor of MEK activity and subsequently inhibited c-Fos activation and AP-1 transactivation and cell transformation (Kang et al. 2007). Importantly, equol specifically bound to MEK noncompetitively with ATP to inhibit MEK activity (Kang et al. 2007). Overall, similar results were observed for the cocoa compound, procyanidin B2 (Figure 13.8e), which also strongly inhibited neoplastic cell transformation by binding directly to MEK1 to suppress its activity (Kang et al. 2008b).

## Myricetin

Myricetin (3,3′,4′,5,5′,7-hexahydroxyflavone) (Figure 13.8f) is a major flavonol found in various foods such as onions, berries, and grapes as well as red wine (German and Walzem 2000; Hakkinen et al. 1999). Myricetin was reported to protect against two-stage skin tumorigenesis (Mukhtar et al. 1988) and inhibit the growth of A549 lung cancer cells (Lu et al. 2006). It also suppresses invasion and both protein expression and enzyme activity of matrix metalloproteinase-2 (MMP-2) in colorectal carcinoma cells (Ko et al. 2005). These accumulated data suggest that myricetin might be an effective chemopreventive agent against carcinogenesis but its specific protein targets still

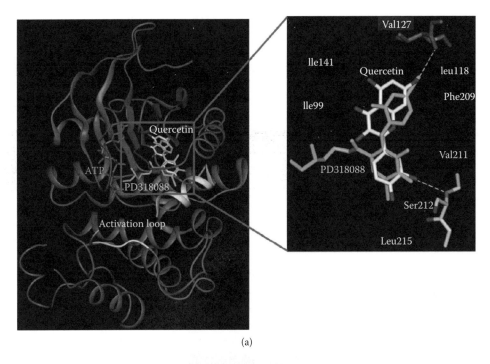

(a)

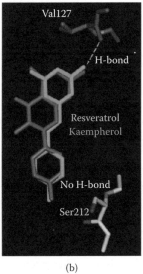

(b)

**FIGURE 13.10** Modeling study of the MEK1 binding of quercetin, resveratrol, or kaempferol. (a) Hypothetical model of MEK1-quercetin complex. Quercetin (white) binds to the pocket adjacent to the ATP-binding site. PD318088 is superimposed on the model structure of MEK1-quercetin complex for comparison. The activation loop is partially disordered. The residues involved in the interactions with quercetin are indicated. Dashed lines are hydrogen bonds. (b) Hypothetical model of MEK1 in complex with resveratrol (light) or kaempferol (dark). Although each of these compounds can retain the hydrogen bond with Val127 and the van der Waals interactions involved in the binding of quercetin to MEK1, neither compound can form a hydrogen bond with the activation loop of MEK1 due to the lack of a hydrogen bond acceptor at the 3′ position of their respective ring adjacent to the activation loop. (From Lee, K. W. et al. 2008. *Cancer Res* 68(3):946–55. With permission.)

need to be identified. Others and we have now reported the direct binding of myricetin with various protein kinases both competitively and noncompetitively with ATP to exert its anticancer and other effects.

We demonstrated that myricetin is an inhibitor of TPA or epidermal growth factor EGF-transformation of JB6 P+ cells corresponding with a decreased activation of c-Fos or AP-1 (Lee et al. 2007). Myricetin strongly inhibited MEK1 kinase activity and suppressed TPA- or EGF-induced phosphorylation of ERK or RSK, downstream targets of MEK. Moreover, myricetin inhibited H-Ras-induced cell transformation more effectively than PD098059, a MEK inhibitor. Notably, myricetin directly bound with MEK1, but did not compete with ATP (Lee et al. 2007). Furthermore, myricetin inhibited UVB-induced wrinkle formation in mouse skin and reduced UVB-induced epidermal thickening of mouse skin, and also suppressed UVB-induced matrix metalloproteinase-9 (MMP-9) protein expression and enzyme activity (Jung et al. 2010b). Myricetin appeared to exert its antiaging effects by suppressing UVB-induced Raf kinase activity and subsequent attenuation of UVB-induced phosphorylation of MEK and ERK in mouse skin and this was confirmed by *in vitro* and *ex vivo* pull-down assays revealing that myricetin bound with Raf in an ATP-noncompetitive manner (Jung et al. 2010b).

Our collaborators reported that myricetin inhibited tumor necrosis factor (TNF)-alpha-induced vascular endothelial growth factor (VEGF) expression in JB6 P+ cells by targeting MAPK kinase 4 (MKK4), as well as MEK1 (Kim et al. 2009). The activation of AP-1 by TNF-alpha was also inhibited by myricetin as was phosphorylation of c-Jun N-terminal kinases (JNKs) and ERKs, but myricetin did not affect the phosphorylation of their upstream kinases MKK4 or MEK1. However, myricetin inhibited TNF-alpha-induced MKK4 activity and, in this case, bound MKK4 competively with ATP. Computer modeling suggested that myricetin docks onto the ATP-binding site in MKK4, which is located between the N- and C-lobes of the kinase domain (Kim et al. 2009). Similarly, myricetin was shown to directly bind with Fyn and suppress Fyn kinase activity also in an ATP-competitive manner (Jung et al. 2008). Docking data (Figure 13.11) suggested that myricetin also docked to the ATP-binding site of Fyn, similarly located between the N and C lobes of the kinase domain. Indeed, the binding of myricetin with Fyn resulted in suppression of UVB-induced COX-2 expression and decreased activation of AP-1 and NF-κB in JB6 P+ cells (Jung et al. 2008). Importantly, myricetin exerted similar inhibitory effects compared with 4-amino-5-(4-chlorophenyl)-7-(t-butyl)pyrazolo[3,4-d]pyrimidine, a well-known pharmacologic inhibitor of Fyn. *In vivo* mouse skin data also revealed that myricetin inhibited Fyn kinase activity directly and subsequently attenuated UVB-induced COX-2 expression and significantly suppressed UVB-induced skin tumor incidence in a dose-dependent manner (Jung et al. 2008).

Akt, a serine/threonine kinase, is a critical regulator in many cellular processes including cell growth, proliferation, and apoptosis. Binding assay data revealed that myricetin bound to Akt directly by competing with ATP resulting in the inhibition of Akt phosphorylation and kinase activity. Molecular modeling suggested that myricetin docks to the ATP-binding site of Akt to inhibit Akt-mediated AP-1 transactivation, cyclin D1 expression, and cell transformation (Kumamoto et al. 2009a). We also reported that topical treatment with myricetin inhibited UVB-induced neovascularization in SKH-1 hairless mouse skin and suppressed UVB-induced VEGF, MMP-9, and MMP-13 expression (Jung et al. 2010a). Myricetin suppressed UVB-induced phosphatidylinositol-3 (PI-3) kinase activity and subsequently attenuated the UVB-induced phosphorylation of Akt/p70$^{S6K}$ in mouse skin by directly binding to PI-3 kinase (Jung et al. 2010a). Finally, the JAK1/STAT3 pathway has been suggested to play a role in cell transformation and carcinogenesis and myricetin was reported to directly bind to JAK1 and STAT3 to inhibit EGF-induced transformation of JB6 P+ cells. Data revealed that myricetin inhibited DNA-binding and transcriptional activity of STAT3 and suppressed phosphorylation of JAK1 and STAT3 at Tyr705 and Ser727 (Kumamoto et al. 2009b).

Thus, similar to EGCG, and probably most natural compounds, myricetin binds to and affects multiple cellular targets. It is a powerful inhibitor of MEK1 kinase activity, resulting in the subsequent inhibition of AP-1 transactivation, c-Fos activation, and EGF- or TPA-induced cell

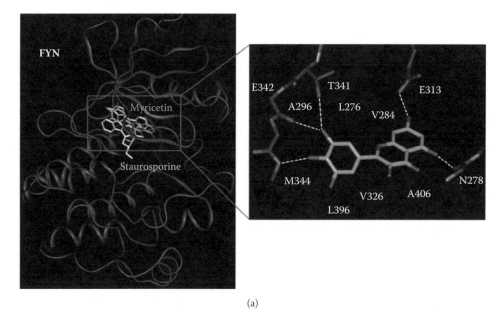

(a)

(b)

**FIGURE 13.11** Modeling study of Fyn binding with myricetin. (a) Myricetin binds to the ATP-binding site in the kinase domain of Fyn. Staurosporine (white) is overlaid on the model structure of the complex for comparison. The close-up view of the inhibitor interaction shows the hydrogen bonds depicted as dashed lines and the hydrophobic contacts as small curves. (b) Myricetin binding to the ATP-binding cleft represented as an electrostatic potential surface. (From Jung, S. K. et al. 2008. *Cancer Res* 68(14):6021–9. With permission.)

transformation (Lee et al. 2007). Myricetin has potent chemopreventive effects against TNF-alpha-related disease, mainly by targeting MKK4 and MEK1 (Kim et al. 2009). It exerts potent anti-photoaging activity by regulating MMP-9 expression through the binding and suppression of Raf kinase activity (Jung et al. 2010b) and suppresses UVB-induced angiogenesis by binding with and regulating PI-3 kinase activity (Jung et al. 2010a). Finally, myricetin binds both JAK1 and STAT3, but has a highest affinity for JAK1 (Kumamoto et al. 2009b).

## Luteolin

Luteolin (Figure 13.8g) is another flavonoid present in various vegetables, including onion and broccoli, that has been reported to possess anticarcinogenic effects. Again, its chemopreventive effect on UV-induced skin cancer and its mechanism are not fully understood. We found that luteolin suppressed UVB-induced COX-2 expression and AP-1 and NF-κB activity in JB6 P+ cells. Immunoblot and kinase assay data showed that luteolin suppressed protein kinase Cε (PKCε) and

Src kinase activities and subsequently inhibited UVB-induced phosphorylation of the Akt signaling pathway (Byun et al. 2010). Notably, pull-down assays revealed that luteolin binds directly to PKCε and Src in an ATP-competitive manner (Byun et al. 2010). Importantly, luteolin suppressed tumor incidence, multiplicity, and overall size in SKH-1 hairless mice. Analysis of the skin by immunohistochemistry and immunoblotting showed that luteolin-treated groups had a substantial reduction in the levels of COX-2, TNF-alpha, and proliferating cell nuclear antigen compared with groups treated with only UVB (Byun et al. 2010). Further analysis using skin lysates confirmed that luteolin inhibited PKCepsilon and Src kinase activities. Together, these data suggest that luteolin exerts potent chemopreventive activity against UVB-induced skin cancer mainly by targeting PKCε and Src (Byun et al. 2010).

## CONCLUSIONS/SUMMARIES

A great deal of scientific data has accumulated elucidating the molecular mechanisms of cancer development and the action of anticancer agents in cancer prevention. This research has provided the basis for the carcinogenic process caused by environmental carcinogens and molecular mechanisms for cancer prevention. Further, such discoveries have identified key molecular targets for screening novel natural anticancer drugs with fewer side effects. Nutritional or dietary factors have attracted a great deal of interest because of their professed ability to act as highly effective chemopreventive agents. They are perceived as being generally safe and may have efficacy as chemopreventive agents by preventing or reversing premalignant lesions and/or reducing second primary tumor incidence. Many of these compounds appear to act on multiple tumor promoter-stimulated cellular pathways and especially the MAP kinase pathways, Raf, MKK4, MEK1, ERK, Akt, and other downstream targets. Some of the most interesting and well documented are components of tea, such as EGCG and other potentially effective dietary compounds including [6]-gingerol, resveratrol, and various flavonoid compounds. A continuing emphasis on obtaining rigorous research data and critical analysis of those data regarding these and other food factors is vital to determine the molecular basis and long-term effectiveness and safety of these compounds as chemopreventive agents. Large-scale animal and molecular biology studies are needed to address the bioavailability, toxicity, molecular target, signal transduction pathways, and side effects of dietary factors. Clinical trials based on clear mechanistic studies are also needed to assess the effectiveness of these dietary factors in the human population.

## ACKNOWLEDGMENTS

We wish to thank Dr. Janos Nadas and Dr. Madhu Mottamal (ZD/AB laboratory) for their contribution in the description of the computer modeling techniques.

## REFERENCES

Afzal, M., D. Al-Hadidi, M. Menon, J. Pesek, and M. S. Dhami. 2001. Ginger: An ethnomedical, chemical and pharmacological review. *Drug Metabol Drug Interact* 18(3-4):159–90.
Aggarwal, B. B., A. B. Kunnumakkara, K. B. Harikumar, S. T. Tharakan, B. Sung, and P. Anand. 2008. Potential of spice-derived phytochemicals for cancer prevention. *Planta Med* 74(13):1560–9.
Aggarwal, B. B., G. Sethi, V. Baladandayuthapani, S. Krishnan, and S. Shishodia. 2007. Targeting cell signaling pathways for drug discovery: An old lock needs a new key. *J Cell Biochem* 102(3):580–92.
Aggarwal, B. B., and S. Shishodia. 2006. Molecular targets of dietary agents for prevention and therapy of cancer. *Biochem Pharmacol* 71(10):1397–421.
Alessi, D. R., A. Cuenda, P. Cohen, D. T. Dudley, and A. R. Saltiel. 1995. PD 098059 is a specific inhibitor of the activation of mitogen-activated protein kinase kinase *in vitro* and *in vivo*. *J Biol Chem* 270(46):27489–94.
Angel, P., K. Hattori, T. Smeal, and M. Karin. 1988. The jun proto-oncogene is positively autoregulated by its product, Jun/AP-1. *Cell* 55(5):875–85.

Arora, A., and E. M. Scholar. 2005. Role of tyrosine kinase inhibitors in cancer therapy. *J Pharmacol Exp Ther* 315(3):971–9.

Athar, M., J. H. Back, L. Kopelovich, D. R. Bickers, and A. L. Kim. 2009. Multiple molecular targets of resveratrol: Anti-carcinogenic mechanisms. *Arch Biochem Biophys* 486(2):95–102.

Baxter, R. A. 2008. Anti-aging properties of resveratrol: Review and report of a potent new antioxidant skin care formulation. *J Cosmet Dermatol* 7(1):2–7.

Bernstein, F. C., T. F. Koetzle, G. J. Williams, E. F. Meyer, Jr., M. D. Brice, J. R. Rodgers, O. Kennard, T. Shimanouchi, and M. Tasumi. 1977. The Protein Data Bank: A computer-based archival file for macromolecular structures. *J Mol Biol* 112(3):535–42.

Bishayee, A. 2009. Cancer prevention and treatment with resveratrol: From rodent studies to clinical trials. *Cancer Prev Res (Phila Pa)* 2(5):409–18.

Bishayee, A., and N. Dhir. 2009. Resveratrol-mediated chemoprevention of diethylnitrosamine-initiated hepatocarcinogenesis: Inhibition of cell proliferation and induction of apoptosis. *Chem Biol Interact* 179(2-3):131–44.

Blume-Jensen, P., and T. Hunter. 2001. Oncogenic kinase signalling. *Nature* 411(6835):355–65.

Bode, A. M., and Z. Dong. 2004a. Cancer prevention by food factors through targeting signal transduction pathways. *Nutrition* 20(1):89–94.

———. 2004b. Targeting signal transduction pathways by chemopreventive agents. *Mutat Res* 555(1-2):33–51.

———. 2005. Signal transduction pathways in cancer development and as targets for cancer prevention. *Prog Nucleic Acid Res Mol Biol* 79:237–97.

———. 2006. Molecular and cellular targets. *Mol Carcinog* 45(6):422–30.

———. 2007. The enigmatic effects of caffeine in cell cycle and cancer. *Cancer Lett* 247(1):26–39.

———. 2009a. Cancer prevention research—then and now. *Nat Rev Cancer* 9(7):508–16.

———. 2009b. Epigallocatechin 3-gallate and green tea catechins: United they work, divided they fail. *Cancer Prev Res (Phila Pa)* 2(6):514–7.

———. 2009c. Signal transduction molecules as targets for cancer prevention. *Sci Signal* 2(59):mr2.

Bode, A. M., W. Y. Ma, Y. J. Surh, and Z. Dong. 2001. Inhibition of epidermal growth factor-induced cell transformation and activator protein 1 activation by [6]-gingerol. *Cancer Res* 61(3):850–3.

Bode, A. M., and Z. Dong. 2004c. Beneficial effects of resveratrol. In Y. Bao and R. Fenwick, eds., *Phytochemicals in Health and Disease*, 257–84. New York: Marcel Dekker, Inc.

———. 2004d. Ginger. In L. Packer, C. N. Ong and B. Halliwell, eds., *Herbal and Traditional Medicine: Molecular Aspects of Health*, 165–77. New York: Marcel Dekker.

———. 2004e. Post-translational modification of p53 in tumorigenesis. *Nature Reviews Cancer* 4:793–805.

Brisdelli, F., G. D'Andrea, and A. Bozzi. 2009. Resveratrol: A natural polyphenol with multiple chemopreventive properties. *Curr Drug Metab* 10(6):530–46.

Byun, S., K. W. Lee, S. K. Jung, E. J. Lee, M. K. Hwang, S. H. Lim, A. M. Bode, H. J. Lee, and Z. Dong. 2010. Luteolin inhibits protein kinase C(epsilon) and c-Src activities and UVB-induced skin cancer. *Cancer Res* 70(6):2415–23.

Cho, Y. Y., K. Yao, H. G. Kim, B. S. Kang, D. Zheng, A. M. Bode, and Z. Dong. 2007. Ribosomal S6 kinase 2 is a key regulator in tumor promoter induced cell transformation. *Cancer Res* 67(17):8104–12.

Cho, Y. Y., K. Yao, A. Pugliese, M. L. Malakhova, A. M. Bode, and Z. Dong. 2009. A regulatory mechanism for RSK2 NH(2)-terminal kinase activity. *Cancer Res* 69(10):4398–406.

Choi, Y. S., B. K. Ryu, H. K. Min, S. W. Lee, and Y. K. Pak. 2005. Analysis of proteome bound to D-loop region of mitochondrial DNA by DNA-linked affinity chromatography and reverse-phase liquid chromatography/tandem mass spectrometry. *Ann N Y Acad Sci* 1042:88–100.

Chung, W. Y., Y. J. Jung, Y. J. Surh, S. S. Lee, and K. K. Park. 2001. Antioxidative and antitumor promoting effects of [6]-paradol and its homologs. *Mutat Res* 496(1-2):199–206.

Cowley, S., H. Paterson, P. Kemp, and C. J. Marshall. 1994. Activation of MAP kinase kinase is necessary and sufficient for PC12 differentiation and for transformation of NIH 3T3 cells. *Cell* 77(6):841–52.

David, G. S., T. H. Chino, and R. A. Reisfeld. 1974. Binding of proteins to CNBr-activated sepharose 4B. *FEBS Lett* 43(3):264–6.

Davis, I. W., A. Leaver-Fay, V. B. Chen, J. N. Block, G. J. Kapral, X. Wang, L. W. Murray, W. B. Arendall, 3rd, J. Snoeyink, J. S. Richardson, and D. C. Richardson. 2007. MolProbity: All-atom contacts and structure validation for proteins and nucleic acids. *Nucleic Acids Res* 35(Web Server issue):W375–83.

Davis, R. J. 1994. MAPKs: New JNK expands the group. *Trends Biochem Sci* 19(11):470–3.

Dixon, S. L., A. M. Smondyrev, E. H. Knoll, S. N. Rao, D. E. Shaw, and R. A. Friesner. 2006. PHASE: A new engine for pharmacophore perception, 3D QSAR model development, and 3D database screening: 1. Methodology and preliminary results. *J Comput Aided Mol Des* 20(10-11):647–71.

Duncia, J. V., J. B. Santella, 3rd, C. A. Higley, W. J. Pitts, J. Wityak, W. E. Frietze, F. W. Rankin, J. H. Sun, R. A. Earl, A. C. Tabaka, C. A. Teleha, K. F. Blom, M. F. Favata, E. J. Manos, A. J. Daulerio, D. A. Stradley, K. Horiuchi, R. A. Copeland, P. A. Scherle, J. M. Trzaskos, R. L. Magolda, G. L. Trainor, R. R. Wexler, F. W. Hobbs, and R. E. Olson. 1998. MEK inhibitors: The chemistry and biological activity of U0126, its analogs, and cyclization products. *Bioorg Med Chem Lett* 8(20):2839–44.

Ermakova, S., B. Y. Choi, H. S. Choi, B. S. Kang, A. M. Bode, and Z. Dong. 2005. The intermediate filament protein vimentin is a new target for epigallocatechin gallate. *J Biol Chem* 280(17):16882–90.

Ermakova, S. P., B. S. Kang, B. Y. Choi, H. S. Choi, T. F. Schuster, W. Y. Ma, A. M. Bode, and Z. Dong. 2006. (-)-Epigallocatechin gallate overcomes resistance to etoposide-induced cell death by targeting the molecular chaperone glucose-regulated protein 78. *Cancer Res* 66(18):9260–9.

Favata, M. F., K. Y. Horiuchi, E. J. Manos, A. J. Daulerio, D. A. Stradley, W. S. Feeser, D. E. Van Dyk, W. J. Pitts, R. A. Earl, F. Hobbs, R. A. Copeland, R. L. Magolda, P. A. Scherle, and J. M. Trzaskos. 1998. Identification of a novel inhibitor of mitogen-activated protein kinase kinase. *J Biol Chem* 273(29):18623–32.

Friesner, R. A., J. L. Banks, R. B. Murphy, T. A. Halgren, J. J. Klicic, D. T. Mainz, M. P. Repasky, E. H. Knoll, M. Shelley, J. K. Perry, D. E. Shaw, P. Francis, and P. S. Shenkin. 2004. Glide: A new approach for rapid, accurate docking and scoring. 1. Method and assessment of docking accuracy. *J Med Chem* 47(7):1739–49.

Gatz, S. A., M. Keimling, C. Baumann, T. Dork, K. M. Debatin, S. Fulda, and L. Wiesmuller. 2008. Resveratrol modulates DNA double-strand break repair pathways in an ATM/ATR-p53- and -Nbs1-dependent manner. *Carcinogenesis* 29(3):519–27.

German, J. B., and R. L. Walzem. 2000. The health benefits of wine. *Annu Rev Nutr* 20:561–93.

Goswami, S. K., and D. K. Das. 2009. Resveratrol and chemoprevention. *Cancer Lett* 284(1):1–6.

Grant, K. L., and R. B. Lutz. 2000. Ginger. *Am J Health Syst Pharm* 57(10):945–7.

Guerrero, R. F., M. C. Garcia-Parrilla, B. Puertas, and E. Cantos-Villar. 2009. Wine, resveratrol and health: A review. *Nat Prod Commun* 4(5):635–58.

Hakkinen, S. H., S. O. Karenlampi, I. M. Heinonen, H. M. Mykkanen, and A. R. Torronen. 1999. Content of the flavonols quercetin, myricetin, and kaempferol in 25 edible berries. *J Agric Food Chem* 47(6):2274–9.

Harikumar, K. B., and B. B. Aggarwal. 2008. Resveratrol: A multitargeted agent for age-associated chronic diseases. *Cell Cycle* 7(8):1020–1035.

Hayakawa, S., K. Saeki, M. Sazuka, Y. Suzuki, Y. Shoji, T. Ohta, K. Kaji, A. Yuo, and M. Isemura. 2001. Apoptosis induction by epigallocatechin gallate involves its binding to Fas. *Biochem Biophys Res Commun* 285(5):1102–6.

He, Z., F. Tang, S. Ermakova, M. Li, Q. Zhao, Y. Y. Cho, W. Y. Ma, H. S. Choi, A. M. Bode, C. S. Yang, and Z. Dong. 2008. Fyn is a novel target of (-)-epigallocatechin gallate in the inhibition of JB6 Cl41 cell transformation. *Mol Carcinog* 47(3):172–83.

Hong, W. K. 2003. General keynote: The impact of cancer chemoprevention. *Gynecol Oncol* 88(1 Pt 2):S56–8.

Huang, C., W. Y. Ma, J. Li, and Z. Dong. 1999. Arsenic induces apoptosis through a c-Jun NH2-terminal kinase-dependent, p53-independent pathway. *Cancer Res* 59(13):3053–8.

Hwang, M. K., N. R. Song, N. J. Kang, K. W. Lee, and H. J. Lee. 2009. Activation of phosphatidylinositol 3-kinase is required for tumor necrosis factor-alpha-induced upregulation of matrix metalloproteinase-9: Its direct inhibition by quercetin. *Int J Biochem Cell Biol* 41(7):1592–600.

Irwin, J. J., and B. K. Shoichet. 2005. ZINC: A free database of commercially available compounds for virtual screening. *J Chem Inf Model* 45(1):177–82.

Jeong, C. H., A. M. Bode, A. Pugliese, Y. Y. Cho, H. G. Kim, J. H. Shim, Y. J. Jeon, H. Li, H. Jiang, and Z. Dong. 2009. [6]-Gingerol suppresses colon cancer growth by targeting leukotriene A4 hydrolase. *Cancer Res* 69(13):5584–91.

Johnson, L. N. 2009. Protein kinase inhibitors: Contributions from structure to clinical compounds. *Q Rev Biophys* 42(1):1–40.

Jones, S. W., E. Erikson, J. Blenis, J. L. Maller, and R. L. Erikson. 1988. A Xenopus ribosomal protein S6 kinase has two apparent kinase domains that are each similar to distinct protein kinases. *Proc Natl Acad Sci USA* 85(10):3377–81.

Jung, S. K., K. W. Lee, S. Byun, N. J. Kang, S. H. Lim, Y. S. Heo, A. M. Bode, G. T. Bowden, H. J. Lee, and Z. Dong. 2008. Myricetin suppresses UVB-induced skin cancer by targeting Fyn. *Cancer Res* 68(14):6021–9.

Jung, S. K., K. W. Lee, S. Byun, E. J. Lee, J. E. Kim, A. M. Bode, Z. Dong, and H. J. Lee. 2010a. Myricetin inhibits UVB-induced angiogenesis by regulating PI-3 kinase *in vivo*. *Carcinogenesis* 31(5):911–7.

Jung, S. K., K. W. Lee, H. Y. Kim, M. H. Oh, S. Byun, S. H. Lim, Y. S. Heo, N. J. Kang, A. M. Bode, Z. Dong, and H. J. Lee. 2010b. Myricetin suppresses UVB-induced wrinkle formation and MMP-9 expression by inhibiting Raf. *Biochem Pharmacol* 79(10):1455–61.

Kallunki, T., B. Su, I. Tsigelny, H. K. Sluss, B. Derijard, G. Moore, R. Davis, and M. Karin. 1994. JNK2 contains a specificity-determining region responsible for efficient c-Jun binding and phosphorylation. *Genes Dev* 8(24):2996–3007.

Kang, N. J., K. W. Lee, J. Y. Kwon, M. K. Hwang, E. A. Rogozin, Y. S. Heo, A. M. Bode, H. J. Lee, and Z. Dong. 2008a. Delphinidin attenuates neoplastic transformation in JB6 Cl41 mouse epidermal cells by blocking Raf/mitogen-activated protein kinase kinase/extracellular signal-regulated kinase signaling. *Cancer Prev Res (Phila Pa)* 1(7):522–31.

Kang, N. J., K. W. Lee, D. E. Lee, E. A. Rogozin, A. M. Bode, H. J. Lee, and Z. Dong. 2008b. Cocoa procyanidins suppress transformation by inhibiting mitogen-activated protein kinase kinase. *J Biol Chem* 283(30):20664–73.

Kang, N. J., K. W. Lee, E. A. Rogozin, Y. Y. Cho, Y. S. Heo, A. M. Bode, H. J. Lee, and Z. Dong. 2007. Equol, a metabolite of the soybean isoflavone daidzein, inhibits neoplastic cell transformation by targeting the MEK/ERK/p90RSK/activator protein-1 pathway. *J Biol Chem* 282(45):32856–66.

Kang, N. J., K. W. Lee, B. J. Shin, S. K. Jung, M. K. Hwang, A. M. Bode, Y. S. Heo, H. J. Lee, and Z. Dong. 2009. Caffeic acid, a phenolic phytochemical in coffee, directly inhibits Fyn kinase activity and UVB-induced COX-2 expression. *Carcinogenesis* 30(2):321–30.

Karlsson, R. 2004. SPR for molecular interaction analysis: A review of emerging application areas. *J Mol Recognit* 17(3):151–61.

Katiyar, S. K., R. Agarwal, and H. Mukhtar. 1996. Inhibition of tumor promotion in SENCAR mouse skin by ethanol extract of *Zingiber officinale* rhizome. *Cancer Res* 56(5):1023–30.

Kim, J. E., J. Y. Kwon, D. E. Lee, N. J. Kang, Y. S. Heo, K. W. Lee, and H. J. Lee. 2009. MKK4 is a novel target for the inhibition of tumor necrosis factor-alpha-induced vascular endothelial growth factor expression by myricetin. *Biochem Pharmacol* 77(3):412–21.

Kim, S. O., K. S. Chun, J. K. Kundu, and Y. J. Surh. 2004. Inhibitory effects of [6]-gingerol on PMA-induced COX-2 expression and activation of NF-kappaB and p38 MAPK in mouse skin. *Biofactors* 21(1-4):27–31.

Kim, S. O., J. K. Kundu, Y. K. Shin, J. H. Park, M. H. Cho, T. Y. Kim, and Y. J. Surh. 2005. [6]-Gingerol inhibits COX-2 expression by blocking the activation of p38 MAP kinase and NF-kappaB in phorbol ester-stimulated mouse skin. *Oncogene* 24(15):2558–67.

Kiuchi, F., S. Iwakami, M. Shibuya, F. Hanaoka, and U. Sankawa. 1992. Inhibition of prostaglandin and leukotriene biosynthesis by gingerols and diarylheptanoids. *Chem Pharm Bull (Tokyo)* 40(2):387–91.

Ko, C. H., S. C. Shen, T. J. Lee, and Y. C. Chen. 2005. Myricetin inhibits matrix metalloproteinase 2 protein expression and enzyme activity in colorectal carcinoma cells. *Mol Cancer Ther* 4(2):281–90.

Kumamoto, T., M. Fujii, and D. X. Hou. 2009a. Akt is a direct target for myricetin to inhibit cell transformation. *Mol Cell Biochem* 332(1-2):33–41.

———. 2009b. Myricetin directly targets JAK1 to inhibit cell transformation. *Cancer Lett* 275(1):17–26.

Kundu, J. K., and Y. J. Surh. 2008. Cancer chemopreventive and therapeutic potential of resveratrol: Mechanistic perspectives. *Cancer Lett* 269(2):243–61.

Langner, E., S. Greifenberg, and J. Gruenwald. 1998. Ginger: History and use. *Adv Ther* 15(1):25–44.

Lawrence, D. S., and J. Niu. 1998. Protein kinase inhibitors: The tyrosine-specific protein kinases. *Pharmacol Ther* 77(2):81–114.

Lee, E., K. K. Park, J. M. Lee, K. S. Chun, J. Y. Kang, S. S. Lee, and Y. J. Surh. 1998. Suppression of mouse skin tumor promotion and induction of apoptosis in HL-60 cells by *Alpinia oxyphylla* Miquel (Zingiberaceae). *Carcinogenesis* 19(8):1377–81.

Lee, E., and Y. J. Surh. 1998. Induction of apoptosis in HL-60 cells by pungent vanilloids, [6]-gingerol and [6]-paradol. *Cancer Lett* 134(2):163–8.

Lee, K. W., N. J. Kang, Y. S. Heo, E. A. Rogozin, A. Pugliese, M. K. Hwang, G. T. Bowden, A. M. Bode, H. J. Lee, and Z. Dong. 2008a. Raf and MEK protein kinases are direct molecular targets for the chemopreventive effect of quercetin, a major flavonol in red wine. *Cancer Res* 68(3):946–55.

Lee, K. W., N. J. Kang, E. A. Rogozin, H. G. Kim, Y. Y. Cho, A. M. Bode, H. J. Lee, Y. J. Surh, G. T. Bowden, and Z. Dong. 2007. Myricetin is a novel natural inhibitor of neoplastic cell transformation and MEK1. *Carcinogenesis* 28(9):1918–27.

Lee, K. W., N. J. Kang, E. A. Rogozin, S. M. Oh, Y. S. Heo, A. Pugliese, A. M. Bode, H. J. Lee, and Z. Dong. 2008b. The resveratrol analogue 3,5,3′,4′,5′-pentahydroxy-trans-stilbene inhibits cell transformation via MEK. *Int J Cancer* 123(11):2487–96.

Li, H., Z. Gao, L. Kang, H. Zhang, K. Yang, K. Yu, X. Luo, W. Zhu, K. Chen, J. Shen, X. Wang, and H. Jiang. 2006. TarFisDock: A web server for identifying drug targets with docking approach. *Nucleic Acids Res* 34(Web Server issue):W219–24.

Li, M., Z. He, S. Ermakova, D. Zheng, F. Tang, Y. Y. Cho, F. Zhu, W. Y. Ma, Y. Sham, E. A. Rogozin, A. M. Bode, Y. Cao, and Z. Dong. 2007. Direct inhibition of insulin-like growth factor-I receptor kinase activity by (-)-epigallocatechin-3-gallate regulates cell transformation. *Cancer Epidemiol Biomarkers Prev* 16(3):598–605.

Lin, H. Y., L. Lansing, J. M. Merillon, F. B. Davis, H. Y. Tang, A. Shih, X. Vitrac, S. Krisa, T. Keating, H. J. Cao, J. Bergh, S. Quackenbush, and P. J. Davis. 2006. Integrin alphaVbeta3 contains a receptor site for resveratrol. *FASEB J* 20(10):1742–4.

Lu, H., and C. Huan. 2007. Transcription factor NFAT, its role in cancer development, and as a potential target for chemoprevention. *Curr Cancer Drug Targets* 7(4):343–53.

Lu, J., L. V. Papp, J. Fang, S. Rodriguez-Nieto, B. Zhivotovsky, and A. Holmgren. 2006. Inhibition of Mammalian thioredoxin reductase by some flavonoids: Implications for myricetin and quercetin anticancer activity. *Cancer Res* 66(8):4410–8.

Manning, G., D. B. Whyte, R. Martinez, T. Hunter, and S. Sudarsanam. 2002. The protein kinase complement of the human genome. *Science* 298(5600):1912–34.

McDonnell, J. M. 2001. Surface plasmon resonance: Towards an understanding of the mechanisms of biological molecular recognition. *Curr Opin Chem Biol* 5(5):572–7.

McWilliam, H., F. Valentin, M. Goujon, W. Li, M. Narayanasamy, J. Martin, T. Miyar, and R. Lopez. 2009. Web services at the European Bioinformatics Institute 2009. *Nucleic Acids Res* 37(Web Server issue):W6–10.

Medyouf, H., and J. Ghysdael. 2008. The calcineurin/NFAT signaling pathway: A novel therapeutic target in leukemia and solid tumors. *Cell Cycle* 7(3):297–303.

Miean, K. H., and S. Mohamed. 2001. Flavonoid (myricetin, quercetin, kaempferol, luteolin, and apigenin) content of edible tropical plants. *J Agric Food Chem* 49(6):3106–12.

Mukhtar, H., M. Das, W. A. Khan, Z. Y. Wang, D. P. Bik, and D. R. Bickers. 1988. Exceptional activity of tannic acid among naturally occurring plant phenols in protecting against 7,12-dimethylbenz(a)anthracene-, benzo(a)pyrene-, 3-methylcholanthrene-, and N-methyl-N-nitrosourea-induced skin tumorigenesis in mice. *Cancer Res* 48(9):2361–5.

Ohren, J. F., H. Chen, A. Pavlovsky, C. Whitehead, E. Zhang, P. Kuffa, C. Yan, P. McConnell, C. Spessard, C. Banotai, W. T. Mueller, A. Delaney, C. Omer, J. Sebolt-Leopold, D. T. Dudley, I. K. Leung, C. Flamme, J. Warmus, M. Kaufman, S. Barrett, H. Tecle, and C. A. Hasemann. 2004. Structures of human MAP kinase kinase 1 (MEK1) and MEK2 describe novel noncompetitive kinase inhibition. *Nat Struct Mol Biol* 11(12):1192–7.

Palermo, C. M., C. A. Westlake, and T. A. Gasiewicz. 2005. Epigallocatechin gallate inhibits aryl hydrocarbon receptor gene transcription through an indirect mechanism involving binding to a 90 kDa heat shock protein. *Biochemistry* 44(13):5041–52.

Park, K. K., K. S. Chun, J. M. Lee, S. S. Lee, and Y. J. Surh. 1998. Inhibitory effects of [6]-gingerol, a major pungent principle of ginger, on phorbol ester-induced inflammation, epidermal ornithine decarboxylase activity and skin tumor promotion in ICR mice. *Cancer Lett* 129(2):139–44.

Pervaiz, S., and A. L. Holme. 2009. Resveratrol: Its biologic targets and functional activity. *Antioxid Redox Signal* 11(11):2851–97.

Phillips, K. S., and Q. Cheng. 2007. Recent advances in surface plasmon resonance based techniques for bioanalysis. *Anal Bioanal Chem* 387(5):1831–40.

Pirola, L., and S. Frojdo. 2008. Resveratrol: one molecule, many targets. *IUBMB Life* 60(5):323–32.

Ramachandran, G. N., C. Ramakrishnan, and V. Sasisekharan. 1963. Stereochemistry of polypeptide chain configurations. *J Mol Biol* 7:95–9.

Reagan-Shaw, S., H. Mukhtar, and N. Ahmad. 2008. Resveratrol imparts photoprotection of normal cells and enhances the efficacy of radiation therapy in cancer cells. *Photochem Photobiol* 84(2):415–21.

Saiko, P., M. Pemberger, Z. Horvath, I. Savinc, M. Grusch, N. Handler, T. Erker, W. Jaeger, M. Fritzer-Szekeres, and T. Szekeres. 2008a. Novel resveratrol analogs induce apoptosis and cause cell cycle arrest in HT29 human colon cancer cells: Inhibition of ribonucleotide reductase activity. *Oncol Rep* 19(6):1621–6.

Saiko, P., A. Szakmary, W. Jaeger, and T. Szekeres. 2008b. Resveratrol and its analogs: Defense against cancer, coronary disease and neurodegenerative maladies or just a fad? *Mutat Res* 658(1-2):68–94.

Sanchez, I., R. T. Hughes, B. J. Mayer, K. Yee, J. R. Woodgett, J. Avruch, J. M. Kyriakis, and L. I. Zon. 1994. Role of SAPK/ERK kinase-1 in the stress-activated pathway regulating transcription factor c-Jun. *Nature* 372(6508):794–8.

Sazuka, M., M. Isemura, and S. Isemura. 1998. Interaction between the carboxyl-terminal heparin-binding domain of fibronectin and (-)-epigallocatechin gallate. *Biosci Biotechnol Biochem* 62(5):1031–2.

Schrödinger Suite. 2006. QM-Polarized Ligand Docking Protocol. Schrödinger, LLC, New York.

Sebolt-Leopold, J. S., D. T. Dudley, R. Herrera, K. Van Becelaere, A. Wiland, R. C. Gowan, H. Tecle, S. D. Barrett, A. Bridges, S. Przybranowski, W. R. Leopold, and A. R. Saltiel. 1999. Blockade of the MAP kinase pathway suppresses growth of colon tumors *in vivo*. *Nat Med* 5(7):810–6.

Shim, J. H., H. S. Choi, A. Pugliese, S. Y. Lee, J. I. Chae, B. Y. Choi, A. M. Bode, and Z. Dong. 2008. (-)-Epigallocatechin gallate regulates CD3-mediated T-cell receptor signaling in leukemia through the inhibition of ZAP-70 kinase. *J Biol Chem* 283(42):28370–9.

Shukla, Y., and M. Singh. 2007. Cancer preventive properties of ginger: a brief review. *Food Chem Toxicol* 45(5):683–90.

Siu, Y. T., and D. Y. Jin. 2007. CREB—a real culprit in oncogenesis. *FEBS J* 274(13):3224–32.

Smyth, LA, and I Collins. 2009. Measuring and interpreting the selectivity of protein kinase inhibitors. *J Chem Biol* 2(3):131–151.

Soleas, G. J., L. Grass, P. D. Josephy, D. M. Goldberg, and E. P. Diamandis. 2002. A comparison of the anticarcinogenic properties of four red wine polyphenols. *Clin Biochem* 35(2):119–24.

Srivastava, K. C., and T. Mustafa. 1992. Ginger (*Zingiber officinale*) in rheumatism and musculoskeletal disorders. *Med Hypotheses* 39(4):342–8.

Steele, V. E. 2003. Current mechanistic approaches to the chemoprevention of cancer. *J Biochem Mol Biol* 36(1):78–81.

Steele, V. E., and G. J. Kelloff. 2005. Development of cancer chemopreventive drugs based on mechanistic approaches. *Mutat Res* 591(1-2):16–23.

Surh, Y. J. 2002. Anti-tumor promoting potential of selected spice ingredients with antioxidative and anti-inflammatory activities: a short review. *Food Chem Toxicol* 40(8):1091–7.

Surh, Y. J., E. Lee, and J. M. Lee. 1998. Chemoprotective properties of some pungent ingredients present in red pepper and ginger. *Mutat Res* 402(1-2):259–67.

Surh, Y. J., K. K. Park, K. S. Chun, L. J. Lee, E. Lee, and S. S. Lee. 1999. Anti-tumor-promoting activities of selected pungent phenolic substances present in ginger. *J Environ Pathol Toxicol Oncol* 18(2):131–9.

Suzuki, Y., and M. Isemura. 2001. Inhibitory effect of epigallocatechin gallate on adhesion of murine melanoma cells to laminin. *Cancer Lett* 173(1):15–20.

Tachibana, H., K. Koga, Y. Fujimura, and K. Yamada. 2004. A receptor for green tea polyphenol EGCG. *Nat Struct Mol Biol* 11(4):380–1.

Tian, W. X. 2006. Inhibition of fatty acid synthase by polyphenols. *Curr Med Chem* 13(8):967–77.

Venter, J. C., M. D. Adams, E. W. Myers, P. W. Li, R. J. Mural, G. G. Sutton, H. O. Smith, M. Yandell, C. A. Evans, R. A. Holt, J. D. Gocayne, P. Amanatides, R. M. Ballew, D. H. Huson, J. R. Wortman, Q. Zhang, C. D. Kodira, X. H. Zheng, L. Chen, M. Skupski, G. Subramanian, P. D. Thomas, J. Zhang, G. L. Gabor Miklos, C. Nelson, S. Broder, A. G. Clark, J. Nadeau, V. A. McKusick, N. Zinder, A. J. Levine, R. J. Roberts, M. Simon, C. Slayman, M. Hunkapiller, R. Bolanos, A. Delcher, I. Dew, D. Fasulo, M. Flanigan, L. Florea, A. Halpern, S. Hannenhalli, S. Kravitz, S. Levy, C. Mobarry, K. Reinert, K. Remington, J.Abu-Threideh, E. Beasley, K. Biddick, V. Bonazzi, R. Brandon, M. Cargill, I. Chandramouliswaran, R. Charlab,K. Chaturvedi, Z. Deng, V. Di Francesco, P. Dunn, K. Eilbeck, C. Evangelista, A. E. Gabrielian, W. Gan, W. Ge, F. Gong, Z. Gu, P. Guan, T. J. Heiman, M. E. Higgins, R. R. Ji, Z. Ke, K. A. Ketchum, Z. Lai, Y. Lei, Z. Li, J. Li, Y. Liang, X. Lin, F. Lu, G. V. Merkulov, N. Milshina, H. M. Moore, A. K. Naik, V. A. Narayan, B. Neelam, D. Nusskern, D. B. Rusch, S. Salzberg, W. Shao, B. Shue, J. Sun, Z. Wang, A. Wang, X. Wang, J. Wang, M. Wei, R. Wides, C. Xiao, C. Yan, A. Yao, J. Ye, M. Zhan, W. Zhang, H. Zhang, Q. Zhao, L. Zheng, F. Zhong, W. Zhong, S. Zhu, S. Zhao, D. Gilbert, S. Baumhueter, G. Spier, C. Carter, A. Cravchik, T. Woodage, F. Ali, H. An, A. Awe, D. Baldwin, H. Baden, M. Barnstead, I. Barrow, K. Beeson, D. Busam, A. Carver, A. Center, M. L. Cheng, L. Curry, S. Danaher, L. Davenport, R. Desilets, S. Dietz, K. Dodson, L. Doup, S. Ferriera, N. Garg, A. Gluecksmann, B. Hart, J. Haynes, C. Haynes, C. Heiner, S. Hladun, D. Hostin, J. Houck, T. Howland, C. Ibegwam, J. Johnson, F. Kalush, L. Kline, S. Koduru, A. Love, F. Mann, D. May, S. McCawley, T. McIntosh, I. McMullen, M. Moy, L. Moy, B. Murphy, K. Nelson, C. Pfannkoch, E. Pratts, V. Puri, H. Qureshi, M. Reardon, R. Rodriguez, Y. H. Rogers, D. Romblad, B. Ruhfel, R. Scott, C. Sitter, M. Smallwood, E. Stewart, R. Strong, E. Suh, R. Thomas, N. N. Tint, S. Tse, C. Vech, G. Wang, J. Wetter, S. Williams, M. Williams, S. Windsor, E. Winn-Deen, K. Wolfe, J. Zaveri, K. Zaveri, J. F. Abril, R. Guigo, M. J. Campbell, K. V. Sjolander, B. Karlak, A. Kejariwal, H. Mi, B. Lazareva, T. Hatton, A. Narechania, K. Diemer, A. Muruganujan, N. Guo, S. Sato, V. Bafna, S. Istrail, R. Lippert, R. Schwartz, B. Walenz, S. Yooseph, D. Allen, A. Basu, J. Baxendale, L. Blick, M. Caminha, J. Carnes-Stine, P. Caulk, Y. H. Chiang, M. Coyne, C. Dahlke,

A. Mays, M. Dombroski, M. Donnelly, D. Ely, S. Esparham, C. Fosler, H. Gire, S. Glanowski, K. Glasser, A. Glodek, M. Gorokhov, K. Graham, B. Gropman, M. Harris, J. Heil, S. Henderson, J. Hoover, D. Jennings, C. Jordan, J. Jordan, J. Kasha, L. Kagan, C. Kraft, A. Levitsky, M. Lewis, X. Liu, J. Lopez, D. Ma, W. Majoros, J. McDaniel, S. Murphy, M. Newman, T. Nguyen, N. Nguyen, M. Nodell, S. Pan, J. Peck, M. Peterson, W. Rowe, R. Sanders, J. Scott, M. Simpson, T. Smith, A. Sprague, T. Stockwell, R. Turner, E. Venter, M. Wang, M. Wen, D. Wu, M. Wu, A. Xia, A. Zandieh, and X. Zhu. 2001. The sequence of the human genome. *Science* 291(5507):1304–51.

Widyarini, S., A. J. Husband, and V. E. Reeve. 2005. Protective effect of the isoflavonoid equol against hairless mouse skin carcinogenesis induced by UV radiation alone or with a chemical cocarcinogen. *Photochem Photobiol* 81(1):32–7.

Williamson, M. P., T. G. McCormick, C. L. Nance, and W. T. Shearer. 2006. Epigallocatechin gallate, the main polyphenol in green tea, binds to the T-cell receptor, CD4: Potential for HIV-1 therapy. *J Allergy Clin Immunol* 118(6):1369–74.

Xiang, Z. 2006. Advances in homology protein structure modeling. *Curr Protein Pept Sci* 7(3):217–27.

Yang, S., K. Irani, S. E. Heffron, F. Jurnak, and F. L. Meyskens, Jr. 2005. Alterations in the expression of the apurinic/apyrimidinic endonuclease-1/redox factor-1 (APE/Ref-1) in human melanoma and identification of the therapeutic potential of resveratrol as an APE/Ref-1 inhibitor. *Mol Cancer Ther* 4(12):1923–35.

Yin, Z., E. C. Henry, and T. A. Gasiewicz. 2009. (-)-Epigallocatechin-3-gallate is a novel Hsp90 inhibitor. *Biochemistry* 48(2):336–45.

Zykova, T. A., F. Zhu, X. Zhai, W. Y. Ma, S. P. Ermakova, K. W. Lee, A. M. Bode, and Z. Dong. 2008. Resveratrol directly targets COX-2 to inhibit carcinogenesis. *Mol Carcinog* 47(10):797–805.

# 14 Green Tea Polyphenols, DNA Repair, and Prevention of Photocarcinogenesis

*Santosh K. Katiyar*

## CONTENTS

## INTRODUCTION

Naturally occurring plant polyphenols constitute a wide variety of phytochemicals which are distributed in plant foods, including fruits, vegetables, seeds, nuts, flowers, and bark. Important dietary sources of polyphenols are grape seeds, grape skin, tea, apples, red wine, onions and cacao, and so on (Manach et al. 2004). These polyphenols possess beneficial health effects of dietary sources. Among the well-known polyphenols, the polyphenols from tea and more specifically from green tea are better known for their health benefits in many organs as well as in general health. In this chapter, we particularly discuss and describe the health benefits of green tea polyphenols (GTPs) with particular emphasis on protection against ultraviolet (UV) radiation-induced skin cancer through repair of damaged DNA.

## GREEN TEA AND GREEN TEA POLYPHENOLS

Beverage-grade commercial tea is manufactured from the leaves of the plant *Camellia sinensis* and is commercially available mainly in three forms: green, black, and oolong tea (Manach et al. 2004; Hara 2001; Katiyar and Elmets 2001). Because of characteristic aroma, flavor, and associated health benefits, tea is consumed as a popular beverage worldwide. Of the total commercial tea production worldwide, about 78% is consumed in the form of black tea, mainly in Western countries and some Asian countries, and 20% is consumed in the form of green tea. Green tea is mainly consumed in some Asian countries such as Japan, China, Korea, parts of India, and a few countries in North Africa and the Middle East (Hara 2001; Katiyar and Elmets 2001). Now, the consumption of green tea is also increasing in Western countries including the United States because of increasingly new investigations on its health benefits and anti-carcinogenic activities in various organs. The remaining 2% of tea is consumed in the form of oolong tea, which is a partially fermented tea product, and mainly consumed in some parts of Southeastern China (Hara 2001; Katiyar and Elmets 2001).

The term *green tea* refers to the product manufactured from the fresh leaves of the tea. The characteristic aroma and health benefits of tea are associated with the presence of catechin/epicatechin molecules and their derivatives, which are commonly called "polyphenols." In general, the production of green tea is characterized by an initial heating process, which inactivates the enzyme polyphenol oxidase, which is responsible for the conversion of the flavanols in the green leaf into the dark polyphenolic compounds of black tea.

The basic steps of manufacturing different tea varieties are similar except to protect and develop their aroma during fermentation process, which also controls the oxidation status of the individual catechin/epicatechin molecules and their derivatives present in fresh tea leaves (Hara 2001). The major polyphenolic constituents present in green tea are (+)-catechins, (−)-epicatechin, (−)-epigallocatechin, (−)-epicatechin-3-gallate, and (−)-epigallocatechin-3-gallate (EGCG) (Manach et al. 2004; Hara 2001; Katiyar and Elmets 2001). The chemical structures of major catechins or polyphenols are shown in Figure 14.1.

These monomeric catechins and their derivatives are important molecules which possess therapeutic properties and anticarcinogenic effects. Studies have shown that polyphenols provide antioxidant and anti-inflammatory activities in nature and have been shown to possess anticarcinogenic

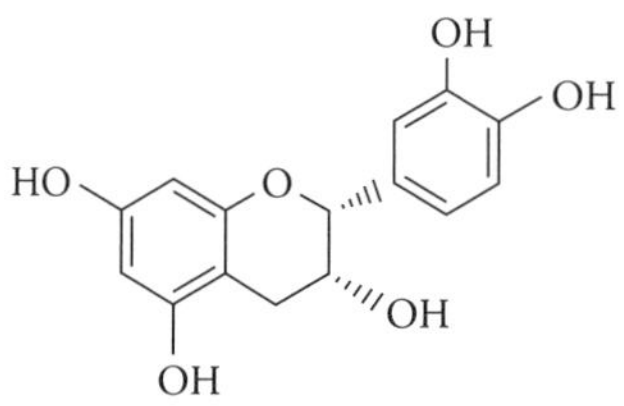

(-)-Epicatechin

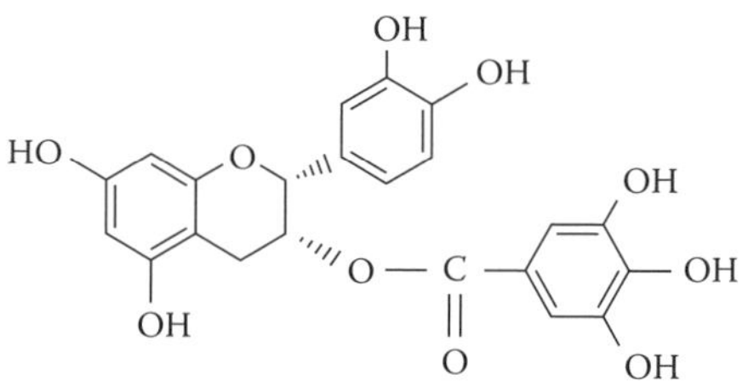

(-)-Epicatechin-3-gallate

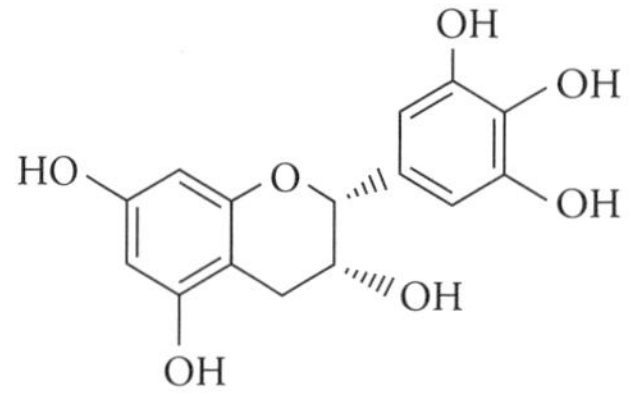

(-)-Epigallocatechin

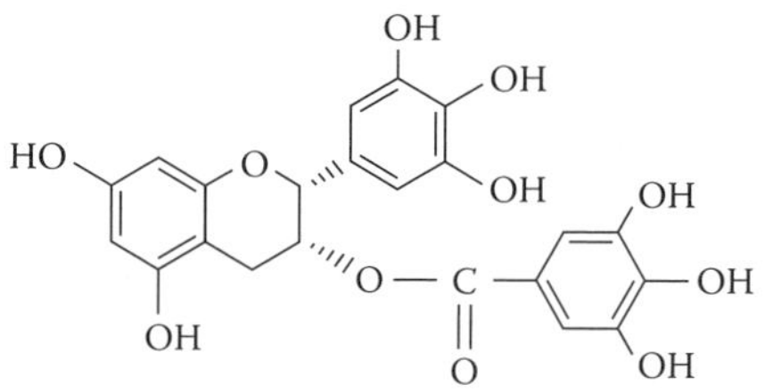

(-)-Epigallocatechin 3-gallate

**FIGURE 14.1** Chemical structures of major epicatechin derivatives present in green tea. These epicatechins are also known as "polyphenols."

activity in several *in vitro* and *in vivo* systems (Baliga and Katiyar 2006; Katiyar et al. 2007). Of these major polyphenols, EGCG is the major and the most effective or active molecule, which has been extensively studied in several disease models including its use in skin photoprotection.

During black tea manufacturing process, the catechin derivatives are oxidized, polymerized during fermentation, and converted into a less well-defined group of compounds known as thearubigens and theaflavins. These compounds with a molecular weight distribution of 1,000–40,000 account for about 15% of dry weight solids of black tea (Hara 2001). Investigations indicate that thearubigins and theaflavins are also anticarcinogenic in nature but their studies are limited compared to the catechins of green tea.

The *in vitro* cell culture and *in vivo* animal experimental studies indicate that GTPs are better chemopreventive agents than those present in black tea. Thus, in this chapter we describe primarily the anti-photocarcinogenic potential of GTPs and their mechanism of actions, with particular emphasis on the DNA repair mechanisms, and its effect on immune system and UV-induced skin tumor development.

## SKIN, SKIN LAYERS, AND CELLS

Skin is much more than a passive physical barrier between the external environment and internal tissues. The skin is the largest organ of the body and comprising a surface area of approximately 1.5–2.0 m$^2$ which protects the internal organs of the body against the detrimental effects of environmental and xenobiotic agents. Thus, skin provides a protective covering at this crucial interface between inside and outside. Morphologically, skin is a composite of a variety of cell types and organellar bodies, each of which has a particular function, and is made up of several different layers, each with particular properties. The major layers include the epidermis, the dermis, and the hypodermis (Figure 14.2). The epidermis is a stratifying layer of epithelial cells that overlies the connective tissue layer, the dermis. The dermis is divided into papillary dermis and reticular dermis. The epidermis and dermis are supported by an internal layer of adipose tissue, called the hypodermis. The major cell type of the epidermis is the keratinocytes. It comprises >90% of the cells of the epidermal layer. Other cell types are Langerhans cells, melanoma cells, and γδ T cells. However, the dermal components of the skin, including dermal fibroblast, microvasculature endothelial cells, dermal dendritic cells, mast cells, and resident perivascular T-cells, also participate in the cutaneous functions. These epidermal and dermal components have evolved into a dynamic network of interacting cells capable of sensing a variety of perturbations including trauma, UV irradiation, toxic environmental chemicals, and pathogenic microorganisms in the cutaneous environment and rapidly sending appropriate signals that alert and recruit other branches of the immune system. In laboratory animals like the mouse, the epidermis is about 2–3 cell layers thick, whereas in human skin, the epidermis is quite thick and comprises of about 8–15 cell layers thick. Among many environmental and xenobiotic factors, the exposure of the skin to solar UV radiation is the

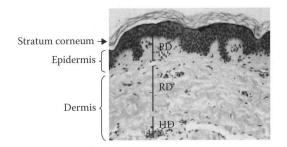

**FIGURE 14.2** Structure of the normal human skin. Stratum corneum is a layer of dead cells and hydrophobic in nature. PD = papillary dermis; RD = reticular dermis; HD = hypodermis.

key factor in the initiation of several skin disorders, such as wrinkling, scaling, dryness, hypopig-mentation and hyperpigmentation, and skin cancer (deGruijl and van der Leun 1994; Ichihashi et al. 2003; Mukhtar and Elmets 1996). Statistically, the average annual UV dose that an average American typically receives in a year is about 2500–3300 mJ/cm$^2$. An average female is exposed to an average of 2200 mJ/cm$^2$ and males 2800 mJ/cm$^2$ each year with an additional exposure of about 800 mJ/cm$^2$ of solar UVB radiation during a conservative vacation period (Godar et al. 2001; Godar 2001).

## SOLAR ULTRAVIOLET RADIATION SPECTRUM

Sunlight is a major source of UV radiation. Solar radiation constitutes approximately 5% of the electromagnetic spectrum that reaches the Earth's surface. The UV spectrum which reaches on the surface of the Earth consists of approximately 5% UVB and 95% UVA. Most of the UVC fraction of solar UV spectrum is blocked by the Earth's stratospheric ozone layer. Solar UV spectrum is divided mainly into three categories based on their wavelengths and biological effects (deGruijl and van der Leun 1994): (1) short-wave UVC (200–290 nm), (2) mid-wave UVB (290–320 nm), and (3) long-wave UVA (320–400 nm).

1. *Short-wave UVC*: UVC radiation normally does not reach the surface of the Earth because it is largely absorbed by the atmospheric ozone layer. UVC radiation has enormous energy and are mutagenic in nature, and therefore comparatively more harmful than UVA or UVB radiation.
2. *Mid-wave UVB*: UVB radiations are also mutagenic and carcinogenic in nature and responsible for a variety of skin diseases. UVB constitutes about 5% of the total solar UV spectrum. UVB radiation can penetrate inside epidermis of the skin and can induce oxida-tive stress, immunosuppression, DNA damage, premature aging of the skin, and skin can-cers including the melanoma and nonmelanoma (deGruijl and van der Leun 1994; Katiyar et al. 2007).
3. *Long-wave UVA*: UVA comprises the largest spectrum of solar UV radiation (≈95%). UVA radiation can penetrate deeper into the epidermis and dermis of the skin. Extensive expo-sure of the skin to UVA can lead to benign tumor formation as well as malignant cancers (Bachelor and Bowden 2004; Wang et al. 2001). Exposure of the skin to UVA induces the generation of singlet oxygen and hydroxyl free radicals, which can cause damage to cel-lular macromolecules like proteins, lipids, and DNA (DiGiovanni 1992). UVA-induced oxidative stress can enhance the process of photoaging in the form of skin sagging rather than wrinkling (Krutmann 2001) and also can suppress some immunological functions (Ullrich 1995).

## SOLAR UV RADIATION IS AN ENVIRONMENTAL CARCINOGEN

UV radiation, particularly UVB, can act as a tumor initiator (Kligman et al. 1980), tumor promoter (Katiyar et al. 1997), and cocarcinogen (Donawho and Kripke 1991; Ziegler et al. 1994). Exposure of the skin to UVB radiation induces a variety of biological effects, including induction of inflamma-tion and inflammatory mediators, immunologic alterations and suppression of the immune system, generation of oxidative stress and DNA damage, and so on. All together, these effects play impor-tant roles in the development of various skin diseases including the development of UV-induced skin cancers (Taylor et al. 1990; Hruza and Pentland 1993; Katiyar and Mukhtar 2001). Although the skin possesses an elaborate defense system consisting of enzymatic and nonenzymatic compo-nents to protect skin from adverse biological effects of UV radiation, the excessive exposure to UV radiation overwhelms and depletes the cutaneous defense system and its ability to protect the skin from deleterious effects (Katiyar et al. 1997, 2001; Katiyar and Mukhtar 2001; Mittal et al. 2003a).

## UV RADIATIONS ARE AN ETIOLOGICAL FACTOR FOR SKIN CANCER INCIDENCE

Chronic exposure to solar UV radiation is the major etiologic agent for over 1 million new non-melanoma skin cancers in the United States each year and is an important factor in the pathogenesis of melanoma, premature aging of the skin, and immunosuppression. Nonmelanoma skin cancer is by far the most common cutaneous malignancy and has a tremendous impact on public health and healthcare expenditures. Skin cancer is mainly divided into two major types of cancers, melanoma and nonmelanoma. Melanoma develops from melanocytes, a kind of skin cell, which contain a pigment called melanin. Depending on the type of cells, nonmelanoma skin cancers are further classified as squamous cell and basal cell carcinomas. Nonmelanoma skin cancers rarely metastasize while melanoma cancer cells can metastasize to other parts of the body, and may lead to death. Chronic exposure of the skin to sunlight, which contains UV radiation, leads to the development of actinic keratoses in human skin, and these actinic keratoses can lead to nonmelanoma skin cancers. Most of the mortality occurs due to melanoma because of its ability to metastasize to other organs. Nonmelanoma skin cancers do not metastasize and can be easily cured, therefore mortality is rare. Protection from melanoma and nonmelanoma skin cancers requires protection from the exposure to solar UV radiation. For the prevention of skin cancers, efforts can be made to educate the people to minimize sun exposure specifically during mid-day (11:00 a.m. to 4:00 p.m.) at which time the intensity of UV radiation is high, wear protective clothing including full sleeve shirt and hat, and adequately use broad-spectrum sunscreen before going outdoors for recreation purposes.

## SUNSCREEN INADEQUATELY PROTECTS AGAINST UV CARCINOGENESIS

Sunscreens are widely advocated as a means of reducing skin cancer risk. This advice is largely based on extrapolation from animal studies, as it is difficult to evaluate long-term protection in humans. Limited data indicate that sunscreens can inhibit actinic keratoses that are regarded as precursors of squamous cell carcinoma (Thompson 1993; Naylor et al. 1995). A study conducted in Australia has shown that daily use of a SPF 16 sunscreen, over a period of 4.5 years, reduced the total number of squamous cell carcinoma by 40% but not the number of people with the tumor. No protective effect was seen for basal cell carcinoma (Green et al. 1999). Some studies even show that sunscreen use is associated with an increased risk of melanoma (Weinstock 1999). Haywood et al. (2003) also showed that sunscreens inadequately protect against the UV-induced free radicals in skin that have been implicated in skin aging and melanoma. Moreover, it is difficult to find an effective sunscreen that can provide full spectral protection against UV light. In addition, sunscreen ingredients may become free radicals themselves when activated by UV irradiation (Xu et al. 2001), and sunscreen chemicals may be absorbed into the skin (Cross et al. 2001) potentially causing harm. For these reasons, the use of antioxidants and immunomodulators, such as the consumption of green tea, may prove highly beneficial for the reduction of UV-induced skin cancer.

## POLYPHENOLS AND SKIN PHOTOPROTECTION

From ancient times people have used plant products for the health benefits of the skin. This interest is continuing to be increased and a considerable human population is constantly using naturally occurring botanicals/phytochemicals for the prevention of sunlight-induced photodamaging effects, including photoaging or premature aging of the skin, wrinkling, and other skin disorders. Botanicals, specifically dietary, possessing anti-inflammatory, immunomodulatory, and antioxidant properties are among the most promising group of phytochemicals that can be exploited as ideal chemopreventive agents for a variety of skin disorders in general. Recent advances in our understanding at the cellular and molecular levels of carcinogenesis have led to the development of promising strategies for the chemoprevention of cancers. Chemoprevention is a means of cancer control by the use of specific natural or synthetic chemical substances which can suppress, retard, or reverse the process of carcinogenesis and other skin diseases. Thus, chemoprevention offers a realistic promise or strategy for controlling the risk of skin diseases. Further, chemopreventive approach appears to have practical implications in reducing skin cancer risk because unlike the carcinogenic or polluted environmental factors that are difficult to control, individuals can modify their dietary habits and lifestyle in combination with a careful use of skin care products to prevent solar UV radiation-induced photodamaging effects. Studies have shown the chemopreventive potential of several naturally occurring phytochemicals, such as GTPs, silymarin, retinoids, grape seed proanthocyanidins, and so on against UV radiation-induced adverse effects including inflammation, oxidative stress, DNA damage, and photocarcinogenesis (Katiyar and Elmets 2001; Baliga and Katiyar 2006; Surh 1999; Pinnell 2003). Here, we will summarize and discuss the recent developments in the area of photoprotective potential of GTPs with emphasis on prevention of photocarcinogenesis through rapid repair of damaged DNA after UVB irradiation of the skin.

## GREEN TEA POLYPHENOLS INHIBIT PHOTOCARCINOGENESIS

Nonmelanoma skin cancers, including basal cell and squamous cell carcinomas, represent the most common malignant neoplasms in humans (Miller and Weinstock 1994; Strom 1996; Urbach 1991). Epidemiological, clinical, and biological studies have indicated that chronic exposure of the skin to solar UV radiation is the major etiological agent for the development of skin cancers (Miller and Weinstock 1994; Strom 1996; Brash et al. 1991; Scotto and Fears 1978). Following standard photocarcinogenesis protocols in SKH-1 hairless mice, it has been found that oral administration of GTPs (a mixture of GTPs) in the drinking water of mice resulted in significant protection against skin tumor development in terms of tumor incidence (percentage of mice with tumors), tumor multiplicity, and tumor size per group compared to non-GTPs-treated UVB-irradiated animals (Mantena et al. 2005) (Figure 14.3). A water extract of green tea leaves, which primarily contained a mixture of polyphenolic ingredients, when given as the sole source of drinking water to mice afforded protection against UVB radiation-induced tumorigenesis (Wang et al. 1992a), and also promoted partial regression of established skin papillomas in mice (Wang et al. 1992b). Topical treatment of GTPs or EGCG was also tested against the risk of photocarcinogenesis. Topical treatment of SKH-1 hairless mouse skin with GTPs or EGCG in a hydrophilic cream-based topical formulation significantly inhibited UVB-induced skin tumor development in terms of tumor incidence, tumor multiplicity, and tumor size (Mittal et al. 2003b). To further illustrate the role of polyphenols in skin photoprotection, mechanism-based studies were performed using both *in vitro* cell culture and *in vivo* animal models.

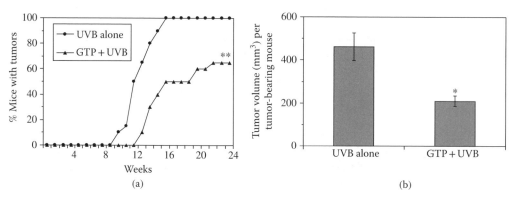

**FIGURE 14.3**  Administration of GTPs (0.2%, w/v) in the drinking water of mice inhibits UVB-induced skin tumorigenesis and tumor growth in SKH-1 hairless mice. (a) The tumor data are presented in terms of the percentage of mice with tumors on a weekly basis. (b) The tumor data are shown in terms of the tumor volume per tumor-bearing mouse as means $\pm$ SD, n = 20. Asterisks indicate different from UVB alone, * $P < 0.005$; ** $P < 0.05$.

# MECHANISTIC STUDIES: DNA AS A MOLECULAR TARGET

## UVB Exposure Induces DNA Damage

Exposure of the skin to UV radiation resulted in DNA damage, which initiates an important cascade of signaling pathways. The DNA photoproducts are altered DNA structures that activate a cascade of responses, beginning with the initiation of cell cycle arrest and activation of DNA repair mechanisms. The biologically harmful effects associated with UV irradiation are largely the result of errors in DNA repair, which can lead to oncogenic mutations (reviewed in Timares et al. 2008). UV-induced DNA damage in the form of thymine dimers or cyclobutane pyrimidine dimers (CPD) is a molecular trigger for the induction of immunosuppression and initiation of photocarcinogenesis (Kripke et al. 1992; Yarosh et al. 1992). Studies have revealed that exposure of the skin to UV radiation results in the formation of CPDs in skin cells (Katiyar et al. 2000). The formation of CPDs occurs immediately after the interaction of photons with the DNA molecule. It has been found that UV exposure of less than one minimal erythema dose is sufficient to damage the DNA in target cells of the skin (Katiyar et al. 2000). The CPDs' formation depends on the penetration ability of the UV radiation inside the skin. The DNA of the majority of epidermal cells got damaged after UVB irradiation.

## GTPs Repair UVB-Induced DNA Damage

Topical treatment of mouse skin with GTPs significantly inhibited UVB-induced DNA damage as assessed using a $^{32}$P-postlabelling technique (Chatterjee et al. 1996). Similarly, topical treatment of human skin with GTPs prior to UV exposure resulted in a dose-dependent inhibition of CPDs formation (Katiyar et al. 2000). It was also found that topical application of green tea or white tea extracts provided human skin protection from solar-simulated UV light (Camouse et al. 2009). These tea extracts were shown to provide protection against the detrimental effects of UV light on cutaneous immunity. The investigators also concluded that these protective effects were not due to direct UV absorption or sunscreen effects. Studies on the effects of GTPs on the DNA repair kinetics and repair mechanisms of UV-induced CPDs have been carried out using *in vitro* cell culture and *in vivo* animal models as well as in human skin. Studies showed that topical treatment of skin with EGCG does not prevent UVB-induced formation of CPDs immediately after UVB irradiation, which indicated that EGCG does not have a significant sunscreen effect. However; in skin samples obtained at 24 hours or 48 hours after UVB exposure, the numbers of CPD-positive cells were significantly reduced (or repaired) in the EGCG-treated mouse skin compared to the control group of mice which were not treated with EGCG (Meeran et al. 2006a). Studies of the DNA repair mechanisms suggested that the rapid repair of UV-induced CPDs by EGCG was mediated through stimulation of a cytokine interleukin-12 (IL-12) on application of the EGCG onto the mouse skin (Meeran et al. 2006a). IL-12 is a 70-kDa heterodimeric protein composed of two disulfide-bonded chains, the p40 and p35 subunits (Wolf et al. 1991; Katiyar 2007b). IL-12 has been shown to possess potent antitumor activity in a wide variety of murine tumor models (Colombo et al. 1996; Brunda et al. 1993; Robertson and Ritz 1996). IL-12 also has been shown to have the capacity to induce DNA repair (Meeran et al. 2006a; Schwarz et al. 2002, 2005) and this concept was verified by testing the effect of EGCG on UV-induced CPD formation in IL-12 knockout mice. EGCG does not remove or repair UV-induced CPDs in the skin of IL-12 knockout mice but repaired in their wild-type counterparts, further confirming the role of IL-12 in rapid repair of DNA damage by this polyphenol (Meeran et al. 2006a). Studies on the effects of oral administration of GTPs in the drinking water of mice on UVB-induced DNA damage also were carried out and it was found that UV-induced DNA damage (i.e., CPDs) was resolved rapidly in the GTPs-treated mice compared to non-GTPs-treated mice (Meeran et al. 2009). Further, the DNA repairing effect of GTPs was less pronounced in the skin of IL-12 knockout mice (Figure 14.4), as was observed in the case of EGCG treatment. Schwarz et al. (2008) observed that treatment of normal human keratinocytes and "human skin equivalent" with GTPs reduced UVB-induced DNA damage in the form of CPDs and that this effect was mediated

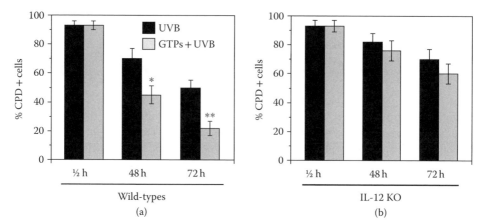

**FIGURE 14.4** Administration of GTPs removes or repairs UVB-induced CPDs more rapidly in the skin of (a) wild-type (IL-12-proficient) mice than in the skin of (b) IL-12 knockout mice. The numbers of CPD-positive cells in each treatment group are summarized in terms of percent positive cells and the data are expressed as the mean ± SD. Significant inhibition versus UVB-exposed alone group, * $P < 0.01$; ** $P < 0.001$.

through the stimulation of IL-12 production. These investigations suggest that the difference in the GTPs-associated DNA repair capacity between IL-12 knockout mice and their wild-type counterparts may be due to the absence of IL-12 in the IL-12 knockout mice. Wei et al. (1998) have shown that water extract of green tea scavenges $H_2O_2$ and inhibits UV-induced oxidative DNA damage in an *in vitro* system. Zhao et al. (1999) demonstrated that application of green tea extract to Epiderm, a reconstituted human skin equivalent, also inhibited psoralen-UVA-induced formation of 8-methoxypsoralen-DNA adducts. Treatment of skin with a green tea extract significantly inhibited DNA damage induced by solar simulator radiation when assessed using a $^{32}$P-postlabeling technique (Chatterjee et al. 1996). These observations demonstrate the potential chemopreventive effects of GTPs against UVB-induced DNA damage.

## GTPs Enhance the Levels of Nucleotide Excision Repair Genes

To identify the DNA repair mechanism in UVB-exposed skin by green tea polyphenols, further studies were conducted in the research laboratory of Dr. Katiyar. It was proposed that a nucleotide excision repair (NER) mechanism is involved in the repair of UVB-induced DNA damage by green tea polyphenols, and that IL-12 has a role in this repairing process (Meeran et al. 2006a,b). To determine whether the NER mechanism is required for the EGCG-induced IL-12-mediated repair of UVB-induced CPDs, NER-deficient fibroblasts from a person suffering from xeroderma pigmentosum complementation group A (*XPA*) and NER-proficient fibroblasts from a healthy person (*XPA*-proficient) were exposed to UVB with or without prior treatment with EGCG. The data analysis of CPD-positive cells revealed that the numbers of CPD-positive cells were significantly lower in EGCG-treated group at 24 hours after UVB exposure in the *XPA*-proficient cells compared to non-EGCG-treated cells. However, EGCG did not significantly remove or repair UVB-induced CPDs in NER-deficient cells. This *in vitro* observation indicated that EGCG-induced DNA repair is mediated through a functional NER mechanism.

## Rapid Repair of UVB-Induced DNA Damage by Green Tea Polyphenols
### Leads to a Reduction in Inflammation

UVB irradiation resulted in inflammation in the skin, and there is increasing evidence that chronic inflammation promotes the initiation of various skin diseases, including the development of skin cancers (DiGiovanni 1992; Mukhtar and Elmets 1996). Both UV-induced inflammatory responses and UV-induced skin tumor development are causally related to UV-induced DNA damage. Therefore, it was of interest to explore the effects of GTPs on DNA repair and their relationship with inflammatory responses in the skin. It is well known that CPDs are formed immediately after the exposure of the skin to UV radiation, and inflammation develops at later stages. Following UV exposure, UV-induced DNA damage in the form of CPDs was repaired or removed more rapidly in the skin of mice that had been treated either with topical application of EGCG or administration of GTPs in drinking water of mice. Subsequently, the levels of UVB-induced inflammation were lowered in the treated mice than the nontreated mice. The levels of inflammation in the mouse skin were assessed through analysis of biomarkers of inflammation, such as cyclooxygenase-2 (COX-2) expression, $PGE_2$ production, and the levels of proinflammatory cytokines, such as tumor necrosis factor-α, interleukin-6, and interleukin-1β. In contrast, this effect of EGCG or GTPs was not observed in IL-12-deficient mice. This may be due to the fact that the treatment of mice with EGCG or GTPs was not able to repair UV-induced DNA damage significantly in the IL-12 knockout mice (Meeran et al. 2009). This new information supports the concept that UV-induced CPDs and inflammatory responses are causally related with the increased risk of photocarcinogenesis. This *in vivo* experimental evidence indicates that the prevention of UVB-induced skin cancer by GTPs or EGCG is mediated through inhibition of UVB-induced inflammation, which in turn is mediated, at least in part, through rapid repair of damaged DNA. The outcome of the studies therefore suggests that regular consumption of green tea or GTPs may be considered as an effective strategy for the prevention of inflammation-associated skin diseases including UV-irradiation-caused skin tumor development.

## UVB-Induced DNA Damage Resulted in Suppression of the Immune System

UV radiation-induced immunosuppression has been implicated in the development of skin cancers, and UVB-induced DNA damage in the form of CPDs has been considered as a molecular trigger for the initiation of immunosuppression and photocarcinogenesis (Katiyar 2007a; Meeran et al. 2006b, 2009). The UV-induced DNA damage also impairs the antigen-presenting capacity of Langerhans cells, which results in a lack of sensitization and the induction of tolerance to contact sensitizers (Vink et al. 1996, 1997). There are evidences suggesting that alterations in the immune system contribute to the pathogenesis of sunlight-induced skin cancer in mice and probably in humans as well (Yoshikawa et al. 1990; Meunier et al. 1998). Chronically immunosuppressed patients living in regions of intense sun exposure experience an exceptionally high rate of skin cancer (reviewed in Meunier et al. 1998; Katiyar 2007a). This observation is consistent with the notion that immune surveillance is an important mechanism designed to prevent the generation and maintenance of neoplastic cells.

## Green Tea Polyphenols Inhibit UVB-Induced Immunosuppression through DNA Repair

As UVB-induced immunosuppression and DNA damage have been implicated in the development of photocarcinogenesis, the effect of GTPs was determined on UVB-induced immunosuppression and whether it is regulated through repair of DNA damage by GTPs. The administration of GTPs in the drinking water of mice inhibited UVB-induced suppression of contact hypersensitivity responses to contact sensitizer and simultaneously the migration of CPD-positive epidermal antigen-presenting cells to the draining lymph nodes of mice was reduced. This indicates that treatment of GTPs might be able to repair UV-induced CPDs in the mice. It is speculated that, as the migrating antigen presenting cells in the epidermis were either not damaged or were repaired in mice, they were able to present antigen to T-cells in the draining lymph nodes resulting in induction of sensitization to contact sensitizer after challenge in contact hypersensitivity reactions. Further, the numbers of CPD-positive cells were significantly higher in the non-GTPs-fed mice in the subcapsular sinus to the paracortical region of the lymph nodes, including the interfollicular areas, which are the sites of T-cell localization (Katiyar et al. 2010). Thus, the damaged DNA in the lymph nodes of non-GTPs-fed mice may adversely affect the ability of the antigen presenting cells to present antigen to T-cells, thus abrogating sensitization after sensitizer treatment. In contrast, the reverse was observed in GTPs-fed mice, and that may be one of the reasons that GTPs prevent UVB-induced immunosuppression in mice (Katiyar et al. 2010).

## INHIBITION OF UVB-INDUCED IMMUNOSUPPRESSION BY GTPS IS MEDIATED THROUGH THE NER MECHANISM

NER is the main mechanism of repair in mammalian cells for the removal of UV radiation-induced DNA damage. Since the treatment of GTPs enhances the removal or repair of UVB-induced DNA damage, it is of interest to examine whether the removal or repair of UV-induced CPDs by GTPs is mediated via induction of NER genes. Subsequent analysis of data reveals that treatment of mice with GTPs in the drinking water of mice increases the levels of some NER genes (e.g., *XPA*, *XPC* and *RPA1*) in UVB-exposed skin sites compared to non-GTPs-fed mice and that may have contributed to the rapid repair of damaged DNA in mouse skin (Katiyar et al. 2010). The role of NER was further confirmed by assessing the effect of GTPs on UVB-induced immunosuppression in xeroderma pigmentosum complementation group A-deficient (*XPA*$^{-/-}$) mice and data were compared with the *XPA*$^{+/+}$ (proficient) mice. That the treatment of mice with GTPs in drinking water prevents UVB-induced suppression of contact hypersensitivity response in *XPA*$^{+/+}$ mice but does not prevent this in *XPA*$^{-/-}$ mice further supports the observations that inhibition of UVB-induced immunosuppression by GTPs requires functional NER genes. This observation was important as the treatment of GTPs does not remove or repair UVB-induced DNA damage in *XPA*$^{-/-}$ (NER-deficient) mice, but repairs in *XPA*$^{+/+}$ (NER-proficient or wild-type) mice which were exposed to UVB (Figure 14.5). These observations were further verified by using NER-deficient cells from *XPA*-patients and NER-proficient cells from healthy persons. Cells derived from patients suffering from xeroderma pigmentosum either lack or have reduced DNA repair capacity due to genetic mutations in several components of the NER. The *XPA* complementation type represents the most severe phenotype, because the *XPA* gene is the most crucial component in the repair process and, thus, cells lacking the *XPA* gene are completely deficient in NER (Carreau et al. 1995; Muotri et al. 2002). Following these experiments, GTPs were able to remove UV-induced CPDs in NER-proficient cells (*XPA*$^{+/+}$) but were not able to remove or repair in NER-deficient (*XPA*$^{-/-}$) human fibroblast cells. These observations indicate that repair of UV-induced DNA damage by GTPs is mediated through the NER mechanism or GTPs-induced DNA repair requires functional NER. These findings have important implications for the chemopreventive mechanism of skin cancer protection by GTPs, and identify a new mechanism by which GTPs prevent UV-induced immunosuppression.

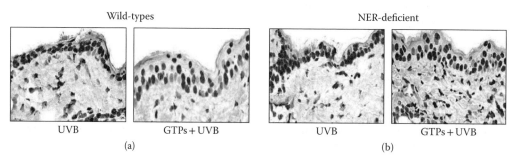

**FIGURE 14.5** GTPs reduced or repaired UVB-induced DNA damage in (a) NER-proficient (wild-types) mice but not in (b) NER-deficient mice. Shaved dorsal skin of the mice was exposed to UVB radiation with or without treatment of GTPs in drinking water. Mice were sacrificed 72 hours after UV irradiation, skin samples were collected and frozen sections were subjected to immunoperoxidase staining of CPD-positive cells, shown as dark color.

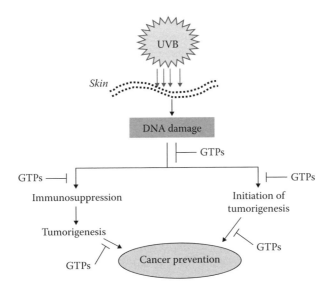

**FIGURE 14.6**    Schematic diagram depicting the mechanism of chemoprotective effect of GTPs on UV radiation-induced skin cancer. GTPs inhibit photocarcinogenesis through rapid repair of UVB-induced DNA damage in the skin.

Together, the studies conducted with GTPs indicate that the prevention of UV radiation-induced immunosuppression and subsequently the prevention of photocarcinogenesis by GTPs either through topical application or in the drinking water of mice are mediated through rapid repair of UVB-induced DNA damage, as summarized in Figure 14.6. As UV-induced DNA damage and immunosuppression play an important role in melanoma and nonmelanoma skin cancers, it is tempting to suggest that drinking green tea should be further investigated as a chemopreventive agent for the prevention of skin cancers in humans, and its possible use in future practice of medicine.

## BIOAVAILABILITY AND METABOLISM OF POLYPHENOLS

The bioavailability and metabolism of polyphenols may influence their effectiveness. The considerable structural diversity among the polyphenols can influence the bioavailability of the individual components. Small molecules, like catechin monomers in green tea, can be easily absorbed through the gut barrier, whereas the large molecular weight polyphenols, such as (−)-epigallocatechin 3-gallate or polyphenols from black tea, are poorly absorbed. Once absorbed, polyphenols are conjugated to glucuronide, sulphate, and methyl groups in the gut mucosa and inner tissues. Nonconjugated polyphenols are virtually not found in plasma. In case of topical application of the polyphenols, the penetration of polyphenols into the skin is limited and successful delivery of plant polyphenols requires cream-based, organic solvent–based or lipid-soluble topical formulations that can enhance the penetration of the polyphenols, and that will result in more efficient skin photoprotection.

## TRANSLATION OF ANIMAL STUDIES TO THE HUMAN SYSTEM

The GTPs which have been discussed have shown significant anti-DNA damaging effects in UVB-irradiated mouse skin. These protective effects of polyphenols on UVB-induced DNA damage contribute to their antiphotocarcinogenic effects and act to abrogate the various biochemical processes induced or mediated by solar UV radiation. Based on the epidemiological evidences and laboratory studies conducted using *in vitro* and *in vivo* systems, it is suggested that routine consumption or topical treatment of GTPs may provide efficient protection against the harmful effects of solar UV radiation in humans. Based on the information obtained in animal models with the use of green tea polyphenols, it can be suggested that the consumption of 5–6 cups (one cup = 150 ml) of green tea (1 g green tea leaves/150 ml) per day by humans may provide the same level of photoprotective effect in human system as was observed in animal models. However, the magnitude of photoprotective effect or UVB-induced immunosuppression and DNA repair by green tea may differ person to person based on the differences in race, and intensity and exposure time of UV radiation. For appropriate conversion of chemopreventive agent doses from animal studies to human system, the body surface area normalization method has been prescribed (reviewed in Reagan-Shaw et al. 2007). Based on this reference, the human equivalent dose of any chemopreventive agent can be calculated using the following formula:

$$\text{Human equivalent dose (mg/kg)} = \text{Animal dose (mg/kg)} \times \frac{\text{Animal } K_m \text{ factor}}{\text{Human } K_m \text{ factor}}$$

Further, the use of polyphenols in combination with sunscreens or skin care lotions may provide an effective strategy for mitigating the effects of UV radiation that will lead to the protection of the skin from various skin diseases caused by excessive sun exposure, including the risk of melanoma and nonmelanoma skin cancers in humans.

## ACKNOWLEDGMENTS

The work reported from Dr. Katiyar's laboratory was supported by the funds from National Institutes of Health (CA104428, AT002536) and the Veteran Affairs Merit Review Award. The content of this article does not necessarily reflect the views or policies of the funding sources. Grateful thanks are also due to our former and current colleagues and postdoctoral fellows for their outstanding contributions.

## REFERENCES

Bachelor, M. A., and G. T. Bowden. 2004. UVA-mediated activation of signaling pathways involved in skin tumor promotion and progression. *Semin Cancer Biol* 14:131–8.
Baliga, M. S., and S. K. Katiyar. 2006. Chemoprevention of photocarcinogenesis by selected dietary botanicals. *Photochem Photobiol Sci* 5:243–53.
Brash, D. E., J. A. Rudolph, J. A. Simon, A. Lin, G. J. McKenna, H. P. Baden, A. J. Halperin, and J. Pontén. 1991. A role for sunlight in skin cancer: UV-induced p53 mutations in squamous cell carcinoma. *Proc Natl Acad Sci USA* 88:10124–8.
Brunda, M. J., L. Luistro, R. R. Warrier, R. B. Wright, B. R. Hubbard, M. Murphy, S. F. Wolf, and M. K. Gately. 1993. Antitumor and antimetastatic activity of interleukin-12 against murine tumors. *J Exp Med* 178:1223–30.
Camouse, M. M., D. S. Domingo, F. R. Swain, E. P. Conrad, M. S. Matsui, D. Maes, L. Declercq, K. D. Cooper, S. R. Stevens, and E. D. Baron. 2009. Topical application of green and white tea extracts provides protection from solar-simulated ultraviolet light in human skin. *Exp Dermatol* 18:522–6.
Carreau, M., E. Eveno, X. Quilliet, O. Chevalier-Lagente, A. Benoit, B. Tanganelli, M. Stefanini, W. Vermeulen, J. H. Hoeijmakers, and A. Sarasin. 1995. Development of a new easy complementation assay for DNA repair deficient human syndromes using cloned repair genes. *Carcinogenesis* 16:1003–9.
Chatterjee, M. L., R. Agarwal, and H. Mukhtar. 1996. Ultraviolet B radiation-induced DNA lesions in mouse epidermis: An assessment using a novel [32]P-postlabelling technique. *Biochem Biophys Res Comm* 229:590–5.
Colombo, M. P., M. Vagliani, F. Spreafico, M. Parenza, C. Chiodoni, C. Melani, and A. Stoppacciaro. 1996. Amount of interleukin 12 available at the tumor site is critical for tumor regression. *Cancer Res* 56:2531–4.
Cross, S. E., R. Y. Jiang, H. A. E. Benson, and M. S. Roberts. 2001. Can increasing the viscosity of formulations be used to reduce the human skin penetration of the sunscreen oxybenzone? *J Invest Dermatol* 117:147–50.

deGruijl, F. R., and J. C. van der Leun. 1994. Estimate of the wavelength dependency of ultraviolet carcinogenesis in humans and its relevance to the risk assessment of stratospheric ozone depletion. *Health Phys* 67:319–59.

DiGiovanni, J. 1992. Multistage carcinogenesis in mouse skin. *Pharmacol Ther* 54:63–128.

Donawho, C. K., and M. L. Kripke. 1991. Evidence that the local effect of ultraviolet radiation on the growth of murine melanomas is immunologically mediated. *Cancer Res* 51:4176–81.

Godar, D. E. 2001. UV doses of American children and adolescents. *Photochem Photobiol* 74:787–93.

Godar, D. E., S. P. Wengraitis, J. Shreffler, and D. H. Sliney. 2001. UV doses of Americans. *Photochem Photobiol* 73:621–9.

Green, A., G. Williams, R. Neale, V. Hart, D. Leslie, P. Parsons, G. C. Marks, P. Gaffney, D. Battistutta, C. Frost, C. Lang, and A. Russell. 1999. Daily sunscreen application and betacarotene supplementation in prevention of basal cell and squamous cell carcinomas of the skin: A randomized controlled trial. *Lancet* 354:723–9.

Hara, Y. 2001. *Green Tea, Health Benefits and Applications*. New York: Marcel Dekker.

Haywood, R., P. Wardman, R. Sanders, and C. Linge. 2003. Sunscreens inadequately protect against ultraviolet-A-induced free radicals in skin: Implications for skin aging and melanoma? *J Invest Dermatol* 121:862–8.

Hruza, L. L., and A. P. Pentland. 1993. Mechanisms of UV-induced inflammation. *J Invest Dermatol* 100:35S–41S.

Ichihashi, M., M. Ueda, A. Budiyanto, T. Bito, M. Oka, M. Fukunaga, K. Tsuru, and T. Horikawa. 2003. UV-induced skin damage. *Toxicology* 189:21–39.

Katiyar, S., C. A. Elmets, and S. K. Katiyar. 2007. Green tea and skin cancer: Photoimmunology, angiogenesis and DNA repair. *J Nutr Biochem* 18:287–96.

Katiyar, S. K. 2007a. UV-induced immune suppression and photocarcinogenesis: Chemoprevention by dietary botanical agents. *Cancer Letts* 255:1–11.

Katiyar, S. K. 2007b. Interleukin-12 and photocarcinogenesis. *Toxicol Appl Pharmacol* 224:220–7.

Katiyar, S. K., F. Afaq, A. Perez, and H. Mukhtar. 2001. Green tea polyphenol (−)-epigallocatechin-3-gallate treatment of human skin inhibits ultraviolet radiation-induced oxidative stress. *Carcinogenesis* 22:287–94.

Katiyar, S. K., N. Ahmad, and H. Mukhtar. 2000. Green tea and skin. *Arch Dermatol* 136:989–94.

Katiyar, S. K., B. M. Bergamo, P. K. Vayalil, and C. A. Elmets. 2001. Green tea polyphenols: DNA photodamage and photoimmunology. *J Photochem Photobiol B: Biology* 65:109–14.

Katiyar, S. K., and C. A. Elmets. 2001. Green tea polyphenolic antioxidants and skin photoprotection. *Int J Oncol* 18:1307–13.

Katiyar, S. K., N. J. Korman, H. Mukhtar, and R. Agarwal. 1997. Protective effects of silymarin against photocarcinogenesis in a mouse skin model. *J Natl Cancer Inst* 89:556–66.

Katiyar, S. K., M. S. Matsui, and H. Mukhtar. 2000. Kinetics of UV light-induced cyclobutane pyrimidine dimers in human skin *in vivo:* An immunohistochemical analysis of both epidermis and dermis. *Photochem Photobiol* 72:788–93.

Katiyar, S. K., and H. Mukhtar. 1997. Tea antioxidants in cancer chemoprevention. *J Cell Biochem*(S) 27:59–67.

Katiyar, S. K., and H. Mukhtar. 2001. Green tea polyphenol (−)-epigallocatechin-3-gallate treatment to mouse skin prevents UVB-induced infiltration of leukocytes, depletion of antigen presenting cells and oxidative stress. *J Leukoc Biol* 69:719–26.

Katiyar, S. K., A. Perez, and H. Mukhtar. 2000. Green tea polyphenol treatment to human skin prevents formation of ultraviolet light B-induced pyrimidine dimers in DNA. *Clinical Cancer Res* 6:3864–9.

Katiyar, S. K., M. Vaid, H. van Steeg, and S. M. Meeran. 2010. Green tea polyphenols prevent UV-induced immunosuppression by rapid repair of DNA damage and enhancement of nucleotide excision repair genes. *Cancer Prev Res* 3:179–89.

Kligman, L. H., F. J. Akin, and A. M. Kligman. 1980. Sunscreens prevent ultraviolet photocarcinogenesis. *J Am Acad Dermatol* 3:30–5.

Kripke, M. L., P. A. Cox, L. G. Alas, and D. B. Yarosh. 1992. Pyrimidine dimers in DNA initiated systemic immunosuppression in UV-irradiated mice. *Proc Natl Acad Sci USA* 89:7516–20.

Krutmann, J. 2001. The role of UVA rays in skin aging. *Eur J Dermatol* 11:170–1.

Manach, C., A. Scalbert, C. Morand, C. Rémésy, and L. Jiménez. 2004. Polyphenols: Food sources and bioavailability. *Am J Clin Nutr* 79:727–30.

Mantena, S. K., S. M. Meeran, C. A. Elmets, and S. K. Katiyar. 2005. Orally administered green tea polyphenols prevent ultraviolet radiation-induced skin cancer in mice through activation of cytotoxic T cells and inhibition of angiogenesis in tumors. *J Nutr* 135:2871–7.

Meeran, S. M., S. Akhtar, and S. K. Katiyar. 2009. Inhibition of UVB-induced skin tumor development by drinking green tea polyphenols is mediated through DNA repair and subsequent inhibition of inflammation. *J Invest Dermatol* 129:1258–70.

Meeran, S. M., S. K. Mantena, C. A. Elmets, and S. K. Katiyar. 2006a. (−)-Epigallocatechin-3-gallate prevents photocarcinogenesis in mice through interleukin-12-dependent DNA repair. *Cancer Res* 66:5512–20.

Meeran, S. M., S. K. Mantena, and S. K. Katiyar. 2006b. Prevention of ultraviolet radiation-induced immunosuppression by (−)-epigallocatechin-3-gallate in mice is mediated through interleukin 12-dependent DNA repair. *Clinical Cancer Res* 12:2272–80.

Meeran, S. M., S. K. Mantena, S. Meleth, and S. K. Katiyar. 2006. Interleukin-12-deficient mice are at greater risk of ultraviolet Radiation-induced skin tumors and malignant transformation of papillomas to carcinomas. *Mol Cancer Ther* 5:825–32.

Meunier, L., N. Raison-Peyron, and J. Meynadier. 1998. UV-induced immunosuppression and skin cancers. *Rev Med Interne* 19:247–54.

Miller, D. L., and M. A. Weinstock. 1994. Nonmelanoma skin cancer in the United States: Incidence. *J Am Acad Dermatol* 30:774–8.

Mittal, A., C. A. Elmets, and S. K. Katiyar. 2003a. Dietary feeding of proanthocyanidins from grape seeds prevents photocarcinogenesis in SKH-1 hairless mice: Relationship to decreased fat and lipid peroxidation. *Carcinogenesis* 24:1379–88.

Mittal, A., C. Piyathilake, Y. Hara, and S. K. Katiyar. 2003b. Exceptionally high protection of photocarcinogenesis by topical application of (−)-epigallocatechin-3-gallate in hydrophilic cream in SKH-1 hairless mouse model: Relationship to inhibition of UVB-induced global DNA hypomethylation. *Neoplasia* 5:555–65.

Mukhtar, H., and C. A. Elmets. 1996. Photocarcinogenesis: Mechanisms, models and human health implications. *Photochem Photobiol* 63:355–447.

Muotri, A. R., M. C. Marchetto, L. F. Zerbini, T. A. Libermann, A. M. Ventura, A. Sarasin, and C. F. Menck. 2002. Complementation of the DNA repair deficiency in human xeroderma pigmentosum group A and C cells by recombinant adenovirus-mediated gene transfer. *Hum Gene Ther* 13:1833–44.

Naylor, M. F., A. Boyd, D. W. Smith, G. S. Cameron, D. Hubbard, and K. H. Neldner. 1995. High sun protection factor sunscreens in the suppression of actinic neoplasia. *Arch Dermatol* 131:170–5.

Pinnell, S. R. 2003. Cutaneous photodamage, oxidative stress, and topical antioxidant protection. *J Am Acad Dermatol* 48:1–19.

Reagan-Shaw, S., M. Nihal, and N. Ahmad. 2007. Dose translation from animal to human studies revisited. *FASEB J* 22:659–61.

Robertson, M. J., and J. Ritz. 1996. Interleukin-12: Basic biology and potential applications in cancer treatment. *Oncologist* 1:88–97.

Schwarz, A., A. Maeda, D. Gan, T. Mammone, M. S. Matsui, and T. Schwarz. 2008. Green tea phenol extracts reduce UVB-induced DNA damage in human cells via interleukin-12. *Photochem Photobiol* 84:350–5.

Schwarz, A., A. Maeda, K. Kernebeck, H. van Steeg, S. Beissert, and T. Schwarz. 2005. Prevention of UV radiation-induced immunosuppression by IL-12 is dependent on DNA repair. *J Exp Med* 201:173–9.

Schwarz, A., S. Stander, M. Berneburg, M. Böhm, D. Kulms, H. van Steeg, K. Grosse-Heitmeyer, J. Krutmann, and T. Schwarz. 2002. Interleukin-12 suppresses ultraviolet radiation-induced apoptosis by inducing DNA repair. *Nat Cell Biol* 4:26–31.

Scotto, J., and T. R. Fears. 1978. Skin cancer epidemiology: Research needs. *Natl Cancer Inst Monogr* 50:169–77.

Strom, S. 1996. Epidemiology of basal and squamous cell carcinomas of the skin. In M. J. Miller, R. S. Weber, and H. Goepfert, eds., *Basal and Squamous Cell Skin Cancers of the Head and Neck*, pp. 1–7. Baltimore: Williams and Wilkins.

Surh, Y-J. 1999. Molecular mechanisms of chemopreventive effects of selected dietary and medicinal phenolic substances. *Mutat Res* 428:305–27.

Taylor, C. R., R. S. Stern, J. J. Leyden, and B. A. Gilchrest. 1990. Photoaging/photodamage and photoprotection. *J Am Acad Dermatol* 22:1–15.

Thompson, S. C., D. Jolley, and R. Marks. 1993. Reduction of solar keratoses by regular sunscreen use. *N Engl J Med* 329:1147–51.

Timares, L., S. K. Katiyar, and C. A. Elmets. 2008. DNA damage, apoptosis and Langerhans cells-activators of UV-induced immune tolerance. *Photochem Photobiol* 84:422–36.

Ullrich, S. E. 1995. Potential for immunotoxicity due to environmental exposure to ultraviolet radiation. *Hum Exp Toxicol* 14:89–91.

Urbach, F. 1991. Incidences of nonmelanoma skin cancer. *Dermatol Clin* 9:751–5.

Vink, A. A., A. M. Moodycliffe, V. Shreedhar, S. E. Ullrich, L. Roza, D. B. Yarosh, and M. L. Kripke. 1997. The inhibition of antigen-presenting activity of dendritic cells resulting from UV irradiation of murine skin is restored by *in vitro* photorepair of cyclobutane pyrimidine dimers. *Proc Natl Acad Sci USA* 94:5255–60.

Vink, A. A., F. M. Strickland, C. Bucana, P. A. Cox, L. Roza, D. B. Yarosh, and M. L. Kripke. 1996. Localization of DNA damage and its role in altered antigen-presenting cell function in ultraviolet-irradiated mice. *J Exp Med* 183:1491–1500.

Wang, Z. Y., M. T. Huang, T. Ferraro, C. Q. Wong, Y. R. Lou, K. Reuhl, M. Iatropoulos, C. S. Yang, and A. H. Conney. 1992a. Inhibitory effect of green tea in the drinking water on tumorigenesis by ultraviolet light and 12-O-tetradecanoylphorbol-13-acetate in the skin of SKH-1 mice. *Cancer Res* 52:1162–70.

Wang, Z. Y., M. T. Huang, C. T. Ho, R. Chang, W. Ma, T. Ferraro, K. R. Reuhl, C. S. Yang, and A. H. Conney. 1992b. Inhibitory effect of green tea on the growth of established skin papillomas in mice. *Cancer Res* 52:6657–65.

Wang, S. Q., R. Setlow, M. Berwick, D. Polsky, A. A. Marghoob, A. W. Kopf, and R. S. Bart. 2001. Ultraviolet A and melanoma: A review. *J Am Acad Dermatol* 44:837–46.

Wei, H., Q. Ca, R. Rahn, X. Zhang, Y. Wang, and M. Lebwohl. 1998. DNA structural integrity and base composition affect ultraviolet light-induced oxidative DNA damage. *Biochemistry* 37:6485–90.

Weinstock, M. A. 1999. Do sunscreens increase or decrease melanoma risk: An epidemiologic evaluation. *J Invest Dermatol Symp Proc* 4:97–100.

Wolf, S. F., P. A. Temple, M. Kobayashi, D. Young, M. Dicig, L. Lowe, R. Dzialo, L. Fitz, C. Ferenz, and R. M. Hewick. 1991. Cloning of cDNA for natural killer cell stimulatory factor, a heterodimeric cytokine with multiple biologic effects on T and natural killer cells. Cloning of cDNA for natural killer cell stimulatory factor, a heterodimeric cytokine with multiple biologic effects on T and natural killer cells. *J Immunol* 146:3074–81.

Xu, C. X., A. Green, A. Parisi, and P. G. Parsons. 2001. Photosensitization of the sunscreen octyl p-dimethyl-aminobenzoate by UVA in human melanocytes but not in keratinocytes. *Photochem Photobiol* 73:600–4.

Yarosh, D., L. G. Alas, V. Yee, A. Oberyszyn, J. T. Kibitel, D. Mitchell, R. Rosenstein, A. Spinowitz, and M. Citron. 1992. Pyrimidine dimer removal enhanced by DNA repair liposomes reduces the incidence of UV skin cancer in mice. *Cancer Res* 52:4227–31.

Yoshikawa, T., V. Rae, W. Bruins-Slot, J. W. vand-den-Berg, J. R. Taylor, and J. W. Streilein. 1990. Susceptibility to effects of UVB radiation on induction of contact hypersensitivity as a risk factor for skin cancer in humans. *J Invest Dermatol* 95:530–6.

Zhao, J. F., Y. J. Zhang, X. H. Jin, M. Athar, R. M. Santella, D. R. Bickers, and Z. Y. Wang. 1999. Green tea protects against psoralen plus ultraviolet A-induced photochemical damage to skin. *J Invest Dermatol* 113:1070–5.

Ziegler, A., A. S. Jonason, D. J. Leffell, J. A. Simon, H. W. Sharma, J. Kimmelman, L. Remington, T. Jacks, and D. E. Brash. 1994. Sunburn and p[53] in the onset of skin cancer. *Nature* 372:773–6.

# 15 Bioactive Food Components and the U-Shaped Health Conundrum

## *Vitamin D and Folate as Examples of Friends and Foes*

John A. Milner

## CONTENTS

## DIET AND CANCER INCIDENCE

Dietary habits are recognized as an important modifiable environmental factor that can have a profound influence on cancer risk and tumor behavior (World Cancer Research Fund 2007; Milner 2008; Mason 2009; Trump et al. 2010). Although some findings have estimated that about 30%–40% of all cancer cases relate to dietary habits, the actual percentage is highly dependent on the foods consumed and the specific type of cancer. Both essential and nonessential food components arising from plants, animals, and mushrooms contain a host of bioactive food components that can influence cancer in a number of model systems (Guarner and Malagelada 2003; Davis and Milner 2007). Even the bacteria in the gastrointestinal tract can influence the metabolism of food components and thus influence the quantities and types of metabolites that are absorbed and potentially bring about a biological response including the risk of developing cancer (Davis and Milner 2009).

Cancer is undeniably a growing concern worldwide. Recent estimates by the World Health Organization (http://globocan.iarc.fr/) suggest that the risk of cancer will increase markedly during the next 30 years. This increase is projected to occur in both developed and developing countries. Part of this increased risk is associated with the aging of populations worldwide and with the increasing frequency of obesity that is occurring globally. It is also clear from a close examination of the data that not all countries are equally susceptible to cancer. Over indulgence is only partially responsible for the risk, and thus inadequacy of the diet must be considered a factor influencing cancer frequency and its complications.

## AN IDEAL DIET

Optimizing the intake of specific foods and/or their bioactive components is a prudent, noninvasive and cost-effective strategy for cancer prevention. Mounting evidence exists that compounds arising from grains, fruits, vegetables, meats and beans, dairy products, and spices can influence risk and/or tumor behavior. Defining "the" ideal food patterns is challenging for multiple reasons including the quantity required to bring about a response is largely unresolved, and numerous nutrient–nutrient and nutrient–gene interactions can occur (Milner 2008; Mariman 2008; Ferguson 2009; Ahmed et al. 2009; Simopoulos 2010). While there is insufficient information to formulate an "ideal diet," there is sufficient knowledge to justify a call for future food-oriented health research. Nevertheless, an "ideal diet" is likely not a reality since it is increasingly apparent that not all individuals respond identically to the foods they consumed and thus a one size prescriptive diet is naïve. Increased attention is being given to strategies that will allow for a personalized or individualized approach in diet and health, especially as it relates to cancer prevention and therapy (Kannan et al. 2008; da Costa et al. 2007; Kaput 2008). This is not a particularly novel approach since classic icons such as the food pyramid are increasingly recognizing individuality in needs (http://www.mypyramid.gov/index.html) as evident by a set of 12 health and nutrition plans. Developing individual risk factors as a function of genetic diversity, the complexity of food supply, culture and lifestyle differences, and the variety of potential processes that relate to disease makes it particularly challenging to develop a personalized approach to health and disease prevention. Nevertheless, the belief in an "ideal diet" is a simplistic approach to nutrition and health and undermines the critical roles that multiple food components have in regulating cellular events and how these are influenced by genetic and epigenetic events.

## BIOMARKERS FOR PREDICTING INDIVIDUAL RESPONSE

Figure 15.1 portrays consumer perspective about the amounts of foods needed to promote health. The first represents a progressive positive response to foods without limits. In biological systems this seldom, if ever, occurs. Nevertheless, consumers frequently have the misconception that if a little is good more is better (Mason 2007). This is often precipitated by the lack of quantitative information about the specific needs for individual bioactive food components and the lack of appreciation that excesses can have ill consequences.

The scientific literature provides evidence that under some circumstances of continues food intake that the benefits plateau. This is frequently observed with the regulation of mineral and vitamin absorption, but likely can occur with multiple food components (Serra et al. 2010). Sodium is one of the most widely studied nutrients that behave in this fashion (Lago et al. 2008). Iron is another example where health status improves up to a point at which additional intake results in diminished well-being (Tzonou et al. 1998). This saturable response is not limited to essential nutrients, but is evident with compounds such as β-carotene (Qiao et al. 2009; Goralczyk 2009).

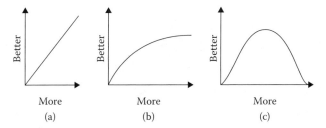

**FIGURE 15.1**  Consumer and scientific perspective about how much is needed to bring about a response. (a) A more is better model. (b) A saturation model. (c) A limitation and excess cause ill consequences model.

While it is unclear how many bioactive food components behave in a maximum capacity fashion, increasingly reports suggest the response is not saturable, but one that brings about ill consequences when intakes are too limited or too excessive (Horl and Ertl 2005; Rouillier et al. 2005; Olafsdottir et al. 2006; Waters et al. 2008). Many examples are present in literature including those associated with inadequate and excess amount of selenium, vitamin A, and iron. Even compounds generally thought to be beneficial to improving health such as flavonoids can, when provided in excess and under specific circumstances, cause gene translocation that has been equated to increased risk of leukemia (Barjesteh et al. 2007).

## DEVELOPING PREDICTIVE MODELS

While foods and food components can have both positive and/or negative health consequences, these are typically evaluated in largely separate evaluation strategies. Palou et al. (2009) has proposed an integrated evaluation of risk and benefit of foods, which likely has broad applications in evaluating all bioactive food components, whether provided as a food or a supplement. They suggest recommended dietary allowance (RDA) and the tolerable upper intake level (UL) fix the boundaries of intakes that should be considered sufficient to prevent deficiency while avoiding toxicity. The difficulty in establishing adequacy comes from the benefits/risk that is being evaluated. This becomes exceedingly challenging when dealing with a disease such as cancer that can take years of exposure to change cancer risk or tumor behavior. Anyway, within these thresholds, two additional thresholds, the lower and upper level of additional benefit (LLAB and ULAB) defines the range of intake that constitutes an additional benefit. Intake within these limits should thus be protective against a specified health or nutritional risk of public health relevance. It is clear families of curve tolerances that include both benefits and risk following exposures to foods occur in the model developed by Palou et al. (2009). These "family of curves" likely reflect the genomics of the consumer and a variety of insults including excess calories, environmental pollutants, viruses, and bacteria, and multiple nutrient–nutrient interactions (Davis and Milner 2007). Unquestionably, a continued system-based approach that builds on sound preclinical and clinical evidence is needed to effectively predict responses that are modified by dietary change. While unraveling the multiple factors influencing benefits and risk will be exceedingly challenging, the societal benefits are unmistakable. How the recognized of individuality in response will shift public health messages is unclear but never and more personalized messages will likely need to be crafted.

## DIFFERENT TYPES OF BIOMARKERS ARE NEEDED

Predictive, validated, and sensitive biomarkers, including those that reliably evaluate "intake/exposure" to a specific food or bioactive component, that assess one or more specific biological "effects" that are linked to cancer, and that effectively predict individual "susceptibility" as a function of nutrient–nutrient interactions and genetics, are fundamental to evaluating who will benefit most or

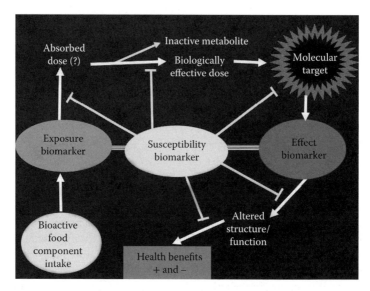

**FIGURE 15.2** Biomarkers for evaluating the biological significance of bioactive food components. Three biomarkers (exposure, effect, and susceptibility) are needed to adequately evaluate the health significance of foods and their components.

be placed at risk due to dietary intervention (Figure 15.2). For widespread use, these biomarkers must be readily accessible, easily and reliably assayed, and predictive of a key process(es) involved with cancer. Since the overall response to a food is determined not only by the effective concentration of the bioactive food component(s), but the quantity reaching the target tissue, all individuals will not be expected to respond identically (Trujillo et al. 2006; Simopoulos and Milner 2010).

Key to understanding a personalized response will be a greater knowledge of nutrigenomics, and related nutritional proteomics and metabolomics (Trujillo et al. 2006; Gorduza et al. 2008; Tseng 2009). Since all of these "omics" can influence the response to foods or their components, each component needs to be evaluated and integrated into a model for health promotion and disease prevention (Figure 15.3). Genetics, as well as nutrient–nutrient interactions, can markedly influence not only absorption, but also the metabolism and elimination of bioactive food components and thus the amount of the agent reaching the molecular target accounting for a biological response (Lampe 2009). The effectiveness of the dose reaching the target is also influenced by the affinity of the receptor for the food component (Cross et al. 2007) and the rate of removal of the active agent (Hwang and Jeffery 2005). While some bioactive food components may be weak ligands, their importance should not be minimized as evident by the ability of food components to inhibit histone deacetylase activity (Dashwood et al. 2006). Since many bioactive food components bring about pleiotropic effects by influencing multiple targets (Calamini et al. 2010), deciphering which site(s) is/are most important is challenging. Regardless, additional research is needed that focuses on the relationship between binding affinity to a particular target and its relationship with a change in a cancer related process. Since most responses to food components appear to be reversible, unless related to apoptosis or differentiation, then a sustained effective concentration is needed (Sundaram and Milner 1993; Bergmann et al. 2009).

The remainder of this chapter will focus on two nutrients, folic acid and vitamin D, which appear to have both beneficial and harmful attributes as proof of principle and to stimulate readers to begin to appreciate that quantity provided is critical and that limitations and excess can have ill-toward circumstances. These two nutrients were selected not only because they have profound implications about appropriate dietary recommendation for the public, but may also assist in defining criteria that can be used to predict benefits or risk from other food components.

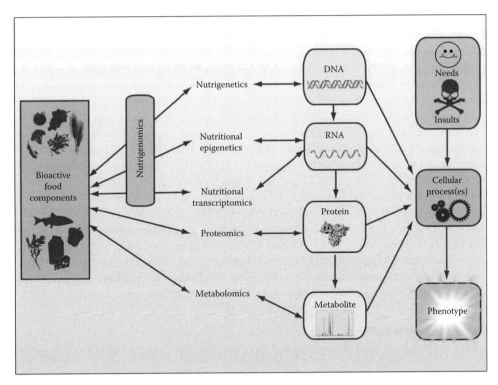

**FIGURE 15.3** Nutrigenomics and beyond. The nutrigenomic response to foods and their components depends on genetics, epigenetics, and transcriptomics and their ultimate effects on nutritional proteomics and metabolomics. Since each of these "omics" operate somewhat independently, information about each is needed to predict the response to foods or their components.

## THE VITAMIN D CONUNDRUM

Much is known about the formation, absorption, and metabolism of vitamin D and its biological impact, especially as related to bone health (Lips 2006; Christakos et al. 2007; Borradale and Kimlin 2009). Vitamin $D_3$ is synthesized in the skin by the sun's ultraviolet light, or obtained from foods, particularly fatty fish, mushrooms, and fortified food products. After hydroxylation in the liver into 25-hydroxyvitamin D (25[OH]D) and kidney into 1,25-dihydroxyvitamin D (1,25[OH]2D), the active metabolite can enter the cell, bind to the vitamin D receptor and subsequently to responsive genes (Campbell et al. 2010). The active metabolite 1,25(OH)2D has an antiproliferative effect and downregulates several genes, including those associated with inflammation, apoptosis, and differentiation (Carlberg and Seuter 2009). Extrarenal synthesis of 1,25(OH)2D occurs under the influence of cytokines and is important for the paracrine regulation of cell differentiation and function (Szatmari and Nagy 2008). Overall, the active metabolite 1,25(OH)2D appears to have pleiotropic effects through the vitamin D receptor (VDR) and vitamin D responsive elements of many genes and on nongenomic effects through a membrane receptor and second messengers. These biological attributes may explain that vitamin D deficiency appears to have a role in the pathogenesis of several diseases, including cancer (Davis et al. 2007).

### EVALUATING VITAMIN D STATUS

A fundamental question remains about how much vitamin D (as well as with all bioactive food components) is needed to bring about a positive health benefit and how best to monitor if the exposure has occurred. A recent systematic review (Seamans and Cashman 2009) confirmed that circulating 25(OH)D is a robust and reliable marker of vitamin D status. The use of other biomarkers or functional endpoints may also assist with the interpretation of variability in human concentrations and biological consequences. Parathyroid hormone (PTH) is a biomarker that may have utility in predicting vitamin D needs (Thomas et al. 1998). A negative relationship exists between serum 25(OH)D and serum PTH. The threshold of serum 25(OH)D, where serum PTH starts to rise is about 75 nmol/l according to most surveys (Lips 2006). Nevertheless, it is evident from several studies that not all patients with hypovitaminosis D develop secondary hyperparathyroidism (Sahota et al. 2004). A distinct group of patients with hypovitaminosis D has been reported to exhibit a blunted PTH response and a disruption in calcium homeostasis but protection against PTH-mediated bone loss (Sahota et al. 2004). Relationships between 25(OH)D and IGF-1 and metabolic syndrome raise additional issues about using various ratios to establish the adequacy of vitamin D (Hypponen et al. 2008). Overall, the use of this type of data may assist in helping define disease subcategories and therefore the optimal amount of vitamin D to provide for health benefits/promotion. It is unfortunate that biomarkers are less readily available to evaluate the status of most bioactive food components. While progress is being made in identifying biologically relevant exposure biomarkers, few have been validated and thus acceptable by the scientific community or regulatory agencies (Kumar and Sabbioni 2010; Loke et al. 2009). Clearly this is a major deficiency and deserves greater scientific attention.

## Vitamin D and Cancer Protection

Suggestions that vitamin D might be a factor in dictating cancer risk surfaced with observations that increased sun exposure was associated initially with a reduction in the risk of colorectal, cancer but the effects may be more widespread (Garland and Garland 1980; Kricker and Armstrong 2006; van der Rhee et al. 2006; Davis et al. 2007). Some of the most compelling evidence linking inadequate vitamin D status with increased cancer risk comes from preclinical studies (Newmark et al. 2009). Providing a Western-type diet high in fat and low in vitamin D and calcium results in the spontaneous formation of intestinal tumors in mice, significantly, in the absence of a carcinogen exposure. While tumors only occurred in some animals in a time-dependent fashion, the addition of vitamin D and calcium reversed the increased risk. Overall, these data are suggestive that long-term inadequate vitamin D status might lead to a progressive increase in colorectal cancers in humans. Additional preclinical evidence points to the ability of vitamin D to influence chemical carcinogenesis may be associated with the promotion phase of carcinogenesis and thus likely proliferation, apoptosis, and differentiation of transformed cells (Jacobson et al. 1989).

Some of the most compelling evidence that vitamin D inadequacy has a role in cancer risk in humans comes from studies involving the colon and rectum. A meta-analysis of epidemiologic studies suggested that individuals with 25(OH)D $\geq$ 33 ng/mL (82 nmol/L) may have a 50% lower risk of colon cancer than individuals with values $\leq$ 12 ng/mL (30 nmol/L) (Gorham et al. 2007). However, there is considerable variation in the individual studies making the overall interpretation of these findings difficult to evaluate. In the recent Women's Health Initiative study, vitamin $D_3$ supplementation (10 mcg [400 IU]) did not affect colon cancer risk (Wactawski-Wende et al. 2006). It is possible that the 10 mcg daily dietary supplement was insufficient to maintain protective concentrations of 25(OH)D, or that higher than average baseline vitamin D intakes may have reduced the differences between controls and cases. Calcium intakes may also have been a confounding factor. It is noteworthy that a nested case-control study within the WHI found that lower baseline serum 25(OH)D concentrations were associated with an increased risk of colorectal cancer (Wactawski-Wende et al. 2006).

## Vitamin D Receptor and Polymorphisms

The VDR is considered a crucial mediator for the cellular effects of vitamin D. Results from previous studies on the association of VDR polymorphisms with different cancer types provide some intriguing yet somewhat equivocal information. Individuals with the FokI polymorphism in the VDR gene have been reported to be particularly sensitive to inadequate calcium in the diet which correlated with an increased risk of colorectal cancer (Wong et al. 2003). Since multiple polymorphisms are known to exist in the VDR gene it is unclear if this is a unique or more common factor influencing cancer risk. The ability of individuals to be influenced by calcium or vitamin D with other polymorphisms remains unknown. Nevertheless, a recent meta-analysis by Raimondi et al. (2009) suggests that VDR FokI and BsmI polymorphisms are not limited to colorectal and might modulate the risk of breast, skin, and prostate cancers in Caucasians. The complexity of understanding genetic polymorphisms is illustrated by the fact that several hundred are known to occur in the VDR gene, thus making it difficult to determine which individual subgroups will be most and least responsive to vitamin D.

## MECHANISMS OF ACTION

Vitamin D is recognized to have pleiotropic effects that go far beyond its role in calcium homeostasis. Literally hundreds of genes with VDR response elements can directly or indirectly influence cell cycling and proliferation, differentiation, and apoptosis (Samuel and Sitrin 2008; Verstuyf et al. 2010). Upon binding, the VDR heterodimerizes with the retinoid X receptor and binds to vitamin D response elements in the promoter region of target genes to induce/repress their expression. The beyond calcium actions of vitamin D can influence normal and pathological cell growth, overall carcinogenesis, and immunocompetence. Despite the rather universal presence of VDRs, different cell types (e.g., keratinocytes, monocytes, bone, placenta) are capable of metabolizing 25(OH)D to 1,25(OH)2D by the enzyme 25(OH)D(3)-1alpha-hydroxylase, encoded by CYP27B1. The combined presence of CYP27B1 and the specific receptor in several tissues introduced the idea of a paracrine/autocrine role for 1,25(OH)2D. 1,25(OH)2D is recognized to induce differentiation and inhibit proliferation of normal and malignant cells (Gocek and Studzinski 2009). It is becoming increasing clear that not all cells are equally effective in producing 1,25(OH)2D (Lou et al. 2004).

## MODIFIERS OF VITAMIN D STATUS

The merits of vitamin D may depend on a host of cellular insults. Thus, the benefits of vitamin D may depend on the number of damaged cells. It is recognized that UV exposure is linked to greater risk of melanoma (Egan 2009), but it is also key to the formation of vitamin D from 7-dehydrocholecalciferol in skin. Thus, DNA strain breaks and therefore, the requirements for cancer protective responses, including apoptosis and differentiation, may call for enhanced vitamin D formation. The balance between these two processes may help explain why vitamin D and associated sunlight exposure has been reported to be protective against melanoma progression (Moan et al. 2008; Egan 2009). Other stressors may reduce the effect of vitamin D exposures. For example, obesity is linked to a reduction in circulating 25(OH)D concentrations (Lee et al. 2009). This would theoretically reduce the amount of 1,25(OH)2D reaching targets and thus potentially promote cancer. It is not clear if the increased cancer risk associated with obesity is in part due to a secondary deficiency of vitamin. Several dietary components can influence vitamin D status, including calcium and genistein (Heaney 2005; Cross et al. 2006). Evidence suggests that genistein enhances CYP27B1 and reduces catabolic CYP24, thereby reducing the catabolism of vitamin D and exaggerating 1,25(OH)2D concentrations (Lechner et al. 2006).

## VITAMIN D AND INCREASED CANCER RISK

It is becoming increasingly apparent that not all individuals benefit equally from enhanced vitamin D and some in fact may experience adverse events, including increased cancer risks (Faupel-Badger et al. 2007; Ahonen et al. 2000). While limited evidence does exist, there is sufficient data to suggest the likelihood of increased risk with exaggerated 25(OH)D concentrations (Davis et al. 2007). It is unclear why this relationship exists, but likely point to a vulnerable subgroup. Dose, timing and duration of exposure, lifestyle factors, and genetics may all contribute to the creation of this vulnerable group.

While epidemiologic studies frequently reveal no association between 25(OH)D concentrations and prostate cancer risk (Faupel-Badger et al. 2007), Ahonen et al. (2000) observed that serum 25(OH)D concentrations were inversely associated with risk of earlier occurrence and more aggressive prostate cancer. Similarly, a "U-shaped" relationship between serum 25(OH)D and prostate cancer risk was detected in a prospective case-control study by Tuohimaa et al. (2004). Increased risk was observed at 25(OH)D concentrations both $\leq$ 8 ng/mL (20 nmol/L) $\geq$ 32 ng/mL (80 nmol/L), compared to a reference range of 16 to 24 ng/mL, suggesting that extremely high concentrations are not needed to observe negative consequences.

Vitamin D status may also influence protection and risk at other cancer sites. While Garland et al. (2007) have asserted that breast cancer incidence may be reduced by 50% with a 25(OH)D concentration of 52 ng/mL (130 nmol/L) compared to 10 ng/mL (25 nmol/L) or less, there is evidence that a "U-shaped" relationship might exist. Goodwin et al. (2009) observed an inverse relationship between 25(OH)D and breast cancer survival. Pancreatic cancer may also provide another case that too much vitamin D might be harmful. In Finnish smokers a high prediagnostic 25(OH)D concentration correlated with a threefold greater risk for pancreatic cancer (Stolzenberg-Solomon et al. 2006a). Finally, Abnet et al. (2007) provide evidence that increased vitamin D exposures likely increase esophageal squamous dysplasia.

Overall, considerable scientific evidence supports a role for vitamin D in decreasing cancer incidence, particularly in the colon. However, it is also becoming increasingly apparent that excess, which is yet to be adequately defined, can cause ill consequences, including possibly an increase in cancer at some sites. Additional studies that focus on the dose–response relation between vitamin D status and cancer-related processes and overall risk, the optimal concentration of 25(OH)D, and individual variation in response as a function of genetics, age, gender, and lifestyle are needed. Additional research is also needed to determine the impact of cellular insults including excess calories and environmental pollutants, bacteria, and viruses on vitamin D needs to optimize health and reduce cancer risk. Likewise, greater attention is needed to identify vulnerable populations through the uses of genomic technologies who are placed at risk due to exaggerated exposures to vitamin D.

## THE FOLATE CONUNDRUM

Folate is a generic term for a group of water-soluble B vitamins that are present in high concentrations in green, leafy vegetables, and are recognized to have a critical role in maintaining DNA stability through their ability to donate one-carbon units. Folate acts as a coenzyme in several single carbon transfers involved in biosynthesis of purine nucleotides and deoxythymidylic acid essential for DNA and RNA synthesis. In addition, they provide one-carbon units for methylation of a wide variety of biological substances including DNA, proteins, phospholipids, and neurotransmitters, thereby influencing multiple cellular function.

### FOLATE AND CANCER PROTECTION

Inadequate intake has been implicated in the development of several cancers, including cancer of the colorectum, breast, ovary, pancreas, brain, lung, and cervix (Prinz-Langenohl et al. 2001). Generally, human studies suggest that people who habitually consume the highest amounts of folate, or with the highest blood folate concentrations, have a reduced risk of developing cancer, particularly colon polyps and colorectal cancer (Mason 2002, 2009). Folate deficiency induces and accelerates carcinogenesis by perturbing multiple cellular processes.

For several decades, evidence from both preclinical and clinical studies has pointed to the benefits of greater dietary folate intake and protection against the development of colorectal cancer.

Collectively, these studies suggest an approximately 40% reduction in the risk of colorectal cancer in individuals with the highest dietary folate intake compared with those with the lowest intake. Less compelling evidence exists about the benefits of folate against cancers of the breast, lung, pancreas, and so on (Mason 2002), but are not as extensively examined. A recent meta-analysis concluded that there is no clear support for an overall relationship between folate intakes or blood folate levels and breast cancer risk (Larsson et al. 2007). However, the authors also concluded that adequate folate intake might reduce the increased risk of breast cancer that has been associated with moderate or high alcohol consumption (Larsson et al. 2007).

Several lines of evidence supports folate inadequacy enhances cancer risk. Controlled preclinical studies provide some of the most convincing evidence in animals treated with chemical carcinogens or with genetic predispositions (Kim 2003, 2004; Liu et al. 2008; Hubner and Houlston 2009). Animal studies suggest the involvement of the Wnt cascade and p53-dependent control of the cell cycle involvement with the response to folate inadequacy (Liu et al. 2008).

## Polymorphisms Influence the Response

Fuchs et al. (2002) provided evidence that the benefits of folate appeared to be greatest in those with a family history of colorectal cancer. This type of finding suggests that genotype may also be a determinant of the needs for folate for cancer prevention. Polymorphism in a key folate-dependent enzyme, methylenetetrahydrofolate reductase (MTHFR), is sometimes associated with a modulation in the risk of developing colorectal cancer (Wolf et al. 2009). Ma et al. (1997) reported a common mutation (677C-->T) in MTHFR reduces enzyme activity, leading to lower levels of 5-methyltetrahydrofolate and presumably increased 5,10-methylenetetrahydrofolate levels for DNA synthesis, which correlated with colorectal cancer risk. Men with the homozygous mutation (TT) had half the risk of colorectal cancer compared with the homozygous normal or heterozygous genotypes. Recently, Taioli et al. (2009) performed both a meta-analysis (29 studies, 11,936 cases, 18,714 controls) of C677T MTHFR polymorphism and colorectal cancer. Overall, the odds ratio for risk for individuals with the TT genotype was 0.83. An inverse association was observed in whites (odds ratio = 0.83, 95% CI: 0.74, 0.94) and Asians (odds ratio = 0.80, 95% CI: 0.67, 0.96) but not in Latinos or blacks. Overall, MTHFR 677TT polymorphism may be associated with a reduced colorectal cancer in some but not all individuals.

## MECHANISMS OF ACTION

The mechanism by which folate insufficiency may increase cancer risk remains an area of active investigation. The synthesis of thymidylate (TMP) occupies a convergence of two critical metabolic pathways: folate metabolism and pyrimidine biosynthesis. Thymidylate is formed from deoxyuridylate (dUMP) using N(5),N(10)-methylene tetrahydrofolate. If folate is persistently limiting, DNA breakage and repair continues to occur and ultimately leads to DNA double-strand breaks, chromosomal aberrations, and malignant transformation (Reidy 1998). Likewise, folate is fundamental for DNA methylation (Duthie 2010). Thus, DNA stability due to shifts in bases or through changes in epigenetic processes may be involved with the ill consequences of folate inadequacy. It should be noted that DNA hypomethylation and site-specific hypermethylation are hallmarks of most cancer genomes. While folate and other methyl donor deficiencies have been observed to decrease methylation in liver, other tissues do not appear to respond in a similar way (Pogribny et al. 2004). Thus, it remains unclear how important methylation shifts are in explaining nonhepatic cancers.

Methyl donor deficiency is often accompanied by an elevation in homocysteine (Selhub 2008). Although there is variability in published findings, an elevation in homocysteine may be a factor in disease conditions since it induces oxidative stress leading to endothelial dysfunction. In addition, homocysteine-induced oxidative stress favors lipid peroxidation and induces production of inflammatory factors (Haijar 2001; Lin et al. 2009). Interestingly, folate treatment of uremic patients has been found to cause a depression in homocysteine, change methylation patterns, and importantly shift allelic expression (Ingrosso et al. 2003). It is possible the benefits of folate, and other methyl donors, will arise from the prevention of allelic expression resulting from secondary metabolic changes.

## FOLATE AND INCREASED CANCER RISK

While considerable evidence points to inadequate folate as a promoter of cancer, it is also clear that by virtue of its role in nucleotide synthesis that it can facilitate the proliferation of cells, a characteristic of dysplastic (precancerous) and malignant (cancerous) neoplasms (Lucock and Yates 2009; Hubner and Houlston 2009; Mason 2009). Thus, overexposures to folate in individuals who harbor a neoplasm may facilitate clonal expansion. Evidence that folate can promote cell proliferation surfaced years ago when Farber noted that folate administrate resulted in an expansion of the leukemic clone (Farber 1949). A recent randomized study provided evidence for a significant 2.3-fold increase in the multiplicity of recurrent adenomas among patients receiving folic acid supplements and 1.7-fold increase in high-risk adenomas (Cole et al. 2007). Admittedly, other supplement studies for shorter durations have not observed this increase in cancer risk (Jaszewski et al. 2008). Nevertheless, there is evidence that fortification of the food supply with folic acid may have increased colorectal cancer risk in the United States (Mason et al. 2007; Mason 2009). While there are concerns with ecological investigations, it should be noted that a similar elevation in colorectal cancer was observed several years later in Canadians and Chileans when their food supply was fortified with folic acid (Mason et al. 2007; Hirsch et al. 2009). Although risk of colorectal cancer is continuing to decrease, the current incidence is higher than would have been expected based on data prior to folic acid fortification. Consistent with this is the observation of the cancer-promoting effects of folic acid in rodents once a preneoplastic lesion is evident (Song et al. 2000 a,b). Overall, preclinical evidence provides compelling evidence of protection when provided early in the life of genetic predisposed animals, but promotes cancer when provided in the life of these vulnerable animals. While most human studies have focused on the impact of folic acid on colorectal cancer risk, there is evidence that exaggerated intakes (greater than 853 mcg) may also increase breast cancer risk (Stolzenberg-Solomon et al. 2006b).

It is not clear if the increased risk following food fortification is because folic acid was added to the food supply. This is not a naturally occurring form of the vitamin, yet it is converted to a natural biological form of the vitamin as it passes through the intestinal wall, with enzymatic reduction and methylation resulting in the circulating form of the vitamin, 5-methyltetrahydrofolate. Nevertheless, folic acid administration can saturate this conversion and result in detectable amounts in the blood, including the quantity being added to the U.S. food supply (Sweeney et al. 2007). Concern exists that this oxidized, nonsubstituted compound might be detrimental because the evidence is not overly compelling. However, Troen et al. (2006) reported that increasing plasma folic acid in elderly women taking folic acid-containing supplements was inversely linked with decreased cytotoxicity of circulating natural killer cells.

## CONCLUSIONS

Evidence continues to mount that a host of food components may serve as deterrents to cancer. However, evidence is also surfacing that these same food components can, when supplied in excess amounts, increase cancer risk. While there is evidence of ill consequences from excessive intake of foods, the chances of these events appear more prevalent when isolated bioactive food components are consumed in excess. Undeniably, multiple factors, including an individual's health status, life stage, nutritional status, and genomic/epigenomics profile, should be taken into consideration when evaluating how much of food components should be consumed. This chapter focused on folic acid and vitamin D as examples of nutrients that can be beneficial and potentially harmful under specific circumstances. These uncertainties in response serve as justification for continued and probing research to determine who is susceptible to exaggerated intakes of food components and under what circumstances. The discovery of susceptibility biomarkers will be useful to identify vulnerability subpopulations. Undeniably, a better understanding of the complex relationships between exposures and health benefits/risk will have profound implications for nutrition guidelines to promote public health and for formulating personalized nutrition intervention strategies.

A dilemma has surfaced from the two example nutrients discussed in this chapter, namely how to identify those whose intake is inadequate and also detect those who will be placed at risk due to exaggerated intakes. There is little question that vitamin D and folate have two faces, in that under certain experimental conditions they can be protective and in others harmful. The challenge before us is to determine how to provide an adequate but not excess to the consumer in order to minimize any ill consequences. New biomarkers that can predict the ill consequences of overindulgence are needed to identify vulnerable individuals to overexposure to these and likely many other bioactive food components. A personalized approach will likely not only be required in the future for these nutrients but for many compounds occurring in the food supply. While the detection of those who benefit and those placed at risk will be scientifically challenging, the societal benefits in terms of growth, development, productivity, and health are indisputable.

## REFERENCES

Abnet, C. C., W. Chen, S. M. Dawsey, W. Wei, M. J. Roth, B. Liu, N. Lu, P. R. Taylor, and Y. Qiao. 2007. Serum 25(OH)-vitamin D concentration and risk of esophageal squamous dysplasia. *Cancer Epidemiol Biomarkers Prev* 16:1889–93.

Ahmed, T., R. Haque, A. M. Shamsir Ahmed, W. A. Petri, and A. Cravioto, Jr. 2009. Use of metagenomics to understand the genetic basis of malnutrition. *Nutr Rev* 67(Suppl 2):S201–6.

Ahonen, M. H., L. Tenkanen, L. Teppo, M. Hakama, and P. Tuohimaa. 2000. Prostate cancer risk and prediagnostic serum 25-hydroxyvitamin D levels (Finland). *Cancer Causes Control* 11:847–52.

Barjesteh van Waalwijk, S. van Doorn-Khosrovani, J. Janssen, L. M. Maas, R. W. Godschalk, J. G. Nijhuis, and F. J. van Schooten. 2007 Dietary flavonoids induce MLL translocations in primary human CD34+ cells. *Carcinogenesis* 28(8):1703–9.

Bergmann H., D. Rogoll, W. Scheppach, R. Melcher, and E. Richling. 2009. The Ussing type chamber model to study the intestinal transport and modulation of specific tight-junction genes using a colonic cell line. *Mol Nutr Food Res* 53(10):1211–25.

Borradale, D., and M. Kimlin. 2009. Vitamin D in health and disease: An insight into traditional functions and new roles for the 'sunshine vitamin.' *Nutr Res Rev* 22(2):118–36.

Calamini B., K. Ratia, M. G. Malkowski, M. Cuendet, J. M. Pezzuto, B. D. Santarsiero, and A. D. Mesecar. 2010. Pleiotropic mechanisms facilitated by resveratrol and its metabolites. *Biochem J* 429(2):273–82.

Campbell, F. C., H. Xu, M. El-Tanani, P. Crowe, and V. Bingham. 2010. The yin and yang of vitamin D receptor (VDR) signaling in neoplastic progression: Operational networks and tissue-specific growth control. *Biochem Pharmacol* 79(1):1–9.

Carlberg, C., and S. Seuter. 2009. A genomic perspective on vitamin D signaling. *Anticancer Res* 29(9):3485–93.

Christakos, S., P. Dhawan, B. Benn, A. Porta, M. Hediger, G. T. Oh, E. B. Jeung, Y. Zhong, D. Ajibade, K. Dhawan, and S. Joshi. 2007. Vitamin D: Molecular mechanism of action. *Ann NY Acad Sci* 1116:340–8.

Cole, B. F., J. A. Baron, R. S. Sandler, R. W. Haile, D. J. Ahnen, R. S. Bresalier, G. McKeown-Eyssen, R. W. Summers, R. I. Rothstein, C. A. Burke, D. C. Snover, T. R. Church, J. I. Allen, D. J. Robertson, G. J. Beck, J. H. Bond, T. Byers, J. S. Mandel, L. A. Mott, L. H. Pearson, E. L. Barry, J. R. Rees, N. Marcon, F. Saibil, P. M. Ueland, and E. R. Greenberg. Polyp Prevention Study Group. 2007. Folic acid for the prevention of colorectal adenomas: A randomized clinical trial. *JAMA* 297(21):2351–9.

Cross, H. S., M. Lipkin, and E. Kállay. 2006. Nutrients regulate the colonic vitamin D system in mice: Relevance for human colon malignancy. *J Nutr* 136(3):561–4.

Cross, J. V., F. W. Foss, J. M. Rady, T. L. Macdonald, and D. J. Templeton. 2007. The isothiocyanate class of bioactive nutrients covalently inhibit the MEKK1 protein kinase. *BMC Cancer* 7:183.

da Costa, E., O. Silva, R. Knöll, and M. Jager. 2007. Personalized nutrition: An integrative process to success. *Genes Nutr* 2(1):23–5.

Dashwood R. H., M. C. Myzak, and E. Ho. 2006. Dietary HDAC inhibitors: Time to rethink weak ligands in cancer chemoprevention? *Carcinogenesis* 27(2):344–9.

Davis, C. D., V. Hartmuller, D. M. Freedman, P. Hartge, M. F. Picciano, C. A. Swanson, and J. A. Milner. 2007. Vitamin D and cancer: Current dilemmas and future needs. *Nutr Rev* 65(8 Pt 2):S71–4.

Davis, C. D., and J. A. Milner. 2007. Biomarkers for diet and cancer prevention research: Potentials and challenges. *Acta Pharmacol Sin* 28(9):1262–73.

Davis, C. D., and J. A. Milner. 2009. Gastrointestinal microflora, food components and colon cancer prevention. *J Nutr Biochem* 20(10):743–52.

Duthie, S. J. 2010. Folate and cancer: How DNA damage, repair and methylation impact on colon carcinogenesis. *J Inherit Metab Dis* 2010 Jun 11.

Egan, K. M. 2009. Vitamin D and melanoma. *Ann Epidemiol* 19(7):455–61.

Farber, S. 1949. Some observations on the effect of folic acid antagonists on acute leukemia and other forms of incurable cancer. *Blood* 4:160–7.

Faupel-Badger, J. M., L. Diaw, D. Albanes, J. Virtamo, K. Woodson, and J. A. Tangrea. 2007. Lack of association between serum levels of 25-hydroxyvitamin D and the subsequent risk of prostate cancer in Finnish men. *Cancer Epidemiol Biomarkers Prev* 16(12):2784–6.

Ferguson, L. R. 2009. Nutrigenomics approaches to functional foods. *J Am Diet Assoc* 109(3):452–8.

Fuchs, C. S., W. C. Willett, G. A. Colditz, D. J. Hunter, M. J. Stampfer, F. E. Speizer, and E. L. Giovannucci. 2002. The influence of folate and multivitamin use on the familial risk of colon cancer in women. *Cancer Epidemiol Biomarkers Prev* 11(3):227–34.

Garland C. F., and F. C. Garland. 1980. Do sunlight and vitamin D reduce the likelihood of colon cancer? *Int J Epidemiol* 9(3):227–31.

Garland, C. F., E. D. Gorham, S. B. Mohr, W. B. Grant, E. L. Giovannucci, M. Lipkin, H. Newmark, M. F. Holick, and F. C. Garland. 2007. Vitamin D and prevention of breast cancer: Pooled analysis. *J Steroid Biochem Mol Biol* 103:708–11.

Gocek, E., and G. P. Studzinski. 2009 Vitamin D and differentiation in cancer. *Crit Rev Clin Lab Sci* 46(4):190–209.

Goodwin, P. J., M. Ennis, K. I. Pritchard, J. Koo, and N. Hood. 2009. Prognostic effects of 25-hydroxyvitamin D levels in early breast cancer. *J Clin Oncol* 27:3757–63.

Goralczyk, R. 2009. Beta-carotene and lung cancer in smokers: review of hypotheses and status of research. *Nutr Cancer* 61(6):767–74.

Gorduza, E.V., L. L. Indrei, and V. M. Gorduza. 2008. Nutrigenomics in postgenomic era. *Rev Med Chir Soc Med Nat Iasi* 112(1):152–64.

Gorham, E. D., C. F. Garland, F. C. Garland, W. B. Grant, S. B. Mohr, M. Lipkin, H. L. Newmark, E. Giovanucci, M. Wei, and M. F. Holick. 2007. Optimal vitamin D status for colorectal cancer prevention: A quantitative meta analysis. *Am J Prev Med* 32:210–6.

Guarner, F., and J. R. Malagelada. 2003. Gut flora in health and disease. *Lancet* 361(9356):512–9.

Hajjar, K. A. 2001. Homocysteine: a sulphrous fire. *J Clin Invest* 107(6):663–4.

Heaney, R. P. 2005. The vitamin D requirement in health and disease. *J Steroid Biochem Mol Biol* 97(1–2):13–9.

Hirsch, S., H. Sanchez, C. Albala, M. P. de la Maza, G. Barrera, L. Leiva, and D. Bunout. 2009. Colon cancer in Chile before and after the start of the flour fortification program with folic acid. *Eur J Gastroenterol Hepatol* 21(4):436–9.

Hörl, W. H., and G. Ertl. 2005. Anaemia and the heart. *Eur J Clin Invest* 35(Suppl 3):20–5.

Hubner, R. A., and R.S. Houlston. 2009. Folate and colorectal cancer prevention. *Br J Cancer* 100(2):233–9.

Hwang, E. S., and E. H. Jeffery. 2005. Induction of quinone reductase by sulforaphane and sulforaphane N-acetylcysteine conjugate in murine hepatoma cells. *J Med Food* 8(2):198–203.

Hyppönen, E., B. J. Boucher, D. J. Berry., and C. Power. 2008. 25-OHhydroxyvitamin D, IGF-1, and metabolic syndrome at 45 years of age: A cross-sectional study in the 1958 British Birth Cohort. *Diabetes* 57:298–305.

Ingrosso, D., A. Cimmino, A. F. Perna, L. Masella, N. G. De Santo, M. L. De Bonis., M. Vacca, M. D'Esposito, M. D'Urso, P. Galletti, and V. Zappia. 2003. Folate treatment and unbalanced methylation and changes of allelic expression induced by hyperhomocysteinaemia in patients with uraemia. *Lancet* 361(9370):1693–9.

Jacobson, E. A., K. A. James, H. L. Newmark, and K. K. Carroll. 1989. Effects of dietary fat, calcium, and vitamin D on growth and mammary tumorigenesis induced by 7,12dimethylbenz(a)anthracene in female Sprague-Dawley rats. *Cancer Res* 49(22):6300–3.

Jaszewski, R., S. Misra, M. Tobi, N. Ullah, J. A. Naumoff, O. Kucuk, E. Levi, B. N. Axelrod, T. B. B. Patel, and A. P. Majumdar. 2008. Folic acid supplementation inhibits recurrence of colorectal adenomas: A randomized chemoprevention trial. *World J Gastroenterol* 14(28):4492–8.

Kannan S., A. Schulz, B. Israel, I. Ayra, S. Weir, T. J. Dvonch, Z. Rowe, P. Miller, and A. Benjamin. 2008. A community-based participatory approach to personalized, computer-generated nutrition feedback reports: The healthy environments partnership. *Prog Community Health Partnersh* 2(1):41–53.

Kaput, J. 2008. Nutrigenomics research for personalized nutrition and medicine. *Curr Opin Biotechnol* 19(2):110–20.

Kim, Y. I. 2003. Role of folate in colon cancer development and progression. *J Nutr* 133(11 Suppl 1):3731S–9S.

Kim, Y. I. 2004. Folate, colorectal carcinogenesis, and DNA methylation: Lessons from animal studies. *Environ Mol Mutagen* 44(1):10–25.

Kricker, A., and B. Armstrong. 2006. Does sunlight have a beneficial influence on certain cancers? *Prog Biophys Mol Biol* 92(1):132–9.

Kumar A., and G. Sabbioni. 2010. New biomarkers for monitoring the levels of isothiocyanates in humans. *Chem Res Toxicol* 23(4):756–65.

Lago, R. M., M. J. Pencina, T. J. Wang, K. J. Lanier, R. B. D'Agostino, W. B. Kannel, and R. S. Vasan. 2008. Interindividual variation in serum sodium and longitudinal blood pressure tracking in the Framingham Heart Study. *J Hypertens* 26(11):2121–5.

Lampe, J. W. 2009 Interindividual differences in response to plant-based diets: Implications for cancer risk. *Am J Clin Nutr* 89(5):1553S–7S.

Larsson, S. C., E. Giovannucci, and A. Wolk. 2007. Folate and risk of breast cancer: a meta-analysis. *J Natl Cancer Inst* 99(1):64–76.

Lechner, D., E. Bajna, H. Adlercreutz, and H. S. Cross. 2006. Genistein and 17beta-estradiol, but not equol, regulate vitamin D synthesis in human colon and breast cancer cells. *Anticancer Res* 26(4A):2597–603.

Lee, P., J. R. Greenfield, M. J. Seibel, J. A. Eisman, and Center, J. R. 2009. Adequacy of vitamin D replacement in severe deficiency is dependent on body mass index. *Am J Med* 122(11):1056–60.

Lin, C. P., Y. H. Chen, H. B. Leu, S. J. Lin, Y. L. Chen, S. L. Huang, and J. W. Chen. 2009. Anti-inflammatory strategies for homocysteine-related cardiovascular disease. *Front Biosci* 14:3836–45.

Lips, P. 2006. Vitamin D physiology. *Prog Biophys Mol Biol* 92(1):4–8.

Liu, Z., S. W. Choi, J. W. Crott, D. E. Smith, and J. B. Mason. 2008. Multiple B-vitamin inadequacy amplifies alterations induced by folate depletion in p53 expression and its downstream effector MDM2. *Int J Cancer* 123(3):519–25.

Loke, W. M., A. M. Jenner, J. M. Proudfoot, A. J. McKinley, J. M. Hodgson, B. Halliwell, and K. D. Croft. 2009. A metabolite profiling approach to identify biomarkers of flavonoid intake in humans. *J Nutr* 139(12):2309–14.

Lou, Y. R., S. Qiao, R. Talonpoika, H. Syvälä, and P. Tuohimaa. 2004. The role of vitamin D3 metabolism in prostate cancer. *J Steroid Biochem Mol Biol* 92(4):317–25.

Lucock, M., and Z. Yates. 2009. Folic acid fortification: a double-edged sword. *Curr Opin Clin Nutr Metab Care* 12(6):555–64.

Ma, J., M. J. Stampfer, E. Giovannucci, C. Artigas, D. J. Hunter, C. Fuchs, W. C. Willett, J. Selhub, C. H. Hennekens, and R. Rozen. 1997. Methylenetetrahydrofolate reductase polymorphism, dietary interactions, and risk of colorectal cancer. *Cancer Res* 57(6):1098–102.

Mariman, E. C. 2008. Epigenetic manifestations in diet-related disorders. *J Nutrigenet Nutrigenomics* 1(5):232–9.

Mason, J. B. 2002. Nutritional chemoprevention of colon cancer. *Semin Gastrointest Dis* 13(3):143–53.

Mason, J. B. 2009. Folate, cancer risk, and the Greek god, Proteus: A tale of two chameleons. *Nutr Rev* 67(4):206–1.

Mason, J. B., A. Dickstein, P. F. Jacques, P. Haggarty, J. Selhub, G. Dallal, and I. H. Rosenberg. 2007. A temporal association between folic acid fortification and an increase in colorectal cancer rates may be illuminating important biological principles: a hypothesis. *Cancer Epidemiol Biomarkers Prev* 16(7):1325–9.

Mason, P. 2007. One is okay, more is better? Pharmacological aspects and safe limits of nutritional supplements. *Proc Nutr Soc* 66(4):493–507.

Milner, J. A. 2008. Nutrition and cancer: Essential elements for a roadmap. *Cancer Lett* 269(2):189–98.

Moan, J., C. A. Porojnicu, A. Dahlback, and R. B. Setlow. 2008. Addressing the health benefits and risks, involving vitamin D or skin cancer, of increased sun exposure. *Proc Natl Acad Sci USA* 105:668–73.

Newmark, H. L., K. Yang, N. Kurihara, K. Fan, L. H. Augenlicht, and M. Lipkin. 2009. Western-style diet-induced colonic tumors and their modulation by calcium and vitamin D in C57Bl/6 mice: A preclinical model for human sporadic colon cancer. *Carcinogenesis* 30(1):88–92.

Olafsdottir, A. S., G. V. Skuladottir, I. Thorsdottir, A. Hauksson, H. Thorgeirsdottir, and L. Steingrimsdottir. 2006. Relationship between high consumption of marine fatty acids in early pregnancy and hypertensive disorders in pregnancy. *BJOG* 113(3):301–9.

Palou, A., C. Pico, and J. Keijer. 2009. Integration of risk and benefit analysis—the window of benefit as a new tool? *Crit Rev Food Sci and Nutr* 49:67–680.

Pogribny, I. P., S. J. James, S. Jernigan, and M. Pogribna. 2004. Genomic hypomethylation is specific for preneoplastic liver in folate/methyl deficient rats and does not occur in non-target tissues. *Mutat Res* 548(1–2):53–9.

Prinz-Langenohl, R., I. Fohr, and K. Pietrzik. 2001. Beneficial role for folate in the prevention of colorectal and breast cancer. *Eur J Nutr* 40(3):98–105.

Qiao, Y. L., S. M. Dawsey, F. Kamangar, J. H. Fan, C. C. Abnet, X. D. Sun, L. L. Johnson, M. H. Gail, Z. W. Dong, B. Yu, S. D. Mark, and P. R. Taylor. 2009. Total and cancer mortality after supplementation with vitamins and minerals: Follow-up of the Linxian General Population Nutrition Intervention Trial. *J Natl Cancer Inst* 101(7):507–18.

Raimondi, S., H. Johansson, P. Maisonneuve, and S. Gandini. 2009. Review and meta-analysis on vitamin D receptor polymorphisms and cancer risk. *Carcinogenesis* 30(7):1170–80.

Reidy, J. A. 1988. Role of deoxyuridine incorporation and DNA repair in the expression of human chromosomal fragile sites. *Mutat Res* 200(1–2):215–20.

Rouillier, P., M. C. Boutron-Ruault, S. Bertrais, N. Arnault, J. J. Daudin, J. N. Bacro, and S. Hercberg. 2005. Alcohol and atherosclerotic vascular disease risk factors in French men: relationships are linear, J-shaped, and U-shaped. *Alcohol Clin Exp Res* 29(1):84–8.

Sahota, O., M. K. Mundey, P. San, I. M. Godber, N. Lawson, and D. J. Hosking. 2004. The relationship between vitamin D and parathyroid hormone: Calcium homeostasis, bone turnover, and bone mineral density in postmenopausal women with established osteoporosis. *Bone* 35(1):312–9.

Samuel, S., and S. D. Sitrin. 2008. Vitamin D's role in cell proliferation and differentiation. *Nutr Rev* 66(10 Suppl 2):S116–24.

Seamans, K. M., and K. D. Cashman. 2009. Existing and potentially novel functional markers of vitamin D status: A systematic review. *Am J Clin Nutr* 89(6):1997S–2008S.

Selhub, J. 2008. Public health significance of elevated homocysteine. *Food Nutr Bull* 29(2 Suppl):S116–25.

Serra, A., A. Macià, M. P. Romero, J. Valls, C. Bladé, L. Arola, and M. J. Motilva. 2010. Bioavailability of procyanidin dimers and trimers and matrix food effects in *in vitro* and *in vivo* models. *Br J Nutr* 103(7):944–52.

Simopoulos, A. P. 2010. Nutrigenetics/Nutrigenomics. *Annu Rev Public Health* 31:53–68.

Simopoulos, A. P., and J. A. Milner. 2010. Research and its translation into medical practice. Preface. *World Rev Nutr Diet* 101:XI–XX.

Song, J., A. Medline, J. B. Mason, S. Gallinger, and Y. I. Kim. 2000a. Effects of dietary folate on intestinal tumorigenesis in the apcMin mouse. *Cancer Res* 60(19):5434–40.

Song, J., K. J. Sohn, A. Medline, C. Ash, S. Gallinger, and Y. I. Kim. 2000b. Chemopreventive effects of dietary folate on intestinal polyps in Apc+/−Msh2−/− mice. *Cancer Res* 60(12):3191–9.

Stolzenberg-Solomon, R. Z., R. Vieth, A. Azad, P. Pietinen, P. R. Taylor, J. Virtamo, and D. Albanes. 2006a. A prospective nested case-control study of vitamin D status and pancreatic cancer risk in male smokers. *Cancer Res* 66:10213–9.

Stolzenberg-Solomon, R. Z., S. C Chang, M. F. Leitzmann, K. A. Johnson, C., Johnson, S. S. Buys, R. N. Hoover, and R. G. Ziegler. 2006b. Folate intake, alcohol use, and postmenopausal breast cancer risk in the Prostate, Lung, Colorectal, and Ovarian Cancer Screening Trial. *Am J Clin Nutr* 83(4):895–904.

Sundaram, S. G., and J. A. Milner. 1993. Impact of organosulfur compounds in garlic on canine mammary tumor cells in culture. *Cancer Lett* 74(1–2):85–90.

Sweeney, M. R., J. McPartlin, and J. Scott. 2007. Folic acid fortification and public health: Report on threshold doses above which unmetabolised folic acid appear in serum. *BMC Public Health* 7:41.

Szatmari, I., and L. Nagy. 2008. Nuclear receptor signalling in dendritic cells connects lipids, the genome and immune function. *EMBO J* 27(18):2353–62.

Taioli, E., M. A. Garza, Y. O. Ahn, D. T. Bishop, J. Bost, B. Budai, K. Chen, F. Gemignani, T. Keku, C. S. Lima, L. Le Marchand, K. Matsuo, V. Moreno, J. Plaschke, M. Pufulete, S. B. Thomas, and G. Toffoli. 2009. Meta- and pooled analyses of the methylenetetrahydrofolate reductase (MTHFR) C677T polymorphism and colorectal cancer: a HuGE-GSEC review. *Am J Epidemiol* 170(10):1207–21.

Thomas, M. K., D. M. Lloyd-Jones, R. I. Thadhani, A. C. Shaw, D. J. Deraska, B. T. Kitch, E. C. Vamvakas, I. M. Dick, R. L. Prince, and J. S. Finkelstein. 1998. Hypovitaminosis D in medical inpatients. *N Engl J Med* 338(12):777–83.

Troen, A. M., B. Mitchell, B. Sorensen, M. H. Wener, A. Johnston, B. Wood, J. Selhub, A. McTiernan, Y. Yasui, E. Oral, J. D. Potter, and C. M. Ulrich. 2006. Unmetabolized folic acid in plasma is associated with reduced natural killer cell cytotoxicity among postmenopausal women. *J Nutr* 136(1):189–94.

Trujillo, E., C. Davis, and J. Milner. 2006. Nutrigenomics, proteomics, metabolomics, and the practice of dietetics. *J Am Diet Assoc* 106(3):403–13.

Trump, D. L., K. K. Deeb, and C. S. Johnson. 2010. Vitamin D: Considerations in the continued development as an agent for cancer prevention and therapy. *Cancer J* 16(1):1–9.

Tseng, M. 2009. Diet, cancer and public health nutrition. *Public Health Nutr* 12(6):737–8.

Tuohimaa, P., L. Tenkanen, M. Ahonen, S. Summe, E. Jellum, G. Hallmans, P. Stattin, S. Harvei, T. Hakulinen, T. Luostarinen, J. Dillner, M. Lehtinen, and M. Hakama. 2004. Both high and low levels of blood vitamin D are associated with a higher prostate cancer risk: A longitudinal, nested case-control study in the Nordic countries. *Int J Cancer* 108:104–8.

Tzonou, A., P. Lagiou, A. Trichopoulou, V. Tsoutsos, and D. Trichopoulos. 1998. Dietary iron and coronary heart disease risk: A study from Greece. *Am J Epidemiol* 147(2):161–6.

van der Rhee, H. J., E. de Vries, and J. W. Coebergh. 2006. Does sunlight prevent cancer? A systematic review. *Eur J Cancer* 42(14):2222–32.

Verstuyf, A., G. Carmeliet, R. Bouillon, and C. Mathieu. 2010. Vitamin D: A pleiotropic hormone. *Kidney Int* 78(2):140–5.

Wactawski-Wende, J., J. M. Kotchen, G. L. Anderson, A. R. Assaf, R. L. Brunner, M. J. O'Sullivan, K. L. Margolis, J. K. Ockene, L. Phillips, L. Pottern, R. L. Prentice, J. Robbins, T. E. Rohan, G. E. Sarto, S. Sharma, M. L. Stefanick, L. Van Horn, R. B. Wallace, E. Whitlock, T. Bassford, S. A. Beresford, H. R. Black, D. E. Bonds, R. G. Brzyski, B. Caan, R. T. Chlebowski, B. Cochrane, C. Garland, M. Gass, J. Hays, G. Heiss, S. L. Hendrix, B. V. Howard, J. Hsia, F. A. Hubbell, R. D. Jackson, K. C. Johnson, H. Judd, C. L. Kooperberg, L. H. Kuller, A. Z. LaCroix, D. S. Lane, R. D. Langer, N. L. Lasser, C. E. Lewis, M. C. Limacher, and J. E. Manson. 2006. Calcium plus vitamin D supplementation and the risk of colorectal cancer. *N Engl J Med* 354:684–696.

Waters, D. J., E. C. Chiang, and D. G. Bostwick. 2008. The art of casting nets: fishing for the prize of personalized cancer prevention. *Nutr Cancer* 60(1):1–6.

Wolf, C. R., C. G. Moore, and J. Little. 2009. Meta- and pooled analyses of the methylenetetrahydrofolate reductase (MTHFR) C677T polymorphism and colorectal cancer: a HuGE-GSEC review. *Am J Epidemiol* 170(10):1207–21.

Wong, H. L., A. Seow, K. Arakawa, H. P. Lee, M. C. Yu, and S. A. Ingles. 2003. Vitamin D receptor start codon polymorphism and colorectal cancer risk: Effect modification by dietary calcium and fat in Singapore Chinese. *Carcinogenesis* 24(6):1091–5.

World Cancer Research Fund/American Institute for Cancer Research. 2007. *Food, Nutrition and the Prevention of Cancer: A Global Perspective*. Washington, DC: American Institute for Cancer Research.

# Section II

---

## *Gene-Linked Networks*

### Intestinal Microbiome-Linked Gene Expression

# 16 Human Intestinal Microbiome

## *Etiology of Inflammation Genomics*

### Roger A. Clemens

## CONTENTS

## INTRODUCTION

The human genome project, completed in 2003, launched the "omics" era with the identification of more than 30,000 genes in the human chromosomal DNA. The subsequent development of analytical molecular tools assisted our understanding of the genetic-health dynamics and can contribute to disease reduction through molecular nutrition. Four of these molecular "omics" tools are genomics, transciptomics, proteomics, and metabolomics. Metabolomics identifies groups of metabolic variables unique to an individual's genetic profile characterized in bodily fluids and provides valuable diagnostic predictors of health and disease. An integrated "omics" approach is one driver leading to our comprehension of the health and disease interplay in which the human microbiome may be central to future strategies for pharmacological developments and nutritional interventions (Kussman and Blum 2007).

The Human Microbiome Project (HMP) is a logical extension of the human genome project. This project provides an interface between medical microbiology and environmental microbiology intended to identify new approaches, including biomarkers and diverse microbial communities, to address health issues relative to an individual's normal physiology. Critical to that physiology is the normal bacterial distribution at major body sites (Table 16.1) (NIH HMP Working Group 2009).

Each of these locations presents a distinct microbial phylotype profile based on 16S rRNA gene-sequence analysis. The four major phylotypes are Firmicutes, Bacteroidetes, Actinobacteria, and Proteobacteria, each of which appears to be consistent among most humans studied to date. Within these phylotypes, there are multiple genera, some of which may be pathogenic and while others are used in food fermentation and considered important sources of probiotics (Table 16.2).

The microbiome has been implicated in several diseases and their inflammatory component. Several examples include obesity and the development of metabolic syndrome may reflect an increased production of bacterial lipopolysaccharide following the consumption of a high-fat diet (Cani et al. 2007). In another example an animal model suggests the onset of type 2 diabetes may be mediated through choline intermediates produced by selected microbial strains. These

**TABLE 16.1**
**Microbial Distribution at Various Body Sites**

| Body Site | Percent Distribution | Dominate Phylotype |
|---|---|---|
| Gastrointestinal tract | 29 | Firmicutes |
| Oral (mouth) | 26 | Firmicutes |
| Skin | 21 | Actinobacteria |
| Airways | 14 | Firmicutes |
| Urogenital | 9 | Firmicutes |
| Blood | 1 | – |
| Eye | 0 | – |

**TABLE 16.2**
**Phylotypes and Their Notable Genera**

| Phylotype | Genera |
|---|---|
| Firmicutes | Clostridia (e.g., Clostridium sp.), Bacilli (e.g., Listeria, Lactobacillus, Staphylococcus, Streptococcus, Leuconostoc) |
| Bacteroidetes | Bacteroides, Flavobacteria, Porphyromonas |
| Actinobacteria | Actinomyces, Bifidobacteria |
| Proteobacteria | Brucella, Rickettsia |

*Source:* Walter, J. and R. Ley 2011. *Annu Rev Microbiol* 65:411–29. With permission.

intermediates, known as methylamines, may contribute to development of nonalcoholic fatty liver disease and insulin resistance (Dumas et al. 2006). Also, inflammatory responses associated with cardiovascular disease and atherosclerosis may be modulated via toll-like receptor 4 (TLR-4) signaling activated by gut microbes. Depending on the TLR-4 polymophism, individuals with either Asp299Gly allele alone or in combination with the Thr399Ile allele, present a diminished inflammatory response to *Chlamydia pneumonia* and *Helicobacter pylori* (Kiechl et al. 2002). These two gram-negative organisms have been implicated in atherogenesis (Triantafilou et al. 2007).

The risk of developing type 1 diabetes (T1D) among the diabetes-prone (BB-DP) rat may be reduced through lower Bacteriodetes levels (Brugman et al. 2006). The microflora profile, including the genera among Bacteriodetes phylum, may be modified through dietary intervention, such as the introduction of hydrolyzed casein. Interestingly, plant-based diets may be more diabetogenic, at least in BB-DP rats, possibly due to the potential antigenic burden of different dietary protein sources (Scott 1996). Oral administration of the antibiotics Bactrimel (suylphamethoxazole and trimethoprim) and Colistin (colistine sulphate) reduced total gram-negative and gram-positive intestinal organisms, significantly decreased the incidence of diabetes, and delayed the onset of the disease. The onset of diabetes was also delayed following the consumption of hydrolyzed casein in the absence of the antibiotics. Importantly, concomitant administration of these antibiotics and hydrolyzed casein was protective against development of diabetes in the BB-DP rat model. These results may be explained, in part, by reduced antigen load, lower inflammatory cytokine production, and decreased microflora, including commensals (Hansen et al. 2005; Bendtzen et al. 1992). These observations suggest variations within the human microbiome at the identified body sites, particularly the gut, may provide therapeutic as well as prophylactic opportunities. These observations also imply microbial manipulation strategies may affect microbial balance, maintenance of epithelial barriers, nutrient metabolism, and inflammatory states (Turnbaugh et al. 2007).

Microbial species and their respective strains within Firmicutes and Actinobacteria dominate probiotics currently incorporated in various foods and utilized in clinical environments. The clinical application of probiotics, even among pre-mature infants that are at risk for developing necrotizing enterocolitis (NEC), is now supported by the American Academy of Pediatrics (Deshpande et al. 2010; Tarnow-Mordi et al. 2010a). NEC is a common serious disease of unknown etiology that primarily affects premature infants. More than one-third of the NEC cases involve the gastrointestinal inflammation (AlFaleh and Bassler 2008). Eleven studies were evaluated by meta-analysis including the probiotics *Bifidobacterium breve, Lactobacillus GG, Saccharomyces boulardii, Bifidobacteria infantis, Streptococcus thermophiles, Bifidobacterium bifidus, Lactobacillus acidophilus, Lactobacillus casei,* and *Bifidobacterium lactis*. These studies, and more recent reports, indicate the incidence of NEC may be reduced by at least 30% through the selective administration of appropriate probiotics at approximately $10^9$ cfu/d through six weeks of discharge when compared with the standard of care (Tarnow-Mordi et al. 2010b). The standard of care for NEC includes, in part, the administration of conventional drugs (antibiotics, corticosteroids) and drip feeds to provide macronutrients necessary for the bowel to heal.

## HUMAN INTESTINAL MICROBIOME: ETIOLOGY OF INFLAMMATION GENOMICS

The human microbiome encompasses approximately 800 different species of microbes represented by more than 7000 genetically unique strains that are integral to our normal ecology. Each of these nearly 100 trillion organisms represent microcommunities that influence genetic expression and metabolic traits characterized in the balance of health and disease (Ley et al. 2006; Majewski et al. 2000). The dynamics of the genetic diversity impacts an array of metabolic systems and provides functional redundancy of some of these systems, including the immune system. The multiple genomes represented by these organisms and an understanding of this biological diversity within individuals and among geographical regions may expose genetic barriers and opportunities for health-promoting interventions (Dethlefsen et al. 2007). Importantly, our immune system may reflect a microbiota partnership or symbiotic relationship in its development, thus allowing us to adapt to environmental stressors, including food challenges and xenobiotics substances (Bäckhed 2007; Dumas et al. 2006).

The human gut is a dynamic organ, and the colon has equally critical functions, beyond just managing fluid balance and excreta. The colon is a complex system with a unique cell composition that prefers short-chain fatty acids for energy. This energy source is typically produced by the anaerobic microbial fermentation of polysaccharides commonly known as dietary fiber. Colonocytes, mucosal epithelia, utilize microbial fermentation products, including short-chain fatty acids, and nutrients which enable these cells to maintain mucosal health (O'Keefe 2008). This relationship with the local microbiota provides a barrier that increases resistance to adverse conditions and disease, such as infection, ulcerative colitis (UC), and colorectal cancer. Disruption of this dependent relationship can lead to destruction of the microbiota barrier and nutrient insufficiency producing local inflammation, which may further lead to systemic inflammatory responses and clinical consequences. This disruption level that contributes to inflammation and disease appears to be dependent upon dietary components (or lack thereof, such as dietary fiber and resistant starch), nutritional status (malnourished at greater risk) and nutrient response, genetics (African descendants at greater risk), and environmental factors.

Evidence suggests that the human gut microflora can play a major role in our health. Early examination by several investigators indicates a balance exists between innate health promoting and pathogenic organisms throughout the intestinal tract (Figure 16.1) (Gibson and Roberfroid 1995). The intestinal tract microflora, particularly in the colon, is a highly metabolically active biomass. Emerging data suggest that fermentation metabolites produced by this biomass within the colon, such as $H_2$ and $H_2S$, can impair mucosal cell function, and possibly contribute to bacterial

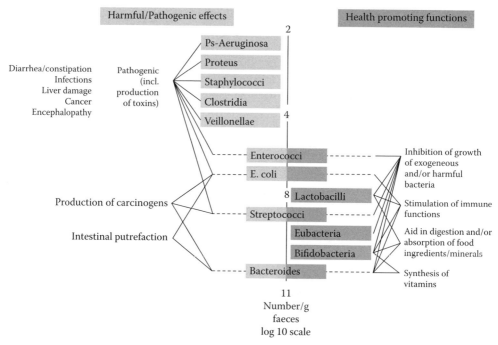

**FIGURE 16.1** Generalized scheme of the composition and health effects of predominant human fecal bacteria. (From Gibson, G., and M. Roberfroid. 1995. *J Nutr* 125:1401–12. With permission.)

overgrowth, stimulate mucosal cell proliferation, and lead to an increased risk of some intestinal diseases (Fujimura et al. 2010). On the other hand, some dietary components are metabolized to short-chain fatty acids, particularly butyrate, that appear to down regulate mucosal cell proliferation and differentiation, thus reducing the risk of some pathogenic processes. Luminal butyrate is the preferred fuel for colonocyte metabolism, and functions as a effective anti-inflammatory, antiproliferative, and potentially antineoplastic agent (Kim and Milner 2007). Mature apical colonocytes may be more responsive to luminal content than less mature cells in the villus crypts, as suggested by increased apical membrane permeability and reduced butyrate uptake by crypt cells (Vidyasagar et al. 2005). Yet, it is the crypt cells that require metabolic fuel and balance as they undergo rapid differentiation and proliferation (Kim and Milner 2007).

There are many facets of the gut–immune function interface that may contribute to improved health. One of those facets includes an improved understanding of the epigenetic regulation of mucosal cell proliferation. Evidence suggests the dynamics of this proliferation differ among populations of high and low risk of certain intestinal disorders. Some cancers and immunological dysfunctions may be modulated by mucosal cell proliferation and products of microflora growth. The ensuing disease-associated inflammation may be attenuated by some anaerobic bacteria by inhibiting NF-κB activation.

NF-κB, activated by many physiological and dietary stimuli, induces the expression of numerous genes, many of which suppress cellular apoptosis. It appears that microflora express substances that inactivate NF-κB by targeting the RelA (reticuloendotheliosis) subunit of NF-κB via a PPAR-γ-dependent pathway. The subsequent decrease in PPAR-γ reduces proinflammatory responses, such as those initiated by pathogens (Kelly et al. 2004). Ultimately, it may be that the synergy among innate microflora and mucosal cells contribute to improved modulation of potential inflammatory responses, or even suppress inflammatory events mediated by NF-κB. These observations indicate decreased gut-associated inflammations and possibly increased food tolerances as well as improved long-term health benefits may be possible through specific dietary interventions and microbe interactions (O'Keefe et al. 2009; Preidis and Versalovic 2009; Aggarwal and Shishodia 2006; Neish et al. 2000).

## MUCOSAL IMMUNE RESPONSE

The importance of the gut and its microbiota in maintaining immunological homeostasis deserves further attention (Mason et al. 2008). There is a delicate balance that protects the body from pathogens, providing a tolerance to commensal organisms, dietary harmful and harmless antigens, and xenobiotics. The apparent immune tolerance involves several mechanisms and unique cellular functions, such as the production of secretory IgA, the synergistic Th2 responses, unique convoluted gut architecture and specialized cells and tissues, an array of dedicated adhesion molecules and chemokine receptors, and the interplay of the indigenous microbiota. The anatomical features of the gut and the mucosal tissues enable the cells to rapidly discriminate and process absorbed antigens. It is also important to note that, as observed in gnotobiotic models, in the absence of bacterial species the animals are more susceptible to opportunistic pathogens and an array of systemic abnormalities (Gordon and Pesti 1971). Those abnormalities include reduced heart size and diminished regional blood flow, decreased weight and surface area of the small intestine, particularly the lamina propria, and an enlarged cecum. This latter observation may reflect a decreased smooth muscle tone.

The intestinal epithelium and its gut-associated lymphoid tissue (GALT) contain M cells, which constantly transport microbiota and antigens from the gut lumen into the lymphoid tissue. This initiates the environmental lamina propria sampling process that is further facilitated by dendritic cells, CD4 T cells, CD8+ T cells, macrophages, and IgA antibody-producing plasma cells (Mowat 2003). This complex process of the mucosal immune system depends, in part, upon the innate commensal microbiota (Figure 16.2) (MacDonald and Monteleone 2005). Controlling the gut immune system and its inflammatory process involves a constellation of events, each of which may be modulated by one's unique microbiome.

The gut immune system is highly regulated and it carefully controls inflammatory processes so that the immune system does not over respond to antigen challenges (Figure 16.3). This unique vigilance allows the gut to remain virtually devoid of disease. In healthy individuals, antigens

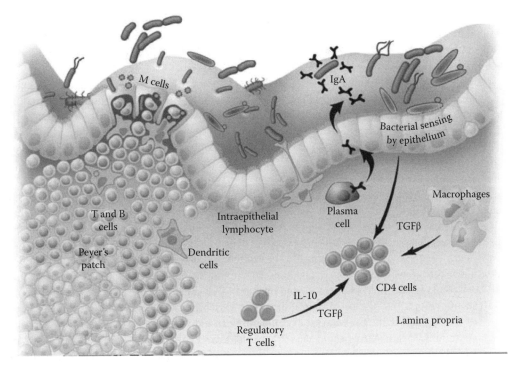

**FIGURE 16.2**  Gut makes a significant investment in maintaining an extensive and highly active immune system. (From McDonald, T. T., and G. Monteleone. 2005. *Science* 307:1920–25. With permission.)

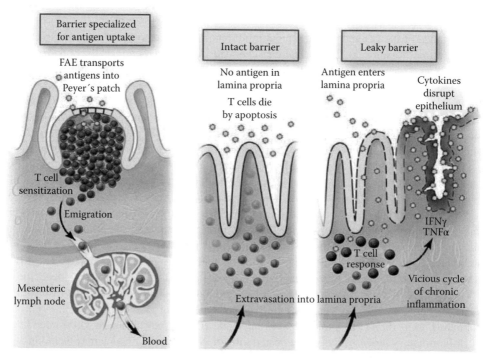

**FIGURE 16.3** Gut makes a significant investment in maintaining an extensive and highly active immune system. (From McDonald, T. T. and G. Monteleone. 2005. *Science* 307:1920–25. With permission.)

are transported into Peyer's patches, where activated CD4 T-cells destroy the antigen or invading pathogens and the respective T cells die by apoptosis. On the other hand, in the presence of a leaky epithelial barrier, as observed in Crohn's and similar genetic diseases, the discrimination between healthy and nonhealthy bacteria is negligible, resulting in continuous T-cell activation and mucosal inflammation.

In addition to a leaky epithelial barrier, Crohn's disease may involve a copy number of β-defensin genes. Beta-defensins are inflammatory mediators implicated in numerous inflammatory/ autoimmune diseases, including, but not limited to multiple sclerosis and psoriasis (Hollox 2010). Importantly, these proteins are also critical in the innate immune system by exerting antimicrobial activity against pathogens such as *E. coli*, *P. aeruginosa*, *S. aureus*, *Candida albicans*, and possibly viruses (Kisich et al. 2007; Vylkova et al. 2007; Hazrati 2006). The copy variations of β-defensins in humans, at least 35 distinct peptides, are expressed by multiple cell types, including Paneth cells in the small intestine and epithelial cells of the female genital tract. The relatively large copy number variations and their diverse tissue location, emphasize their importance as signaling molecules in inflammatory diseases, and augment the challenges in therapeutic approaches in resolving immune-related diseases (Hollox 2008).

In this era of genomics, the emergence of the human microbiome as a key component in human health, nutrient responsiveness, innate and adaptive immunity, and gut development is not surprising. The diversity of the microbiome core and its manipulation can impact basic human physiology, overall health status, and disease susceptibility. The variability of the microbiome appears to be central to and dependent upon several factors, including an individual's lifestyle, genotype, environment, and immune system (Preidis and Versalovic 2009). For example, many studies with select and genetically characterized strains of probiotics suggest that inflammatory processes and signaling pathways can be suppressed by reducing proinflammatory cytokine production by intestinal epithelial cells, and blocking the NF-κB pathway. Modeling systems indicate antibiotics, while having

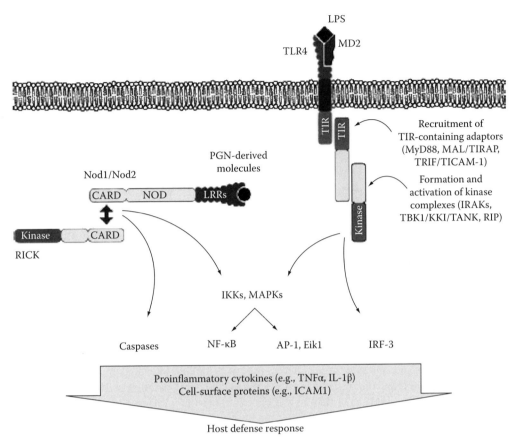

**FIGURE 16.4** Nod1, Nod2, and TLR4 signaling pathways. (From Inohara, N. et al. 2005. *Ann Rev Biochem* 74:355–83. With permission.)

considerable therapeutic benefits, disrupt the balanced microbial community thereby decreasing the diversity of organisms, alter TLR signaling, and possibly lead to dysregulation of the mucosal immune system (Figure 16.4) (Inohara et al. 2005).

This disruption may contribute to the onset of antibiotic associated diarrhea (AAD) and increased risk of UC, inflammatory bowel disease (IBD), irritable bowel syndrome (IBS), and related disorders including Celiac disease, possibly necrotize enterocolitis, and even be an associative factor to the pathogenesis related to obesity (MacDonald and Monteleone 2005; Brook 2008; Turnbaugh et al. 2006). AAD reflects alterations in intestinal microbial profile, and often contributes to *Clostridium difficile*-associated diarrhea (Fujimura et al. 2010). Numerous intervention studies suggest dietary supplementation with various strains of probiotics, or combinations of probiotics, may suppress or eliminate pathogens, while "replenishing" the commensal microbiota (Wolvers et al. 2010). These observations reflect, in part, the production of bacteriocins, inhibition of enteropathogen growth and mucosal adhesion, or modulation of the innate immune system (Naidu et al. 1999). On the other hand, select strains of probiotics may modulate gut immunity through factors such as immunomodulins, select nutrients, and production of short-chain fatty acids that can lead to anti-inflammatory responses by immune cells and the intestinal epithelium (Inohara et al. 2005).

This immunomodulation suggests there is mucosal cell-microbe cross-talk (Figure 16.5) (Freitas et al. 2003). As the mucosal cells and microbiota interact at the cellular surface interface, a cascade of events is initiated. One important event is the production of soluble modulins by the bacteria. These compounds trigger specific cell surface glycosylation patterns with galactose, sialic acid, or

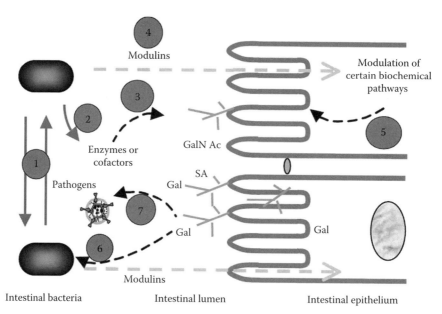

**FIGURE 16.5**  General scheme of the intestinal bacterium-eukaryote cells "cross-talk." (From Freitas, M. et al. 2003. *Biol Cell* 95:503–06. With permission.)

N-acetyl-galactosamine. These unique patterns, in turn, stimulate intestinal epithelial cell defense mechanisms that increase our resistance to microbial and viral pathogens. In the neonate, this apparent microbial cross-talk is critical as the once intrauterine germ-free newborn enters highly contaminated extrauterine environment. Evidence now suggests inadequate colonization with commensal organisms delays the development of normal mucosal immune function and may present several consequences of clinical significance (Walker 2008).

Importantly, immature dendritic cells in the lamina propria extend an appendage between enterocytes into the intestinal lumen. Through toll-like receptor 2 (TLR-2) and TL-4 on this appendage, the cells sample the luminal commensal bacteria, and through this interaction the dendritic cells mature and release cytokines creating a microenvironment enabling naïve T-helper cells (Th0) to mature into balanced Th1 and Th2/Tr1 helper subsets. Colonization with non-pathogens leads to the maturation of dendritic cells. The release of anti-inflammatory cytokines contributes to the maturation of naïve T-helper cells that are important components of disease resistance (Inohara et al. 2005).

The microbiome responds to different gastrointestinal conditions that are impacted by many host factors, such as genotype, immune system, and transient community composition. The two classes of "pattern recognition receptors," namely TLRs and NOD molecules, through several cell signaling mechanisms, instruct the innate immune system to detect and clear pathogens. This apparently conferred disease resistance also involves the activation of macrophage-like cells with specific pattern recognition receptors (Inohara et al. 2005).

Using a mouse model, modulation of the intestinal microbial environment can interact with components of the innate immune system, such as TLRs and the MyD88 protein, that may be critical epigenetic factors that can modify the predisposition to developing type 1 diabetes. Similarly, Crohn's disease susceptibility reflects several genetic aberrants, including frameshift mutations, specifically around the leucine-rich repeat (LRR) domain of the TLR in their extracellular domain, a Toll/Interleukin-1 (TIR) domain in the intracellular domain, and the required recognition peptide, bacterial muramyl dipeptide (MDP), and variants in nucleotide-binding domain 2 (NOD2) (Giradin et al. 2003a).

These physical characteristics generate specific recognition patterns, also termed pathogen-associated molecular patterns, which are unique to microbial pathogens. TLRs actively participate in the regulation and activation of both the innate and adaptive immune system. LRRs are required for recognition of bacterial muramyl dipeptides (MDP) and nucleotide-binding domains (NOD). MDP occurs naturally in the cell walls of pathogens, such as gram-positive bacteria. NODs are unique cytosolic proteins found in humans as well as plants that function as key regulators of apoptosis and pathogen resistance. Thus, they have important roles in the innate and acquired immune system by "sensing" bacterial components, such as bacterial peptidoglycan-derived molecules. This sensing leads to the recruitment of several downstream kinase proteins and the production of proinflammatory cytokines and cell-surface molecules. Many pathogens secrete aminases that disrupt NODs, thus decreasing the anti-inflammatory processes while further inducing pathogenesis. Polymorphisms of NOD-LRR proteins have been implicated in several inflammatory diseases, including Crohn's disease and UC (Giradin et al. 2003b; Inohara et al. 2005).

## CONCLUSION

Inflammation, induced through various processes that involve NF-κB, is the cornerstone of many human diseases. Importantly, individual inflammatory responses genetically mediated through the innate immune system, alterations in leukocyte activity, and perturbations in the human microbiome.

Through extensive microarray data analysis, there appear to be broad transcriptional changes when individuals are exposed to bacterial-derived endotoxins. Computerized modeling of the complex immune system evaluated in humans trials and rodent studies, indicate more than 300 genes are involved in the inflammatory process and innate immunity. Typically within 24 hours post endotoxin exposure, normal homeostasis of the immune system is restored. This model illustrates the importance of functional networks and the genome-wide interactions necessary for appropriate inflammatory response and the normalization of the immune system. Disruption or functional decline of any aspect of this immune network can lead to chronic inflammation and enhance a significant disease sequelae in otherwise normally, healthy individuals.

Inflammation is also a fundamental response to normal injury or insult. Within the gut, a local acute inflammation can lead to chronic inflammatory disorders, such as IBD, IBS, and food allergies, and inflammatory diseases, including autoimmune arthritis, cardiovascular disease, and possibly some cancers (NIH HMP Working Group 2009). The innate immunity involves TLRs, epithelial cells, macrophages, and an array of proteins and cell types, which if compromised through genetic expression, contribute to the production of proinflammatory cytokines, oxidative stress, and tissue damage. This potential compromise is readily observed in the gut which is continuously challenged by dietary components, environmental substances, and microbial exposure.

The gut and the human microbiome reflect a significant barrier to protect the body. Select bacterial colonization or transient introduction can alter intestinal physiology by gene modulation that impacts nutrient absorption, mucosal defenses, and xenobiotics metabolism. The cascade of events following these exposures impact amino acid homeostasis, lipid metabolism, protein and carbohydrate fermentation, and the balance within the gut microbial ecology. Metabolic profiles within the microbially-laden gastrointestinal tract emphasize the importance of bacterial fermentation of dietary carbohydrates to produce short-chain fatty acids in the proximal bowel to enrich the mucosal barrier and improve cellular health. In addition, the cross-talk of commensal bacteria, mucosal cells, and innate and adaptive immune systems highlight the significance of the human microbiome on intestinal homeostasis and overall health, including appropriate immune responses and tissue functions (Martin et al. 2009).

## REFERENCES

Aggarawal, B. B., and S. Shishodia. 2006. Molecular targets of dietary agents for prevention and therapy of cancer. *Biochem Pharmacol* 71:1397–1421.

AlFaleh, K. M., and D. Bassler. 2008. Probiotics for prevention of necrotizing enterocolitis in preterm infants. *Cochrane Database of Systematic Reviews* DOI: 10.1002/14651858.CD005496.pub2.

Bäckhed, F. 2010. 99th Dahlem conference on infection, inflammation and chronic inflammatory disorders: the normal gut microbiota in health and disease. *Clin Exptl Immunol* 160:80–4.

Bendtzen, K., M. Diamant, T. Horn, C. Pedersen, and K. Buschard. 1992. Effect of fusidic acid on interleukin-1 (IL-1)- and IL-6-induced pancreatic beta-cell functions in rats. *J Endocrinol* 132:345–52.

Brook, I. 2008. Microbiology and management of neonatal necrotizing enterocolitis. *Am J Perinatol* 25(2):111–8.

Brugman, S., F. A. Klatter, J. T. Visser, A. C. Wildeboer-Veloo, H. J. Harmsen, J. Rozing, and N. A. Bos. 2006. Antibiotic treatment partially protects against type 1 diabetes in the bio-breeding diabetes-prone rat. Is the gut flora involved in the development of type 1 diabetes? *Diabetologia* 49:2105–8.

Cani, P. D., J. Amar, M. A. Iglesias, M. Poggi, C. Knauf, D. Bastelica, A. M. Neyrinck, F. Fava, K. M. Tuohy, C. Chabo, A. Waget, E. Delmée, B. Cousin, T. Sulpice, B. Chamontin, J. Ferrières, J. F. Tanti, G. R. Gibson, L. Casteilla, N. M. Delzenne, M. C. Alessi, and R. Burcelin. 2007. Metabolic endotoxemia initiates obesity and insulin resistance. *Diabetes* 56:1761–72.

Deshpande, G., S. Rao, S. Patole, and M. Bulsara. 2010. Updated meta-analysis of probiotics for preventing necrotizing enterocolitis in preterm neonates. *Pediatr* 125:921–30.

Dethlefsen, L., M. McFall-Ngai, D. A. Relman. 2007. An ecological and evolutionary perspective on human-microbe mutualism and disease. *Nature* 449:811–8.

Dumas, M. E., R. H. Barton, A. Toye, O. Cloarec, C. Blancher, A. Rothwell, J. Fearnside, R. Tatoud, V. Blanc, J. C. Lindon, S. C. Mitchell, E. Holmes, M. I. McCarthy, J. Scott, D. Gauguier, and J. K. Nicholson. 2006. Metabolic profiling reveals a contribution of gut microflora to fatty liver phenotype in insulin-resistant mice. *Proc Natl Acad Sci USA* 103:12511–6.

Freitas, M., E. Tavan, C. Cayuela, L. Diop, C. Sapin, and G. Trugnan. 2003. A Network-based analysis of systemic inflammation in humans. *Biol Cell* 95:503–06.

Fujimura, K. E. , N. A. Slusher, M. D. Cabana, and S. V. Lunch. 2010. Role of the gut microbiota in defining human health. *Expert Rev Anti Infect Ther* 8:435–54.

Gibson, G., and M. Roberfroid. 1995. Dietary Modulation of the Human Colon Microbiota: Introducing the Concept of Prebiotics. *J Nutr* 125:1401–12.

Girardin, S. E., I. G. Boneca, J. Viala, M. Chamaillard, A. Labigne, G. Thomas, D. J. Philpott, and P. J. Sansonetti. 2003a. Nod2 is a general sensor of peptidoglycan through Muramyl dipeptide (MDP) detection. *J Biol Chem* 278:8869–72.

Girardin, S. E., L. H. Travassos, M. Herve, D. Blanot, I. G. Boneca, D. J. Philpott, P. J. Sansonetti, and D. Mengin-Lecreulx. 2003b. Peptidoglycan molecular requirements allowing detection by Nod1 and Nod2. *J Biol Chem* 278:41702–08.

Gordon, H. A., and L. Pesti. 1971. The gnotobiotic animal as a tool in the study of host microbial relationships. *Bacteriol Rev* 35:390–429.

Hansen, A. K., F. Ling, A. Kaas, D. P. Funda, H. Farlov, and K. Buschard. 2005. Diabetes preventive gluten-free diet decreases the number of caecal bacteria in non-obese diabetic mice. *Diabetes Metab Res Rev* 22:220–5.

Hazrati, E., B. Galen, W. Lu, W. Wang, Y. Ouyang, M. J. Keller, R. I. Lehrer, and B. C. Herold. 2006. Human alpha- and beta-defensins block multiple steps in herpes simplex virus infection. *J Immunol* 177:8658–66.

Hollox, E. J. 2008. Copy number variation of beta-defensins and relevance to disease. *Cytogenet Genome Res* 123:148–55.

Hollox, E. J. 2010. β-Defensins and Crohn's disease: Confusion from counting copies. *Am J Gastroenterol* 105:360–2.

Inohara, N., M. Chamaillard, C. McDonald, and G. Nuñez. 2005. NOD-LRR proteins: Role in host microbial interactions and inflammatory disease. *Annu Rev Biochem* 74:355–83.

Kelly, D., J. I. Campbell, T. P. King, G. Grant, E. A. Jansson, A. G. P. Coutts, S. Pettersson, and S. Conway. 2004. Commensal anaerobic gut bacteria attenuate inflammation by regulating nuclear-cytoplasmic shuttling of PPAR-γ and RelA. *Nat Immunol* 5:104–12.

Kiechl, S., E. Lorenz, M. Reindl, C. J. Wiedermann, F. Oberhollenzer, E. Bonora, J. Willeit, and D. A. Schwartz. 2002. Toll-like receptor 4 polymorphisms and atherogenesis. *N Eng J Med* 347:185–92.

Kim, Y. S., and J. A. Milner. 2007. Dietary modulation of colon cancer risk. *J Nutr* 137:S2576–9.

Kisich, K. O., M. D. Howel, M. Boguniewicz, H. R. Heizer, N. U. Watson, and D. Y. Leung. 2007. The constitutive capacity of human keratinocytes to kill *Staphylococcus aureus* is dependent on beta-definsin 3. *J Invest Dermatol* 127:2368–80.

Kussman, M., and S. Blum. 2007. OMICS-derived targets for inflammatory gut disorders: Opportunities for the development of nutrition related biomarkers. *Endocr Metab Immune Disord Drug Targets* 7(4):271–87.

Ley, R. E., D. A. Peterson, and J. I. Gordon. 2006. Ecological and evolutionary forces shaping microbial diversity in the human intestine. *Cell* 124:837–48.

MacDonald, T. T., and G. Monteleone. 2005. Immunity, inflammation, and allergy in the gut. *Science* 307:1920–25.

Majewski, J., P. Zawadski, P. Pickerill, F. M. Cohan, and C. G. Dowson. 2000. Barriers to genetic exchange between bacterial species: *Streptococcus pneumoniae* transformation. *J Bacteriol* 182(4):1016–23.

Mason, K. L., G. B. Huffnagle, M. C. Noverr, and J. Y. Kao. 2008. Overview of gut immunology. *Adv Exp Med Biol* 635:1–14.

Mowat, A. M. 2003. Anatomical basis of tolerance and immunity to intestinal antigens. *Nat Rev Immunol* 3:331–41.

Naidu, A. S., W. R. Bidlack, and R. A. Clemens. 1999. Probiotic spectra of lactic acid bacteria (LAB). *CRC Crit Rev Food Sci Nutr* 39:13–126.

Neish, A. S., A. T. Gewirtz, H. Zeng, A. N. Young, M. E. Hobert, V. Karmali, A. S. Rao, and J. L. Madara. 2000. Prokaryotic regulation of epithelial responses by inhibition of IkappaB-alpha ubiquitination. *Science* 289:1560–3.

NIH HMP Working Group. 2009. The NIH Human Microbiome Project. *Genome Res* 19:2317–23.

O'Keefe, S. J. D. 2008. Nutrition and colonic health: The critical role of the microbiota. *Curr Opin Gastroenterol* 24:51–8.

O'Keefe, S. J. D., J. Ou, S. Aufreiter, D. O'Connor, S. Sharma, J. Sepulveda, T. Fukuwatari, K. Shibata, and T. Mawhinnery. 2009. Products of the colonic microbiota mediate the effects of diet on colon cancer risk. *J Nutr* 139(11):2044–8.

Preidis, G. A., and J. Versalovic. 2009. Targeting the human microbiome with antibiotics, probiotics, and prebiotics: Gastroenterology enters the metagenomics era. *Gastroenterol* 136(6):2015–31.

Scott, F. W. 1996. Food-induced type 1 diabetes in the BB rat. *Diabetes Metab Rev* 12:341–59.

Tarnow-Mordi, W. O., D. Wilkinson, A. Trivedi, J. Sinn, S. Dutta, T. Parikh, and H. C. Lin. 2010a. Re: ESPGHAN commentary and education that probiotics substantially reduce all-cause mortality and necrotizing enterocolitis in preterm infants. *J Pediatr Gastroenterol Nutr* 50:694.

Tarnow-Mordi, W. O., D. Wilkinson, A. Trivedi, and J. Brok. 2010b. Probiotics reduce all-cause mortality and necrotizing enterocolitis: It is time to change practice. *Pediatr* 125:1068–70.

Triantafilou, M., F. G. J. Gamper, P. M. Lepper, M. A. Mouratis, C. Schumann, E. Harokopakis, R. E. Schifferle, G. Hajishengallis, and K. Triantafilou. 2007. Lipopolysaccharides from atherosclerosis-associated bacteria antagonize TLR4, induce formation of TLR2/1/CD36 complexes in lipid rafts and trigger TLR2-induced inflammatory responses in human vascular endothelial cells. *Cell Microbiol* 9:2030–9.

Turnbaugh, P. J., R. E. Ley, M. Hamady, C. M. Fraser-Liggett, R. Knight, and J. I. Gordon. 2007. The human microbiome project. *Nature* 449:804–10.

Turnbaugh, P. J., R. E. Ley, M. A. Mahowald, V. Magrini, E. R. Mardis, and J. I. Gordon. 2006. An obesity-associated gut microbiome with increased capacity for energy harvest. *Nature* 444:21–8.

Vidyasagar, S., C. Barmeyer, J. Geibel, H. J. Binder, and V. M. Rajendran. 2005. Role of short-chain fatty acids in colonic $HCO_3$ secretion. *Am J Physiol Gastrointest Liver Physiol* 288:G1217–26.

Vylkova, S., N. Nayyar, W. Li, and M. Edgerton. 2007. Human beta-defensins kill *Candida albicans* in an energy-dependent and salt-sensitive manner without causing membrane disruption. *Antimicrob Agents Chemother* 51:154–61.

Walker, M. A. 2008. Mechanism of Action pf Probiotics. *Clin Infect Disease* 46:587–91.

Walter, J. and R. Ley. 2011. The Human Gut Microbiome: Ecology and recent evolutionary changes. *Annu Rev Microbiol* 65:411–29.

Wolvers, D., J-M. Antoine, E. Myllyluoma, J. Schrezenmeir, H. Szajewska, and G. T. Rijkers. 2010. Guidance for substantiating the evidence for beneficial effects of probiotics: prevention and management of infections by probiotics. *J Nutr* 140:698S–712S.

# Section III

## Food Needs to Meet Nutrigenomic Health Needs

### Agriculture's Ability to Enhance Food Quality

# 17 Calcium Biofortification of Crops

*Kendal D. Hirschi and Sean M. Thompson*

## CONTENTS

## INTRODUCTION

More than half of the world's population is deficient in calcium (Ca), iron (Fe), iodine (I), magnesium (Mg), selenium (Se), or zinc (Zn) (Graham et al. 2007; White and Broadley 2009). The consumption of plants, directly or via livestock, containing inadequate concentrations of particular minerals causes these deficiencies. Agronomic and genetic strategies can increase the delivery of bioavailable minerals (Figure 17.1) (Morris et al. 2008; White and Broadley 2009). Although the focus is predominately on Ca, the framework discussed here should be generally applicable to boosting the levels of other elements in agriculturally important crops.

A recent panel of the world's foremost economists deemed plant biofortification, the process of increasing the bioavailable concentration of an element in foods, as one of the preeminent global challenges (www.copenhagenconsensus.com). Furthermore, the economists predicted tremendous benefits compared to costs associated with developing this technology. Indeed, genome projects are providing novel approaches for identifying plant genes of nutritional importance. The term "nutritional genomics" has been coined to describe this work at the interface of plant genomics and human nutrition (DellaPenna 1999; Mayer et al. 2008; Zhu et al. 2007; Zhu and Shimamoto 2007). However, work done to measure the nutrient value of these engineered plant foods to date is minimal (Adams et al. 2002; Hambidge et al. 2005; Hirschi 2008; Mendoza et al. 1998).

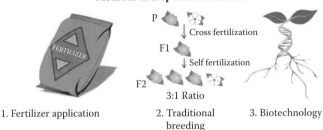

**Methods of crop biofortification**

P → Cross fertilization

F1 → Self fertilization

F2 → 3:1 Ratio

1. Fertilizer application       2. Traditional        3. Biotechnology
                                   breeding

**FIGURE 17.1** Methods of developing nutrient dense crops. (1) Application of fertilizers during plant growth can be used to increase specific nutrients in crops. (2) Traditional breeding is an effective and proven approach to increasing specific nutrients in closely related crops. (3) Biotechnology can be used to increase the nutrient content of crops by expressing genes from unrelated organisms.

## GENETICALLY MODIFIED FOODS: FARMERS AND CONSUMERS

The debate over genetically modified foods is being decided on the ground. Global demands for both food and fuel have sold farmers on transgenic technologies. In America, despite some consumer concerns, we eat more genetically modified foods than any other country. The possibilities associated with transgenic approaches keep plant biologists and production agriculture optimistic despite the current political and economic landscape that is not completely receptive to this technology (Freese and Schubert 2004; ISLI 2008; Johnson et al. 2007; Powell 2007; Weil 2005). Even with these current limitations, the potential for genetic modifications to alleviate hunger and nutrient deficiencies warrants advocacy of this technology among both scientists and citizens. Furthermore, despite the hurdles, if you have been to the grocery store lately, odds are you've eaten genetically modified foods.

The integration of genetically modified foods into modern agriculture and our current food supply can be exemplified by the work of the St. Louis-based biotechnology company, Monsanto (Hindo 2007). Although humans do not directly consume the majority of Monsanto crops, there are plenty of ways for people to ingest them indirectly. For example, the majority of Monsanto's genetically modified corn is used to make animal feed and ethanol, but a portion goes into the food supply as corn syrup and cornstarch, as well as helping to make some corn tortilla chips. Also, the bulk of the biotech cotton feeds the textile industry. However, cotton by-products do end up in the food chain as cottonseed oil, which can be used to make mayonnaise and margarine.

Another illuminating example is papaya. An estimated 80% of the papayas from Hawaii are genetically engineered. This technology is publically licensed and sold to farmers through a not-for-profit group. Although these are just a few examples, in the United States about 70% of all "formulated foods"—processed food with more than one ingredient—contain GMOs (Figure 17.2).

The work outlined here may promote positive viewpoints among consumers regarding genetically modified foods. Our studies detailed below addressing the bioavailability of Ca were well received by the media, including news articles by the BBC and NPR (press articles can be found at http://www.bcm.edu/cnrc/faculty/kendalh.htm). An article in the *London Telegraph* talking about our work contained the subheading "Europe 'will be forced to re-think on GM crops'" (Highfield 2008).

## BIOFORTIFICATION AND NUTRITIONAL STUDIES

Biofortified foods offer a potentially powerful intervention tool that targets the most vulnerable people (resource-poor women, infants, and children). However, few studies have measured the most important parameter to determine the eventual successes of conventionally bred foods or genetically modified lines; namely, are these foods actually functional foods (Powell 2007)?

The most notable example of this gap between the technology of the transgenic plants and measuring nutritional efficacy is the case of "Golden Rice," engineered to produce beta-carotene in the

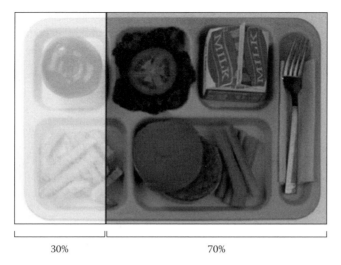

30%                              70%

**FIGURE 17.2** A biotechnologist's view of a school lunch. Genetically modified ingredients have been introduced into nearly 70% of processed foods available in the United States; including snacks, cereals, and hot dogs. (Data from Bren, L. 2003. *FDA Consumer Magazine*, November–December 2003.)

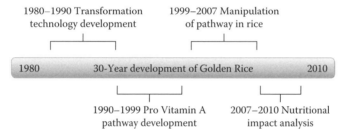

**FIGURE 17.3** Vitamin A fortification of Golden Rice. Technological developments in the 1980s allowed for different genes to be expressed in rice, a process termed transformation. The carotenoid pathway was characterized and the genes identified during the 1990s. The engineering of different genes responsible for carotenoid biosynthesis in rice was optimized by 2007. In the last several years nutritional studies have been initiated to assay the effectiveness of the food. The ecological impact of Golden Rice is currently under study.

edible portion of the grain (Ye et al. 2000). Seven plus years after the introduction of Golden Rice, scientists are still pondering its nutritional benefits; however, some recent preliminary findings are encouraging (Krawinkel 2007; Tang et al. 2009). While the first human bioavailability studies have just been completed, we are left to consider the timing of these events (Figure 17.3). From one vantage point, it was prudent to wait on the bioavailability studies, since the initially engineered versions of Golden Rice did, in fact, have very little beta-carotene (Nestle 2001). On the other hand, how can so much time and effort be invested in this technology without performing fundamental studies regarding nutritional benefits? Unfortunately, the lack of nutritional assessment in biofortified foods is the norm rather than the exception. Different carotenoid-enriched foods (DellaPenna 2007) as well as crops enriched with other micronutrients such as vitamin E (Ajjawi and Shintani 2004) and folate (Bekaert et al. 2008) are further examples where the genetically modified foods have not been adequately assessed at the nutritional level.

A notable exception to the lack of nutritional studies is the work manipulating phytic acid content in crops (Adams et al. 2002; Hambidge et al. 2005; Raboy 2001). As detailed next, phytic acid is the storage form of phosphorus in seeds. Human feeding studies were done that compared the absorption of Ca from tortilla meals prepared from low-phytate maize with that from meals prepared from maize with typical phytate content. Mean fractional absorption of Ca from tortillas prepared from the low-phytate maize was significantly greater than that from tortillas prepared from the control maize. This type of approach serves as a paradigm for the integration of plant biology and human nutritional studies.

## TARGETS FOR Ca BIOFORTIFICATION

Popular vegetables among American consumers, namely potatoes, carrots, and lettuce, are excellent targets for enhancing Ca content (Table 17.1, Figure 17.4). On average, Americans consume only 3.3 servings of vegetables a day (CDC 2011a,b). Currently, less than 25% of the U.S. population reaches the adequate daily consumption of five daily portions of fruits and vegetables. These statistics argue for the development of strategies to increase the levels of health promoting compounds, such as Ca in the vegetables that people consume in substantial amounts.

The dark green vegetables (broccoli, bok choy, collard greens, kale, etc.), which may be high in dietary Ca, represent only 0.2 of these daily servings. The number one vegetable eaten by Americans is the potato followed by iceberg lettuce. In terms of Ca content, these vegetables are not as desirable as

## TABLE 17.1
### Comparison of Absorbable Dietary Ca in Traditional and Fortified Foods

| Food | Serving Volume | Serving Weight (g) | Ca Content (mg)/Serving | Ca Fractional Absorption (%) | Ca Absorbed (mg)/Serving |
|------|----------------|--------------------|-----------------------|------------------------------|--------------------------|
| Milk | 1 cup | 245 | 285 | 32 | 91 |
| Kale | 1 cup | 118.5 | 47 | 49 | 23 |
| Spinach | 1 cup | 118.5 | 244 | 5 | 12 |
| Spinach (–Ca OX) | 1 cup | 118.5 | 244 | 59* | 144* |
| Carrots | 1 cup | 128 | 42 | 53 | 22.26 |
| Ca fortified carrots | 1 cup | 128 | 420 | 26 | 109.2 |
| Lettuce | 1 cup | 72 | 13 | 49 | 6.37 |
| Ca fortified lettuce | 1 cup | 72 | 130 | 25 | 32.5 |
| Potato | 1 med | 173 | 26 | 22 | 5.72 |
| Ca fortified potato | 1 med | 173 | 260 | 11 | 28.6 |

*Notes:* Ca content and bioavailability varies among foods. Fortified vegetables have the potential to positively impact consumers whose current diet is deficient in Ca (Titchenal and Dobbs 2007).

*Potential change in Ca absorption if Ca oxalate were removed from spinach.

(a)                                          (b)                                          (c)

**FIGURE 17.4** Targets for Ca biofortification. (a) Potatoes, (b) carrots, and (c) lettuce are targets for Ca biofortification because they are currently low in Ca content and are popular among American consumers.

some of the dark leafy greens (Table 17.1). However, rather than trying to alter consumer preferences, our work here will enhance the Ca content of vegetables that are already popular among consumers.

## POTATOES

Traditionally, potatoes have been grown and consumed in Europe and North America. Recently, there has been a dramatic increase in potato production and demand in Asia, Africa, and Latin America. In fact, Asia now consumes almost half of the world's potato supply, but its huge population means that consumption per person is approximately 25 kg. The heartiest potato eaters remain Europeans (85 kg) and North Americans (60 kg). Per capita consumption is lowest, but increasing, in Africa and Latin America.

Potatoes are eaten all over the world in an assortment of different dishes. Potatoes are high in carbohydrate content (19%), mostly in the form of starch, which makes them such a good source of energy, but the potato also provides a fair amount of protein (2%). In addition, a medium-sized potato can provide about a third of the recommended daily allowance (RDA) of vitamin C, around a fifth of the RDA of vitamin B6 and about 30% of the RDA of Fe, as well as small amounts of thiamin, riboflavin, folate, niacin, Mg, phosphorus, and Zn. An average-sized baked potato (including skin) contains approximately 26 mg of Ca and is not a good source of Ca (Table 17.1).

## CARROTS

Carrots are one of the most popular vegetables in the United States (Simon and Goldman 2007). Although consumption varies by ethnicity, age, and income, the average American eats 4.98 kg of carrots per year. Carrots can be eaten in a variety of ways and raw carrots have become a popular ready-to-eat snack food. Since the 1980s, baby carrots (carrots that have been peeled and cut into uniform cylinders) have transformed the way people eat carrots. Carrots are an excellent source of beta-carotene as pro vitamin A, which can be converted by the human body into an active form. In addition, they are a very good source of vitamin C, vitamin K, dietary fiber, and potassium. Ca content in carrots is approximately 42 mg/cup and is currently a fair source of Ca (Table 17.1).

## LETTUCE

Lettuce is an attractive dietary option for enhancing consumption of dietary Ca. Like carrots, lettuce can be eaten raw (FoodReference.com 1990). Per-capita consumption of all lettuce varieties has been increasing since 1960. In 2004 total lettuce consumption reached a record high of 15.65 kg per capita. Some lettuces (especially iceberg) have been specifically bred to remove the bitterness from their leaves. These lettuces have a high water content with very little nutrient value. However, lettuce is rich in vitamin K (as much as 167 µg/head, but is currently low in Ca (13 mg/1 cup shredded) and is not a good source of Ca (Table 17.1).

In our studies modified potatoes, carrots, and lettuce are being used to translate our knowledge from model systems to crops that can benefit consumers. Alternative vegetables we have considered modifying (after discussions with dieticians) include tomatoes and corn; however, these plants have proven recalcitrant to the approaches used in our studies (Park et al. 2005). Preliminary studies suggest Ca biofortification efforts using potatoes; carrots and lettuce should concentrate on fertilizer applications and genetic engineering approaches. There is limited genetic variation among potato, carrot, and lettuce varieties in terms of Ca content but these plants can be easily transformed to facilitate genetic engineering approaches. Additionally, we have working relationships in place with carrot and potato breeders at Texas A&M University (Park et al. 2004). These advantages have already allowed us to design and implement an experimental pipeline to genetically alter these crops and measure bioavailability (Morris et al. 2008). The inability to transform many crops, such as spinach, makes these plants recalcitrant to our experimental approaches.

## ANTINUTRIENTS

Ca present in plant foods typically exists as a complex with oxalate, phytate, fiber, lactate, fatty acid, protein, and other anions (Linder 1991; Mendoza et al. 1998; Wilson and Clifford 1990). Many plants are particularly high in total Ca, but it is often sequestered by an antinutrient. Antinutrients, although not toxic, are plant compounds that decrease the nutritional value of a food, usually by making an essential nutrient unavailable or indigestible.

### PHYTIC ACID

Grains and legumes are foods often rich in phytic acid (myo-inositol-1,2,3,4,5,6-hexakis-phosphate), the storage form of phosphorus in seeds. This compound binds to mineral nutrients such as I and Zn, forming salts that are excreted. This can contribute to mineral depletion and deficiency. To solve this problem, the U.S. Department of Agriculture and others have isolated cereal and legume low-phytic acid mutations and have used these to breed low-phytate hybrids, cultivars and lines of maize (*Zea mays*), barley (*Hordeum vulgare*), rice (*Oryza sativa*), and soybean (*Glycine max*). Seed phytic acid is reduced in these crops by 50%–90%.

A maize mutant in the synthesis of phytic acid during seed maturation has been used to study the consequences of the lack of this important reserve substance on seed survival. Data on germination, free Fe levels, free radical relative abundance, protein carbonylation level, damage to DNA, and other parameters were recorded on seeds of maize and of an isogenic low phytic acid mutant (lpa1-241), either unaged or incubated for 7 days in accelerated aging conditions (46°C and 100% relative humidity). The lpa1-241 mutant, compared to the control showed a lower germination capacity, which decreased further after accelerated aging. Whole lpa1-241 mutant kernels contained about 50% more free or weakly bound Fe than control ones and showed a higher content of free radicals, mainly concentrated in embryos. These findings suggest antioxidant activity for phytic acid through Fe complexation. Therefore, a role in plant seed development can be assigned to phytic acid, that is, protection against oxidative stress during maturation. These studies suggest that removal of phytic acid may have negative consequences for plant development.

### OXALATE

Oxalate is also an "antinutrient"; it sequesters Ca in a state that may render it unavailable for nutritional absorption (Figure 17.5) (Weaver et al. 1987). In support of this, Ca absorption appears to be inversely proportional to the oxalic acid content in food (Table 17.1) (Weaver et al. 1987). While spinach contains between 23.8 and 26.7 mg/g Ca, oxalate content is high and drastically reduces Ca bioavailability; however, kale which contains between 26.3 and 27.6 mg/g Ca has low oxalate levels and much higher bioavailable concentrations of Ca. (Weaver et al. 1987); a notable exception to this correlation is soybeans, where oxalate levels are high (35mg/g) (Massey et al. 2001) but bioavailable Ca is also high (total Ca 27.7 mg/g) (Massey et al. 2001).

Previous studies have strongly suggested that Ca when sequestered in the form of Ca oxalate is unavailable for nutritional absorption (Franceschi and Nakata 2005; Heaney et al. 1988; Weaver 1990). These studies lacked precision because the genetic mechanisms underlying the inherent differences in the plants were too numerous to accurately determine the antinutrient(s). In order to clarify these nutritional studies, comparisons were made in the Ca absorption from isogenic lines of a plant that differed in a single gene, which mediates oxalate crystal content (Morris et al. 2007; Nakata and McConn 2000). Using genetic analysis and mice fed labeled diets, this experimental platform shows that plants lacking Ca oxalate crystals are better sources of bioavailable Ca (Figure 17.5). However, oxalate crystals are also an important factor in deterring insects from eating the plants (Korth et al. 2006). Recent work has shown that the size and shape of oxalate crystals are important factors in determining effects on insect growth (Korth et al. 2006). If manipulation of Ca oxalate is to

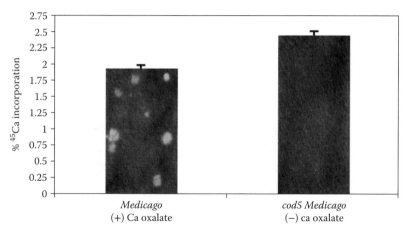

**FIGURE 17.5** Ca oxalate crystals impair Ca absorption in mice feeding studies. A representative portion of a leaf is shown using partially polarized light. Spinach cleared of chlorophyll displays Ca oxalate crystals (bright crystal structures) that are not present in the kale. Mice given diets containing the *cod5* mutant lacking Ca oxalate crystals (Nakata and McConn 2000) have a higher percent of Ca incorporation into bones (Morris et al. 2007). (From Nakata, P. A., and M. M. McConn. 2000. *Plant Physiology* 124:1097–104. With permission.)

be used in developing improved nutritional qualities and insect resistance in plants, then controlling not only the overall amount, but also the size and shape of crystals, could be valuable traits.

## GOT VEGGIES? ENHANCING Ca CONTENT IN POPULAR VEGETABLES

We will concentrate on the tools that we have developed and outline the approaches we used to combine these technologies to produce potatoes, carrots, and lettuce that have high levels of dietary Ca.

### BIOFORTIFICATION: INCREASING Ca TRANSPORT ACTIVITY

Heightened activity of a plant Ca transporter doubles Ca content in potatoes, carrots, and lettuce. In its simplest form, this strategy can be compared to nutrient mining; Ca is transported from the soil into the edible portions of plants. Specifically, one approach is to manipulate plant endomembrane transporters to increase Ca transport. In animal cells and in yeast, capacitative Ca entry (CCE) mechanisms are activated when vacuolar Ca transporters are highly expressed (Hirschi 2001). In plants, there is evidence to suggest that expression of a gene from a different plant species may be able to bypass some of the endogenous regulatory elements in the plant. This may produce higher activity of a given gene product (Diener and Hirschi 2000). With this in mind, we have expressed an *Arabidopsis* Ca transporter (CAX) in lettuce, carrots, and potatoes to increase Ca content (Figure 17.6). In each case, these modifications did not appear to alter the levels of antinutrients or perturb growth, development, or fertility (Park et al. 2004).

In the potatoes, carrots, and lettuce we have verified the increased Ca content is caused by expression of a single copy of the CAX cassette. In carrots, we have used genetic crosses to verify this is a heritable trait. Using both biochemical analysis and microscopic studies, we noted no alteration in oxalate levels, a potential antinutrient.

### Ca ABSORPTION MEASUREMENTS USING MOUSE FEEDING STUDIES

The use of radioisotopes in mice feeding studies is a rapid and economically feasible method of getting preliminary results regarding a net gain in total Ca absorbed from biofortified foods (Figure 17.7). The mouse has become the preeminent mammalian model animal because of the

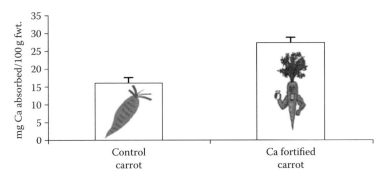

**FIGURE 17.6** Ca absorption in human subjects fed control and Ca fortified carrots. Increase in total Ca absorbed by human subjects from control carrot lines and those fortified using biotechnology. Increased expression of a Ca transporter from *Arabidopsis* increases total bioavailable Ca. Increased Ca in these fortified carrots demonstrates the potential of this technology to address Ca deficiencies in today's diets.

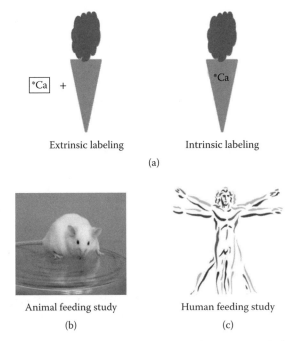

**FIGURE 17.7** Design for testing bioavailability in biofortified foods. (a) Extrinsic and intrinsic labeling of food. Extrinsic labeling adds the tracer (*) to the food whereas intrinsic labeling incorporates the label into the plant matrix. (b) Initially, modified plants are labeled with inexpensive radioactive tracers and fed to animals. (c) Human feeding studies using safe, but expensive, stable isotopes are used if the animal work shows promise. (Adapted from Hirschi, K. D. 2008. *Trends in Plant Science* 13:459–63.)

underlying biological similarity to humans, the emerging genomic sequence data, and the ability to manipulate the mouse genome in a targeted or random fashion (Nguyen and Xu 2008). Some of the most elegant mice genetics has been directed at the study of bone development (Provot et al. 2008; Wagner and Karsenty 2001). For example, studies using mice mutants elucidated the role of leptin as a potent inhibitor of bone formation. Long-term nutrition studies that are directed at understanding gene for gene interactions between the food and the "consumer" must utilize malleable genetic systems on both ends of the equation. For example, if a plant is expressing high levels of a Ca transporter, in the near future it will be interesting to observe how mice defective in Ca bone deposition respond to diets containing genetically altered food.

We have modified protocols previously used in rat feeding studies for analyzing Ca bioavailability in mice (Morris et al. 2007). Our data on Ca absorption with the $CaCl_2$, $CaC_2O_4$ and spinach (both extrinsic and intrinsic) compare favorably to similar studies in rats and thus establish the equivalence between the two *in vivo* models (Weaver et al. 1987). For example, the published rat studies have absorption values around 1.8% for both $CaCl_2$ and $CaCO_3$ diets and only 0.2% from $CaC_2O_4$. In our work, we showed absorption of $^{45}Ca$ from $CaCl_2$ and $CaCO_3$ diets at 1.21% and 1.17%, compared to only 0.07% from $CaC_2O_4$. While the precise absorption values are not identical to the rat studies (Weaver et al. 1987), the findings from both studies suggest that oxalic acid binds Ca, rendering it unavailable for absorption.

As mentioned in the introduction, modifying plants to reduce the concentration of antinutrients have been shown to increase bioavailable mineral content (Adams et al. 2002; Hambidge et al. 2005; Mendoza et al. 1998; Morris et al. 2007). We have performed studies looking at the role of Ca oxalate crystals in *Medicago truncatula* variants fed to mice (Morris et al. 2007). *M. truncatula* has been chosen as a model organism for legume biology. It has a small diploid genome, is self-fertile, has a rapid generation time and prolific seed production, and is also amenable to genetic transformation. The Ca oxalate deficient (cod) mutant, *cod5*, of *M. truncatula* contains identical Ca concentrations to wild-type, but scanning electron micrographs show the mutants contain no oxalate crystals (Morris et al. 2007). For our feeding studies, diets using wild type *M. truncatula* (oxalate crystals) were compared to the *cod5* mutant based diets (no oxalate crystals). We fed equal numbers of male and female mice four Ca-labeled diets: *M. truncatula* extrinsically or intrinsically labeled, and *cod5* extrinsically or intrinsically labeled by radioisotope $^{45}Ca$ (*Ca). Extrinsic labeling adds the tracer to the food before ingestion, whereas for intrinsic labeling plants are grown in the presence of the label to incorporate the tracer into the plant matrix (Figure 17.7). Using both diets, absorption of the tracer was determined one day after consumption. In the intrinsically labeled diets, Ca absorption was 22.87% (p < 0.001) higher in mice fed *cod5* (Morris et al. 2007). We postulate from our findings that during the formation of crystals in Medicago, much like oxalate crystals in spinach, Ca is bound in a nonbioavailable form. To our knowledge this study presents the first genetic evidence to demonstrate the nutritional impact of removing oxalate crystals from food. This work also demonstrates the usefulness of the mouse model for Ca absorption studies.

## Ca Absorption in Humans

The ultimate demonstration of our hypothesis regarding nutritional benefits to alterations in Ca content requires human studies. The use of stable isotopes makes this feasible and safe as a method for determining a net gain in total Ca absorbed. Although animal models provide evidence related to bioavailability, there are fundamental differences in the mechanism of Ca absorption between humans and small animals. In particular, humans utilize a greater proportion of Ca absorption in the upper small intestine than small animals (Abrams 2003). Recently, we have analyzed the bioavailability of the Ca from modified carrots in a human study (Morris et al. 2008). We chose young adults who were healthy and represent a typical population that might utilize vegetable sources to obtain a substantial amount of their Ca intake. In addition, unlike studies done with children, we could more easily obtain their informed consent.

Through use of both mouse and human feeding studies we have demonstrated that *sCAX1*-expressing carrots have increased Ca bioavailability (Figure 17.8). Although there is a 10% reduction in absorbed Ca from the *sCAX1*-expressing carrots, the total concentration of Ca absorbed from the *sCAX1*-expressing carrots is 42 ± 2% higher compared to an equal quantity of control carrots. In the human feeding trials, our working hypothesis to explain the differences between controls and *sCAX1* lines is that not all the Ca sequestered in the vacuole by ectopic expression of *sCAX1* is bioavailable; it may be conjugated to phytates, phosphates, or other antinutrients within the edible carrot.

To the best of our knowledge, our findings represent the first report to directly evaluate the nutritional consequence of transgenic foods in both animal and human feeding studies. We establish

**FIGURE 17.8** Serving sizes of carrots, Ca-fortified carrots, and milk required to obtain 300 mg of Ca. One 8 oz serving (240 mL) of milk is comparable to 10 (650 g) Ca fortified carrots or 15 (1000 g) control carrots. (Adapted from Hawthorne, K. M., J. Morris, T. Holtze, K. D. Hirschi, and S. A. Abrams. 2009. *J Bioequiv Availab* 1:34–8.)

unequivocally that modifying a single plant Ca transporter improves plant Ca bioavailability. Additionally, we can now use this approach to further boost Ca levels in a variety of different foods.

## IMPACT OF THIS RESEARCH

Single servings of our biofortified vegetables could significantly contribute to dietary Ca intake. These improved vegetables could be widely consumed, making them a substantial component of the Ca intake of Americans. Historically, nutritional problems have centered on the inadequate intake of certain vitamins and minerals resulting in nutritional deficiencies (Frazao 1999). Currently, nutritional problems in America are driven by the discovery of strong links between nutrition and chronic diseases.

A majority of Americans are interested in improving their diets. Evidence suggests that many are changing their diets and attempting to move closer to established dietary recommendations (Hornick et al. 2008). However, the direction and magnitude of these changes vary considerably, both among individuals and among food groups. For example, survey data suggest a trend toward lower fat diets. However, the same data show that individuals are not increasing their consumption of fruits and vegetables as recommended and that the prevalence of obesity is rising (CDC 2008; Wareham et al. 2005).

Government assistance can also have some influence on consumer dietary choices (Ralston et al. 2008). Food assistance programs can affect the amount and the types of foods consumed by low-income populations. However, the long-term changes on the consumer diet, especially after leaving the program, have so far been uncertain. Certainly, greater food expenditure does not necessarily imply a more healthful diet. For the more narrowly targeted programs, such as the School Lunch Program, nutrient intake is typically increased (at least while the recipient remains in the program). For example, children in schools with restricted snack availability had significantly higher frequency of fruit and vegetable consumption than children in schools without restricted snack availability. These recent findings suggest that a restrictive snack policy should be part of a multi-faceted approach to improve children's diet quality (Gonzalez et al. 2009).

Here we are seeking to provide substantial benefit to the public at no inconvenience and minimal cost to the consumer. This passive method does not require consumer knowledge, understanding, or commitment to change food consumption behavior. Increasing the dietary Ca levels five-fold in potatoes, lettuce, and carrots could have a significant affect on total Ca consumption in the United States without requiring altering dietary habits. The U.S. Dietary Reference Intake (DRI) for Ca is linked to a person's age and stage of life, so among adolescents these vegetables contribute even less to their daily Ca requirements. The DRI for Ca for 19- to 30-year-old adults is an adequate intake (AI) level of 1000 mg/day. If Americans continue to eat 60 kg of potatoes a year (approximately 150 medium size potatoes) they would receive a net benefit of approximately 3432 mg of absorbed Ca from these alterations (4290 mg in the biofortified potatoes versus 858 mg in the varieties found today (Table 17.1). For lettuce, if Americans continue to eat 14.4 kg of lettuce per year (approximately 200 servings of 1 shredded cup), the net gain in Ca would be 5226 mg from this food (6500 mg in the biofortified versus 1274 in the standard variety) (Table 17.1). For carrots, if Americans consume 4.989 kg of carrots per year (39 servings of chopped carrots) that would increase Ca consumption by 3390 mg (4258 mg from the biofortified carrots compared with the 868 mg from standard carrots) (Table 17.1). These back-of-the-envelope calculations and predictions assume optimal preparation of the vegetables to ensure Ca bioavailability. While bearing in mind these caveats, these Ca biofortified vegetables could contribute approximately 3%–5% of the DRI of Ca for a wide array of Americans.

## FUTURE STUDIES

The ability to substantially increase bioavailability of Ca in vegetables through a combination of biofortification efforts will be pursued. Vegetables, namely potatoes, lettuce, and carrots, containing single gene alterations in Ca transport activity will be grown in hydroponic conditions to accumulate significantly higher levels (>10X) of Ca. The benefit associated with increasing nutrient bioavailability in foods already regularly consumed is in the potential to impact those most in need. Increasing the Ca content in vegetables provides potential increased calcium in the diet without changing the diet.

We anticipate that in pilot mice feeding studies, the biofortified vegetables will have at least fivefold more bioavailable Ca than the respective control vegetables. In humans, we also assume that total Ca absorption from a serving of biofortified vegetables will be increased at least fivefold compared to controls. In sensory analysis tests, the biofortified vegetables should be equivalent in taste and texture to the nonbiofortified vegetables. We have developed nutritional readouts that can be used to quantify differences among plant-based diets that differ in the location of Ca within the plant matrices. This positions us to rapidly iterate between nutrient partitioning within the plant matrices and bioavailability to devise strategies to further improve the nutritional status of a multitude of agriculturally important crops.

## CLOSING REMARKS

Changing metabolic functions may impact plant growth and productivity. For example, changes in metal content could alter various enzyme and protein functions. It is important to establish if an alteration in plant metabolism is cost effective. A useful plant improvement should increase nutrient content while keeping cultivation and production costs affordable.

The tests used to analyze genetically modified foods should mimic clinical trials with a novel pharmacological agent. With the novel foods, after the nutritional efficacy of the food is proven comes the difficult task of determining the collateral effects. Interactions with nutrients in the plant matrices, allergic responses to the consumer, and altering plant-stress responses are some of the measurements that need to be performed. As mentioned in the *Oxalate* section, removal of the antinutrient Ca oxalate crystals causes these modified plants to be more nutritious but reduces the plant's defense to insect chewing (Korth et al. 2006). For consumer confidence, the most important thing the scientific community must do is be careful in our analysis of these foods before they become available to consumers.

Demand for any biofortified food must drive the product through developmental stages and to offset associated cost increases (Freese and Schubert 2004; ISLI 2008; Johnson et al. 2007; Powell 2007; Weil 2005). For this to occur, the health benefits must be apparent to the consumer. The first steps in this process include nutritional studies in both animal and human feeding trials.

Breeding approaches can be used to enhance the nutritional qualities of foods (Bouis 2000; Grusak and Cakmak 2005). However, breeding alone will not be an adequate approach because of the limitations of particular plant species (Jeong and Guerinot 2008). Breeding and molecular genetics were elegantly combined to characterize an important wheat gene associated with grain

protein, Zn, and Fe content (Uauy et al. 2006). Wild durum (pasta) wheat has shorter grain maturation periods and higher protein, Zn, and Fe contents than domesticated wheats. The version of the
gene found in a wild ancestor of durum wheat was isolated, sequenced, and compared to the version
in modern domesticated wheats. The domesticated wheats have an inactive form of the gene thus
explaining their lower nutrient content relative to the wild durum wheat. The active version can now
be incorporated by breeding or genetic engineering to increase the protein, Zn, and Fe contents of
domesticated wheats. This is likely an example of the optimal experimental approach in biofortification, where research is directed primarily at breeding and genetic modifications will occur only
when necessary.

## REFERENCES

Abrams, S. A. 2003. Using stable isotopes to assess the bioavailability of minerals in food-fortification programs. *Forum of Nutrition* 56:312–3.

Adams, C. L., M. Hambidge, V. Raboy, J. A. Dorsch, L. Sian, J. L. Westcott, and N. F. Krebs. 2002. Zinc
absorption from a low-phytic acid maize. *American Journal of Clinical Nutrition* 76:556–9.

Ajjawi, I., and D. Shintani. 2004. Engineered plants with elevated vitamin E: A nutraceutical success story.
*Trends in Biotechnology* 22:104–7.

Bekaert, S., S. Storozhenko, P. Mehrshahi, M. J. Bennett, W. Lambert, J. F. Gregory III, K. Schubert,
J. Hugenholtz, D. Van Der Straeten, and A. D. Hanson. 2008. Folate biofortification in food plants.
*Trends in Plant Science* 13:28–35.

Bouis, H. E. 2000. Enrichment of food staples through plant breeding: A new strategy for fighting micronutrient malnutrition. *Nutrition* 16:701–4.

Bren, L. 2003. Genetic engineering: The future of foods? *FDA Consumer Magazine*, November–December
2003.

CDC. 2008. State-Specific Prevalence of Obesity Among Adults—United States, 2007. Atlanta, GA: Morbidity
and Mortality Weekly Report.

CDC. 2011a. Fruit and vegetable consumption data and statistics. http://apps.nccd.cdc.gov/5ADaySurveillance/
(accessed July 4, 2011).

CDC. 2011b. How many fruits and veggies do you need? www.fruitandveggiesmatter.gov/ (accessed July 4,
2011).

DellaPenna, D. 1999. Nutritional genomics: Manipulating plant micronutrients to improve human health.
*Science* 285:375–9.

DellaPenna, D. 2007. Biofortification of plant-based food: Enhancing folate levels by metabolic engineering.
*Proceedings of the National Academy of Sciences USA* 104:3675–6.

Diener, A., and K. Hirschi. 2000. Heterologous expression for dominant-like gene activity. *Trends in Plant
Science* 5:10–1.

FoodReference.com. 1990. Lettuce. http://www.foodreference.com/html/artlettuce.html (accessed July 4,
2011).

Franceschi, V. R., and P. A. Nakata. 2005. Calcium oxalate in plants: Formation and function. *Annual Review
of Plant Biology* 56:41–71.

Frazao, E. 1999. America's Eating Habits: Changes and Consequences, T.E.o.F. U.S. Department of Agriculture
Economic Research Service, Farming, Natural Resources, and Rural America. Washington, DC: U.S.
Department of Agriculture Economic Research Service.

Freese, W., and D. Schubert. 2004. Safety testing and regulation of genetically engineered foods. *Biotechnology
and Genetic Engineering Reviews* 21:299–324.

Gonzalez, W., S. J. Jones, and E. A. Frongillo. 2009. Restricting snacks in U.S. elementary schools is associated
with higher frequency of fruit and vegetable consumption. *Journal of Nutrition* 139:142–4.

Graham, R. M., J. W. Thompson, J. Wei, N. H. Bishopric, and K. A. Webster. 2007. Regulation of Bnip3 death
pathways by calcium, phosphorylation, and hypoxia-reoxygenation. *Antioxid Redox Signal* 9:1309–15.

Grusak, M. A., and I. Cakmak. 2005. Methods to improve the crop delivery of minerals to humans and livestock. In M. R. Broadley, and P. J. White, eds., *Plant Nutritional Genomics*, pp. 265–86. Oxford, UK:
Blackwell.

Hambidge, K. M., N. F. Krebs, J. L. Westcott, L. Sian, L. V. Miller, K. L. Peterson, and V. Raboy. 2005.
Absorption of calcium from tortilla meals prepared from low-phytate maize. *American Journal of
Clinical Nutrition* 82:84–7.

Hawthorne, K. M., J. Morris, T. Holtze, K. D. Hirschi, and S. A. Abrams. 2009. Biotechnologically-modified carrots: Calcium absorption relative to milk. *J Bioequiv Availab* 1:34–8.

Heaney, R. P., R. R. Recker, and S. M. Hinders. 1988. Variability of calcium absorption. *American Journal of Clinical Nutrition* 47:262–4.

Highfield, R. 2008. GM carrot may help treat osteoporosis. *The London Telegraph*, January 14, 2008. http://www.telegraph.co.uk/scienceandtechnology/science/sciencenews/3321694/GM-carrot-may-help-treat-osteoporosis.html (accessed July 4, 2011).

Hindo, B. 2007. Monsanto: Winning the ground war. BusinessWeek.

Hirschi, K. D. 2001. Vacuolar H+/Ca2+ transport: Who's directing the traffic? *Trends in Plant Science* 6:100–104.

Hirschi, K. D. 2008. Nutritional improvements in plants: Time to bite on biofortified foods. *Trends in Plant Science* 13:459–63.

Hornick, B. A., A. J. Krester, and T. A. Nicklas. 2008. Menu modeling with MyPyramid food patterns: Incremental dietary changes lead to dramatic improvements in diet quality of menus. *Journal of the American Dietetic Assocociation* 108:2077–83.

ILSI. 2008. Nutritional and safety assessments of foods and feeds nutritionally improved through biotechnology: Case studies. *Comprehensive Reviews in Food Science and Food Safety, International Life Sciences Institute* 7:50–99.

Jeong, J., and M. L. Guerinot. 2008. Biofortified and bioavailable: The gold standard for plant-based diets. *Proceedings of the National Academy of Sciences USA* 105:1777–8.

Johnson, K. L., A. F. Raybould, M. D. Hudson, and G. M. Poppy. 2007. How does scientific risk assessment of GM crops fit within the wider risk analysis? *Trends in Plant Science* 12:1–5.

Korth, K. L., S. J. Doege, S. H. Park, F. L. Goggin, Q. Wang, S. K. Gomez, G. Liu, L. Jia, and P. A. Nakata. 2006. Medicago truncatula mutants demonstrate the role of plant calcium oxalate crystals as an effective defense against chewing insects. *Plant Physiol* 141:188–95.

Krawinkel, M. B. 2007. What we know and don't know about Golden Rice. *Nature Biotechnology* 25:623; author reply 624.

Linder, M. C. 1991. *Nutritional Biochemistry and Metabolism: With Clinical Applications*. New York: Elsevier.

Massey, L. K., R. G. Palmer, and H. T. Horner. 2001. Oxalate content of soybean seeds (Glycine max: Leguminosae), soyfoods, and other edible legumes. *J Agric Food Chem* 49:4262–6.

Mayer, J. E., W. H. Pfeiffer, and P. Beyer. 2008. Biofortified crops to alleviate micronutrient malnutrition. *Curr Opin Plant Biol* 11:166–70.

Mendoza, C., F. E. Viteri, B. Lonnerdal, K. A. Young, V. Raboy, and K. H. Brown. 1998. Effect of genetically modified, low-phytic acid maize on absorption of iron from tortillas. *Am J Clin Nutr* 68:1123–7.

Morris, J., K. M. Hawthorne, T. Hotze, S. A. Abrams, and K. D. Hirschi. 2008. Nutritional impact of elevated calcium transport activity in carrots. *Proceedings of the National Academy of Sciences USA* 105:1431–5.

Morris, J., P. Nakata, M. McConn, A. Brock, and K. D. Hirschi. 2007. Increased calcium bioavailability in mice fed genetically engineered plants lacking calcium oxalate. *Plant Molecular Biology* 64:613–18.

Nakata, P. A., and M. M. McConn. 2000. Isolation of medicago truncatula mutants defective in calcium oxalate crystal formation. *Plant Physiology* 124:1097–104.

Nestle, M. 2001. Genetically engineered Golden Rice unlikely to overcome vitamin A deficiency. *Journal of the American Dietetic Association* 101:289–90.

Nguyen, D., and T. Xu. 2008. The expanding role of mouse genetics for understanding human biology and disease. *Disease Models & Mechanisms* 1:56–66.

Park, S., N. H. Cheng, J. K. Pittman, K. S. Yoo, J. Park, R. H. Smith, and K. D. Hirschi. 2005. Increased calcium levels and prolonged shelf life in tomatoes expressing *Arabidopsis* H+/Ca2+ transporters. *Plant Physiol* 139:1194–206.

Park, S., C. K. Kim, L. M. Pike, R. H. Smith, and K. D. Hirschi. 2004. Increased calcium in carrots by expression of an Arabidopsis H+/Ca2+ transporter. *Molecular Breeding* 14:275–82.

Powell, K. 2007. Functional foods from biotech—an unappetizing prospect? *Nature Biotechnology* 25:525–31.

Provot, S., E. Schipani, J. Wu, and Kronenberg. 2008. Development of the Skeleton. In R. Marcus, D. Feldman, D. A. Nelson, and C. J. Rosen, eds., Osteoporosis, pp. 241–69. New York: Academic Press.

Raboy, V. 2001. Seeds for a better future: 'Low phytate' grains help to overcome malnutrition and reduce pollution. *Trends in Plant Science* 6:458–62.

Ralston, K., C. Newman, A. Clauson, J. Guthrie, and J. Buzby. 2008. The National School Lunch Program Background, Trends, and Issues, t.E.o.F. U.S. Department of Agriculture Economics Research Service, Farming, Natural Resources, and Rural America. Washington, DC: U.S. Department of Agriculture Economics Research Service.

Simon, P. W., and I. L. Goldman. 2007. Carrot. In R. J. Singh, ed., *Genetic Resources, Chromosome Engineering, and Crop Improvement: Vegetable Crops*, pp. 497–517. Boca Raton, FL: CRC Press/Taylor & Francis Group.

Tang, G., J. Qin, G. G. Dolnikowski R. M. Russell, and M. A. Grusak. 2009. Golden Rice is an effective source of vitamin A. *Am J Clin Nutr* 89:1776–83.

Tichenal, C. A., and J. Dobbs. 2007. A system to assess the quality of food sources of calcium. *J Food Composition and Analysis* 20:717–24.

Uauy, C., A. Distelfeld, T. Fahima, A. Blechl, and J. Dubcovsky. 2006. A NAC Gene regulating senescence improves grain protein, zinc, and iron content in wheat. *Science* 314:1298–301.

Wagner, E. F., and G. Karsenty. 2001. Genetic control of skeletal development. *Curr Opin Genet Dev* 11:527–32.

Wareham, N. J., E. M. van Sluijs, and U. Ekelund. 2005. Symposium on "Prevention of Obesity." Physical activity and obesity prevention: A review of the current evidence. *Proceedings of the Nutrition Society* 64:229–47.

Weaver, C. M. 1990. Assessing calcium status and metabolism. *J Nutr* 120:1470–3.

Weaver, C. M., B. R. Martin, J. S. Ebner, and C. A. Krueger. 1987. Oxalic acid decreases calcium absorption in rats. *Journal of Nutrition* 117:1903–06.

Weil, J. H. 2005. Are genetically modified plants useful and safe? *IUBMB Life* 57:311–4.

White, P. J., and M. R. Broadley. 2009. Biofortification of crops with seven mineral elements often lacking in human diets—iron, zinc, copper, calcium, magnesium, selenium and iodine. *New Phytol* 182:49–84.

Wilson, D. S., and A. J. Clifford. 1990. Bioavailability: How the nutrients in food become available to our bodies. In D. T. Smith, ed., *Nutrition: Eating for Good Health*, Bulletin 685, pp. 72–7. Washington, DC: The U.S. Department of Agriculture.

Ye, X., S. Al-Babili, A. Klöti, J. Zhang, P. Lucca, P. Beyer, and I. Potrykus. 2000. Engineering the provitamin A (beta-carotene) biosynthetic pathway into (carotenoid-free) rice endosperm. *Science* 287:303–05.

Zhu, C., S. Naqvi, S. Gomez-Galera, A. M. Pelacho, T. Capell, and P. Christou. 2007. Transgenic strategies for the nutritional enhancement of plants. *Trends Plant Sci* 12:548–55.

Zhu, J. K., and K. Shimamoto. 2007. Cell signaling and gene regulation: When will signaling meet small RNAs and epigenetic regulation? *Current Opinion in Plant Biology* 10:433–5.

# 18 Use of Genomics-Aided Breeding to Improve the Nutritional Content of Lettuce

*David W. Still*

## CONTENTS

## INTRODUCTION

The FAO estimates that worldwide, 1.02 billion people were undernourished in 2009 (FAO 2009). The number of undernourished people has been increasing since the mid-1990s, following a period during the 1970s and 1980s in which the number of undernourished dramatically decreased. That decrease was largely due to the global food crisis which precipitated a large investment in government sponsored investment into the agriculture sector. Today, there is recognition that governments have neglected the agricultural sector, believing that global food security had been achieved and could be maintained with decreasing investments (Godfray et al. 2010; Lele 2010). Ironically, in the United States about 68% of all adult men and women are overweight, and 27%–32% of adults are obese, which is defined as a body mass index of 30 or greater (CDC 2010; Flegal et al. 2010). The undernourished population is a result from being priced out of the market, while the over-nourished group suffers from the consumption of too many foods with low nutritional content and a lack of physical activity (CDC 2010; FAO 2009). Both groups would benefit from the availability of foods that have higher nutritional content. The development of more nutritious food can be accomplished either through conventional breeding or may be engineered to synthesize compounds that are lacking or produced in low quantity. Either approach is facilitated having a full genomic sequence of the plant available. It is important to note that literally every plant that has been grown for food, ornament, or turf, has gone through a domestication and breeding process. Because most people are not familiar with how their food supply was developed, a brief overview of how crops were developed and how breeding has evolved in parallel with technology has been included.

## DOMESTICATION OF CROPS

The genetic improvement of plants began when man first started gathering seeds from the wild, planting them, and harvesting these seeds. This repetitive cycle played out over hundreds to thousands of years and led to the domestication of most of the foods commonly eaten today. Cereals provide most of the food calories consumed worldwide, and they appear to be the first in the domestication process pipeline (Balter 2007). The domestication of cereals, and likely most other crops, included a gathering stage followed by a predomestication cultivation stage, and finally the fixation of genes responsible for domestication (Allaby 2010; Allaby et al. 2008). These domestication genes control a suite of traits that are advantageous for the plant to survive and evolve in the wild, but hamper its usefulness and therefore adoption, as a food source for humans. Almost without exception the traits affected include the loss of seed dormancy, loss of seed and fruit dehiscence, and the redirection of assimilates to reproductive structures (Hancock 2004). Archaeological evidence indicates cereal gathering as early as 23,000 years ago, predomestication cultivation approximately 13,000 years ago, and the fixation of the traits occurring over long periods of time during the fixation stage (Balter 2007). One of the key factors contributing to the length of time between gathering a plant species and the fixation of an allele was to interrupt gene flow between the cultivated field and the wild plants growing adjacent to the fields (Fuller et al. 2009; Kovach et al. 2007).

Interestingly, all major crops, that is, those that provide the vast majority of our calories, were domesticated thousands of years ago. By comparison, modern domesticates include relatively minor horticultural crops such as strawberries, pecans, macadamia nuts, and blueberries (Diamond 2002). While there are an estimated 230,000 to 250,000 species of angiosperms (Takhtajan 1997; Wikstrom et al. 2001), the majority of calories consumed by humans are provided from the grass and legume plant families and only nine plant families provide the 154 or so foods commonly consumed worldwide (Harlan 1992). Ethnobotanists have reported indigenous cultures gathered and consumed between 1100 and 1400 different plant species while Freedman has published online a database of approximately 1200 species that have served as "famine foods" (http://www.hort. purdue.edu/newcrop/FamineFoods/faminefoods.html). Given the large number of potential food sources, one may ask why some species were domesticated while others were not, and further, are these all the species capable of being domesticated? Clearly, Neolithic civilizations did not have knowledge of genetics nor was it necessary to successfully domesticate a species. Domestication occurred as a result of restricting the gene flow between wild plants and the seeds that the early hunters and gatherers planted. By necessity, this resulted in a genetic bottleneck, with the plants undergoing domestication gradually losing genetic (allelic) diversity (Tanksley and McCouch 1997). The domestication of rice, for example, occurred over thousands of years, apparently because villagers continued to harvest seeds from wild plants to supplement their cultivated rice harvests (Fuller 2007; Fuller et al. 2009). Once the number of individual plants that are harvested from, or the number of plants that serve as pollen donors is restricted, then domestication of plants, and the fixation of traits, will occur very quickly. This is not unique to plants; indeed, elite lines of thoroughbred horses and most dog breeds are highly inbred and have had very few sires (Lindgren et al. 2004; Sundqvist et al. 2006). These are the tenants of breeding, whose first scientific underpinnings were described by Mendel.

## EARLY CROP IMPROVEMENT

Gregor Mendel is credited with publishing the first results and most importantly, the interpretation of experiments from crosses of inbred lines of peas (Mendel 1866). It is well known that pollinations (i.e., genetic crosses) had been made for centuries, and $F_1$ hybrids were also being sold in Europe (e.g., Fairchild's Sweet William dianthus). Why then, has Mendel gotten credit while we do not remember anybody else's work on discovering the "units of heredity"? His genius appears to have been the ability to work with a model plant (peas) and tractable traits (for example, seed shape, seed color, flower color, stem length) whose phenotypes were easily observed and minimally affected by the environment. Further, because each of the $F_1$ plants were allowed to self-pollinate, several phenotypic classes were discovered. These experiments were carefully crafted and the data correctly interpreted which established the principles of segregation of genes and the independent assortment of alleles. Some have speculated that Mendel's data are "too good" (Allchin 2005; Cleophas et al. 2009; Trocchio 1991), but the impressive fact remains of the careful planning Mendel expended in setting up and carrying out the experiments. In Mendel's own words, "The value and utility of any experiment are determined by the fitness of the material to the purpose for which it is used, and thus in the case before us it cannot be immaterial what plants are subjected to experiment and in what manner such experiments are conducted (Mendel 1866, p. 4)." These experiments established the basis for genetics and provided the scientific foundation by which evolutionary and breeding theory operate. Establishing the genetic basis by which populations change over time, is, in fact, what separates domestication from breeding. It is interesting to note that it has taken over 100 years to identify the genes underlying the traits described in Mendel's experiments. (See, e.g., Bhattacharyya et al. 1990; Lester et al. 1997).

Conventional plant breeding has been practiced for at least 100 years. The earliest plant breeding textbooks on my shelf date from the early 1900s. The fact that crops have been developed with increased yields, improved resistance to disease and insects, and that many ornamental species have been developed from their wild progenitors, indicates that conventional plant breeding has been a very successful enterprise. Yet the genetic basis underlying most traits remains unknown. Thus, breeding progress can be made without identifying specific genes, but knowing which genes influence traits would make the breeding process more efficient. In essence, breeding is a numbers game, and can be quite inefficient. An example will illustrate this: if two parents which differ by 10 heterozygous loci are crossed, there will be 59,049 different genotypes in the $F_2$ and one would have to grow a minimum of 1,048,576 plants to ensure the chance presence of every genotype. Breeders are typically faced with the requirement to grow very large populations in hopes of obtaining the genotype they wish to recover. However, breeders are not able to identify the desired genotype without laboratory methods, but instead will walk the fields or greenhouses examining and selecting superior plants based on the phenotype observed under field conditions. Considering this, the fact that breeders have managed to improve the performance of plants is quite remarkable. Again, plant improvement is a gradual process. Writing about maize, Hayes and Garber (1927, p. 2) remarked that "Squaw Flint from the Indian reservations in Minnesota has averaged as large a yield … as the more carefully selected varieties" that they had been breeding and selecting. Thus, the domestication and repeated cultivation methods by which Native Americans over the millennia had employed resulted in maize landraces whose yield was not increased until the genetics of increased yield was teased apart through scientific approaches. The process of increasing yield in maize required decades of research and hundreds of thousands (if not millions) of crosses. The first few decades of maize production utilized open pollinated varieties and little to no yield increases were observed. From the early 1920s through the early 1940s maize was bred first by essentially combining four genetic lines into a single line through a two-step process, the "double cross" method because it required a total of four parents and two crosses. A very large number of crosses were typically made to determine which parental combinations resulted in the highest yields. Once the best parental combinations were determined, the maize breeders were able to make crosses between a single pair of superior parents, known as the "single cross" method. Thus, by examining yield data, breeders identified the best yielding parents and used a relatively small number of elite parental lines to increase yields at an even greater rate. It took over 40 years of empirical observation to learn which genetic lines when combined would be the best performers. This process worked, but it convincingly illustrates that conventional breeding can be both inefficient and unpredictable.

## THE ROAD TO NEXT-GENERATION PLANT BREEDING

It would be very useful to predict, a priori, the outcome of a cross between two plants. At present this is not possible and the process of choosing superior plants in reality is very inefficient. The breeder, after making a cross, depends on heterosis to improve the agronomic or horticultural traits of the plant. Heterosis is defined as the occurrence of phenotypes that are beyond the parental values. Strictly speaking, this would include values that are higher or lower than that observed in the parents. When making a hybrid, this is referred to a "hybrid vigor." If the trait is conditioned by a single gene, then there should be minimal environmental influence on the phenotype and the improved phenotype would be easily recognized. Unfortunately, most traits we are interested in improving are genetically complex and involve multiple genes. Complex traits would include yield, drought tolerance and nutritional content, to name a few examples. A breeder using conventional approaches typically begins the plant improvement process by making a cross between two parents, planting a large segregating population ($F_2$ or backcross, for example) in the field, and hope to identify plants with superior phenotypes. Although the selection is made on the phenotype, it is the genotype that is controlling the trait and ultimately the genotype is what the breeder has selected or culled. The task of finding the plants with the superior phenotype and superior genotype is a classical statistical exercise in terms of limiting your type I or type II errors. A type I error occurs when the breeder throws away the plant that in reality had an allelic combination that actually confers a superior phenotype; the type II error occurs when the plant is kept based on its promising phenotype but in reality does not have the combination of favorable alleles that ultimately are needed. The type I error is fatal in that it requires the breeder to go back to previous generations or make a new cross and start over. The consequence of a type II error is that more plants than are needed are carried along through the next stages, making the breeding process more inefficient and costly.

Breeders develop genetic maps which use recombination frequencies to determine the order of genes (or markers) in a segregating population following a cross between two parents. The first genetic map was created by Sturtevant in 1913 (Sturtevant 1913) in which the linear order of sex-linked traits were mapped in the fruit fly *Drosophila*. The traits he mapped were simple and each was controlled by a single gene. For the next few decades geneticists were able to map traits in a wide variety of model organisms as well as crops. Early genetic maps were based on morphological characters in which the recombinant phenotypes could be easily identified as being different from parental phenotypes. For example, a tomato genetic map in 1952 contained 35 phenotypic markers across 12 linkage groups (Butler 1952). This early tomato map included easily discernible traits like fruit color, leaf color, fruit shape, locule numbers, flower color, fruit fasciation, and so on. Easily categorized traits were mapped first because more complex traits were controlled by multiple genes and the progeny displayed a continuum of phenotypic values instead of easily discernible classes. It was argued at the time that if simple traits could be mapped, the more complex traits (quantitative traits) could likewise be mapped using similar approaches because the underlying genetics were the same (Thoday 1961). Those involved in the early mapping of quantitative traits were faced with two practical limitations. The availability of suitable markers was still limited by the requirement of having at least one of the parents possessing an easily distinguished phenotypic trait and the development of a genetic map required that the parents differ in the expression of that trait. Second, the large number of recombinant classes required considerable effort to discover and locate the relevant loci (Thoday 1961). The usefulness of morphological markers are further limited because they normally behave in a dominant/recessive manner and they interact epistatically (Tanksley et al. 1989). Taken together, these requirements greatly limited the development of genetic maps comprised of quantitative traits.

The development of molecular markers replaced the use of morphological markers and facilitated the mapping of quantitative traits. The detection of polymorphisms between the parental genotypes is accomplished by various biochemical or molecular methods. The first molecular mapping projects employed biochemical methods and capitalized on the development of gel electrophoresis (Smithies 1955) to assess isozyme variation between parental genotypes. Linkage maps based on isozymes were first applied to crops as diverse as barley, maize, and tomatoes (Kahler and Allard 1970; Ott and Scandalios 1978; Tanksley and Rick 1980). Isozymes are different forms of the same protein and are caused by different amino acids as a consequence of a genetic mutation within the gene encoding that protein. Although it was known that that isozyme variation had a genetic basis, the techniques had not yet been developed to directly detect the nucleotide polymorphism causing the coding change. The discovery of restriction enzymes (Linn and Arber 1968) and the demonstration that these enzymes recognize specific DNA sequences in double-stranded DNA (Kelly and Smith 1970; Smith and Wilcox 1970) was quickly followed by their application to resolving genetic differences among individuals. The use of restriction enzymes coupled with the advent of Sanger sequencing (Sanger et al. 1977) enabled the development of a wide variety of molecular markers including RFLP, RAPD, AFLP, and SSR, to name a few. Scientists used Sanger sequencing for almost three decades before the second-generation sequencing platforms were developed and became commercially available. The second-generation sequencing platforms (Roche 454, Illumina and SOLiD platforms) featured much lower sequencing costs and increased throughput, which led to the sequencing of more individuals, which in turn led to the development and widespread use of the ultimate polymorphic DNA-based marker, the single nucleotide polymorphism (SNP). With the coevolution of molecular markers and software capable of working with large datasets, a solution gradually took shape that for the breeder would minimize both type I and type II errors. A seminal paper was published some 20 years ago by Eric Lander and David Botstein that used statistical likelihoods to create linkage maps based on molecular markers to identify and map quantitative trait loci and calculate the effects of quantitative trait loci on the expression of the trait being measured (Lander and Botstein 1989).

As mentioned earlier in this section, detecting polymorphisms among individuals requires that the parents differ at that locus and the ability to identify and locate that locus. Resolving the location of the polymorphism requires creating recombination in the region of the chromosome containing that locus. One method by which this is accomplished in plant breeding is to develop a population of related individuals from a cross between two dissimilar parents and develop a large number of related families, called recombinant inbred lines (RIL). These RIL populations are known as "immortal" mapping populations because once created they can be used ad infinitum to study a new trait and establish its genetic basis. A RIL population is typically created by single seed descent in which two parents are crossed to produce an $F_1$ plant which is then allowed to self-pollinate to produce seed for the $F_2$ population. Based on theoretical considerations, the $F_2$ population should contain as many plants as possible; having a larger number of plants should increase the number of recombinant events during meiosis, and a larger number of RIL families in the $F_8$ generation or later will improve mapping resolution and provide greater accuracy in determining the value of a locus in the expression of that trait (Beavis 1998; Melchinger et al. 1998; Xu 2003). From the $F_2$ generation forward, a single seed is taken from each plant to establish the next filial generation until the $F_8$ population, at which time a theoretical 99.32% of all loci in the genome are fixed. These RIL populations are used primarily for mapping because ideally, no selection (natural or artificial) would have taken place during the development of the population. Because all variation existing in the population was derived from the loci of only two parents, the disadvantage of this approach is that the variation in a given RIL population represents only a fraction of the total variation present in the species.

Another very powerful method by which to resolve loci comes from association mapping, also known as linkage disequilibrium mapping. This is a favored approach in human studies to associate loci with phenotypic variation. As sequencing prices have dropped, this approach is increasingly used by plant breeders. The power of association mapping results from capturing recombination events from very large populations of unrelated individuals. The approach is the same as that used in the RIL, namely detecting nonrandom associations between a particular locus, usually identified as a single nucleotide polymorphism, and the trait of interest. For a review of association mapping in plants see Myles et al. (2009) or in humans see Weir (2008). A variation of association mapping, called nesting association mapping (NAM), was developed in maize (Flint-Garcia et al. 2005; McMullen et al. 2009). The maize NAM population was created by crossing a single reference genetic line with 25 diverse inbred maize lines to create about 5000 RILs. The key to all mapping projects is to create and detect recombination events, and the maize NAM population captured approximately 136,000 crossover events with an average of about three crossover events per gene (McMullen et al. 2009). The advantage of the NAM approach is that it can resolve loci to extremely small intervals (an average of 1.3 cM in the maize NAM population). For example, Buckler and colleagues (2009) detected loci involved in flowering in maize and resolved these loci to 0.5 cM regions, but on the downside, this required phenotyping over 40,000 plants. Nonetheless, when compared to the typical mapping resolution of a RIL family of 10–20 cM, the effort required to create and phenotype a NAM population is well worthwhile. Note that one cM region in the model plant Arabidopsis corresponds to about 250 kb and an average of 44 genes while in maize a cM covers 1.2 Mb and approximately 31 genes (Lukowitz et al. 2000; Salvi and Tuberosa 2005).

Identification of a locus through QTL or association mapping does not in itself reveal the genetic cause. Further work typically requires fine mapping and cloning of the putative gene, in which another set of crosses are made between those genotypes that contain the mapped locus and those that do not contain the locus. The resulting progeny are again phenotyped and genotyped. Those carrying the molecular marker and the phenotype are sequenced to locate and identify the genes underlying this locus. It is often necessary to produce and screen a very large number of progeny before enough recombination events occur and the putative gene can be identified. This is simply because recombination frequency is inversely proportional to the physical proximity of two loci. For example, while tracking down the gene responsible for seed shattering in rice, a key domestication gene, over 12,000 progeny containing the shattering locus were screened for recombinants to narrow the genetic interval region and find the causative gene (Li et al. 2006). The next step is to demonstrate the function of this gene by cloning and then silencing or over-expressing the gene. Excellent examples of this approach have been reported in rice, maize, and tomato (Frary et al. 2000; Li et al. 2006; Yano et al. 2000).

Traditional breeding depends on natural variation being present in the germplasm. A breeder assembles a wide diversity of genetic materials including breeding lines and unimproved germplasm. To minimize genotype × environment interactions, which will confound phenotyping, all genetic lines should be grown on a common nursery. Because plant morphology and biochemistry is extraordinarily sensitive to environmental stimuli (light, temperature, soil conditions, water and nutrient availability, insect and disease pressure), growing all plant materials simultaneously on a single site minimizes environment–genetic interactions that are likely to influence plant phenotypes. The vast majority of plant improvement has occurred through domestication followed by and traditional breeding methods, and by all accounts, this has been a spectacular success. The future of breeding, however, will depend and benefit from having the entire genomic sequence of every crop and at some point, every plant in the breeder's materials.

## WHY LETTUCE?

Lettuce is the most popular fresh market crop in the United States (USDA, 2009), and it is grown and consumed on every continent except Antarctica (USDA 2009; FAO 2010). Over 23.5 million metric tons were produced on a little more than 1 million ha (Table 18.1). Although lettuce has been part of the Western diet for decades, most in the United States and Europe would be surprised to know the world's largest producer and consumer of lettuce is China (Table 18.1). In fact, although yield is only about 64% of that in the United States on a per hectare basis, China produces 300% more lettuce than the United States. Of the top ten lettuce producing countries, in terms of yield per hectare, production efficiency is by far the highest in the United States. The yield differential is likely not due to genetics but rather cultural practices. Lettuce production in the United States is a high-input enterprise in which all lettuce is regularly irrigated on tight schedules, pesticides are applied as needed, and high-nitrogen fertilizers are regularly used (Kerns et al. 1999). Typical of horticultural crops, intensively managed lettuce results in high yields. Lettuce yields of 481,481 Hg/ha have been reported for Kuwait, but note that total production area was only 135 ha (FAO 2010).

## NUTRITIONAL VALUE OF LETTUCE

Although lettuce does not rank as the most nutritious vegetable, there are appreciable amounts of iron, folate, beta carotene, and vitamins A and K. Spinach, on the other hand, is a rich source of a wide variety of essential minerals and vitamins but its per capita consumption in the United States

**TABLE 18.1**

**Top Ten Lettuce Producing Countries and World Lettuce Production for 2008, Ranked by Production**

| Country | Production (metric tons) | Ha harvested | Yield (Hg/ha) |
|---|---|---|---|
| China | 12,505,500 | 542,265 | 229,347 |
| United States | 4,110,160 | 114,290 | 359,625 |
| Spain | 1,107,700 | 37,000 | 299,378 |
| Italy | 847,666 | 44,003 | 192,638 |
| India | 790,000 | 120,000 | 65,833 |
| Turkey | 439,641 | 23,000 | 191,148 |
| Japan | 560,000 | 22,000 | 245,545 |
| France | 420,400 | 15,898 | 264,435 |
| Germany | 316,741 | 14,859 | 213,164 |
| Mexico | 284,709 | 13,309 | 213,922 |
| World | 23,531,961 | 1,061,929 | 188,050 (average) |

*Source:* FAO. 2009. The state of food insecurity in the world: Economic crises—impacts and lessons learned. Rome: Electronic Publishing and Support Branch, Communication Division, FAO.

is relatively low. Of the five different lettuce types consumed, romaine lettuce has the most nutritious profile while iceberg lettuce has the least amount of nutrition (Figure 18.1). Thus, it is ironic that iceberg lettuce remains the most consumed lettuce in the United States. Song et al. (2010) indicated that because of its high consumption lettuce ranks fourth in providing antioxidants in U.S. diets. It is important to realize that genetic variation exists in nutritional content both within and between lettuce types (Mou 2005), which makes lettuce amenable to improvement through breeding.

Breeders must always begin a project by deciding what traits to improve. Any trait for which there exists natural genetic variation can be improved using traditional breeding approaches. Traits that ensure a successful yield are often the first priority of seed companies. For example, breeding for disease resistance after a new race of a pathogen has been discovered. In such a situation, one may have no choice—either they find resistance or be faced with the prospect of not growing the crop in that area. Because the caloric needs of the world have largely been met (FAO 2009), plant breeders have recently focused on improving the nutritional content of specific crops (Godfray et al. 2010; Graham et al. 1999). Breeding for increased content of essential minerals, vitamins, or amino acids is backed by long-standing evidence that these components contribute to maintenance of human health. However, these components make up only a small proportion of the tens of thousands of phytochemicals a plant may synthesize. Most of the phytochemicals a plant produces are secondary metabolites and their role in human health or plant health has largely not been established. It is widely accepted though, that these secondary compounds evolved as mechanisms as an adaptive response to biotic and abiotic stresses (Penuelas and Munne-Bosch 2005; Pourcel et al. 2007; Rice-Evans et al. 1997). Many plant compounds have been shown to have antitumor, anticancer, anti-inflammatory, or other therapeutic effects in cells or model animal systems. Their role in humans is supported with epidemiological evidence (DellaPenna 1999). Diets rich in fruits and vegetables are associated with reduced risks of chronic and degenerative diseases, certainly as a result of these foods containing essential vitamins and minerals, but also as a result of the plethora of secondary compounds whose role in human nutrition and health has yet to be elucidated.

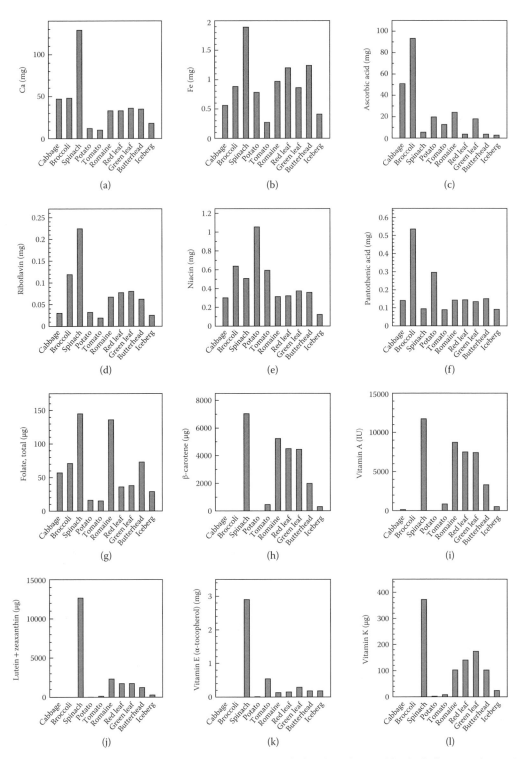

**FIGURE 18.1** Nutritional content of essential minerals and vitamins of vegetables including romaine, red leaf, green leaf, butterhead, and iceberg lettuce types. All values are per 100 grams fresh weight. Common cabbage, broccoli florets, spinach leaves, potato (with skin, uncooked), tomato (raw, uncooked), and trimmed lettuce leaves were used. (Data from USDA. 2010. Nutrient availability data. USDA Economic Research Service.)

## IMPROVING THE ANTIOXIDANT CONTENT IN LETTUCE

Our laboratory is actively engaged in improving the nutritional content of lettuce. We have initially focused on increasing the antioxidant (vitamin) content in lettuce using traditional breeding aided by mapping phenotypic, biochemical, and gene expression traits. Antioxidant content is known to increase in the plant in response to a wide variety of stimuli, which include excess light, smog, wounding, high osmotic potential, chilling, and drought. For example, controlled dehydration was used to increase antioxidant content in lettuce grown in a controlled environment (Oh et al. 2010). Perhaps most critical to a plant's survival, antioxidants provide protection for their photosystems. Plant photosystems are dynamic complexes of carotenoids, chlorophylls, and protein, and evolved to harvest light energy. Carotenoids have at least five functions in plants, including (1) light harvesting via singlet state energy transfer, (2) photoprotection via the quenching of chlorophyll triplet states, (3) singlet oxygen scavenging, (4) excess energy dissipation, and (5) structure stabilization (Frank and Cogdell 1996; Pascal et al. 2005). In humans, antioxidants are thought to play largely the same role, that is, to protect against reactive oxygen species and reestablish a healthy cellular redox potential (Demmig-Adams and Adams 2002).

Our approach to improve the nutritional content of lettuce required the development of a RIL population for mapping and the availability of genomic sequence to develop molecular markers to construct high-density maps, identify genes along metabolic pathways, and to develop primers to assess gene expression of genes encoding the enzymes along these pathways. We developed an RIL population from a cross between two commercial cultivars, "Diplomat" and "Margarita," a iceberg and butterhead, respectively, hereafter referred to as the D × M RIL. A genetic map for the RIL population was constructed using amplified fragment length polymorphic markers (Hayashi et al. 2008). A second-generation map is being constructed from single nucleotide polymorphisms identified from sequence data from the SOLiD and Illumina sequencing platforms and Illumina GoldenGate assays. In plants, several different metabolic pathways contribute antioxidant activity, including the phenylpropanoid, carotenoid, vitamin E, and ascorbic acid pathways. Using genes encoding enzymes along these pathways in the model plant *Arabidopsis*, we identified lettuce (*Lactuca sativa*) homologs in the expressed sequence tag database developed from the Composite Genome Project (http://compgenomics.ucdavis.edu/).

Because antioxidant content is a complex trait influenced by many genes, the environmental influence is quite high. Progress can only be made by breaking down the trait into individual components using quantitative trait loci (QTL) analysis. This approach in effect "Mendelizes" a complex trait and allows the breeder to identify genomic regions associated with a trait (Tanksley et al. 1989). The ability to improve the antioxidant content through breeding depends on the existence of natural variation in the species which our initial screening of genetic materials confirmed existed. The nutritional values presented in Figure 18.1 for lettuce and other vegetables are only typical values and do not represent the range that exists within any of these crop species. We have observed wide variation among lettuce cultivars in our lettuce D × M RIL population when grown under commercial production conditions (Figure 18.2). We have grown the RIL population in multiple production sites and years and assessed the antioxidant content of each RIL family and approximately 30 commercial cultivars using an oxygen radical absorbance capacity (ORAC) assay. "Diplomat," the iceberg lettuce type, typically has lower antioxidant content than "Margarita," the butterhead type parent of the D × M RIL, as shown by the arrows in the distribution curve in Figure 18.2. Breeding progress also depends on the presence of transgressive segregation, which we have observed in every production site and year this population has been grown. The transgressive segregants are RIL families which have phenotypic values beyond either parent. When the expression of the trait is favorable this is more commonly called hybrid vigor. However, it should be pointed out that lettuce, and all plants which normally self-pollinate, are not sold as hybrids—they are inbred lines. Using the phenotypic values from the ORAC assay, we identified QTL along linkage group (LG) 1 and LG 4 associated with antioxidant content when grown in the field (LG 1) or in the greenhouse (LG 4) (Figure 18.3).

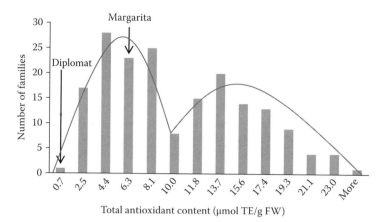

**FIGURE 18.2** Frequency distribution of antioxidant content evaluated in an $F_8$ RIL derived from a cross between cultivars "Diplomat" and "Margarita," an iceberg and butterhead type, respectively. Total antioxidant content was assessed of three biological replicates of each RIL family using an ORAC assay. The plants were grown under commercial conditions at the Yuma Agricultural Center, Yuma, Arizona, and the plants harvested and assayed at market maturity. Note the bimodal distribution of the RIL population and the presence of many transgressive segregants families.

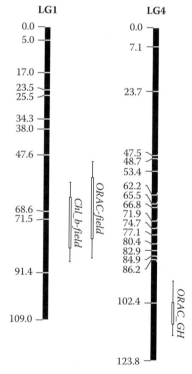

**FIGURE 18.3** QTL associated with antioxidant and chlorophyll content were identified on LG one and four on a linkage map created from the D × M RIL using amplified fragment length polymorphism markers. Total antioxidant content was assayed using an ORAC assay and two QTL were identified, with the LG1 locus identified on field-grown plants at market maturity while the LG4 locus was only identified in greenhouse-grown plants and before plants had formed heads. A QTL associated with chlorophyll was identified from the field-grown population and colocated with an ORAC QTL. Chlorophyll was measured as a proxy for lutein and β-carotene. (Mou 2005). (Data from unpublished data of Y. You, G. Wan, and D. W. Still, collected between 2006–2011.)

Because the synthesis of chlorophyll and carotenoids are intricately entwined, chlorophyll can be measured as a proxy for lutein and β-carotene (Mou 2005). At the same time we measured ORAC content, we measured chlorophyll content of these RIL families and identified a QTL associated with chlorophyll *b* (Chl *b*) content on LG 1 (Figure 18.3). This suggests this genomic region contains genes encoding chlorophyll or carotenoids which contribute to antioxidant content, or an upstream gene or transcription factor that controls these pathways.

Although a trait can be improved without understanding its genetic basis, the efficiency of breeding can be greatly improved if the gene(s) underlying the trait is (are) identified. Further, if a biotechnology approach is to be used, it is essential that key genes of these pathways are identified and the tissues and conditions under which they are expressed are understood. To that end, we developed primers of key genes along the antioxidant-contributing pathways and developed primers to assess transcript abundance. The phenylpropanoid pathway and associated enzymes in lettuce are shown in Figure 18.4. We next harvested plants from each RIL family at market maturity and used quantitative PCR to determine the relative abundance of transcripts of about 50 antioxidant-contributing genes. The distribution frequency of the relative expression of phenylalanine ammonia lyase (*LsPal*), which converts L-phenylalanine to trans-cinnamate, is given for each RIL family in Figure 18.5. Again, a wide range in expression values and a large number of transgressive segregants were observed. These data

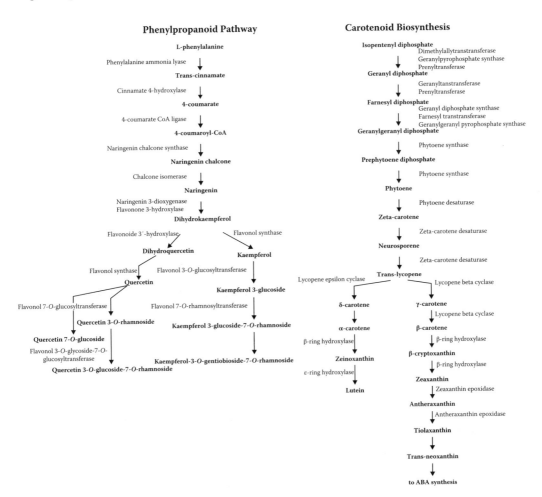

**FIGURE 18.4** Pathways associated with synthesis of antioxidant producing compounds in lettuce. The phenylpropanoid and carotenoid pathways contribute to antioxidants content and genes encoding these and other pathways were assayed using quantitative PCR.

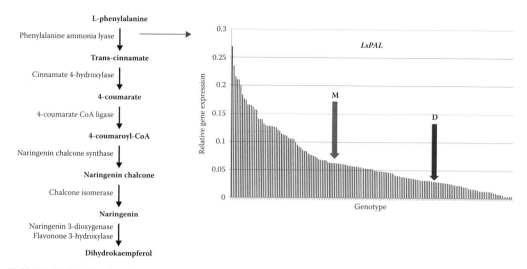

**FIGURE 18.5** The development of expression level polymorphisms (eQTL). Plants were grown in the field and RNA was extracted from three biological replications of each recombinant inbred line (RIL) family (n = 169). In this example, the transcript level of the gene encoding the enzyme to convert L-phenylalanine to trans-cinnamate was assayed and the distribution frequency of each RIL family (X-axis) of the relative expression level (Y-axis) for *LsPal* is shown. The parental values of "Diplomat" and "Margarita" are indicated by the arrows. Note the wide distribution of expression level and the great number of families displaying transgressive segregation. This specific gene mapped as an eQTL. (Data from unpublished data of Y. You, G. Wan, and D. W. Still 2006–2011.)

were then used to map gene expression as expression level polymorphisms (eQTL). The value of this approach is that when the eQTL colocates with the QTL for a trait of interest, it provides very strong evidence this gene is involved in the expression of the trait. When growing the D × M RIL population in two different production sites a large number of eQTL were identified of which five colocated with either the ORAC or Chl QTL (unpublished data from Y. You, G. Wan and D. W. Still 2006–2011).

Gene expression patterns of these genes for all genotypes can be shown as heat maps, in which the relative expression is represented by colors that indicate up- or downregulation of a gene relative to a base level. Using a hierarchical grouping of genotypes, distinct blocking patterns were apparent, an example of which is shown in Figure 18.6. We suspect these patterns reflect expression haplotypes and are working to identify the allelic composition of these genes for each genotype. There is no direct relationship between the gene expression of these genotypes and antioxidant content, but because of the multiple genes contributing to the antioxidant content this is not surprising. Interestingly, using the approach I have just outlined above, the gene underlying a β-carotene QTL in maize was identified and the allelic combination serving to increase this enzyme was recently reported (Yan et al. 2010).

## FUTURE PROSPECTS

Breeding, and indeed most aspects of biology, are focused on identifying and understanding the genetic basis of a given disease or phenotype. This was, in fact, largely the rationale for obtaining funding which supported the sequencing and annotation of the complete human genome. The data presented in the chapters of this book have illustrated both the promise and progress of how genomics can and will be applied to society and ultimately to individuals. The degree to which we realize this goal, and the amount of time before widespread applications of these objectives materialize, is very much dependent on the development of technology and the ability to make sense of the almost unfathomable amount of data that will be generated. First-generation sequencing (Sanger sequencing) established the first genomic sequence for humans and plants over a decade ago. The total cost

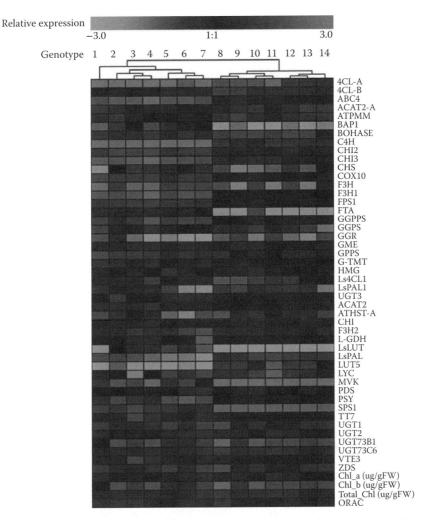

**FIGURE 18.6** Heat map showing hierarchical clustering of genotypes of the expression of antioxidant-encoding genes, chlorophyll, and antioxidant content of the D × M RIL population of lettuce. Two distinct clades are apparent, one containing genotypes 1–7 and the other containing genotypes 8–14. The biological basis for this clustering is not known but it is suggestive of alleles within certain genes that are similar among genotypes within a clade that differ between clades. (Data from unpublished data of Y. You, G. Wan, and D. W. Still 2006–2011.)

for the working "draft" of the first publically funded human genome was approximately $350 million, involved scientists from 16 institutions worldwide, and took about 10 years (NIH 2000). Often, a $3 billion figure is associated with the cost (see, e.g., Dolgin 2009) but this is the original estimate of the total funding for genome-related projects funded over a 15-year period (1990–2005) which included other organisms and development of technologies and bioinformatic support (NIH 2000). The working draft was defined as coverage of 90% of the genome with some gaps and ambiguities, while the sequence would be declared finished once there were no gaps and was 99.99% accurate. The finished sequence was published in 2004 and contained some 341 gaps (International Human Genome Sequencing Consortium 2004). The cost from inception to finished sequence was estimated at $1 billion (Bentley 2006).

Second-generation sequencing (454, Illumina, and SOLiD platforms) increased throughput and decreased the price of obtaining a complete genome, which in turn allowed scientists to sequence

**TABLE 18.2**

**Complete Human Genome Sequencing Costs Associated with First- (Sanger), Second-(Roche 454 and Illumina GA), and Third-Generation (Complete Genomics and Helicos) Sequencing Platforms**

| Year | Reference[a] | Technology | Sample[b] | Coverage Depth | Consumable Cost | Estimated Cost per 40-Fold Coverage |
|------|-----------|------------|--------|----------------|-----------------|-------------------------------------|
| 2000 | 1 | Sanger (ABI) | Composite | 1 | NA | NA |
| 2007 | 2 | Sanger (ABI) | JCV | 7 | $10,000,000 | $57,000,000 |
| 2008 | 3 | 454 (Roche) | JDW | 7 | $1,000,000 | $5,700,000 |
| 2009 | 4 | GA (Illumina) | NA18507 | 30 | $250,000 | $330,000 |
| 2009 | 5 | Helicos | SRQ | 28 | $48,000 | $69,000 |
| 2009 | 6 | DNBs (Complete Genomics) | NA20431 | 45 | $1,726 | $1,500 |

[a] References: 1) International Human Genome Sequencing Consortium. 2001. *Nature* 409(6822):860–921. 2) Levy, S. et al. 2007. *PLoS Biol* 5(10):e254. 3) Wheeler, D. A. et al. 2008. *Nature* 452(7189):872–6. 4) Bentley, D. R. et al. 2008. *Nature* 456(7218):53–59. 5) Pushkarev, D., N. F. Neff, and S. R. Quake. 2009. *Nat Biotech* 27(9):847–50. 6) Drmanac, R. et al. 2010. *Science* 327(5961):78–81.

[b] Sample: Composite of pooled DNA from several individuals: JCV = J. Craig Venter; JDW = James D. Watson; NA18507 = Yoruba individual; SRQ = Stephen R. Quake; NA20431 = Caucasian male sample.

*Notes:* Second-generation platforms produce shorter reads but require greater coverage depth to construct contigs and scaffolds. The first human draft sequence was estimated to cost approximately $350 million; as this project pioneered the tools of genomics and bioinformatics the consumable cost and cost per 40-fold coverage are not estimable. Note that not all major sequencing platforms are represented in this table.

the genomes of more species and many more individuals within a species (Table 18.2). The key to establishing genotype–phenotype links is dependent on capturing the wide spectrum of allelic diversity that exists in the species. Given the expected wide diversity of loci of certain genes, the concept of a consensus or "normal" sequence is debatable. Although SNP chips have been profitably employed as an economically efficient method to survey allelic diversity in a wide variety of species, the gold standard is the complete genome sequence of individuals. The amount of information a complete genome provides in terms of providing a basis of comparison is invaluable. A number of recent reports illustrates this quite dramatically. These projects included obtaining the complete sequence of monozygotic twins discordant for multiple sclerosis and the sequence of a family of four, thus ushering in the age of family-based sequencing (Baranzini et al. 2010; Roach et al. 2010). The sequencing of a family (both parents and two siblings) resulted in an unparalleled resolution of recombination, established a family-specific mutation rate, and detected rare loci associated with diseases (Roach et al. 2010). In plants, current technology and prices are approaching the economic threshold that will make this technology affordable to all plant species and individuals in a mapping population. The commercial release of the third-generation sequencing platforms promises to lower prices and increase coverage depth even further, which will allow plant breeders to obtain the full sequence of every individual in the mapping population, giving extraordinary resolution at reasonable costs. It is well recognized that the bottleneck has shifted from obtaining genome sequence to the ability to accurately phenotype large populations and also, how to efficiently analyze the massive amounts of data generated from the new platforms.

The latest sequencing platforms will allow geneticists and breeders to simultaneously obtain the genome, transcriptome, and epigenome from individuals and most importantly, from every individual in a mapping population. As the cost per data point spirals downward, this will allow greater participation by small research laboratories and will allow scientists to apply this technology to study organisms other than humans, models species, and major crops. The epigenome will allow breeders to evaluate what has been called soft inheritance (Mayr 1982; Richards 2006). Identifying alleles whose epigenetic status may be environmentally conditioned or unconditionally environmentally dependent will allow the identification of epialleles and the subsequent development of epimaps. Our lab has recently mapped as a QTL this environmentally induced variation and we are exploring its biological basis (unpublished data of D. W. Still 2010–2011). Mapping of epialleles will provide another layer of information to be overlaid on to the existing genetic, QTL, and eQTL maps. This approach will finally begin to establish a basis by which to breed for the heretofore intractable genotype x environment interaction.

## REFERENCES

Allaby, R. 2010. Integrating the processes in the evolutionary system of domestication. *J Exp Bot* 61:935–44.

Allaby, R., D. Q. Fuller, and T. A. Brown. 2008. The genetic expectations of a protracted model for the origins of domesticated crops. *Proc Natl Acad Sci USA* 105:13982–6.

Allchin, D. 2005. The dilemma of dominance. *Biol Phil* 20:427–51.

Balter, M. 2007. Seeking agriculture's ancient roots. *Science* 316(5833):1830–5.

Baranzini, S. E., J. Mudge, J. C. Van Velkinburgh, P. Khankhanian, I. Khrebtukova, N. A. Miller, L. Zhang, A. D. Farmer, C. J. Bell, R. W. Kim, G. D. May, J. E. Woodward, S. J. Caillier, J. P. Mcelroy, R. Gomez, M. J. Pando, L. E. Clendenen, E. E. Ganusova, F. D. Schilkey, T. Ramaraj, O. A. Khan, J. J. Huntley, S.Luo, P.-Y. Kwok, T. D. Wu, G. P. Schroth, J. R. Oksenberg, S. L. Hauser, and S. F. Kingsmore. 2010. Genome, epigenome and RNA sequences of monozygotic twins discordant for multiple sclerosis. *Nature* 464(7293):1351–6.

Beavis, W. D. 1998. QTL analyses: Power, precision, and accuracy. In *Proceedings of the Forty-Ninth Annual Corn and Sorghum Industry Research Conference*, 250–66. Washington, DC: American Seed Trade Association.

Bentley, D. R. 2006. Whole-genome re-sequencing. *Curr Opin Genet Dev* 16(6):545–52.

Bentley, D. R., S. Balasubramanian, H. P. Swerdlow, G. P. Smith, J. Milton, C. G. Brown, K. P. Hall, D. J. Evers, C. L. Barnes, H. R. Bignell, J. M. Boutell, J. Bryant, R. J. Carter, R. Keira Cheetham, A. J. Cox, D. J. Ellis, M. R. Flatbush, N. A. Gormley, S. J. Humphray, L. J. Irving, M. S. Karbelashvili, S. M. Kirk, H. Li, X. Liu, K. S. Maisinger, L. J. Murray, B. Obradovic, T. Ost, M. L. Parkinson, M. R. Pratt, I. M. J. Rasolonjatovo, M. T. Reed, R. Rigatti, C. Rodighiero, M. T. Ross, A. Sabot, S. V. Sankar, A. Scally, G. P. Schroth, M. E. Smith, V. P. Smith, A. Spiridou, P. E. Torrance, S. S. Tzonev, E. H. Vermaas, K. Walter, X. Wu, L. Zhang, M. D. Alam, C. Anastasi, I. C. Aniebo, D. M. D. Bailey, I. R. Bancarz, S. Banerjee, S. G. Barbour, P. A. Baybayan, V. A. Benoit, K. F. Benson, C. Bevis, P. J. Black, A. Boodhun, J. S. Brennan, J. A. Bridgham, R. C. Brown, A. A. Brown, D. H. Buermann, A. A. Bundu, J. C. Burrows, N. P. Carter, N. Castillo, M. Chiara, E. Catenazzi, S. Chang, R. Neil Cooley, N. R. Crake, O. O. Dada, K.D. Diakoumakos, B. Dominguez-Fernandez, D. J. Earnshaw, U. C. Egbujor, D. W. Elmore, S. S. Etchin, M. R. Ewan, M. Fedurco, L. J. Fraser, K. V. Fuentes Fajardo, W. Scott Furey, D. George, K. J. Gietzen, C. P. Goddard, G. S. Golda, P. A. Granieri, D. E. Green, D. L. Gustafson, N. F. Hansen, K. Harnish, C. D. Haudenschild, N. I. Heyer, M. M. Hims, J. T. Ho, A. M. Horgan et al. 2008. Accurate whole human genome sequencing using reversible terminator chemistry. *Nature* 456(7218):53–9.

Bhattacharyya, M. K., A. M. Smith, T. H. N. Ellis, C. Hedley, and C. Martin. 1990. The wrinkled-seed character of pea described by Mendel is caused by a transposon-like insertion in a gene encoding starch-branching enzyme. *Cell* 60(1):115–22.

Buckler, E. S., J. B. Holland, P. J. Bradbury, C. B. Acharya, P. J. Brown, C. Browne, E. Ersoz, S. Flint-Garcia, A. Garcia, J. C. Glaubitz, M. M. Goodman, C. Harjes, K. Guill, D. E. Kroon, S. Larsson, N. K. Lepak, H. Li, S. E. Mitchell, G. Pressoir, J. A. Peiffer, M. O. Rosas, T. R. Rocheford, M. C. Romay, S. Romero, S. Salvo, H. S. Villeda, H. Sofia Da Silva, Q. Sun, F. Tian, N. Upadyayula, D. Ware, H. Yates, J. Yu, Z. Zhang, S. Kresovich, and M. D. Mcmullen. 2009. The genetic architecture of maize flowering time. *Science* 325(5941):714–8.

Butler, L. 1952. The linkage map of the tomato. *J Hered* 43(1):25–36.

CDC. 2010. U.S. obesity trends: Center for Disease Control and Prevention. http://www.cdc.gov/obesity/data/trends.html. (accessed August 15, 2011).

Cleophas, T. J., A. H. Zwinderman, T. F. Cleophas, and E. P. Cleophas. 2009. Statistical analysis of genetic data. In *Statistics Applied to Clinical Trials*, 4th ed. pp. 331–9. New York: Springer.

Dellapenna, D. 1999. Nutritional genomics: Manipulating plant micronutrients to improve human health. *Science* 285(5426):375–9.

Demmig-Adams, B. and W. W. Adams, III. 2002. Antioxidants in photosynthesis and human nutrition. *Science* 298(5601):2149–53.

Diamond, J. 2002. Evolution, consequences and future of plant and animal domestication. *Nature* 418:700–7.

Dolgin, E. 2009. Human genomics: the genome finishers. *Nature* 462:843–5.

Drmanac, R., A. B. Sparks, M. J. Callow, A. L. Halpern, N. L. Burns, B. G. Kermani, P. Carnevali, I. Nazarenko, G. B. Nilsen, G. Yeung, F. Dahl, A. Fernandez, B. Staker, K. P. Pant, J. Baccash, A. P. Borcherding, A. Brownley, R. Cedeno, L. Chen, D. Chernikoff, A. Cheung, R. Chirita, B. Curson, J.C. Ebert, C. R. Hacker, R. Hartlage, B. Hauser, S. Huang, Y. Jiang, V. Karpinchyk, M. Koenig, C. Kong, T. Landers, C. Le, J. Liu, C. E. Mcbride, M. Morenzoni, R. E. Morey, K. Mutch, H. Perazich, K. Perry, B. A. Peters, J. Peterson, C. L. Pethiyagoda, K. Pothuraju, C. Richter, A. M. Rosenbaum, S. Roy, J. Shafto, U. Sharanhovich, K. W. Shannon, C. G. Sheppy, M. Sun, J. V. Thakuria, A. Tran, D. Vu, A. W. Zaranek, X. Wu, S. Drmanac, A. R. Oliphant, W. C. Banyai, B. Martin, D. G. Ballinger, G. M. Church, and C. A. Reid. 2010. Human genome sequencing using unchained base reads on self-assembling DNA nanoarrays. *Science* 327(5961):78–81.

FAO. 2009. The state of food insecurity in the world: Economic crises—impacts and lessons learned. Rome: Electronic Publishing and Support Branch, Communication Division, FAO.

Flegal, K. M., M. D. Carroll, C. L. Ogden, and L. R. Curtin. 2010. Prevalence and trends in obesity among US adults, 1999-2008. *JAMA* 303:235–41.

Flint-Garcia, S. A., A. C. Thuillet, J. Yu, G. Pressoir, S. M. Romero, S. E. Mitchell, J. Doebley, S. Kresovich, M. M. Goodman, and E. S. Buckler. 2005. Maize association population: A high-resolution platform for quantitative trait locus dissection. *Plant J* 44(6):1054–64.

Frank, H. A. and R. J. Cogdell. 1996. Carotenoids in photosynthesis. *Photochem Photobiol* 63:257–64.

Frary, A., T. C. Nesbitt, A. Frary, S. Grandillo, E. V. D. Knaap, B. Cong, J. Liu, J. Meller, R. Elber, K. B. Alpert, and S. D. Tanksley. 2000. fw2.2: A quantitative trait locus key to the evolution of tomato fruit size. *Science* 289(5476):85–8.

Fuller, D. Q. 2007. Contrasting patterns in crop domestication and domestication rates: Recent archaeobotanical insights from the Old World. *Ann Bot* 100(5):903–24.

Fuller, D. Q., L. Qin, Y. Zheng, Z. Zhao, X. Chen, L. A. Hosoya, and G.-P. Sun. 2009. The domestication process and domestication rate in rice: Spikelet bases from the Lower Yangtze. *Science* 323(5921):1607–10.

Godfray, H. C. J., J. R. Beddington, I. R. Crute, L. Haddad, D. Lawrence, J. F. Muir, J. Pretty, S. Robinson, S. M. Thomas, and C. Toulmin. 2010. Food security: The challenge of feeding 9 billion people. *Science* 327(5967):812–8.

Graham, R., D. Senadhira, S. Beebe, C. Iglesias, and I. Monasterio. 1999. Breeding for micronutrient density in edible portions of staple food crops: Conventional approaches. *Field Crops Res* 60(1-2):57–80.

Hancock, J. F. 2004. *Plant Evolution and the Origin of Crop Species*. Cambridge, MA: CABI Publishing.

Harlan, J. R. 1992. *Crops and Man*. 2nd ed. Madison, WI: American Society of Agronomy.

Hayashi, E., N. Aoyama, and D. W. Still. 2008. Quantitative trait loci associated with lettuce seed germination under different temperature and light environments. *Genome* 51(11):928–47.

Hayes, H. K. and R. J. Garber. 1927. *Breeding Crop Plants*. New York: McGraw-Hill.

International Human Genome Sequencing Consortium. 2001. Initial sequencing and analysis of the human genome. *Nature* 409(6822):860–921.

International Human Genome Sequencing Consortium. 2004. Finishing the euchromatic sequence of the human genome. *Nature* 431(7011):931–45.

Kahler, A. L. and R. W. Allard. 1970. Genetics of isozyme variants in barley. I. esterases1. *Crop Sci* 10(4):444–8.

Kelly, T. J. J. and H. O. Smith. 1970. A restriction enzyme from *Hemophilus influenzae*. II. Base sequence of the recognition site. *J Mol Biol* 51:393–409.

Kovach, M. J., M. T. Sweeney, and S. R. Mccouch. 2007. New insights into the history of rice domestication. *Trends Genet* 23:578–87.

Lander, E. S. and D. Botstein. 1989. Mapping Mendelian factors underlying quantitative traits using RFLP linkage maps. *Genetics* 121(1):185–99.

Lele, U. 2010. Food security for a billion poor. *Science* 327(5973):1554.

Lester, D. R., J. J. Ross, P. J. Davies, and J. B. Reid. 1997. Mendel's stem length gene (Le) encodes a gibberellin 3-hydroxylase. *Plant Cell* 9(8):1435–43.

Levy, S., G. Sutton, P. C. Ng, L. Feuk, A. L. Halpern, B. P. Walenz, N. Axelrod, J. Huang, E. F. Kirkness, G. Denisov, Y. Lin, J. R. Macdonald, A. W. C. Pang, M. Shago, T. B. Stockwell, A. Tsiamouri, V. Bafna, V. Bansal, S. A. Kravitz, D. A. Busam, K. Y. Beeson, T. C. Mcintosh, K. A. Remington, J. F. Abril, J. Gill,

J. Borman, Y.-H. Rogers, M. E. Frazier, S. W. Scherer, R. L. Strausberg, and J. C. Venter. 2007. The diploid genome sequence of an individual human. *PLoS Biol* 5(10):e254.

Li, C., A. Zhou, and T. Sang. 2006. Rice domestication by reducing shattering. *Science* 311(5769):1936–9.

Lindgren, G., N. Backstrom, J. Swinburne, L. Hellborg, A. Einarsson, K. Sandberg, G. Cothran, C. Vila, M. Binns, and H. Ellegren. 2004. Limited number of patrilines in horse domestication. *Nat Genet* 36(4):335–6.

Linn, S. and W. Arber. 1968. Host specificity of DNA produced by *Escherichia coli*, X. *In vitro* restriction of phage fd replicative form. *Proc Natl Acad Sci USA* 59:1300–6.

Lukowitz, W., C. S. Gillmor, and W. R. Scheible. 2000. Positional cloning in *Arabidopsis*. Why it feels good to have a genome initiative working for you. *Plant Physiol* 123:795–805.

Mayr, E. 1982. *The Growth of Biological Thought*. Cambridge, MA: The Belknap Press of Harvard University Press.

Mcmullen, M. D., S. Kresovich, H. S. Villeda, P. Bradbury, H. Li, Q. Sun, S. Flint-Garcia, J. Thornsberry, C. Acharya, C. Bottoms, P. Brown, C. Browne, M. Eller, K. Guill, C. Harjes, D. Kroon, N. Lepak, S. E. Mitchell, B. Peterson, G. Pressoir, S. Romero, M. O. Rosas, S. Salvo, H. Yates, M. Hanson, E. Jones, S. Smith, J. C. Glaubitz, M. Goodman, D. Ware, J. B. Holland, and E. S. Buckler. 2009. Genetic properties of the maize nested association mapping population. *Science* 325(5941):737–40.

Melchinger, A. E., H. F. Utz, and C. C. Schon. 1998. Quantitative trait locus (QTL) mapping using different testers and independent population samples in maize reveals low power of QTL detection and large bias in estimates of QTL effects. *Genetics* 149(1):383–403.

Mendel, G. 1866. Versuche über Plflanzen-hybriden. Verhandlungen des naturforschenden Ver-eines in Brünn, Bd. IV für das Jahr 1865 (Experiments in plant hybridization. Read at the February 8th, and March 8th, 1865, meetings of the Brünn Natural History Society). Abhand-lungen: 3–47.

Mou, B. 2005. Genetic variation of beta-carotene and lutein contents in lettuce. *J Am Soc Hort Sci* 130(6):870–76.

Myles, S., J. Peiffer, P. J. Brown, E. S. Ersoz, Z. Zhang, D. E. Costich, and E. S. Buckler. 2009. Association mapping: Critical considerations shift from genotyping to experimental design. *Plant Cell* 21(8):2194–202.

Oh, M.-M., E. E. Carey, and C. B. Rajashekar. 2010. Regulated water deficits improve phytochemical concentration in lettuce. *J Am Soc Hort Sci* 135(3):223–9.

Ott, L. and J. G. Scandalios. 1978. Genetic control and linkage relationships among aminopeptidases in maize. *Genetics* 89(1):137–46.

Pascal, A. A., Z. Liu, K. Broess, B. Van Oort, H. Van Amerongen, C. Wang, P. Horton, B. Robert, W. Chang, and A. Ruban. 2005. Molecular basis of photoprotection and control of photosynthetic light-harvesting. *Nature* 436(7047):134–7.

Penuelas, J. and S. Munne-Bosch. 2005. Isoprenoids: an evolutionary pool for photoprotection. *Trends Plant Sci* 10(4):166–9.

Pourcel, L., J. M. Routaboul, V. Cheynier, L. Lepiniec, and I. Debeaujon. 2007. Flavonoid oxidation in plants: from biochemical properties to physiological functions. *Trends Plant Sci* 12(1):29–36.

Pushkarev, D., N. F. Neff, and S. R. Quake. 2009. Single-molecule sequencing of an individual human genome. *Nat Biotech* 27(9):847–50.

Rice-Evans, C., N. Miller, and G. Paganga. 1997. Antioxidant properties of phenolic compounds. *Trends Plant Sci* 2(4):152–9.

Richards, E. J. 2006. Inherited epigenetic variation—revisiting soft inheritance. *Nat Rev Genet* 7(5):395–401.

Roach, J. C., G. Glusman, A. F. A. Smit, C. D. Huff, R. Hubley, P. T. Shannon, L. Rowen, K. P. Pant, N.Goodman, M. Bamshad, J. Shendure, R. Drmanac, L. B. Jorde, L. Hood, and D. J. Galas. 2010. Analysis of genetic inheritance in a family quartet by whole-genome sequencing. *Science* 328(5978):636–9.

Salvi, S. and R. Tuberosa. 2005. To clone or not to clone plant QTLs: Present and future challenges. *Trends Plant Sci* 10(6):297–304.

Sanger, F., S. Nicklen, and A. R. Coulson. 1977. DNA sequencing with chain-terminating inhibitors. *Proc Natl Acad Sci USA* 74(12):5463–7.

Smith, H. O. and K. W. Wilcox. 1970. A restriction enzyme from Hemophilus influenzae. I. Purification and general properties. *J Mol Biol* 51:370–91.

Smithies, O. 1955. Zone electrophoresis in starch gels: Group variations in the serum proteins of normal human adults. *Biochem J* 61:629–41.

Sturtevant, A. H. 1913. The linear arrangement of six sex-linked factors in Drosophila, as shown by their mode of association. *J Exp Zoo* 14:43–59.

Sundqvist, A.-K., S. Bjornerfeldt, J. A. Leonard, F. Hailer, A. Hedhammar, H. Ellegren, and C. Vila. 2006. Unequal contribution of sexes in the origin of dog breeds. *Genetics* 172(2):1121–8.

Takhtajan, A. 1997. *Diveristy and Classification of Flowering Plants*. New York: Columbia University Press.

Tanksley, S. D. and S. R. McCouch. 1997. Seed banks and molecular maps: Unlocking genetic potential from the wild. *Science* 277(5329):1063–6.

Tanksley, S. D. and C. M. Rick. 1980. Isozymic gene linkage map of the tomato: Applications in genetics and breeding. *Theor Appl Genet* 58(2):161–70.

Tanksley, S. D., N. D. Young, A. H. Paterson, and M. W. Bonierbale. 1989. RFLP mapping in plant breeding: new tools for an old science. *Nat Biotech* 7(3):257–64.

Thoday, J. M. 1961. Location of polygenes. *Nature* 191(4786):368–70.

Trocchio, F. 1991. Mendel's experiments: A reinterpretation. *J Hist Biol* 24(3):485–519.

USDA. 2010. Nutrient availability data. USDA Economic Research Service. http://www.ers.usda.gov/Data/FoodConsumption/NutrientAvailIndex.htm. (accessed August 15, 2011).

Weir, B. S. 2008. Linkage disequilibrium and association mapping. *Annu Rev Genom Hum Genet* 9(1):129–42.

Wheeler, D. A., M. Srinivasan, M. Egholm, Y. Shen, L. Chen, A. Mcguire, W. He, Y.-J. Chen, V. Makhijani, G. T. Roth, X. Gomes, K. Tartaro, F. Niazi, C. L. Turcotte, G. P. Irzyk, J. R. Lupski, C. Chinault, X.-Z. Song, Y. Liu, Y. Yuan, L. Nazareth, X. Qin, D. M. Muzny, M. Margulies, G. M. Weinstock, R. A. Gibbs, and J. M. Rothberg. 2008. The complete genome of an individual by massively parallel DNA sequencing. *Nature* 452(7189):872–6.

Wikstrom, N., V. Savolainen, and M. W. Chase. 2001. Evolution of the angiosperms: Calibrating the family tree. *Proc R Soc Lond B* 268:2211–20.

Xu, S. 2003. Theoretical basis of the Beavis effect. *Genetics* 165(4):2259–68.

Yan, J., C. B. Kandianis, C. E. Harjes, L. Bai, E.-H. Kim, X. Yang, D. J. Skinner, Z. Fu, S. Mitchell, Q. Li, M. G. S. Fernandez, M. Zaharieva, R. Babu, Y. Fu, N. Palacios, J. Li, D. Dellapenna, T. Brutnell, E. S. Buckler, M. L. Warburton, and T. Rocheford. 2010. Rare genetic variation at *Zea mays* crtRB1 increases [beta]-carotene in maize grain. *Nat Genet* 42(4):322–7.

Yano, M., Y. Katayose, M. Ashikari, U. Yamanouchi, L. Monna, T. Fuse, T. Baba, K. Yamamoto, Y. Umehara, Y. Nagamura, and T. Sasaki. 2000. *Hd1*, a major photoperiod sensitivity quantitative trait locus in rice, is closely related to the *Arabidopsis* flowering time gene CONSTANS. *Plant Cell* 12(12):2473–84.

# Section III

**Food Needs to Meet Nutrigenomic Health Needs**

*Food Science and Technology Enhancement of Diet Quality*

# 19 Natural Colorants as Bioactive Agents in Functional Foods

*Ann Marie D. Craig*

## CONTENTS

## INTRODUCTION

The focus of this monograph is to provide an introduction to the world of natural food colorant compounds, specifically carotenoids and anthocyanins. These two classes of colorants are readily available from several plant sources and are often recognized by the vibrant colors they produce. Consumers often select food by what is appealing to the eye and frequently color is the deciding factor. Regardless of whether the color indicates a certain flavor or serves as an indicator of quality to the item, consumer choice is dependent on this visual property.

Both natural and synthetic colorants are used in the food industry in processed foods. The concerns of consumers about synthetic food colorants as "chemicals" are based on historical use of petroleum-based colorants that proved to cause or increase the risk for cancer. This and a general fear of chemicals have caused them to move to more natural selections. However, the incorporation of natural colorants as a replacement for artificial colorants is a challenge requiring specific processing conditions related to the structural sensitivity of the natural compounds. Aside from the aspect of replacing synthetics, natural colorant compounds typically possess potential health-promoting benefits. Therefore, it is essential for food scientists to advance the field of natural colorants by determining the conditions that enhance processing stability, as well as establishing ways to maximize and maintain the biological activity of these substances.

## BIOACTIVE COMPOUNDS IN FUNCTIONAL FOODS

When a discussion begins on bioactives, phytochemicals, and functional foods, it is often a good idea to ensure the meaning of the terms as they are used for different reasons. In general, compounds derived from plants are typically known as *phytochemicals*, and more specifically referred to as any naturally occurring compound that may possess a health-protective or biologic medicinal value (Duke et al. 2000; Craig 1997). A resurrection in the interest of phytochemicals has occurred in the past few decades leading researchers to study them for a myriad of reasons. Food scientists want to incorporate them into beverages and other foods creating the industry niche known as "functional foods," while nutrition scientists desire a deeper understanding into their bioactivity. Hence, the term *bioactive* goes one step further to be defined as a phytochemical with known specific biologic activity.

Why the renaissance for natural compounds? It is likely due to several things, such as the growing knowledge that all botanical compounds have not been discovered, nor have those discovered all been extensively examined, or the fact that nature has a way of selecting for its own biological activity, or lastly, scientific methods of analysis have continued to vastly improve. The idea of exploring such "medicinal foods" is something to be more deeply investigated. An ethnopharmacologic point of view is just what food and nutrition scientists need to keep this field of bioactives and functional foods alive (Etkin 2006; Shu 1998).

Both natural and manufactured food products can have bioactive compounds present that can influence human function. According to the Institute of Medicine's (www.iom.edu) Food and Nutrition Board, a functional food is defined as "any food or food ingredient that may provide a health benefit beyond the traditional nutrients it contains." The development of functional foods is a topic of interest considering that there is little to no doubt that nutritional habits are closely connected to the state of one's health. The common quote by Hippocrates is to "let food be thy medicine and medicine be thy food." This centuries-old philosophy still holds true today as many researchers study various aspects of food from post-harvest effects, processing concerns, and nutritional value to the body's ability to metabolize, absorb, and effectively utilize the food components. Numerous foods are linked to health promotion and disease prevention, as many have been reported to modify redox potential of cells which can ultimately determine one's own health and well-being (Milner 2000). Table 19.1 provides a brief selected list of functional foods with the corresponding proposed health benefit.

A centralized database for specific classes of bioactive compounds including carotenoids, anthocyanins, flavonoids, and plant sterols is under development (Day et al. 2009). To date, extensive literature searches and professional conference proceedings have served as the source of this scientific information. Specialized databases are in need of more detailed information that would include a specific description of the foods and types of samples analyzed, the number of samples used in specific studies, and measures of unpredictability for all samples. This data would be helpful for food composition databases and for clinical and epidemiological studies on bioactive compounds (Pennington 2002). The incorporation of functional ingredients into foods is also a concern. Beyond

---

**TABLE 19.1**

**List of Selected Functional Foods, Both Natural and Manufactured, and Their Benefit to Overall Health**

| Functional Food | Health Benefit | Source |
|---|---|---|
| Beta-glucan in oats | Heart health | Williams et al. 2006 |
| Plant sterols incorporated into margarines | Lowering cholesterol | Williams et al. 2006 |
| Chewing gum sweetened with sugar alcohols | Prevention of tooth cavities | Williams et al. 2006 |
| Golden rice | Pro-vitamin A | Williams et al. 2006 |
| Kale and spinach containing lutein and zeaxanthin | Protective antioxidants in macular region of the human retina; protection against cataract formation | Maiani et al. 2009 |
| Astaxanthin found in salmon, shrimp, and other "red/pink" seafood or crustaceans | Potent antioxidant properties | Maiani et al. 2009 |
| Epigallocatechin gallate (EGCG) found in green tea | Reduces risk of certain cancers and reduced heart disease | Bidlack and Wang 2000 |

the ability to isolate a bioactive component, the considerations of regulatory issues include dosage, bioavailability, stability, and the effect on the food product itself (Day et al. 2009).

## CONSUMER ACCEPTANCE AND COLOR ADDITIVES

The addition of color to food products is an important factor in consumer acceptance. Colors must be pleasing to the eye and safe for the body since it is ranked alongside freshness as one of the key criteria for selection. Color is certainly a notable trait in food since it often predetermines the expectation and acts as a predictor of flavor intensity. Color can also act as an indicator of quality (Francis 1995). Since food quality is often measured by general organoleptic properties such as taste and texture, it is likely the color that attracted the consumer to the food in the first place. Color can also be considered a manufacturing necessity in the food industry. Color additives emphasize colors that are already present in the food or guarantee consistency of a product from batch to batch (Griffiths 2005). It is the responsibility of the food scientist to ensure that food meets the necessary criteria to maintain consumer acceptability. Also noted, color additives should never be used to cover up any negative aspects of food developing at any phase of food production. Color can also be an indication of food safety since foods of an unsuitable or non-uniform color may be associated with spoilage or bad processing. (Delgado-Vargas and Paredes-Lopez 2003a). Thus, color can assure the consumer of food quality within the regulatory uses of food colorants and not to mask poor product quality.

A color additive is defined by the United States Food and Drug Administration (FDA) as any dye, pigment, or substance that can impart color when added or applied to food, drugs, and cosmetics or to the body. Food colorants are chemicals that can be naturally occurring or synthetically derived. Today, in the United States, all food color additives are regulated by the FDA to ensure safety of the products. According to the FDA's Code of Federal Regulations (www.fda.gov), Title 21 parts 73 and 74 (subpart A), all color additives used in food are divided into two categories, "exempt from certification" and "certified," respectively. Certified colors are man-made color additives derived primarily from petroleum and coal sources, while those exempt from certification include pigments that are derived from natural sources such as minerals, animals, vegetables, and man-made counterparts from the natural derivatives. Certification has no bearing on their overall safety; however, exempt colors are perceived as less of a health hazard than those derived synthetically from petroleum sources. Although there is no such thing as "absolute safety," the FDA works diligently to determine if there is "a reasonable certainty of no harm" under the prescribed use of a color according to the "How Safe are Color Additives?" publication (U.S. Food and Drug Administration 2010a). Table 19.2 lists examples of colors from both classifications. Since safety is always a concern of consumers regarding any food "chemicals," the FDA does ensure that the toxicological properties of the synthetic colorants are understood (i.e., well studied by standard research criteria). Several steps are in place to confirm that there are no adverse effects on human health through laboratory testing that includes the assay of toxicity in rats, a teratology study, and a multigenerational reproductive study in mice. Nonsynthetic colorant compounds simply have to comply with identity and purity specifications since they are assumed safe by historical consumption of naturally produced foods containing those colorants.

Colorants exempt from certification come from natural sources. A majority of natural colorants belong to two specific groups, the carotenoids and the anthocyanins. The interesting thing about these two classes of chemicals is their color distinction. The carotenoids are best known for their vibrant yellow, orange, and some red hues found in carrots, squash, and tomatoes. The anthocyanins are responsible for the deep red, violet, purple, and blue hues of several berries, grapes, and flowers. The color a person observes is dependent on the interaction of light with the chemical structure of the carotenoid or anthocyanin. Color perception also depends on the type and intensity of light, the chemical and physical characteristics of the food, and an individual's ability to distinguish color (Hutchings 2002). The specific chemistry of the carotenoid and anthocyanin structures will be discussed later in their respective sections.

**TABLE 19.2**

**Colors Listed under the U.S. Food and Drug Administration's Code of Federal Regulation for "Certified" and "Exempt from Certification" Additives**

| Color Additive | Regulatory Status | 21 CFR Section |
|---|---|---|
| Blue No. 1 | Certified | 74.101 |
| Blue No. 2 | Certified | 74.102 |
| Green No. 3 | Certified | 74.203 |
| Red No. 3 | Certified | 74.303 |
| Red No. 40 | Certified | 74.340 |
| Yellow No. 5 | Certified | 74.705 |
| Yellow No. 6 | Certified | 74.706 |
| Annatto | Exempt | 73.30 |
| Beta-carotene | Exempt | 73.35 |
| Lycopene | Exempt | 73.585 |
| Fruit juice | Exempt | 73.250 |
| Vegetable juice | Exempt | 73.260 |
| Paprika | Exempt | 73.340 |
| Tumeric | Exempt | 73.600 |

*Note:* Certified color additives are analyzed and batch certified before use in a regulated product while certification is not necessary for the protection of the public health for those exempt (not an inclusive list of "exempt" colors).

## CAROTENOIDS

The carotenoids are one of the largest and most important classes of plant pigments. They play a crucial role in defining quality parameters of many fruits and vegetables. There are approximately 600 isolated carotenoids, of which about 50 are present in a typical diet (Krinsky 1994). Chemically they are sorted by structure into two classes, the carotenes and the xanthophylls, each demonstrating unique biological activity. It is the structure of a carotenoid that determines its biological function and coloring capability (Rodriguez-Amaya 2001; Delgado-Vargas et al. 2000).

Food carotenoids are built from eight $C_5$ isoprenoid units resulting in the usual $C_{40}$ backbone. This basic structure can be linear or cyclized at one or both ends with lateral methyl groups along the chain. Other modifications include hydrogenation, dehydrogenation, double-bond migration, isomerization, chain shortening, and inclusion of oxygen. The carotenoids void of any oxygen species are referred to as the carotenes, while those containing oxygen are known as the xanthophylls. Examples of each class of carotenoids are shown in Figure 19.1.

Sources of dietary carotenoids for humans come from fruits, vegetables, fish, and other meats. It is thought that a diet high in these foods can protect against diseases such as atherosclerosis, certain types of cancers, and age-related macular degeneration (Hinds et al. 1997). Carotenoids are known for their pro-vitamin-A activity. Vitamin A can be produced within the body from certain carotenoids, notably β-carotene. This activity serves many vital systemic functions in humans, for example, it is essential for vision, immune response, epithelial cell growth and repair, bone growth, reproduction, maintenance of the surface linings of the eyes, and epithelial integrity of respiratory, urinary, and intestinal tracts (Rodriguez-Amaya 1996). However, not all carotenoids have pro-vitamin-A activity. Only those carotenes with an unsubstituted β-ring and a $C_{11}$ polyene chain exhibit the ability to convert to retinol, or vitamin A (Rodriguez-Amaya 1996, 2001; Delgado-Vargas et al. 2000).

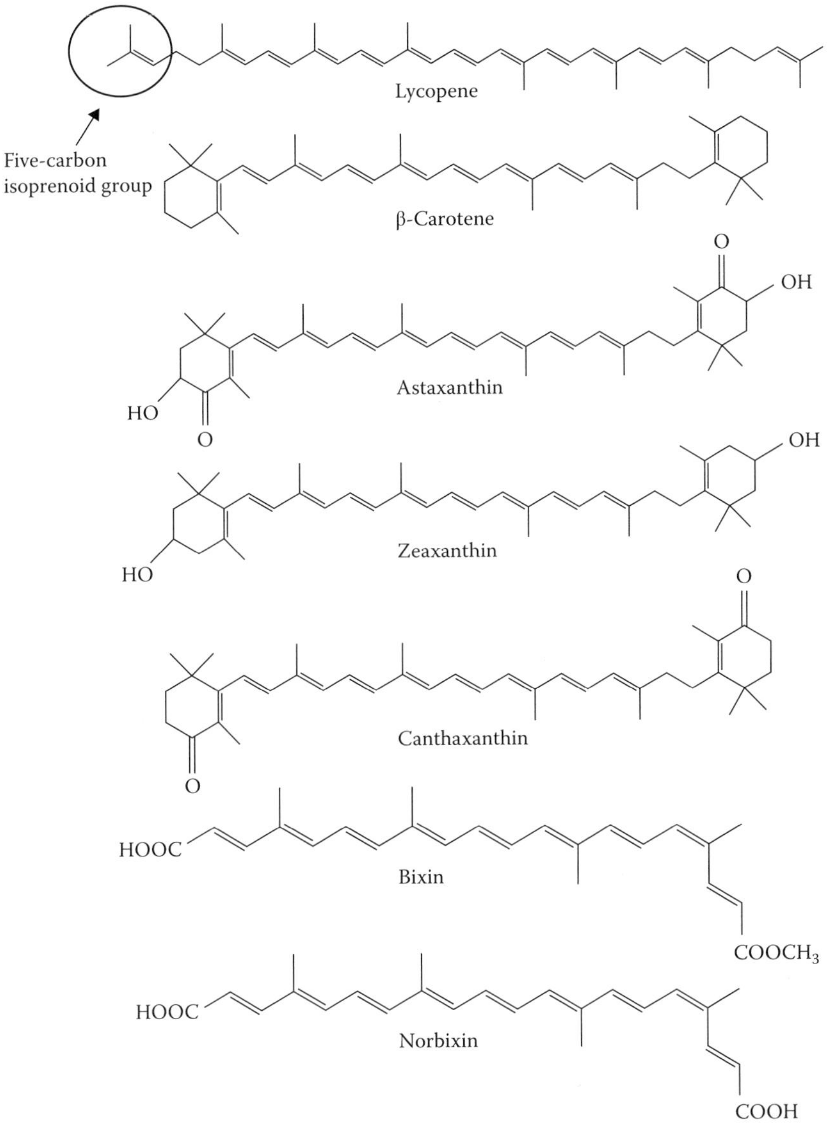

**FIGURE 19.1** Selected examples of the two classes of carotenoids, including the depiction of the isoprenoid group, which is the integral part of the structural backbone.

Aside from their brilliant pigment capability and potential health benefits, carotenoids also exhibit strong antioxidant properties (Paiva and Russell 1999), which accounts for their important function in food and biological systems (El-Agamey et al. 2004; Krinsky 1994; van den Berg et al. 2000). They are considered as potential membrane antioxidants due to their strong interaction with reactive oxygen species and free radicals. By intercepting these damaging chemical reactants, carotenoids essentially prevent damage to membrane lipids. Their distinct chemical structure is what allows for energy absorption from the reactive species. Figure 19.2 depicts several physical (bonding pattern of structural backbone) and chemical properties (significance of the bonding pattern) of carotenoids.

The colorant capacity of a carotenoid is imparted by its conjugated double-bond system that resides mainly in a *trans* configuration. Any extensive conjugated system is highly delocalized around the pi-bonds, forming the double bonds, resulting in a low-energy excited state. Essentially, this accounts for the color that we see. When samples are exposed to light energy that matches a

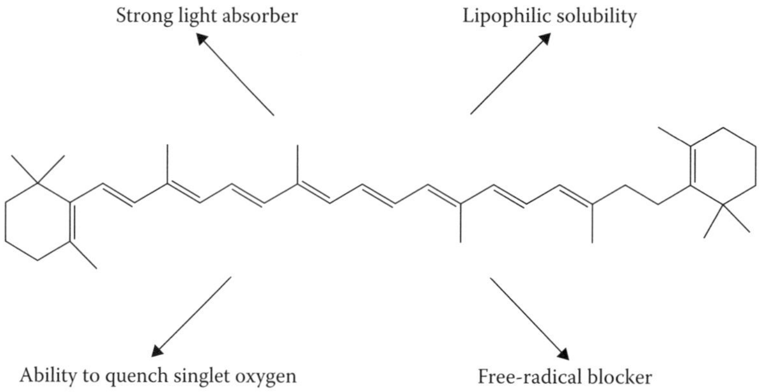

Strong light absorber                          Lipophilic solubility

Ability to quench singlet oxygen                          Free-radical blocker

FIGURE 19.2   General physical structure depicting the extensive conjugated bond system that allows for the characteristic colors exhibited by the carotenoids, as well as the chemical properties as a result of the isoprenoid backbone of carotenoids. (Adapted from Krinsky, N. I. 1994. *Pure Appl Chem* 66(5):1003–1010.)

possible electronic transition within a molecule, part of the energy will be absorbed as the electrons are promoted to a higher energy orbital (Melendez-Martinez et al. 2007). The transition energy occurs in the visible region of the electromagnetic spectrum, approximately 380–800 nm, but more specifically in the 400–500 nm range for the carotenoids, which is where yellow, orange, and red hues are observed.

In addition, carotenoids are capable of energy transfers to their excited triplet state. This ability allows the structure to accept the transfer of energy from other highly reactive species; the most notable species is singlet oxygen ($^1O_2$). β-carotene was the first carotenoid used to characterize the reaction between carotenoids and singlet oxygen (Foote and Denny 1968). The ability to quench unpaired electrons is based on the number of conjugated bonds within the carotenoid structure; thus, lycopene is more effective than β-carotene (Krinsky 1994). Research by Cantrell et al. (2003) suggested that singlet oxygen quenching is dependent upon the type of carotenoid and the environment in which they are incorporated, for example, a lipid membrane. Their results indicate lycopene and β-carotene are the fastest to quench an unpaired electron, while lutein was the slowest; astaxanthin and canthaxanthin were intermediate quenchers. The quenching ability of the xanthophylls may be hindered by the longer delocalized pi-bond system present in their structure.

Carotenoids are a very diverse group of compounds given their physicochemical properties. The structure determines the unique functionality that each carotenoid possesses. Depending on the molecular arrangement and length of the conjugated bond system, carotenoids can act as singlet oxygen quenchers or free radical blockers. Current research has established the efficacy of carotenoids in intercepting such detrimental molecules, thus adding to the vital diversity of these compounds and also assigning them as antioxidants.

Carotenoids are biologically active beyond those that have pro-vitamin-A activity and their general antioxidant capacity. Carotenoids have also been studied in cell signaling pathways where there is an indication that they or their metabolites influence expression of certain genes or inhibit certain regulatory enzymes associated with cancer pathways (Stahl and Sies 2005). Other studies conducted *in vitro* or using animal systems have provided sufficient evidence that carotenoids have the ability to protect against several kinds of cancer (Mayne 1996; Maiani et al. 2009).

## ANTHOCYANINS

Anthocyanins are another large class of natural colorant compounds. In contrast to the carotenoids, anthocyanins are water soluble. These compounds are responsible for the red, purple, and blue colors found in many fruits, vegetables, and flowers aptly named after the Greek words *anthos* (flower) and *kyanos* (dark blue). Anthocyanins are sensitive to changes in pH which allows for the observed array of colors produced. This class of compounds is not only of interest for its coloring ability, but also for the potential health effects which characterize anthocyanins as another *bioactive* species.

Anthocyanins are unique in structural qualities (Figure 19.3). Their core structure is built on a $C_{15}$ skeleton consisting of a $C_6$-$C_3$-$C_6$ construction. The first ring is a chromane moiety with a second aromatic ring attached (known as ring B). On C-2 of ring B, a phenyl ring is attached that can have its own amount of substitution. The multiple R-groups of all three rings can be substituted with hydrogens, hydroxyl, or sugars. Structures not substituted with sugar molecules are known as anthocyanidins or aglycones. The unique substitution of the R-groups at a given pH is what allows for the myriad of colors produced by these molecules. In Table 19.3, the names of dominant anthocyanins and their substitution patterns are shown.

Aside from the substitution patterns, pH is the other factor that dictates the expressed color of the anthocyanin. Like the carotenoids, anthocyanins are colored because of the extensive double bond system throughout the $C_6$-$C_3$-$C_6$ skeleton. The delocalized pi-bonds are in a low energy state, but when excited by light energy equal to the energy of the electron state, a transition occurs. Part of this energy is absorbed as electrons move to a higher orbital. The energy involved in this transition is equivalent to energy in the visible region of the electromagnetic spectrum. For the anthocyanins, this energy can be in several areas because they are red (490–560 nm), violet (560–580 nm), and

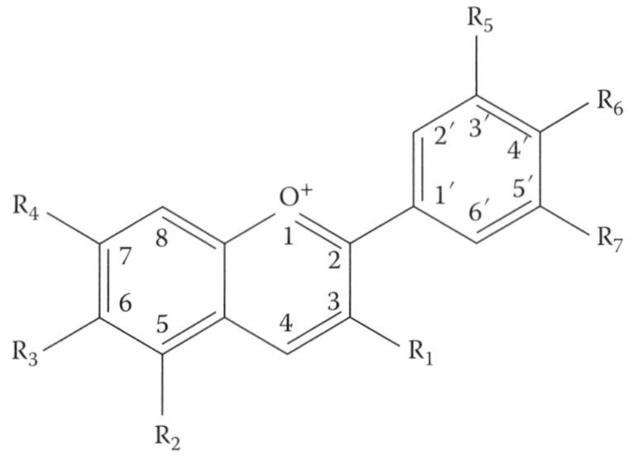

**FIGURE 19.3** Basic anthocyanidin structure indicating several substitution locations. Glycosylation at R1 deems it an "anthocyanin."

## TABLE 19.3
## Substitution Pattern for Selected Anthocyanidins and Their Respective Color for General Structure Shown in Figure 19.3

| Aglycone | Hydroxyl Group | Color |
|---|---|---|
| Cyanidin | 3,5,7,3',4' | Magenta and crimson |
| Delphynidin | 3,4,7,3',4',5' | Purple, mauve, blue |
| Pelargonidin | 3,5,7,4' | Salmon |
| | **Methoxy Group** | |
| Malvidin | 3,5' | Purple |
| Peonidin | 3' | Magenta |
| Rosinidin | 7 | Red |

*Source:* Adapted from Delgado-Vargas, F., and O. Paredes-Lopez, In *Natural Colorants for Food and Nutraceutical Uses*, 167–191, CRC Press, Boca Raton, FL, 2003.

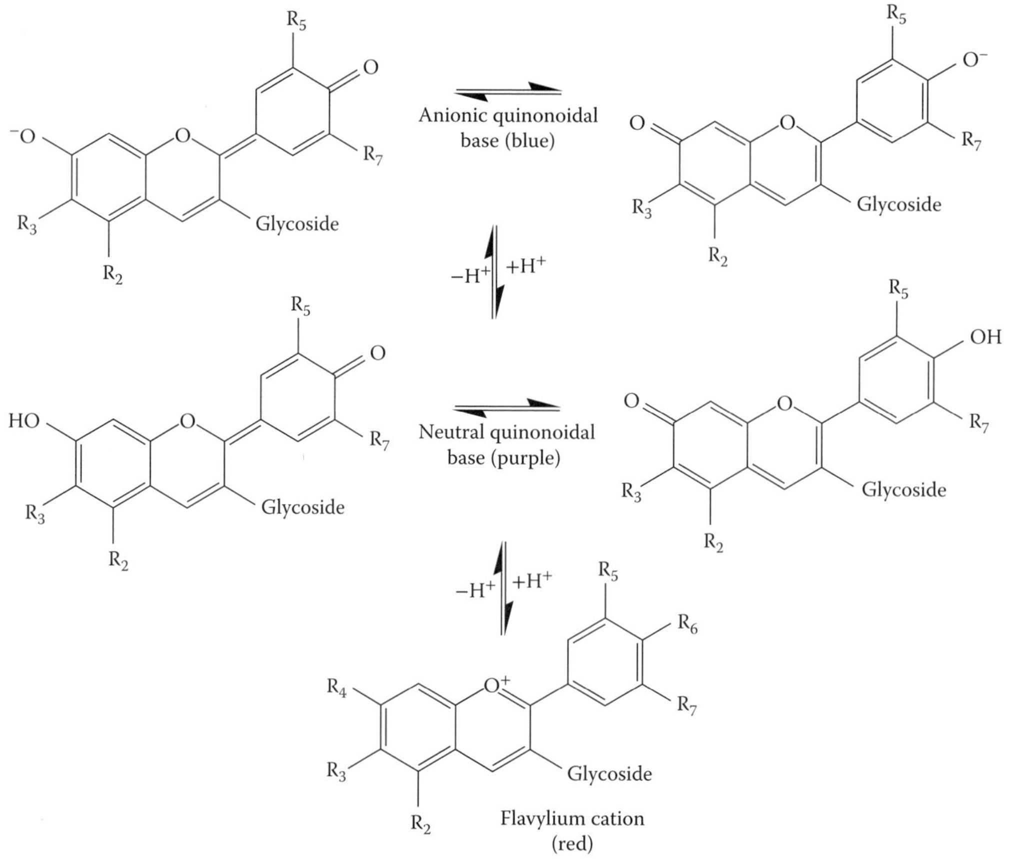

**FIGURE 19.4** Transformation of anthocyanin structure as a result of addition or removal of acid. As pH increases, the anthocyanins exist in the quinonoidal resonance form, and in acidic conditions, the flavilium cation. (Adapted from Delgado-Vargas, F., and O. Paredes-Lopez, In *Natural Colorants for Food and Nutraceutical Uses*, 167–191, CRC Press, Boca Raton, FL, 2003; and Stintzing, F. C., and R. Carle, *Trends Food Sci Tech* 15:19–38, 2004.)

blue (560–620 nm). Figure 19.4 shows the quinonoidal resonance form of the anthocyanin structure as in basic conditions and the flavilium ion favored in acidic conditions. In the cationic form, anthocyanins are red and as the pH increases, or becomes more alkaline, shades of violet and blue result (quinonoidal anion). It has been reported that as the number of sugar substitutions increase, a hypsochromic shift occurs in the visible spectrum resulting in shorter wavelengths that produce deeply red colored pigments (Giusti and Wrolstadt 2003). Understanding the behavior of anthocyanins as they are exposed to different pH values is important in the preparation of manufactured foods, establishing shelf-life stability, and passage through the gastrointestinal tract, where the bioactive properties can vary as they are exposed to different pH and temperature systems.

Beyond their magnificent color display, anthocyanins have also been touted for their potential biological activity and health-promoting benefits. Anthocyanins are the most abundant flavonoid found in fruits and vegetables. It is estimated that the daily intake of these compounds in the United States is between 180–215 mg as part of a balanced diet (Wang and Stoner 2008). Epidemiological studies suggest that consuming anthocyanins regularly can lower the risk of cardiovascular disease, arthritis, diabetes, and cancer due to their antioxidant and anti-inflammatory capabilities (Prior and Wu 2006). The phenolic structures laden with conjugated double bonds are responsible for the antioxidant capacity of the anthocyanins. The double bond network is able to scavenge reactive oxygen species (ROS) such as superoxide ($O_2^-$), singlet oxygen ($^1O_2$), peroxides (ROO–), and hydroxyl radicals (OH•). In several *in vitro* studies using cell culture systems from colon, endothelial, liver, breast, and leukemic cell lines, the anthocyanins displayed numerous antitoxic and anticarcinogenic effects based on their ability to scavenge ROS, which stimulated the expression of phase II enzymes and ultimately reducing oxidative adducts in DNA and hindering mutagenesis (Wang and Stoner 2008).

## STABILITY IN PROCESSING

Both the carotenoids and the anthocyanins are of interest to the food industry to serve as a "natural" replacement for the currently used synthetic dyes and lakes. However, the synthetic colorants are more chemically stabile, making them an attractive and easy-to-use food additive. As consumers push away from the use of artificial or synthetic "chemicals" in food, the food industry has to continuously keep up with these trends and seek out alternative replacements. Some of the currently used carotenoids include β-carotene, annatto (norbixin and bixin blend), lycopene, and paprika. While the use of anthocyanins is practiced, they will be listed as "vegetable juice" or "fruit juice" on the label. The challenge to have a widespread acceptance of these natural alternatives by the food industry resides in their instability during many of the processing scenarios currently practiced. Therefore, it is the responsibility of food scientists to understand ways to maximize the stability of carotenoids and anthocyanins by designing and evaluating the effects in model food systems.

For carotenoids, β-carotene has been the primary test compound used to establish their processing stability (Delgado-Vargas et al. 2000). While β-carotene is a reasonable representative of the carotenoids, others carotenoids may possess additional stability benefits due to differences in their chemical structural backbone. As discussed in the section *Carotenoid* (Figure 19.2), the carotenoid backbone is capable of a certain amount of stabilization from deleterious effects of light, heat, and oxygen (pH is not a factor). Preliminary results obtained from a photochemical degradation study that surveyed five carotenoid colorants (Craig and Zivanovic 2006) determined that after 30 days exposure to fluorescent light, annatto was least stable and astaxanthin was most stable (β-carotene, paprika oleoresin, and apo-8'-carotenal were the others studied). Annatto consists of two carotenoids, bixin and norbix, and is an example of one of the shorter backbones, while astaxanthin resembles β-carotene, the conjugated double-bond system is much more extensive, therefore stabilizes itself upon attack by the light energy. Maiani et al. (2009) reviewed the effects of food processing on carotenoid stability and bioavailability during postharvest, thermal applications, and product storage. For the thermal treatments, cooking and canning of tomato products showed an increase in the amount of carotenoid recovered, while hot air drying decreased tomato carotenoid, and pasteurization had no effect. In storage below 0°C, a variety of products had a decrease in carotenoid content, as did the 0°–10°C conditions. In Maiani's study, all carotenoids compared were non-pro-vitamin-A active. Several studies have been conducted in food systems to determine the stability of several other fruit and vegetable sourced carotenoids (Minguez-Mosquera et al. 1993; Boileau and Erdman 2004; Lin and Chen 2005; Barret and Anthon 2008; Schieber and Reinhold 2008).

When studying anthocyanins, the same factors that degrade the carotenoids (light, heat, and oxygen) will alter the color of these compounds as well. However, pH is an added factor to consider in terms of color stabilization (Figure 19.4). Also, the rate of anthocyanin degradation can be increased when in the presence of increase sugar, ascorbic acid, other phenolics, peroxides, and sulfites (Wrolstadt et al. 2008; Delgado-Vargas and Paredes-Lopez 2003b). Because of their heightened sensitivity, anthocyanins have been proposed as an indicator of good manufacturing practice. For example, if red raspberries are badly processed, the formation of brown compounds (polymeric anthocyanins) can be seen and is subsequently considered poor quality by a consumer (Garcia-Viguera et al. 1998). Since anthocyanins from fruit and vegetable sources are either processed as fresh product or extracted for use as a colorant in other products, researchers are trying to develop strategies to minimize effects of postharvest treatments, improve extraction technologies, and reduce processing effects (Wrolstadt et al. 2008; Giusti and Wrolstadt 2008; Garcia-Viguera et al. 1997).

## FOOD SCIENCE MEETS NUTRIGENOMICS

The way that food molecules are ingested and ultimately interact with our individual human matrix leads to the discussion of "nutrigenomics." The concept that food components, like those with bioactive ability, interact with our genetic makeup has become a growing focus of nutrition scientists. When considering bioactive colorant compounds and their potential gene interaction, the question needs to be asked: "Can the carotenoids and anthocyanins contribute to more than just color?" The search for natural bioactive compounds able to increase one's own well-being is a topic of research among food and nutritional scientists. As consumers become savvier and their demand for "better" food rises, science needs to work at meeting these demands. The move towards healthy nutritious food with balanced calories and additional health functions (functional foods), can only be developed through innovative multidisciplinary research programs (Day et al. 2009). A new focus in nutritional science is the ongoing personalization of diet for an individual. The "personalized nutrition" approach is just one aspect of the field of nutrigenomics (Gillies 2003). The understanding of what, how, and when food is consumed by an individual plays a role in how it is digested, absorbed, and metabolized; not to mention how a nutrient, non-nutrient, or their metabolites potentially interact with genes. Carotenoids may be able to decrease malignant transformation of cells as indicated by one study (Krinsky 1994). This research also indicated that beta-carotene, lycopene, and canthaxanthin enhanced gap junction communication in certain cells due to up-regulation of a specific gene, connexin43. This gene is the most commonly expressed constituent of the connexin family of molecules (Burri 2000) and is essential for the development of normal cardiac and ventricular structure. A recent study on anthocyanins isolated from bilberries and black currants indicated they suppressed the expression of numerous genes associated with inflammatory processes (Karlsen et al. 2007). A review of berry phytochemicals by Duthie (2007) also discussed data from numerous cell and animal model studies that indicate anthocyanins are protective against gene instability at several sites in pathways involving carcinogenesis.

The complexities of nutrient–gene interaction should not be underestimated; it is more intricate than diet alone and adds a level of sophistication to select personal dietary components to enhance health. Environmental factors, overall cellular equilibrium, and an individual's metabolism all contribute to these gene interactions. The challenge to assess the regulation of this interaction requires analytical methodologies sensitive enough to account for genetic differences and exposure diversity. Another complex area of nutrigenomics is metabolomics. The objective of metabolomics in this capacity is the investigation of metabolic alterations produced by the effect of different nutrients or bioactive food components (Garcia-Canas et al. 2010). Two basic approaches are target analysis and metabolic profiling. In target analysis, selected biomarkers are quantified, whereas metabolic profiling studies a group of metabolites that may characterize specific pathways. This is in an effort to understand the metabolic profile of a cell and provide an accurate description of a specific phenotype (Wai-Nang and Go 2005). Most bioactives are over-categorized as antioxidants. Thus, further analytical and screening techniques are necessary to assess specific effects on cellular biologic or gene expression activities. When metabolic patterns are standardized under normal conditions, insight into specific disease characteristics can then be established. The advantage of adding metabolomics under the umbrella of nutrigenomics is that metabolic outcomes can be correlated directly to the regulation of gene(s) expression in cells and tissue function.

The concept of target analysis of biomarkers offer food and nutrition scientists the ability to acquire a comprehensive "molecular" picture of food including analysis of food component, determination of food quality/authenticity, and monitoring of food consumption data (Wishart 2008). Wishart (2008) also points out that a food metabolomics researcher views food as a complex matrix consisting of several metabolites. There are over 3600 substances allowed in processed food, according to the FDA's "Everything Added to Food in the United States" (U.S. Food and Drug

Administration 2010b). These, along with over 8000 non-nutrient phytochemicals (Ehrman et al. 2007), can be referred to as the nutrition metabolome (Gibney et al. 2005). Gibney continues with the discussion of how extrinsic factors (diet, drugs, exercise, and stress) and intrinsic factors (body composition, age, metabolic rate, genotype, and health) will affect the metabolome (sum of all metabolites).

The marriage of food science and nutrigenomics has created a new scientific model described as "nutridynamics" (deVos et al. 2006), characterized as the study of how food components are affected while in the food matrix and while in the human body. According to deVos (2006, page 218):

It addresses the fate, kinetics and interactions of food components embedded in a food matrix and their physiological effects upon consumption, including the mechanism of action, the effect of concentration and interaction with other food components in the diet. In this way, nutridynamics couples food characteristics with processes in the body in a systematic way, including both quantitative and kinetic aspects, and hence links aspects of food and biotechnology on the product side with target delivery of nutrigenomics on the consumer side.

While the molecular form of food consumed is generally known, how it is delivered throughout the body upon consumption is often speculated, especially when discussing the fate of bioactives. In the case of bioactive colorant compounds, challenges exist in understanding the bioavailability of anthocyanins (McGhie and Walton 2007) and carotenoids (Faulks and Southon 2005; Elliot 2005). Figure 19.5 proposes a pathway showing the interconnection of food science, nutrigenomics, and metabolomics.

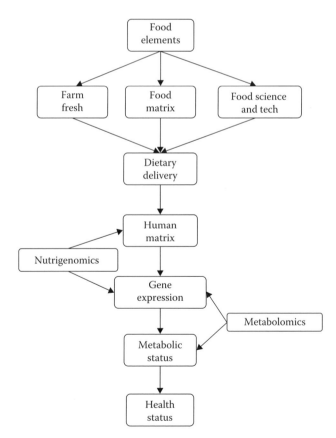

**FIGURE 19.5** General concept of nutridynamics: the collaboration of food science, nutrigenomics, and metabolomics.

## FUTURE OF NATURAL COLORANTS

The quality of the color perceived by the human eye is related to the chemical stability of that compound. The stability of natural colorants faces challenges all the way from postharvest storage to processing techniques. If color is degraded or is lost, then the structural integrity of that compound is lost, which ultimately affects the possible bioactive features it once possessed. The metabolic fate of food colorants is difficult to discern. It is estimated that under regular food and beverage processing, a 50–100 ppm concentration of colorant is added (Parkinson and Brown 1981), an amount less than that consumed as part of a healthy balanced diet. Aside from maintaining certain color integrity, chemical stability will maintain any health-promoting benefits as well. Several natural colorants belonging to the classes of carotenoids and anthocyanins have bioactive properties. Understanding the ways in which to protect the fundamental structure as an ingredient in a functional food system is the first step in knowing what a consumer actually ingests. Without knowing this, it will be difficult to track the colorants absorption, metabolism, and potential gene interactions on the human system.

## SUMMARY

Natural color compounds are known for their potential health benefits. The yellow, orange, and red carotenoids are best known for the pro-vitamin-A activity of beta-carotene, while lutein, lycopene, and zeaxanthin are better recognized for prevention of certain types of cancer. The anthocyanins, having red, blue, and purple hues, have been linked to such health benefits as anti-inflammatory and anticancer agents, as well as the potential to reduce the risk of coronary heart disease. Health proclamations need continued research despite the fact that carotenoids and anthocyanins are relatively harmful. After all, the consumption of these natural compounds has been practiced since evolution's early stages. The consideration must be made that high doses or concentrated amounts could be toxic. However, if being used as a natural color replacement to synthetic dyes, the quantity consumed would be limited by the desired color of the product at hand.

Under nutrigenomics, the notion of metabolomics is paramount to the target analysis and metabolic profiling needed to understand how the intake of food components affects the human genome. With advances in food and nutritional sciences, research to explore how food metabolites behave within a food matrix or human target can only propel the knowledge of food's interaction with specific disease characteristics. "Nutridynamics" is the next step for these two important scientific fields to merge resources and knowledge to better serve humankind. Some may even refer to it as "foodomics" (Cifuenes 2009).

## REFERENCES

Barrett, D. M., and G. E. Anthon. 2008. Color quality of tomato products. In C. A. Culver, and R. E. Wrolstadt, eds., *Color Quality of Fresh and Processed Foods*, 131–139. ACS Symposium Series 983. Cary, NC: Oxford University Press.

Bidlack, W. R., and W. Wang. 2000. Designing functional foods to enhance health. In W. R. Bidlack, S. T. Omaye, M. S. Meskin, and D. K. W. Tophan, eds., *Phytochemicals As Bioactive Agents*, 241–266. Boca Raton, FL: CRC Press.

Boileau, A. C., and J. W. Erdman. 2004. Impact of food processing on content and bioavailability of carotenoids. In N. I. Krinsky, S. T. Mayne, and H. Sies, eds., *Carotenoids in Health and Disease*, 209–228. New York: Marcel Dekker.

Burri, B. J. 2000. Carotenoids and gene expression. *Nutr* 16(7/8):577–578.

Cantrell, A., D. J. McGarvey, T. G. Truscott, F. Rancan, and F. Bohm. 2003. Singlet oxygen quenching by dietary carotenoids in model membrane environment. *Arch Biochem Biophys* 412:47–54.

Cifuentes, A. 2009. Food analysis and foodomics. Foreword. *J Chrom A* 1216:7109.

Craig, A. D., and L. Zivanovic. 2006. Photochemical modification of carotenoid food colorants. Paper presented at the Color Quality of Processed Food Symposium, 231st American Chemical Society Meeting, Atlanta, Georgia, March 26–30, 2006.

Craig, W. J. 1997. Phytochemicals: Guardians of our health. *J Am Diet Assoc* 97(10):S199–S207.

Day, L., R. B. Seymour, K. F. Pitts, I. Konczak, and L. Lundin. 2009. Incorporation of functional ingredients into foods. *Trends Food Sci Tech* 20:388–395.

Delgado-Vargas, F., A. R. Jimenez, and O. Paredes-Lopez. 2000. Natural pigments: Carotenoids, anthocyanins, and betalaines-characteristics, biosynthesis, processing, and stability. *Crit Rev Food Sci Nutr* 40(30):173–289.

Delgado-Vargas, F., and O. Paredes-Lopez. 2003a. Pigments as food colorants. In *Natural Colorants for Food and Nutraceutical Uses*, 35–62. Boca Raton, FL: CRC Press.

Delgado-Vargas, F., and O. Paredes-Lopez. 2003b. Anthocyanins and betalains. In *Natural Colorants for Food and Nutraceutical Uses*, 167–191. Boca Raton, FL: CRC Press.

deVos, W. M., J. J. M. Castenmiller, R. J. Hamer, and R. J. Brummer. 2006. Nutridynamics-studying the dynamics of food components in products and in the consumer. *Curr Opin Biotech* 17:217–225.

Duke, S. O., A. M. Rimando, F. E. Dayan, C. Canel, D. E. Wedge, M. R. Tellez, K. K. Schrader, L. A. Weston, T. J. Smillie, R. N. Paul, and M. V. Duke. 2000. Strategies for the discovery of bioactive phytochemicals. In W. R. Bidlack, S. T. Omaye, M. S. Meskin, and D. K. W. Tophan, eds., *Phytochemicals as Bioactive Agents*, 1–20. Boca Raton, FL: CRC Press.

Duthie, S. J. 2007. Berry phytochemicals, genomic stability and cancer: Evidence for chemoprotection at several stages in the carcinogenic process. *Mol Nutr Food Res* 51:665–674.

Ehrman, T. M, D. J. Barlow, and P. J. Hylands. 2007. Phytochemical databases of Chinese herbal constituents and bioactive plant compounds with known target specificities. *J Chem Inf Model* 47:254–263.

El-Agamey, A., G. M. Lowe, D. J. McGarvey, A. Mortensen, D. M. Phillip, T. G. Truscott, and A. J. Young. 2004. Carotenoid radical chemistry and antioxidant/pro-oxidant properties. *Arch Biochem Biophys* 430:37–48.

Elliott, R. 2005. Mechanisms of genomic and non-genomic actions of carotenoids. *Biochem Biophys Acta* 1740:147–154.

Etkin, N. L. 2006. *Edible Medicines: An Ethnopharmacology of Food*. Tucson, AZ: University of Arizona Press.

Faulks, R. M., and S. Southon. 2005. Challenges to understanding and measuring carotenoid bioavailability. *Biochem Biophys Acta* 1740:95–100.

Foote, C. S., and R. W. Denny. 1968. Chemistry of singlet oxygen: Quenching β-carotene. *J Am Chem Soc* 90:6233–6235.

Francis, F. J. 1995. Quality as influenced by color. *Food Qual Prefer* 6:149–155.

Garcia-Canas, V., C. Simo, C. Leon, and A. Cifuentes. 2010. Advances in nutrigenomics research: Novel and future analytical approaches to investigate the biological activity of natural compounds and food functions. *J Pharm Bio Anal* 51:290–304.

Garcia-Viguera, C., P. Zafrilla, and F. A. Tomas-Barberan. 1997. Determination of authenticity of fruit jams by HPLC analysis of anthocyanins. *J Sci Food Agri* 73:207–213

Garcia-Viguera, C., P. Zafrilla, F. Artes, F. Romero, P. Abellan, and F. A. Tomas-Barberan. 1998. Colour and anthocyanin stability of red raspberry jam. *J Sci Food Agri* 78:565–573.

Gibney, M. J., M. Walsh, L. Brennan, H. M. Roche, B. German, and B. vanOmmen. 2005. *Am J Clin Nutr* 82(3):497–503.

Gillies, P. J. 2003. Nutrigenomics: the rubicon of molecular nutrition. *J Am Diet Assoc* 103(12):S50–S55.

Giusti, M. M., and R. E. Wrolstadt. 2003. Acylated anthocyanins from edible sources and their application in food systems. *Biochem Eng J* 14:217–225.

Giusti, M. M., and R. E. Wrolstadt. 2008. Color quality of maraschino cherries. In C. A. Culver and R. E. Wrolstadt, eds., *Color Quality of Fresh and Processed Foods*, 43–54. ACS Symposium Series 983. Cary, NC: Oxford University Press.

Griffiths, J. C. 2005. Coloring food and beverages. *Food Tech* 59(5):38–44.

Hinds, T. S., W. L. West, and E. M. Knight. 1997. Carotenoids and retinoids: Review of research, clinical, and public health applications. *J Clin Pharm* 37:551–558.

Hutchings, J. 2002. The perception and sensory assessment of colour. In D. B. MacDougall, ed., *Colour in Food*, 9–32. Boca Raton, FL: CRC Press.

Karlsen, A., L. Retterstol, P. Laake, I. Paur, S. Kjolsrud-Bohn, L. Sandvik, and R. Blomhoff. 2007. Anthocyanins inhibit nuclear factor-kB activation in monocytes and reduce plasma concentrations of pro-inflammatory mediators in healthy adults. *J Nutr* 137(8):1951–1954.

Krinsky, N. I. 1994. The biological properties of carotenoids. *Pure Appl Chem* 66(5):1003–1010.

Lin, C. H., and B. H. Chen. 2005. Stability of carotenoids in tomato juice during storage. *Food Chem* 90:837–846.

Maiani, G., M. J. P. Caston, G. Catasta, E. Toti, I. G. Cambrodon, A. Bysted, F. Granado-Lorencio, B. Olmedilla-Alonso, P. Knuthsen, M. Valoti, V. Bohm, D. Mayer-Miebach, E. Behsnilian, and U. Schlemmer. 2009. Carotenoids: Actual knowledge on food source intakes, stability and bioavailability and their protective role in humans. *Mol Nutr Food Res* 53:S194–S218.

Mayne, S. J. 1996. Beta-carotene, carotenoids, and disease prevention in humans. *FASEB J* 10:690–701.

McGhie, T. K., and M. C. Walton. 2007. The bioavailability and absorption of anthocyanins: Towards a better understanding. *Mol Nutr Food Res* 51:702–713.

Melendez-Martinez, A. J., G. Britton, I. M. Vicario, and F. J. Heredia. 2007. Relationship between the color and the chemical structure of carotenoid pigments. *Food Chem* 101:1145–1150.

Milner, J. A. 2002. Functional foods and health: A U.S. perspective. *Brit J Nutr* 88(2):S151–S158.

Minguez-Mosquera, M. I., M. Jaren-Galen, and J. Garrido-Fernandez. 1993. Effect of processing of paprika on the main carotenes and esterfied xanthophylls present in the fresh fruit. *J Agri Food Chem* 41:2120–2124.

Paiva, S. A. R., and R. M. Russell. 1999. Beta-carotene and other carotenoids as antioxidants. *J Am Coll Nutr* 18(5):426–433.

Parkinson, T. M., and J. P. Brown. 1981 Metabolic fate of food colorants. *Ann Rev Nutr* 1:175–205.

Pennington, J. A. T. 2002. Food composition database for bioactive food components. 15(4):419–434.

Prior, R. L., and X. Wu. 2006. Anthocyanins: Structural characteristics that result in unique metabolic patters and biological activities. *Free Radic Res* 40:1014–1028.

Rodriguez-Amaya, D. B. 1996. Assessment of the provitamin A content in foods—a Brazilian experience. *J Food Comp Anal* 9:196–230.

Rodriguez-Amaya, D. B. 2001. *A Guide to Carotenoid Analysis in Food*. Washington, DC: OMNI Research.

Schieber, A., and C. Reinhold. 2008. Stability of carotenoids in vegetables, fruits, functional foods, and dietary supplements with particular reference to trans-cis-isomeration. In C. A. Culver, and R. E. Wrolstadt, eds., *Color Quality of Fresh and Processed Foods*, 140–150. ACS Symposium Series 983. Cary, NC: Oxford University Press.

Shu, Y. Z. 1998. Recent natural products based drug developments: A pharmaceutical industry perspective. *J Nat Prod* 61:1053–1071.

Stahl, W., and H. Sies. 2005. Bioactivity and protective effects of natural carotenoids. *Biochem Biophys Acta* 1740:101–107.

Stintzing, F. C., and R. Carle. 2004. Functional properties of anthocyanins and betalains in plants, food, and in human nutrition. *Trends Food Sci Tech* 15:19–38.

U.S. Food and Drug Administration. 2010a. Consumer update: How safe are color additives? http://www.fda .gov/ForConsumers/ConsumerUpdates/ucm048951.htm (accessed March 10, 2010).

U.S. Food and Drug Administration. 2010b. Everything added to food in the United States. http://www .accessdata.fda.gov/scripts/fcn/fcnNavigation.cfm?rpt=eafusListing (accessed March 3, 2010).

Van den Berg, H., R. Faulks, H. F. Granado, J. Hirschberg, B. Olmedilla, G. Sandman, S. Southon, and W. Stahl. 2000. The potential for the improvement of carotenoid levels in foods and the likely systemic effects. *J Sci Food Agric* 80:880–912.

Wai-Nang, P. L., and V. L. W. Go. 2005. Nutrient-gene interaction: Tracer-based metabolomics. *J Nutr* 135(12): 3027S–3032S.

Wang, L.-S., and G. D. Stoner. 2008. Anthocyanins and their role in cancer prevention. *Canc Let* 269:281–290.

Williams, M., E. Pehu, and C. Ragasa. 2006. Functional foods: Opportunities and challenges for developing countries. *Agri Rural Dev Notes* 19(9).

Wishart, D. S. 2008. Metabolomics: Applications to food science and nutrition research. *Trends Food Sci Tech* 19:482–493.

Wrolstadt, R. E., T. Ngo, C. E. Finn, and Y. Zhao. 2008. Color quality of fresh and processed strawberries. In C. A. Culver, and R. E. Wrolstadt, eds., *Color Quality of Fresh and Processed Foods*, 18–42. ACS Symposium Series 983. Cary, NC: Oxford University Press.

# 20 Manufacturing Functional Foods

## *Effects on Quality and Bioavailability*

*Ryan J. Elias and John D. Floros*

## CONTENTS

## INTRODUCTION

In the past, the main purpose of food preservation was to achieve safety and extend the shelf life of food products. However, more recently, attention has been given to maintaining the fresh-like characteristics of certain foods, retaining important organoleptic attributes such as color, texture, taste, and appearance, and improving the nutritional quality and functional properties of foods. With the recent introduction of functional foods, nutraceuticals, and bioactives in the marketplace, the retention of nutrients such as vitamins, the bioavailability of functional food components such as antioxidants, and the preservation of other beneficial food compounds have become increasingly important.

To obtain high retention of heat-sensitive nutrients and functional components, reduced thermal processes or nonthermal processes have been developed. Terms such as "minimally processed" or "fresh-like" products have become commonplace in the food industry (Mertens and Knorr 1992). Manvell (1997) defines minimal processing as the least possible treatment to achieve a purpose. Under this definition, minimal processing does not specify the quality to be retained. Therefore, a more appropriate term that covers both safety and quality could be "optimal processing." Optimal processing would thus be defined as the process needed to obtain safe products with the highest possible quality (Dock and Floros 1999).

Historically, the justification for processing foods has been to reduce or eliminate microbial spoilage organisms and pathogens, to prolong physical and chemical stability, and to achieve certain organoleptic properties deemed beneficial by consumers. Both thermal and nonthermal processing methods have been used to carry out these objectives, and in many cases these methods have resulted in the loss of key microcomponents. Indeed, many consumers still assume that "processing" can only result in foods of diminished nutritive value. While this is true in some cases (e.g., vitamin losses during heating), many processing operations actually improve the nutritional (e.g., enzymatic hydrolysis of proteins during fermentations) and biological activity (e.g., improved bioavailability of lycopene during thermal processing) of food components. Furthermore, research carried out over the past decade has demonstrated that improved functionality and biological activity can be engineered into processed foods at the nano-, micro-, and macro-levels. Such an approach is based on the implementation of structural design principles and delivery system assembly to create functional foods, rather than on utilization of processing operations to create these structures (McClements et al. 2009). By understanding the physicochemical environments of foods, and by implementing the principles of material science, it is possible for the food scientist to engineer specific design features into foods and food ingredients for the purpose of stabilizing bioactives or even augmenting their functionality.

The chemical nature of the functional component, as well as its intended food system and desired performance attributes, determines the most appropriate design strategy. For example, lipophilic bioactive compounds destined for an aqueous food require emulsification, whereas a bioactive peptide intended for targeted delivery in the human intestines requires encapsulation. The objectives of this chapter are to first present a short commentary on the history of food processing, then briefly review appropriate processing methods for functional foods, and finally discuss the major types of bioactive components and some novel bioactive delivery systems.

## PROCESSING AND MANUFACTURING FOR FUNCTIONAL FOODS

According to Richard Wrangham, a Harvard University biological anthropologist, humanity was launched about 2 million years ago by a distant ancestor that discovered cooking, the original form of food processing (Wrangham 2009). Later, but still during prehistoric times, cooking was augmented by slaughtering, fermenting, drying, preserving with salt, and other primitive forms of food processing (Table 20.1), which allowed groups and communities to form and survive. Thus, humans first learned how to cook food, then how to transform, preserve, and store it safely. This experience-based technology led to food processing (Hall 1989; Floros 2008). Much later, the domestication of plants became widespread, and at the end of the last ice age, humans revolutionized eating meat by domesticating animals for food. As a result, land cultivation and plant and animal agriculture contributed to improving the human condition.

The study of every ancient civilization clearly shows that throughout history humans overcame hunger and disease, not only by harvesting food from a cultivated land, but also by processing it with sophisticated methods. For example, the three most important foods in Ancient Greece—bread, olive oil, and wine—were all products of complicated and sophisticated processing that transformed perishable, unpalatable, or hardly edible raw materials into safe, tasty, nutritious, stable, and enjoyable foods (Floros 2004).

As time passed, an explosion in the human population took place about 200 years ago. This population explosion necessitated rapid progress in agricultural sciences, which brought about new and more efficient ways to cultivate the land and raise animals. It also necessitated progress in food science and technology that resulted in more sophisticated methodologies to process and preserve food (Table 20.1). Although we continued to use most of the traditional methods of processing food—fermenting, pickling, drying, salting, smoking, and so on—we developed many new technologies and more sophisticated processes such as pasteurization, canning, aseptic processing and packaging, modified and controlled packaging, and many others (Tables 20.1 and 20.2). Once again, food processing improved the safety, nutrition, convenience, affordability, and availability of foods. It changed the perception of foods and beverages, helped society overcome hunger and disease, and made food the basis of a healthy civilization.

Today, food processing is a sophisticated form of handling, cooking, transforming, preserving, packaging, and finally, delivering foods to the consumer. Food processing is the basis for the food industry and has contributed immensely to the development of a healthy human civilization and helped society prosper and flourish (Lund 1989). In developed countries, modern food processing methods and applications of food science and technology are responsible for practically eliminating food losses, while helping to extend shelf life, improve quality, increase seasonal availability of many foods, preserve sensitive nutrients, add beneficial ingredients such as vitamins, remove toxins, increase food safety, and facilitate distribution (Floros 2008). On the other hand, as much

**TABLE 20.1**
**Early Interventions versus Modern Processes**

| Early Interventions | Modern Processes |
|---|---|
| • Fermentation and pickling | • Canning |
|   • Wine, yogurt, vegetables | |
| • Drying | • Pasteurization |
|   • Meat, fruits | |
| • Salting | • Refrigeration/freezing |
|   • Meat | |
| • Smoking | • Freeze drying |
|   • Meat | |
| | • Packaging and storage conditions |
| | • Aseptic processing |
| | • Hazard analysis |
| | • Critical control point |

**TABLE 20.2**
**New Processing and Packaging Technologies**

| | |
|---|---|
| • Aseptic processing and packaging | • MAP |
| • Microwave heating | • Active packaging |
| • Ohmic heating | • Intelligent packaging |
| • Irradiation | • Edible coatings |
| • High pressure | • Surface modification |
| • Pulsed electric fields | • Sensors/indicators |
| • Oscillating magnetic fields | • Nanocomposites |
| • Oscillating magnetic fields | • $O_3$/ClO |
| • Ultrasonication | • Bio-based materials |

as half of the food grown and harvested in underdeveloped and developing countries is never consumed, partly because proper handling, processing, packaging, and distribution methods are lacking (Floros 2008).

Both conventional and nonconventional, or novel, processing operations are discussed in the next sections. Attention is given to the effect of these operations on the stability and fate of nutrients and bioactive compounds.

## CONVENTIONAL THERMAL PROCESSING

Conventional thermal processes, such as cooking, blanching, pasteurization, and sterilization have different objectives. Cooking aims at producing a more palatable food, blanching is primarily aimed at destroying enzymes, pasteurization is aimed at destroying some, but not all vegetative cells, while sterilization is the destruction of all microorganisms (Lund 1975). Sterilization used for food processing prevents growth of microorganisms at some expected storage conditions, and thus the term "commercial sterility" should be used (Lund 1975). For example, vegetables meant for frozen storage must first be blanched, while conventionally pasteurized milk, yogurt, or other functional foods containing prebiotics and probiotics must be refrigerated. Thermal treatments needed to produce commercially sterile and shelf-stable foods, such as infant formulas and some weight-control products, depend on product characteristics (e.g., pH, viscosity, and water activity), type, heat resistance and the initial load of target microorganism(s), and the heat transfer characteristics of the food.

The amount of heat needed for commercial sterilization depends on the organisms that survive and grow in the product. Most sterile containers are hermetically sealed and contain a low oxygen concentration that inhibits the growth of aerobic microorganisms. Hence, the microorganisms of concern are the anaerobes or facultative anaerobes. Spore forming organisms are of particular concern, because spores are more heat resistant than vegetative cells. Consequently, anaerobic spore forming pathogens, which produce lethal toxins, set the constraints needed for heat sterilization, and in many cases, the organism of concern is *Clostridium botulinum*. Products with pH higher than 4.6 (low acid) must receive a treatment resulting in a 12-log reduction of *C. botulinum*. At pH 4.6 or lower, *C. botulinum* does not grow, and for these products microbial destruction is based on other target microorganisms. Many other organisms are more heat resistant than *C. botulinum* and are used to calculate the extent of thermal treatments (Lund 1975).

Thermal destruction of microorganisms, most nutrients, proteins and enzymes, and other quality factors such as bioactive components, flavors, colors, and texture obey first-order reaction kinetics (Lund 1975). This means that the reaction rate depends on the concentration of the substance of interest. Two parameters are important in thermal destruction kinetics: the destruction rate at a given temperature, and the dependence of the destruction rate to temperature. Both parameters form the basis for thermal process calculations and are the most common methods used in industry to design manufacturing processes (Lund 1975).

In certain cases, thermal processing results in positive effects for functional ingredients and bioactive components. For example, lycopene, a carotenoid found primarily in tomatoes and suspected to lower the risk of several types of cancer, becomes more bioavailable when tomatoes are thermally processed. Lycopene absorption increases 2.5 times in tomato paste as compared to the absorption from raw tomatoes (Broihier 1999). Similarly, heating blueberries by a mild steam treatment increases the levels of available antioxidants such as anthocyanins (Zind 1999). Lastly, phytochemicals present in rice were more active in rice bran oil rather than in rice bran, inferring that processing improves the health-promoting action of functional ingredients such as phytochemicals by probably increasing their bioavailability (Zind 1999).

Food proteins are denatured by mild heat (60–90°C) resulting in insolubility and decreased functional properties. However, protein denaturation also improves digestibility and biological availability of essential amino acids. For example, several raw plant proteins exhibit poor digestibility, but moderate heating improves their digestibility without toxin production (Damodaran 1996). In other cases, plant proteins contain antinutritional factors. For example, trypsin and chymotrypsin inhibitors are present in many plant proteins and impair efficient digestion. Similarly, lectins agglutinate red blood cells, impair protein digestion, and cause malabsorption of other nutrients (Damodaran 1996). Fortunately, both protease inhibitors and lectins are thermo-labile. Moderate heat also inactivates proteinaceous toxins such as the *Clostridium botulinum* toxin, which is deactivated at 100°C (Damodaran 1996).

Similarly, most oxidative and hydrolytic enzymes (i.e. proteases, lipases, lipoxygenases, amylases, and polyphenoloxidases) are inactivated by moderate heat treatments. The treatment, termed blanching, consists of heating the product to approximately 70–105°C and holding it there for a few minutes. However, some heat resistant isozymes exist, which may cause off flavors and colors (Lund 1975).

## Aseptic Processing

Aseptic products are free of pathogenic and spoilage bacteria and packaged under aseptic conditions into a pre-sterilized container (Holdsworth 1992). The aseptic process is different from conventional retorting in that the product is continuously heated and then continuously cooled before packaging. This results in several advantages including better nutritional and organoleptic quality retention, and utilization of flexible, plastic/paper laminated containers, because the product is cooled prior to packaging (Floros 1993). Another advantage of aseptic processing technology is its ability to package products in large aseptic bulk storage tanks. Containers as large as one million gallons (about four million liters) are used to store and ship aseptic products around the world.

An important part of aseptic processing is the use of high-temperature-short-time (HTST) processing. One of the reasons that aseptic processing is effective in retaining quality can be explained by the activation energy of important food ingredients. For example, oxidation and enzyme catalyzed reactions have lower activation energies ($E_a$=2–15 kcal/mole) as compared to other quality degradation reactions involving color, flavor, texture, and nutrients ($E_a$=15–30 kcal/mole), while microbial and some enzymatic destruction reactions have the highest $E_a$ at 50–100 kcal/mole (Schwartz 1992). The activation energy reflects the change in the rate of the reaction with temperature. High $E_a$ indicates that reaction rates increase rapidly with temperature, while processes with low $E_a$ are minimally influenced by temperature. Therefore, high $E_a$ reactions are favored during HTST processes, and thus microbial destruction proceeds rapidly, while degradation of nutrients, bioactives, and other quality factors is minimized (Schwartz 1992).

Enzyme renaturation or reversible denaturation sometimes occurs in aseptically processed foods. In some aseptic products, enzyme activity was undetectable immediately after processing, but slowly enzymes regained their activity throughout storage. For example, peroxidase activity was found to regenerate during storage of aseptically processed peas in starch (Schwartz 1992). Enzymes encapsulated in particulates, such as peas, may be somewhat protected within the tissue, resulting in the need for additional heating or pre-processing in order to properly inactivate them. An enzymatic problem associated with aseptically processed orange juice not-from-concentrate is settling and separation of pectin caused by pectinesterase. Therefore, aseptically processed orange juice is distributed and stored under refrigerated temperature. Amylase also exhibits high thermal resistance, and thinning of starch based puddings is caused by bacterial α-amylase. Similarly, lipase and protease activities in UHT milk have been associated with growth of pcychrophilic bacteria in raw milk prior to processing (Anderson et al. 1983). Therefore, it is important to assure the quality of raw materials, because although bacteria are killed during processing, the enzymes they produced may not be completely inactivated. In an aseptically processed enriched soymilk product marketed by Vitasoy USA, Inc. (San Francisco, CA) unwanted beany flavor was minimized by using a UHT process. The process deactivated the enzymes, but it did not affect other important nutrients added to soymilk such as calcium, riboflavin, and vitamins A, D, and B$_{12}$ (Ohr 1999).

## OHMIC HEATING

Ohmic heating is also called resistance or electro heating (Reznick 1996). The process consists of passing an electric current through the product to be treated. The product acts like a resistor and heats up. The concept of this technology is not new and several patents from the nineteenth century have described the use of electrical energy for heating flowable materials (Sastry and Palaniappan 1992). More recently, ohmic heating gained interest for processing particulate foods, especially in combination with aseptic processing. Examples include potato soups with chunks of potato, vegetable soups with whole peas and other sizeable vegetable pieces, and other liquid food products containing solid pieces of size up to 1–2 cm. Ohmic heating is capable of fast, uniform heating, and can heat particulates as fast or faster than the surrounding liquid media (Sastry and Palaniappan 1992). It is this rapid and uniform heating of food throughout that makes ohmic heating different than traditional thermal processing, which depends on heat transfer from liquid to particulate surface and eventually into the particulate's center (Kim et al. 1996). Furthermore, ohmic heating offers better uniformity of heating than microwave heating (Sastry and Palaniappan 1992).

In relation to bioactive components, nutraceuticals, and functional foods, ohmic heating presents a good alternative to thermal heating, because of its ability to heat particles as fast as or faster than liquid. Vitamins, proteins, and other functional ingredients that are sensitive to heat can be processed by ohmic heating in a more tender fashion, providing better retention of heat sensitive ingredients as compared to processing by traditional retorting.

## MICROWAVE HEATING

Microwave energy has been used for heating food since the first microwave ovens were developed in 1947. Consumer microwave ovens were introduced by Tappan in 1955, and today they are present in more than 90% of American homes (Schiffmann 1992). Microwaves kill microorganisms by generating heat. However, a shorter time, up to fivefold less (Harlfinger 1992), is needed to heat food by microwaves as compared to conventional heating. Therefore, microwaves yield foods with better texture, taste, and appearance (Harlfinger 1992).

Microwaves carry far too little energy to break any chemical bonds, and are thus a form of non-ionizing radiation (Heddelson and Doores 1994). Microwaves consist of a magnetic and an electrical field oriented perpendicular to each other. The electrical field is the primary cause of heating by promoting rotation of polar (dielectric) molecules, and resulting in heat caused by molecular friction (Heddelson and Doores 1994).

In a study of microwave heating of breast milk, the effect on vitamins $B_2$ (thiamin) and E (α-tocopherol), polyunsaturated fatty acids (linoleic and linolenic), and specific immunoglobulins was evaluated (Lassen and Ovesen 1995). The study showed that vitamins and fatty acids were stable over the temperature range tested (37–77°C), while immunoglobulins were stable up to 57°C, but completely destroyed at 77°C. Loss of vitamin C during microwave and conventional heating for various vegetables was studied (Schnepf and Driskell 1994). Steaming of vegetables in a microwave showed the lowest loss of vitamin C, which depended on the amount of water used. Similarly, a study on inactivation of pectin methyl esterase in orange juice by microwave and conventional heating showed that inactivation during microwave heating was significantly faster, indicating that some non-thermal effects may be present during microwave heating (Tajchakavit and Ramaswamy 1996). A study of α-tocopherol stability in virgin olive oil cooked in a frying pan or in a microwave oven for 8 minutes, showed that the retention of α-tocopherol was 38% and 51%, respectively (Ruiz-Lopêz et al. 1995).

## IRRADIATION

Food irradiation is one of the most effective nonthermal methods for microbial destruction while retaining food quality. However, the public has not accepted irradiation as a safe technology. In the early 1990s the U.S. regulatory agencies classified irradiation as an "additive" (Pauli 1991). Thus, scientists had to prove that irradiation did not "add" or produce any dangerous substances in the food. This obstacle was unique to irradiation, unlike other new processing techniques that were never considered to be or produce "additives." All irradiated foods must be labeled with a radura (the international symbol of irradiation), and the words "treated by irradiation" or "treated with radiation" (ADA 1996).

Radiation destroys microorganisms and other contaminants such as insects and parasites by partial or total inactivation of their genetic material, or in other words, by directly attacking DNA or by producing radicals and ions that attack DNA (WHO 1994). The breaking of chemical bonds by irradiation, called radiolysis, produces unstable reactive products which are subsequently converted to stable final products (WHO 1994). However, these radiolytic products have been shown to be virtually identical to thermolytic products produced by conventional heating (WHO 1994).

Irradiation can destroy microorganisms at levels as low as 1–3 kGy. The resistance of microorganisms to irradiation is influenced by several factors, including the pH, temperature, and chemical composition of the food. Because irradiation affects the cell's DNA, the cell's ability to repair this damage is important. For example, the vegetative cells of some bacteria have a resistance to irradiation similar to that of spores, because they are able to repair large numbers of breaks in their DNA (Moseley 1984). However, spores are generally more resistant than vegetative cells of bacteria, yeasts, and molds, while viruses are even more resistant (Monk et al. 1995), and toxins are not affected (WHO 1994).

All nonthermal preservation techniques have the potential of retaining superior quality and nutritional value of food as compared to conventional methods, like retorting. Studies on the effect of irradiation on the three major food components, carbohydrates, lipids, and proteins, showed that doses as high as 50 kGy induce only very small changes (WHO 1994). Carbohydrates react with hydroxyl radicals to form ketones, aldehydes, or acids as end products, leading to a drop in pH (ADA 1996). Starch degrades to dextrins, maltose, and glucose, resulting in a decrease in the viscosity of polysaccharides in solution (ADA 1996). Glucose alone produces at least 34 radiolytic products (von Sonntag 1980). In general, the nature and number of products formed depends on the amount of water present (WHO 1994). Furthermore, pure carbohydrates are more susceptible to irradiation effects than when they are present in a food (WHO 1994).

Irradiation of fruits and vegetables often causes tissue softening due to breakdown of pectin (Kilcast 1994). The effect of irradiation on lipids is production of cation radicals and molecules in an excited state, which leads to deprotonation, followed by dimerization or disproportionation (WHO 1994). Excited triglycerides may undergo several reactions producing many different products, such as fatty acids, propanediol esters, aldehydes, ketones, diglycerides, diesters, alkanes, methyl esters, and shorter-chain triglycerides (WHO 1994). However, production of potential carcinogenic substances, such as aromatic or heterocyclic rings, has not been proven (WHO 1994).

Some antioxidative effects of protein-carbohydrate interaction products increase with irradiation dose, and may mitigate oxidative changes in lipids (WHO 1994). Usually, irradiation of foods sensitive to oxidation is done under anaerobic conditions, because in the presence of oxygen auto-oxidation may be rapid (WHO 1994). Proteins present in a food matrix are not greatly accessible, and are therefore less susceptible to attack. The radicals formed are mostly immobile, and prone to recombination rather than to reaction with other components in the food matrix (WHO 1994). Splitting of carbon-nitrogen bonds and disulfide bridges may occur, degrading proteins into smaller polypeptides. On the other hand, globular proteins may aggregate, producing solutions with higher viscosity (WHO 1994). Enzymes are not affected by irradiation to any great extent. Therefore, enzyme activity must be prevented by other means. Similarly, irradiation of amino acids in proteins with up to 50 kGy showed limited changes in amino acid composition (WHO 1994). Although, some off flavors can be formed from breakdown products of sulfur containing amino acids (Kilcast 1994).

Most vitamins are not affected severely by irradiation up to 10 kGy, which is the limit for food irradiation. In general, the most thermo-liable vitamins are also the ones most sensitive to irradiation (ADA 1996). Vitamin C and thiamin (vitamin $B_1$) are the most sensitive of the water-soluble vitamins, and their losses usually increase with increased irradiation dose and temperature (Kilcast 1994). Nevertheless, thiamin is more sensitive to heat than to irradiation (Josephson et al. 1978), while a study on oranges showed no significant loss of vitamin C at irradiation levels up to 1 kGy through 6 weeks of storage (Nagai and Moy 1985). From all fat-soluble vitamins, α-tocopherol (vitamin E) is the most sensitive. However, foods containing significant amounts of vitamin E (i.e. oils and dairy products) are not suitable for irradiation due to off-flavor production.

In conclusion, irradiation processes have been used for many decades now, and have been refined to employ appropriate dose-time exposures that result in the formation of far less secondary products. Although irradiation causes some changes in foods, in general, they are small and without significant health effects. The safety of irradiated foods has been studied for longer than 40 years in multispecies animal studies that showed no toxic effects (Thayer 1994). In a human study, volunteers were fed 100% irradiated foods and no ill effects were observed (Diehl 1995).

## High Hydrostatic Pressure

The basis for applying high pressure to foods is to compress the medium surrounding the food, usually water. The effects of high pressure on microorganisms are multiple, including biochemical, genetic, and morphological changes, as well as changes in cell wall and membrane (Hoover et al. 1989). At high pressure, microbial death is probably caused by permeabilization of the cell membrane (Morita 1975). However, it is also believed that pressure affects a combination of processes and does not inhibit or destroy only one specific function (Hoover et al. 1989). Pressure acts in two ways on the cell biosystem: the available molecular space is decreased and interchain reactions are increased (Hoover et al. 1989). Reactions that produce a volume increase tend to be inhibited, while the opposite is true for volume decreasing reactions (Farr 1990).

Variables that influence the effect of pressure on microorganisms include the magnitude and duration of compression, the stage of microbial growth, the composition of the growth medium, and environmental parameters such as temperature, pH, and water activity ($a_w$) during processing (Hoover et al. 1989; Knorr 1993). Normally, an increase in duration and magnitude of the pressure treatment increases vegetative cell death (Hoover et al. 1989). Cells from the log phase were more sensitive to pressure as compared to stationary, dormant, or death-phase cells (ZoBell 1970). Bacterial spores are resistant to several treatments, such as temperature, homogenization, irradiation, and pressure (Lechowich 1993). Bacterial spores may survive pressures higher than 1200 MPa in part due to the thickness and structure of their coat (Larson et al. 1918).

High-pressure treatments may cause either reversible or irreversible changes to different enzymes and may result in complete or partial inactivation or activation. The effect depends on the level and duration of the treatment, the chemical nature of the enzyme, and the substrate, temperature, and other environmental parameters (Cheftel 1992). Enzymatic browning is usually accelerated in pressure treated foods. For example, enzymatic activity remained in jams treated with 600 MPa, and refrigerated storage was required to reduce color and flavor deterioration (Horie et al. 1991). On the contrary, high-pressure treatment of meat showed that proteolytic activity was modified to improve meat quality (Ohmori et al. 1991).

The effect of pressure treatment on polyunsaturated fatty acids was studied in fish products. Extracted sardine oil treated with 600 MPa for 60 minutes did not show any changes in peroxide values (POV) or thiobarbituric acid values (Ohshima et al. 1993). However, in cod muscle POV increased with increasing pressure and process time (Ohshima et al. 1992). This indicated that extracted oils may be more resistant to oxidation than oils existing in the muscle during high-pressure treatments (Ohshima et al. 1993).

High pressure can be used to extend the shelf life of foods by destroying microorganisms and inactivating some undesirable enzymes, while retaining desirable quality characteristics. Several applications of high pressure are already used in the food industry. Examples include application to prolong the shelf life of milk, guacamole, and fruit juices, and to reduce microbial loads of meats, eggs, fish, seafood, pickles, cheese, and spices (Pothakamury et al. 1995).

## HIGH-INTENSITY PULSED LIGHT

High-intensity pulsed light (HIPL) is a novel technique for sterilization, and little research has been done to confirm its effectiveness. A few reports show significant reduction of both microorganisms and spores (Dunn and Clark 1995). High-intensity light inactivates microorganisms through a combination of photochemical and photothermal reactions, whereas ultraviolet (UV) light commonly used to sterilize packaging material acts primarily in a photochemical way (Manvell 1997). High-intensity light is a simple way to quickly transfer large amounts of energy onto the surface of a material in order to heat up a very thin layer of that material (Mertens and Knorr 1992).

HIPL treatments of food with 30 J/cm$^2$ had a very severe effect on microorganisms, but they did not show any difference in nutrient quality when compared to untreated controls (Dunn et al. 1995). Some of the components measured were protein, riboflavin, nitrosamine, benzpyrene, and vitamin C. However, it is well known that many nutrients, especially vitamins, unsaturated fatty acids, certain amino acids, and pigments, are sensitive to light (Gravani 1985).

## HIGH-INTENSITY PULSED ELECTRIC FIELDS

High-intensity pulsed electric field (PEF) pasteurization involves application of a short burst of high voltage to foods placed between two electrodes (Qin et al. 1996). When a critical electric field strength is applied across a cell or microorganism, pores are formed in its membrane. These pores may be reversible or irreversible depending on field strength, number, and duration of pulses (Vega-Mercado et al. 1997.

The effect of PEF on vegetative microbial cells has been proven in several studies (Zhang et al. 1995; Grahl and Märkl 1996; Martín et al. 1997. The effect of PEF on spores has also been addressed in a few studies (Grahl and Märkl 1996), which concluded that it is impossible to sterilize spore-containing media with PEF.

Enzymes are stabilized by weak noncovalent forces, such as hydrogen bonds and hydrophobic interactions, and they must maintain their three-dimensional structure to remain active. The effect of PEF on enzymes varies. Some enzymes are activated by PEF, while others are partly or completely inactivated (Ho et al. 1997; Vega-Mercado et al. 1997, 1995).

## ULTRASONICS

Most applications for ultrasonics (sonication) are nonmicrobial. To mention a few, ultrasound has been utilized to enhance drying and dewatering (Fairbanks 1974; Ensminger 1988), accelerate extraction processes (Kim and Zayas 1989), assist heat transfer (Lima and Sastry 1990), and facilitate diffusion through membranes and biomaterials (Floros and Liang 1994). Ultrasonic equipment has been developed for cleaning, emulsification, and cell disruption (Gunasekaran and Chyung 1994). Development of ultrasonics for microbial destruction has been studied with some success.

Ultrasound can cause disruption of biological structures (Manvell 1997), and an array of other acoustic effects. For example, ultrasound has been shown to cause cavitation, which is the formation, growth, and collapse of small voids in liquids; localized heating, especially at membrane interfaces; and free radical formation (Floros and Liang 1994). Cavitation is recognized as the mechanical effect responsible for the destruction of bacterial cells (Shukla 1992). The effect of ultrasound is amplified when used in combination with prolonged treatment time, heat, pH, or chlorine (Earnshaw et al. 1995; Lillard 1993; Wrigley and Llorca 1992).

## OSCILLATING MAGNETIC FIELDS

Magnetic fields have long been known to influence living organisms. For example, magnetotactic bacteria are capable of synthesizing magnetosomes composed of magnetite, which help the bacteria migrate in certain directions following the Earth's magnetic field (Pothakamury et al. 1993). Growth of bacteria may also be inhibited if subjected to a certain magnetic field. The use of oscillating magnetic fields (OMF) as a pasteurization technique is a new and rather unexplored area. Certain magnetic fields may either stimulate or inhibit microbial growth, but only reductions of about two-log cycles have been noted (Pothakamury et al. 1993). Furthermore, reports on the effect of OMF on spores, important microbial pathogens, and nutritional quality are not available.

## OZONE

Ozone ($O_3$) is a naturally occurring gas with strong oxidizing proporties. It is readily soluble in water, and its solubility increases with increasing temperature. $O_3$ may be produced by several chemical methods, but commercially $O_3$ is produced by passing dry air or dry oxygen through an electric arc (EPRI 1997). During autodecomposition, $O_3$ produces numerous free radical species, among which the hydroxyl free radical ($OH^-$) is the most prominent (Graham 1997). The main reaction byproduct of $O_3$ decomposition is oxygen; thus, no harmful products are produced. The rate of decomposition of $O_3$ increases with temperature and pH (Graham 1997). The half-life of gaseous $O_3$ is about 12 hours, while in water its half-life is a matter of seconds and depends on the amount of $O_3$ demanding material present (Graham 1997). Chemically, $O_3$ is a powerful oxidizer that reacts up to 300 times faster than chlorine with organic material.

Ozone has been used as a disinfectant of potable water in Europe for almost a century, while the first potable treatment with $O_3$ in the United States did not occur until 1940 (Graham 1997). Several studies (Liyange et al. 1997; Restaino et al. 1995; Finch et al. 1993; Rickloff et al. 1987) investigated the effect of aqueous $O_3$ on bacteria and spores. Ozone destroys bacteria by rupturing their cellular membrane (EPRI 1997) and dispersing their cytoplasm in the water. The effects on microbial destruction from gaseous ozone have also been investigated (Liew and Prange 1994; Naitoh 1992; Isizaki et al. 1986). Ozone is generally recognized as safe (GRAS) for treatment of bottled water (FDA 1995), and it was declared GRAS by an expert panel for use in food processing. This may allow U.S. processors to use $O_3$ in processes such as fish processing, control of fruit ripening, minimal processing of fruits and vegetables, sanitation of bottles, and disinfection of poultry, beef, and cereal products.

## CHLORINE DIOXIDE

In the past, chlorine compounds have been used extensively in the food industry as sanitizing agents on equipment and surfaces due to their strong antimicrobial effects. More recently, chlorine dioxide ($ClO_2$), both as gas and as liquid, has been recognized as an effective antimicrobial substitute for chlorine. Compared to chlorine, $ClO_2$ has 2.5 times the oxidation capacity (Benarde 1965), and is five times as soluble in water (White 1972). It also maintains its bactericidal activity longer, because it is not affected by alkaline pH or high concentration of organic matter (White 1972). Furthermore, unlike chlorine, $ClO_2$ does not react with ammonia or nitrogenous compounds, thus it does not form chloramine compounds which may be carcinogenic (Aieta et al. 1984; Alliger 1980). Destruction of bacteria in poultry processing chill water showed that $ClO_2$ was seven times more effective as compared to chlorine (Lillard 1979).

The antimicrobial effect of $ClO_2$ has been demonstrated against several bacteria including *Escherichia coli*, *Listeria monocytogenes*, *Salmonella* sp., and spores of *Bacillus cereus* and *Clostridium perfringens* (Lin et al. 1996; Villarreal et al. 1990; Foegeding et al. 1986; Thiessen et al. 1984), fungi (Roberts and Reymond 1994), and viruses (Chen and Vaughn 1990). Applications of $ClO_2$ in the food industry have been presented in several studies, but no readily available information exists on its effects on nutritional quality.

## OTHER CHEMICAL TREATMENTS

Early attempts to preserve food consisted of treatments of heat, cold, drying, and fermentation. Chemical preservation methods were limited to salting of fish and meat, and pickling of fruit and vegetables, although the mechanism of preservation was poorly understood. It is now known that manipulation of the chemical characteristics of food, such as pH, water activity, and redox potential, prevents growth of spoilage organisms, and/or inhibits activity of harmful enzymes.

Over time many organic and inorganic chemicals have been examined for their antimicrobial and antienzymatic properties. The most important of these chemicals were discussed by Dock and Floros (1999): benzoates, sorbates, organic acids, trisodium phosphate, and bacteriocins.

# EFFECT OF PROCESSING ON BIOACTIVE COMPONENTS

## VITAMINS

Vitamin losses are not caused by processing alone. Instead, loss of vitamins starts from the time of harvest for most foods (Gregory 1996). For example, if a food is a major source of a specific vitamin, or a claim is made, then the food processor must maintain and prove the specific vitamin's bioavailability and concentration. Inactivation of enzymes often has a stabilizing effect on vitamins during storage. Heat is not the primary factor of vitamin loss. Instead, losses occur primarily by oxidation and leaching (aqueous extraction) (Gregory 1996). The extent of thermally induced vitamin loss depends on the chemical nature of the food (e.g., pH, $a_w$, dissolved $O_2$) and the likelihood of leaching (Gregory 1996). HTST processing is very effective in increasing retention of heat-labile vitamins and other nutrients. The reason is that the z-values for nutrients and quality factors (e.g., color, flavor) generally are larger than those for microorganisms and heat-labile enzymes (Gregory 1996). Therefore, HTST has a greater effect on microorganisms than on most nutrients.

## MINERALS

Most research shows that heat, pH, oxidation, and other factors affecting organic nutrients do not affect minerals. These findings are usually related to ash values, which are largely unaffected by heating (Watzke 1998). The only way to significantly decrease the mineral content of a food is by removing/loosing it through leaching, milling, or other physical means (Miller 1996). The most extensive loss of minerals occurs during milling. For example, during milling of wheat a significant amount of minerals is lost—up to 86% Mn, 76% Fe, 78% Zn, and 68% Cu (Rotruck 1982). Loss of calcium during draining of whey in low pH cheeses has also been shown (Miller 1996). However, instead of measuring ash values, the bioavailability of minerals was examined before and after processing, and some significant differences were found. Some minerals may become unavailable or less available for digestion due to their interaction with conutrients or nonfood components during processing. On the contrary, increased mineral bioavailability may also occur. For example, thermal destruction of binding ligands, such as phytates, increases mineral bioavailability (Watzke 1998). Fortification after processing usually includes iron only (Miller 1996), but some cereal products are now fortified with several minerals and vitamins.

## NOVEL DELIVERY SYSTEMS OF BIOACTIVE LIPIDS

A large number of bioactive food components are lipids. Some examples include omega-3 fatty acids, conjugated linoleic acid (CLA), phytosterols, carotenoids, and lipophilic antioxidants (e.g., polyphenols). Operationally speaking, these compounds, like all lipids, are soluble in organic solvent and insoluble or only sparingly soluble in water. This property presents both challenges and opportunities to the food scientist interested in incorporating bioactive lipids within heterogeneous foods.

### Conventional Emulsions

Most dietary lipids are encountered and consumed as dispersions in foods, which is to say that they are seldom consumed are bulk fats and oils (McClements 2005). Food lipids are therefore often incorporated within processed foods as emulsions—thermodynamically unstable dispersions of nano- or microscale lipid droplets within an aqueous continuum made kinetically stable through the adsorption of surface active compounds (i.e., emulsifiers) to the lipid-water interface. A consequence of emulsification is a substantial increase in the surface area of the lipid in terms of exposure to the aqueous phase, which often contains pro-oxidants (e.g., transition metal catalysts) and water-soluble radical initiators. This can lead to the acceleration of deleterious oxidation reactions resulting in the chemical degradation of the dispersed bioactive lipid. However, many researchers have demonstrated that emulsion droplet interface can be manipulated or designed in a way that protects lipids from deterioration.

It has been demonstrated that the electrostatic properties of emulsion droplets can be modified in order to repel pro-oxidative metals for the purpose of inhibiting lipid oxidation (McClements and Decker 2000). For example, the oxidation of an omega-3-enriched oil-in-water (O/W) emulsion can be retarded by conferring cationic properties to emulsion droplets (Donnelly et al. 1998). This was done by lowering the pH of the aqueous phase below that of the emulsifying protein's isoelectric point (pI), thus establishing an area of positive charge on the droplet interface. In this system, cationic transition metals are electrostatically repelled from the oxidatively labile lipid, and a net reduction in oxidation is observed. The opposite trend was observed at pH value above the pI of the emulsifier, in which case the droplets hold a net anionic charge that attracts pro-oxidant metals to the interface, and lipid oxidation proceeds at a high rate. Other studies have demonstrated the importance that droplet charge has on lipid oxidation kinetics in O/W emulsions (Hu et al. 2003a,b, Klinkesorn et al. 2005b; McClements and Decker 2000).

The interfacial thickness of emulsion droplets has been shown to affect the rate of oxidation reactions. By increasing the thickness of the interfacial membrane, it is possible to physically hinder the interaction between emulsified lipid and aqueous phase pro-oxidants (Chaiyasit et al. 2000, Silvestre et al. 2000). For example, it has been demonstrated that lipid oxidation proceeds at a higher rate in emulsions stabilized by Brij 76 (a nonionic surfactant) versus emulsions stabilized by Brij 700. The authors suggested that the large polar head groups of Brij 700 resulted in a considerably thicker interfacial membrane, which in turn made it difficult for aqueous oxidants to access the lipid.

The application of antioxidants in food emulsions is perhaps the most common strategy for inhibiting lipid oxidation in foods. In general, antioxidant function by binding, or chelating, transition metals or interrupting free radical chain reactions (i.e., chain-breaking antioxidants). Enhanced oxidative stability is often seen when both metals chelators and chain-breaking antioxidants are used in O/W emulsions. The choice of which antioxidant to add to which food product is often not a straight-forward matter. Antioxidants vary greatly in their polarity, as some are water-soluble and others lipid-soluble, which forms the basis of their physical location within a food system. In the case of O/W emulsions, lipid-soluble antioxidants (e.g., α-tocopherol) are often more effective than water-soluble antioxidants (e.g., ascorbic acid), because they aggregate at lipid droplet surfaces where oxidation reactions are thought to occur.

Antioxidants can be used with great effectiveness to extend emulsion shelf life, the extent of which varies as a function of the food matrix (e.g., levels of endogenous trace metal catalysts), the instrinsic properties of the food (e.g., the fatty acid profile of the omega-3 oil), and the storage conditions (e.g., temperature). Although it is impossible to predict with any specificity the shelf life of a given antioxidant stabilized food product, it appears that several months of shelf stability are possible under refrigeration temperature. For example, Djordjevic and coworkers found that a 25% menhaden O/W emulsion with 500 ppm mixed tocopherols (chain-breaking antioxidant) and 100 mM EDTA (chelator) was virtually free of propanal (a secondary lipid oxidation product associated with rancidity) after 16 weeks at 4°C (Djordjevic et al. 2004). The same emulsion system without added tocopherol or EDTA had significant ($p > 0.05$) accumulation of propanol after just two weeks. The effectiveness of EDTA (100 mM) versus a control was observed at elevated temperature (37°C), however the time before secondary lipid oxidation products were detected was reduced to just 3 weeks. In another study, Lee et al. reported that a 25% (wt) algal O/W emulsion (pH 3) containing mixed tocopherols and EDTA could be added to restructured ham and stored under vacuum packaging at 4°C for at least 21 days (the length of the study) before the onset of lipid oxidation (Lee et al. 2006).

Bioavailability, or the fraction of biologically active components that is released from a food and is taken up in the gastrointestinal tract, is dependent not only the chemical properties of the compound, but on the properties of the associated food matrix. Depending on the application, colloid scientists may be interested in designing emulsion systems that reduce lipid bioavailability (e.g., for reduced calorie formulations) or enhance bioavailability (e.g., for bioactive lipid delivery systems). This is currently a major area of research in food emulsion research.

The bioavailability of emulsified bioactive compounds can potentially be manipulated by a variety of means. For example, if calorie reduction is desired, specific emulsifiers or surfactants can be employed in order to impede the activity of gastric and/or pancreatic enzymes (e.g., lipases) required for digestion and absorption (McClements et al. 2008). Another strategy may involve the use of highly surface-active emulsifiers in order to retard absorption by competing with the body's endogenous enzymes. Some have proposed in recent years the use of cross-linked emulsion interfaces, or even enzyme-resistant interfacial layers, as a means of inhibiting lipid bioavailability. From functional food perspective, reducing lipid (e.g., omega-3) bioavailability would be counterproductive; however, research in this area may lead to technologies that allow for increased absorption of bioactive compounds through emulsion interfacial engineering.

## Multiple Emulsions

A proven and effective means of controlling the interfacial characteristics of lipid droplets is the layer-by-layer (LBL) electrostatic deposition method (Aoki et al. 2005; Gu et al. 2007; Klinkesorn et al. 2005a; Ogawa et al. 2004; Park et al. 2007). The charge, thickness, permeability, and environmental responsiveness of interfacial layers surrounding lipid droplets can be controlled using this method (McClements et al. 2008). The LBL approach (Figure 20.1) involves successive deposition

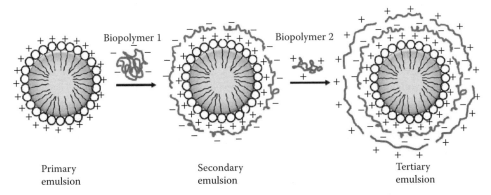

Primary emulsion    Secondary emulsion    Tertiary emulsion

**FIGURE 20.1** Schematic demonstrating the preparation of a multilayer O/W emulsion using layer-by-layer deposition.

of oppositely charged polymers onto a droplet surface, which may then be cross-linked using a suitable method (e.g., physically, enzymatically, or chemically). By building thick, impermeable interfacial layers from biopolymers (proteins or polysaccharides) that are resistant to enzymatic degradation and displacement, it may be possible to reduce lipid bioavailability. It may also be possible to encapsulate a lipid and release it at a particular site of action (e.g., the stomach, small intestine, or large intestine) by controlling the environmental responsiveness of the interfacial layers (e.g., to pH, ionic strength of specific enzymes).

Such an approach would have great advantages for a functional food processor, as much research in recent years has emphasized that the bioavailability of active compounds should be weighed over and above their final concentration in the food system. Moreover, the targeted release of omega-3 lipids to the lower GI would potentially be an effective means of reducing "burp-back," an undesirable condition often associated with omega-3 supplements resulting in the reflux of off-aromas from the stomach to the upper esophagus. This could be accomplished if the lipid remained emulsified through transit through the upper GI.

### Solid Lipid Particle Emulsions

Solid lipid particle emulsions consist of partially or completed solidified lipid droplets that are coated with an emulsifier and dispersed within an aqueous phase (McClements et al. 2009). These emulsions are prepared by homogenizing an oil and water phase together in the presence of a hydrophilic emulsifier at a temperature above the melting point of the lipid phase, a process known as "hot emulsification." The most common application of solid lipid particles is the delivery of lipophilic drugs, which the pharmaceutical industry has been using for years (Muller and Keck 2004); however, extension of this technology to food processing has yet to occur.

The purported advantages of delivering lipophilic bioactive compounds within solid lipid particles compared to conventional emulsions related mostly to increases in physical and chemical stability. With respect to physical stability, many of the mechanisms leading to droplet coalescence and growth are avoided when the liquid lipid phase is converted to solid. The location of bioactive lipids within solid lipid particle emulsion droplets is potentially easy to control, as well. For example, it is possible to restrict bioactive compounds within the grain boundaries of the crystallized bulk lipid phase, in essence restricting these labile compounds from the droplet interface where oxidation reactions are thought to occur.

Solid lipid particles may have other advantages over conventional emulsions, particularly with respect to their release and delivery attributes of lipophilic materials. The loading capacity and efficiency of solid lipid particle emulsions is greater than conventional systems in many cases, and the release kinetics of their functional constituents can be easy to predict and control (McClements et al. 2009).

## CONCLUSIONS AND FUTURE DIRECTIONS

Food processing encompasses a wide range of technologies and operations that transform raw food materials and ingredients into consumer food products. The processing of a food or beverage begins with harvesting and postharvest handling of raw materials, continues through several complex manufacturing operations, and ends with packaging and delivery to consumers. The primary purpose of processing is to ensure food safety and extend product shelf life. However, in recent years, this term has taken on an almost pejorative meaning, and has been associated with foods of depleted quality and nutrition. This is clearly not the case, especially when one considers the role of processing in preserving sensitive endogenous bioactive components, augmenting bioavailability of many ingredients, increasing safety with respect to human food-borne pathogens, and eliminating toxins. In most situations, processed foods and beverages are similar, if not superior, in nutritive value to their raw counterparts or products prepared by traditional cooking methods.

Furthermore, appropriate processing technologies can be used to modulate macronutrient conversion to calories, while novel processing methods (i.e., high-pressure processing, pulsed electric fields, etc.) can be used to make more bioavailable the micronutrients bound up in the cellular structure of foods. Novel emerging processes not totally relying on heat seem to offer the potential to increase bioavailability of classic micronutrients and to spare many of the labile phytochemicals that are a major advantage in fresh fruits and vegetables.

Finally, the future of our food system will be largely determined by three major trends: population growth; availability of energy, land, water, and other natural resources; and climate change. However, the science and technology deployed will also be important, not only as it pertains to problem solving and issue resolution, but also as a matter of public understanding, public policy, consumer attitudes, consumer acceptance, and fiscal resources. In addition to the novel processing methods briefly described in this chapter, a number of new technologies are being developed with promising benefits to our food system. They include biotechnology, molecular biology, and many other evolving "omics" tools; microbial ecology and the microbiology of the human gut; personalized nutrition and its associated targeted methods for disease prevention or better health and wellness; nanoscale science, engineering, and technology—referred to as nanotechnology—and other science and engineering approaches; and information technologies.

Our modern food system is very complex and changes continuously in time and space. As Albert Einstein once said, "We need a new way of thinking to solve the problems caused by the old way of thinking." Food science and technology can help us advance the food system, minimize risks, maximize benefits, and deliver a safe, nutritious, and abundant food supply to all people around the world. With science and technology solutions available to address specific issues throughout the food system, our ability to feed a growing population in a sustainable way, while safeguarding both human and planet health, looks not only possible, but promising. We must however, remain steadfast and rational about our approach to help both humanity and nature.

## REFERENCES

ADA.1996. Position of the American Dietetic Association: Food irradiation. *Journal of the American Dietetic Association* 96(1):69–72.

Aieta, E. M., P. V. Roberts, and M. Hernandez. 1984. Determination of chlorine dioxide, chlorine, and chlorate in water. *J Am Water Works Ass* 76:64–70.

Alliger, H. 1980. *An Overall View of Chlorine Dioxide*. Farmingdale, NY: The Alcide Corp.

Anderson, J. E., D. M. Adams, and W. M. Walter. 1983. Conditions under which bacterial amylases survive ultrahigh temperature sterilization. *J Food Sci* 48(6):1622–25.

Aoki, T., E. A. Decker, and D. J. McClements. 2005. Influence of environmental stresses on stability of O/W emulsions containing droplets stabilized by multilayered membranes produced by a layer-by-layer electrostatic deposition technique. *Food Hydrocolloids* 19(2):209–220.

Benarde, M. A., B. M. Israel, V. P. Olivieri, and M. L. Granstrom. 1965. Efficiency of chlorine dioxide as a bactericide. *J App Microbiol* 13(5):776–80.

Broihier, K. 1999. A tomato a day keeps cancer away. *Food Processing* 4:58.

Chaiyasit, W., M. P. Silvestre, D. J. McClements, and E. A. Decker. 2000. Ability of surfactant hydrophobic tail group size to alter lipid oxidation in oil-in-water emulsions. *J Agric Food Chem* 48(8):3077–80.

Cheftel, J. C. 1992. Effect of high hydrostatic pressure on food constituents and overview. In C. Balny, R.Hayashi, K. Hermans, and P. Masson, eds., *High Pressure Technology*, vol. 224, pp. 195–209. Montrouge, France: John Libby Eurotext Ltd.

Chen, Y. S. and J. M. Vaughn. 1990. Inactivation of human simian rotaviruses by chlorine dioxide. *Appl Environ Microbiol* 56(5):1363–66.

Damodaran, S. 1996. Amino acids, peptides, and proteins. In O. R. Fennema, ed., *Food Chemistry* 321–429. New York: Marcel Dekker, Inc.

Diehl, J. F. 1995. *Safety of Irradiated Foods*. New York: Marcel Dekker, Inc.

Djordjevic, D., D. J. McClements, and E. A. Decker. 2004. Oxidative stability of whey protein-stabilized oil-in-water emulsions at pH 3: Potential omega-3 fatty acid delivery systems (Part B). *J Food Sci* 69(5):C356–62.

Dock, L. L., and J. D Floros. 1999. Thermal and non-thermal processing techniques. In M. Schmidl and T. Labuza, eds., *Science and Technology of Functional Foods* 49–88. Gaithersburg, MD: Aspen Publishers.

Donnelly, J. L., E. A. Decker, and D. J. McClements. 1998. Iron-catalyzed oxidation of Menhaden oil as affected by emulsifiers. *J Food Sci* 63(6):997–1000.

Dunn, J., T. Ott, and W. Clark. 1995. Pulsed-light treatment of food and packaging. *Food Technol* 49(9):95–98.

Earnshaw, R. G., J. Appleyard, and R. M. Hurst. 1995. Understanding physical inactivation processes: Combined preservation opportunities using heat, ultrasound and pressure. *Int J Food Microbiol* 28(2):197–219.

Ensminger, D. 1988. Acoustic and electroacoustic methods of dewatering and drying. *Dry Technol* 6(3):473–99.

EPRI (Electrical Power Research Institute). 1997. Ozone-GRAS affirmation for use in food. *Food Industry Currents* 1(1):1–6.

Fairbanks, H. V. 1974. Ultrasonically assisted drying of fine particles. *Ultrasonics* 12(6):260–62.

Farr, D. 1990. High pressure technology in the food industry. *Trends Food Sci Technol* 1(1):14–16.

FDA. 1995. Beverages: Bottled water, final rule. *Federal Register* 60:57075–130.

Finch, G. R., E. K. Black, L. L. Gyürék, and M. Belosevic. 1993. Ozone inactivation of *Cryptosporidium parvum* in demandfree phosphate buffer determined by *in vitro* excystation and animal inefectivity. *Appl Environ Microbiol* 59(12):4203–10.

Floros, J. 2004. Food and diet in Greece from ancient to present times. Paper presented at the Indigenous Knowledge Conference Proceedings May 27–28, 2004, The Pennsylvania State University, State College, Pennsylvania. Available at http://www.ed.psu.edu/ICIK/2004Proceedings/section2-floros-withpics.pdf (accessed March 21, 2010).

Floros, J. 2008. Food science: Feeding the world. *Food Technol* 62(5):11.

Floros, J. D. 1993. Aseptic packaging technology. In J. V. Chambers and P. E. Nelson, eds., *Principles of Aseptic Processing and Packaging*, pp.115–48. Washington, DC: The Food Processors Institute.

Floros, J. D. and H. Liang. 1994. Acustically assisted diffusion through membranes and biomaterials. *Food Technol* 48(12):79–84.

Foegeding, P. M., V. Hemstapat, and F. G. Giesbrecht. 1986. Chlorine dioxide inactivation of Bacillus and Clostridium spores. *J Food Sci* 51(1):197–201.

Graham, D. M. 1997. Use of ozone for food processing. *Food Technol* 51(6):72–75.

Grahl, T., and H. Märkl. 1996. Killing of microorganisms by pulsed electric fields. *Appl Microbiol Biotechnol* 45(1-2):148–57.

Gravani, R. B. 1985. Food deterioration and spoilage caused by light. *Dairy Food Sanitation* 5(10):386–87.

Gregory, J. F. 1996. Vitamins. In O. R. Fennema, ed., *Food Chemistry* 531–616. New York: Marcel Dekker, Inc.

Gu, Y. S., E. A. Decker, and D. J. McClements. 2007. Application of multi-component biopolymer layers to improve the freeze-thaw stability of oil-in-water emulsions: Beta-lactoglobulin-iota-carrageenan-gelatin. *J Food Eng* 80(4):1246–54.

Gunasekaran, S., and A. Chyung. 1994. Evaluating milk coagulation with ultrasonics. *Food Technol* 48(12):74–78.

Hall, R. L. 1989. Pioneers in food science and technology: Giants in the earth. *Food Technol* 43(9):186–95.

Harlfinger, L. 1992. Microwave sterilization. *Food Technol* 46:57–61.

Heddelson, R. A. and S. Doores. 1994. Factors affecting microwave heating of foods and microwave induced destruction of foodborne pathogens—a review. *J Food Prot* 57(11):1025–37.

Ho, S. Y., G. S. Mittal, and J. D. Cross. 1997. Effects of high field electric pulses on the activity of selected enzymes. *J Food Eng Nutrition Food Science* 31(1):69–84.

Holdsworth, S. D. 1992. *Aseptic Processing and Packaging of Food Products*. New York: Elsevier.

Hoover, D. G., C. Metrick, A. M. Papineau, D. F. Farkas, and D. Knorr. 1989. Biological effects of high hydrostatic pressure on food microorganisms. *Food Technol* 43(3):99–107.

Horie, Y., K. Kimura, M. Ida, Y. Yoshida, and K. Ohki. 1991. Jam preparation by pressurization. *Nippon Nogeikagaku Kaushi* (*Journal of the Japan Society for Bioscience Biote*) 65(6):975–80.

Hu, M., D. J. McClements, and E. A. Decker. 2003a. Impact of whey protein emulsifiers on the oxidative stability of salmon oil-in-water emulsions. *J Agric Food Chem* 51(5):1435–39.

Hu, M., D. J. McClements, and E. A. Decker. 2003b. Lipid oxidation in corn oil-in-water emulsions stabilized by casein, whey protein isolate, and soy protein isolate. *J Agric Food Chem* 51(6):1696–1700.

Isizaki, K., N. Shinriki, and H. Matsuyama.1986. Inactivation of *Bacillus* spores by gaseous ozone. *J Appl Bacteriol* 60(1):67–72.

Josephson, E. S., M. H. Thomas, and W. K. Calhoun 1978. Nutritional aspects of food irradiation: An overview. *J Food Process Preserv* 2(4):299–313.

Kilcast, D. 1994. Effect of irradiation on vitamins. *Food Chem* 49(2):157–64.

Kim, H. J., Y. M. Choi, A. P. P. Yang, I. A. Taub, J. Giles, C. Ditusa, S. Chall, and P. Zoltai. 1996a. Microbiological and chemical investigation of ohmic heating of particulate foods using a 5kW ohmic system. *J Food Process Preserv* 20(1):41–58.

Kim, S. M., and J. F. Zayas. 1989. Processing parameters of chymosin extraction by ultrasound. *J Food Sci* 54(3):700–03.

Klinkesorn, U., P. Sophanodora, P. Chinachoti, E. A. Decker, and D. J. McClements. 2005a. Encapsulation of emulsified tuna oil in two-layered interfacial membranes prepared using electrostatic layer-by-layer deposition. *Food Hydrocolloids* 19(6):1044–53.

Klinkesorn, U., P. Sophanodora, P. Chinachoti, E. A. Decker, and D. J. McClements. 2005b. Increasing the oxidative stability of liquid and dried tuna oil-in-water emulsions with electrostatic layer-by-layer deposition technology. *J Agric Food Chem* 53(11):4561–66.

Knorr, D. 1993. Effects of high-hydrostatic-pressure processes on food safety and quality. *Food Technol* 47(6):156–61.

Larson, W. P., T. B. Hartzell, and H. S. Diehl. 1918. The effects of high pressure on bacteria. *J Infect Dis* 22(2):271–79.

Lassen, A., and L. Ovesen. 1995. Nutritional effects of microwave-cooking. *Nutr & Food Sci* 4:8–10.

Lechowich, R. V. 1993. Food safety implications in high hydrostatic pressure as a food processing method. *Food Technol* 47(6):170–72.

Lee, S., P. Hernandez, D. Djordjevic, H. Faraji, R. Hollender, C. Faustman, and E. A. Decker. 2006. Effect of antioxidants and cooking on stability of n-3 fatty acids in fortified meat products. *J Food Sci* 71(3):C233–38.

Liew, C. V., and R. K. Prange. 1994. Effect of ozone and storage temperature on postharvest diseases and physiology of carrots. *J Am Soc Hortic Sci* 119(3):563–67.

Lillard, H. S. 1979. Levels of chlorine and chlorine dioxide of equivalent bactericidal effect in poultry processing water. *J Food Sci* 44(6):1594–97.

Lillard, H. S. 1993. Bactericidal effect of chlorine on attached Salmonellae with and without sonification. *J Food Prot* 56(8):716–17.

Lima, M., and S. K. Sastry. 1990. Influence of fluid rheological properties and particle location on ultrasound-assisted heat transfer between liquid and particles. *J Food Sci* 55(4):1112–15.

Lin, W. F., T. S. Huang, J. A. Cornell, C. M. Lin, and C. I. Wei. 1996. Bactericidal activity aqueous chlorine and chlorine dioxide solutions in a fish model system. *J Food Sci* 61(5):1030–34.

Liyange, L. R. J., G. R. Finch, and M. Belosevic. 1997. Sequential disinfection of *Cryptosporidium parvum* by ozone and chlorine dioxide. *Ozone Sci Eng* 19(5):409–23.

Lund, D. 1975. Heat transfer in foods. Ch. 3. In O. R. Fennema, ed., *Physical Principles of Food Preservation*, chapter 3. New York: Marcel Dekker, Inc.

Lund, D. 1989. Food processing: From art to engineering. *Food Technol* 43(9):242–308.

Manvell, C. 1997. Minimal processing of food. *Food Science and Technology Today* 11(2):107–11.

Martín, O., B. L. Qin, F. J. Chang, G. V. Barbosa-Canovas, and B. Swanson. 1997. Inactivation of Escherichia coli in skim milk by high intensity pulsed electric fields. *J Food Proc Eng* 20(4):317–36.

McClements, D. J. 2005. *Food Emulsions: Principles, Practice, and Techniques*. Boca Raton, FL: CRC Press.

McClements, D. J., and E. A. Decker. 2000. Lipid oxidation in oil-in-water emulsions: Impact of molecular environment on chemical reactions in heterogeneous food systems. *J Food Sci* 65(8):1270–82.

McClements, D. J., E. A. Decker, Y. Park, and J. Weiss. 2008. Designing food structure to control stability, digestion, release and absorption of lipophilic food components. *Food Biophys* 3(2):219–28.

McClements, D. J., E. A. Decker, Y. Park, and J. Weiss. 2009. Structural design principles for delivery of bioactive components in nutraceuticals and functional foods. *Crit Rev Food Sci Nutr* 49(6):577–606.

Mertens, B., and D. Knorr. 1992. Developments of nonthermal processes for food preservation. *Food Technol* 46(5):124–33.

Miller, D. D. 1996. Minerals. In O.R. Fennema, ed., *Food Chemistry*, chapter 9. New York: Marcel Dekker, Inc.

Monk, J. D., L. R. Beuchat, and M. P. Doyle. 1995. Irradiation inactivation of food-borne microorganisms. *J Food Prot* 58(2):197–208.

Morita, R. Y. 1975. Psychrotrophic bacteria. *Bacteriological Reviews* 39(2):144–67.

Moseley, B. E. B. 1984. Radiation damage and its repair in non-sporulating bacteria. In M. E. H. Andrew and A. D. Russel, eds., *The Revival of Injured Microbes*, pp. 147–74. London: Academic Press.

Muller, R. H., and C. M. Keck. 2004. Challenges and solutions for the delivery of biotech drugs—a review of drug nanocrystal technology and lipid nanoparticles. *J Biotech* 113(1-3):151–70.

Nagai, N. Y., and J. H. Moy. 1985. Quality of gamma irradiated California valencia oranges. *J Food Sci* 50(1):215–19.

Naitoh, S. 1992. Studies of the application of ozone in food preservation—ozone inhibition of yeast. *Journal of Antibacterial and Antifungal Agents—Japan* 21(6):341–46.

Ogawa, S., Decker, E. A. Decker, and D. J. McClements. 2004. Production and characterization of O/W emulsions containing droplets stabilized by lecithin-chitosan-pectin mutilayered membranes. *J Agric Food Chem* 52(11):3595–600.

Ohmori, T., T. Shigehisa, S. Taji, and R. Hayashi. 1991. Effect of high pressure on the protease activities in meat. *Agric Biol Chem* 55(2):357–61.

Ohr, L. M. 1999. Functional foods put flavor up front. *Prepared Foods* 2:59–64.

Ohshima, T., T. Nakagawa, and C. Koizumi. 1992. In E. G. Bligh, ed., Effects of high pressure on the enzymatic degradation of phospholipids in fish muscle during storage. *Seafood Science and Technology*, pp. 64–75. Oxford, UK: Fishing New Books.

Ohshima, T., H. Ushio, and C. Koizumi. 1993. High-pressure processing of fish products. *Trends Food Sci Tech* 4(11):370–75.

Park, G. Y., S. Mun, Y. Park, S. Rhee, E. A. Decker, J. Weiss, and D. J. McClements. 2007. Influence of encapsulation of emulsified lipids with chitosan on their *in vivo* digestibility. *Food Chem* 104(2):761–67.

Pauli, G. H. 1991. Food irradiation in the United States. In S. Thorne, ed., *Food Irradiation*, pp. 235–59. New York: Elsevier.

Pothakamury, U. R., G. V. Barbosa-Cánovas, and B. G. Swanson. 1993. Magnetic field inactivation of microorganisms and generation of biological changes. *Food Technol* 47(12):85–93.

Pothakamury, U. R., G. V. Barbosa-Cánovas, B. G. Swanson, and R. S. Meyer. 1995. The pressure builds for better food-processing. *Chem Eng Prog* 91(3):45–53.

Qin, B. L., Pothakamury, U. R., G. V. Barbosa-Cánovas, and B. G. Swanson. 1996. Nonthermal pasteurization of liquid foods using high-intensity pulsed electric fields. *Crit Rev Food Sci Nutr* 36(6):603–27.

Restaino, L., E. W. Frampton, J. B. Hemphill, and P. Palnikar. 1995. Efficacy of ozonated water against food related microorganisms. *Appl Environ Microbiol* 61(9):3471–75.

Reznick, D. 1996. Ohmic heating of fluid foods. *Food Technol* 50(5):250–51.

Rickloff, J. R. 1987. An evaluation of the sporicidal activity of ozone. *Appl Environ Microbiol* 53(4):683–86.

Roberts, R. G. and S. T. Reymond. 1994. Chlorine dioxide for reduction of postharvest pathogen inoculum during handling of tree fruits. *Applied and Environmental Microbiology* 60(8):2864–68.

Rotruck, J.T. 1982. Effect of processing on nutritive value of food: Trace elements. In M. Rechcigl, ed., *Handbook of Nutritive Value of Processed Food*, pp. 521–28. Boca Raton, FL: CRC Press.

Ruiz-Lopêz, M. D., R. Artacho, M. A. Pineda, H. Lopez Garcia de la Serrana, and M. C. Lopez Martinez. 1995. Stability of a-tocopherol in virgin olive oil during microwave heating. *Lebensmittel-Wissenschaft & Technologie (LWT-Food Science and Technology)* 28(6):644–46.

Sastry, S. K., and S. Palaniappan. 1992. Ohmic heating of liquid-particle mixtures. *Food Technol* 46(12):64–7.

Schiffmann, R. F. 1992. Microwave processing in the U.S. food industry. *Food Technol* 46(12):50–7.

Schnepf, M., and J. Driskell. 1994. Sensory attributes and nutrient retention in selected vegetables prepared by conventional and microwave methods. *J Food Qual* 17(2):87–99.

Schwartz, J. S. 1992. Quality considerations during aseptic processing of foods. In R. K. Singh and P. E. Nelson, eds., *Advances in Aseptic Processing Technologies*, pp. 245–59. New York: Elsevier.

Shukla, T. P. 1992. Microwave ultrasonics in food processing. *Cereal Foods World* 37(4):332–33.

Silvestre, M. P., W. Chaiyasit, R. G. Brannan, D. J. McClements, and E. A. Decker. 2000. Ability of surfactant headgroup size to alter lipid and antioxidant oxidation in oil-in-water emulsions. *J Agric Food Chem* 48(6):2057–61.

Tajchakavit, S., and H. S. Ramaswamy 1996. Thermal vs. microwave inactivation kinetics of pectin methylesterase in orange juice under batch mode heating conditions. *Lebensmittel-Wissenschaft & Technologie (LWT-Food Science and Technology)* 30(1):85–93.

Thayer, D. W. 1994. Wholesomeness of irradiated foods. *Food Technol* 48(5):132–35.

Thiessen, G. P., W. R. Usborne, and H. L. Orr. 1984. The efficacy of chlorine dioxide in controlling Salmonella contamination and its effect on product quality of chicken broiler carcasses. *Poult Sci* 63(4):647–53.

Vega-Mercado, H., O. Martín-Belloso., B. L. Qin, F. J. Chang, M. Marcela Gongora-Nieto, G. V. Barbosa-Canovas, and B. G. Swanson. 1997. Non-thermal food preservation: Pulsed electric fields. *Trends Food Sci Technol* 8(5):151–57.

Vega-Mercado, H., J. R. Powers, G. V. Barbosa-Canovas, and B. G. Swanson. 1995. Plasmin inactivation with pulsed electric fields. *J Food Sci* 60(5):1143–46.

Villarreal, M. E., R. C. Baker, and J. M. Regenstein. 1990. The incidence of Salmonella on poultry carcasses following the use of slow release chlorine dioxide (Alcide). *J Food Prot* 53(6):465–67.

von Sonntag, C. 1980. Free-radical reactions of carbohydrates as studied by irradiation techniques. *Adv Carbohydr Chem Biochem* 37(1):7–77.

Watzke, H. J. 1998. Impact of processing on bioavailability examples of minerals in foods. *Trends Food Sci Technol* 9:320–27.

White, G. C. 1972. *Handbook of Chlorination*. New York: Van Nostrand Reinhold, Co.

WHO. 1994. *Safety and Nutritional Adequacy of Irradiated Food*. Geneva: World Health Organization.

Wrangham, R. 2009. *Catching Fire: How Cooking Made Us Human*. New York: Basic Books.

Wrigley, D. M., and N. Llorca. 1992. Decrease of *Salmonella typhimurium* in skim milk and egg by hat and ultrasonication wave treatment. *J Food Prot* 55(9):678–80.

Zhang, Q., B. L. Qin, G. V. Barbosa-Canovas, and B. G. Swanson. 1995. Inactivation of E. coli for food pasteurization by high-strength pulsed electric fields. *J Food Process Preserv* 19(2):103–18.

Zind, T. 1999. The functional foods frontier. *Food Processing* 4:45–50.

ZoBell, C. E. 1970. Pressure effects on morphology and life processes of bacteria. In A. M. Zimmerman, ed., *High Pressure Effects on Cellular Processes* 85–130. New York: Academic Press.

# 21 New Whole Foods Designed to Deliver Bioactive Components

*Cheryl R. Mitchell*

## CONTENTS

## INTRODUCTION: THE ROLE OF WHOLE FOODS IN NUTRITIONAL GENOMIC HEALTH

In the future, more individuals based on their genomic profile will be instructed to consume or avoid different foods. As such, the food industry must be prepared to provide for these consumer demands. Consumers have access to a variety of both fresh and processed foods. While fresh foods should constitute the majority of our diet, the reality is that for most Americans, it is processed foods that predominate. It should be understood that the category of "processed foods" refers to both "simple" and "complex" processed foods. The food industry refers to simple processed foods as those foods that have been processed specifically for microbiological and product quality stability. This process provides a greater distribution and variety of safe and convenient basic foods inclusive of dairy, juices, flours, sweeteners, canned fruits, and vegetables. These products have simple ingredient declarations and usually include only those ingredients necessary for the preservation or processing of the basic food product itself. Complex processed food products include those consumer products that have been formulated with many ingredients, such as breakfast cereals, breads, granola bars, baking mixes, and confectionary.

Today, there are more choices in processed foods as the food industry has responded to nutritional needs with bioactive functional foods that contribute to healthier diets. Consumers are reading ingredient labels and looking for processed foods that contain ingredients that may be more compatible with their nutrition and health needs. The latter is supported by the extraordinary growth in the Natural Foods Industry (Natural Product Companies 2008). To meet individual health needs related to gene-linked diseases, consumers are looking for alternatives based upon genomic knowledge, implied genomics (heredity), or symptomatic genomics (allergies, high cholesterol,

**385**

bone deficiency). The food industry needs to provide simple and complex processed foods that have been nutritionally (genomically) improved by processing or by improved formulations. Another significant change includes consumers' recognition of the importance of the nutrient density of foods. Consumers are beginning to select those processed foods that are nutrient dense, and preferably, contain the *whole food* and not just a conglomerate of refined ingredients.

During the twentieth century, the objective of the food scientist was extraction and refinement. Many food components, such as proteins, carbohydrates, fibers, and oils, were separated into a variety of refined ingredients. This was done to produce a series of ingredients that were stable, consistent in taste and functionality, and could be easily employed in the formulation of complex processed foods. Unfortunately, during these processes minor constituents were separated from the major components, removed, and in many cases discarded as animal feed. These constituents are now recognized as the critical bioactive components that contribute to the nutritional value of whole food. The food scientist now understands that the refined ingredients were lacking the essential bioactive components that contribute to "nutrient density." Effectively, it has been realized that the sum of the refined parts are *not* equal to the whole food. The importance of bioactives such as soluble fibers, minerals, vitamins, antioxidants, phospholipids, sterols, and so on, while not necessarily present in large amounts, are being recognized as significant, causing the food scientist to rethink how whole foods should be processed to retain and maintain both the bioactive components and "nutrient density" of the whole food.

Starting with whole foods, the food scientist has developed food products and extracts where the food, if not whole food, at least retained most of its nutrient density. For example, instead of refining sugar, "cane extracts" containing the minerals and vitamins associated with the sugar started to emerge and become very popular. Brown rice syrup, containing some of the oil, minerals, and vitamins of brown rice, was developed to replace refined glucose syrup. Various sea salts, as opposed to refined or iodized salt, have been gaining in popularity. Likewise, nutrient-dense extracts of grains and seeds were being used to create more nutrient dense beverages that were suitable alternatives to dairy milk but had a different balance of bioactive components and nutritional value. While these beverages from grains or seeds contained some of the nutrient density and genomic functionality of the food source, they were only the beginning of a new revolution in the food industry. The food scientist was beginning to work on the "transformation" of the whole food into food products that retain *all* of the nutritional value of the whole food while substantially improving the stability, functionality, and application of the whole food. This transformation of the whole food allows for the use of whole foods in broader applications and diverse food categories. This chapter provides a brief discussion of the general process and formulation improvements that have been utilized with whole foods; the trend to provide foods with higher nutrient density, such as the grain and seed beverage extracts; and an introduction to the revolutionary concept of transformation of whole grains. A comparison of the compositions of currently available "milk" beverages from grain and seed extracts with "milks" resulting from the transformation of whole grains is provided as well as a discussion of the potential benefits and application of these transformed whole grains in both simple and more complex processed foods.

## NUTRITIONAL IMPROVEMENTS IN WHOLE FOODS

Once bioactive agents are identified in a food source, it is up to the food scientists to work with these food sources to convert them into stable foods or ingredients that retain most, if not all, of the bioactive and nutritional components of the food. To improve the nutritional value of processed whole foods, the food scientist can make changes in process and packaging technology, as well as make changes in formulations of simple and complex whole food products.

### PROCESS IMPROVEMENTS

The method of processing and handling of bioactive whole foods is critical not only to preserve the bioactive agents in the food, but to create a form of the whole food that has sufficient application diversity for consumers. To preserve whole foods, processing and packaging methods such as dehydration, pickling, canning, aseptic, and recently high-pressure technologies are frequently employed. The primary objective of processing has been to preserve the basic form of the whole food while stabilizing it against microbiological degradation. Recent methods such as high-temperature-short-time (HTST) processing (Heldman and Hartel 1997) reduces the exposure time to high temperatures, resulting in better flavor, longer stability, and better retention of bioactives in whole foods such as juices and dairy milks. High-pressure processing of avocados has been used to retain the fresh flavor, color, and texture of avocados for months in refrigerated or frozen storage (IFT 2008). Different types of radiation from ultraviolet (UV) to gamma have also been successfully used to reduce the microbiological load in fresh foods thereby extending the shelf life and distribution of these foods (Ragheb 2009). Tomatoes and strawberries have been successfully irradiated with extension of their shelf life. In the case of mushrooms, the UV exposure has also resulted in a conversion of the vitamin D precursors to vitamin D making fresh mushrooms now an excellent source of vitamin D (Roberts et al. 2008). In general, process improvements of whole foods have been utilized to improve the taste, functionality, texture, microbiological stability, and shelf life, and simultaneously enhance or preserve the bioactive components of the whole food.

## FORMULATION IMPROVEMENTS

During the last 20 years, food scientists have also improved processed whole food by either the removal of negative ingredients utilized in the processing of the whole food, or the addition of functional ingredients to replace or enhance the nutritional density of the whole food. Many simple and complex whole food products contain not only the whole food, but additional ingredients such as salt, sugar, acids, or preservatives that are sometimes a necessity in preserving the food. Additionally, complex whole food products may contain ingredients that would be considered undesirable especially for individuals having a genomic predisposition. Consequently, whole food products containing high level of salts, sugars, saturated fats, trans fats, cholesterol, or allergens like gluten, soy, dairy, or nuts, have been reformulated to remove or replace these ingredients. Currently, it is easy for the consumer to find traditional whole foods such as soups that are low in sodium, dried fruits without added sugar or preservatives, fruit juices with no added sugar, lowfat and nonfat milks, low cholesterol eggs, wheat-free breads, and granola bars that are selectively free of soy, dairy, or nuts.

Whole foods can also be improved by the use of beneficial additives or functional ingredients that are believed to improve the nutritional value of the whole food products. Addition of bioactive ingredients such as antioxidants, omega-3 fatty acids, soluble fibers, vitamins, amino acids or peptides, minerals, and even probiotics are frequently added to enrich whole foods. Milk and eggs have been enriched with omega-3 fatty acids through enhancing the diet of cows and chickens with algae, grains, and seeds rich in omega-3 fatty acids. Vitamins A and D have been added to fortify dairy milk to assure the nutritional health of our children. Inulin, a soluble fiber and prebiotic, has been added to yogurts to enhance whole food dairy products. Iodine has been added to salt to assure thyroid health, and vitamins added to "enrich" flours, replacing essential vitamins and iron lost in removing the outer coat of the grains. Other foods such as juices have been enriched with vitamin C or calcium to improve the nutritional value of these whole foods. Overall, formulation improvements have provided a variety of improved simple and complex whole food products. However, even with these improvements in processing, packaging and formulation, the actual form and subsequent applications of the whole food has not changed significantly.

Juices pressed from fruits or vegetables, or grains that have been ground into flours, have been the extent of changes in whole foods for centuries. The flour, of course, increased applications in cereals, breads, and pastas. However, because the protein and fiber remain linked together even in the smallest of particles, during the cooking process, the chewy texture of the protein and fiber aggregate remains. Texturally, this is great for cereals and breads, but truly limits the applications of the whole grain for other uses in foods.

## NUTRIENT-DENSE FOODS

### GRAIN EXTRACTS

Extracts of grains in the form of "milk-like" beverage products have been known for over 1000 years and have more recently gained in popularity, due to flavor and emulsion palatability and consumer availability during the last 30 years. The grain milks, or "flour-based" milk products, are prepared by blending grain flours with water, cooking, and sometimes enzymatically hydrolyzing these flours. Various flour milks made from oat, wheat, and rice (horchata) have been commercialized. Upon the preparation of these flour milks, the interaction of the oil, protein, and fiber produces undesirable insoluble particles of oil/protein/fiber aggregates. Additionally, if whole grain flours are used, the natural oils contained in the whole grain are easily oxidized, resulting in a bitter taste or rancid flavor. Consequently, in the final consumer product, either small concentrations of the whole grain flour are used or the protein/fiber aggregates are removed so as to make smooth milk without particulates or rancid flavor, or something other than whole grains are used as a starting material (white rice versus brown rice, for example). In any case, these milks made from grain flours either do not contain the whole grain, or contain whole grains at nutritionally insignificant levels. In rice beverage extracts, after the removal of the oil/protein/fiber aggregate, it is necessary to actually add an outside source of oil to interact with the carbohydrate from the rice in order to make a stable milk-like emulsion. The latter is the case of a very popular rice-based beverage known as Rice Dream. Again, the popularity of this beverage, while not nutritionally whole-grain significant, indicates a strong consumer desire and need for alternatives to dairy milk. However, the values of whole grains in the human diet are of growing importance. Consequently, while these grain extracts had a higher nutrient density as compared with a sugar-based beverage, it was still desirable to develop a complete whole grain milk beverage and food product maximizing natural nutrient density. The primary use of the rice extracts are as a substitute for dairy milk. The fact that rice is hypoallergenic, contains no lactose, cholesterol, or casein, and, in some cases is low in protein, provides positive benefits for the consumer. However, the low protein content also limits the use of rice milk as a protein substitute for milk, as well as its use in baking applications where the protein has certain functionality.

### SOY, ALMOND, AND HEMP SEED EXTRACTS

High-nutrient-density seed milks made from soy, almonds, and hemp are currently popular and are readily available throughout the United States (Grace 2008). In general, these milks are made by grinding the dehulled seed with water, heating and cooking the slurry. During the cooking process, an oil, protein, and fiber aggregate is formed that is readily removed by sifting or centrifugation to produce a smooth milk-like emulsion. Consequently, like the grain extracts above, the seed milks start with the whole seed, however, the seed milk resulting from this process does not contain the same nutritional composition as the whole seed. The composition of these seed extracts as compared to the whole seed for soy, almond, and hemp are provided in Tables 21.1 through 21.3, respectively. These tables provide a normalized comparison of the label declaration for a commercial "milk" beverages made from soy, almond, and hemp, as compared to the seed from which the milk is extracted. Since these seeds are naturally rich in oil, no additional oil needs to be added, and hence it was easiest to normalize the seed content of the milk based upon the oil content found in the milk. The significance or benefits of the seed milks are discussed individually next and provide an insight into their nutrient density advantage as well as the applications and use of these seed extracts.

Soybean milk extract is a significant source of protein and oil and is a suitable replacement for the protein in dairy milk as well as providing the functionality in many food applications including baking applications. In spite of the availability of dairy milk in the United States and Europe, the use of soy milk has significantly increased as a substitute for dairy milk for reasons associated with philosophical avoidance of mammalian milk, as well as the avoidance of casein, lactose, cholesterol, and saturated fats. More recently, the benefits of the phytosterols (estrogen) contained in soy have caused an upsurge in women drinking soy milk (Messina and Hughes 2003). Soy milks, which are naturally high in protein, have also been used as the basis for other food applications including frozen desserts, ice creams, meat analogues/substitutes, and cheeses. While there are broader uses of the soy milk extracts as compared with rice, the negative aspects of soy include the fact that it is an allergenic material, and the taste of the soy is distinctive and not as widely accepted as rice. According to Table 21.1, soy milks do not differ substantially from the soybean itself. Most of the popular soy milks have added sugar and salt to improve the taste, and added carrageenan (a soluble fiber) to improve the stability of the milk emulsion. For the most part, only some of the fiber is removed, making it a high density extract of soybeans. More recently, seed milks made from almonds and hemp seeds have also entered the market place. These milks, while also serving as an alternative to dairy milk, are being used because of a preferred flavor as compared with soy, or because of the nutritional genomic value of the bioactive components known to be contained in the seed source.

## TABLE 21.1
## Nutrients in Soybeans and a Commercial Soymilk Beverage

| | Soybeans[a] | Soymilk[b] Normalized[c] | Soymilk[b] Label[c] |
|---|---|---|---|
| Equivalent soybean solids, g[c] | 100 | **18.4** | – |
| Calories | 486 | 89 | 100 |
| Calories from fat | 196[d] | 36 | 35 |
| Total fat, g | **21.7** | **4** | **4** |
| Saturated fat, g | 3.1 | 0.6 | 0.5 |
| *Trans* fat, g | – | – | 0 |
| Polyunsaturated fat, g | 12.3 | 2.3 | 2.5 |
| Monounsaturated fat, g | 4.8 | 0.9 | 1 |
| Cholesterol, g | 0 | 0 | 0 |
| Sodium, mg | 2 | 0 | 135 |
| Potassium, mg | 1959 | 361 | 250 |
| Total carbohydrate, g | 33 | 6 | 8 |
| Dietary fiber, g | 10 | 2 | 2 |
| Sugars, g | 8 | 1.5 | 4 |
| Nonfiber carbohydrate, g[e] | 15 | 3 | 6 |
| Protein, g | 40 | 7 | 7 |
| INGREDIENTS: | [Whole soybean solids] | Filtered water, organic whole soybeans, organic evaporated cane juice, calcium carbonate, sea salt, carrageenan, vitamin E, vitamin A palmitate, vitamin $B_{12}$, vitamin $D_2$. | |

[a] USDA-NAL Database: Soybeans, mature seeds, raw; data for 100 grams soybean solids (109 grams "as is").

[b] Soy Dream® Soymilk (Organic) Original Enriched—package label nutrition facts as shown for serving size of 8 fl oz (240 mL). Soy Dream is a registered trademark of Westbrae Natural, Inc.

[c] Normalized to total fat.

[d] Not listed, calculated: Total fat × 9.

[e] Calculated by difference: Total carbohydrates less dietary fiber.

**TABLE 21.2**

**Nutrients in Almonds and a Commercial Almond Beverage**

| | Almonds[a] | Almond Beverage[b] | |
|---|---|---|---|
| | | Normalized[c] | Label |
| Equivalent almond solids, g[c] | 100 | **4.7** | |
| Calories | 610 | 31 | 60 |
| Calories from fat | 478[d] | 23 | 25 |
| Total fat, g | **53** | **2.5** | **2.5** |
| Saturated fat, g | 4 | 0.2 | 0 |
| *Trans* fat, g | – | – | 0 |
| Polyunsaturated fat, g | 13 | 0.6 | – |
| Monounsaturated fat, g | 34 | 1.6 | – |
| Cholesterol, g | 0 | 0 | 0 |
| Sodium, mg | 29 | 1 | 150 |
| Potassium, mg | 721 | 34 | 180 |
| Total carbohydrate, g | 21 | 1 | 8 |
| Dietary fiber, g | 11 | 0.5 | 1 |
| Sugars, g | 5 | 0.2 | 7 |
| Nonfiber carbohydrate, g[e] | 10 | 0.5 | 7 |
| Protein, g | 23 | 1.1 | 1 |
| INGREDIENTS: | [Almond solids] | Purified water, evaporated cane juice, almonds, tricalcium phosphate, sea salt, potassium citrate, carrageenan, soy lecithin, vitamin A palmitate, vitamin $D_2$, D-alpha-tocopherol (natural vitamin E). | |

[a] USDA-NAL Database: Nuts, almonds, blanched; data for 100 grams almond solids (105 grams "as is").

[b] Blue Diamond Natural® Almond Breeze® Original Non-Dairy Beverage—Package label nutrition facts as shown for serving size of 8 fl oz (240 mL). Blue Diamond Natural and Almond Breeze are registered trademarks of Blue Diamond Growers.

[c] Normalized to total fat.

[d] Not listed, calculated: Total fat × 9.

[e] Calculated by difference: Total carbohydrates less dietary fiber.

Seed milks from almond, are rich in vitamin E, manganese, and magnesium. While the taste is significantly different from soy, like soy, the most preferred almond milk contains added sugar. It should be noted from Table 21.2 that the amount of almonds contained in the almond milk itself is very low as compared to the soy and rice so that the overall gain from nutrient density of the almond milk is limited. Almond milks contain the equivalent of about 5 g of almonds per serving. This equates to about 5 almonds per serving of the milk. Comparing the 5 grams of almonds with the 18 grams of soybean solids or 32 grams of rice solids per serving of the equivalent serving of milk products, it is obvious the nutrient density per serving of almond milk is considerably less. Almonds, like soy, while a suitable substitute for dairy milk, are also in the list of the top eight food allergens (Food Allergies Labeling and Consumer Protection Act of 2004). As a cautionary note, designation as allergenic depends on the number of consumers exposed, frequency, and duration of data reported, so nonallergenic in one group may prove allergenic in another.

Seed milk from hemp seed is a relatively new entry into the seed milk extracts. Hemp nuts are naturally rich in omega-3 fatty acids and the milks provide a good source of high-quality fat and protein. The hemp itself is not allergenic and has a flavor and taste that is pleasant and mild. Hemp milks are rich in polyunsaturates and have a unique protein, making the hemp milk more similar in

**TABLE 21.3**
**Nutrients in Hemp Seeds and a Commercial Hemp Milk Beverage**

| | Hemp Seeds[a] | Hemp Milk[b] | |
| --- | --- | --- | --- |
| | | Normalized[c] | Label |
| Equivalent hempseed solids, g[c] | 100 | **13.0** | – |
| Calories | 590 | 79 | 100 |
| Calories from fat | 417[d] | 54 | 50 |
| Total fat, g | **46** | **6** | **6** |
|     Saturated fat, g | 16 | 2 | 0.5 |
|     *Trans* fat, g | – | – | 0 |
|     Polyunsaturated fat, g | 41 | 5 | 4.5 |
|     Monounsaturated fat, g | 0.3 | 0 | 1 |
| Cholesterol, g | 0 | 0 | 0 |
| Sodium, mg | 2 | 0 | 25 |
| Potassium, mg | – | – | – |
| Total carbohydrate, g | 13 | 2 | 10 |
|     Dietary fiber, g | 7 | 1 | 0 |
|     Sugars, g | 3 | 0 | 6 |
| Nonfiber carbohydrate, g[e] | 5 | 1 | 10 |
| Protein, g | 35 | 5 | 2 |
| INGREDIENTS: | [Hempseed solids] | Hemp nut base (filtered water, hemp nut [shelled hemp seed]), brown rice syrup, organic evaporated cane juice, tricalcium phosphate, carrageenan, sea salt, vitamin A palmitate, vitamin $D_2$, riboflavin, vitamin $B_{12}$. | |

[a] http//earthfriendlygoods.com—Hulled hemp seed; data from web page calculated for 100 grams hempseed solids (105.3 grams "as is").
[b] Tempt™ Original Living Harvest® Hempmilk Creamy Non-Dairy Beverage—Package label nutrition facts as shown for serving size of 1 cup (240 mL). Tempt is a trademark of Living Harvest Foods, Inc.
[c] Normalized to total fat.
[d] Not listed, calculated: Total fat × 9.
[e] Calculated by difference: Total carbohydrates less dietary fiber.

appearance, texture, and functionality to that of dairy milk. From Table 21.3, it can be seen that a serving of hemp milk actually contains about 13 grams of hemp nuts. While some of the protein and fiber have been removed, the amount of the polyunsaturated oils (and hence omega-3 fatty acids) is significant. While not a complete source of the whole food, the enrichment of the fatty acids may be beneficial to some individuals.

## NEW WHOLE FOODS RICH IN BIOACTIVE COMPONENTS

### TRANSFORMING WHOLE GRAINS INTO "MILKS"

During the last few years, the food scientist has examined the methodology of handling whole foods and has started to create totally new ways of processing whole foods. The revolution has begun to transform whole foods into new forms that would have wider applications across all food categories and yet retain the nutrient density of the complete whole food. To mankind, this transformation of whole grain ingredients is as significant as the transformation of sand into glass. The latter provided man with a plethora of applications for the glass that could not have been realized with sand alone.

Similarly, the transformation of whole foods such as whole grains and seeds into their corresponding "milks" has provided a new class of high-nutrient-density ingredients suitable for broader applications across all food categories.

The consumer and food industry have come to realize that there is considerable benefit and more food value in the consumption of the whole grain (Rajasree Pai and Raghesh 2007; Jones et al. 2010; Fardet et al. 2008), than the consumption of the isolated and refined parts. Consuming high fructose corn syrup is not the same as eating a whole kernel corn. Consuming filtered fruit juice does not benefit the individual the same way as the whole fruit. Eating refined fiber is not a substitute for consuming whole grains. What is important about whole grain, fruit, or other whole foods is the nutrient density and the balance of nutritional and bioactive composition. Preserving this balance of the whole food and yet changing the *form* of the whole food such that it can be easily used in a variety of food categories is the new challenge for the food scientist.

Because of the versatility of these transformed whole foods, they may be used as a replacement for low nutritional density ingredients such as sugars, maltodextrins, and oils. Because these new whole food milks are casein and lactose free, and derived from a primary agricultural resource, they also may be readily used to replace allergenic and environmentally challenging ingredients such as dairy milk.

Recent technology has been used to transform whole grains into whole grain "milks." This patented technology (Mitchell 2005) has been appropriately identified with the trademarked phrase, "We 'Milk' the Grain and Not the Cow" (U.S. Trademark Registered 2009). This transformation of whole grains is significantly different than previous technology which effectively cooked whole grains or seeds, and then removed the coagulated protein and fiber to produce a milk-like product. In the case of transformed whole grains, or whole grain milks, the bioactive fractions such as the protein, fiber, oil, antioxidants, vitamins, and minerals are first liberated from each other prior to any heat process, such that in particular the oil is free to form an emulsion with water and the fiber and protein are no longer in close proximity. Consequently, when transformed grain milks are further heat processed, they do not form the typical protein/fiber/oil aggregate which is the main component of cooked grains or seeds and which are removed in the making of typical grain and seed milks. In making the transformed grains, no other ingredient is used other than purified water. Using a combination of physical mechanical processes, heat, and enzymes, the various components of the whole grain are first liberated, heat processed, and then recombined. The transformed whole grain milk can then be utilized as a milk, as illustrated in Figure 21.1, for the transformation of whole grain corn; concentrated to a whole grain syrup, as illustrated in Figure 21.2, for the transformation of whole grain rice; or dried to a powder via spray or drum drying methodology for further use in simple or complex processed foods.

The new transformed whole grains most importantly contain all the nutritional components of the whole grain. The powders in particular have an appearance similar to dried dairy milk powder. They are microbiologically and oxidatively stable, making them convenient for storage and ease of use. When dissolved in water, they produce a milky appearance similar to cow's milk. The concentrates or dried powder forms of these whole grain milks can subsequently be used by manufacturers and consumers to create new food products containing whole grains and having improved nutrient density. Table 21.4 provides a comparison of the nutrients in brown rice versus the nutrients as

**FIGURE 21.1**  Transforming whole grain corn to whole grain corn milk.

**FIGURE 21.2** Transforming whole grain rice to whole grain rice syrup.

**TABLE 21.4**
**Nutrients in Brown Rice and Commercial Rice Beverages**

| | Brown Rice[a] | Rice Whole-Grain Beverage[b] | | Rice Extract Beverage[c] | |
|---|---|---|---|---|---|
| | | Normalized[d] | Adj. Label[b] | Normalized[d] | Label |
| Equivalent rice solids, g[d] | 100 | **32.4** | – | **28.6** | – |
| Calories | 414 | 133 | 125 | 117 | 120 |
| Calories from fat | 28[e] | 9 | 10[e] | 8 | 20 |
| Total fat, g | 3.1 | 1.0 | 1.2 | 0.9 | 2.5 |
| Saturated fat, g | 0.6 | 0.2 | 0.3 | 0.2 | 0 |
| *Trans* fat, g | – | – | 0 | – | 0 |
| Polyunsaturated fat, g | 1.1 | 0.4 | 0.5 | 0.3 | 0.5 |
| Monounsaturated fat, g | 1.1 | 0.4 | 0.4 | 0.3 | 1.5 |
| Cholesterol, g | 0 | 0 | 0 | 0 | 0 |
| Sodium, mg | 4 | 1 | 125 | 1 | 100 |
| Potassium, mg | 292 | 95 | 86 | 84 | – |
| Total carbohydrate, g | 87 | 28 | 27 | 25 | 23 |
| Dietary fiber, g | 6.7 | 2 | 1 | 2 | 0 |
| Sugars, g | – | – | 12 | – | 10 |
| Nonfiber carbohydrate, g[f] | **80** | **26** | **26** | **23** | **23** |
| Protein, g | 8.6 | 2.8 | 2 | 2.5 | 1 |
| INGREDIENTS: | [Whole brown rice solids] | Filtered water, organic whole brown rice, calcium phosphate, sea salt, carrageenan, vitamins and minerals (zinc gluconate, riboflavin, vitamin A, vitamin $D_2$, vitamin $B_{12}$). | | Filtered water, organic brown rice (partially milled), organic expeller pressed safflower and/or sunflower oil and/or canola oil, tricalcium phosphate, sea salt, vitamin A palmitate, vitamin $D_2$, vitamin $B_{12}$. | |

[a] USDA-NAL Database: Medium-grain cooked brown rice; data for 100 grams rice solids (370 grams "as is").
[b] Ryza™ Organic Fortified Rice Beverage—Calculated to 240 mL from package label nutrition facts as shown for serving size of 1 cup (British) (250 mL). Ryza is a trademark of the Australian Conference Association Limited.
[c] Rice Dream® Organic Rice Drink—Package label nutrition facts as shown for serving size of 1 cup, 8 fl oz (240 mL). Rice Dream is a registered trademark of Westbrae Natural, Inc.
[d] Normalized to non-fiber carbohydrate.
[e] Not listed on label, calculated: Total fat × 9.
[f] Calculated by difference: Total carbohydrates less dietary fiber.

found in two commercial rice beverages: one containing whole brown rice and another containing an extract of brown rice. The data in Table 21.4 was normalized against nonfiber carbohydrates since it is known that in these rice beverages, no added source of nonfiber carbohydrates, such as sucrose, has been added. As such, the whole grain beverage was found to have been made using approximately 32.4 grams of brown rice solids as compared with 28.6 grams in the rice extract beverage. Based on the content of brown rice that was used to make the beverage, the Normalized column provides the total amount of nutrients that would exist in the beverage if 100% of the whole grain existed in the beverage. The column adjacent to the Normalized column reflects the nutrients as stated in the label declaration. In the Rice Extract Beverages column, there are significant differences in the fat content indicating that a significant amount of fat was added to the product as declared in the ingredients label. Specifically, safflower or sunflower oil has been added. This was done to provide a more "milk-like" emulsion since most of the natural fat from the rice had been removed during the extraction process. Additionally, it is quite evident that most of the protein and substantially all of the fiber had been removed. Consequently, the natural nutrient density of the brown rice extract as found in this rice extract beverage product is substantially less than that found in the whole grain beverage product. Very importantly, it should be recognized that since there is an absence of fiber, this would indicate that the final product does not contain any whole grain in the finished consumer product, even though according to the ingredient label, it is made with brown rice, which is a whole grain. In the Rice Whole Grain Beverage column, the slightly higher level of fat and slightly lower fiber and protein value would indicate that the beverage, while made from 100% brown rice, apparently does not contain 100% whole grain in the finished product, but does contain as much as 70% whole grain in the finished product. One should note that in the ingredient declaration, there is no added oil or fat, which confirms the fact that the resulting milk is indeed made from the whole grain. This raises an interesting dilemma for the consumer with regard to understanding if the finished product they are consuming actually contains whole grains or is simply made with the extract of whole grains.

To remedy this potential confusion between products that contain the whole grain from those that do not contain the whole grain in the finished product, a nonprofit organization has recently been established to enable consumers to know immediately whether or not the product they are purchasing or consuming actually contains the whole grain. This organization, known as the Whole Grains Council (Whole Grain Council 2010) has created a stamp, as shown in Figure 21.3, which is currently being used by manufacturers to identify those consumer products that actually contain whole grains and the number of grams of whole grains per serving. Since the current Codes of Federal Regulations state that the ingredient used in making the product must be declared, but does not necessarily identify as to whether that whole ingredient is actually in the finished product, this

**FIGURE 21.3**   Identifying whole grain formulated foods with the whole grain stamp.

**FIGURE 21.4** Foods formulated with whole grains.

whole grain stamp is essential to identify products actually containing whole grains, as it may not be immediately evident from the label. As such, in the two examples provided, that for Rice Dream, which is a brown rice extract, and that for Ryza, which actually contains whole grains, only the Ryza product can be identified with the whole grain stamp.

As more and more manufacturers start to utilize the whole grain milk products, the uses and food categories of use will continue to be expanded. Since the food category may be expanded into unusual areas such as confectionary, frozen desserts such as nondairy ice cream, and even cultured product such as nondairy yogurt, these products will depend on identifying their whole grain content by use of the whole grain stamp. An example of this is in the whole grain rice product known as Ricera. As shown in Figure 21.4 which identifies some commercial food products that are formulated with whole grains, the products are readily identified with the whole grain stamp along with how much whole grain the finished product actually contains.

Recently, a California company, Creative Research Management, has started to make whole grain milks containing sucralose, or a high-intensity natural sweetener derived from stevia. Incredibly, this sweetened whole grain product, in syrup or powdered form, can be used in bakery applications as a 1:1 replacement for sugar, corn syrup, maltodextrin, and high fructose corn syrup. Basically, in consumer baked goods, whole grains can now easily be used to replace refined sugars! Additionally, this replacement can occur without increasing the calories. The concept is to use more nutrient-dense ingredients to replace refined empty calories. For the very first time in history of food formulation, these sweetened whole grain products, because of their unique functionality in baking and other food applications, allow for a replacement of refined sugars with whole grains.

These transformed whole grain milks and sweetened whole grain milks can be used in formulating foods across a variety of food categories from beverages to confectionary. Using whole grain milks, whole grains need not be limited to their use in breads and cereals. Additionally, these transformed whole grains have greatly improved microbiological and oxidative stability, allowing for convenient usage by both manufacturers and consumers. Transformed whole grains are high-density food ingredients having versatility in simple foods as an alternative to dairy milk, or in more complex foods such as confectionary and baking mixes. It is known that the transformation of whole grains has been done with rice, corn, oats, wheat, barley, and quinoa, and it is anticipated that the market of each of these transformed whole grains will be driven in part by the genomic requirements of individuals.

## FUTURE TRENDS

As we continue to unravel the human genome, recommendations of foods based on the individual's genome will be inevitable. Processed foods have the distinction of being convenient, stable, and in a form that is easy for the consumer to utilize. Based upon the understanding of nutritional genomics, we are starting to see new products and categories in processed foods that utilize improved ingredients and new process technologies. However, the objective of the food scientist is no longer to improve just the process and packaging of the whole food, or the ingredients used in making whole food containing products, but actually transforming the whole food into high-nutrient-density ingredients that are convenient, safe, stable, and suitable for use in simple and complex foods in a greater diversity of food products beyond what would have normally been conceived for that food source. Whole grains, transformed beyond the simplicity of grinding to a fine flour, now exist as "milks" that are not limited to breads and cereals but may be used in a greater variety of food categories such as frozen desserts, confectionary, cultured foods (such as yogurts), and beverages. The transformed whole food source can more readily meet the industrial need for whole food ingredients and the subsequent consumer need for simple and complex food products while providing high nutrient density and specific nutritional genomic requirements.

## REFERENCES

Fardet, A., E. Rock, and C. Remesy. 2008. Is the *in vitro* antioxidant potential of whole grain cereals and cereal products well reflected *in vivo*? *J Cereal Sci* 48(2):258–6.

Food Allergen Labeling and Consumer Protection Act of 2004. U.S. Food and Drug Administration Center for Food Safety and Applied Nutrition. http://www.CfSan.FDA.gov/~dms/alrgact.html (accessed March 10, 2010).

Grace, V. 2008. Milk alternatives and dairy-free beverages. http://foodallergies.about.com/od/dairy/tp/milkalternatives.htm (accessed March 10, 2010).

Heldman, D. R. and R. W. Hartel. 1997. *Principles of Food Processing*. New York: Chapman & Hall.

IFT. 2008. Innovative food packaging solutions: A scientific status summary of the Institute of Food Technologists. Eds., Brody, A. L., B. Bugusu, J. H. Han, C. K. Sand, and T. McHugh. *J Food Science* 73(8):R107–R116.

Jones, J. M., and J. Engleson. 2010. Whole grains: Benefits and challenges. *Annu Rev Food Sci Technol* 1:19–40.

Messina, M., and C. Hughes. 2003. Efficiency of soyfoods and soybean isoflavone supplements for alleviating Menopausal symptoms is positively related to initial hot flash frequency. *J Med Foods* 6:1 11.

Mitchell, C. R. 2005. US Patent # 7,678,403, 2005.

Natural Product Companies. 2008. http://www.24-7pressrelease.com/press-release/ground-breaking-natural-prodcut-cpmpanies-propel-mainstream-growth-110808.php (accessed March 10, 2010).

Ragheb, M. 2009. *Food Preservation by Radiation*, pp. 1–19. Urbana-Champaign, IL: University of Illinois Urbana-Champaign.

Rajasree Pai, R. and V. K. Raghesh. 2007. Health benefits of whole grains: A literature review. *Internet Journal of Nutrition and Wellness* 4(2):36. http://.isub.com/journal/the_internet_journal_of_nutrition_and_wellness/vol4_2_31/article/health benefits of whole grains:a literature review.html (accessed August 16, 2011).

Roberts, J. S., A. Teichert, and T. H. McHugh. 2008. Vitamin D2 formation from postharvest UV-B treatment of mushrooms (*Agaricus bisporus*) and retention during storage. *J Agric Food Chem* 56(12):4541–4.

Whole Grain Council. 2010. http://www.WholeGrainsCouncil.org (accessed March 10, 2010).

# Section III

Food Needs to Meet
Nutrigenomic Health Needs

*Regulatory Oversight*

# 22 Nutritional Genomics and the Future of Food Labeling in the United States

*Evelyn D. Cadman*

## CONTENTS

## INTRODUCTION

Advancements in nutrigenomic research may lead us to a time when genetic testing may enable each of us to know which foods would help us achieve optimal health. In the meantime, nutritional genomic research such as that discussed in the other chapters of this book may provide the basis for understanding specific biochemical constituents of food that may promote health. Yet, how will such information be conveyed to consumers? Could the data discussed throughout this book be used to make claims on food products or at least lay the foundation for such claims? This chapter evaluates the current state of science-based food labeling and then describes the potential for future changes as data from nutritional genomics continues to come forth.

Most Americans encounter food and beverage labels on a daily basis. According to U.S. Department of Agriculture (USDA) food expenditure data, Americans spent over $500 billion ($565,284,000,000) on food products in 2008 (USDA ERS 2009a), the equivalent of $2141.00 per person per year (USDA ERS 2009a). In the year 2000, U.S. consumers had about 40,000 food products to choose from in the typical supermarket (Harris et al. 2002). Over the past several years over 20,000 new food products were introduced into the U.S. market (USDA ERS 2009b) and several thousand others were retired. In 2008, over 5.5%, or 1389, of these new products included the claims "high vitamin" or "low trans fats" (USDA ERS 2009b).

The Nutrition Labeling and Education Act of 1990 (NLEA 1990) required most food products to bear labels that included nutrition information beginning in 1994. While a recent Food and Drug Administration (FDA) health and diet survey (USDHHS 2008) found that 61% of the people surveyed reported that nutrition information was important to them and another 35% reported that nutrition information is somewhat important when shopping for food, a study by the Economic Research Service of the USDA found that consumer use of food labels for making purchasing decisions declined by 10% from 1995 to 2006 (Todd and Variyam 2008). Among adults over 20 years of age the decline was greatest (17.2%) for the use of health claims and was significant for use of the ingredient list, health claims, and all nutrients except for fiber and sugars. Interestingly, the FDA released results of a 2008 survey showing a 10% increase in respondents who reported reading a food label the first time they buy a product but that 56% of those surveyed believed that only some or none of the nutrient content claims on labels are accurate (FDA 2008, 2010a; 21 CFR 1.3 (a)).

Common sense tells us that what we eat affects our health and that whole foods and a quality diet can provide nutrients and bioactive components for good health. We have many common expressions about food and diet: "You are what you eat"; "An apple a day keeps the doctor away"; and "Through the lips and on the hips." The effects of diet on gene expression were the topics of discussion presented at the nutritional genomics conference held at Cal Poly Pomona in November 2009. From these discussions it is clear that the potential exists for the understanding of these interactions to one day enable good health and prevent the onset of disease. If current research results further support this potential and individual genotypic susceptibilities to diet related diseases are elucidated, what will food manufacturers be allowed to say on food labels?

Most of us know people with medical conditions such as diabetes or high blood pressure that are responsive to diet. Hippocrates said, "Let food be your medicine and your medicine be your food." Consumers report that reading food labels could be a tool in helping them improve their health (Borra 2006). Yet, in spite of common knowledge and this sage advice, food labeling regulations generally do not permit foods to be marketed as products that may treat, cure, or prevent disease. In fact, had there been an FDA and if Hippocrates had offered food products for sale to his patients, the agency would likely have responded with a warning letter including a statement similar to "Because of these intended uses, these products are drugs within the meaning of the Food, Drug and Cosmetic Act." Recently, the FDA's National Center for Toxicological Research created an initiative to look at personalized nutrition and medicine, appointing Jim Kaput, PhD, as director. He has an established career in the rapidly evolving field of nutritional genomics.

## LABELING

Labeling is the intersection between the nutritional genomic scientists and the world of consumers. Labeling refers not only to information affixed to the immediate container of a product but also "includes all written, printed, or graphic matter accompanying [a product]…in interstate commerce" (21 CFR 1.3(a)) and so extends to any promotional material or advertising. Federal regulations allow for several types of claims on food product labels, including nutrient content claims, health claims, qualified health claims, and Food and Drug Administration Modernization Act (FDAMA) claims. Each of these types of claims must comply with FDA regulations and must be truthful, not misleading, and supported by scientific evidence.

## NUTRIENT CONTENT CLAIMS

Nutrient content claims are statements that discuss the level or range of a particular nutrient in a food product (21 CFR 101.13). Examples of nutrient content claims include "contains vitamin C," "high fiber," "low fat," and "healthy." FDA regulations spell out the requirements that must be met for foods to bear particular nutrient content claims in sections 101.54 through 101.67 of title 21 of the Code of Federal Regulations (Table 22.1).

A main dish food product bearing the nutrient content claim, "healthy" must meet the following nutritional criteria:

1. Low in fat, containing less than 3 g of fat per 100 g of product and no more than 30% of the calories may be from fat
2. Low in saturated fat; containing no more than 1 g of saturated fat per 100 g of the food and no more than 10% of the calories may be from saturated fats
3. No more than 90 mg of cholesterol per serving
4. Must contain at least 10% of the recommended daily intake (RDI) per serving of two nutrients from the list of vitamin A, vitamin C, calcium, iron, protein, or fiber
5. May not exceed 480 mg of sodium per serving

It should be noted that the brand name "Healthy Choice" is exempted from this definition since the name was in use prior to promulgation of the regulation in October of 1989 (21 CFR 101.13 (q)(1)).

## ALLOWABLE HEALTH CLAIMS

FDA regulations permit the use of certain health claims that characterize the relationship between a food and disease. Such claims are the subject of petitions submitted by interested parties and are based on strong supporting evidence in publically available information from numerous, well-controlled studies. Allowable health claims are authorized by the FDA when there is significant scientific agreement over the interpretation of the data supporting the claim. In a guidance document the FDA states, "The assessment of significant scientific agreement then derives from the conclusion that there is a sufficient body of sound, relevant scientific evidence that shows

---

## TABLE 22.1
## Selected Nutrient Content Claims

| Claim | Eligible Nutrients | Level Required to Use Claim |
|---|---|---|
| Contains, good source, provides* | Nutrients with RDI/DRV | 10% to 19% of the RDI or DRV |
| Rich in, high, excellent source* | Nutrients with RDI/DRV | 20% or more of the RDI or DRV |
| Fortified, enriched, added, extra, plus | Nutrients with RDI/DRV | 10% or more of the RDI or DRV |
| Sodium free, no sodium, zero sodium, without sodium** | Sodium | The food contains less than 5 mg of sodium per labeled serving |
| Low in saturated fat; low saturated fat; little saturated fat*** | Saturated fat | The food contains 1 g or less of saturated fatty acids per reference amount customarily consumed and not more than 15% of calories from saturated fatty acids |

This table contains a partial listing of the requirements for the selected nutrient content claims.

Code of Federal Regulations, 21CFR, Nutrient Content Claims.

*21 CFR 101.54

**21 CFR 101.61

***21 CFR 101.62

**TABLE 22.2**
**Allowable Health Claims**

| Defined in 21 CFR | Title of Health Claim |
| --- | --- |
| 101.72 | Calcium and osteoporosis |
| 101.73 | Dietary lipids and cancer |
| 101.74 | Sodium and hypertension |
| 101.75 | Dietary saturated fat and cholesterol and risk of coronary heart disease |
| 101.76 | Fiber containing grain products, fruits, and vegetables and cancer |
| 101.77 | Fruits, vegetables, and grain products that contain fiber, particularly soluble fiber, and risk of coronary heart disease |
| 101.78 | Fruits and vegetables and cancer |
| 101.79 | Folate and neural tube defects |
| 101.80 | Dietary noncariogenic carbohydrate sweeteners and dental caries |
| 101.81 | Soluble fiber from certain foods and risk of coronary heart disease |
| 101.82 | Soy protein and risk of coronary heart disease |
| 101.83 | Plant sterol/stanol esters and risk of coronary heart disease |

*Source:* Code of Federal Register, 21 CFR 101, Health Claims.

consistency across different studies and among different researchers and permits the key determination of whether a change in the dietary intake of the substance will result in a change in a disease endpoint" (FDA 1999).

Once authorized, the health claims are codified in regulation that describes the relationship between the nutrient and disease, the requirements for and limitations on use of the claim, as well as model claim language. The model language for two authorized health claims are given below and a list of all allowable health claims is provided in Table 22.2.

- Calcium and vitamin D for osteoporosis: Adequate calcium and vitamin D throughout life, as part of a well-balanced diet, may reduce the risk of osteoporosis (21 CFR 101.72).
- Folic acid and neural tube defects: Healthful diets with adequate folate may reduce a woman's risk of having a child with a brain or spinal cord birth defect. (21 CFR 101.79).

On March 22, 1995, the Quaker Oats Co. submitted a health claim petition to the FDA requesting that the agency authorize a health claim on the relationship between the consumption of oat products and the risk of coronary heart disease. Thirty-seven human studies on oat bran and oatmeal and serum cholesterol were included in the petition. In 2001 a Quaker Oats petition to expand the oats and cholesterol claim to include extracted beta glucan included 80 references (Quaker Oats and Rhodia, Inc 2001). The resulting allowable health claim is shown below.

- Oats and cholesterol: Diets low in saturated fat and cholesterol that include 3 grams or more per day of [beta]-glucan soluble fiber from whole oats may reduce the risk of heart disease. One serving of [name of food] provides _____ grams of this soluble fiber.

The health claim regarding the relationship between oats and cholesterol was in the news in May of 2009 after the FDA issued a warning letter to General Mills regarding the use of unauthorized language discussing oats and cholesterol on boxes of Cheerios (Figure 22.1). The letter stated in part,

Based on claims made on your product's label, we have determined that your Cheerios® Toasted Whole Grain Oat Cereal is promoted for conditions that cause it to be a drug because the product is intended

**FIGURE 22.1** Cheerios label, which was the subject of a 2009 warning letter.

for use in the prevention, mitigation, and treatment of disease. Specifically, your Cheerios® product bears the following claims on its label:

- "you can Lower Your Cholesterol 4% in 6 weeks"
- "Did you know that in just 6 weeks Cheerios can reduce bad cholesterol by an average of 4 percent? Cheerios is ... clinically proven to lower cholesterol. A clinical study showed that eating two 1.5 cup servings daily of Cheerios cereal reduced bad cholesterol when eaten as part of a diet low in saturated fat and cholesterol."

These claims indicate that Cheerios® is intended for use in lowering cholesterol, and therefore in preventing, mitigating, and treating the disease hypercholesterolemia. Additionally, the claims indicate that Cheerios® is intended for use in the treatment, mitigation, and prevention of coronary heart disease through, lowering total and "bad" (LDL) cholesterol. Elevated levels of total and LDL cholesterol are a risk factor for coronary heart disease and can be a sign of coronary heart disease. Because of these intended uses, the product is a drug within the meaning of section 201(g)(1)(B) of the Act [21 U.S.C. § 321 (g)(P)(B)]. The product is also a new drug under section 201(p) of the Act [21 U.S.C. § 321(p)] because it is not generally recognized as safe and effective for use in preventing or treating hypercholesterolemia or coronary heart disease. Therefore, under section 505(a) of the Act [21 U.S.C. § 355(a)], it may not be legally marketed with the above claims in the United States without an approved new drug application.

**(FDA Minneapolis Office 2009)**

General Mills responded to this letter by submitting studies to support the cholesterol-lowering claims. Dr. Barbara O. Schneeman, director of the Office of Nutrition, Labeling and Dietary Supplements at the Center for Food Safety and Applied Nutrition, responded after assessing the newly submitted studies and concluded that,

> In determining whether to amend 21 C.F.R. § 101.81 to authorize the soluble fiber/coronary heart disease claim to include a statement about a specific percentage reduction in LDL cholesterol, FDA would need to consider the totality of the publicly available scientific evidence to support such a statement and also how to convey information in a way that is not misleading (Schneeman 2009a).

This remains an active issue and only time will tell what happens here. General Mills has stated they have not misrepresented the data. As of this writing, claims regarding Cheerios® and cholesterol reduction remain on the company website (General Mills 2010). In a private conversation a former FDA staff member stated that the only court cases FDA has lost have been based on First Amendment principles. It will be interesting to see whether such arguments are made in this case.

## QUALIFIED HEALTH CLAIMS

Qualified health claims represent a class of claims that are also petitioned for by interested parties but for which the supporting evidence is weaker and does not meet the criteria required for allowable health claims. Accordingly, the claims are "qualified" by inclusion of a disclaimer. Rather than being the subject of regulations, enforcement discretion letters are issued for qualified health claims. These letters describe the FDA's assessment of the data submitted in the health claim petition and also include the language that may be used for the claim. The qualified claim for omega-3 fatty acids serves as an illustration (see Table 22.3) for a list of health conditions that are the subject of qualified health claims:

> Supportive but not conclusive research shows that consumption of EPA and DHA omega-3 fatty acids may reduce the risk of coronary heart disease. One serving of [Name of the food] provides [ ] gram of EPA and DHA omega-3 fatty acids (Hubbard 2004).

The provisions for qualified health claims were created as a result of court decisions in *Pearson v. Shalala*, which challenged the FDA's general health claims regulations and the agency's decision not to authorize health claims for four specific substance/disease relationships. A district court first ruled for the FDA (14 F. Supp. 2d 10 (D.D.C. 1998)). However, on appeal, the U.S. Court of Appeals for the DC Circuit reversed the lower court's decision (164 F.3d 650 (D.C. Cir. 1999)). The appeals court ruled that the First Amendment does not permit the FDA to reject health claims that are deemed potentially misleading unless the agency also determines that no disclaimer would eliminate the potential deception. In general, qualified health claims are not supported by data that meet the standard of significant scientific agreement (Hubbard 2004) that is required for health claims and are therefore "qualified" by inclusion of a statement regarding the strength of the data. In many cases the disclaimer is stronger than the claim, as is illustrated by the qualified health claim for selenium and prostate cancer.

> Two weak studies suggest that selenium intake may reduce the risk of prostate cancer. However, four stronger studies and three weak studies showed no reduction in risk. Based on these studies, FDA concludes that it is highly unlikely that selenium supplements reduce the risk of prostate cancer (Scheenman 2009c).

**TABLE 22.3**
**Qualified Health Claims Describing Nutrient Disease Relationships**

| Nutrient | Disease Relationship |
|---|---|
| 1. Omega 3 fatty acids | Coronary heart disease |
| 2. B-vitamins | Vascular disease |
| 3. Selenium | Cancer |
| 4. Antioxidant vitamins | Cancer |
| 5. Phosphatidylserine | Cognitive dysfunction and dementia |
| 6. Nuts | Heart disease |
| 7. Walnuts | Heart disease |
| 8. Monounsaturated fatty acids from olive oil | Coronary heart disease |
| 9. Green tea | Cancer |
| 10. Chromium picolinate | Diabetes |
| 11. Calcium | Colon/rectal cancer and Recurrent colon/rectal polyps |
| 12. Calcium | Hypertension Pregnancy-induced hypertension Preeclampsia |
| 13. Tomatoes and/or tomato sauce | Prostate, ovarian, gastric, and pancreatic cancers |
| 14. Unsaturated fatty acids from canola oil | Reduced risk of coronary heart disease |
| 15. Corn oil and corn oil-containing product | Reduced risk of heart disease |
| 16. Folic acid | Neural tube birth defects |

*Note:* Qualified health claims: Letters of enforcement discretion available at http://www.fda.gov/
Food/LabelingNutrition/LabelClaims/QualifiedHealthClaims/ucm072756.htm.

Appeals of *Pearson v. Shalala* and new court cases regarding qualified health claims continued through 2009. In May of 2010, the U.S. District Court for the District of Columbia ruled that the FDA had failed to comply with *Pearson v. Shalala* and the agency's own guidance document regarding evaluation of scientific evidence for qualified health claims. The Court remanded the case back to the FDA (Alliance for Natural Health U.S. v. Sebelius, 2010).

for the purpose of 1) drafting one or more disclaimers to accompany plaintiffs' certain cancers, anti-carcinogenic, and prostate claims, or, alternatively, setting forth empirical evidence that any disclaimer would fail to correct the claims' purported misleadingness; 2) determining an appropriate disclaimer to accompany plaintiffs' lung and respiratory tract claim in light of the SU.VI.MAX study; and 3) reevaluating plaintiffs' colon and digestive tract claim and drafting one or more disclaimers.

## FDAMA CLAIMS

FDAMA claims were created by passage of the Food and Drug Administration Modernization Act of 1997 (21 U.S.C. § 343). These claims are authorized based on authoritative statements from "a scientific body of the United States Government with official responsibility for public health protection or research directly relating to human nutrition (such as the National Institutes of Health or the Centers for Disease Control and Prevention) or the National Academy of Sciences or any of its subdivisions." The statement must be "currently in effect" and "[identify] the nutrient level to which the claim refers" (21 U.S.C. § 343). Interested parties are permitted to notify the FDA regarding their intention to make a claim based on an authoritative statement. According to an FDA guidance document the notification must include the language of the intended claims, a description of how it

**TABLE 22.4**
**Seven FDAMA Claims**

| Claim | Source of Statement |
|---|---|
| Nutrient content claim choline containing foods | Food and Nutrition Board, Institute of Medicine (IOM), National Academy of Sciences |
| Fluoride and the risk of dental caries | Centers for Disease Control, Surgeon General, Public Health Service |
| Potassium and the risk of high blood pressure and stroke | National Academy of Sciences |
| Saturated fat, cholesterol, and trans fat, and the risk of heart disease | Department of Health and Human Services, Department of Agriculture |
| Substitution of saturated fat with unsaturated fatty acids and risk of heart disease | National Academy of Sciences |
| Whole grain foods and the risk of heart disease and certain cancers | National Academy of Sciences |
| Health claim notification for whole grain foods with moderate fat content and the risk of heart disease and certain cancers | National Academy of Sciences |

*Source:* FDAMA, 21 U.S.C. § 343.

was determined that the requirements for an authoritative statement have been satisfied, a copy of the authoritative statement, and a balanced representation of the scientific literature regarding nutrient disease relationship that is the subject of the claim (FDA 1998). Some sample FDAMA claims are as follows:

- Diets rich in whole grain foods and other plant foods and low in total fat, saturated fat, and cholesterol may reduce the risk of heart disease and some cancers.
- Diets containing foods that are a good source of potassium and that are low in sodium may reduce the risk of high blood pressure and stroke.

Of the seven FDAMA claims (Table 22.4) that are authorized, most derive from statements made by various offices at the National Academy of Sciences. Other agencies that are considered authorities for the purpose of FDAMA claims include the National Institutes of Health (NIH), the Centers for Disease Control and Prevention (CDC), which are both part of the Department of Public Health; the surgeon general as well as the Food and Nutrition Service (FNS), Food Safety and Inspection Service (FSIS), and Agricultural Research Service (ARS), which are part of the Department of Agriculture.

## U.S. FOOD INDUSTRY USE OF FOOD LABEL CLAIMS

The food industry in the United States has tried to creatively apply the various regulations covering food label claims to come up with ways to inform customers about the quality or benefits of their products (Figure 22.2). Whether this is simply public service or a means to differentiate their products and drive sales, the result has been a plethora of symbols and systems that vary greatly. Ultimately, consumers probably remain confused.

### Guiding Stars Program

The Guiding Stars program, begun in 2006, is described as "objective, based on consumer research, and not influenced by price, brand or manufacturer trade groups" (Guiding Stars 2010). This program was developed by a panel of experts and uses a proprietary algorithm to evaluate data in

**Guiding Stars®**
Nutritious choices made simple®

**FIGURE 22.2**   Various front of package (FOP) symbols. Left: Smart Choices logo appearing with nutrition at a glance information. Right: NuVal and Guiding Stars logos.

the nutrition facts panel to determine the amount of essential vitamins, minerals, fiber, and whole grains per 100 calories for positive value. In addition, the levels of saturated fat, trans fat, cholesterol, added sodium, and added sugars are evaluated and detract from the product's score and potential rating from zero to three stars. In 2009 the Guiding Stars program modified their algorithm and criteria twice. One adjustment was to account for the fact that naturally occurring trans fat are found in dairy and some meat products—originally *any* trans fat reduced a food's eligibility for the program (Guiding Stars 2009a). Another adjustment was made to accommodate the American Heart Association's recommendation to reduce added sugars (Guiding Stars 2009b). The program lists grocery store chains, food service organizations, and insurance companies among the groups using their system.

## NuVal Program

The NuVal program provides composite scores—you get positive number scores for good nutrients and "bad" nutrients lower the score by being subtracted from the number.

This program utilized a panel of experts led by Dr. David Katz of the Yale Prevention Research Center to develop an algorithm that converts nutritional information into a numerical score ranging from 1 to 100. The development was funded by the nonprofit Griffin Hospital and Yale University School of Medicine affiliate. According to company information, the algorithm evaluates over 30 nutrients and nutrition factors including protein, calcium, vitamins, sugar, sodium, and cholesterol. Similar to the Guiding Stars program, some nutrients increase the NuVal score and others such as fat and cholesterol lower the score. The company website lists three grocery store partners (http://www.nuval.com/).

## Nutrition at a Glance

In an unusual twist, Kellogg's Nutrition at a Glance front of package (FOP) labeling, which lists the amounts of calories and major nutrients such as fat and carbohydrates as well as some vitamins and minerals, uses the United Kingdom's guideline daily amounts (GDAs) to also present the percentage of each nutrient that that product provides in each serving. It is interesting that the GDAs for fat and sugar are higher than the recommended daily intake (RDI) and percent daily values (DV) used by the FDA for nutritional labeling, while the GDA for fiber is lower than the RDI and DV for this nutrient. It will be interesting to see if this company, beleaguered by reaction to the use of immunity enhancing claims on several sugary cereals marketed for children, will draw the FDA's attention for the use of the GDAs as well as the noncompliant nutrient content claims the manner in which the amounts are stated represents. This may well have been one of the programs that FDA views as using "symbols that either expressly or by implication are nutrient content claims" and promised to assess and compare with regulatory criteria in the agency's letter on point of purchase food labeling (Schneeman 2009b).

## SMART CHOICES PROGRAM

The Smart Choices Program made 2009 a very interesting year for nutrition claims on food packages. Created by a group of scientists, academicians, health and research organizations, food and beverage manufacturers, and retailers, this complicated program specified criteria for 19 different product categories, including beverages, cereals, meats, dairy, and snacks.

The program was launched in August of 2009 and quickly came under scrutiny. On September 4, 2009 an article in *The New York Times* discussed the inclusion of Froot Loops cereal as a Smart Choice (Neuman 2009). The article cited nutritionists who were not part of the Smart Choice program as being appalled by some of the choices, while nutritionists who were part of the program defended standards stating that they were based on government dietary guidelines and widely accepted nutritional standards and influenced by research showing that consumers did not want negative messages or the sense that food choices were being dictated to them. The designation of Froot Loops as a Smart Choice was justified as being a better choice than a donut for a child's breakfast.

The public outcry over Froot Loops being designated a Smart Choice grew and included a letter to the FDA from Representative Rosa L. DeLauro, chairwoman of the House Appropriations Subcommittee on Agriculture Rural Development, Food and Drug Administration, and Related Agencies. An October 23, 2009 news release from Smart Choice announced that it would "voluntarily postpone active operations and not encourage wider use of the logo ... by either new or currently enrolled companies" (Smart Choices 2009). By October 30, companies, including Kellogg's, Kraft Foods, ConAgra Foods, Unilever, General Mills, PepsiCo, and Tyson Foods, that had been using the Smart Choices logo stated they would cease doing so.

The action to curtail the Smart Choices program was not only in response to public outcry about Froot Loops being denoted as a Smart Choice but primarily in response to an October 20, 2009 FDA letter to industry regarding point of purchase and FOP labeling. The letter stated in part that criteria used by the various programs to label the front of packages and store shelves with nutrition information were not consistent and that consumers would be less likely to check the nutrition facts box when such symbols were used. The letter also stated that all label claims and symbols are subject to FDA regulations and that the agency "[was] analyzing FOP labels that appear to be misleading" and "developing a proposed regulation to define the nutritional criteria" for making broad FOP or shelf label claims concerning the nutritional quality of a food. The letter further stated that it is the FDA's "intent is to provide standardized, science based criteria on which FOP nutrition labeling must be based" (Schneeman 2009a).

On November 3, 2009 Smart Choices Chairman of the Board Michael Hughes published an opinion piece in *USA Today* in which he stated that leading experts in nutrition, public health, and food science, as well as food industry representatives participated in the process. Government agencies sent observers. "We will gladly participate in the FDA's process, just as the agency participated with us" (Hughes 2009).

In May of 2009 the American Dietetic Association released a review of various front of package labeling systems (American Dietetic Association 2009) and published a chart comparing the ratings various foods would receive under several of these labeling paradigms. As shown in the portion of the chart depicted in Figure 22.3, Planters® Lightly Salted Cocktail Peanuts would receive disparate ratings from Guiding Stars, NuVal and Smart Choices.

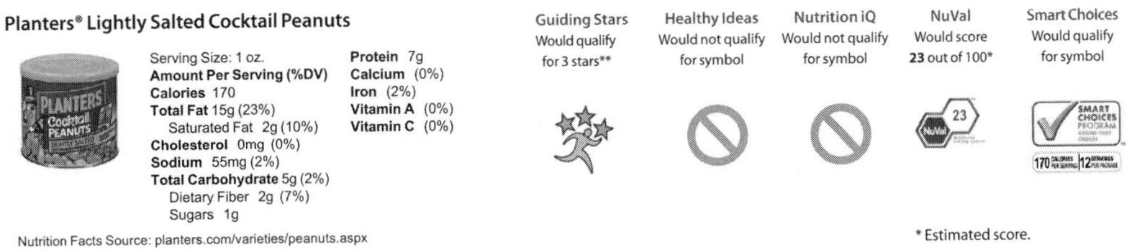

Planters® Lightly Salted Cocktail Peanuts
Serving Size: 1 oz.
Amount Per Serving (%DV)
Calories 170
Total Fat 15g (23%)
Saturated Fat 2g (10%)
Cholesterol 0mg (0%)
Sodium 55mg (2%)
Total Carbohydrate 5g (2%)
Dietary Fiber 2g (7%)
Sugars 1g
Protein 7g
Calcium (0%)
Iron (2%)
Vitamin A (0%)
Vitamin C (0%)
Nutrition Facts Source: planters.com/varieties/peanuts.aspx
Guiding Stars Would qualify for 3 stars**
Healthy Ideas Would not qualify for symbol
Nutrition iQ Would not qualify for symbol
NuVal Would score 23 out of 100*
Smart Choices Would qualify for symbol
23
* Estimated score.

**FIGURE 22.3** The differing ratings a single product would receive from various front-of-package labeling programs. (Excerpted from American Dietetic Association, 2009, used by permission.)

## Traffic Light Labeling

The FDA's letter to industry on point of purchase food labeling also discussed the United Kingdom's traffic light FOP labeling program:

> The recent experience with FOP labeling in the United Kingdom demonstrates the potential of voluntary initiatives to provide consumers helpful FOP labeling. In that instance, the government set certain criteria for the use of such labeling, and retailers took the initiative to implement FOP labeling in their stores. The agency wants to explore the potential of that approach (Schneeman 2009b).

It should be noted that, unlike in the United States, nutrition labeling is not compulsory on UK food products. In order to help consumers make nutritionally sound food purchase choices, the British government has encouraged the use of a standardized traffic light symbol on the front of food product labels. The levels of various nutrients present in the product are rated as green, yellow, or red (Figure 22.4). According to the UK's Food Standards Agency, a red light symbol indicates that "the food is high in something we should be trying to cut down on. It's fine to have the food occasionally, or as a treat, but try to keep an eye on how often you choose these foods, or try eating them in smaller amounts." The yellow light is neutral and green indicates "the healthier choice" (Food Standards Agency 2010). The number of grams of fat, saturated fat, sugars, and salt in what the manufacturer or retailer suggests as a "serving" of the food is also listed in the graphic.

The traffic light system has not been embraced by industry. According to the thesis of Debra Van Camp at Ohio State University, its adoption on products introduced from January 2002 through December 2008 ranged from 27% among foods not targeted by the agency to 55% of hot cereal, cold cereal, pastry dishes, pizzas, prepared meals, and sandwiches, which were targeted by the Food

Low fat
Low sat fat
High sugar
Med salt
7.7g Per serve
2.0g Per serve
42.2g Per serve
2.0g Per serve
Per serving
Fat 7.7g
Saturates 2.0g
Sugars 42.2g
Salt 2.0g
High Medium Low
Low Fat 7.7g per serving
Low Saturates 2.0g per serving
High Sugars 42.2g per serving
Med Salt 2.0g per serving

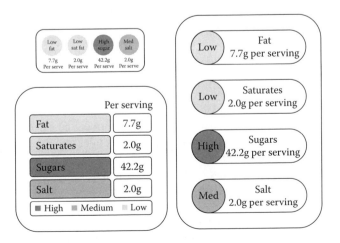

**FIGURE 22.4** UK traffic light symbols that may be used voluntarily on food product labels.

Standards Agency (Van Camp et al. 2009). The primary adopters were private label products; none of the major brands used the system (Food Standards Agency 2010).

Even though nutritional labeling is required on food product in the United States, the FDA appears to be intent on adding FOP labeling. In the October 2009 letter to industry on point of purchase labeling the agency stated (Scheenman 2009b),

> If voluntary action by the food industry does not result in a common, credible approach to FOP and shelf labeling, we will consider using our regulatory tools toward that end. This effort will include research to assess through consumer studies the likely effects of FOP symbols on information search behavior related to the Nutrition Facts label, which in turn can affect consumer understanding of the full nutrition profile of a product. The foundation of that approach should be a common set of mandatory nutritional criteria that consumers can rely on when they view FOP labels, even if no one symbol is ultimately selected as superior.

In March of 2010, FDA Commissioner Margaret Hamburg published an open letter to industry regarding FOP labeling in which she stated,

> I believe we now have a wonderful opportunity to make a significant advancement in public health if we can devise a front-of-pack labeling system that consumers can understand and use. We intend to work closely with food manufacturers, retailers, and others in the design process, and I hope that every food processor will contribute its views on how we can do this in the best way possible. In the meantime, FDA will soon issue new draft guidance relating to front-of-pack calorie and nutrient labeling (Hamburg 2010).

The letter was followed by the April 29, 2010 publication of Docket FDA-2010-N-0210, "Front-of-Pack and Shelf Tag Nutrition Symbols" for comment (FDA 2010b).

Whether such government-mandated FOP labeling would be helpful to consumers or would simply negate product differentiation via food label claims is an open question. It will be important for anyone with an interest in communicating nutrition information on food labels to participate in the policy development process. In fact, the FDA's statement that it is their "intent is to provide standardized, science based criteria on which FOP nutrition labeling must be based" (Schneeman 2009b), the March 2010 letter (Hamburg 2010) and the April 29, 2010 publication of Docket FDA-2010-N-0210 "Front-of-Pack and Shelf Tag Nutrition Symbols" are calls to action. Nutrigenomic scientists, the food industry, and other interested parties may start their participation by commenting on this and other proposals outlined in the letter to industry. It is imperative to be part of the discussion and shape the future of food labels with an eye on the growing field of nutritional genomics. Otherwise, what the FDA eventually decides could be detrimental to product labeling and the public's access to important nutritional information.

## OTHER COMMENTS ON NUTRITIONAL GENOMICS AND FOOD LABELING FROM THE FDA

The FDA's letter to industry regarding FOP labeling gives us a good understanding of the agency's view on the subject. Before that letter was published several former and current FDA staff members were interviewed and asked their thoughts on nutritional genomics and the future of food labeling. Their comments ranged widely and are summarized next.

The idea that FDA draws a bright line separating food and drug was stated several times along with the idea that the agency would be very wary about the sorts of claims that potentially could arise if a person could have their DNA assayed and learn what nutrients could be particularly beneficial to them. One interviewee commented, "The agency would not want to allow the line between foods and drugs to be blurred." The concept of claims needing to meet the criteria of significant scientific agreement was stated by several interviewees.

One person interviewed stated that, "From a regulatory standpoint and FDA; there is none more conservative in regard to being pessimistic or wary of new developments in nutrition than nutritionists and people at CFSAN" (Center for Food Safety and Applied Nutrition). He went on to explain that CFSAN staff are very skeptical because "there is a history of fraud and that charlatans are always there before the science." He said, "If companies wanted to discuss links between genomics and food, FDA would be very wary and would like the evidence to be very strong due to history of fraud."

One interviewee revealed that a former FDA commissioner said, "There are people who do molecular biology and there are people that feed animals and look for lumps and bumps—which attracts the smartest people?" The interviewee went on to explain, "Nutrition and toxicology studies are done in a black box. The studies involve broad groups and vague findings. Nutritional genomics may change that. Studies on pharmacology and genomics of disease have provided lots of evidence that people respond to drugs differently and the future may lead us to similar knowledge about food."

Other interviewees noted that regulations change to accommodate the science and that regulations are not prescriptive but are permissive and are changed via petition, some by legislation or via legislative authority given to the agency by Congress. Other interviewees noted that drugs must go through an approval process to show safety and effectiveness and suggested that in a similar fashion, the industry would need to show validity for nutrigenomic claims. The potential need for a process similar to generally regarded as safe (GRAS) notifications or new dietary ingredient (NDI) notifications was noted. These notifications must include safety data and exposure estimates in order to determine whether the use of the ingredient would pose health risks to consumers.

Another interviewee suggested that the best approach would be to make claims about obtaining a benefit versus avoiding harm, supporting health versus disease prevention, and went on to state,

> The idea that you have information that someone with a [genetic] tendency to develop a certain disease if they consume your product leads to ethical questions. Should you note on the label that certain folks should avoid this food? (Alcoholic beverage labels include warnings to pregnant women.) The idea that you would have material information regarding the effect of your product when used as intended gets right to the heart of the concept of misbranding. One aspect of misbranding is when the information presented is misleading in a particular. But if you have certain information and do not reveal it; that could also be misbranding.*

The most interesting comment from those interviewed was "FDA certainly will not lead the way [to nutrigenomic claims on food labels] but may be dragged kicking and screaming." Ultimately, knowledge changes and people's choices change, but the FDA is charged with protecting consumers and this responsibility will figure greatly in the process of responding to the science and industry attempts to inform consumer of the benefits of certain foods. Will future food labels guide individuals to select foods for optimizing their health? Will we have a series of color codes for those susceptible to diseases? Perhaps we will end up with a "eat right for your type" system. Or perhaps only general positive statements will be permitted, such as, "Great for people who require plenty of omega-3s," or "Formulated for people who require high levels of antioxidants."

## Possible Pathways to Nutritional Genomics Claims

So far, various types of claims that food products may qualify for have been discussed along with the applications of these claims and the FDA's response to some of them. So far all the claims discussed—nutrient content claims, allowable health claims, qualified health claims, and FDAMA claims—are authorized via the petition processes and are based on the level of scientific agreement in publically available information about the nutrient and disease states. In order to take advantage of

---

\* Misbranding is prohibited by the Federal Food, Drug, and Cosmetic Act1 (FD&C Act), which states, "The introduction or delivery for introduction into interstate commerce of any food, drug, device, or cosmetic that is adulterated or misbranded." A product is misbranded if it is incorrectly labeled either by the inclusion of false or misleading information or it fails to include information required by law.

these regulatory mechanisms as applied to nutritional genomic advancements, industry must not only watch the literature in order to gather publicly available information to support these petition processes but must also fund, conduct, and publish studies to add to the body of data. It is also important for scientists and industry to comment on FDA rulemakings. The FDA's letter on FOP labeling asked for comments, and these can be provided from a group or anonymously through attorneys or regulatory consultants. It is important for parties interested in influencing food labeling regulations to understand the regulations and to use real science and sound principles of nutrition when commenting on FDA guidance documents or call for comments. In order to effectuate future claims relevant to nutritional genomics, it will also be helpful to watch for authoritative statements from federal experts.

## OTHER ROUTES TO NUTRITIONAL GENOMICS CLAIMS

We have reviewed the various petition processes that may be used to gain approval for various types of claims that discuss the relationship between diet and health. There are also mechanisms that rely on public information that are not subject to premarket FDA review or approval but that may be useful for disseminating information about nutritional genomic advancements. It may be possible to market medical foods or foods for special dietary uses based on advancing science; it may also be possible to market products using structure function claims.

### Medical Foods

Medical foods are formulated for the dietary management of a disease or condition with distinctive, established nutritional requirements (Table 22.5). The definition of medical foods as extracted from the Orphan Drug Act can be found in 21 CFR 101.9(j)(8):

> A medical food is a food which is formulated to be consumed or administered enterally under the supervision of a physician and which is intended for the specific dietary management of a disease or condition for which distinctive nutritional requirements, based on recognized scientific principles, are established by medical evaluation.

The regulation goes on to stipulate that a medical food

- Is a specially formulated and processed product, not a naturally occurring food used in its natural state
- Provides nutritional support specifically modified for the management of the unique nutrient needs that result from the specific disease or condition, as determined by medical evaluation

The regulation describes the patient medical foods as formulated for

- A patient who has limited or impaired capacity to ingest, digest, absorb, or metabolize food or certain nutrients because of therapeutic or chronic medical needs
- A patient who has other special medically determined nutrient requirements which cannot be met by the modification of the diet alone
- A patient who requires medical care on a recurring basis and is receiving active and ongoing medical supervision

Medical foods are exempt from nutrition labeling regulations but are subject to other basic labeling requirements, as well as the food GMP (Good Manufacturing Practice) regulations. In 1996, the FDA published an Advanced Notice of Proposed Rulemaking (ANPR) (FDA 1996) with the intent to regulate medical foods more rigorously. However, after a number of years of inaction, the agency withdrew the proposed rule while stating that the proposed rule could still reflect the FDA's view on the subject. Chatter in trade publications indicates that some expect the FDA to take up the issues

**TABLE 22.5**
**Examples of Medical Foods**

| Target Disease | Medical Food | Purpose |
| --- | --- | --- |
| Kidney disease | Type and quality of protein | Provide adequate nutrition that is minimally taxing for diseased kidneys due to moderate protein levels, low fluid and electrolyte levels, and high caloric content |
| Liver disease | Type and quality of protein | Provide adequate protein without inducing or exacerbating hepatic encephalopathy and to overcome malnutrition |
| Hypermetabolic states (severe burns, trauma or infection) | Type and quality of protein; added amino acids; elevated levels of specific vitamins or minerals | Provide high calories for increased energy needs; glutamine to aid intestinal cell proliferation to overcome impaired GI function |
| Lung disease (chronic obstructive pulmonary disease, acute respiratory distress syndrome, cystic fibrosis | High fat, low carbohydrate content | Reduce load of carbon dioxide the lungs must clear |
| Compromised immune function, human immunodeficiency virus infection, acquired immune deficiency syndrome (AIDS) | Enriched with specific amino acids; fortified with increased levels of vitamins | To support immune function and increase $CD_4/CD_s$ ratio |
| Oral rehydration solutions | Solutions of water electrolytes and carbohydrate source | To quickly restore fluids and minerals lost in diarrhea and vomiting in infants and children |
| Phenylketonuria (PKU) | Restricts dietary phenylalanine | Phenylalanine-ree to allow greater intake of complete protein |
| Malabsorption as found in inflammatory bowel disease (ulcerative colitis, Crohn's disease); radiation enteritis; short bowel syndrome | Pre-digested macronutrients; altered type or quantity of fat | To reduce the severity and incidence of diarrhea and abdominal discomfort in individuals with fat malabsorption |
| Diabetes mellitus | Type and quantity of carbohydrate; high fiber | To enhance blood sugar control |

*Source:* FDA, 1996, Regulation of Medical Foods.

once again, while others see the overburdened and under-funded agency as unable to address medical foods in light of more pressing concerns.

While the formulation for the medical food must be based on "recognized scientific principles," under current regulations, there is no explicit requirement that clinical studies be conducted on the medical foods nor must the product be reviewed by FDA prior to market introduction. As stated in the 1996 ANPR, medical foods "may bear claims that have not been evaluated under the 1990 amendments to ensure that they are scientifically valid. Moreover, there is no assurance that the formulation of a medical food has been evaluated prior to sale to ensure that it is suitable for the intended patient population." However, the requirement that claims must be supported by scientific evidence does necessitate the availability of supporting studies. Under the guidelines of the 1996 ANPR, medical foods would have to be clinically tested and would very likely have to be reviewed and approved by FDA; however, testing is not presently required by FDA regulation.

It is possible that nutritional genomic research could lead the discovery of nutritional dysfunction that may be occurring in many diseases currently treated with pharmaceuticals. If so, these findings could be applied to the production of medical foods that could potentially lead to great improvements in patient care and outcomes. If supported by scientific data, medical foods may currently be marketed without prior FDA approval of the formula or the claims made about the product providing it conforms to the regulations discussed above. Table 22.5 provides information from the 1996 ANPR on medical foods that were on the market at that time.

## Foods for Special Dietary Uses

If a food is marketed for a particular dietary need, it is classified as a food for special dietary uses. Special dietary uses are defined in 21 CFR 105.3 as

  (i) Uses for supplying particular dietary needs which exist by reason of a physical, physiological, pathological or other condition, including but not limited to the conditions of diseases, convalescence, pregnancy, lactation, allergic hypersensitivity to food, underweight, and overweight;
 (ii) Uses for supplying particular dietary needs which exist by reason of age, including but not limited to the ages of infancy and childhood;
(iii) Uses for supplementing or fortifying the ordinary or usual diet with any vitamin, mineral, or other dietary property. Any such particular use of a food is a special dietary use, regardless of whether such food also purports to be or is represented for general use.

The regulation also notes that the use of an artificial sweetener in a food in order to effect caloric value or available carbohydrates, or to make the food suitable for use by diabetics is a special dietary use. Food sold by weight loss programs qualify as foods for special dietary uses as do the ubiquitous low calorie beverages. The Nutrisystem program offers a line of products with double special dietary uses, Nutrisystem D, which are formulated for weight loss and diabetes control.

In the 1996 ANPR on medical foods, the FDA discussed the differences between medical foods and food for special dietary uses (FDA 1996):

> The statutory definitions of medical foods and foods for special dietary use overlap to the extent that both categories encompass foods that are intended for use by sick people. The differences in the statutory definitions evidence, however, that Congress intended foods for special dietary use under section 411(c)(3)(A) of the act to be a broader category of foods for use by people with special dietary needs or desires, while it intended medical foods to be a narrower category of foods for use by people with particular diseases or conditions that have distinctive nutritional requirements. Since a medical food must address the "distinctive nutritional requirements" of a disease or condition, a medical food is suitable only for use by patients with that disease or condition. Of course, it is possible for more than one disease or condition to create the same distinctive nutritional requirements. A product that is intended to address the distinctive nutritional requirements of a particular disease is a medical food, even though some of those requirements may also be created by other diseases. A product that is designed to address a problem that is common to several diseases, but not the full range of requirements of any specific disease, would be a food for special dietary use. For example, the distinctive nutritional requirements of burn patients include a greater energy requirement due to hypermetabolism and a requirement for dietary glutamine because endogenous synthesis of this amino acid does not meet the metabolic requirement. Thus, a product formulated to meet the higher energy requirement due to the hypermetabolic state, but which does not meet the requirement for glutamine, would be a food for special dietary use and not a medical food because it does not meet the full range of distinctive nutritional requirements in patients with burn injuries.

Foods for special dietary uses most likely represent the most straightforward route for incorporating nutritional genomic findings into food labeling because of the provisions from the regulations defining the use of foods "for supplying particular dietary needs which exist by reason of a … physiological condition" and of "foods for supplementing or fortifying the ordinary or usual diet with any vitamin, mineral or other dietary property." Nutritional genomics will likely lead us to be able to formulate and market foods that contain a particular bioactive nutrient or "dietary property" for particular physiological conditions.

**FIGURE 22.5**    Structure function claims appearing on Minute Maid enhanced juice label.

## Structure Function Claims

Although originally devised for dietary supplements, in recent years, the FDA has allowed structure function claims to be made about conventional foods. Structure function claims describe the effects that a food or food ingredient has on the normal, healthy structure or normal, healthy function of the body and its systems (21 CFR 101.93). Such claims may not discuss or imply drug actions or abnormal/ disease states, must be truthful, not misleading, and must be supported by scientific data. Structure function claims for conventional foods are based on nutritive value such as "fiber supports healthy digestion." When used on dietary supplement products, structure function claims must be accompanied by the disclaimer, "These statements have not been evaluated by the Food and Drug Administration. This product is not intended to diagnose, treat, cure or prevent any disease." In addition, dietary supplement manufacturers must notify the FDA of the structure function claims in use within 30 days of market introduction. Food products bearing structure function claims are not required to include the disclaimer and manufacturers do not have to notify the FDA of the claims in use.

Minute Maid recently introduced a series of enhanced juices which are marketed with a number of structure function claims. The pomegranate blueberry variety is marketed for brain nourishment and the front label includes the structure function claims "Omega-3/DHA* Help Nourish Your Brain" and "5 Nutrients to Support Brain and Body" (Figure 22.5). The back of the package includes several other structure function claims:

- DHA is a key building block in the brain.
- Choline and B12 play a role in brain and nervous system signaling.
- Antioxidant vitamin E helps shield the omega-3s in the brain from free radicals.
- Antioxidant vitamin C is highly concentrated in brain nerve endings.

In October of 2009, the National Advertising Division of the Council of Better Business Bureaus announced that they had reviewed the "help nourish your brain" claim and determined that Minute Maid had sufficient data to support the claim. However, the body recommended that a television commercial depicting immediate memory improvement after drinking the product be discontinued due to its implication that people who drink the juice will experience immediately improved memory capabilities (NAD News 2009). Nutritional genomic findings that demonstrate how a particular

---

* Docosahexaenoic acid.

food or food component supports the normal, healthy structure or normal, healthy function of the body or its systems can be communicated to consumers through structure function claims.

## Dietary Guidance Statements

Unlike health claims, which were discussed in the section *Allowable Health Claims*, dietary guidance statements do not require FDA authorization and do not link a particular nutrient to a disease but are general statements regarding health and diet. The distinction between the two types of statements was exemplified in a 1993 federal register docket on food labeling and health claims (FDA 1993):

> The following illustration using the National Cancer Institute's (NCI's) Five-a-Day Program … exemplifies how the context of the label will determine whether a statement is a health claim or dietary guidance. A cereal label that says "The National Cancer Institute recommends that you eat five servings daily of fruits and vegetables" is not a health claim because the information cannot be reasonably understood to be about a substance. There is neither a nutrient nor a product-specific element in the claim, and there is therefore no characterization between a substance and the disease included in the name or the organization. However, if the statement said, "The National Cancer Institute recommends that you eat five servings daily of fruits and vegetables to increase your intake of fiber," it would be a health claim because of the reference to a specific nutrient, fiber, and to a disease, cancer.

These statements were succinctly described in a January 2007 guidance document (Schneeman 2007):

> Dietary Guidance statements tend to focus on general dietary patterns, practices and recommendations that promote health. Typically "dietary guidance" statements make reference to a category of foods and not a specific substance. An example of a dietary guidance statement is: "Carrots are good for your health." Dietary guidance statements can be made without FDA review or authorization before use but the statements must be truthful and non-misleading.

## CONCLUSION

This review has outlined the current regulations governing food product claims, the position of the FDA regarding such claims and the need to watch the science and to participate in FDA discussions on food labeling in order to influence the future of food labels. Emerging nutritional genomic science looks promising and several potential pathways exist for making use of nutritional genomic data. The application of the regulations authorizing the promotion of foods for special dietary uses looks particularly promising as a way to utilize nutritional genomic data for the promotion of public health. The concept that labeling that fails to reveal material facts about a product causes the product to be misbranded, coupled with assertions of First Amendment rights to free speech may provide a path for making nutritional genomic claims.

Whether the regulatory environment is shaped by and keeps pace with nutritional genomic findings will depend largely upon the participation of interesting parties in FDA regulatory processes. While there are current regulatory frameworks that would permit the use of some claims that may be made possible through nutritional genomic research, it may be necessary to completely revamp labeling regulations via petitions to the FDA and to Congress in order to fully communicate useful health messages to consumers. Or, it may be necessary to petition for a mechanism similar to the GRAS process to deal with some nutrigenomic claims, as suggested by a former FDA staff member.

Not only will the future of food labeling as it relates to nutritional genomic data require awareness of the emerging science, it will also require vigilance to ensure that the increasingly enforcement driven FDA does not narrow the use of the claims and food classifications through guidance document and enforcement precedence. Ultimately, we cannot predict the future of food labeling but with knowledge, planning, and effort we can perhaps shape it.

# REFERENCES

21 CFR 1.3 (a) General Enforcement Regulations, Definitions, labeling, 2002 CFR Title 21.

21 CFR 101.9(j)(8) Nutrition labeling of food, Foods Exempt or Subject to Special Labeling Requirements, Medical Foods, CFR Title 21.

21 CFR 101.13 Nutrient Content Claims, General Principles, CFR Title 21.

21 CFR 101.13 (q)(1) Nutrient Content Claims, General Principles, Exemptions, CFR Title 21.

21 CFR 101.72 Health Claims: Calcium and osteoporosis.

21 CFR 101.79 Health Claims: Folate and neural tube defects.

21 CFR 101.93 Certain types of statements for dietary supplements.

21 CFR 105.3 Foods for Special Dietary Use, Definitions and Interpretations, CFR Title 21.

21 U.S.C.§ 343 Food and Drug Administration Modernization Act of 1997 (FDAMA) (Public Law 105–115).

Alliance for Natural Health U.S. v. Sebelius, 714 F.Supp.2d 48 (2010), United States District Court, District of Columbia., May 27, 2010.

American Dietetic Association. 2009. Supermarkets get in the game of nutrient profiling: Shelf tags could be displaying more than prices as store chains launch nutrition symbols to help shoppers identify healthy foods. http://www.eatright.org/Media/content.aspx?id=1250&terms=guiding+stars (accessed June 10, 2010).

Borra, S. 2006. Consumer perspectives on food labels. *Am J Clin Nutr* 20683(suppl):1235S.

FDA. 1993. Food and Drug Administration, (January 6, 1993). Food labeling: General requirements for health claims for food. *Federal Register* 58(3):2478.

FDA. 1996. Food and Drug Administration (November 29, 1996). Regulation of medical foods: Notice of proposed rulemaking. *Federal Register* 61(231):60661–71.

FDA. 1998. Office of Nutrition, Labeling, and Dietary Supplements, Center for Food Safety and Applied Nutrition (June 1998). Guidance for industry: Notification of a health claim or nutrient content claim based on an authoritative statement of a scientific body. http://www.fda.gov/Food/GuidanceComplianceRegulatoryInformation/GuidanceDocuments/FoodLabelingNutrition/ucm056975.htm (accessed March 31, 2010).

FDA. 1999. Guidance for industry: Significant scientific agreement in the review of health claims for conventional foods and dietary supplements. http://www.fda.gov/Food/GuidanceComplianceRegulatoryInformation/GuidanceDocuments/FoodLabelingNutrition/UCM059132 (accessed March 31, 2010).

FDA. 2008. Health and diet survey, March 2, 2010. http://www.fda.gov/Food/ScienceResearch/ResearchAreas/ConsumerResearch/ucm193895.htm#FOODLABELUSEALL (accessed March 31, 2010).

FDA. 2010a. FDA consumer health information survey shows gains in food-label use, health/diet awareness. http://www.fda.gov/ForConsumers/ConsumerUpdates/ucm202611.htm (accessed March 31, 2010).

FDA. 2010b. Front-of-pack and shelf tag nutrition symbols: Establishment of docket, request for comments and information. *Federal Register* 75(82):22602–6.

FDA Minneapolis Office. 2009. Warning letter to General Mills, Inc. http://www.fda.gov/ICECI/EnforcementActions/WarningLetters/ucm162943.htm (accessed March 31, 2010).

Food Standards Agency. 2010. http://www.eatwell.gov.uk/foodlabels/trafficlights/ (accessed February 18, 2010).

General Mills. 2010. Get the power of oats working for you. http://www.cheerios.com/forAdults/cholesterol/PowerOfOats.aspx (accessed February 13, 2010).

Guiding Stars. 2009a. Guiding Stars to implement minor adjustments to proprietary algorithm. http://www.guidingstars.com/news/guiding-stars-to-implement-minor-adjustments-to-proprietary-algorithm/ (accessed February 14, 2010).

Guiding Stars. 2009b. Guiding Stars adjusts proprietary algorithm. http://www.guidingstars.com/news/guiding-stars-adjusts-proprietary-algorithm/ (accessed February 14, 2010).

Guiding Stars. 2010. http://www.guidingstars.com/ (accessed February 14, 2010).

Hamburg, M. 2010. Food and Drug Administration open letter to industry from Dr. Hamburg on front of package labeling. http://www.fda.gov/Food/LabelingNutrition/ucm202733.htm (accessed March 31, 2010).

Harris, J. M., P. R. Kaiufman, S. W. Martinez, and C. Price. 2002. The U.S. Food Marketing System, 2002, Competition, Coordination, and Technological Innovations Into the 21st Century, U.S. Food Marketing System, AER 811, Economic Research Service USDA. http://www.ers.usda.gov/publications/aer811/aer811.pdf (accessed February 13, 2010).

Hubbard, W. K. 2004. Letter responding to health claim petition dated June 23, 2003 (wellness petition): Omega-3 fatty acids and reduced risk of coronary heart disease (Docket No. 2003Q-0401). http://www.fda.gov/Food/LabelingNutrition/LabelClaims/QualifiedHealthClaims/ucm072936.htm (accessed February 7, 2010).

Hughes, M. 2009. We have learned a lot. USA Today, Final Edition, News Section, page 8A.

NAD News. 2009. NAD finds Coca-Cola can support certain claims for Minute Maid enhanced juice. NAD recommends advertiser discontinue certain broadcast claims. http://www.nadreview.org/NewsRoom.aspx (accessed February 24, 2010).

NCTR FDA, National Center for Toxicological Research, Food and Drug Administration, Personalized Nutrition and Medicine, 2007 http://www.fda.gov/AboutFDA/CentersOffices/NCTR/WhatWeDo/ResearchDivisions/ucm079056.htm (accessed March 31, 2010).

Neuman, W. 2009. For your health, Froot Loops. New York Times, September 5, 2009, Section B, page 1.

NLEA. 1990. Nutrition Labeling Education Act of 1990 (Public Law 101-535, 101st Congress).

Quaker Oats and Rhodia, Inc. 2001. Petition to expand the oats soluble fiber and coronary heart disease health claim, (21 CFR 101.81: soluble fiber from certain foods and risk of CVD). http://www.fda.gov/ohrms/dockets/dailys/01/Jul01/072001/hcn0001.pdf (accessed February 6, 2010).

Schneeman, B. O. 2007. Guidance for industry and FDA: Dear manufacturer letter regarding food labeling. http://www.fda.gov/Food/GuidanceComplianceRegulatoryInformation/GuidanceDocuments/FoodLabelingNutrition/ucm053425.htm (accessed February 24, 2010).

Schneeman, B. O. 2009a. FDA response to General Mills addressing the scientific basis for Cheerios cholesterol claims. http://www.fda.gov/Food/LabelingNutrition/FoodLabelingGuidanceRegulatoryInformation/InspectionCompliance/WarningOtherLetters/ucm190492.htm (accessed February 17, 2010).

Schneeman, B. O. 2009b. Guidance for industry: Letter regarding point of purchase food labeling. http://www.fda.gov/Food/GuidanceComplianceRegulatoryInformation/GuidanceDocuments/FoodLabelingNutrition/ucm187208.htm (accessed February 17, 2010).

Scheenman, B. O. 2009c. Selenium and a reduced risk of site-specific cancers., FDA-2008-Q-0323, http://www.fda.gov/Food/LabelingNutrition/LabelClaims/QualifiedHealthClaims/ucm168527.htm (accessed August 19, 2011).

Smart Choices, 2009. Smart Choices program postpones active operations. PRNewswire, http://www.prnewswire.com/news-releases/consumer-products-retail-latest-news/smart-choices-program-postpones-active-operations-65809752.html (accessed August 19, 2011).

Todd, J. E., and J. N. Variyam. 2008. Decline in consumer use of food nutrition labels, 1995–2006, Economic Research Report 63, Economic Research Service, United States Department of Agriculture. http://www.ers.usda.gov/Publications/ERR63/ERR63.pdf (accessed February 13, 2010).

USDA ERS. 2009a. Economic Research Service, United States Department of Agriculture, Food CPI and Expenditures, Tables 1 and 13. http://www.ers.usda.gov/Briefing/CPIFoodAndExpenditures/Data/Expenditures_tables.htm (accessed February 9, 2010).

USDA ERS. 2009b. Economic Research Services. United States Department of Agriculture, Food marketing system in the U.S.: New product introductions of consumer packaged goods. http://www.ers.usda.gov/Briefing/FoodMarketingSystem/new_product.htm (accessed March 31, 2010).

USDHHS. 2008. United States Department of Health and Human Services, Office of Disease Prevention and Health Promotion, Food and Drug Administration, 2008 Health and Diet Survey: Dietary guidelines supplement report of findings (2004 and 2005). http://www.fda.gov/downloads/Food/ScienceResearch/ResearchAreas/ConsumerResearch/ucm080413.pdf (accessed February 13, 2010).

Van Camp, D. J., N. H. Hooker, and D. M. Souza Monteiro. 2009. UK food manufacturer responses to voluntary front of package nutrition schemes, Department of Agricultural, Environmental and Development Economics, The Ohio State University, Columbus, Ohio. http://hdl.handle.net/1811/37149 (accessed February 18, 2010).

# Index

UNIVERSITIES AT MEDWAY LIBRARY